Progress in Inorganic Chemistry

Volume 51

Advisory Board

JACQUELINE K. BARTON
   CALIFORNIA INSTITUTE OF TECHNOLOGY, PASADENA, CALIFORNIA
THEODORE J. BROWN
   UNIVERSITY OF ILLINOIS, URBANA, ILLINOIS
JAMES P. COLLMAN
   STANFORD UNIVERSITY, STANFORD, CALIFORNIA
F. ALBERT COTTON
   TEXAS A & M UNIVERSITY, COLLEGE STATION, TEXAS
ALAN H. COWLEY
   UNIVERSITY OF TEXAS, AUSTIN, TEXAS
RICHARD H. HOLM
   HARVARD UNIVERSITY, CAMBRIDGE, MASSACHUSETTS
EIICHI KIMURA
   HIROSHIMA UNIVERSITY, HIROSHIMA, JAPAN
NATHAN S. LEWIS
   CALIFORNIA INSITITUTE OF TECHNOLOGY, PASADENA, CALIFORNIA
STEPHEN J. LIPPARD
   MASSACHUSETTS INSTITUTE OF TECHNOLOGY, CAMBRIDGE, MASSACHUSETTS
TOBIN J. MARKS
   NORTHWESTERN UNIVERSITY, EVANSTON, ILLINOIS
EDWARD I. STIEFEL
   EXXON MOBIL RESEARCH & ENGINEERING CO., ANNANDALE, NEW JERSEY
KARL WIEGHARDT
   MAX-PLANCK-INSTITUT, MÜLHEIM, GERMANY

PROGRESS IN INORGANIC CHEMISTRY

Edited by

KENNETH D. KARLIN

DEPARTMENT OF CHEMISTRY
JOHNS HOPKINS UNIVERSITY
BALTIMORE, MARYLAND

VOLUME 51

AN INTERSCIENCE PUBLICATION
JOHN WILEY & SONS, INC.

Cover Illustration or "a molecular ferric wheel" was adapted from Taft. K. L. and Lippard. S. J., *J. Am. Chem. Soc.*, **1990**. 112, 9629.

Copyright © 2003 by John Wiley & Sons, Inc. All rights reserved.

Published by John Wiley & Sons, Inc., Hoboken, New Jersey.
Published simultaneously in Canada.

No part of this publication may be reproduced, stored in a retrieval system, or transmitted in any form or by any means, electronic, mechanical, photocopying, recording, scanning, or otherwise, except as permitted under Section 107 or 108 of the 1976 United States Copyright Act, without either the prior written permission of the Publisher, or authorization through payment of the appropriate per-copy fee to the Copyright Clearance Center, Inc., 222 Rosewood Drive, Danvers, MA 01923, 978-750-8400, fax 978-750-4470, or on the web at www.copyright.com. Requests to the Publisher for permission should be addressed to the Permissions Department, John Wiley & Sons, Inc., 111 River Street, Hoboken, NJ 07030, (201) 748-6011, fax (201) 748-6008, e-mail: permreq@wiley.com.

Limit of Liability/Disclaimer of Warranty: While the publisher and author have used their best efforts in preparing this book, they make no representations or warranties with respect to the accuracy or completeness of the contents of this book and specifically disclaim any implied warranties of merchantability or fitness for a particular purpose. No warranty may be created or extended by sales representatives or written sales materials. The advice and strategies contained herein may not be suitable for your situation. You should consult with a professional where appropriate. Neither the publisher nor author shall be liable for any loss of profit or any other commercial damages, including but not limited to special, incidental, consequential, or other damages.

For general information on our other products and services please contact our Customer Care Department within the U.S. at 877-762-2974, outside the U.S. at 317-572-3993 or fax 317-572-4002.

Wiley also publishes its books in a variety of electronic formats. Some content that appears in print, however, may not be available in electronic format.

Library of Congress Catalog Card Number 59-13035
ISBN 0-471-26534-9

Printed in the United States of America

10 9 8 7 6 5 4 3 2 1

Contents

Chapter 1	Fundamental Coordination Chemistry, Environmental Chemistry, and Biochemistry of Lead(II) ELIZABETH S. CLAUDIO, HILARY ARNOLD GOLDWIN, AND JOHN S. MAGYAR	1
Chapter 2	Chromium in Biology: Toxicology and Nutritional Aspects AVIVA LEVINA, RACHEL CODD, CAROLYN T. DILLON, AND PETER A. LAY	145
Chapter 3	Laterally Nonsymmetric Aza-Cryptands PARIMAL K. BHARADWAJ	251
Chapter 4	Coordination Complexes in Sol–Gel Silica Materials STEPHEN P. WATTON, COLLEEN M. TAYLOR, GRANT M. KLOSTER, AND STEPHANIE C. BOWMAN	333
Chapter 5	Crystal Chemistry of Organically Templated Vanadium Phosphates and Organophosphonates ROBERT C. FINN, ROBERT C. HAUSHALTER, AND JON ZUBIETA	421
Subject Index		603
Cumulative Index, Volumes 1–51		625

Progress in Inorganic Chemistry

Volume 51

Fundamental Coordination Chemistry, Environmental Chemistry, and Biochemistry of Lead(II)

ELIZABETH S. CLAUDIO, HILARY ARNOLD GODWIN,[*] **and JOHN S. MAGYAR**

Department of Chemistry
Northwestern University
Evanston, IL

CONTENTS

I. INTRODUCTION
   A. Why Study Lead?
   B. General Properties of Lead and Lead(II)
      1. Isotopes of Lead
      2. Oxidation States
      3. Electronic Properties
      4. Solubility and Water-Exchange Rate of Lead(II)
   C. Scope of This Chapter

II. SPECTROSCOPIC STUDIES OF LEAD(II) COMPLEXES
   A. Introduction
   B. Questions of Interest
   C. Relativistic Effects
   D. Absorption Spectroscopy
      1. Introduction to Possible Electronic Transitions for Lead
      2. Optical Electronegativities
      3. Solid-State Absorption Spectra
   E. Photoelectron Spectroscopy
      1. Relativistic Effects and the Myth of the $6s^2$ "Inert" Pair
      2. Lead(II) Oxides
      3. Lead(II) Halides
      4. Lead(II) Chalcogenides

---

[*] Author to whom correspondence should be addressed.

Progress in Inorganic Chemistry, Vol. 51, Edited by Kenneth D. Karlin.
ISBN 0-471-26534-9   © 2003 John Wiley & Sons, Inc.

F. Vibrational Spectroscopy (Infrared and Raman)
G. Nuclear Magnetic Resonance Spectroscopy
   1. Lead-207 Parameters: Chemical Shifts and Coupling Constants
   2. Lead-207 NMR Spectroscopy of Proteins
H. Spectroscopy Conclusions

III. STRUCTURAL STUDIES ON LEAD(II) COMPLEXES

   A. Introduction
   B. Questions of Interest
   C. X-Ray Crystal Structures of Lead(II) Small Molecule Complexes
      1. Commonly Observed Donor Groups in Lead(II) Structures
      2. Commonly Observed Coordination Numbers and Geometries in Lead(II) Structures
      3. The Structural Effects of the Lead(II) $6s^2$ Electron Pair
      4. Structural Insights into the Rational Design of Chelation Therapy Agents
      5. New Types of Lead(II) Structures
   D. X-Ray Crystal Structures of Lead(II) Biomolecules
      1. Complexes of Lead(II) with Small Biomolecules
      2. Use of Lead as a Heavy Atom Derivative in Proteins and Nucleic Acids
   E. EXAFS Studies on Lead(II) Compounds
   F. Structural Conclusions

IV. KINETICS AND THERMODYNAMICS OF LEAD–LIGAND INTERACTIONS

   A. Introduction
   B. Questions of Interest
   C. Kinetics of Lead–Ligand Interactions
      1. Simple Ligand-Exchange Constants
      2. Mechanisms and Rates of Lead–Chelate Association and Substitution Reactions
      3. Kinetic Studies with Macrocyclic Ligands
      4. Implications of Kinetic Studies for Lead(II)–Protein Interactions and Other Complex Systems
   D. Thermodynamics of Lead–Ligand Interactions
      1. Methods for Determining Lead(II) Ligand Binding Constants
      2. Thermodynamics of Lead–Small Molecule Interactions
      3. Thermodynamic Stability of Lead–Protein Interactions
   E. Conclusions about Lead–Ligand Thermodynamics and Kinetics

V. LEAD IN THE ENVIRONMENT

   A. Introduction
   B. Questions of Interest
   C. Lead Minerals
      1. PbS (Galena)
      2. $PbCO_3$ (Cerussite)
      3. PbO (Litharge and Massicot)
      4. Pb_3O_4 (Minium)
      5. $PbHPO_4$ (Schulterite) and Other Lead Phosphates
   D. Lead in Soil
   E. Lead in Aquatic Systems
   F. Environmental Contamination with Lead: The Historical Record and Geochemistry

VI. BIOLOGICAL CHEMISTRY OF LEAD

    A. Symptoms of Lead Poisoning
    B. Questions of Interest
    C. Biodistribution of Lead
        1. Lead Body Burden and Uptake
        2. Lead Distribution in the Body
        3. Fluorescent Sensors
        4. Toxicokinetics
    D. Molecular Targets for Lead
        1. Interactions with Calcium Proteins
        2. Interactions with Zinc Proteins
        3. Interactions with DNA and RNA
        4. Other Possible Targets for Lead
    E. Biomarkers for Lead Poisoning
        1. Blood Lead Level
        2. Alterations in Heme Biosynthesis
        3. Bone Lead Levels
        4. Other Potential Markers for Lead Poisoning
    F. Chelation Therapy
        1. Chelation Therapy Agents Currently in Use
        2. Challenges in Developing New Chelation Therapy Agents

VII. CONCLUSIONS

ACKNOWLEDGMENTS

ABBREVIATIONS

REFERENCES

I. INTRODUCTION

A. Why Study Lead?

"Hot lead can be almost as effective coming from a linotype as from a firearm." John O'Hara (The Portable F. Scott Fitzgerald, 1945, Introduction) (1)

Lead, often referred to as the "useful" metal (2), has been both exploited extensively and feared as a toxin since antiquity (3). Because metallic lead is readily—albeit slowly—oxidized to Pb(II) under atmospheric conditions, most lead on the earth's surface is naturally found in the form of ores, such as galena (PbS). Galena is widely available and can be converted to metallic lead at low temperatures (e.g., "roasting" in a campfire) in the presence of simple reducing agents (e.g., "smelting" with charcoal) (4). Consequently, lead was one of the first metals to be used by humans: The earliest samples of metallic lead date back to 6500 B.C. (3).

Roasting (conversion to oxide)

$$2\,\underset{\text{(galena)}}{PbS} + 3\,O_2 \xrightarrow{800°C} 2\,\underset{\text{(litharge)}}{PbO} + 2\,SO_2$$

Smelting (reduction of metal)

$$PbO + C \xrightarrow{heat} Pb + CO$$
$$PbS + 2\,PbO \xrightarrow{heat} 3\,Pb + SO_2$$

Because lead has a low melting temperature (327.5 °C) (5), it was relatively simple for early humans to manipulate lead into useful shapes and tools. Consequently, lead was used for a wide variety of objects—ranging from axes to tumblers to coins and figurines (3). Because it is a major component in alloys such as bronze ($\sim$5–60% Pb), lead also figured heavily in the development of the Bronze Age ($\sim$3000–1000 B.C.) (3).

It was not lead's utility, however, that was originally responsible for widespread contamination in the environment by humans. Rather, humans first produced large quantities of lead in their hungry search for the more valuable metal silver. Not surprisingly, given its similar affinity for sulfur, silver is frequently codeposited with lead and is present in large quantities in galena ($\sim$1–8000 µg Ag g^{-1} of PbS) (3). The first real boom in world lead production occurred $\sim$5000 years ago with the discovery of cupellation, the process for extracting silver from lead ores (Fig. 1) (3, 6–8). [It is interesting to note that this colocalization of silver and lead ores may also be responsible for the myth later promulgated by alchemists that lead could be transmuted into noble metals. In fact, when mixtures of ores are reduced and heated, the more volatile metallic lead (bp 1740 °C) (5) is vaporized, often leaving behind a small sample of silver or gold (9)]. Consequently, lead emissions increased from 160 tons/year 4000 years ago to $\sim$10,000 tons/year $\sim$2700 years ago, when silver began to be used extensively for coins (6).

Lead production underwent a further explosive increase during the Roman Empire $\sim$2000 years ago: Roman production of lead topped out at 50,000–80,000 tons/year (Fig. 1) (3, 6, 7). Lead was used during Roman times for a wide range of applications: from drinking vessels, to pipes, to coins and coffins (3). Thus it is not surprising that lead is frequently referred to as the "Roman metal" (3). Indeed, our chemical name and symbols for lead also come from the Roman word for lead, "plumbum", as do our modern words "plumbing" and

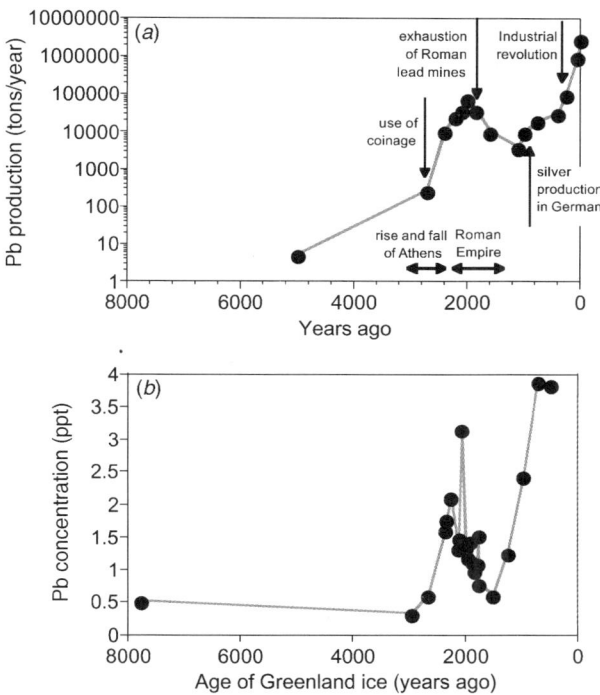

Figure 1. Anthropogenic production of lead over time [(a), logarithmic plot] is reflected in deposition of lead in Greenland ice (b), suggesting that lead pollution is a global issue. Large increases in lead production occurred with the discovery of cupellation around 5000 years ago, when silver began to be used in coins ~2700 years ago, and during the Industrial Revolution. [Reprinted with permission from S. Hong, J.-P. Candelone, C. C. Patterson, and C. F. Boutron, *Science*, 265, 1841–1843 (1994). Copyright © 1994 American Association for the Advancement of Science.]

"plumber" (10). This extensive use of lead came at a cost: Lead levels in human remains from the Roman empire are typically 10–100-fold higher than "natural" levels of lead in bones from pre-Roman sites, and the adverse health effects of lead were noted even by early Greeks and Romans (3). The extensive use of lead by the Romans—and the concomitant contamination of the environment and population—led to the widely cited hypothesis that lead poisoning contributed to the fall of Rome (3, 11, 12). However, this "myth" has been largely debunked (3, 13), and most classicists today agree that economic factors, including those arising from overexpansion and decentralization, coupled with a decline of the power of the Roman aristocracy, were the major contributors to the transformation and gradual decline of the Roman Empire (14).

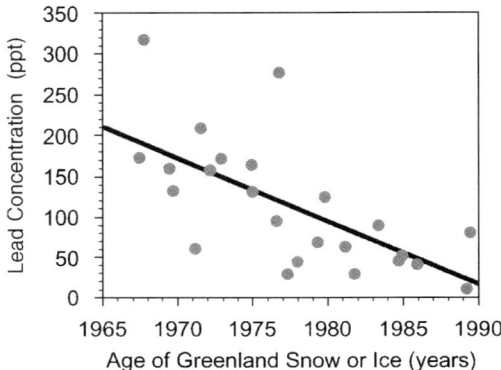

Figure 2. Lead emissions into the environment (as measured by deposition in Greenland ice) decreased dramatically after the introduction of catalytic converters in the 1970s. Lead levels are given in picograms of lead per gram of ice. [Reprinted with permission from C. F. Boutron, U. Gorlach, J.-P. Candelone, M. A. Bol'shov, and R. J. Delmas, *Nature (London)*, *353*, 153–156 (1991). Copyright © 1991 Macmillan Magazines Limited.]

In addition to surging during Roman times, anthropogenic emissions of lead increased again dramatically during the industrial revolution (Fig. 1) and have continued to increase for much of the twentieth century. In modern times, lead has been used extensively in paint, batteries, building materials, ammunition, and solder (5). In addition, one of the largest contributions to lead in the atmosphere came from the use of Pb(IV) compounds (e.g., tetraethyllead) as a gasoline additive and antiknocking agent. Lead emissions into the environment did not decrease significantly until the introduction of the catalytic converter in the 1970s (Fig. 2) (15). Although catalytic converters were not introduced with the goal of decreasing atmospheric lead per se, phasing out leaded gasoline had an immediate and profound effect upon public health in the United States: Declines in the average blood lead level (BLL, given in $\mu g\ dL^{-1}$) in the United States mirror decreases in the total amount of lead used in gasoline over the period 1976–1980 (Fig. 3) (10, 16). Critically, the average blood lead level in the United States fell *below* the level that is currently considered "high risk" for lead poisoning by the Centers for Disease Control (CDC) (10 $\mu g\ dL^{-1}$) in 1980 and has continued to gradually decrease over time (17, 18). Although leaded gasoline has been phased out much more gradually in developing countries, studies to date suggest that populations worldwide are benefiting from reductions in lead levels in gasoline (19, 20).

Despite these impressive advances, lead poisoning is still a major health concern worldwide. [For recent popular books on this subject, see (2, 21, 22)]. Lead is still used in a wide variety of applications; and as a result, ~3.4 million

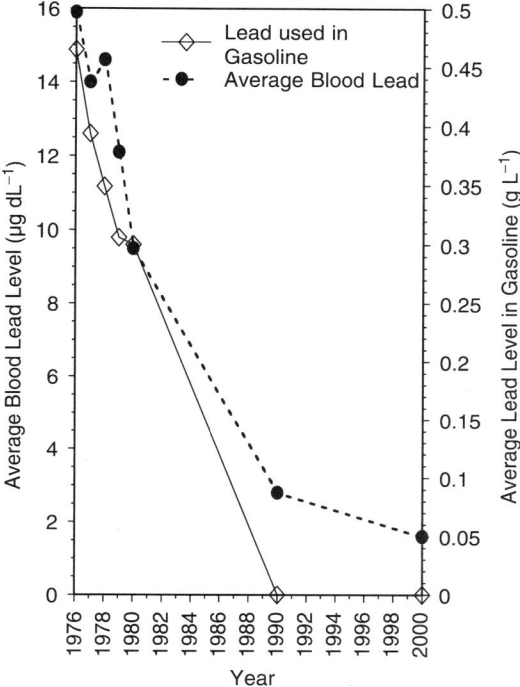

Figure 3. Average blood lead levels in the United States dropped dramatically in the late 1970s when catalytic converters were introduced and leaded gasoline started to be phased out of use (16, 20). Unfortunately, although leaded gasoline is no longer used in the United States today, there is still residual lead contamination in the environment and in our blood.

metric tons of lead are produced each year worldwide, resulting in the release of ~1.6 million metric tons of lead into the environment each year (10). In addition, most of the 300 million metric tons of lead that have been produced to date (Fig. 4) are still circulating in the environment, principally in soil and dust (10, 23). Lead is not only a common contaminant in industrial sites—it is a top 10 contaminant at Superfund sites (24)—but also in our cities and homes. Soil lead levels in many inner city areas (up to 7900 ppm in some sites in Chicago) rival that of Superfund sites (25, 26), and 74% of houses in the United States that were built before 1980 are contaminated with lead paint (22). As a result, the average lead body burden in the United States is still ~100–1000 times that of prehistoric levels (10, 27), lead was ranked as number 2 on the national hazard priority list for 1999 (28), and lead poisoning remains the most common environmentally caused disease in the United States (29, 30).

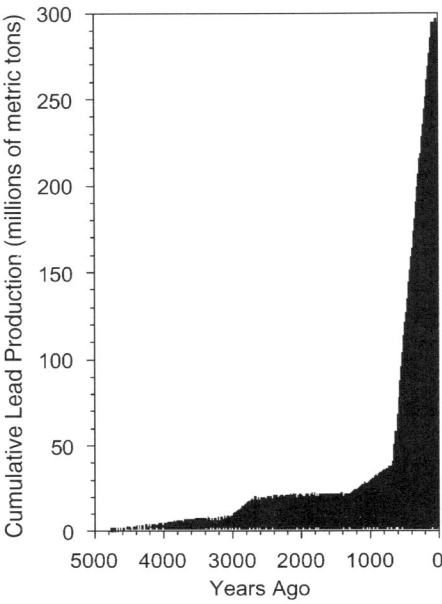

Figure 4. Cumulative production of lead (in millions of metric tons) as a function of time. [Adapted from (23).]

What can be done to address this pressing problem? The following recommendations and goals were put forth to the National Research Council in 1993 by the Committee on Measuring Lead in Critical Populations and remain largely relevant today (10):

- To determine how lead affects targeted organ systems (e.g., the reproductive system and the nervous system).
- To determine the toxicokinetic behavior of lead (especially lead in bone).
- To develop new methods for characterizing how lead affects calcium signaling in living systems.
- To determine the mechanism of low-level lead toxicity, particularly how lead affects "gene expression, calcium signaling, heme biosynthesis, and cellular energy stores" (10).
- To determine which of the effects of lead poisoning are reversible and which are permanent.
- To develop standard reference materials for analysis of biological and environmental samples.

- To use mass spectrometry (MS) of stable isotopes to investigate human lead metabolism.
- To develop more sensitive techniques to quantify human lead body burden, particularly lead in bone.

Chemists can contribute to this cause through a variety of mechanisms, including developing new analytical techniques for quantifying lead and studying lead speciation both in vitro and in situ, and by studying the fundamental aqueous coordination chemistry, biochemistry, and environmental chemistry of lead. Our progress as a community toward each of these goals—and the important questions that remain to be addressed—are the primary focus of this chapter.

B. General Properties of Lead and Lead(II)

Before delving into the fundamental coordination chemistry of lead, we briefly summarize the critical properties of the element lead and how these properties affect our ability to investigate the chemistry of Pb(II).

1. Isotopes of Lead

Lead, with an atomic weight of 207.2 and an atomic number of 82, is widely distributed on Earth (5, 31). Lead has four common stable isotopes: ^{204}Pb (1.48%), ^{206}Pb (23.6%), ^{207}Pb (22.6%), and ^{208}Pb (52.3%) (5). The ratio of the stable isotopes of lead varies from one source to another in the environment, depending on the radioactive source (and hence geologic formation) from which the lead was derived and the relative decay rates of the radioactive elements: ^{206}Pb is formed from ^{238}U ($t_{1/2} = 4.5 \times 10^9$ years), ^{207}Pb is formed from ^{235}U ($t_{1/2} = 0.70 \times 10^9$ years), and ^{208}Pb is formed from ^{232}Th ($t_{1/2} = 1.40 \times 10^{10}$ years). By contrast, ^{204}Pb has no long-lived radioactive parent isotope (10, 31, 32). The ^{206}Pb/^{207}Pb ratios (and ratios of other stable lead isotopes) have proven particularly useful for identifying the primary source of lead contamination in environmental and human samples (see Section V.F) (10, 31, 33). The stable isotope ^{207}Pb ($I = \frac{1}{2}$) has also proved to be useful for studying the lead nucleus using nuclear magnetic resonance (NMR) spectroscopy (see below).

Lead also has several radioactive isotopes (notably ^{203}Pb, ^{210}Pb, and ^{212}Pb). Lead-210 ($t_{1/2} = 22.5$ years) has proven useful for dating relatively "young" environmental samples (33), much the way that ^{14}C dating ($t_{1/2} = 5730$ years) is used to assess the age of older samples of biotic origin (31). Radioactive isotopes of lead have also been used to study lead metabolism, but these studies have been relatively limited both because the half-lives of the common radioactive isotopes of lead are not ideal for metabolic studies ($t_{1/2}$ of ^{203}Pb $= 51.88$ h, $t_{1/2}$ of ^{210}Pb $= 22.5$ years, and $t_{1/2}$ of ^{212}Pb $= 10.64$ h) and because they are either negative β or orbital electron emitters, and hence require extensive safety precautions (10, 34). Fortunately, recent advances in mass spectrometry (MS), particularly in high-resolution isotope dilution mass spectrometry (IDMS) and inductively coupled plasma–mass spectrometry (ICP–MS) have obviated the need to use radioactive elements to study metal metabolism. Because the sensitivity of these techniques and the ability to differentiate between different isotopes of the same element using these techniques are so great, metal distribution and metabolism can now be monitored using stable isotopes, even when the concentration of lead in the samples is quite low (10, 34).

2. Oxidation States

Lead has three readily accessible oxidation states: Pb(0), Pb(II), and Pb(IV) (5, 35) (Table I). Metallic lead is readily oxidized by oxygen to yield soluble Pb(II) (34, 36):

$$2\,Pb + O_2 + 4\,H^+ \rightarrow 2\,Pb^{2+} + 2\,H_2O \qquad E° = 1.355\,V$$

As a result, lead used for pipes or solder in municipal water supply lines is slowly leached into the water supply, providing a low-level chronic source of lead exposure in many urban areas. The rate of this leaching is highly dependent on pH (with more lead being reduced and swept into the water supply at lower pH), as is the speciation and solubility of the resulting Pb(II) (see below) (3, 10, 36). In addition, Pb(IV) compounds are strong oxidizing agents, and hence Pb(IV) compounds are readily reduced to Pb(II) in the environment: By the time that gasoline additives such as PbEt$_4$ make it through a car engine, they have been reduced in high percentages to Pb(II) compounds (e.g., PbBrCl) (5, 36). Because Pb(II) is the predominant form of lead in the environment and the most common form of lead to which humans are currently exposed, we have chosen to focus on the chemistry of Pb(II) both in our own work and in the remainder of this chapter (37–41).

TABLE I

Properties and Applications of the Common Oxidation States of Lead (35, 36)

Oxidation State	Electronic Structure	Radius (Å)	Industrial Uses
Pb(0)	[Xe]$4f^{14}5d^{10}6s^26p^2$	1.44	Lead pipes, solder
Pb(II)	[Xe]$4f^{14}5d^{10}6s^2$ 6s lone pair can be stereochemically active	1.19	Lead paint [Pb$_3$(OH)$_2$(CO$_3$)$_2$] lead batteries
Pb(IV)	[Xe]$4f^{14}5d^{10}$	0.775	Gasoline additives (e.g., PbEt$_4$)

3. Electronic Properties

Lead has an electronegativity of 1.8 (35), and as such has a high affinity for very electronegative elements such as S, O, N, Cl, and Br. The ionization enthalpies of lead are (in kJ mol^{-1}): 715.3, 1450, 3080, 4082 (35). The electronic structure of Pb(II) (group 14 (IVA), [Xe]$4f^{14}5d^{10}6s^2$) has been studied fairly extensively, with one of the most critical questions being the extent to which the $6s^2$ "lone pair" is mixed with other (lead and/or ligand) orbitals versus the extent to which the $6s$ orbital is energetically isolated from other orbitals, and hence "inert". In addition, the effect that the lone pair has on geometry (i.e., when it is "stereochemically active" and why) is a critical question in Pb(II) coordination chemistry. These issues are discussed extensively in the sections on the spectroscopic and structural properties of Pb(II) complexes (Sections II and III, respectively).

4. Solubility and Water-Exchange Rate of Lead(II)

Inorganic complexes of Pb(II) are renowned for their low solubility (Table II): Lead chloride is only moderately soluble and only lead nitrate, lead acetate, and lead citrate are reasonably soluble (5, 34, 42–46). This low solubility necessarily has a profound effect upon both the environmental and biological chemistries of

TABLE II

Solubility of Common Inorganic and Organic Complexes of Lead(II)

Compound	Solubility (g/100 cm^3 of cold H$_2$O)	K_{sp} (25°C)	Reference
Pb$_3$(C$_6$H$_5$O$_7$)$_2$ (citrate)	140		42
Pb(C$_2$H$_3$O$_2$)$_2$ (acetate)	44.3		34
Pb(NO$_3$)$_2$	37.56		34
Pb(SCN)$_2$		2.11 × 10^{-5}	44
PbCl$_2$		1.70 × 10^{-5}	34
PbBr$_2$		6.6 × 10^{-6}	45
PbSeO$_4$		1.37 × 10^{-7}	45
PbF$_2$		3.3 × 10^{-8}	45
PbSO$_4$		2.53 × 10^{-8}	45
PbI$_2$		9.8 × 10^{-9}	45
PbC$_2$O$_4$ (oxalate)		8.5 × 10^{-10}	44
PbCrO$_4$		2.8 × 10^{-13}	46
Pb(IO$_3$)$_2$		3.69 × 10^{-13}	45
PbCO$_3$		7.4 × 10^{-14}	44
Pb(OH)$_2$		1.40 × 10^{-20}	34
PbS		8.81 × 10^{-29}	34
Pb$_3$(PO$_4$)$_2$		1 × 10^{-54}	43

lead. In its native form as ores in the environment [e.g., PbS (galena), Table II], lead tends to be extremely insoluble, and hence is not very mobile (Section V). It is interesting to speculate that this insolubility may play a role in the susceptibility of living organisms to lead poisoning. Although lead is relatively abundant compared to other heavy metals [with an average abundance of ~20 ppm in the earth's crust (47)], most naturally occurring lead compounds are so insoluble that most biological systems (including people) have been exposed to relatively little lead on an evolutionarily relevant timescale. Consequently, it is not surprising that living organisms have generally not evolved extensive mechanisms for coping with lead intoxication (27). (The obvious exceptions to this generalization are bacteria, which have a short generation time and evolve on a rapid timescale.) The solubility of lead compounds also plays a critical role in accessibility from different sources. For example, lead tends to form insoluble hydroxides and carbonates when these ions are present in solution, and hence the risk of lead exposure from lead pipes is much lower in municipal areas with hard water than those with soft water. Many municipalities with high levels of soluble lead in their water supplies have successfully combated this problem by adding ions that complex the lead (e.g., phosphates) and passivate the surface of lead pipes (48) (Section V.E). Because the solubility of lead is strongly pH dependent, once lead in the environment is disturbed (e.g., by mining and smelting) and taken up into people (e.g., via airborne emissions and dust), it is fairly readily absorbed and mobilized (e.g., in the stomach, which is highly acidic). The speciation and distribution of lead inside the body presumably also reflects this propensity of lead to complex tightly to a variety of ligands (Section IV). Our ability to characterize the aqueous coordination chemistry of lead is necessarily dependent on our ability to accurately quantify these effects and the speciation of lead in complex solutions.

The kinetics of the reactions of lead complexes (Section IV) in aqueous solutions and mixed media is of equal importance to understanding the fundamental environmental (Section V) and biological chemistry (Section VI) of lead. It is important to note that although lead forms very thermodynamically stable complexes, ligands bound to lead tend to be extremely labile in aqueous solutions: The water-exchange rate for lead is $7 \times 10^9 \text{ s}^{-1}$ (49).

C. Scope of This Chapter

As the title indicates, this chapter covers the fundamental coordination chemistry, environmental chemistry (Section V), and biochemistry (Section VI) of Pb(II), with particular emphasis on spectroscopic properties (Section II), structures (Section III), and thermodynamics and kinetics (Section IV) of Pb(II) complexes. We have intentionally placed the section on the spectroscopic properties of Pb(II) at the beginning of the chapter (Section II) because we feel that many of the principles described therein—and the insights that these studies provide into the electronic structure of Pb(II)—will help to clarify many of the properties described in later sections. However, we also recognize that the spectroscopy section is the most technical section and encourage those individuals who are primarily interested in an empirical discussion of the properties of Pb(II) (including environmental chemistry and biochemistry) to skip directly to subsequent sections. Throughout, we have attempted to identify the critical questions that are of importance in this field (outlined at the beginning of each section), and to provide insights into the answers (or partial answers) that have been obtained to these questions at the time this chapter was written (2002).

II. SPECTROSCOPIC STUDIES OF LEAD(II) COMPLEXES

A. Introduction

Lead, with an electron configuration $[Xe]4f^{14}5d^{10}6s^{2}6p^{2}$, has been widely perceived to be "spectroscopically silent" because it has a d^{10} electronic configuration in all three common oxidation states (0, +2, and +4). However, nothing could be farther from the truth: Although lead compounds do not exhibit d–d transitions, they do exhibit both intraatomic and charge-transfer (CT) transitions. As early as 1952, Klotz et al. (50) reported absorption spectra attributed to lead [Pb(II)] binding to proteins (37, 51–53). Even earlier, the absorption spectra of both aqueous (54–60) and solid-state (61–68) lead-doped alkali halides had been reported, as well as the absorption spectra of some Pb(II) doped alkaline earth compounds (69–73).

In addition, the presence of the $6s^2$ lone pair and the relativistic splitting of the empty $6p$ orbitals are responsible for a range of fascinating electronic as well as structural effects. A wide variety of spectroscopic techniques, including absorption, photoelectron, infrared (IR), and Raman spectroscopies, along with associated theory and modeling, have been used to examine the electronic structure and coordination environment of divalent lead complexes. In this section, we will examine how each of these methods can increase our understanding of the electronic structure of Pb(II) and how this electronic structure affects the structure and bonding of Pb(II) compounds. In addition, we will discuss the ^{207}Pb NMR spectroscopy of lead complexes and the insights into lead coordination chemistry that NMR spectroscopy provides.

B. Questions of Interest

In this section, we summarize what is known about the spectroscopic properties of Pb(II) compounds, both in the solid state and in solution. In particular, we focus on addressing the following questions:

- What is the electronic origin of the absorption spectrum of Pb–thiolate complexes?
- What do the photoelectron spectroscopy of lead halides, oxides, and sulfides tell us about the makeup of the valence band in each case?
- To what extent is an understanding of relativistic effects necessary for interpretation of lead photoelectron spectroscopy (PES)?
- What do the parameters obtained from photoelectron spectroscopy tell us about the chemical–physical properties of Pb(II) compounds?
- What can be learned about the covalency of lead–ligand interactions from vibrational spectroscopy?
- How do trends in NMR parameters depend on the lead coordination environment?
- Can changes in coordination number and in the identities of ligand atoms be observed by NMR spectroscopy?
- Can structural effects of the lead lone pair of electrons be observed by NMR spectroscopy?

C. Relativistic Effects

For much of chemistry, theoretical treatment with the nonrelativistic Schrödinger equation is sufficient. However, in some instances—particularly for the heavy elements—the relativistic approach to quantum mechanics developed by Dirac is required to accurately describe electronic properties (74, 75). The Dirac equation provides four quantum numbers, just as the Schrödinger equation does. However, with the exception of the principal quantum number n, the Dirac quantum numbers are not exactly the same four quantum numbers as in nonrelativistic quantum mechanics. One of the results of this change in quantum numbers is that although orbital type (s, p, d, f, etc.) is still determined by the quantum number ℓ, orbital shape and energy are now determined by the angular momentum quantum number j and the magnetic quantum number m. Thus, instead of three p orbitals of equal energy and shape (p_x, p_y, p_z), there are instead three p orbitals with different symmetries ($p_{1/2(m=1/2)}$, $p_{3/2(m=1/2)}$, and $p_{3/2(m=3/2)}$) in the relativistic model. The $p_{1/2}$ orbital holds two electrons and is spherically symmetric (and hence able to mix easily with s orbitals). There are two types of $p_{3/2}$ orbitals, each holding two electrons; both are at higher energy than the $p_{1/2}$ orbital due to spin–orbit splitting (76). The $p_{3/2(m=1/2)}$ has an angular distribution similar to a "dogbone", whereas the $p_{3/2(m=3/2)}$ orbital is a "doughnut"; none of the relativistic orbitals have any nodes (77, 78). Spin–orbit interaction, although often included as an addition to nonrelativistic quantum theory because of its importance to electronic structure, is a relativistic effect that appears naturally from the Dirac equation.

Relativistic effects, although they are negligible for lighter elements, greatly increase in importance for the heavy elements. As inner-shell s electrons approach the nucleus, they accelerate noticeably, their velocity relative to the fixed speed of light cannot be ignored, and they experience a significant increase in mass. That mass increase leads to a greater attraction to the nucleus, the inner s shells are pulled in, and the outer s and p electrons follow. A side effect of the contraction of the s and p orbitals is increased shielding of the d and f shells from the nuclear charge; the d and f orbitals actually expand. The overall relativistic effect is generally an atomic radius contracted more than would be expected based solely on periodic trends (76, 79, 80). Since relativistic effects increase with Z^2, they are most important for the heavy elements; for example, the relativistic contraction becomes important somewhere near $Z = 60-70$ (80). Several reviews (76, 79, 81) as well as the text by Huheey et al. (80), provide a good introduction to relativistic effects. Greater mathematical detail is included in two other reviews by Pyykkö (82, 83). Because of the greatly increased complexity of relativistic quantum mechanics compared to nonrelativistic theory, the relativistic approach is only invoked when absolutely necessary.

Relativistic effects are indeed necessary, however, for explaining several of the characteristics of Pb(II). In particular, relativistic effects account for the so-called "inert pair effect" observed for lead (76, 79, 80). The "inert pair effect" describes the observation that elements in groups 14–17 (IVA, VA, VIA, VIIA), plus Tl in group 13 (IIIA), form ions with oxidation numbers two less than the noble gas configuration; for example, Pb(II) and Tl(I) are preferred over Pb(IV) and Tl(III). This characteristic has been termed the "inert pair effect" because early descriptions of this phenomenon remarked that it appeared as if two of the valence electrons (the $6s^2$ pair) had become inert, as if absorbed into the core (84). Later work has shown that the inert pair effect is likely due to relativistic effects and can be explained in terms of the binding energies of the s electrons: $6s$ electrons have larger binding energies than $5s$ electrons. As a result of this phenomenon, Pb(II) is significantly stabilized over Pb(IV) (76, 79, 84, 85, 86). Although some have argued that the hybridization of $6s$ and $6p$ orbitals is energetically unfavored, calculations by Fricke (87) and by Pyykkö and Desclaux (88) indicate that the $6s$ orbitals are not isolated from bond formation. These calculations suggest that the $6s$ orbital can mix with the $6p_{1/2}$ orbital, but not with the higher, empty $6p_{3/2}$ orbital (76, 82). Since recent studies on the spectroscopy of lead oxides also suggest that the $6s$ and $6p$ orbitals of Pb(II) mix (see below), we recommend that the term "inert pair" *only* be used when referring to the differential stabilization of Pb(II) over Pb(IV) and *not* be used to rationalize geometries.

D. Absorption Spectroscopy

Lead bound to cysteine residues in proteins results in the appearance of several intense absorption bands in the ultraviolet (UV) region of the absorption spectrum (37, 50–53, 89, 90). These absorption bands have been used successfully in our laboratory and elsewhere to monitor the stability of lead–protein interactions (37, 53) (discussed in detail in Sections IV and VI) but there has not yet been a detailed study of the electronic origin of these bands. Here, we review previous work on the theory of CT transitions in related systems and then apply those ideas to the spectra observed for lead in proteins.

1. Introduction to Possible Electronic Transitions for Lead

The absorption of visible (vis) or UV light by most inorganic complexes results in electronic transitions. These transitions may be between orbitals primarily associated with the metal (e.g., the familiar, weak d–d transitions), between orbitals primarily associated with the ligand (e.g., π–π^* transitions in aromatic ligands), or between a predominantly ligand orbital and an orbital that is primarily metallic in character (91). The latter transitions are referred to as ligand-to-metal charge-transfer (LMCT) or metal-to-ligand charge-transfer (MLCT). Charge-transfer transitions tend to be much more intense than ligand field transitions, as CT bands are typically symmetry allowed (92).

Most work in electronic spectroscopy has focused on the transition metals and their partially filled d orbitals, which can result in both d–d ligand field transitions and LMCT or MLCT bands to or from the metal d orbitals (93, 94). By contrast, relatively little attention has been focused on the spectroscopy of the posttransition metals [such as Pb(II), Tl(I), etc.]. In these elements, the d electrons are all part of the unreactive core; the outer shell consists of a lone pair of electrons in an s orbital [$6s^2$, in the case of Tl(I) and Pb(II)]. Both intraatomic transitions ($s^2 \rightarrow sp$) and interatomic CT transitions have been observed in posttransition metal complexes, including complexes of Pb(II) (95).

2. Optical Electronegativities

We can learn much about the electronic structure of Pb(II) complexes by comparing their spectra with those of other related systems. Optical electronegativities have been used in several cases to analyze the absorption spectra of metal–thiolate complexes. These studies provide qualitative (but critical) insights into the origin of the electronic transitions. The concept of optical electronegativity (χ_{opt}) was originally developed by Jørgensen (96–101) to describe electronic transitions in CT spectra and is analogous to the concept of bond energy electronegativity developed by Pauling (102, 103).

Optical electronegativities have been used extensively in transition metal complexes to determine the direction of charge transfer (i.e., LMCT vs MLCT) responsible for absorption bands. The frequency of the CT transition is proportional to the difference in optical electronegativities of the ligand and metal (104). Typically, the frequency of the putative CT transition and the optical electronegativity of the metal [$\chi_{opt}(M)$] are known and the optical electronegativity of the ligand [$\chi_{opt}(L)$] is calculated for complexes of the ligand with several different metal ions. Consistency in the values of $\chi_{opt}(L)$ obtained for each of these calculations strongly supports assignment of a transition as LMCT. By using this approach, McMillin (105) described the intense absorption of Cu(II)–, Co(II)–, and Ni(II)–azurin as a LMCT transition and calculated χ_{opt}(thiolate) = 2.60. He also used spectral data for oxidized spinach ferredoxin (106) and calculated χ_{opt}(thiolate) = 2.64; together these results suggest a general χ_{opt}(thiolate) ≈ 2.6 (105). In an unrelated study, Kennedy and Lever (107) examined complexes of Co(II), Ni(II), Zn(II), Cd(II), Hg(II), Pt(II), Bi(III), and Sn(IV) with mercaptopyridines. In order to determine the direction of CT, they used spectral data and the known optical electronegativities of mercury, tin, nickel, and cobalt to calculate the ligand optical electronegativities under both the assumption that the transfer is metal to ligand and that it is ligand to metal. They found that when using the MLCT assumption, the calculated ligand optical electronegativities were widely scattered and ranged from 0.58–0.85. This inconsistency, coupled with the fact that the accepted electronegativity of sulfur is three-to-four times higher than their calculated ligand electronegativities, suggested immediately that the MLCT assumption was incorrect (107). Under the LMCT assumption, however, the calculated ligand optical electronegativities were within the narrow range 2.35–2.50 and averaged 2.43. Jørgensen also reported previously that the optical electronegativity of the sulfide ion is near 2.4–2.6 (101). Since the data are much more consistent with the LMCT assumption, Kennedy and Lever concluded that the CT in this system is indeed ligand to metal (i.e., sulfur to metal) in origin.

What do optical electronegativities tell us about d^{10} metal–thiolate CT bands in proteins? By comparing the spectra of metalloproteins with those of Zn(II), Cd(II), and Hg(II) tetrahalides using Jørgensen's method (101), Vasák et al. (108) unambiguously identified the first resolved absorption band of Zn(II), Cd(II), and Hg(II) metallothioneins as an LMCT transition and suggested that there is also a second LMCT transition at higher energy. Additionally, Vasák et al. (108) reported the calculated metal optical electronegativities $\chi_{opt}(M)$ for Zn(II) (1.15), Cd(II) (1.27), and Hg(II) (1.5). The trend of increasing $\chi_{opt}(M)$ down the group of the periodic table agrees well with the trends of recalculated Pauling electronegativities (109) and is interpreted as corresponding to the increase in covalency of ligand–metal bonds (108). By using the value reported by McMillin (105) for the optical electronegativity of the cysteine thiolate

TABLE III

Observed Absorption Maxima (λ_{max}) for Metal–Thiolate CT Bands in Zinc Proteins (37, 52, 53, 108, 110, 111) and the Corresponding Calculated Optical Electronegativities [$\chi_{opt}(M)$][a]

	Zn(II)		Cd(II)		Hg(II)		Pb(II)		References
	λ_{max}(nm)	χ_{opt}(M)	λ_{max}(nm)	χ_{opt}(M)	λ_{max}(nm)	χ_{opt}(M)	λ_{max}(nm)	χ_{opt}(M)	
Metallothionein	231	1.15[b]	250	1.27[b]	303	1.5[b]	255	1.3	52, 108
Zinc finger consensus peptide	N.R.[c]		230	1.2	N.R.[c]		255	1.3	37, 111
Human immunodeficiency virus (HIV) nucleocapsid protein	N.R.[c]		240[d], 255[e]	1.2[d], 1.3[e]	N.R.[c]		255[d]	1.3[d]	37, 110, 111
CadC	N.R.[c]		240	1.2	N.R.[c]		240	1.2	53

[a] Optical electronegativities χ_{opt}(M) were calculated by Jørgensen's method (101) from the energy of the first absorption band and the reported value (105) of χ_{opt}(thiolate) = 2.6. The absorption bands in all cases result from metal–thiolate (cysteine) coordination with tetrahedral (or pseudotetrahedral) geometry. The trend in optical electronegativity Hg(II) > Cd(II) ≈ Pb(II) > Zn(II) corresponds to the relative covalency of the metal–ligand bonds and is in agreement with the scale of recalculated Pauling electronegativities (109).
[b] Calculated by Vasák et al. (108).
[c] Not reported = N.R.
[d] The N-terminal structural zinc-binding domain.
[e] Full length protein.

ligand [$\chi_{opt}(L) = 2.6$] and reported spectral data (37, 52, 53, 108, 110, 111), we have calculated the optical electronegativities for metals bound to cysteines in a series of zinc proteins. The results of this calculation are shown in Table III. In all cases, the absorption bands result from the coordination of a metal ion to cysteine(s) in a tetrahedral (or pseudotetrahedral) environment. The calculated metal optical electronegativities compare favorably with those of Vasák et al. (108) and the relevant Pauling electronegativities (109), and the trend in the calculated $\chi_{opt}(M)$ for the series of metals $Hg(II) > Pb(II) \approx Cd(II) > Zn(II)$ corresponds to the expected relative covalency of the metal–ligand bonds (112). These trends all point to the explanation that the absorption spectrum observed when d^{10} metals [such as Pb(II)] bind to sulfur in proteins is likely due—at least in part—to S $\rightarrow$ M charge transfer.

3. Solid-State Absorption Spectra

Although we can conclude from the optical electronegativities that LMCT transitions play a role in the absorption spectra of lead thiolates, it is not currently known to what extent intraatomic transitions may also contribute to these bands, or what orbitals are involved in the transitions. There have been no detailed theoretical studies of lead–thiolate CT bands, and very little work in general has been reported on the absorption spectroscopy of Pb(II) coordination complexes. Although the absorption spectra have been reported for solid-state Pb(II) alkaline earth sulfides (69–71), these data have not been analyzed extensively. In addition, much of the data that have been reported (69–71) are ambiguous due to insufficient information on sample preparation and composition (95).

However, detailed insights into the electronic transitions of lead–thiolate complexes can be gained from studies on Tl(I) [which is isoelectronic with Pb(II)] and Pb(II)-doped alkali halides in the solid state (Fig. 5) (37, 50, 54, 113). The details of the electronic transitions in Tl(I) doped alkali halides and related compounds have been studied extensively (both theoretically and experimentally) because these compounds have interesting luminescent properties and are useful in phosphors. [We will not discuss the emission spectra of these compounds, as they are not relevant to our discussion of lead-thiolate CT in coordination complexes; rather, the reader is directed to several extensive reviews of luminescence in doped alkali halide systems (95, 113, 114).] The characteristic absorption spectra of alkali halides doped with a Tl(I) type ion consist of four bands, known as the A, B, C, and D bands. The A band is at lowest energy, followed by B, C, and D respectively; the extinction coefficients of the bands follow the general trend $D > C > A > B$ (Fig. 6) (115, 116). Two weaker bands labeled D' and D'' are also shown in Fig. 6, which are attributed to the same CT transitions as the main D band (116). In Section II.E, we will

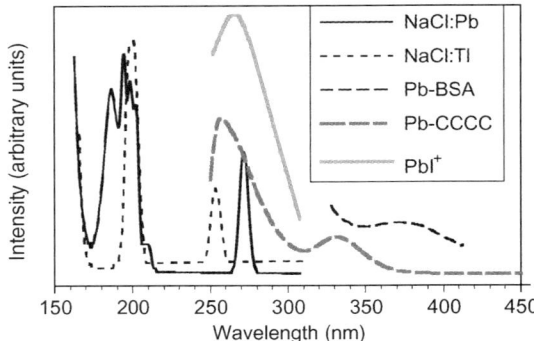

Figure 5. Representative absorption spectra of solid-state Tl(I) and Pb(II) doped alkali halides (NaCl:Tl and NaCl:Pb) and solution spectra of both PbI$^+$ and Pb^{II}–protein complexes (Pb–BSA (BSA = bovine serum albumin) and Pb–CCCC) reveal the similarities between the absorption bands in the different systems. Spectra used to assemble this figure were originally reported in references (37, 50, 54, 113). [Reprinted with permission from J. C. Payne, M. A. ter Horst, and H. A. Godwin, *J. Am. Chem. Soc., 121*, 6850 (1999), Copyright © 1999 American Chemical Society; I. M. Klotz, J. M. Urquhart, and H. A. Fiess, *J. Am. Chem. Soc., 74*, 5537 (1952), Copyright © 1952 American Chemical Society; and H. Fromherz and K.-H. Lih, *Z. Physik. Chem., A153*, 321–375 (1931), Oldenbourg Wissenschaftsverlag GmbH.]

concentrate on the details of the results obtained from absorption studies on these systems, the interpretation of these data, and insights that they provide into possible assignments of lead-thiolate CT bands.

The earliest theoretical treatment of thallium-doped alkali halides is due to Seitz (115), who interpreted the absorption spectra in terms of a substitutional

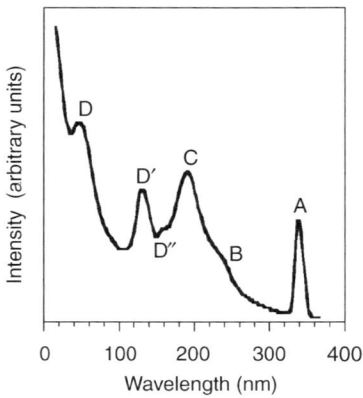

Figure 6. Absorption spectrum of Pb(II) doped KI, showing the relative positions and intensities of the A, B, C, and D absorption bands. The D' and D'' bands result from the same CT transitions as the more intense D band. [Adapted from (116).]

model. He assumed that the Tl(I) ions are distributed regularly throughout the alkali halide crystal and substitute for alkali metal ions in the crystal lattice. Seitz presented two possibilities for the source of the observed absorbance associated with thallium: either that the peaks are due to absorption by the thallous ion (i.e., intraatomic transitions) or that the absorption bands result from transfer of an electron from a halogen ion to the thallous ion (i.e., LMCT) (115). Although Seitz favored the first explanation, subsequent investigators have shown that the absorption is likely due to a combination of both intraatomic and CT transitions (57, 72, 95, 117, 118).

The refinement of the Seitz substitutional model took place over a number of years as further experiments were conducted and as the modern ideas of molecular orbitals developed. Williams provided the first quantitative application of the model to a specific case, Tl(I) doped KCl (119). Yuster and Delbecq examined Tl(I) doped KI at cryogenic temperatures and observed additional absorption bands not visible at room temperature. They explained this fine structure in terms of crystal field splitting of the Tl(I) absorbance bands in an extension of the Seitz model (120). Knox and Dexter (121) extended the Williams treatment of the Seitz model (119) to include predictions of oscillator strengths in the crystal, but their results do not correspond well with experimental results. These discrepancies were soon addressed by Williams et al. (122), who expanded the theory to include the effects of crystalline interactions and spin–orbit coupling. Including spin–orbit interactions was extremely important to improving the model. Then, in 1959, Knox showed that considering only the excited-state wave functions of free activator ions in an explanation of the properties of doped alkali halides is insufficient. Rather, it is necessary to allow interaction between different types of excitation states, for example, between an excited activator ion and a CT state (117). Knox concluded that the Seitz model is still valid but that more weight should be given to CT transitions relative to the intraatomic effects preferred by Seitz (115, 117).

In 1962, Sugano showed that the Seitz model (115) could be interpreted as a molecular orbital model (123), an interpretation that clarifies analysis of these systems. In this interpretation, the absorption bands observed in the Tl(I) doped alkali halide system come from the electronic transition $(a_{1g})^2 \rightarrow (a_{1g})(t_{1u})$, but the excited states are still calculated assuming an ionic interaction between the metal and the ligand. Since the thallium–chlorine bond is actually largely covalent, Bramanti et al. (118) modified the approach and used a semiempirical molecular orbital (MO) calculation to describe the energy levels of Tl(I) doped KCl. Molecular orbitals were constructed by the linear combination of atomic orbitals (LCAO) method from the $6s$ and $6p$ metal orbitals and the $3p$ chlorine orbitals. Initial calculations were conducted with the one-electron approximation; the method was then expanded to include Coulomb and spin–orbit interactions. The results of Bramanti et al. were consistent with experimental

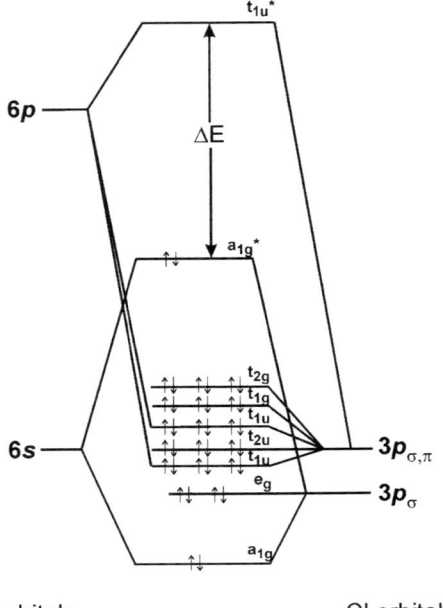

Figure 7. Molecular orbital energy-level diagram developed by Bramanti et al. (118) for thallium sites in Tl(I) doped KCl; both intraatomic and Cl–Tl charge-transfer processes are involved in the transitions observed in the absorption spectrum. [Adapted from reference (118).]

observations of the absorption spectrum and with the transition $(a_{1g}^*)^2 \rightarrow (a_{1g}^*)(t_{1u}^*)$, as seen in Fig. 7 (118). The excited Tl(I) ion has an electron population of $6s^{1.45}6p^{1.12}$, as opposed to the $6s^2 6p^0$ ground-state electron configuration (118). These results indicate that both intraatomic and Cl–Tl CT processes are involved in the observed absorption spectra. Overall, the covalent approximation of Bramanti et al. (118) is more satisfying conceptually than the previous ionic approximations (117, 123).

Although most of the work on this series of doped alkali halides has dealt with Tl(I), several studies have been conducted on Pb(II) doped systems (63–68). From this work, it is clear that for the most part, the same trends that have been observed and explained in the Tl(I) case are relevant to Pb(II) as well: that is, the same basic series of bands is observed in the absorption spectrum, and these bands are due to a combination of intraatomic and CT transitions. The C band is a multiplet that has been interpreted in terms of the Jahn–Teller effect (63); the Jahn–Teller effect may also lead to asymmetry in the A band (65, 66). Additionally, a fourth absorption band, known as the D band, is particularly apparent in Pb doped alkali halides (68). The D band has been interpreted both as a perturbation of the fundamental absorption band of the alkali halide crystal (114), and as a CT transition (67, 68, 116). Although it may have some character of a perturbed exciton (113), Tsuboi (67, 116, 124) shows by magnetic circular dichroism (MCD) spectroscopy and MO calculations that the D band is likely a LMCT band. Tsuboi's MCD studies also confirm the (at least partial) CT character of the A, B, and C bands (67). Taken together, these studies on solid-state Tl and Pb halides suggest that the observed absorption spectra are due to a combination of Pb intraatomic transitions and strong LMCT transitions.

In summary, the A, B, C, and D bands of alkali halides doped with Tl(I) type ions have been assigned to specific electronic transitions. The A, B, and C bands are all due to $(a_{1g}^*)^2 \to (a_{1g}^*)(t_{1u}^*)$ transitions. These transitions are from a molecular orbital with slightly more ligand character to one with slightly more metal character, that is, a LMCT. Multiple absorption bands arise because of the effects of spin–orbit coupling; both singlet–singlet (spin allowed) and singlet–triplet (spin forbidden) transitions take place. Transitions between these orbitals are also consistent with intraatomic transitions, but it is likely that the a_{1g}^* orbital is only partially metallic in character. The D bands are LMCT bands; the transition is $(e_g)^4 \to (e_g)^3 (t_{1u}^*)$. Using molecular term symbols, the electronic transitions in cubic symmetry are as follows (114):

A $^1A_{1g} \to {}^3T_{1u}$ (spin–orbit allowed)
B $^1A_{1g} \to ({}^3E_u + {}^3T_{2u})$ (spin–orbit and vibration allowed)
C $^1A_{1g} \to {}^1T_{1u}$ (dipole allowed, multiplet due to Jahn–Teller effect)
D $^1A_{1g} \to {}^1T_{1u}$, (dipole allowed); $^1A_{1g} \to {}^3T_{1u}, {}^1A_{1g} \to {}^3T_{2u}$ (spin–orbit allowed)

The interpretation used to explain the absorption spectra of Tl(I) and Pb(II) solid-state complexes can also be extended to aqueous solution spectra of Pb(II) complexes (Fig. 5) (54–60, 62). Although aqueous lead halide absorption spectra reported by Fromherz et al. (54, 55) were not originally interpreted in the context of CT spectroscopy (56, 125), in retrospect it is clear that the bands reported are the same as those observed in doped alkali halide crystals, as are later spectra reported by Bendiab et al. (59, 60). For example, in solution a spectral shift to longer wavelengths is observed for increasing atomic number of the halogen ligand (54, 55), which is consistent with a CT process (126).

The generality of these assignments can also be seen from the similar spectroscopic assignments that have been made for solid-state Pb doped alkaline earth fluorohalides and carbonates (72, 73). Two separate transitions have been observed in the absorption spectra, one of which has been assigned to a Pb(II) $s^2 \to sp$ intraatomic transition, the other of which appears to be a CT band.

These reports further indicate that the detailed explanations of electronic transitions in thallium and lead alkali halides are general and can be applied to a wide variety of related systems, such as the sulfides and thiolates, both in the solid state and in aqueous solution.

A similar interpretation is also likely to apply to the CT bands observed for lead bound to thiol groups in proteins. Although the symmetries of the molecular orbitals would be slightly different in tetrahedral or pseudotetrahedral symmetry, for example, than found in structural zinc-binding domains in proteins, the nature of the electronic transitions is expected to be the same. The parallels between the solid state and aqueous absorption spectra are dramatic (Fig. 5), and they indicate that the theory developed for solids is applicable to the extremely similar spectra observed in aqueous solution. Based on these parallels, and on the calculations of optical electronegativities, we feel that it is likely that the CT bands observed for lead bound to cysteine-rich sites in proteins are in the ligand-to-metal direction but may contain some intraatomic character. Further rigorous theoretical studies must be conducted, however, to confirm this hypothesis. The typical absorption spectrum for a lead-thiolate complex in a protein consists of two bands, an intense peak near 250 nm and a smaller peak near 340 nm. These bands are characteristic of CT transitions both in position and in their extinction coefficients. Based on comparisons to the doped alkali halides, these bands are likely derived both from intraatomic transitions from the Pb(II) $6s$ orbital to the Pb(II) $6p$ orbital and from interatomic CT transitions between mixed s and p orbitals with both metal and ligand character. There should be large spin–orbit coupling effects on the splittings of the energy levels. The energy levels derived from the lead and from the sulfur are likely very close, and it will require a detailed study of the lead-thiolate system using density functional theory, taking into account spin–orbit coupling, to know the ordering of the energy levels.

E. Photoelectron Spectroscopy

Complementary insights into the electronic structure of Pb(II) compounds can be obtained from studies by PES. These studies are particularly useful for assessing the degree to which the $6s^2$ orbital may (or may not) be mixed with p or d orbitals. This sort of orbital hybridization, or lack of it, would profoundly affect the geometry of Pb(II) complexes. In addition, the character of the valence orbitals should critically influence the reactivity of the ion. The extent to which relativistic effects are needed to qualitatively explain Pb(II) PES spectra is of particular interest.

Photoelectron spectroscopy is especially valuable when coupled with band structure theory for making assignments of molecular orbitals to the observed peaks in the spectra and determining orbital energies (127). When a sample is irradiated by a beam of energetic photons (generally X-rays or UV light), ionization occurs and electrons are ejected. From the known energy ($h\nu$) of the incident photons, the kinetic energy (E_k) of the ejected electron, and Einstein's photoelectric equation $h\nu = \text{BE} + E_k$, the binding energies (BE) of all the electrons in a molecule can be determined (128).

In a 1965 paper, Anderson and Gold (129) reported a detailed description of the Fermi surface and band structure of lead based on de Haas–van Alphen measurements and pseudopotential interpolations. They concluded that including spin–orbit interaction in the model is extremely important for accurately describing the Fermi surface. Later ab initio calculations (130) and experiments (131, 132) confirmed the necessity of taking into account relativistic effects, including spin–orbit interactions, when analyzing the band structure of lead. Studies by other investigators on a variety of simple lead compounds have confirmed the importance of considering relativistic effects in the interpretation of photoelectron spectra and band structure calculations (127, 133–140).

1. Relativistic Effects and the Myth of the $6s^2$ "Inert" Pair

Photoelectron spectroscopy has been particularly useful for providing insights into the role that the lead $6s$ lone pair of electrons plays in the solid-state structures of simple lead compounds. One question that arose from inspection of lead salts and ores is why PbS has the cubic rock salt structure, whereas PbI_2 has the hexagonal CdI_2 structure, and PbO has two distorted polymorphic forms, tetragonal α-PbO and orthorhombic β-PbO (141–143). Insights into the reasons for these different structures can be obtained by comparing the ordering of energy levels and the extent of orbital mixing in these compounds. The extent of orbital mixing is also of inherent interest, since it provides insights into whether the $6s$ pair of electrons is energetically isolated or whether this orbital in fact mixes significantly with the $6p$ orbitals. Each of these parameters is available from PES.

Although there have been a number of studies of the PES of simple Pb(II) compounds, especially the halides, oxides, and chalcogenides, there is still debate over the proper interpretation of these results, particularly in the assignment of PES peaks to atomic and molecular orbitals. Most of this debate centers over whether it is necessary to include relativistic effects to accurately interpret PES spectra of Pb(II) compounds. Although Breeze (144) suggested in 1974 that spectra based on the density of states, such as PES, could be interpreted with little reference to relativistic effects, other authors have found that taking such effects into account noticeably improves the agreement between theory and experiment (127, 133–140), and hence are more likely to provide accurate assignments of PES peaks to atomic and molecular orbitals.

2. Lead(II) Oxides

The PES of lead(II) oxides have been studied in detail. Whether or not relativistic effects are taken into consideration, however, has a critical influence on the interpretation of the photoelectron spectra of Pb(II) oxide as well as the interpretation of the effects of orbital mixing on the solid-state structure of this compound.

Terpstra et al. (143) reported photoelectron spectra and band structure calculations for the lead monoxides, but they did not take relativistic effects into account. They calculate that in the valence bands of both α-PbO and β-PbO, there is significant mixing of the O $2p$ orbitals with both Pb $6s$ and Pb $6p$ orbitals. From these results, they conclude that the Pb(II) $6s^2$ lone pair actually plays an active role in the chemical bonding and is not truly an "inert" pair; the distortions in solid-state PbO from high-symmetry cubic structure to the observed lower symmetry tetragonal (α-PbO) or orthorhombic (β-PbO) structures are then attributed to these hybridization effects. Likewise, Watson et al. (145) concluded based on nonrelativistic ab initio calculations that the distortion in α-PbO (with an idealized CsCl-type structure) does not arise from hybridization of the lead $6s$ and $6p$ orbitals; rather, it comes from mixing of the lead $6s$ with the oxygen $2p$ orbitals. These are intriguing conclusions; but since neither Terpstra et al. (143) nor Watson et al. (145) included relativistic effects in their treatment, one must be skeptical of their conclusions regarding energy levels and orbital overlaps.

On the other hand, Dybowski et al. (140) recently used ab initio calculations, taking into account correlation and relativistic effects, to explain the ^{207}Pb NMR and X-ray absorption near edge spectroscopy (XANES) spectra of Pb(II) oxides. These calculations show that $6s$–$6p$ mixing not only accounts for the distortions of β-PbO from cubic symmetry, it can explain the *different* distortion in α-PbO. In β-PbO, the Pb(II) ion is coordinated to two O^{2-} ions, and the term nearest in energy to the $6s_{1/2}$ orbital is the $6p_{3/2(m\,=\,3/2)}$ orbital; in α-PbO, by contrast, the Pb(II) is coordinated to four O^{2-} ions in a square-planar geometry, and the closest energy term to $6s_{1/2}$ is the $6p_{3/2(m\,=\,1/2)}$ orbital (140). Hybridization of the Pb $6s$ orbital with one or the other of these $6p$ orbitals with different angular dependences (the $6p_{3/2(m\,=\,1/2)}$ "dog-bone" vs the $6p_{3/2(m\,=\,3/2)}$ "doughnut") leads to the stereochemical activity of the Pb valence pair of electrons and the distortions from high symmetry observed in the solid state. The good agreement between theory and experiment seen in this study (140), coupled with the earlier work that showed the importance of relativistic effects in calculations for lead (129–132), support the calculations and conclusions of Dybowski et al. (140) over the nonrelativistic approaches of both Watson et al. (145) and Terpstra et al. (143).

3. Lead(II) Halides

The PES and band structure of PbI_2 and the other Pb(II) halides have also been well studied (133, 135, 136, 138, 139, 146–148). Even so, there continues to be controversy in the literature regarding the *detailed* interpretation of the nature of the valence band (139). Itoh et al. (139) recently reported UV photoelectron spectra of PbF_2 and $PbCl_2$ using a synchrotron light source. Based on these high-resolution data, Itoh et al. (139) suggest that the central part of the valence band has primarily Cl^- $3p$ character, while the upper and lower parts are mainly due to the Pb(II) $6s$ orbitals mixed with a small amount of Cl^- $3p$ character. Previous studies of the PES and band structure of lead halides are in good agreement with these qualitative results; several investigators (136, 138, 146, 148) proposed a valence band structure that is constructed from a mixture of Pb(II) $6s$ and halide p orbitals. The calculations of Nizam et al. (138) also agree well with their ^{207}Pb NMR chemical shift data. It is clear based on the results from studies of lead(II) oxides, however, that additional theoretical work (including relativistic effects) needs to be performed on lead halide valence bands in order to clarify the degree of orbital mixing and to quantify the ordering of the orbitals in the valence band.

4. Lead(II) Chalcogenides

Results of studies of the Pb(II) chalcogenides suggest that orbital mixing is not as pronounced in PbS as in PbO and PbI_2 and indicate a strong link between the degree of orbital mixing and the structural distortion of these Pb(II) compounds (i.e., those compounds that have a higher degree of orbital mixing exhibit more distorted structures) (127, 134, 137, 142, 149, 150). McFeely et al. (127) used high-resolution X-ray PES to examine the valence band density of states for PbS, PbSe, and PbTe. They determined that the experimental results agreed well with results from several theoretical methods but that the best agreement came from relativistic orthogonalized-plane-wave calculations (127). Kohn et al. (134) and Grandke et al. (137) also concluded that calculations including relativistic effects agree best with the experimental PES data for lead chalcogenides. From the combination of theory and experiment, McFeely et al. (127) concluded that the valence band of PbS is composed of three p-like bands and two lower energy s-like bands. Although McFeely et al. (127) did not explicitly indicate what degree of mixing might occur between orbitals, no mixing was implied. Ettema et al. (142) more recently reported, based on scanning tunneling microscopy (STM) studies of the PbS (001) surface, that the PbS valence band is primarily composed of S $3p$ states and the conduction band of Pb $6p$ states. Taken together, these studies suggest that the most likely assignment for the CT transition(s) observed for lead-thiolate compounds is S

$3p \to$ Pb $6p$, possibly with some mixing of Pb $6s$ character into the Pb $6p$ orbitals. However, detailed experimental and theoretical studies on lead-thiolate PES and absorption spectra are needed to test this hypothesis.

Since the orbital energies and shapes for a heavy element like lead depend on relativistic effects, and the photoelectron spectra are based on orbital binding energies, it should not be surprising that theories that take relativistic effects into account agree better with experiment than those that do not. Indeed, PES studies and calculations reveal that relativistic effects are required to explain the properties of PbO [and presumably other Pb(II) compounds as well]. In addition, recent PES studies clearly demonstrate that the distortions observed for PbO (and the stereochemical activity of the lone pair) arise from mixing of $6s$ and $6p$ orbitals, again suggesting that the term "inert pair" should not be used in reference to orbital mixing in Pb(II).

F. Vibrational Spectroscopy (IR and Raman)

Infrared spectroscopy provides a valuable qualitative indication of the degree of covalency of a given Pb–ligand interaction and also has been used (with much less success) to determine the coordination number of Pb(II). There have been limited studies of Pb(II) complexes by IR (149, 151–162) and Raman (163–166) spectroscopies. The interpretation of these results is most successful when coupled with structural data from X-ray crystallography, X-ray absorption spectroscopy [extended X-ray absorption fine structure (EXAFS) or X-ray absorption near edge spectroscopy (XANES)], PES, or NMR spectroscopy (149, 154, 155, 157, 158, 160–162).

Most vibrational studies on Pb(II) complexes to date have focused on carboxylate compounds [e.g., Pb(II) ethylenediamine-N,N,N',N',-tetraacetic acid, Pb^{II}–EDTA, and Pb^{II}–nitrilo-2,2',2''-triacetic acid, Pb^{II}–NTA]. In general, in carboxylate compounds the carbonyl stretch shifts to higher energy as the carboxylate oxygen–metal interaction becomes more covalent (151, 167, 168). This characteristic has been observed in several Pb(II) complexes (151, 154, 159–161). These studies indicate that in both Pb^{II}–EDTA and Pb^{II}–NTA chelates, the Pb(II) ion is bound fairly ionically to oxygen and covalently to nitrogen (159, 160). In Pb^{II}–iminodiacetate complexes, the Pb–O bonds are also mostly ionic (161). Tajmir-Riahi concluded from a study of a series of related carboxylate complexes that D-glucurono-6,3-lactone, D-glucono-1,5-lactone, and D-glucuronic acid bind Pb(II) covalently, but that D-glucuronate and D-gluconate form ionic salts with Pb(II) (154).

Several authors have also attempted to report the coordination number of lead bound to EDTA based on solely on IR spectra (151, 159, 169). Their interpretations, however, conflict with the reported crystal structure of Pb^{II}–EDTA, in which the Pb(II) ion is either seven- or eight-coordinate depending on whether a monomeric or a dimeric species is formed (170). For dimeric Pb^{II}–EDTA in the solid state, the Pb(II) ion is coordinated to a hexadentate EDTA ligand, a water molecule, and a carboxylate oxygen from a second EDTA molecule. The monomeric structure is similar to that of the dimer except lacking the extra carbonyl oxygen. In its place, in the eighth coordinate position, appears to be the Pb(II) stereochemically active lone pair. By contrast, the lone pair is *not* stereochemically active in the dimeric structure (170). From earlier IR studies, however, Langer concluded that Pb(II) bound to EDTA is four coordinate in the solid state (powdered sample) and six coordinate in solution (169). Dragan and Fitch (159) agreed with Langer's interpretation and also suggest an alternate five-coordinate solid-state structure in which the fifth position is occupied by the Pb(II) lone pair. Although it is possible that the presence of solvent molecules in the crystal lattice leads to a different structure in a single crystal than in a powdered sample, it seems much more likely that determinations of coordination number based on the IR spectra are simply inaccurate.

Raman spectra of Pb^{II}–EDTA are consistent with strong ionic Pb—O interactions in the complex (163, 164), but provide much more accurate predictions of lead coordination number. From Raman spectra of Pb^{II}–EDTA, Krishnan and Plane (163) suggest that the Pb(II) is six coordinate in solution; and McConnell and Nuttall (164) conclude that $Na_2PbEDTA \cdot 2H_2O$ has a seven-coordinate Pb(II) ion, with six positions taken up by ligand bonds and the seventh filled by the lone pair. These results agree well with the structural conclusions from NMR spectroscopy (171) and X-ray crystallography (170). These results further emphasize that the information in the IR spectra alone cannot be interpreted sufficiently well to make conclusions about coordination number but are better interpreted along with structural information from other techniques (e.g., NMR spectroscopy, EXAFS, XANES, X-ray crystallography).

G. Nuclear Magnetic Resonance Spectroscopy

Lead-207 NMR spectroscopy does not provide the details of electronic structure that can be gleaned from PES or atomic transitions, but can—at least in theory—provide complementary information about the Pb(II) coordination environment in solution. Lead-207 NMR spectroscopy is an excellent probe for the lead coordination environment due to the characteristics of the ^{207}Pb isotope ($I = \frac{1}{2}$): Excellent receptivity (11.7 times greater than that of ^{13}C), high natural abundance (22.6%), and large chemical shift range (>16,000 ppm) (172, 173). Although ^{207}Pb NMR spectroscopy has been used extensively to characterize alkyl Pb(IV) derivatives and solid-state Pb(II) compounds, relatively few studies have been conducted of soluble Pb(II) coordination compounds (171–183). Nonetheless, those studies that have been conducted to date on Pb(II) coordination compounds in solution reveal that the ^{207}Pb chemical shift is particularly

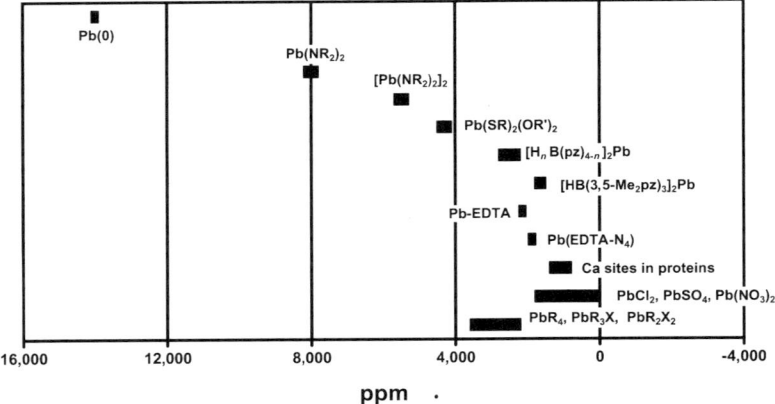

Figure 8. Lead-207 NMR chemical shifts of selected lead compounds referenced to $Pb(NO_3)_2$ (0 ppm). Lead chemical shifts can cover a range of ~16,000 ppm and are very sensitive to the lead coordination environment. The lead resonance tends to shift *upfield* as the electronegativity of the ligand atom increases (S > N > O). [Reprinted with permission from E. S. Claudio, M. A. ter Horst, C. E. Forde, C. L. Stern, M. K. Zart, and H. A. Godwin, *Inorg. Chem.*, 39, 1391–1397 (2000). Copyright © 2000 American Chemical Society.]

sensitive to the type of donor atoms surrounding the lead ion (Fig. 8) (39). In addition, coupling constants and two-dimensional (2D) spectroscopies can provide information about the lead coordination environment and may eventually prove useful for deducing lead coordination environments in complex samples (e.g., proteins).

1. Lead-207 Parameters: Chemical Shifts and Coupling Constants

The lead chemical shift provides an excellent measure of the type of donor atoms bound to lead. As in other heavy metal NMR spectroscopies (e. g., ^{113}Cd, ^{199}Hg) (184, 185), the ^{207}Pb resonances become more deshielded and shift downfield as the electronegativity of the donor atom decreases—opposite the effect observed in ^1H NMR spectroscopy (186, 187). This trend results from mixing of states in lead due to orbital angular momentum, which results in a paramagnetic term in the Ramsey equation (187). Consequently, the ^{207}Pb resonance shifts downfield as donor atoms are changed from oxygen to nitrogen to sulfur (Table IV) (39, 178, 181, 182, 188–190). Although the Pb(II) $6s$ lone pair of electrons is expected to also have a profound effect on the ^{207}Pb chemical shift, only one report to date addresses this question. Reger et al. (178) showed that in a series of similar compounds, the presence of a stereochemically active lone pair shifts the lead resonance downfield (see Table IV). However, in the absence of other examples, it is not clear whether or not this effect is a general one.

TABLE IV

Lead-207 Chemical Shifts for Selected Lead(II) Compounds, Referenced to the Chemical Shift of $Pb(NO_3)_2$[a]

Compound	Coordination Number	Donor Atoms	^{207}Pb Chemical Shift (ppm)	References
$Pb(R)C_6H_3$-2,6-$Trip_2$[b]	2	1 C, 1 Br	6657–7420	220
$Pb(SR)_2(OR')_2$	4	2 O, 2 S	4516–4454 (1555–1493)	181, 182
$Pb(R,R'$-dithiocarbamate$)_2$[c]	4	2 N, 2 S	3367–3575 (406–614)	188
$[B(pz)_4]_2Pb$	4	4 N	2427 (−534)	177
$[H_nB(pz)_{4-n}]_2Pb$[d] ($n = 1, 2$) with active lone pair	6	6 N	$n = 1$:2066 (−895) $n = 2$:2427 (−139)	177
$[HB(Me_2pz)_3]_2Pb$[d] with inactive lone pair	6	6 N	1480 (−1481)	177
$Pb[EDTA]^{2-}$	7–8 (1–2·water)	2 N, 5–6 O	2441	39
$Pb[EDTA-N_2]$	6 (No. water unknown)	2 N, 5–6 O	2189	39
$Pb[EDTA-N_4]^{2+}$	6	2 N, 4 O	1764	39
$Pb-NtnOtnH_4$[e] (16-member macrocycle)	4	2 N, 2 O	230	189
$PbNenOdienH_4$ and $PbNtnOdienH_4$[f] (17- and 18-member macrocycles)	5	2 N, 3 O	206–222	189

[a] If original referencing was to Me_4Pb, then the shift relative to Me_4Pb is also given in parentheses.
[b] R = Me, t-Bu or Ph; Trip = C_6H_2-2,4,6-i-Pr_3.
[c] R = Et, n-Pr, i-Pr, n-Bu, i-Bu, n-Pen, c-Hx.
[d] Pyrazol = pz.
[e] $NtnOtnH_4$ = 1,13-diaza-3,4:10,11-dibenzo-5,9-dioxacyclohexadecane.
[f] $NenOdienH_4$ = 1,15-diaza-3,4:12,13-dibenzo-5,8,11-trioxacycloheptadecane, $NtnOdienH_4$ = 1,15-diaza-3,4:12,13-dibenzo-5,8,11-trioxacyclooctadecane.

In theory, valuable information about the Pb(II) coordination environment can be determined from the Pb—X coupling constant (J). Typically, coupling constants provide a measure of bond strength and sterics; however, insufficient reports have appeared in the literature to date to observe trends for Pb(II) coordination compounds. Constants have been reported for coupling of ^{207}Pb to ^{1}H and ^{13}C through up to three bonds in Pb(II) compounds: Pb^{II}–EDTA and the lead complexes of amido derivatives of EDTA exhibit $^{3}J(^{207}\text{Pb}-^{1}\text{H})$ of 15–17 Hz (39), lead–salen (salen = bis(salicylidene)ethylenediamine) complexes exhibit $J(^{207}\text{Pb}-^{1}\text{H}) = 14$ Hz and $^{3}J(^{207}\text{Pb}-^{1}\text{H}) = 46$ Hz (191). In addition, Power and co-workers (192) reported coupling constants for diorganolead(II) compounds [$J(^{207}\text{Pb}-^{13}\text{C}) = 248$ Hz and $J(^{207}\text{Pb}-^{1}\text{H}) = 40$ Hz] that are comparable to the coupling observed in the Pb(IV) analogue, tetramethyllead coupling [$J(^{207}\text{Pb}-^{13}\text{C}) = 250$ Hz and $J(^{207}\text{Pb}-^{1}\text{H}) = 60$ Hz] (193). Although these

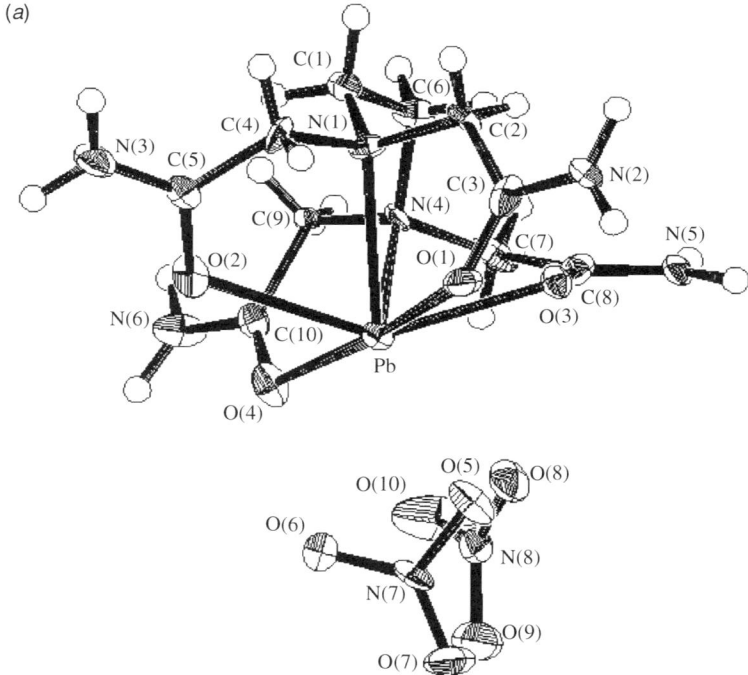

Figure 9. The Pb(II) complex [Pb^{II}–EDTA–N$_4$]$^{2+}$ has been characterized both (a) crystallographically and (b) by two-dimensional ^{207}Pb–^{1}H heteronuclear multiple quantum (HMQC) NMR spectroscopy. The technique is sensitive enough to detect Pb(II) coupling to methylene protons through three bonds to the protons on the acetoamido arms (–CH$_2$CONH$_2$) and ethylene backbone (–NCH$_2$). [Reprinted with permission from E. S. Claudio, M. A. ter Horst, C. E. Forde, C. L. Stern, M. K. Zart, and H. A. Godwin, *Inorg. Chem.*, 39, 1391–1397 (2000). Copyright © 2000 American Chemical Society.]

data are insufficient in number to determine general trends or correlations to structural properties, they are encouraging in that they suggest that a more comprehensive survey of coupling constants in Pb(II) coordination compounds might allow for such parallels to be drawn in the future.

In addition, the first 2D ^{207}Pb–^{1}H NMR spectrum was recently reported (Fig. 9) (39). In this study, the ^{207}Pb–^{1}H HMQC NMR spectrum (194) was used to determine the sources of couplings observed in the one-dimensional (1D) ^{1}H NMR spectrum of Pb(II) ethylenediaminetetraacetamide [Pb^{II}–EDTA–N$_4$] and confirmed that ^{207}Pb–^{1}H coupling can be observed through at least three bonds. In addition, the 2D spectrum assisted in predicting the coordination of Pb(II)— which was then confirmed by X-ray crystallography (Fig. 9) (39)—suggesting that this technique may provide a potentially useful new method for characterizing lead-binding sites in complex samples (e.g., proteins).

3. Lead-207 NMR Spectroscopy of Proteins

The first application of ^{207}Pb NMR spectroscopy to metalloproteins was reported relatively recently by Vogel and co-workers (173), who used this

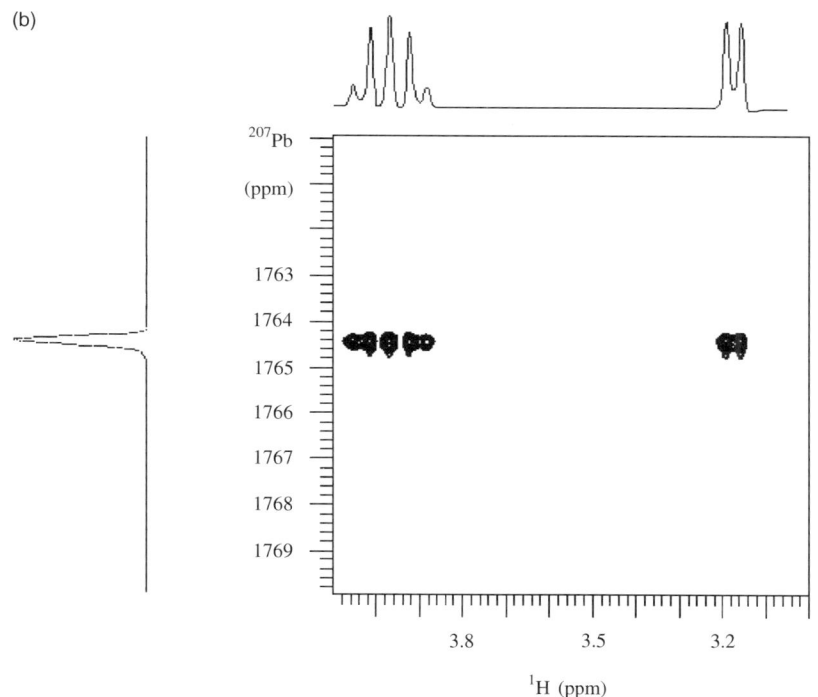

Figure 9. (Continued)

technique to study lead binding to a series of calcium proteins, including calmodulin. The sensitivity of ^{207}Pb chemical shifts to the local chemical environment can be seen from the addition of ^{207}Pb(II) to apo-calmodulin: Although the chemical environments of the four metal-binding sites in calmodulin are extremely similar, four distinct ^{207}Pb resonances are observed (Fig. 10). The other calcium-binding proteins reported in this study also exhibited different ^{207}Pb chemical shifts (747–1262 ppm). In all cases, the ^{207}Pb resonances were significantly upfield from resonances that had been previously reported for small molecules containing donor nitrogen or sulfur, reflecting the hard, oxygen-rich coordination environments found in calcium proteins (Fig. 8) (173).

This study suggests that ^{207}Pb NMR spectroscopy should be a useful technique for characterizing lead-binding sites in proteins; however, no other reports of protein ^{207}Pb NMR spectra have appeared to date. In particular, it would be very interesting to characterize the nature of the lead coordination environment in zinc proteins (see Section IV) using NMR spectroscopy. Because these sites tend to be cysteine- (thiol-) rich, we would expect ^{207}Pb bound to zinc proteins to exhibit resonances that are downfield of those for lead bound to calcium proteins. Ideally, the sensitivity provided by the indirect detection of the 1D HMQC experiment [analogous to the 2D experiment reported by Claudio et al. (39)] should allow for more rapid sweeping of the sizable ^{207}Pb chemical shift window when searching for the resonances of these sites.

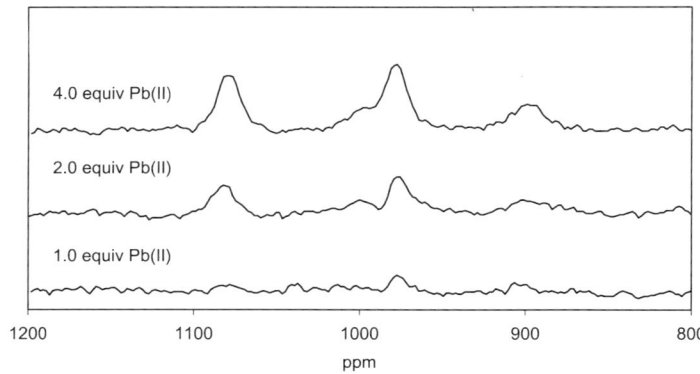

Figure 10. Lead-207 NMR spectra (104.435 MHz) of calmodulin with one, two, or four equivalents of ^{207}Pb(II) reveal that lead binds to all four sites in calmodulin simultaneously. Despite the similarity of the chemical environments in each of the four binding sites, the lead ion bound to each site exhibits a distinct ^{207}Pb chemical shift. [Reprinted with permission from J. M. Aramini, T. Hiraoki, M. Yazawa, T. Yuan, M. Zhang, and H. J. Vogel, *J. Biol. Inorg. Chem.*, 1, 39–48 (1996). Copyright © 1996 Society of Biological Inorganic Chemistry.]

H. Spectroscopy Conclusions

Taken together, the studies that have been conducted to date on the absorption, photoelectron, vibrational, and NMR spectra of Pb(II) compounds suggest that divalent lead is far from "spectroscopically silent". These studies have provided useful insights into the electronic structure of Pb(II) compounds and the role that relativistic effects play in determining the properties of Pb(II). In addition, they provide useful tools not only for characterizing the fundamental coordination chemistry of lead (Section IV), but also for determining lead coordination environments in complex environmental (Section V) and biological (Section VI) samples.

III. STRUCTURAL STUDIES ON LEAD(II) COMPLEXES

A. Introduction

Compared to those of many transition metal ions, the structures of Pb(II) compounds have received relatively little attention. However, pioneering structural studies by Raymond, Reger, and others have revealed that Pb(II) exhibits a rich and interesting coordination chemistry (39, 170, 176–183, 195–198): lead(II) binds to a broad range of donor groups, assumes a wide range of coordination numbers, and can exhibit a stereochemically active 6s lone pair. A survey of the Cambridge Structural Database (CSD) (199, 200) reveals 295 Pb(II) structures for which coordinates have been deposited, including both extended structures and discrete complexes. In addition, lead has been used extensively to prepare heavy atom derivatives to aid in the solution of three-dimensional structures of protein molecules; the environments occupied by lead in these structures provide important insights into the fundamental coordination chemistry and biochemistry of Pb(II). Other diffraction techniques, such as EXAFS spectroscopy, have also provided critical insights into Pb(II) coordination chemistry, particularly in complex samples.

B. Questions of Interest

Here, we review the structural studies that have been conducted to date on Pb(II) coordination compounds, with particular emphasis on addressing the following questions:

- What are commonly observed donor ligands for Pb(II)?
- What are the most commonly observed coordination numbers and geometries for Pb(II) compounds?

- How do coordination number and geometry correlate with donor type?
- When is the lone pair of electrons on Pb(II) stereochemically active?
- How can these insights be used to rationally design chelation therapy agents for lead?
- Are the trends exhibited by Pb(II) in small molecule structures also seen in biological molecules?
- What do lead-binding sites in biological molecules tell us about the biological chemistry of lead?
- What insights into lead coordination chemistry can be gained from EXAFS spectroscopy that are not available from X-ray crystal structures?

C. X-Ray Crystal Structures of Lead(II) Small Molecule Complexes

Recent reviews by Holloway and Melnick (201) and Glusker and co-workers (197) survey both Pb(IV) and Pb(II) coordination compounds with the goal of understanding the preferred stereochemistry and structure of the compounds. Here, we focus on the properties of Pb(II); the previously reported results are updated by our own survey of the recent literature and search of the CSD (199, 200) (Table V).

1. Commonly Observed Donor Groups in Lead(II) Structures

According to Pearson's classifications of hard and soft acids and bases (202), Pb(II) is considered to be an intermediate acid, and as such is expected to bind to a broad range of donor atoms. In general, common ligands for Pb(II) have donor atoms of O, N, S, P, Cl, Br, or I, or some combination thereof (Table V) (201). Despite the fact that most inorganic chemists think of lead's chemistry as being dominated by its thiophilicity, by far the most common donor atom observed is oxygen, which is present in the coordination sphere of 48% of all Pb(II) compounds deposited in the CSD. By contrast, 33% of all Pb(II) compounds contain one or more nitrogens as donor atoms and only 16% of all Pb(II) compounds contain one or more sulfurs as donor atoms. Average Pb—X bond lengths for each donor atom type as a function of denticity of the ligand and coordination number of lead are listed in Table VI (201).

2. Commonly Observed Coordination Numbers and Geometries in Lead(II) Structures

Unlike Pb(IV) compounds, which tend fairly universally to be four-coordinate and tetrahedral, Pb(II) compounds exist as both discrete and polynuclear complexes and exhibit a broad range of coordination numbers, from 2 to 12

TABLE V

Number of Lead(II) Complexes Found in the CSD, Grouped According to Coordination Number and Common Donor Ligands

Donor Atom(s)	Coordination Number											
	2	3	4	5	6	7	8	9	10	11	12	Total
N only	1	2	3	1	2	1	a	1	a	a	a	11
O only	22	5	8	6	6	4	2	a	7	1	a	61
S only	3	1	5	2	1	1	a	a	1	a	a	14
X (= F, Cl, Br, I) only	14	a	a	a	1	a	a	a	a	a	a	15
C only	6	5	8	a	a	a	a	a	5	a	a	24
N and O	a	1	3	2	3	9	10	10	3	a	a	41
N and S	a	a	1	a	1	3	1	a	a	a	a	6
N and C	a	2	6	3	a	a	a	a	a	a	3	14
N and X	a	a	2	a	17	2	a	a	a	a	a	21
N and Other	a	a	1	a	1	3	a	a	a	a	a	5
N and O and S	a	a	1	1	1	a	2	a	a	a	a	5
N and O and X	a	a	a	1	1	2	a	a	a	a	a	4
N and O and other[b]	a	a	a	1	a	a	a	a	a	a	a	1
N and S and X	a	a	a	1	a	a	a	a	a	a	a	1
O and S	a	a	7	a	2	1	2	a	a	a	a	12
O and C	a	a	a	2	1	1	a	a	a	a	a	4
O and X	a	a	a	2	4	1	2	a	a	a	a	9
O and other[b]	a	a	3	a	1	1	a	a	a	a	a	5
S and C	a	a	6	a	a	a	a	a	a	a	a	6
S and X	a	1	a	a	a	1	a	a	a	a	a	2
S and other[b]	a	a	a	2	a	a	a	a	a	a	a	2
Other[b]	1	5	17	5	1	3	a	a	a	a	a	32
Total number of Pb(II) compounds	47	22	71	29	43	33	19	11	16	1	3	295
% of Total	16	7	24	10	15	11	6	4	5	0	1	100

[a] This combination of donor atom type(s) and coordination number was not found in the CSD.
[b] Where "other" indicates, for example, P, B, a metal, or some combination thereof.

(Table V) (197, 201). The most common coordination numbers (Table V) are two [16% of all Pb(II) compounds reported in the CSD], four [24% of all Pb(II) compounds], and six [15% of all Pb(II) compounds]. Interestingly, Pb(II) compounds with coordination numbers of 4 and 6 do not typically exhibit "standard" tetrahedral or octahedral geometries. High coordination numbers are occasionally seen [10% of all Pb(II) compounds in the CSD exhibit coordination numbers >8] but are highly unusual if sulfur is present in the Pb(II) coordination sphere. Only one compound with a coordination number >8 has sulfur in the first coordination sphere and 80% (24 out of 30) of the reported structures that contain a sulfur donor have a coordination number <6 (Table V). Of course, the

TABLE VI

Average Pb—X Distances in Angstroms (Å) in Lead(II) Coordination Compounds According to Atom, Denticity of Ligand, and Lead Coordination Number[a]

Donor Atom	Denticity of Ligand	Coordination Number						
		2	3	4	5	6	7	8
O	Mono	b	2.40	2.52	2.53	2.53	2.65	2.76
	Bi	b	b	b	2.64	2.55	2.77	b
	Penta	b	b	b	b	2.53	b	2.72
	Hexa	b	b	b	b	b	2.75	2.76
	Octa	b	b	b	b	b	b	2.54
N	Mono	2.24	2.40	2.44	2.58	2.85	2.82	2.55
	Bi	2.55	2.45	2.49	2.49	b	2.60	b
	Tri	b	2.43	b	2.63	2.57	b	b
	Tetra	b	b	2.46	2.54	b	b	2.75
	Penta	b	b	b	b	2.53	b	b
	Hexa	b	b	b	b	2.64	2.75	b
	Octa	b	b	b	b	b	b	2.76
S	Mono	b	2.56	2.705	2.92	3.08	2.97	3.00
	Bi	b	b	2.86	2.88	2.93	2.96	b
C	Mono	2.37	2.37	b	b	b	b	b
P	Bi	b	2.77	2.83	b	b	b	b
Cl	Mono	b	b	b	b	2.87	2.69	2.78
Br	Mono	b	2.82	2.96	b	b	b	b
I	Mono	b	b	3.13	b	3.22	2.90	b

[a] Adapted from (201).
[b] This combination was not found in a search of the CSD.

"coordination number" is somewhat subjective: not surprisingly, bond lengths tend to increase as the coordination number increases (Table VI) and in many cases some of the atomic interactions that have traditionally been considered to be "bonds" in high coordination number compounds (197) have lengths that are longer than the sum of the van der Waals radii for lead and the respective donor atom. Nonetheless, it is clear from this survey that lead adopts a wide range of coordination numbers, especially when oxygen and nitrogen serve as donor groups, and that the geometries of the compounds tend to be irregular.

The unusual coordination chemistry of Pb(II) is illustrated by the complex Pb(II) forms with EDTA. Although EDTA is typically considered to be the prototypical hexacoordinate ligand, two molecules of Pb^{II}–EDTA are observed in the crystal structure—one monomeric and one dimeric—neither of which are octahedral, or even six coordinate (Fig. 11) (170, 203). The monomeric form exhibits a coordination number of 7, with the seventh ligand provided by a coordinated water molecule. Each lead atom in the dimeric form is *eight* coordinate, with the seventh donor atom provided by water and the eighth provided by a carboxylate group from an adjacent EDTA molecule. Furthermore, whereas the eight donor atoms surrounding each Pb(II) in the dimeric form are distributed roughly evenly (spherically, or "holodirected") around the central lead ion, the seven donor groups on the monomeric form are bunched to one side (hemidirected), leaving an "open" coordination site, the hallmark of a stereochemically active 6s lone pair.

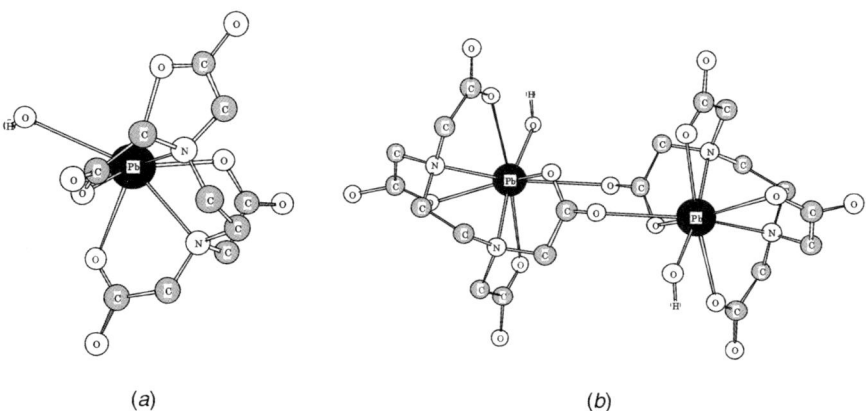

Figure 11. The crystal structure of Pb^{II}–EDTA contains both a monomeric form and a dimeric form of the compound (170, 203). The monomeric form (*a*) has coordination number seven with oxygen from a coordinated water providing the seventh donor ligand. The dimeric form (*b*) has a coordination number eight with the eighth ligand provided by an oxygen from a carboxylate group from an adjoining molecule. Coordinates downloaded from the CSD.

Why are Pb(II) structures so interesting and irregular? Lead's large size [Pb(II) has an ionic radius of 1.19 Å (35)] allows high coordination numbers without the steric effects encountered by smaller metals. In addition, a heavy atom such as lead is subject to relativistic effects, which change the size, shape, and energetics of the s and p orbitals. In addition, the $6s^2$ "lone pair" can be stereochemically active, thus drastically distorting expected geometries (see Section II.E).

3. *The Structural Effects of the Lead(II) $6s^2$ Electron Pair*

Whether or not the $6s^2$ electron pair is "stereochemically active" has a profound influence on the geometry observed for Pb(II) coordination compounds, as was first observed in 1972 by Lawton and Kokotailo (204). Unfortunately, the terminology used to describe the behavior of the $6s^2$ pair of electrons is confusing. Although the terms "inert pair" and "stereochemically active lone pair" have both been used to describe the same structural effects, the term "inert pair effect" should *only* be used to refer to the relativistic stabilization of Pb(II) versus Pb(IV) (Section II) and *not* to account for the observed geometries of Pb(II). The pair of electrons has an effect on geometry, but *not* because it is inert. Whether the $6s^2$ electron pair on Pb(II) is "stereochemically active" (the ligands are "hemidirected", i.e., clustered on one side of the Pb ion with a void on the other) or "stereochemically inactive" (the "holodirected" case, in which the ligands are evenly distributed around the central atom in close to spherically symmetric fashion) (197) is more appropriately explained using simple (*not* relativistic) hybridization/valence shell electron-pair repulsion (VSEPR) arguments (205, 206). The reason for the *activity* of the Pb $6s^2$ lone pair is no different than that for the activity of any other lone pair of electrons: similar effects are seen in SF_4, for example, in which an electron pair stabilizes a "seesaw" shape (a distortion of the trigonal bipyramid, with the electron pair axial) rather than tetrahedral geometry (206). The geometries of Pb complexes can also be interpreted using an orbital hybridization model (206, 207). Calculations indicate that in "hemidirected" Pb complexes, there is both s and p character in the lead lone pair orbital, whereas in "holodirected" complexes, only s character is observed (197).

These observations are consistent with a model in which the Pb s and p orbitals hybridize under normal energetic conditions to yield a stereochemically active lone pair and hemidirected structures. [The symmetries of either the nonrelativistic or the relativistic sets of orbitals are consistent with this hybridization (77, 78).] In the holodirected case, on the other hand, no hybridization takes place. The $6s^2$ electrons remain in the spherically symmetric s orbital and do not affect the molecular geometry. Holodirected complexes only appear with very large or bulky ligands, in which steric effects outweigh the energetic benefits of hybridization. In this case, the lone pair is both "inert" and "stereochemically inactive"; but its inactivity is a result of external (i.e., steric) effects rather than of any inherent quality of the electron pair itself.

How does the stereoactivity of the lone pair correlate with coordination number? When the number of donor atoms is high (>8), then sterics outweigh hybridization and the structures are holodirected (lone pair is stereochemically inactive). By contrast, all Pb(II) coordination compounds with coordination numbers <6 are hemidirected and have a stereochemically active lone pair. Both types of structures are found for coordination numbers 6–8, depending on the size of the ligand and the nature of the donating group. Ligand-to-metal distances are usually shortened on the side of the metal opposite the active lone pair. Ligand–metal distances flanking the "empty" coordination site are usually longer than average (197).

For intermediate coordination numbers (6–8), the nature of the donor atoms plays a major role in determining whether or not the lone pair is stereochemically active. Hemidirected structures tend to have harder donor ligands (O, N) that are connected to one another (e.g., in chelators, bidentates, or ring systems). By contrast, holodirected structures tend to have relatively softer donor atoms (S, Cl, Br, I). Why would this be so? For relatively electronegative donor atoms such as O and N, hemidirected structures were calculated to have more net positive charge on the Pb(II) ion than holodirected structures, and hence are energetically favored. Molecular orbital calculations suggest that repulsive interactions between very polarizable (soft) ligands as a result of their effective charge may result in holodirected structures (197).

Given these correlations, it seems that it should be possible to predictably "turn on or off" the stereochemical activity of the lead lone pair. To test this idea, Reger and co-workers prepared a series of lead complexes of polypyrazolyl borate and polypyrazolyl methane ligands in which the number of donor atoms and the size of the ligand was systematically varied (177, 180, 183, 208). For a constant coordination number (6), only the bulkier ligand is holodirected (Fig. 12). This observation is consistent with the idea that the "natural" (i.e., energetically favored) state for Pb(II) compounds is one in which the $6s$ and $6p$ orbitals are hybridized and the lone pair is stereochemically active, but that sufficient steric repulsion can overcome this effect and make the holodirected structure more stable.

4. Structural Insights into the Rational Design of Chelation Therapy Agents

One of the greatest goals of this field is to develop a sufficiently good understanding of lead coordination chemistry to be able to rationally design improved ligands for lead that could be used as chelation therapy agents to treat

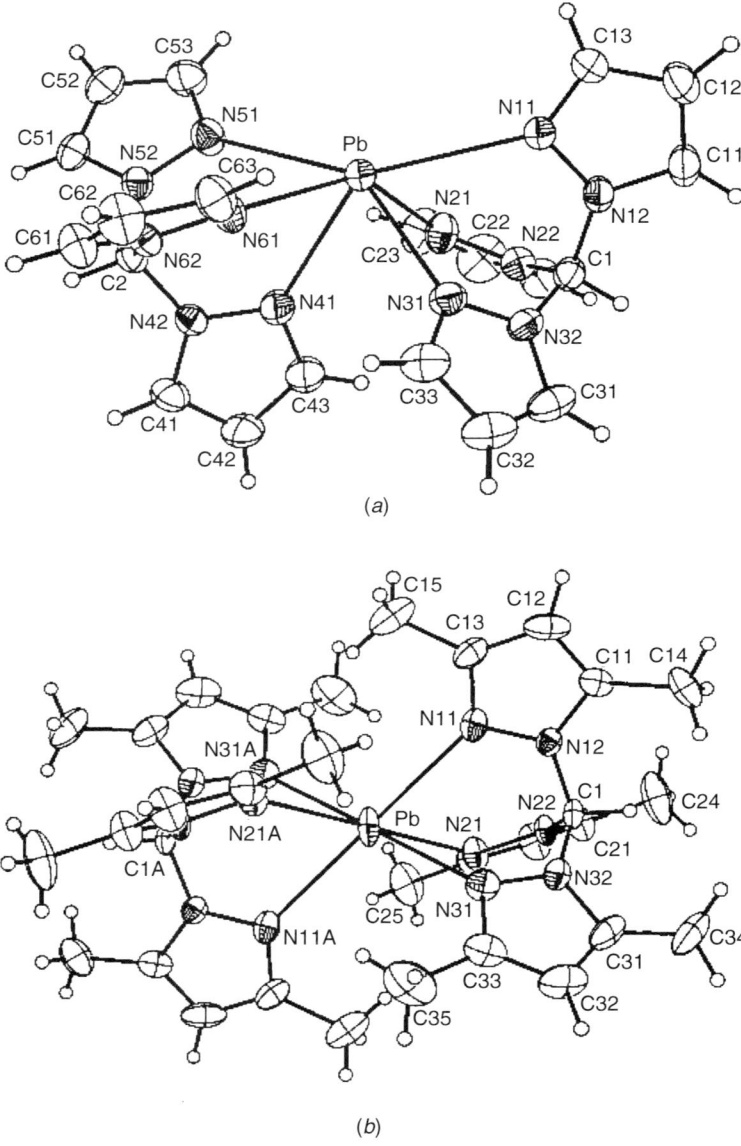

Figure 12. Six-coordinate Pb(II) compounds can either be (a) hemidirected (with a stereochemically active lone pair) or (b) holodirected (with a stereochemically inactive lone pair). Here, in two six-coordinate compounds with similar ligand sets, the steric repulsions in the bulkier ligand (b) trigger the stereoinactivity of the lone pair. [Reprinted with permission from D. L. Reger, T. D. Wright, C. A. Little, J. J. S. Lambda, and M. D. Smith, *Inorg. Chem.*, **40**, 3810–3814 (2001). Copyright © 2001 American Chemical Society.]

lead poisoning. With this goal in mind, we review the structures of known and potential chelation therapy agents.

The most commonly used chelation therapy agents for Pb(II) are *meso*-2,3-dimercaptosuccinic acid (DMSA) and CaNa$_2$EDTA (see Section IV) (209). At the time this chapter was written, no crystal structure had been reported for Pb^{II}–DMSA. By contrast, the structure of Pb^{II}–EDTA has been thoroughly characterized. [See Fig. 11 and Section III.C.2 for a detailed description of the structure. (170)] The observation that lead recruits additional donor groups (from water and neighboring molecules) in the solid state beyond the six groups provided by a single EDTA molecule suggests that Pb(II) prefers high coordination numbers (≥ 6), if possible. In addition, the structure indicates that it is important to consider that the actual structure of Pb^{II}–EDTA in solution may include coordinated water or other small molecules.

In addition to EDTA and DMSA, penicillamine and 2,3-dimercaptopropanol (British anti-Lewisite or BAL) have also been used as chelation therapy agents for lead. The crystal structure for D-penicillaminatolead(II) (195) [Fig. 13(*a*)] reveals that multiple ligand molecules are bound to each lead ion in the solid state, resulting in an extended structure. The donor atoms to lead include three sulfurs and two oxygens, coordinated in a distorted pentagonal bipyramid. The structure around the lead ion is hemidirected with a stereochemically active lone pair. Although no crystal structure is available for Pb(II)–BAL, the structure of Pb^{II}–ethanedithiol [Fig. 13(*b*)] serves as a useful model system (209) and recapitulates many of the properties seen for Pb^{II}–penicillamine. Although the net stoichiometry is 1:1 metal to ligand, each Pb(II) is associated with an additional four thiolates from bridging ethanedithiols (see Fig. 13). The primary Pb—S distances within the molecule are 2.660 and 2.665 Å. The Pb—S distances to the four additional thiols are much longer (3.056–3.584 Å), but are all within the sum of the van der Waals radii of lead and sulfur (3.8 Å) (211). If the more distant thiolates are considered "ligands," the complex has a distorted octahedral geometry with a stereochemically active lone pair. The structures of Pb^{II}–penicillamine and Pb^{II}–ethanedithiol suggest that lead will recruit additional ligands to fill up its coordination sphere if an individual ligand does not provide a sufficient number of donor atoms. In addition, the extended structures observed for these compounds in the solid state offer an explanation for the low solubility exhibited by each of these complexes. Given that under physiologically relevant conditions these chelators are very dilute and unlikely to form dimeric or multimeric structures, these structures lead us to conclude that when lead is complexed to penicillamine or BAL in the body, Pb(II) probably recruits water or biomolecules to fill up its coordination sphere.

To systematically test whether increasing denticity and rigidity and "preorganizing" the chelating agent provide discrete complexes with improved solubility, selectivity, and affinity for lead, Raymond and co-workers rationally

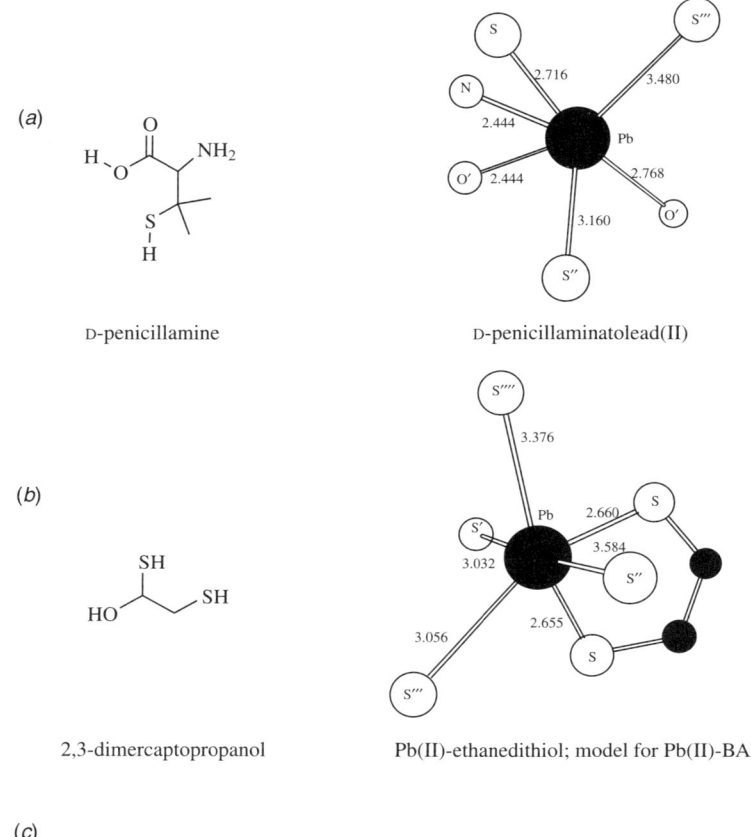

Figure 13. Crystal structures have been reported for lead complexed to the chelation therapy agents Pb^{II}–EDTA (170, 203) (Fig. 11) and (a) Pb^{II}–penicillamine (195), as well as for the potential agents (b) Pb^{II}–ethanedithiol (210) (an analogue of the chelating agent BAL), (c) bis(N-methyl-3-pyridothiohydroxamato–lead(II) (181), and (d) Pb^{II}–EDTA–N$_4$ (39). The lead coordination environments in these systems vary dramatically: in (a), (b), and (c), additional donor atoms are recruited from adjacent molecules and metal–ligand bond lengths vary from averages depending on whether coordination is intraligand or interligand atoms. Coordinates for (b), (c), and (d) were downloaded from the CSD.

(d)

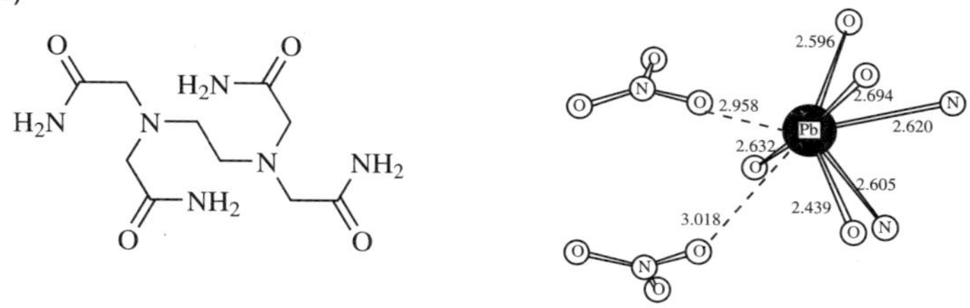

ethylenediaminetetraacetamide
(EDTA-N$_4$)

Pb(II)-EDTA-N$_4$

Figure 13 (Continued)

designed and structurally characterized a series of Pb^{II}–thiohydroxamic and Pb^{II}–hydroxypyridinethione compounds [Fig. 13(c)] (176, 179, 181, 182). Thiohydroxamic acids are the sulfurous equivalent of the hydroxamic functionality frequently found in naturally occurring iron chelators (siderophores) (212, 213) and were chosen as a key component of potential chelating agents because of their high affinity for Pb(II) and their excellent selectivity over Zn(II) and Ca(II) (176, 214). The hydroxypyridinethione ligands are constrained analogs of the thiohydroxamic acids, in which the oxygen and sulfur groups are held in the (Z) configuration desired for metal coordination (179). The lead complexes of the hydroxamic acid and hydroxypyridinethione ligands exhibit pseudo-trigonal-bipyramidal geometries about lead with two oxygen donor atoms, two sulfur donor atoms, and the fifth coordination site occupied by a stereochemically active lone pair [Fig. 13(c)] (176, 179, 181, 182). To decrease the propensity of these ligands to form extended structures with lead in the solid state, the steric bulk of the ligand was increased, resulting in a discrete 2:1 ligand complex for bis(N-methyl-3-pyridothiohydroxamato) lead(II) [Fig. 13(c)] (181). To improve on the thermodynamic stability and solubility of lead complexes of monothiohydroxamic acids, Raymond and co-workers attached two acids to make one ligand, yielding bis(hydroxamic) acids; the rigidity and the lengths of the linking portion between the two acids were systematically varied. Unfortunately, they found that ligands with very rigid linking portions polymerized in solution and were therefore insoluble. The most soluble tetradentate ligand in this series that has been reported to date has a two-carbon backbone between ether oxygens but a poor affinity for lead relative to that of EDTA (see Section IV) (182).

To test whether it is possible to design an improved chelating agent for lead *starting* from the EDTA framework, we prepared a series of lead complexes of amido derivatives of EDTA, with the goal of "softening" the donor set available to lead (39). Amido functional groups were selected because of earlier reports by Battistizzi et al. (215) that this functionality confers selective binding for Pb(II) over other metal ions and because related amido-functionalized ligands have been successfully employed in coordination compounds used for magnetic resonance imaging spectroscopy (216–218). Out of the series of complexes that were prepared, the most soluble was the tetraamido analogue, Pb^{II}–EDTA–N_4, which exhibits a discrete 1:1 ligand/metal complex in the solid state. Interestingly, both solution and solid-state characterization of this compound reveal that the amido functionality binds to lead via the carbonyl oxygens [Fig. 13(d)] (39). Binding constants for these compounds have yet to be reported, so the effect of the amido group on the affinity and selectivity for lead in this particular system is not known.

Taken together, the structural characteristics of reported Pb(II) chelation agents demonstrate several key points about the coordination environment of Pb(II) that determine how effective a potential ligand will be as a chelating agent. It is clear from the structures reported to date that if a single ligand does not have sufficient denticity, Pb(II) seeks higher coordination numbers from additional ligand molecules or solvent. This preference for high coordination numbers can lead to extended structures in the solid state for these complexes, which reduces the solubility of the complex. However, it is likely that under the dilute conditions that exist in vivo, lead recruits water or biomolecules to fill its coordination sphere. The effect that this tendency to recruit additional ligands has on solubility and biodistribution of the complexes of lead with chelating agents in vivo is not known. In addition, all the Pb–chelate complexes that have been reported to date have a stereochemically active lone pair, with the exception of the dimeric form of Pb^{II}–EDTA. Whether or not ligands that have idealized geometries that accommodate this lone pair exhibit improved selectivity for Pb(II) over other metal ions is not known. Given the limited success of the rational design approaches taken to date, however, switching to higher through-put methods for synthesizing and screening ligands may be the most promising approach for developing better chelating agents in the future (38).

5. *New Types of Lead(II) Compounds*

One of the most exciting advances in lead coordination chemistry in the last couple of years is the report of structurally well-characterized lead compounds containing lead–lead multiple bonds. Until recently, lead was the only atom in group 14 (VIA) for which an analogue of ethylene ($H_2C=CH_2$) had not been reported. However, the first molecule with a lead–lead double bond was recently reported (Fig. 14) (219). This and subsequent examples of diplumbenes are characterized by Pb—Pb bond lengths of 3.0515–3.537 Å, which are longer than the predicted length for a Pb=Pb double bond in the calculated structure of $H_2Pb=PbH_2$ (2.95–3.00 Å) (219, 220). Sturmann et al. (220) suggest that the

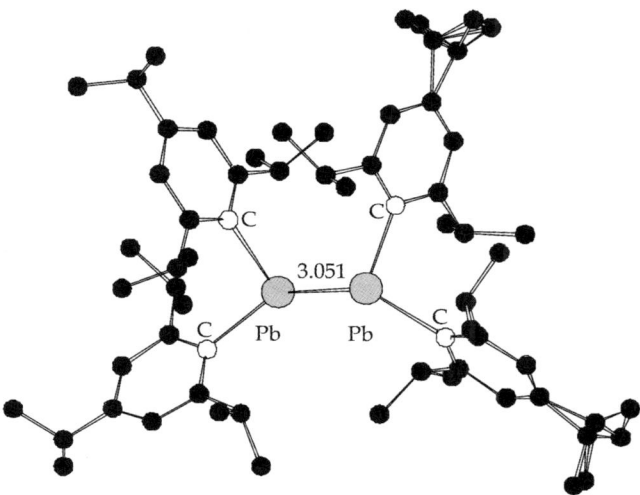

Figure 14. The crystal structure of tetrakis(2,4,6-triisopropylphenyl)diplumbene, one of the first reported Pb(II) molecules with a Pb=Pb double bond (219). Unlike a C=C double bond, the Pb=Pb double bond is formed by electron donation from the doubly occupied 6s orbital of one lead atom to empty 6p orbitals of the adjacent lead atom. The Pb=Pb double bond is 0.05–0.10 Å longer than the calculated distance for $H_2Pb=PbH_2$. Coordinates downloaded from the CSD. A compound with a Pb≡Pb triple bond has also been reported by Pu et al. (190) (2,6-$Trip_2H_3C_6$Pb–PbC_6H_3-2,6-$Trip_2$, where $Trip_2 = C_6H_2$-2,4,6-i-Pr_3) with a Pb–Pb bond distance of 3.188 Å.

diplumbene bond is weaker than would be predicted from calculations, possibly due to relativistic effects that contract the 6s electron pair. In addition, the first compound containing a Pb≡Pb triple bond was recently reported by Power and co-workers (190). The Pb–Pb bond distance in this compound is 3.188 Å, which is considerably shorter than the lead–lead distance in metallic Pb (3.49 Å). This molecule exhibits a bent geometry with a Pb–Pb–C angle of 94.3° and exhibits a stereochemically active lone pair on each metal. The Pb≡Pb triply bonded compound is particularly interesting because it constitutes the first example of a heavier group 14 (VIA) analogue of acetylene (192).

D. X-Ray Crystal Structures of Lead(II) Biomolecules

1. Complexes of Lead(II) with Small Biomolecules

From studies on the structures of lead complexed to small biomolecules, we hope to gain insights into the possible coordination chemistry and speciation of lead in a biological milieu. Unfortunately, only a few structures of Pb(II) with small biomolecules have been deposited in the CSD to date. The structures that

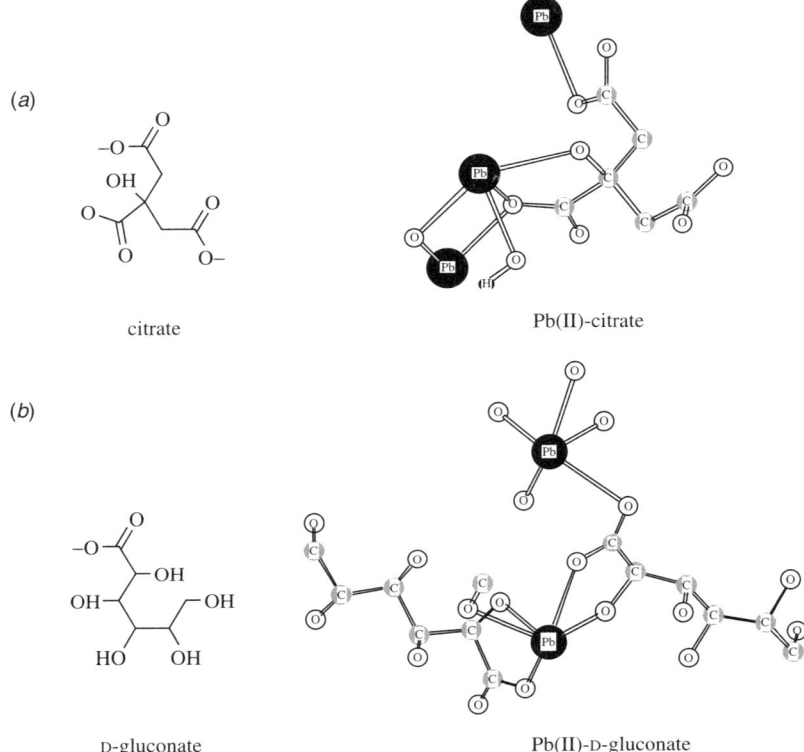

Figure 15. Lead complexes to small biomolecules that have been characterized crystallographically include Pb^{II}–citrate (221) and Pb^{II}–D-gluconate (222). In both cases, lead forms an extended structure and is bound to oxygen-rich sites. Coordinates downloaded from the CSD.

have been reported to date include Pb^{II}–citrate (221), Pb^{II}–D-gluconate (222), and a flavin metal complex, bis(10-methylisoalloxazine) lead(II) perchlorate tetrahydrate (223) (Fig. 15). In all three structures, Pb(II) has a stereochemically active lone pair and forms multimeric structures, which presumably has a profound effect on the solubility of these complexes.

The structure of Pb(II) citrate was solved only recently (221). The complex displays a distorted trigonal-bipyramidal geometry indicative of an active lone pair [Fig. 15(a)]. The uncoordinated terminal carboxylate binds a lead atom in an adjoining molecule, resulting in dimeric units. This complex is distinguished by extensive hydrogen bonding with water and with adjoining complex molecules through the terminal carboxylate, resulting in a very highly interconnected structure.

The only report to date of a structure of a lead–saccharide complex is that of Pb–D-gluconate (222). This sugar–lead complex has a coordination number of 6 and a distorted octahedral geometry. Each gluconate is tridentate with three donor oxygen atoms coordinating to lead. Like the Pb^{II}–citrate complex, Pb^{II}–D-gluconate forms multimeric units where each carboxylate links two lead atoms [see Fig. 15(b)]. Although an active lone pair was not explicitly identified by the authors, an examination of the three Pb—O distances show two short lengths and one length longer than average, suggesting a stereochemically active lone pair.

Lead appears to be able to interact with complex small biomolecules as well, such as flavins; for example, bis(10-methylisoalloxazine) perchlorate tetrahydrate (223). Isoalloxazine is a planar three-ringed heterocyclic amino cofactor associated with riboflavin and is active in oxidation–reduction reactions with metals such as Mo and Fe. Lead binds to bis(10-methylisoalloxazine) in a 1:1 metal–ligand complex, with two additional waters bound resulting in a four coordinate molecule with a total of four oxygen donors. An active lone pair results in a distorted square-pyramidal structure. As is the case for citrate, extensive hydrogen bonding was observed in the crystal lattice.

Although there are few reported structures of Pb(II) complexes with small biomolecules, making it difficult to report trends, a few observations can be made. Coordination numbers ranged from 4 to 6 with the lone pair of electrons stereochemically active. Because all of the reported ligands have relatively low denticities (2–4), the lead coordination sphere is completed by a water molecule, an additional ligand molecule, or both. In each of the small biomolecule structures reported to date, the donor atoms were all oxygens and the complexes exhibited extended structures, with two complexes showing extensive hydrogen-bonding networks attached through the donor oxygens. These structures suggest that lead speciation in biological systems is apt to be extremely complex: It is unlikely that lead will exist in solution as an aquated inorganic ion; rather, it will likely be bound by a variety of small—and large—biomolecules within the cell.

2. *Use of Lead as a Heavy Atom Derivative in Proteins and Nucleic Acids*

Lead has also been used as a heavy atom to prepare isomorphous derivatives of proteins and nucleic acids that can be used to solve the "phase problem" that is inherent to solving the crystal structures of large molecules (224, 225). Although the oxidation state and form of lead used to prepare these samples varies [e.g., Pb(IV) compounds such as trimethyllead acetate are commonly used], the binding sites that lead occupies in the resulting derivatives provide some insights into the types of sites in large biomolecules that are likely to be occupied by lead under physiologically relevant conditions. These include sites on RNA molecules, in addition to sites normally occupied by calcium or zinc in proteins.

One of the earliest reports in which lead was used to prepare a heavy atom derivative was the crystal structure of the transfer RNA for phenylalanine (tRNA^{Phe}) [Fig. 16(a)] (226, 227). This structure was a landmark in the field of RNA biochemistry both because it constituted the first crystal structure of a transfer RNA molecule (226) and because the observation that the Pb(II) soaked into the structure catalyzed cleavage of the sugar–phosphate backbone [Fig. 16(b)] helped to lay the foundation for the field of catalytic RNA (227). Although the observation that lead is able to catalyze autohydrolysis of RNA is widely cited as evidence that nucleic acids are a target for Pb(II), more detailed studies (228–238) suggest that the concentrations of lead required for this reaction are too high for this reaction to be physiologically relevant (see Section VI).

Lead has also been used as a heavy atom derivative to solve crystal structures of proteins, and inspection of these structures provides useful insights into the biological chemistry of Pb(II). Although considerably larger than both Ca(II) (0.99 Å) and Zn(II) (0.74 Å), Pb(II) (1.19 Å) (35) can bind to both calcium- and zinc-binding sites in proteins (Fig. 17). The structures of two critical calcium-binding proteins, calmodulin (239) and synaptotagmin (240–242), have been solved using lead as a heavy atom derivative. In both cases, lead binds a high coordination number site that is made up of carboxylate groups that is usually occupied by calcium. In the case of calmodulin, additional adventitious sites of lead binding are also observed, but the occupancies of these sites are relatively low (240). In synaptotagmin, lead binds only to the high-affinity calcium-binding site in the multinuclear calcium-binding site of the C2A domain (240–242). In the latter case, the inability of lead to reproduce the stoichiometry of calcium binding to this site may be an important factor in the different activities observed for lead- and calcium-bound forms of synaptotagmin (see Section VI).

Lead has also been used to solve the crystal structure of the zinc enzyme, δ-aminolevulinic acid dehydratase (ALAD), a known target for lead in vivo (Section VI) (243, 244). Lead occupies more than one site in the ALAD crystal structure, but shows the highest occupancy in the cysteine-rich catalytic site that is occupied by zinc in the active form of the protein (244, 245). Lead's ability to substitute for zinc in this site explains why lead is able to inhibit the protein at very low concentrations (246). By contrast, the site occupied by lead does not overlap well with the native zinc-binding site in the crystal structure of another zinc enzyme, carboxypeptidase A, which is not known to be inhibited by lead (247). These studies suggest that the ability of lead to target metalloproteins in vivo may be dependent on whether or not the native metal-binding site of the protein constitutes a "good" site for lead (see Section VI).

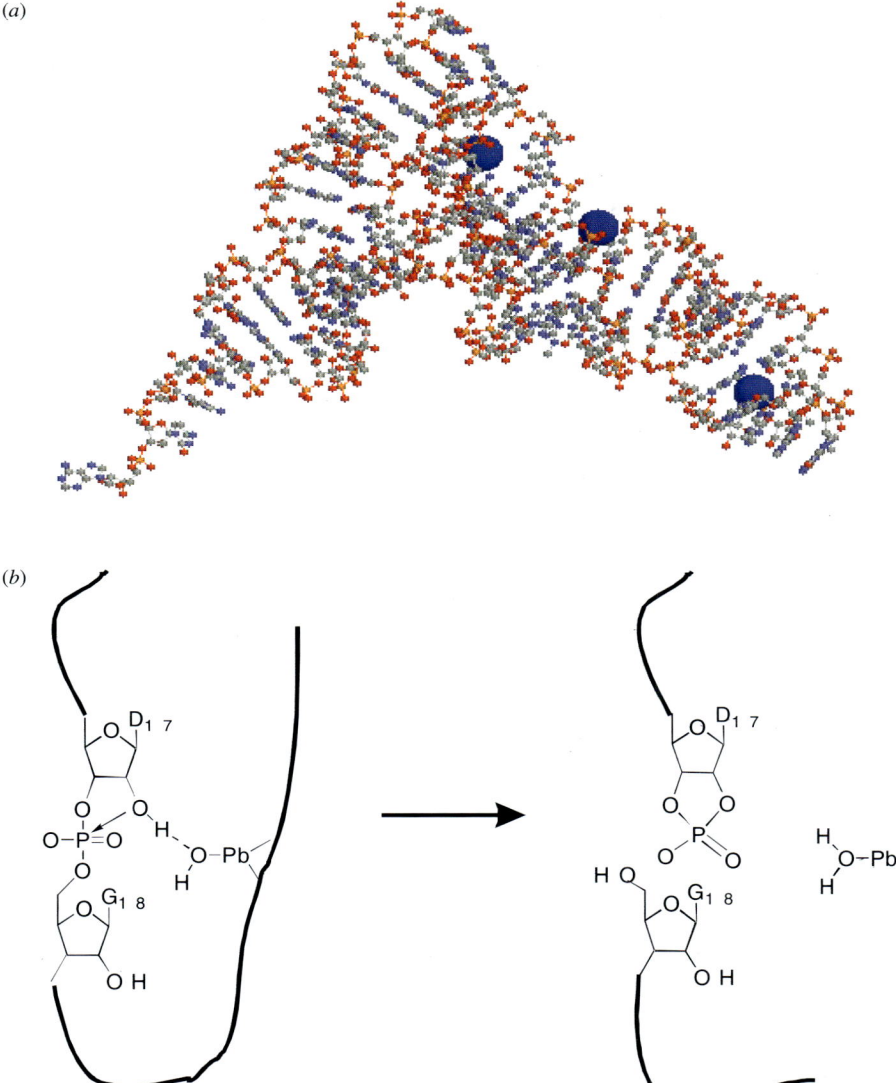

Figure 16. Lead was used as a heavy atom derivative to solve the structure of (*a*) tRNA[Phe] (lead shown as large space-filling blue spheres) (224, 225). The observation that (*b*) Pb(II) catalyzed cleavage of the sugar-phosphate backbone in this structure was a critical development in the field of catalytic RNA and lead to the widely cited hypothesis that Pb(II) targets nucleic acids. The coordinates for the Ph$_3$tRNA[Phe] structure (*a*) were downloaded from the protein databank (1TN2); [Part (*b*) was adapted from (227).]

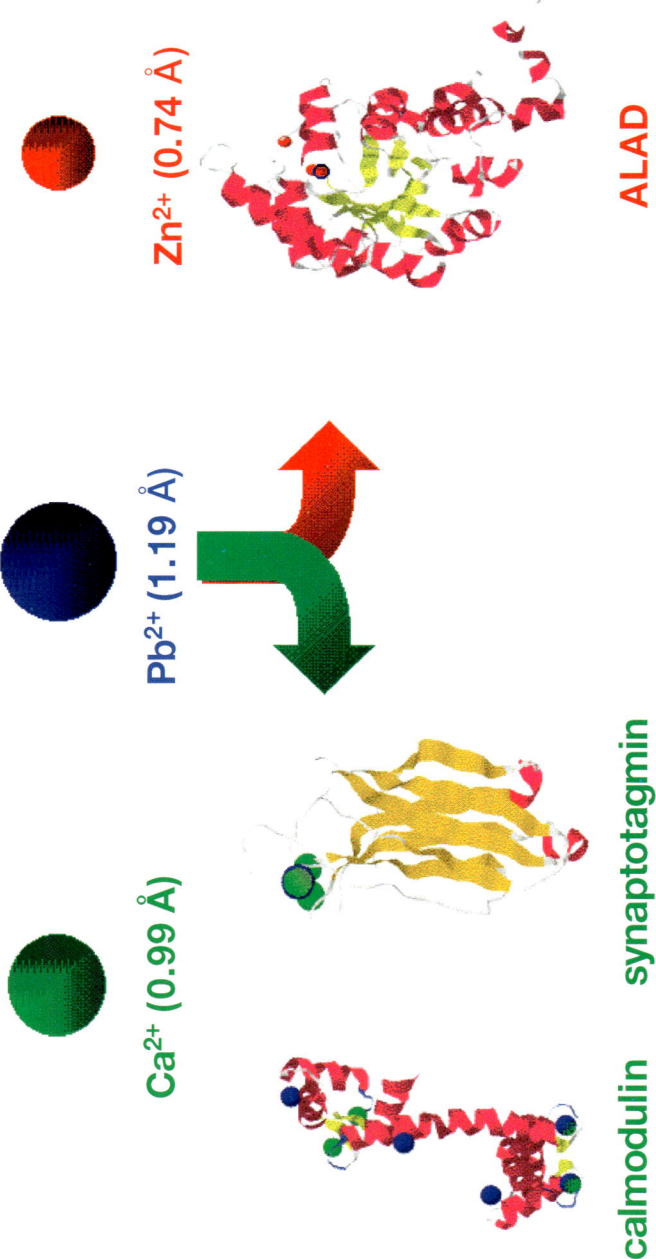

Figure 17. Lead has been used as a heavy atom derivative to solve the phase problem in the crystal structures of a variety of protein including the calcium proteins calmodulin and synaptotagmin and the zinc protein δ-aminolevulinic acid dehydratase, ALAD. These structures provide useful insights into the coordination environments preferred by lead, which can bind both to carboxylate-rich calcium sites and thiol-rich zinc sites. Structures downloaded from the protein databank (3CLN, 1RSY, 1AW5, 1QNV); where necessary, lead was added to the figure based upon coordinates provided by the authors (240, 241, 243, 246). Figure adapted from *Curr. Opin. Chem. Biol.*, Vol. 5, H.A. Godwin, "The biological chemistry of lead," pp. 223–227, Copyright © 2001, with permission from Elsevier Science.

E. EXAFS Studies on Lead(II) Compounds

For complex systems that are difficult or impossible to crystallize (e.g., biological or environmental samples), other diffraction techniques such as EXAFS spectroscopy can often provide details about the Pb coordination environment. The information about a metal complex that is available from EXAFS spectroscopy includes the coordination number of the metal ion and bond distances. Bond angles and geometry are difficult to determine directly but can be inferred from careful comparisons to model complexes of known geometry (248, 249). The identity of coordinated atoms can also be determined from the EXAFS spectrum, although it is often difficult to differentiate between atoms of similar atomic number (e.g., N vs O) (250). Only a few Pb(II) systems have been examined by EXAFS, but the following studies provide excellent examples of the kinds of questions EXAFS is uniquely suited to answer (157, 158, 162, 251–259).

In particular, EXAFS spectroscopy has great promise for examining the coordination environment of Pb(II) in proteins that could not be crystallized and has been used to characterize the metal-binding sites of metallothionein and bovine ALAD (251, 252). Hasnain et al. (251) used EXAFS spectroscopy to examine the metal environment of Pb, Hg, Ag, Zn, Cu, and Cd bound to metallothionein and found metal–sulfur coordination in all cases, with a Pb—S bond distance of 2.65 Å in Pb–metallothionein. Prior to the report of the crystal structure of ALAD, Dent et al. (252) reported EXAFS data for only the zinc sites in Zn_4Pb_4–ALAD. They noted slight changes in the spectrum of Zn_4Pb_4–ALAD compared to that of Zn_4–ALAD, which they attributed to slight changes in the geometrical arrangement of the ligands upon binding of Pb. X-ray absorption spectroscopy has also been used to examine the coordination environment of lead in nonbiological inorganic complexes (157, 158, 162). Studies of this sort are particularly useful for complicated systems such as the lead hydro sodalite $[Pb_2(OH)(H_2O)_3]_2[Al_3Si_3O_{12}]_2$ reported by Eiden-Assmann et al. (162), in which the lead is enclosed in a polyhedral cage.

Another application of EXAFS spectroscopy is the determination of lead speciation in natural systems. The EXAFS spectra of mine tailings (257), soils (253–255, 258), plants (259), and fungal cell walls (256) have all been observed directly. The resulting spectra are fit using linear combinations of model compound spectra to determine the identity of lead compounds in these environmental samples (253, 257). Manceau et al. (253) caution that because of the high variability of the Pb(II) coordination sphere, recording high-quality EXAFS spectra and then interpreting them quantitatively is extremely difficult. Indeed, the paucity of lead EXAFS studies is due in part to the experimental difficulties that exist, but Manceau et al. (253) demonstrate that lead EXAFS spectra can be used as a "powerful structural fingerprint of [lead's] coordination environment in matrices as complex as soils."

F. Structural Conclusions

Taken together, the structures that have been reported for Pb(II) coordination compounds and Pb(II) bound to molecules provide the following insights into the coordination chemistry of Pb(II):

- Lead binds to a wide range of donor atoms (principally O, N, and S) and assumes a wide range of geometries and coordination numbers (from 2 to 12).
- Most Pb(II) compounds exhibit a stereochemically active lone pair of electrons, which arises from mixing of $6s$ and $6p$ orbitals and results in a "hemidirected" structure.
- Sterically demanding ligands and high coordination numbers can shift the lead geometry to a more spherically symmetric, "holodirected" structure, in which the lone pair of electrons is stereochemically inactive.
- Lead tends to form extended structures in the solid state, which may explain the poor solubility of many Pb(II) complexes in aqueous solutions. These structures also suggest that the solution speciation of lead in biological and environmental samples is likely to be complex.
- Lead can substitute for both calcium and zinc in proteins, consistent with its ability to bind to both O- and S-rich small molecule ligands.

As with any technique, crystallography has its limits: Crystallography cannot provide insights into the factors that contribute to tight and selective binding of lead and what the coordination and speciation of lead is in aqueous solutions. To answer these questions, we turn to studies on the thermodynamics and kinetics of Pb(II) complexes (Section IV) and studies on the environmental (Section V) and biological (Section VI) chemistry of lead.

IV. KINETICS AND THERMODYNAMICS OF LEAD–LIGAND INTERACTIONS

"Lead is the least favoured base metal of the experimentalist. The complexity and low solubility of many of its most common compounds in aqueous solutions make it a relatively unattractive target for laboratory research...A cynic would suggest, with good reason, that if there is no uncertainty about the stability of any particular lead complex, it has not been measured enough times."

—*Rickard and Nriagu (1978)* (260)

A. Introduction

To fully understand how lead behaves in biological and environmental systems, it is critical to understand not only the structure of lead complexes in the solid state, but also the nature of lead–ligand interactions in solution. Both the dynamics of lead–ligand binding and substitution and the stability of lead–ligand complexes affect the bioavailability of lead and thus its role as a toxin. By understanding the mechanisms of ligand exchange and the thermodynamic ligand preferences of Pb(II), we can also better design molecules for the immobilization of lead and treatment of lead poisoning.

The pioneering work of Basolo and Pearson showed that metal binding and substitution reactions could be categorized, like organic reactions, into three main mechanistic categories: associative (S_N2 or S_E2), dissociative (S_N1 or S_E1), or concerted interchange (I) (261–263). In the dissociative case, the rate-limiting step involves only bond breaking, whereas equal bond making and breaking are involved in the rate-limiting step of an associative model.

Associative mechanism (S_N2)

$$Y + M-X \rightleftharpoons Y \cdots M \cdots X \rightleftharpoons Y-M + X$$

Associative mechanism (S_E2)

$$M-X + M' \rightleftharpoons M \cdots X \cdots M' \rightleftharpoons M + X-M'$$

Dissociative mechanism (S_N1)

$$M-X \rightleftharpoons M + X$$
$$M + Y \rightleftharpoons M-Y$$

Dissociative mechanism (S_E1)

$$M-X \rightleftharpoons M + X$$
$$X + M' \rightleftharpoons M'-X$$

Which mechanistic type prevails depends on many factors, including electrostatic, steric, and solvent effects (263). The mechanistic type, in turn, has physiological relevance; for example, if the mechanism is dissociative and rate of dissociation of a natural metal from a protein is very slow, then the affinity of lead for the site is immaterial because lead may never have the opportunity to bind.

In addition to the mechanism of metal–ligand exchange, the affinity of lead for different types of ligands is of interest. Lead's toxicity and the variety of coordination environments to which lead can bind prompted several studies that investigated the thermodynamics of lead–ligand interactions in aqueous solutions, either with an eye toward designing better (i.e., strong binding, yet selective) chelation therapy agents or as part of understanding likely targets of Pb(II) in an aqueous and/or biological environment (37, 41, 182, 264–269). In several of these cases, size and intermediate hard–soft acid character of Pb(II) have put a twist in the conventional wisdom of ligand design guidelines suggested by Pearson's hard–soft rules or macrocyclic–cryptand effects. For example, the large size (1.19 Å) (35) of Pb(II) allows it to accommodate high coordination numbers and as a result Pb(II) often forms multimeric complexes. This tendency in turn both reduces solubility and the effectiveness of low coordinate ligands in sequestering lead. Furthermore, despite lead's reputation for being thiophilic, lead exhibits as large a binding constant for the relatively hard set of donor groups in EDTA (log K_{PbL} = 18.10) (270), as it does for the softer, sulfur-rich environment found in DMSA (log K_{PbL} = 17.4) (271). These examples indicate that a multitude of factors is important in determining the affinity and selectivity for lead and/or that there is no "one" preferred coordination environment for Pb(II). Instead, there may be several "good" environments that can be adopted by lead under a given set of conditions (272).

B. Questions of Interest

Here, we review the studies that have been conducted to date on the kinetics and thermodynamics of lead coordination compounds, with particular focus on addressing the following questions:

- What are the mechanisms of lead binding and substitution? Does lead binding proceed by an associative or a dissociative mechanism?
- What sorts of intermediates are formed in these mechanisms?
- What other effects have an impact on the rate of binding/substitution (e.g., buffer, sterics)?
- What is the preferred coordination environment, chelate ring size, and coordination number for lead?
- Do macrocycles confer selectivity or improved affinity for Pb over nonmacrocyclic ligands?
- Does the ability of a ligand to accommodate a stereochemically active Pb(II) lone pair affect selectivity or affinity of that ligand for Pb(II)?
- What is the physiological and environmental relevance of the thermodynamic preferences and kinetic lability of different ligands for Pb(II)?

C. Kinetics of Lead–Ligand Interactions

1. Simple Ligand-Exchange Constants

The simplest reaction of a solvated metal ion is the exchange of solvent molecules in aqueous solution, that is, water exchange. The water-exchange rate for Pb(II) is relatively fast: 7×10^9 s^{-1} (Table VII) (49). The characteristic water-exchange rate often is mirrored by the rates of binding and substitution of other ligands in aqueous solution (273); hence, exchange rates for other monodentate ligands bound to lead in aqueous solution are also expected to be rapid. Likewise, in the simplest case ligand-exchange rates are also expected to be rapid in multidentate ligand-exchange reactions involving lead. However, it is important to realize that steric effects in ligands can slow down the inherent ligand-exchange rate, just as some outer sphere interactions can hasten it (273).

In general, the diffusion rate can also be a limiting factor in metal–ligand exchange rates. For the reaction of metal ions with neutral or negatively charged ligands at low ionic strength, the diffusion controlled rate constants range from

TABLE VII

Characteristic Water-Exchange Rate Constants [a]

Metal Ion	k_{-w} (s^{-1})
Pb^{2+}	7×10^9
Hg^{2+}	2×10^9
Cu^{2+}	1×10^9
Ca^{2+}	6×10^8
Cd^{2+}	3×10^8
La^{3+}	1×10^8
Zn^{2+}	7×10^7
Mn^{2+}	3×10^7
Fe^{2+}	4×10^6
Co^{2+}	2×10^6
Mg^{2+}	3×10^5
Ni^{2+}	3×10^4
Fe^{3+}	2×10^2
Ga^{3+}	8×10^2
Al^{3+}	1
Cr^{3+}	5×10^{-7}
Rh^{3+}	3×10^{-8}

[a] Reference 46. [Reproduced from W. Stumm and J. J. Morgan, *Aquatic Chemistry: Chemical Equilibria and Rates in Natural Waters*, pp. 294–298, Copyright © 1996 John Wiley & Sons, Inc. Reprinted by permission of John Wiley & Sons, Inc.]

10^9–10^{11} $M^{-1}s^{-1}$ (273). For most Pb(II) compounds, the complexation reactions are sufficiently slow that the diffusion-controlled limit is not a factor; but a key exception is the reaction of Pb(II) with aminocarboxylate chelating agents such as NTA, for which the formation rate constants have been determined to be 2×10^{10}–2.4×10^{11} $M^{-1}s^{-1}$ (273, 274). (For structures of chelate compounds for which kinetic studies are discussed, see Table VIII.)

TABLE VIII

Structures of Chelators and Macrocycles for which Kinetic and Thermodynamic Studies Are Discussed in Text

Compound Name	Structure
trans-1,2-Cyclohexanediaminetetraacetic acid (CyDTA)	
2,3-Diaminobutane-N,N,N',N'-tetraacetic acid (DBTA)	
2,3-Dimercaptosuccinic acid (DMSA)	
1,4,7,10-Tetrakis(acetamido)-1,4,7,10-Tetraazacyclododecane (DOTAM)	
Diethylenetriamine-N,N,N',N'',N''-pentaacetic acid (DTPA)	

TABLE VIII (Continued)

Compound Name	Structure
Ethylenediamine-N,N'-diacetic-N,N'-dipropionic acid (EDDDA)	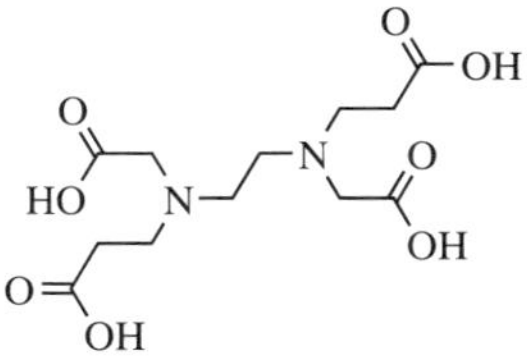
Ethylenediamine-N,N'-bis(2-hydroxyphenylacetic acid) (EDDHA)	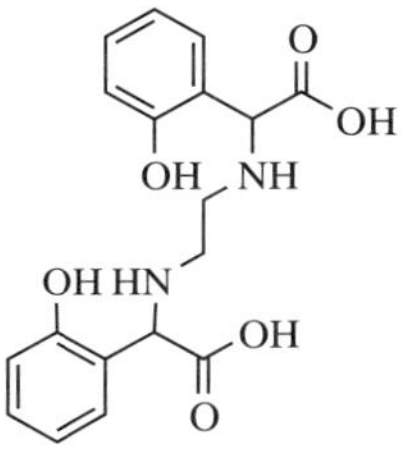
Ethylenediamine-N,N,N',N'-tetraacetic acid (EDTA)	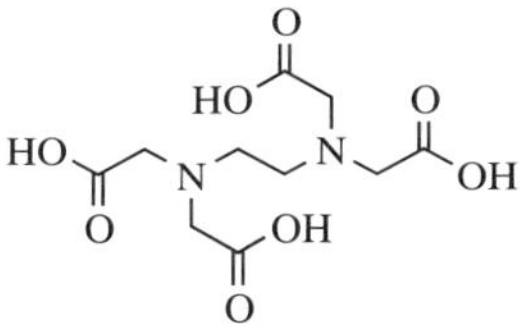
1,2-Bis(2-aminoethoxyethane)-N,N,N',N'-tetraacetic acid (EGTA)	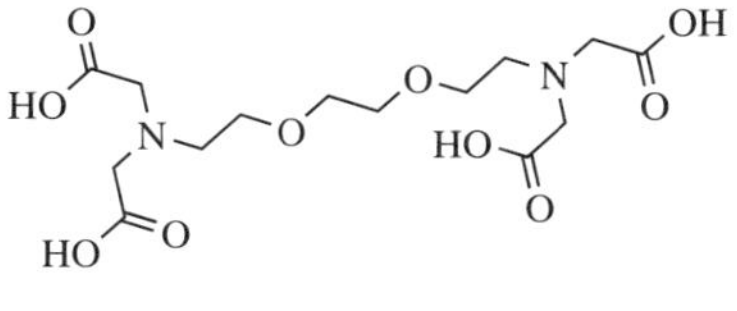
N-(2-Hydroxyethyl)ethylenediamine-N,N',N'-triacetic acid (HEDTA)	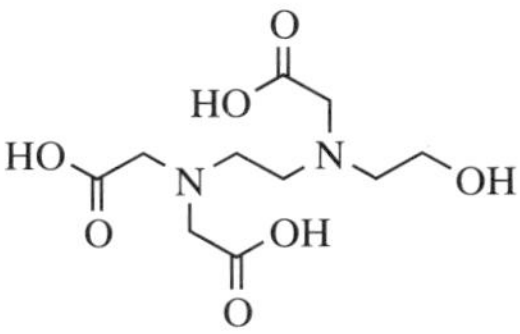
N-(2-Hydroxyethyl)iminodiacetic acid (HEIDA)	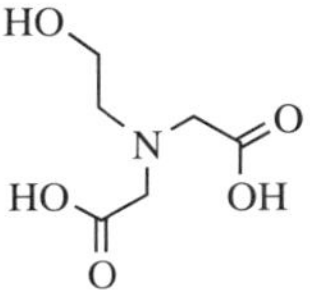

(*continues*)

TABLE VIII (Continued)

Compound Name	Structure
Nitrilo-2,2′,2″-triacetic acid (NTA)	
1,2-Propylenediaminetetraacetic acid (1,2-PDTA)	
1,3-Propylenediaminetetraacetic acid (1,3-PDTA)	
1,4,7,10-Tetra-azacyclododecane	
1,4,10,13-Tetraoxa-7,16-diazacyclooctadecane-N,N'-diacetic acid	
Tetraphenylporphyrin (TPP)	

2. Mechanisms and Rates of Lead–Chelate Dissociation and Substitution Reactions

With the exception of solvent exchange, replacement reactions of one monodentate ligand with another have not been extensively studied for Pb(II) compounds (273). The kinetics and reaction mechanisms of Pb(II) complexes with a variety of multidentate ligands have been examined, however; these are reviewed below (275–293). For a more general review of metal–chelate kinetics, we recommend reference (273).

a. Association Reactions of Lead with Chelating Ligands. The simplest picture of lead–ligand interactions can be gained from studies in which a single ligand type is used and no other metal ions are involved. The rates for these lead-association reactions tend to mirror the water-exchange rate for lead, as would be predicted: Most of the association rates reported to date for Pb(II) and chelating ligands are in the range of 10^6–$10^{11}\,M^{-1}\,s^{-1}$. The formation and dissociation of Pb^{II}–NTA provides an excellent example of this type of system (285, 289, 290):

$$Pb^{2+} + HNTA^{2-} \rightleftharpoons Pb(NTA)^- + H^+$$

Changes in the protonation state of the ligand due to coordination with the metal ion have been monitored by ^{1}H NMR line-broadening techniques and by observing the collapse of ^{207}Pb–^{1}H spin–spin coupling (285, 289). From these experiments, Rabenstein and Kula (Fig. 18) proposed that the rate of NTA binding depends on the relatively slow migration of a proton from nitrogen (where it resides in the normal monoprotonated form of the ligand) to a carboxylate group (285). In this mechanism, the reaction goes through an intermediate carboxylate-protonated Pb^{II}–HNTA species on the way to formation of the final Pb^{II}–NTA complex (273, 285). Margerum et al. (274) pointed out, however, that if the ligand intramolecular proton transfer from the nitrogen to the carboxylate group occurs prior to metal binding, then the metal-binding rate will not be affected by the proton-transfer rate (Fig. 19). Based on comparisons with the formation rates for Zn^{II}–, Cd^{II}–, Cu^{II}–, and Ni^{II}–NTA, they conclude that the proton transfer step is likely not limiting and prefer a mechanism that starts with carboxylate-protonated $HNTA^{2-}$.

Small structural changes in the ligand can have a profound effect on the reaction rate, even when the overall mechanism remains the same. This effect can be seen by comparing the lead-association rate for protonated 1,3-PDTA, $6.8 \times 10^{10}\,M^{-1}\,s^{-1}$, with that of protonated EDDDA, $2.5 \times 10^{10}\,M^{-1}\,s^{-1}$ (288). The differences in the lead-association rates for these two ligands can be readily understood if the rate of complex formation is assumed to be governed by the

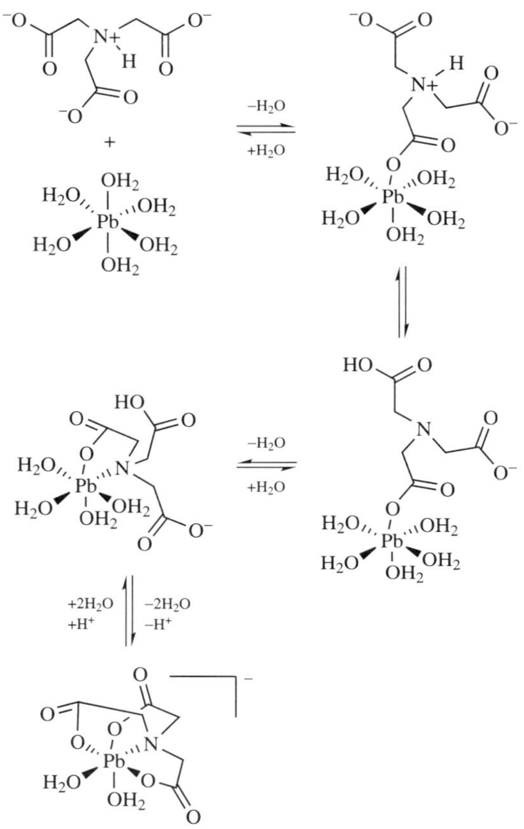

Figure 18. Proposed mechanism for the binding of NTA to aquated Pb(II). The rate depends on the slow intramolecular proton transfer from the nitrogen to the carboxylate group. [Adapted from (285).]

rate of formation of an individual metal–ligand bond. In the case of EDDDA, the rate is limited by formation of one of the six-membered chelate rings (Table VIII). Because formation of five-membered chelate rings tends to be faster than formation of six-membered rings, 1,3-PDTA (which *only* forms five-membered chelate rings) binds Pb(II) more rapidly (288).

b. Substitution Reactions of Lead(II) with Multidentate Ligands. Metal–Ligand substitution reactions fall into two main categories: (1) those in which metal ions are replaced in a given ligand,

$$M_1L + M_2 \rightleftharpoons M_2L + M_1$$

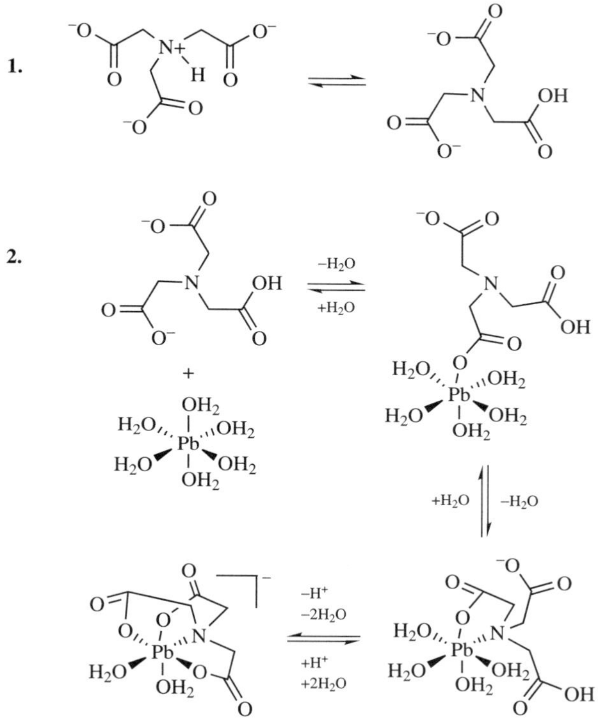

FUNDAMENTAL CHEMISTRY OF LEAD(II)   61

Figure 19. Alternate proposed mechanism for the binding of aquated Pb(II) to NTA. Here, the intramolecular proton transfer from the nitrogen to the carboxylate occurs first and thus does not affect the rate of metal binding. Step two is the coordination of carboxylate-protonated NTA to Pb(II) (273).

and (2) those in which different ligands compete for the same metal, that is,

$$ML_1 + L_2 \rightleftharpoons ML_2 + L_1$$

In general, the rates of multidentate ligand exchange with lead tend to be significantly slower than those predicted for mechanisms limited only by water-exchange rates (289). The mechanisms for complex ligand systems also are typically more complicated than a simple exchange picture might suggest.

i. Metal Substitution Reactions. The complexity involved in metal substitution reactions with chelates is illustrated by the mechanism by which lead substitutes for other metals that are bound to EDTA (275–282, 284, 292). The general mechanism of metal exchange is believed to involve three simultaneous reaction paths:

$$Pb^{2+} + ML^{2-} \underset{k_{-1}}{\overset{k_{+1}}{\rightleftharpoons}} PbL^{2-} + M^{2+} \qquad (1)$$

$$ML^{2-} + H^+ \rightleftharpoons MHL^- \qquad (2)$$

$$Pb^{2+} + MHL^- \underset{k_{-2}}{\overset{k_{+2}}{\rightleftharpoons}} PbHL^- + M^{2+}$$

$$PbHL^- \rightleftharpoons PbL^{2-} + H^+$$

$$ML^{2-} + H^+ \rightleftharpoons M^{2+} + HL^{3-} \qquad (3)$$

$$Pb^{2+} + HL^{3-} \underset{k_{-3}}{\overset{k_{+3}}{\rightleftharpoons}} PbL^{2-} + H^+$$

This mechanism of lead–metal exchange on EDTA is believed to proceed by an associative (S_E2) process, with the formation of dinuclear intermediates in which each metal is coordinated to one of the iminodiacetate halves of the EDTA molecule (273). Tanaka et al. (277, 278) showed that the presence of buffer molecules in solution slows the rate of metal exchange by limiting Pb accessibility but proposed that the buffer does not alter the overall mechanism. Consequently, it is necessary to include the kinetics of metal-buffer coordination in calculations in order to produce meaningful rate constants (275, 276, 278, 280). Studies of Pb(II) exchange with Co^{II}–EDTA confirm that both hydrated and monoacetato Pb(II) ions take part in the substitution reaction, but substitution reactions with monoacetato lead complexes occur more slowly than those with hydrated lead ions (280, 281).

Despite the apparent generality of this mechanism, the lead–metal exchange *rates* observed for different chelating ligands are highly dependent on the structure of the individual ligands. For example, lead exchanges more rapidly with Co^{II}–NTA ($k = 1.65\ M^{-1}\ s^{-1}$) than with Co^{II}–EDTA ($k = 0.10\ M^{-1}\ s^{-1}$) (291), presumably because the lower denticity of the NTA ligand (Table VIII) allows for more rapid collapse into the fully chelated Pb complex. By contrast, Pb(II) exchanges more slowly with Zn^{II}–1,2-PDTA ($k = 0.66\ M^{-1}\ s^{-1}$) than with Zn^{II}–EDTA ($k = 8.9\ M^{-1}\ s^{-1}$), presumably because the sterics of the methyl group of 1,2-PDTA slow the formation of the final lead complex (283, 284).

Steric effects play such a large role in exchange reactions between lead and metal complexes of *trans*-1,2-cyclohexanediaminetetraacetic acid (CyDTA), that the mechanism of metal exchange is altered. In this case, by contrast with EDTA, direct replacement of one metal by another is impossible because of the great steric bulk of the cyclohexane ring on the ligand (see Table VIII). The rate of reaction depends not on the metal concentration but on the concentration of H^+, which can act as a catalyst for exchange. Additionally, because metal–CyDTA complexes are extremely stable, the CyDTA dissociation rate tends to be slow and rate limiting. In general, the mechanism of metal exchange for

CyDTA is purely dissociative (S_E1); the replacing metal only reacts with free CyDTA and cannot actively aid in the dissociation of the initial metal–CyDTA complex. It is thought that protons may be involved in assisting the dissociation reaction via penetration into the chelate "cage" without totally displacing the metal ion (273). This theory is supported by the work of Laurenczy et al. (293), who studied Pb(II) exchange with *meso-* and *rac*-Ce^{III}-2,3-diaminobutane-N,N,N',N'-tetraacetate (DBTA) and concluded that the mechanism can be best described as a proton-catalyzed dissociation reaction. Dramatic steric effects from the ligand methyl groups also served to slow this reaction compared to that of Pb(II) with Ce^{III}–EDTA (292, 293). In the case of Pb(II) exchange with M–CyDTA complexes, the situation is even more complex: Lead can form a mixed-metal species that inhibits the proton-assisted dissociation of M–CyDTA and as a result can actually inhibit its own exchange reaction (273).

ii. Ligand Substitution Reactions. Ligand substitution reactions can also proceed via either associative or dissociative mechanisms, and the stereochemistry of products and reactants in a ligand substitution process can be very informative about the mechanism of the reaction (286, 287). The use of optically active multidentate ligands not only makes it easier to monitor the course of a reaction by spectropolarimetry, but also means that potential mechanisms can be clearly ruled out on account of reactant stereospecificity (286). For the reaction,

$$Pb-EDTA^{2-} + D\text{-}(-)CyDTA^{4-} \rightleftharpoons L^*(+)\text{-}Pb\text{-}D\text{-}(-)CyDTA^{2-} + EDTA^{4-}$$

L* indicates the absolute (*) stereochemistry of the Pb complex. Several mechanisms have been proposed but all share the following features: Pb^{II}–EDTA is aquated stepwise and D-(−)CyDTA attacks the aquated species. Because of steric hindrance of the D-(−)CyDTA nitrogen, the acetato group on CyDTA must bind to lead first. Subsequent steps involve breaking the five-membered rings with the dissociating ligand and forming five-membered rings with the incoming ligand, one ring at a time. Other possible intermediates, in which both EDTA and D-(−)CyDTA are bound through one nitrogen each in a cis configuration, are sterically unfavorable (286). Carr and Baker (287) observed that although a sterically bulky leaving ligand effectively slows the substitution reaction by slowing the aquation of the metal ion, an identically bulky entering ligand does not affect the rate. They suggest that one of the earlier steps in the multistep reaction mechanism must be rate limiting; that way, the bulky substituents of the entering ligand do not interact with the coordination sphere of the initial complex ion and destabilize early intermediates (287). However, the rate-determining step is dissociative for CyDTA only because this ligand is so sterically encumbered; in general the rate-determining step for chelate-exchange reactions is an associative step, and hence the incoming ligand usually has a greater effect on the rate constant than the leaving ligand (273).

3. Kinetic Studies with Macrocyclic Ligands

The effect of steric hindrance on the rate of lead association and exchange is particularly pronounced for macrocyclic ligands. To be inserted into a macrocycle, a metal must be extensively desolvated. Ideally, this desolvation occurs stepwise as for flexible open-chain multidentate ligands, with the macrocyclic donors replacing coordinated solvent molecules one by one. For example, rates of insertion for lead into the monoprotonated forms of the two macrocycles 1,4,7,10-tetra-azacyclododecane and 1,4,10,13-tetraoxa-7,16-diazacyclooctadecane-N,N'-diacetic acid are 8.3×10^5 and $2.3 \times 10^8\ M^{-2}\ s^{-1}$, respectively (294, 295), which are at the slow end of the range observed for open chelating ligands (10^6–$10^{11}\ M^{-1}\ s^{-1}$) (273). [See (296, 297) for reviews of the kinetics of metal–macrocycle interactions.] However, not all macrocycles are sufficiently flexible to maintain a stepwise desolvation pathway. In the more rigid cases, multiple desolvation is necessary, which considerably slows the rate of metal association or exchange (273). For example, the substitution of metals in rigid porphyrins [e.g., Zn(II) into Pb(II) tetraphenylporphyrin (TPP)] tend to be relatively slow (e.g., $k = 4.0 \times 10^{-4}\ M^{-2}\ s^{-1}$ (273, 298).

4. Implications of Kinetic Studies for Lead(II)–Protein Interactions and Other Complex Systems

To date, very few—if any—detailed studies have been conducted on the kinetics of lead–protein interactions. However, the kinetic studies that have been conducted to date on the binding of Pb(II) complexes to small molecules provide insights into the likely behavior of lead in the body. Because the rates of multidentate ligand exchange are affected by, but do not solely depend on, the characteristic water-exchange rates (289), ligand-exchange rates for lead tend to be rapid; even the slowest exchange reactions discussed above still occur on a timescale that is physiologically relevant. In addition, the observation that small molecules (e.g., buffer) can profoundly affect the accessibility of lead and slow down lead–ligand association rates is likely to be pertinent to conditions inside cells, where a wide variety of small biomolecules are present in substantial concentrations (e.g., glutathione, which is present in millimolar concentrations in the cytosol) (264). In small molecules, substitution reactions of Pb(II) often occur through associative mechanisms, but the mechanism and rate of Pb substitution can be altered dramatically by outside effects, particularly sterics. The relevance of these mechanisms to the kinetics and mechanism of lead–protein binding is unknown but suggest that ligand- and metal-exchange rates in these systems will tend to be rapid as well.

Some insights into the timescale of metal exchange in biological systems can be gained from examination of thermodynamic studies of lead binding to sites in proteins (37). Metal-exchange reactions between lead and zinc in structural zinc-binding domains reveal that lead and zinc rapidly equilibrate in competition experiments and that the ratio of lead to zinc bound in a given site depends directly on the relative affinities of the two metals for that site. These results indicate that lead binding to structural zinc sites is under thermodynamic, rather than kinetic, control (37). In addition, when children with lead poisoning are treated with zinc, the lead-induced inhibition of ALAD is reversed (299), suggesting that zinc is able to displace lead from the enzyme on a physiologically relevant timescale. Given that the complex environment of proteins and cells is likely to have a dramatic effect on the lead-binding rate and mechanism of metal exchange, it would be interesting to examine the actual rates of metal binding and substitution and to determine the mechanism of metal exchange in each of these systems.

D. Thermodynamics of Lead–Ligand Interactions

The thermodynamics of lead–ligand interactions provide key insights both into the probable mechanisms of lead's toxicity and into the design of improved chelation therapy agents. Various constants can be used to describe the affinity of a metal ion for a given ligand described by the net equilibrium (300, 301):

$$M^{n+} + xL \rightleftharpoons ML_x^{n+}$$

with the stepwise equilibria:

$$M^{n+} + L \rightleftharpoons ML^{n+}$$
$$ML^{n+} + L \rightleftharpoons ML_2^{n+}$$
$$\vdots$$
$$ML_{x-1}^{n+} + L \rightleftharpoons ML_x^{n+}$$

$$\beta = \text{overall stability constant} = \frac{[ML_x^{n+}]}{[M^{n+}][L]^x} \qquad \beta_1 = \frac{[ML^{n+}]}{[M^{n+}][L]}$$

$$\beta_2 = \frac{[ML_2^{n+}]}{[M^{n+}][L]^2} \quad \text{etc.}$$

$$K_b = \text{binding constant} = \frac{[ML_x^{n+}]}{[ML_{x-1}^{n+}][L]} \qquad K_{b1} = \frac{[ML^{n+}]}{[M^{n+}][L]}$$

$$K_{b2} = \frac{[ML_2^{n+}]}{[ML^{n+}][L]} \quad \text{etc.}$$

$$K_d = \text{dissociation constant} = \frac{1}{K_b} = \frac{[ML_{x-1}^{n+}][L]}{[ML_x^{n+}]} \qquad K_{d1} = \frac{[M^{n+}][L]}{[ML^{n+}]}$$

$$K_{d2} = \frac{[ML^{n+}][L]}{[ML_2^{n+}]} \quad \text{etc.}$$

where

$$\beta_1 = K_{b1} \quad \beta_2 = K_{b1} \times K_{b2}, \text{ etc.}$$

In addition, the stepwise stability constant K_{MLH} is frequently used when protonation of the metal complex is examined (301):

$$K_{MLH} = \frac{[MLH^{n+1}]}{[ML^{n+}][H^+]}$$

When discussing the affinity of a ligand or comparing the affinities of a set of ligands for a particular metal ion (or set of metal ions), each of these constants has its own advantages and disadvantages. Although stability constants are commonly reported (especially in the chemical literature), the units and magnitude of β are only the same for two complexes of the same stoichiometry. If complexes with different stoichiometries are to be compared, it is more useful to compare the binding constants, K_b. Dissociation constants are frequently used when discussing the affinity of the metal for a biomolecule, because, for a 1:1 complex, $K_d = [M^{n+}]$, the concentration of "free" metal in solution that half saturates that biomolecule with metal. [Alternatively, some authors use $pM = -\log[M]$, where $[M]$ is the free metal concentration, for this value is independent of the stoichiometry of the complex and takes into account all of the simultaneous equilibria in a given solution. However, pM must necessarily be reported for a specific set of conditions (e.g., pH), total ligand and metal concentrations, and ionic strength (302).] Here, because we are primarily comparing the stabilities of 1:1 lead/ligand complexes (for which $K_{ML} = K_b = \beta_1$), we primarily use either K_{ML} (for small molecules) or K_d (for proteins).

1. Methods for Determining Lead(II)–Ligand Binding Constants

Several techniques have been employed to examine the thermodynamics of Pb(II)–ligand interactions and determine consistent values of log K_{PbL}. By far, the most common method used to date is a direct determination by potentiometry in aqueous solution. However, in cases where the reagents or the complex are insoluble, other techniques have been employed (303). Generally, one can use any method that can measure the concentration of at least one of the species involved in the formation of a metal complex at equilibrium (301), provided that the concentration of the species and the stoichiometry of the reaction provide enough information to account for all of the species present in solution.

a. Potentiometry. Approximately 80% of all small molecule metal-binding constants have been determined using potentiometric titrations (301, 304, 305).

In a potentiometric titration, the pH is determined at every point after an aliquot of base is added to a solution containing a solution of the metal–ligand complex. The constant reported is for a particular temperature and ionic strength [see reviews by Harris, Martell, and Motekaitis for a complete description of the technique (306, 307)]. Potentiometry is most useful for determining stability constants at pH values that range from two to twelve. If the metal forms a complex that is too tight for the K_{ML} to be determined by direct titration, then the binding constants must be determined indirectly either by ligand–ligand or metal–metal competition experiments.

b. Spectrophotometry. For situations where potentiometry is not ideal (such as for conditions requiring a pH > 12 or the maintenance of a stable pH), absorption spectroscopy can be useful. For ligands containing thiol groups, the lead-binding constant can be determined directly by monitoring lead-thiolate CT bands in the UV (37). Otherwise, competition titrations with a metal ion having a visible signature [e.g., Co(II)] can be used (37, 308–310).

c. Polarography. Polarography can measure very low concentrations of metal ions ($\geq 10^{-6}$ M), and hence tends to be useful for compounds that are only sparingly soluble, such as those of posttransition metals such as Pb(II) (301). The technique involves measuring peaks of a polarographic wave of current as a function of applied potential and decreasing pH. The stability constant for a complex is determined from a plot of peak potential versus pH. [For a more complete description of this technique, see (301, 303).] Polarography has been used to determine binding constants for Pb(II) to a wide variety of ligands, including EDTA and NTA (311), fulvic acids in river sediments (312, 313), and macrocycles (314).

d. Nuclear Magnetic Resonance Spectroscopy. This spectroscopy has not been used extensively to determine stability constants for Pb(II) complexes to small molecules, primarily because most systems to date could be studied by the methods outlined above. However, NMR spectroscopy has attractive features for studying metal binding to peptides, which tend to be difficult to study using potentiometry or polarography because they cannot tolerate fluctuations in pH, and for systems that are not soluble in water (171, 198).

2. Thermodynamics of Lead–Small Molecule Interactions

A summary of known binding constants for Pb(II) to common functional groups, biological molecules, macrocycles, and chelation therapy agents is provided in Tables IX–XI (264, 270, 306, 315–329). Structures of chelating ligands and macrocycles discussed in the text are presented in Table VIII.

TABLE IX

Protonation and Lead(II)-Binding Constants in Aqueous Solution for Specific Functional Groups, Amino Acids, Nucleic Acids, and Buffers[a]

Compound	pK_a1	pK_a2	$\log K_{PbL}$	$\log K_{PbLH}$	$\log K_{PbL} - \log K_{ML}$		References
					$M = Zn^{2+}$	$M = Ca^{2+}$	
Functional Groups							
OH	13.78		6.00		1.30	5.00	270
NH	9.5		2.00				270, 315
NH_3	9.28		1.60[b]		−0.61[b]	1.8	270, 315
Ethylenediamine	9.89	7.08	5.04		−0.66		270, 315
Acetate	4.6		2.15				270, 315
Imidazole	7.01		2.30		−0.26	2.4	270, 315
Thiol	9.4		7.30		1.6	7.85	270, 315
Phosphate	12.35[c]	7.2[c]	3.20				270, 315
Chloride			1.08		1.08[b]	1.0	270, 315
Amino Acids							
Glycine	9.57	2.36	5.50		0.12	4.11	264
Glutamic acid	9.59	4.20	4.70		0.21	3.27	270
Histidine	9.08	6.02	6.80		−0.59		264
Mercaptoethylamine	10.71	8.21	11.10		1.2	8.89	264
Cysteine	10.29	8.15	12.20		3.03	9.7	326
Methionine	9.05	2.20	4.38		0.01		270
Glutathione	9.53	8.64	10.60[d]	6.91[e]	2.03[d]		264, 270
Guanosine	9.15	2.15	3.50		0.94	3.6	264, 270
Diphosphate	9.40[c]	6.70[c]	6.40				264

Nucleotides and Phosphonates

Methyl phosphonate (MeP^{2-})	7.53		3.60	327, 329
Ethyl phosphonate (EtP^{2-})	7.77		3.69	327, 329
n-Butyl phosphate (BuP^{2-})	6.72		3.27	327, 329
Phenyl phosphate (PhP^{2-})	5.85		2.84	327, 329
4-Nitrophenyl phosphate (NPhP^{2-})	5.05		2.36	327, 329
D-Ribose 5′-monophosphate (RibMP^{2-})	6.24		3.01	327, 329
Uridine 5′-monophosphate (UMP^{2-})	6.15		2.80	327, 329
Cytidine 5′ monophosphate (CMP^{2-})	6.19	4.33	2.93	327, 329
Adenosine 5′-monophosphate (AMP^{2-})	6.21	3.84	2.92	328, 329
Guanosine 5′-monophosphate (GMP^{2-})	9.49	6.25	3.23	324, 328, 329
Inosine 5′-monophosphate (IMP^{2-})	9.02	6.22	3.06	328, 329
1-(2′-Deoxy β-D-ribofuranosyl) thymine 5′-Monophosphate (dTMP^{2-})	6.36		2.93	327, 329

Buffers

Bis–tris	6.56		4.32	1.94	317
Tris	8.31		2.70	0.76	317
Citrate	5.67	4.35	4.44^f	2.98^f	−0.94	270, 315

^{a} Ionic strength $I = 0.10$ and $T = 25°C$, unless otherwise noted.
^{b} $I = 0.0$.
^{c} $I = 1.0$.
^{d} $I = 3.0$.
^{e} $T = 37°C$.
^{f} $I = 0.15$, $T = 37°C$.

TABLE X

Protonation and Lead(II)-Binding Constants in Aqueous Solution for Chelating Agents[a]

| Compound | pK_a1 | pK_a2 | log K_{PbL} | log K_{PbLH} | log K_{PbL} − log K_{ML} | | Reference |
					M = Zn^{2+}	M = Ca^{2+}	
Nitrilo-2,2′,2″-triacetic acid (NTA)	9.65	2.48	11.34		0.68		270
Iminodiacetic acid (IDA)	9.34	2.61	7.41		0.17		270
Ethylenediamine-$N,N,N′,N′$-tetraacetic acid (EDTA)	10.17	6.11	18.10		1.66	7.49	270
1,2-Propylenediaminetetraacetic acid (1,2-PDTA)	10.85	6.23	18.92		1.6		270
DL-(Methylene)dinitrilodiacetic acid $N,N′$ diamide[b]	7.56	3.80	16.89		5.67	10.21	318
meso-2,3-Dimercaptosuccinic acid (DMSA)	11.15	9.62	17.40				270
2,3-Dimercaptopropanol (BAL)	10.68	8.65					270
D-Penicillamine (DPA)	10.60	7.95	12.3	4.05	2.9		270
$N,N′$ Ethylene-L-cysteine (EC)	11.14	9.88	19.86	27.25	−1.12		325

[a] Ionic strength $I = 0.10$ and $T = 25°C$, unless otherwise noted.
[b] $T = 37°C$.

Table XI

Protonation and Lead(II)-Binding Constants in Aqueous Solution for Macrocyclic Ligands[a]

Compound	pK_a1	pK_a2	log K_{PbL}	log K_{PbL} − log K_{ML}		References
				M = Zn^{II}	M = Ca^{II}	
1,4,7-Triazacyclononane (9-ane N_3)	10.44	6.81	11.0	−0.6		270
1-Oxa-4,7-diazacyclononane (9-ane N_2O)			5.17	−1.19		320, 321
1,4,7-Trioxa-10-thiacyclododecane (12-ane O_3S)			0.94[b]			270
1,4,7,10-Tetraoxa cyclododecane (12-crown-4 ether)			2.00			270
12-ane N_3O	10.18	8.56	11.54	1.0		320, 321
1,4,7,10-Tetraazacyclodecane (12-ane N_4)	10.60	9.60	15.90	−0.3	12.8	331
DOTA (12-ane N_4 with 4 pendant acetic arms)			22.7	4.0	6.3	316, 322
DOTAM (12-ane N_4 with 4 pendant amide arms)			19.00	8.5		332
1,4,7,10-Tetra-(2-hydroxypropyl)-1,4,7,10-tetraazacyclododecane (THP-12 ane N_4)	9.98	8.26	15.07	1.62	9.4	331
4,7,10-Tri-(2-hydroxypropyl)-1-oxa-4,7,10-triaazacyclododecane (THP-12 ane N_3O)	10.14	6.46	12.17	2.3	7.6	331
1,4,7,10,13-Pentaoxa cyclopentadecane (15-crown-ether)			1.85			270
1,4,7,10,13-Pentaaza cyclopentadecane (15-ane N_5)	10.73	9.53	17.3	−1.8		270
1,4,7,10-Tetraoxa-13-thiacyclopentadecane (15-ane O_4S)			1.65			270
1-Oxa-4,13-dithia-7,10-diazacyclopentadecane (15-ane ON_2S_2)	8.90	5.21	6.80	2.4		270
18-crown-6-ether			4.27			270
1,4,10,13-Tetraoxa-7,16-diazacyclooctadecane (18-ane O_4N_2)	9.08	7.94	6.80			270
1,4,10,13-Tetraoxa-7,16-dithiacyclooctadecane (18-ane O_4S)			3.13			270
1,4,10,13-Tetraoxa-7,16-diazacyclooctadecane-7,16-diacetic acid (DAK-22)	9.04	7.81	14.54	5.64	5.94	330

[a] Ionic strength $I = 0.10$ and $T = 25°C$, unless otherwise noted.
[b] $I = 0.0$.

a. **Thermodynamic Preferences of Lead(II): Donor Atom.** When the affinity of lead for different *isolated* functional groups is examined, Pb(II) exhibits a clear preference for thiols over harder oxygen- and nitrogen-containing functionalities, as would be predicted from Pearson's hard–soft rules (202, 265). (See Table IX, "Functional Groups.") Lead's preference for thiols can also be observed from comparison of the affinities of lead for a series of simple ligand systems, such as amino acids (see Table IX, "Amino Acids"): The affinity of lead for cysteine ($\log K_{PbL} = 12.20$) is considerably greater than that of lead for glutamic acid ($\log K_{PbL} = 4.70$) or histidine ($\log K_{PbL} = 6.80$) However, when Pb(II) binds to more complex molecules, other factors, such as steric effects and chelate effects necessarily contribute to affinity and selectivity for lead. When the affinities of lead for different chelating agents are examined (Table X), it is clear that lead can have at least as high affinity for molecules that contain only oxygen and nitrogen donor groups (e.g., EDTA, $\log K_{PbL} = 18.10$) as for molecules that contain a mixture of sulfur and oxygen groups (e.g., DMSA, $\log K_{PbL} = 17.40$).

b. **Thermodynamic Preferences of Lead(II): Coordination Number.** Relatively few systematic studies have been reported in which both the stability and the structure of Pb(II) complexes were investigated. Notable exceptions include studies by Parkin and co-workers on models for the zinc protein ALAD, studies by Raymond and co-workers on thiohydroxamic acid compounds, and a series of studies on Pb(II) and macrocycles (182, 198, 265–267, 314, 325, 330, 331).

From these studies, several trends emerge. We learn that lead does not exhibit any single preferred coordination number. Tight binding is seen for both tetra- and hexadentate ligands (Table X): the best thiohydroxamic acid is likely to be four coordinate and exhibits a $\log K_{PbL}$ of 12.53, DMSA is likely to be four coordinate and exhibits a $\log K_{PbL}$ of 17.4, and EDTA, which is hexadentate, exhibits a $\log K_{PbL} = 18.1$. However, in the case of macrocyclic compounds, four-, five-, and six-coordinate structures with mostly nitrogen donors tend to have higher affinities for Pb(II) than do oxygen-rich or lower coordination number macrocycles (see Table XI). A general observation can be made that across categories, small molecules with more nitrogen donors tend to have higher binding constants than do comparable ligands with fewer nitrogen donors.

In order for a chelating agent to be therapeutically useful, it must bind lead not only tightly, but also selectively. Good selectivity over Ca(II) is fairly easy to come by, as is witnessed by the relatively large number of compounds that exhibit large, positive values for ($\log K_{PbL} - \log K_{CaL}$) (see Tables X and XI). By contrast, it is relatively difficult to design a ligand that is selective for lead over Zn(II), another common ion in biological systems. The best selectivity for Pb(II) over Zn(II) is provided by the macrocycle DOTAM [1,4,7,10-tetrakis (acetamido)-1,4,7,10-tetraazacyclododecane], which exhibits not only a $\log K_{PbL}$

of 19.0, but also a (log K_{PbL} − log K_{ZnL}) of 8.5 (332). Increasing the number of sulfur donors seems to improve selectivity over zinc: A novel tridentate ligand with three sulfur groups [tris(2-mercapto-1-phenylimidazolyl)hydroborate] exhibits a clear preference for Pb(II) over Zn(II) by a difference of 2.5 log units (198).

c. Thermodynamic Preferences of Lead(II): Chelate Ring Size. Five-membered chelate rings are clearly preferred by large metal ions such as Pb(II) over chelate rings of other sizes. The ability of a ligand to form this preferred ring size is typically more important than macrocyclic or chelate effects. Multidentate structures that maximize the number of five-membered rings (e.g., EDTA, which can form five five-membered rings) tend to have very high affinities for lead (265, 267, 301, 331, 333–335).

d. Thermodynamic Preferences of Lead(II): Macrocycles versus Non-macrocyclic Ligands. Macrocycles do not appear to have an improved affinity or selectivity for Pb(II) compared to equivalent nonmacrocyclic ligands (265, 267, 330, 331). Other factors, such as the ability to form five-membered chelate rings and the presence of sufficient nitrogen donors to fill lead's coordination sphere, seem to dominate effects on the binding affinity of a given macrocycle for Pb(II). In addition, macrocycles are not necessarily desirable for use as chelation therapy or lead-sequestration agents because of their relatively slow rates of lead association (see Section IV.C.3) (265, 336).

e. Thermodynamic Preferences of Lead: Stereochemical Activity of the Lone Pair. From the studies conducted to date, it is difficult to deconvolute the ability of a ligand to confer a stereochemically lone pair from other factors that influence affinity or selectivity for Pb(II). Hancock and Martell (265) and Hancock et al. (330) suggested that increasing the number of donor nitrogens provided by a ligand improves the affinity for Pb(II) by "activating" the lone pair. However, the selectivity for Pb(II) over Zn(II) is worse when more than three donor nitrogens are present in the macrocycle. Hancock and co-workers (330) argue that this lack of selective binding arises because the smaller Pb(II) with the stereochemically active lone pair behaves more like Zn(II). However, this observation is not general: Some Pb(II) complexes that exhibit a stereochemically active lone pair also exhibit reasonable selectivity for lead over zinc (e.g., PbEDTA, where log K_{Pb} − log K_{Zn} = 1.6 (270).

3. Thermodynamic Stability of Lead–Protein Interactions

Many of the symptoms associated with lead poisoning are believed to arise due to the interactions between lead and proteins that naturally bind either calcium or zinc (see Section VI) (41, 244, 337–340). This hypothesis raises several questions about the thermodynamic stability of lead–protein interactions:

- What is the affinity of lead for zinc- and calcium-binding sites in proteins?
- Does lead bind to any of the proposed target proteins more tightly than the native metal ion?
- How is it possible that lead might target both thiol-rich zinc-binding sites and carboxylate-rich calcium-binding sites in vivo?

Three methods have been used to determine virtually all of the lead–protein dissociation constants reported in the literature: equilibrium dialysis (52), radiotracer studies (341–344), and spectroscopic titrations (37, 53, 89, 344, 345). Of these, the most rigorous quantitative data to date have been provided by spectoscopic studies for which the binding constants are evaluated using a rigorous fitting procedure that takes into consideration multiple equilibria in solution. By contrast, the dissociation constants calculated from equilibrium dialysis and radiolabeling studies that have been analyzed using graphical methods provide much less rigorous estimates for binding constants. [For discussion of the proper construction and analysis of Scatchard plots—as well as of common myths and pitfalls—the reader is directed to the recent book by Klotz (346).] Here, we focus on the results of the spectroscopic studies that have been conducted to date (see Table XII) (37, 52, 53, 89, 341–345, 347–349); details of how these experiments were performed and information about the relevance of the interactions to lead toxicity are discussed in Section VI.

Spectroscopic studies of lead binding to zinc proteins have reported a wide range of dissociation constants, from $>10^{-4}$ M for carboxypeptidase A (345) to 3.9×10^{-14} M for the four cysteine zinc finger consensus peptide, CP–CCCC (37) (see Table XII). By far, the tightest lead binding occurs for sites rich in cysteine sulfur coordination. The weakest binding sites are those where the lead is bound to carboxylate oxygens and histidine residues, such as that found in carboxypeptidase A. Lead is also able to compete with zinc best for the cysteine-rich sites; lead binds more tightly than zinc to a four cysteine site in a consensus zinc finger domain. These results indicate that although lead does bind to zinc enzymes such as carboxypeptidase A, these proteins are unlikely to be a target of lead poisoning. More likely targets are the cysteine-rich proteins, such as structural zinc-binding sites that contain Cys_4 zinc-binding sites (e.g., estrogen receptor), and the zinc enzyme ALAD, which has a novel Cys_3 catalytic site (244–246).

Fewer studies have been conducted to directly measure the thermodynamics of lead binding to calcium proteins; most of the studies performed to date have instead focused on activity assays for the proteins (see Section VI) (338, 340,

TABLE XII

Dissociation Constants for Lead (K_d^{Pb}) and Selectivity Over Other Metal Ions (K_d^M/K_d^{Pb}) for Proteins (37, 52, 53, 89, 341–345, 347–349)

Protein	K_d^{Pb}	K_d^M	K_d^M/K_d^{Pb}	M	Ligand Atoms	Notes	References
CP–CCCC	3.9×10^{-14}	1.1×10^{-12}	31	Zn	S	a	37
CP–CCHC	8×10^{-11}	3.2×10^{-12}	0.04	Zn	S, N	a	37
CP–CCHH	5×10^{-11}	5.7×10^{-12}	0.11	Zn	S, N	a	37
HIV–CCHC	3.0×10^{-10}	7.0×10^{-11}	0.23	Zn	S, N	a	37
CadC	$\leq 9.1 \times 10^{-8}$	2.3×10^{-13}	2.6×10^{-6}	Cd	S	a	53
Metallothionein	8.7×10^{-7}	5×10^{-13}	5.7×10^{-7}	Zn	S	b	52, 348, 349
HP2	1.0×10^{-6}	6.3×10^{-8}	0.063	Zn	Unknown, believed to be S	a	89
α-Lactalbumin	1.6×10^{-5}	5.0×10^{-6}	0.31	Zn			
	1.0×10^{-5}	2×10^{-6}	0.20	Zn	O	a	345
Carboxypeptidase A	$>10^{-4}$	3×10^{-11}	3×10^{-7}	Zn	O, N	a	344, 347
Rat kidney cytosolic proteins	5×10^{-8}				Unknown	c	341
Human renal cytosolic proteins	1.4×10^{-8}				Unknown	c	343
Human brain cytosolic proteins	1×10^{-9}				Unknown	c	342

[a] The parameters K_d^{Pb} and K_d^M were determined by spectroscopic titration (37, 53, 89, 344, 345, 347).
[b] The parameter K_d^{Zn} was determined by spectroscopic titration (348, 349); K_d^{Pb} was determined by equilibrium dialysis and Scatchard plot analysis (52).
[c] The parameter K_d^{Pb} was determined by radiolabel studies and Scatchard plot analysis (341–343).

350–355). Spectroscopic studies in competition with the fluorescent metal ion Tb(III) suggest that Pb(II) binds more tightly than calcium to the calcium sensing protein synaptotagmin (40). In theory, it is possible that lead could bind as tightly to these sites as to the zinc sites. After all, lead does bind as tightly to EDTA (a carboxylate-rich chelator) as to DMSA (a thiol-rich chelator). However, the studies conducted on synaptotagmin suggest that the absolute affinity of lead for this protein is many orders of magnitude weaker than the affinity of lead for the "best" zinc-binding sites. Whether or not lead targets both calcium and zinc proteins—or either one—is likely to depend both on the concentration of lead in different cell types and on the subcellular distribution of lead and speciation of lead within cells (see Section VI).

E. Conclusions about Lead–Ligand Kinetics and Thermodynamics

From the broad range of studies that have been conducted to date on both the kinetics and thermodynamics of lead–ligand interactions, we draw the following conclusions:

- Lead(II) is, in general, kinetically labile and undergoes ligand-association and exchange reactions that are rapid on a physiological or environmental timescale.
- Small molecules (e.g., buffer) and pH can have a pronounced effect on lead–ligand exchange rates and are expected to have a profound effect on the kinetics of lead complexes in biological and environmental systems.
- Lead(II) can bind tightly to a wide range of ligands but has the highest affinity for multidentate ligands that can form five-membered chelate rings.
- More studies are needed on the kinetics and thermodynamics of lead–ligand interactions in complex systems (including mixed-phase systems) so that we can better understand lead speciation in vivo and in the environment.

V. LEAD IN THE ENVIRONMENT

A. Introduction

Lead is prevalent in nature and is found in a wide variety of geochemical environments. Not surprisingly, many parallels can be found between the environmental chemistry of lead and the fundamental coordination chemistry of Pb(II), including preferences in donor atoms, structural motifs, and thermo-

TABLE XIII
Lead Concentrations in Common Rock-Forming Minerals (47, 356)[a]

Mineral Name	Composition	Mean Pb Conc. (ppm)	Range (ppm)	No. of Analyses
Potassium feldspar, pegmatitic	$KAlSi_3O_8$	98	0.9 to >10,000	219
Potassium feldspar, granitic	$KAlSi_3O_8$	53	3–700	419
Plagioclase feldspar	$NaAlSi_3O_8$	20	1–75	61
Amphiboles	$(Ca, Mg, Fe, Mn)SiO_3$; $Na(Al, Fe)(SiO_3)_2$	15	1–70	85
Biotites	$HK(Mg,Fe)_2Al_2(SiO_4)_3$	21	4–95	259
Muscovite	$H_2KAl_3(SiO_4)_3$	26	5–77	32
Olivines	$(Ca, Mg, Mn)_2SiO_4$	1.9	0.2–8.5	15
Pyroxenes	$(Li_2, Na_2, Ca, Mg, Mn, Fe, Zn)SiO_3$	6	0.3–20	20
Quartz	SiO_2	<1	0–5	15
Chalcopyrite	$CuFeS$	22	10–37	32
Pyrites, sedimentary	FeS_2	62	2 to >1000	297
Pyrites, metamorphic	FeS_2	56	4 to >500	20
Arsenopyrite	$FeSAs$	118	78–160	3
Pyrrhotite	FeS	18	4–28	7
Cinnabar	HgS	150		2
Sphalerite	ZnS	88	2–800	35
Barite	$BaSO_4$	40	5–200	15
Apatite	$Ca_5(PO_4)_3F$	102	5 to >2000	58
Tetrahedrite	Cu_3SbS_3	750	80–2100	8
Magnetite	Fe_3O_4	6	1.5–27	17
Halite	$NaCl$	0.2	0–1.7	109
Sylvite	KCl	0.6	0.04–2.5	27
Limonite	$2Fe_2O_3 \cdot 3H_2O$	370	0 to >1000	71
Manganite	$MnO(OH)$	290	100 to >3000	290
Braunite	Mn_2O_3	390	90 to >10,000	7
Hausmanite	Mn_3O_4	1500		2

[a] Adapted from (47).

dynamic and kinetic trends. As is the case for small molecules, lead in minerals is both chalcophilic (exhibits sulfur affinity) and lithophilic (exhibits oxygen affinity) (47). Lead(II) easily forms sulfides, selenides, tellurides, oxides, sulfates, arsenates, phosphates, and carbonates, among other compounds. In mineral lattices, Pb(II) will replace ions with sufficiently similar ionic radii, including Na^+, K^+, Ca^{2+}, Sr^{2+}, and Ba^{2+}. High concentrations of Pb(II) are also found in potassium feldspars (silicates), apatites, and iron and manganese oxides (Table XIII) (47, 356).

TABLE XIV

Lead Levels in Water, Diet, and Air[a]

Source of Lead	Typical Level in United States Today	Legal Limit	References
Drinking water	0.5 μmol L^{-1}	15 ppb (72 nM)	5, 357
Overall diet	0.5–0.15 μmol/day		5
Lead in air (Urban)	2.5–50 nmol m^{-3}	0.15 mg m^{-3}	358
Lead in air (Remote areas)	0.04 nmol m^{-3}	0.15 mg m^{-3}	358
Daily intake		0.007 mg kg^{-1}	358

[a] By comparison, blood lead levels in the United States average 2 μg(dL^{-1}) (0.1 μmol L^{-1}) and onset of symptoms occurs at 10 μg(dL^{-1}) (0.5 μmol L^{-1}) or lower (10, 19).

However, not all of the lead currently in the environment is in a "natural" state. Lead's usefulness and natural abundance have led to its current position as a ubiquitous pollutant: Some 300 million tons of Pb from anthropogenic sources is currently dispersed worldwide (Fig. 4 and Table XIV) (5, 10, 18, 23, 357, 358). Here, we discuss the properties of some of the major minerals of lead and then turn to a discussion of anthropogenic lead in soil, water, and air. For a more in-depth discussion, the reader is directed to several books that discuss the environmental biogeochemistry of lead in greater detail (3, 47, 359, 360).

B. Questions of Interest

It would be impossible to describe all of the studies that have been conducted on lead and its role in the environment here—that is a topic for another review. Rather, our goal here is simply to discuss the major forms that lead takes in the environment and to address the following questions:

- What is the speciation of lead in natural systems?
- How does the speciation of lead affect its solubility and bioavailability?
- What can we learn about historical lead levels from geochemistry?

C. Lead Minerals

World lead mine production currently averages ∼3 million metric tons per year (361–363). Major lead producing countries in 2000 included Australia (630,000 metric tons), China (560,000 metric tons), the United States (480,000 metric tons), Peru (270,000 metric tons), Canada (140,000 metric tons), and Mexico (140,000 metric tons) (362). Perhaps even more impressive is that in the United States in 2000, an additional 1.1 million metric tons of lead, or

~66% of annual domestic consumption, were obtained from recycled lead, nearly all from postconsumer scrap. Used batteries alone accounted for ~1 million metric tons of recycled lead (362). Recycling efficiency for lead is currently estimated at 95% (364).

Currently in the United States, most of the lead produced comes from mines in Missouri, Alaska, Idaho, and Montana, primarily from lead–zinc and lead ores (361, 362). Worldwide, major lead deposits exist in association with zinc, silver, and/or copper (362). There are five major geological types of lead deposits: volcanic-hosted massive sulfide deposits [Canada, Cyprus, Japan, Australia (Tasmania), Turkey]; sediment-hosted deposits of sulfides interbedded with shales, and so on, formed in an anaerobic marine environment [Australia, Canada, Germany, United States (Alaska)]; strata-bound carbonate deposits containing sulfide minerals [United States (Mississippi Valley), southern European Alps, Canada, Poland]; sandstone-hosted deposits of finely crystalline sulfides (Canada, France, Morocco, Sweden); and vein deposits of coarsely crystalline sulfide aggregates (western United States, Germany, Japan, Mexico, Peru) (364). The wide variety of compositions seen for lead minerals is illustrated by the representative lead minerals listed in Table XV (3, 47). Below, we discuss the lead minerals that are most prevalent in nature in more detail.

1. *PbS (Galena)*

Lead sulfide (PbS, galena) is the primary lead ore (47) and was undoubtedly the first to be used by humans (3). Galena is often found alongside silver ore and is unearthed in the process of mining silver; the concentration of silver in galena ranges from ~1–8000 μg Ag/g PbS (3). Lead sulfide is black and opaque with a metallic luster (365). It crystallizes with a perfect cubic rock salt structure and is the only lead sulfide mineral with octahedral symmetry around the metal ion. By contrast, in other minerals such as bournonite ($PbCuSbS_3$), seligmannite ($PbCuAsS_3$), and marrite ($PbAgAsS_3$), lead exhibits an irregular polyhedral coordination geometry that is generally considered a result of the stereochemically active $6s^2$ lone pair (47).

2. *$PbCO_3$ (Cerussite)*

Cerussite ($PbCO_3$) is also naturally abundant and crystallizes in an orthorhombic form isomorphous with strontianite ($SrCO_3$) and aragonite ($CaCO_3$). $PbCO_3$ is often found with galena in ore veins because it forms when carbonate-rich waters interact with PbS in the common lead–zinc–limestone ore association (47). The derivative $PbCO_3 \cdot Pb(OH)_2$, known as ceruse or "white lead", was the most important artificial white pigment in antiquity (3). Ceruse is made by treating naturally occurring lead compounds [mixtures of $PbCO_3$, PbO, and

TABLE XV

Names and Composition of Representative Lead Minerals (3, 47, 360)[a]

Mineral	Composition
Galena	PbS
Geocronite	$Pb_5(Sb,As)_2S_8$
Beegerite	$Pb_6Bi_2S_9$
Bournonite	$PbCuSbS_3$
Meneghinite	$CuPb_{13}Sb_7S_{24}$
Boulangerite	$Pb_{2-5}Sb_{2-4}S_{5-11}$
Cosalite	$Pb_2Bi_2S_5$
Selenocosalite	$Pb_2Bi_2(S,Se)_5$
Kobellite	$Pb_2(Bi,Sb)_2S_5$
Selenokobellite	$Pb_2(Bi,Sb)_2(S,Se)_5$
Franckeite	$Pb_5Sn_3Sb_2S_{14}$
Cylindrite	$Pb_3Sn_4Sb_2S_{14}$
Jamesonite	$Pb_4FeSb_6S_{14}$
Semseyite	$Pb_9Sb_8S_{21}$
Zinkenite	$Pb_6Sb_{14}S_{27}$
Plagionite	$Pb_5Sb_8S_{17}$
Altaite	PbTe
Nagyagite	$Pb_5Au(Te,Sb)_4S_{5-8}$
Wittite	$Bi_6Pb_5(Se,S)_{14}$
Penroseite	$(Ni,Cu,Pb)Se_2$
Litharge	PbO
Massicot	PbO
Minium	Pb_3O_4
Pyromorphite	$Pb_5(PO_4)_3Cl$
Plumbogummite	$PbAl_3(PO_4)_2(OH)_5 \cdot H_2O$
Tsumebite	$Pb_2Cu(OH)_3(PO_4) \cdot 3H_2O$
Percyclite	$Pb_3(CO_3)Cl_2$
Cerussite	$PbCO_3$
Anglesite	$PbSO_4$
Boleite	$Pb(Cu,Ag)Cl_2(OH)_2 \cdot H_2O$
Argentian plumbojarosite	$(Pb,Ag)Fe_{3-6}(SO_4)_{2-4}(OH)_{6-12}$
Benjaminite	$Pb(Cu,Ag)Bi_2S_4$
Fizelyite	$Pb_5Ag_2Sb_8S_{18}$
Ramdohrite	$Pb_3Ag_2Sb_3S_{13}$
Andorite	$PbAgSb_3S_6$
Hutchinsonite	$(Pb,Tl)_2(Cu,Ag)As_5S_{10}$
Seligmannite	$PbCuAsS_3$
Marrite	$AgPbAsS_3$
Lengenbachite	$Pb_6(Ag,Cu)_2As_4S_{13}$
Diaphorite	$Pb_2Ag_3Sb_8S$
Freieslebenite	$Pb_3Ag_5Sb_5S_{12}$
Owyheenite	$Pb_5Ag_2Sb_6S_{15}$
Schirmerite	$PbAg_4Bi_4S_9$

[a] Adapted from (3).

Pb(OH)$_2$] with carbon dioxide to form a species in which the ratio of PbCO$_3$ to Pb(OH)$_2$ ranges from 1.5 to 3 (3).

3. *PbO (Litharge and Massicot)*

Lead oxide exhibits two common forms: tetragonal red α-PbO (litharge) and orthorhombic yellow β-PbO (massicot) (47). Litharge exhibits a layered structure. Lead is bound on one side by four oxygen atoms arranged in a square; the Pb(II) lone pair is presumed to occupy the other side, forming the apex of a tetragonal pyramid (366). Between the layers, there is only Pb—Pb contact. The structure of massicot (β-PbO) is similar to that of α-PbO, but instead of four Pb—O bonds of equal length, two are shortened and two lengthened compared to those in litharge (366). In their pure forms, litharge is the more stable compound; but the presence of common impurities (e.g., Si, Ge, P, As, Sb, Se, Te, Mo, and W) stabilizes massicot over litharge. Therefore, PbO in the environment is often yellow or orange (i.e., impure massicot, or a mixture of massicot and litharge), not red (pure litharge) (47).

4. *Pb$_3$O$_4$ (Minium)*

Minium (Pb$_3$O$_4$) is commonly known as "red lead" and has been used as a pigment for millennia (3). Minium is a mixed-valence lead oxide, effectively 2Pb^{II}O • Pb^{IV}O$_2$, or Pb(II) plumbate. In the structure, chains of Pb^{IV}O$_6$ octahedra sharing opposite edges are linked by Pb(II) atoms, each of which has three nearest neighbor O atoms in a pyramidal arrangement (366). The tendency of lead to form mixed-valence oxides—whereas naturally occurring Si, Ge, and Sn oxides contain only Si(IV), Ge(IV), and Sn(IV)—highlights the relativistic effect-induced stability of Pb(II) over Pb(IV) (35, 47).

5. *PbHPO$_4$ (Schulterite) and Other Lead Phosphates*

Lead phosphates are extremely insoluble and are believed to be the most stable form of lead in soil (47, 367). The most important phosphate minerals are schulterite, (PbHPO$_4$); the pyromorphites (or lead apatites), [Pb$_5$(PO$_4$)$_3$X, where X = halogen or hydroxyl group]; plumbogummite, [PbAl$_3$(OH)$_5$(PO$_4$)$_2$ • H$_2$O]; and tsumebite, [Pb$_2$Cu(OH)$_3$(PO$_4$) • 3H$_2$O] (47).

D. Lead in Soil

Lead occurs naturally in all soils at levels ranging from <1 ppm to >10% (in ore bodies) (47). The average concentration of lead in uncontaminated soils is ~20 ppm (47, 368). Soil with anthropogenic lead contamination, however, is a major health risk and a main source of human lead exposure (25, 369). Urban areas are particularly likely to have high concentrations of lead in soil [as high as 7900 ppm in a Chicago residential neighborhood (25)] because of the effects of industrial emissions, leaded-gasoline emissions, and dust from lead-based paint (25, 26).

There have been a number of studies on the spatial distribution of lead in soils, which often yield critical insights into the probable pollution source (25, 370). Many factors affect the local concentration of lead in soil. For example, soil lead concentrations at urban intersections depend not only on distance from the traffic lights but also on the light cycling time (371). This pattern suggests that a primary source of lead contamination in soil is leaded gasoline emissions: In areas where leaded gasoline was in use, the concentration of lead from automobile emissions is greatest within 100 m of a road and near roads with higher speeds (372, 373). These trends in spatial distribution of lead levels are valuable for identifying high-risk areas and exploring potential historic sources of lead emissions (25, 370).

Of critical interest is the *speciation* of lead in soil, which provides a measure of the bioavailability of the heavy metal; the risks to living organisms depend more on the solubility of lead than on the total lead concentration (253). Speciation of lead in soil samples has been examined directly using spectroscopic methods (see Section III.E) (253, 254, 257). Sequential extraction methods have also been used succesfully to examine the speciation of lead in environmental systems (374–377).

Speciation of lead in soil also has a major effect on its mobility in the environment. Miller and Friedland (378) have shown that the large amount of anthropogenic lead deposited in forests in the northeastern United States between the 1950s and the 1980s is being actively redistributed through the soil profile. This migration of lead through the soil suggests that the high levels of lead deposited from atmospheric pollution after 1960 will begin to appear in upland streams within the next 50 years (378). Although other researchers have not necessarily observed the same degrees of migration, it is clear that the speciation of lead is important to metal ion mobilization (379) and that changes in metal speciation can have a significant environmental impact (380). Changes in metal speciation can be caused by a variety of factors, both natural and anthropogenic. Li and Shuman (381, 382) showed that unlike zinc and cadmium, lead is not mobilized from soils significantly by application of poultry litter extract. Addition of a chelating agent (EDTA), however, did tend to mobilize soil lead (381–383). Once the lead is solubilized, it can readily migrate to groundwater sources or streams.

One goal of speciation and mobilization studies is to apply the insights gained from these studies to develop better methods to remediate environmental lead contamination (253). Nonetheless, the remediation of lead-contaminated soil remains a complex problem. Although the lead can be solubilized and washed away by treatment with an extracting agent such as EDTA (383), this method just moves the lead (and hence the problem) out of the soil and into water systems. Often, a preferable alternate remediation technique is the *immobilization* of lead in soil (367, 384–389). Although a variety of compounds have been found to bind lead tightly in soils [e.g., fulvic acids (386) and iron oxides (385)], several recent studies have indicated that treating contaminated soils with phosphates is the most effective immobilization technique (367, 387, 388, 390–395), because lead phosphates are so insoluble (47, 396). A common method of phosphate application is treatment with apatite [$Ca_{10}(PO_4)_6OH_2$], which is readily available, inexpensive, and easy to apply (e.g., as bonemeal) (367, 388). It has been shown that in the presence of phosphate, pyromorphites [$Pb_5(PO_4)_3X$, X = halogen or hydroxyl group] form readily *in situ* (367, 390, 391, 394, 395). The pyromorphites are among the *least* soluble of the lead phosphates (47, 397), and research by Ruby et al. (367, 398, 399) indicates that these compounds have extremely limited bioavailability even under acid conditions similar to those in the human stomach. Plant uptake of lead from contaminated soil is also decreased dramatically when the soil is treated with apatite (400). Recent work by Hettiarachchi et al. (387) suggests that soil amendments composed of manganese oxides as well as phosphorus are even more effective at forming pyromorphites and immobilizing lead than is remediation with phosphorus alone.

Similar immobilization appears to take place in naturally lead-rich soils (e.g., soils contaminated with lead from natural weathering of bedrock) on a geological timescale (258). Most of the lead in the soil profile is found as plumbogummite [$PbAl_3(PO_4)_2(OH)_5 \cdot H_2O$], a highly insoluble phosphate mineral. Lower in the soil profile, a fraction of Pb forms surface complexes with Mn oxides, but these complexes are less resistant to breakdown than plumbogummite. These results indicate the importance of the role of phosphate minerals in the long-term immobilization of lead in soils, and they suggest that the presence of Al can have an important effect on the nature of Pb-phosphate formation (258). As a result, immobilization of lead *in situ* is likely to be a viable long-term solution for the remediation of contaminated soils.

Another potential treatment for lead-contaminated soils is phytoremediation (389, 400–403). Plants are grown on a contaminated site, they uptake lead, and the plants are then harvested and disposed of safely. There are basically two approaches to choosing plants for phytoremediation. A species may be chosen that hyperaccumulates metals; unfortunately, most such plants have very low biomass so the overall uptake of lead is low (404–406). Alternatively, high biomass plants can be used, with uptake enhanced by soil treatments to mobilize the lead compounds (401–403, 406). The latter technique seems to be generally more effective, as lead uptake by plants is considerably enhanced when the soil is treated with a chelating agent such as EDTA. A variety of chelating agents have been examined, and the order of effectiveness for some of these in phytoremediation is EDTA > HEDTA > CyDTA > DTPA > EGTA > HEIDA > EDDHA ≈ NTA (see Table VIII) (401, 403). Additionally, EDTA was most effective at mobilizing lead from the plant roots to the shoots, which eases harvesting (401). Phytoremediation shows great promise; but much work remains to optimize the techniques, as the amounts of lead taken up by plants even under the most favorable conditions are still very low (0.4 kg Pb ha^{-1}) (403). Depending on the site, plants may be used for phytostabilization (to minimize the amount of bioavailable Pb in the soil) or for phytoextraction, such that the plants are harvested and incinerated. The ash from incineration may be disposed of as hazardous waste or treated to recover/recycle the metal. However, estimates suggest that recovery becomes economically feasible only if plants can accumulate >1% Pb in the shoots or 10,000 mg Pb kg^{-1} (401). Unfortunately, the most efficient plants known (at least as of 1997) accumulate only 1000 mg Pb kg^{-1} in untreated soils (401).

E. Lead in Aquatic Systems

Aquatic lead pollution has been an environmental and public health issue since ancient times (3). Widespread use of lead pipes, lead-lined storage tanks, and pipes joined with lead-based solder led to contamination of drinking water (3). In addition, natural waters have been polluted through atmospheric deposition and industrial effluent, including leaching from mine tailings (3, 407). The speciation of lead in these aqueous systems has important implications, both for lead bioavailability and for the treatment of lead-contaminated systems.

The speciation of lead—and hence the amount of lead leached in water—can vary dramatically depending on environmental factors. Although PbS (the primary lead ore found in nature) is extremely stable and insoluble when it is buried deep in the earth under reducing conditions, the sulfide is readily oxidized in air and oxygen-rich surface water, to yield PbSO$_4$ and H$_2$SO$_4$. These products are highly soluble and not only cause aquatic heavy metal contamination but also contribute to acid mine drainage (407). The solubilization of PbS can also occur directly in the presence of highly acidic waters, as can be seen in the following reactions (48):

$$PbS\ (s) \rightleftharpoons Pb^{2+}\ (aq) + S^{2-}\ (aq)$$
$$S^{2-}\ (aq) + H^+\ (aq) \rightleftharpoons HS^-\ (aq)$$
$$HS^-\ (aq) + H^+\ (aq) \rightleftharpoons H_2S\ (aq)$$
$$PbS\ (s) + 2\,H^+\ (aq) \rightleftharpoons Pb^{2+}\ (aq) + H_2S\ (aq)$$

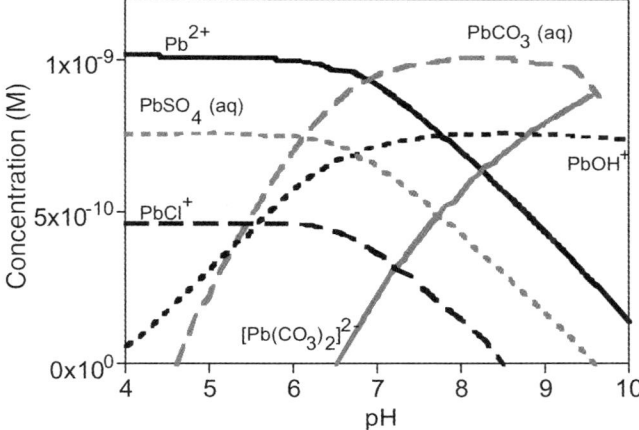

Figure 20. Speciation of Pb(II) in natural waters. $[Pb]_T = 10^{-9} M$; [total carbonates] = $2 \times 10^{-3} M$. [Adapted from (49).]

As a result, the solubility of PbS increases linearly with the H^+ concentration (48) and significant dissolution of PbS can occur in the environment in the presence of highly acidic waters. By contrast, under alkaline conditions, insoluble lead carbonate complexes are stabilized (Fig. 20) (49).

The composition and pH of water also affect the oxidation of lead metal. Modern drinking water treatment is highly effective at removing lead from water before it is pumped to the consumer; most water contamination comes from corrosion in distribution systems, that is, lead pipes and lead-based solder (408). Under acidic or neutral conditions in the presence of oxygen, metallic lead is readily oxidized to Pb^{2+}, as can be seen from the equations below and from Figure 21 (48, 408, 409).

$$2\,Pb\,(s) + O_2\,(g) + 4\,H^+\,(aq) \rightarrow 2\,Pb^{2+}\,(aq) + 2\,H_2O\,(\ell)$$
$$PbO\,(s) + 2\,H^+\,(aq) \rightarrow Pb^{2+}\,(aq) + H_2O\,(\ell)$$
$$Pb^{2+}\,(aq) + H_2O\,(\ell) + CO_2\,(g) \rightarrow PbCO_3\,(s) + 2\,H^+\,(aq)$$
$$3\,Pb^{2+}\,(aq) + 3\,CO_3^{2-}\,(aq) + 2\,H_2O\,(\ell) \rightarrow Pb_3(CO_3)_3(OH)_2\,(s) + 2\,H^+\,(aq)$$

At higher pH, the formation of PbO and Pb_3O_4 is favored. In the presence of CO_2, these Pb oxides are relatively soluble but readily convert to the more insoluble $PbCO_3$ and $Pb_3(CO_3)_3(OH)_2$ (408). From the Pourbaix diagram for Pb (Fig. 21), it is clear that soft water [alkalinity = 0–25 mg L^{-1} $CaCO_3$ (3)] with pH < 6.5 has a much greater corrosive effect on lead pipes than hard, basic water. Hard water actually has dual protective effects: Not only is lead less

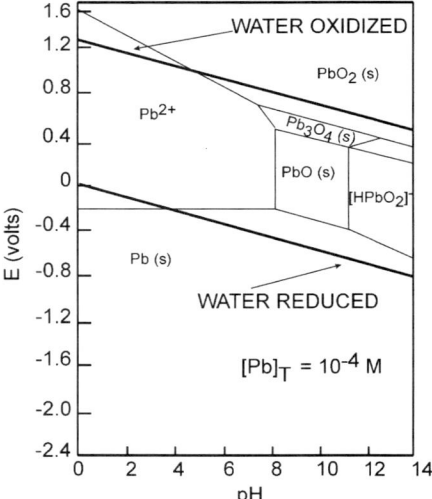

Figure 21. Pourbaix (potential-pH) diagram for the speciation of Pb in water, 25°C. $[Pb]_T$ is the total lead concentration. [Adapted from (409).]

soluble under basic conditions, but the scaly lead carbonate precipitate that forms on the inner walls of the pipes prevents further corrosion of the base metal (49, 408).

Lead is generally removed from drinking water by exploiting the conditions under which insoluble lead compounds form. For example, water treatment with lime ($CaCO_3$) leads to the precipitation of lead carbonates at the water treatment plant. As a result, the buildup of insoluble carbonates in the pipes is decreased, resulting in fewer pipe blockages from scale, but increasing the risk of solubilizing lead in the distribution system (408). In areas with particularly soft water, phosphates are often added to the water supply to decrease lead solubility and passivate the surface of lead pipes (48). Another common water treatment method is adsorption of lead by ferric sulfates or hydrous iron or manganese oxides. Precipitation or coagulation of particulate matter and its subsequent removal serves to remove the lead (408). Similar effects have been observed in natural systems; Taillefert et al. (410, 411) recently reported that lead in a meromictic lake is adsorbed by particles formed by aggregation of hydrous iron oxides and organic matter at the oxic-anoxic transition. Lead is bound tightly enough to these organic–mineral clusters that PbS does not precipitate, even in the anoxic zone (411). It is calculated that 98% of Pb in the lake studied by Taillefert et al. is complexed by organic matter, illustrating the important role that organic ligands can play in the speciation of lead in natural waters. Organic acids, such as humic and fulvic acids, are prevalent in aquatic systems. Although it has been shown that fulvic acids bind Pb(II) ($K_d \approx 10^{-2}$ M) (386, 412) and that the rate of Pb^{II}–fulvic acid dissociation is slow ($k \approx 10^{-5}$ s^{-1}) (413), the complete role of fulvic and humic acids in aquatic metal speciation remain to be elucidated (49, 386, 412–415). It is clear from these studies, however, that both pH and ligand concentrations have critical influences on the aquatic speciation of Pb. As in soil, controlling the metal speciation (e.g., limiting solubility) is a viable method for reducing the bioaccessibility of lead contaminants.

F. Environmental Contamination with Lead: The Historical Record and Geochemistry

How do we know the approximate amount of lead produced during different points in history? Soils and sediments are a valuable source of such information (416). In addition to analysis of archeological remains, scientists have turned to lead deposits in peat bogs and Greenland ice, which are highly stratified and have remained undisturbed for thousands of years (7, 8, 15, 33, 417). Pioneering work by Patterson and co-workers (7) on lead deposits in Greenland ice revealed that widespread environmental pollution with lead occurred during ancient Greek and Roman times and that deposition of lead closely maps anthropogenic production (Figs. 1 and 2). These assertions have been supported by studies on Swedish lake sediments (417) and recent studies that focus on samples of Swedish forest soils (418) and samples from a peat bog in Switzerland (33). In the latter study, the authors not only quantify the amount of lead present in peat samples, but also use the ratio of ^{206}Pb/^{207}Pb and the lead enrichment factor (Pb EF) relative to scandium [Pb EF = (Pb/Sc)$_{sample}$/(Pb/Sc)$_{background}$; where (Pb/Sc)$_{background}$ = 3.9 ± 1.5, the ratio of lead to scandium in preanthropogenic background samples] to determine whether the lead deposited in the peat bog at a given period in time is from "natural" or anthropogenic sources (33). High levels of Pb at a constant Pb/Sc ratio of ~3.9 indicate lead emitted from natural sources, such as storms, volcanic eruptions, forest fires, and natural dust (33). Since 3000 B.C., not only has total lead deposited increased; but the values of ^{206}Pb/^{207}Pb decreased and Pb EF increased (to >2), suggesting that most lead emitted into the atmosphere since that time has been due to anthropogenic (nonnatural) sources (Fig. 22).

Recent patterns of Pb deposition can be monitored through sediment analysis as well. Sample ages within the past 100–200 years can be determined by ^{210}Pb dating ($t_{1/2}$ = 22.26 years) (31). The source of the lead can be determined either by ratios of stable isotopes as described above (33), or by comparing the level of ^{210}Pb to the levels of ^{214}Bi and ^{214}Pb, with which it should be in secular equilibrium (419). The "excess ^{210}Pb" is assumed to be due to atmospheric

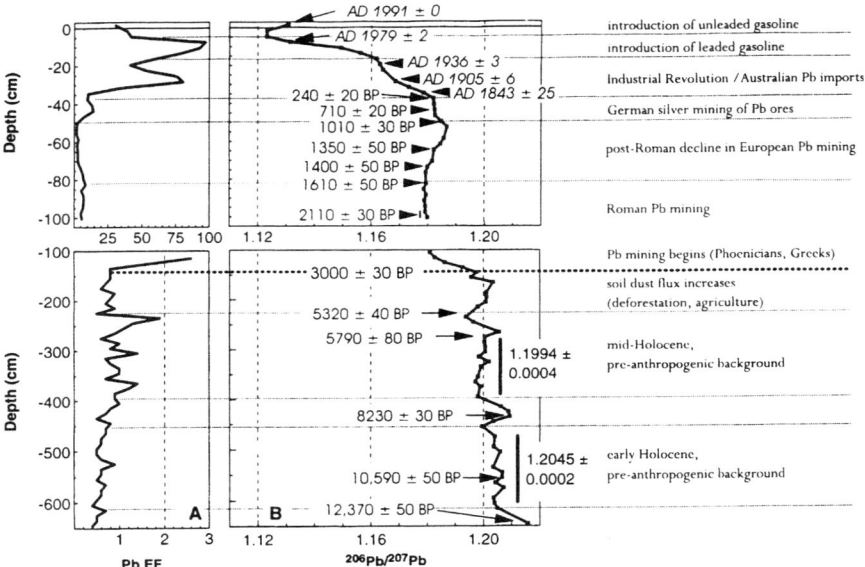

Figure 22. Lead deposition in the environment is determined from the Pb/Sc ratio (Pb enrichment factor or "Pb EF") in peat samples from the Jura Mountains in Switzerland (**A**), where EF values >2 correlate with anthropogenic sources of lead (note different scale for recent dates). Likewise, declining ^{206}Pb/^{207}Pb ratios (**B**) in modern times suggest that the majority of the lead deposited in peat is due to nonnatural sources. The layers of peat were dated using ^{210}Pb analysis (for more recent samples) or ^{14}C dating (for older samples). A large increase in the amount of anthropogenic lead released in the atmosphere occurred ~3000 years ago, coincident with the introduction of lead mining by the Phoenicians and Greeks, and levels have remained high until the present. Levels increased again dramatically in the 1850s, during the industrial revolution and continued to increase for much of the twentieth century, until leaded gasoline was phased out of use in the 1970s. [Reprinted with permission from W. Shotyk, D. Weiss, P. G. Appleby, A. K. Cheburkin, R. Frei, M. Gloor, J. D. Kramers, S. Reese, and W. O. Van Der Knaap, *Science*, *281*, 1635–1640 (1998). Copyright © 1998 American Association for the Advancement of Science.]

deposition as opposed to natural *in situ* decay (419). Some recent studies by these methods have examined lead distribution in sediments from Central Park Lake (New York City) (419), Lake Geneva (France–Switzerland) (420), Lake Tahoe (California–Nevada) (421), sites throughout the Czech Republic (422), and the Sea of Galilee (423). The results show that atmospheric Pb deposition over the past 100–200 years comes from two major sources: emissions from the combustion of leaded gasoline and emissions from industrial sources, for example, coal burning. Lead contamination of Lake Geneva, assorted sites in the Czech Republic, and Central Park Lake is largely from industrial sources (419, 420, 422). Various methods were used to reach these conclusions. Lead stable isotope ratios clearly point to industrial sources of contamination for the

Lake Geneva site (420). At the Czech sites, there is a strong correlation between levels of Czech coal production and Pb deposition; but without stable isotope studies, the role of leaded gasoline emissions is unknown (422). Municipal solid waste incineration is indicated as the primary source of Pb contamination of Central Park Lake based on good correlation between trends of lead, zinc, and tin deposition and solid waste incineration history (419). By contrast, lead deposition at Lake Tahoe and the Sea of Galilee has been attributed to leaded gasoline combustion; peak lead levels in the late 1970s–early 1980s are consistent with that conclusion (421, 423). From a wide variety of studies, it is clear that the decrease in gasoline lead levels worldwide has led to a concomitant decrease in air lead levels (20). From an analysis of data from 19 studies on six continents, Thomas et al. (20) concluded that reductions in gasoline lead levels have also led directly to decreased population blood lead levels. Studies of historical lead levels have also been conducted in trees, determining contaminant source by stable isotope ratios and using tree rings to date samples (424–428). Obviously, this method documents historical levels of *bioavailable* lead (427). A study of a small sample of English trees found industrial emissions to be the major lead source; but it was possible to observe a post-1987 decrease in the fraction of lead attributed to gasoline emissions (427). There is debate in the literature over the reliability of tree-ring dating techniques for reconstructing historical lead levels (424, 425); but it seems likely that with proper sample preparation and rigorous analytical techniques, valuable information can be gained by this method (425).

Several recent studies have used lead levels in wines as a means of determining localized historical lead levels. Initial comparisons of the concentrations of inorganic and organic lead in wines of known vintage, made of grapes picked in exactly the same area each year, correlated well with patterns of leaded gasoline use (429). Subsequent Pb stable isotope studies, however, indicate that the lead concentration in wines is not necessarily a good indication of historical atmospheric lead levels due to significant additional lead contamination during the winemaking process (430).

The result common to all the studies discussed above, both the historical sediment records and the studies examined by Thomas et al. (20), is the dramatic decrease in lead deposition in the past 15 years (7, 8, 15, 33, 417, 419–430). Whether the drop is attributable mainly to decreased gasoline emissions or to a combination of decreases in automobile, industrial, and other emissions, the *existence* of the trend is unambiguous. The large quantities of anthropogenic lead already in the environment, though, require continued attention. The remediation of contaminated soil and water, the treatment of lead poisoning, and the elucidation of the mechanisms of lead toxicity are all areas in which a firm understanding of the fundamental chemistry of lead complexes can contribute dramatically.

VI. BIOLOGICAL CHEMISTRY OF LEAD

A. Symptoms of Lead Poisoning

"The file-cutter never lives the span of life allotted to man. After many small warnings his thumb weakens. He neglects that; and he gets touches of paralysis in the thumb, the arm, and the nerves of the stomach; can't digest; can't sweat; at last, can't work; goes to the hospital: there they galvanize him, which does him no harm; and boil him, which does him a deal of good. He comes back to work, resumes his dirty habits, takes in fresh doses of lead, turns dirty white or sallow, gets a blue line round his teeth, a dropped wrist, and to the hospital again, or on to the file-cutter's box; and so he goes miserably on and off, till he drops into a premature grave, with as much lead in his body as would lap a hundred-weight of tea."

—*Charles Reade (1870)*, cited in (3)

Many people are surprised to find that lead poisoning is still the most common environmentally caused disease in the United States today (3, 29, 30, 431): As recently as 1997, an estimated 890,000 U. S. children had elevated blood lead levels (defined as ≥ 10 μg dL^{-1}) (10, 17, 432). Most people assume that this problem should have disappeared when leaded gasoline was phased out of use and lead paint was banned in the United States. Indeed, BLLs have dropped considerably in the last 20 years, mirroring the dramatic decrease in airborne lead levels worldwide since the catalytic converter was introduced (Fig. 3) (20). However, despite the decrease in airborne lead and the *rate* of deposition of lead in the environment in recent years, the *total amount* of lead contamination in the environment has continued to increase slowly (10, 23). This finding is particularly a problem given that soil lead (see Section V.D) has recently been found to be a major contributor to lead poisoning in many urban areas (25, 26). In addition, even though lead paint was banned for use in housing in 1978, 74% of houses in the United States that were built before 1980 are still contaminated with lead paint (22). Unfortunately, leaded paint, which gets incorporated into household dust and is then inhaled or ingested, is also a major source of exposure to lead in small children. As a result, age and condition of housing are excellent predictors for whether a child is likely to be lead poisoned (17, 22). As a result of these factors, lead poisoning disproportionately affects children who live in inner city neighborhoods: Low-income children who live in older housing are 30 times more likely to have elevated blood lead levels than middle class children who live in newer housing (433).

The adverse effects of exposure to lead on human health have been known since antiquity, but it was not until the last century that lead poisoning was recognized to be a serious health threat to children (3, 431). Children are not only more likely to put things in their mouths and hence to ingest lead, but also experience symptoms of lead poisoning at much lower BLLs than do adults (29,

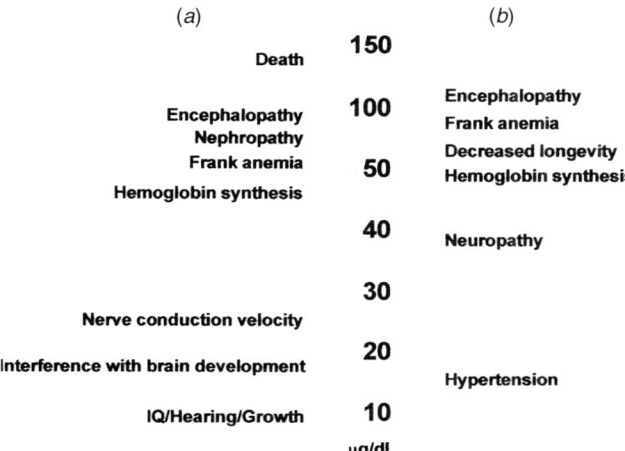

Figure 23. Symptoms associated with lead poisoning in children (a) and adults (b), and the corresponding BLL (in μg dL^{-1}) at which onset of the symptoms typically occurs. Baseline BLLs in the United States are typically 2–3 μg dL^{-1} (~0.1 μM); onset of lead poisoning in children is usually considered to occur at 10 μg dL^{-1} (~0.5 μM). Onset of lead poisoning occurs at much higher BLLs in adults (~40 μg dL^{-1}). [Adapted from (22a).]

30). Whereas adults do not not typically exhibit symptoms of lead poisoning unless they have BLLs of 40 μg dL^{-1} (2.0 μM) or higher, children start to exhibit symptoms at BLLs as low as 10 μg dL^{-1} (0.5 μM) (Fig. 23) (22). Consequently, adults today do not typically experience symptoms of lead poisoning unless they have been exposed occupationally to lead, whereas children frequently become lead poisoned through casual exposure to environmental contaminants (e.g., dust from leaded paint and contaminated soil) (17). (For a detailed discussion of routes of exposure and uptake, see Section VI.C.1.) Symptoms in children also tend to be more severe. Adults with moderate BLLs (40–60 μg dL^{-1}, 2.0–3.0 μM) develop hypertension, anemia, and neurological problems, but these symptoms tend to be reversible. By contrast, children with even slightly elevated BLLs (10–20 μg dL^{-1}, 0.5–1.0 μM) experience delayed development, behavioral disorders, and neurological problems; these problems tend to persist even once the source of exposure has been removed (Fig. 23) (22, 434). By studying the fundamental biological chemistry of lead, we and others hope to provide insights into what molecular targets are responsible for the symptoms associated with developmental problems in children. These insights should lay the foundation for the rational design of improved methods for preventing and treating lead poisoning.

B. Questions of Interest

Despite the prevalence of lead poisoning and the seriousness of the health problems associated with lead poisoning, many fundamental questions about the mechanism of lead poisoning still remain unanswered (29). These include

- What is the biological and subcellular localization of lead; how does this localization depend on organism and cell type?
- What is the speciation of lead within cells?
- What is the distribution of bioavailable lead in different organs and cell types?
- What are the molecular targets for lead, and how does lead interact with these biomolecules?
- How does lead affect gene expression and developmental processes?
- Are there specific genotypes that account for increased susceptibility to lead poisoning?
- Why are children different than adults in terms of onset and permanence of symptoms?
- Is lead binding to biomolecules under thermodynamic or kinetic control?

The extent to which these questions have been addressed to date and the challenges that remain in this field are discussed below.

C. Biodistribution of Lead

1. Lead Body Burden and Uptake

Total lead body burden is typically estimated from the amount of lead that is present in an individual's bone (see below). Although bone lead in living people must be determined using X-ray fluorescence spectroscopy (XRF, see Section VI.E.3), lead levels in bone from autopsy samples can be determined more accurately using techniques such as thermal ionization IDMS, in which a known amount of a stable isotope (e.g., ^{208}Pb, ^{42}Ca, or ^{136}Ba) is added to the sample to be analyzed as an internal standard (27, 31). Additionally, ^{206}Pb/^{207}Pb and Pb/Ca (or Pb/Sr) ratios can be used to distinguish between "natural" and industrial sources of lead (27, 33). These data reveal that the average adult today in the United States has a lead body burden that is $\sim$0.6 mg kg^{-1}, 100–1000 times greater than that of prehistoric humans, and that the majority of lead to which we are currently exposed is industrial in origin (6, 10, 27, 33).

Studies on autopsy samples and on volunteers exposed to radioactive lead (conducted primarily at a time when "informed consent" was not required or meant something different than it does today) have also provided critical insights into the routes by which lead is taken up into the body (10). Lead is taken up by humans through inhaled air and dust, drinking water, and diet (5). In countries in which leaded gasoline is still in use, absorption via the lungs is a major contributor to daily lead uptake. Organic Pb(IV) compounds, which tend to be extremely volatile, are readily taken up into the lungs (5); particulate inorganic Pb(II) (e.g., in automobile exhaust or household dust) can also be absorbed by the lungs (5, 10). In addition, whereas organic Pb(IV) compounds are rapidly absorbed via the skin, uptake of inorganic Pb(II) compounds via the skin is relatively uncommon for individuals with nonoccupational exposure (5, 10). Lead uptake into the gastrointestinal tract is also a major concern: The average person in the United States ingests ~0.5 µmol of Pb/day, 10% of which is absorbed (5). Infants and children, who tend to put nonfood items in their mouths (e.g., contaminated toys, soil, paint chips), are particularly at risk for oral exposure to lead (10). Absorption of lead from the intestines can be minimized by maintaining sufficient iron, calcium, zinc, and phosphate in the diet (10). By contrast, ascorbic acid and citric acid (which both chelate and solubilize lead) increase dietary lead absorption, as do alcohol and ulcers (10).

2. Lead Distribution in the Body

Once absorbed from the intestines, lead enters the blood stream and is rapidly distributed throughout the body, to the erythrocytes, bones, and soft tissue (10). In adults, 90–95% of the total lead body burden is found in bone (5, 435, 436), resulting in a mean level of 14 µg Pb/g of skeletal bone ash (14 ppm) in middle-aged adults in the United States today (27, 437) (Table XVI). In children, the percent of total body lead that resides in bone is closer to 70–80% (436). By contrast, lead concentrations in soft tissues are typically 0.5 ppm (438) and lead concentrations in brain are usually <0.2 ppm, with the highest levels being found in the hippocampus and frontal cortex (10). The average concentration of lead in whole blood for people in the United States in 1999 was 1.6 µg/dL^{-1} (16 ppb) (18); 94–99% of blood lead is found in the erythrocytes and only 1–6% is in the plasma (10, 27). (Methods for the analysis of lead content in blood are discussed in Section VI.E.)

3. Fluorescent Sensors

Information about the subcellular distribution and speciation of lead is absolutely critical if we wish to understand what molecules are targeted by lead. Important questions in this field include

- What is the concentration of lead in specific cell types and organelles?

TABLE XVI

Lead Levels in Human Tissues

Location	"Natural" Level	Level Today	Legal Limit	References
Mean body burden (adults)	40 µg/70 kg adult (0.6 ppb)	40 mg/70 kg adult (0.6 ppm)		27
Mean bone Lead (adults)	13 ng Pb/g of bone ash (13 ppb)	14 µg Pb g^{-1} of bone ash (14 ppm)		27, 437
Teeth		0.1–300 ppm		438
Mean (whole) blood lead levels (BLLs)		1.6 µg dL^{-1} (0.1 µmol L^{-1}, 16 ppb)	10 µg dL^{-1} [a] (0.5 µmol L^{-1}, 100 ppb)	10, 18
Plasma		1–6% of BLL		5, 10
Erythrocytes		94–99% of BLL		5
Soft tissues		0.5 ppm		10
Liver		0.2–0.8 ppm		438
Kidney		0.1–0.4 ppm		438
Hair		0.1–5 ppm		438
Brain		0.2 ppm		10
Urine		2–8 ppb		438

[a] U.S. Centers for Disease Control guideline, 1991.

- How much of this lead is "bioavailable" (i.e., thermodynamically and kinetically accessible and hence able to bind to biomolecules)?

To answer these questions, fluorescent sensors are needed that can be used to quantitate the concentration of bioavailable lead in cells, much the way that fluorescent calcium sensors have been used to study calcium signaling in real time (439–441). The ideal fluorescent sensor for lead would exhibit the following properties (442):

- Bind lead selectively over other metals commonly found in vivo [especially Ca(II), Zn(II), Mg(II), and Fe(II/III)].
- Bind lead tightly enough to be useful for monitoring lead concentrations typically seen in cells ($K_d \approx 10^{-9}$–10^{-15}).
- Respond ratiometrically (i.e., the fluorescence spectrum shifts in the presence of lead, so that the relative amounts of bound and free dye can be determined using a calibration curve developed in vitro).
- Be nontoxic.
- Absorb and emit light in regions of the electromagnetic spectrum that are not obscured by cellular components (excitation at a wavelength >340 nm; emission at a wavelength >500 nm).

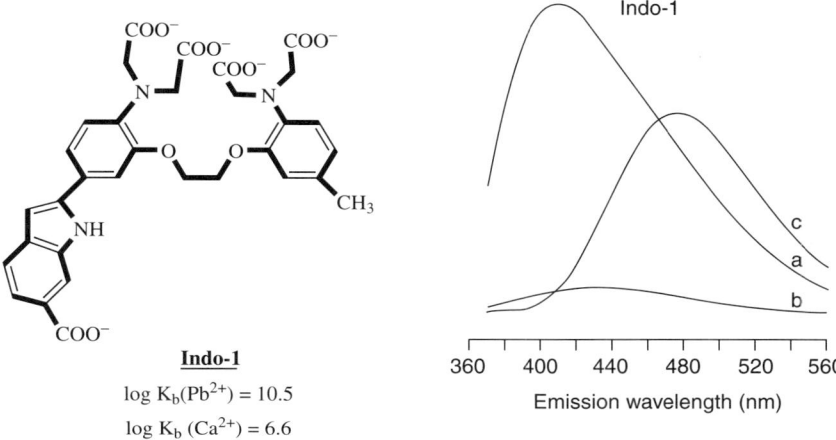

Figure 24. The fluorescent sensor Indo-1 has been used to monitor lead levels in living cells using fluorescence microscopy. The spectrum of Indo-1 is given in the presence of 1 mM CaCl$_2$ (a), 10 μM Pb(NO$_3$)$_2$ (b), or 10 mM EGTA (c). Although originally designed as a calcium sensor, Indo-1 binds Pb^{2+} preferentially; and the Pb- and Ca-bound forms of Indo-1 are readily distinguishable, which allows intracellular lead levels to be quantitated even in the presence of calcium. [Adapted from (443).]

- Absorb readily into cells or easily delivered into cells.
- Exhibit a reversible response (i.e., on- and off-rates of lead binding are fast, so equilibrium with cellular pool of lead is achieved rapidly).

Unfortunately, no lead sensors have been developed to date that meet all of these criteria. The system that has been used most extensively to quantitate lead levels in vivo is the fluorescent sensor Indo-1 (Fig. 24) (443–445). Although originally developed as a calcium dye (446) {log K_b[Ca(II)] = 6.6}, Indo-1 (2-[4-[bis(carboxymethyl)amino]-3-[2-[2-[bis(carboxymethyl)amino]-5-methyl-phenoxy]ethoxy]phenyl]-1H-indole-6-carboxylic acid) binds lead quite tightly {log K_b[Pb(II)] = 10.5} and exhibits a very different fluorescence emission spectrum when bound to lead than when free in solution or bound to calcium (Fig. 24). As a result, Indo-1 can be used to determine whether lead is present in cells, even in the presence of excess calcium, provided that the fluorescence spectrum is deconvoluted to account for calcium interference and all of the possible equilibria are taken into consideration (443, 445). The main drawback of the Indo-1 detection system is that the dye is almost completely quenched when bound to lead (Fig. 24, spectrum b), making it difficult to quantitate the amount of free lead present.

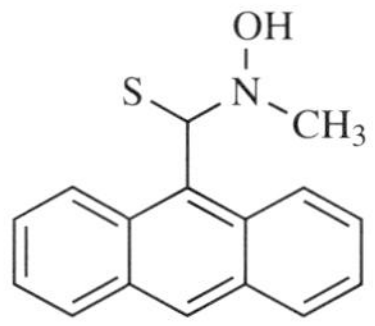

(a)

N-methyl-9-anthrylthiohydroxamic acid

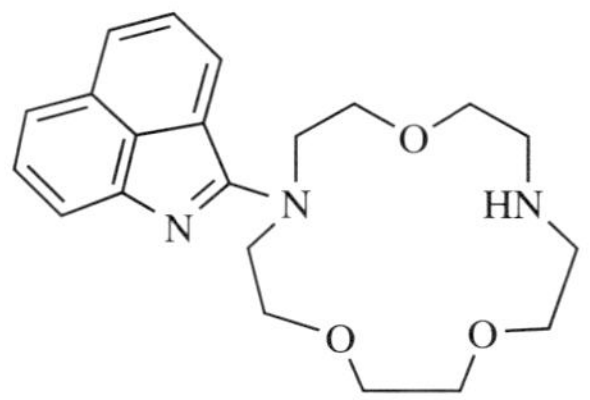

(b)

Benz(c,d)indole-1,7-diaza-15,crown-5 ("BI-crown")

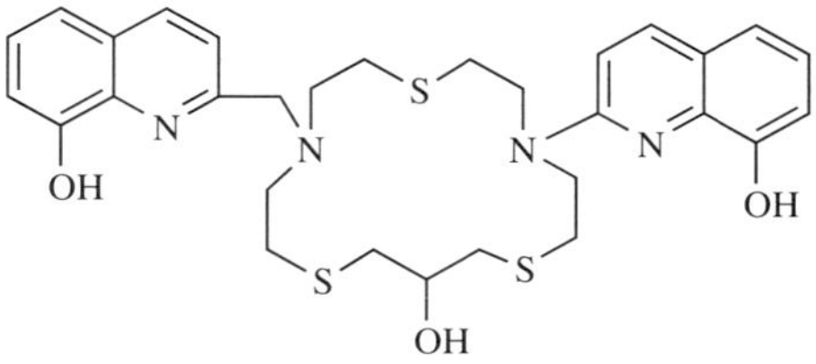

(c)

3-hydroxy-8,14-bis(8-hydroxyquinolin-2-ylmethyl)
-8,14-diaza-1,5,11-trithiacyclohexadecane

Figure 25. Two small molecule fluorescent sensors have been reported recently that exhibit an increase in fluorescence emission in the presence of lead: (*a*) *N*-methyl-9-anthrylthiohydroxamic acid (448) and (*b*) benz(*c*,*d*)indole-1,7-diaza-15,crown-5 (BI-crown) (447); the BI crown is also ratiometric. In addition, Bradshaw and co-workers recently reported a series of bis(hydroxyquinoline) derivatives of diazatrithia-15-crown-5 [e.g., (*c*) 3-hydroxy-8,14-bis(8-hydroxyquinolin-2-ylmethyl)-8,14-diaza-1,5,11-trithiacyclohexadecane] that exhibit shifted fluorescence emission maxima upon binding of Pb^{2+} (449).

Some progress has been made in the last couple of years toward designing new fluorescent sensors that bind lead but are not quenched in the bound state. Two small molecule sensors have been reported recently that exhibit *increased* fluorescence in the presence of lead (relative to the free ligand): a benz(c, d) indole-conjugated crown (BI crown) (447) and a 9-*N*-methylthiohydroxamate derivative of anthracene (*N*-methyl-9-anthrylthiohydroxamic acid) (448) (Fig. 25). The BI-crown looks extremely promising—not only does the fluorescence emission increase in intensity when lead binds, but the dye is also ratiometric and absorbs and emits in an excellent range (Table XVII) (38, 442, 443, 447–451)—but the affinity for Pb(II) and selectivity over other cations have not been reported. The anthracene compound absorbs too far in the ultraviolet to be useful in cells (Table XVII) and no binding constant for Pb(II) (or other metal ions) has been reported for the dye (448). In addition,

TABLE XVII

Advantages and Limitations of Fluorescent Sensors Currently Available for Lead(II)

Sensor	Affinity for Lead[a]	Selective for Lead Over Other Metals?[a]	λ_{max} Absorption (nm)		λ_{max} Emission (nm)		Ratiometric?	Quenched by Lead?	References
			Free Dye	Pb + Sensor	Free Dye	Pb + Sensor			
"Ideal" sensor	$K_d \sim$ fM–nM	Yes (by at least 10^3)	>340	>340	500–800	500–800	Yes	No	442
Indo-1	$K_d \sim$ 30 pM	Yes (over Ca^{2+} by $\sim 10^4$)	336	336	~ 400	~ 440	Yes	Yes	443
Benz(c,d)indole-1,7-diaza-15.crown-5 (BI crown)	N.R.	N.R.	428	381	580	508	Yes	No	447
N-Methyl-9-anthrylthio hydroxamic acid	N.R.	N.R.	258	272	420	420	Yes	No	448
3-Hydroxy-8,14-bis(8-hydroxyquinolin-2-ylmethyl)-8,14-diaza-1,5,11-trithiacyclohexadecane	$K_d <$ 0.1 μM	No	310	390	N.R.	530	Yes	No	451, 450
dns-ECEE	$EC_{50} \sim$ 120 μM[b]	Yes (over Ca^{2+}, Zn^{2+}, and Cd^{2+})	337	337	557	510	Yes	No	38
Catalytic DNA biosensor	$K_d \sim$ 13.5 μM	Yes (over Zn^{2+}, Ca^{2+}, and Mg^{2+}, by $\sim$80-fold)	560	560	~ 580	~ 580	No	No	451

[a] N.R., not reported.
[b] Where EC_{50} is the concentration of total lead required to obtain half-maximal response.

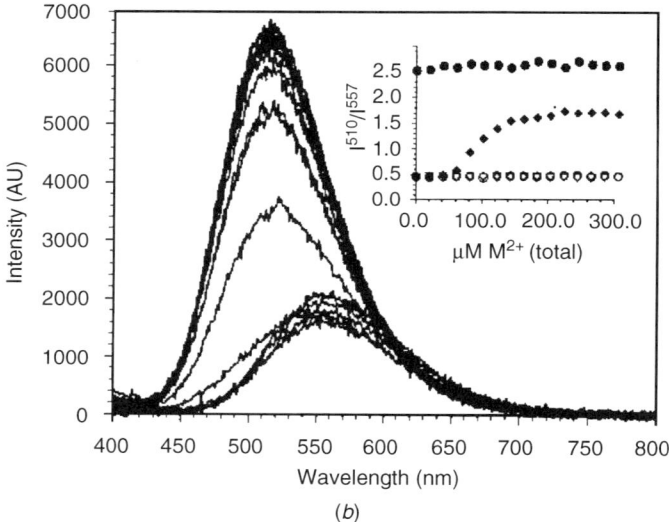

Figure 26. (a) Chemical formula of the fluorescent sensor dns-ECEE. (b) The fluorescence emission spectrum of the sensor shifts to shorter wavelengths and increases in intensity upon binding of Pb^{2+}. The response of the probe (inset) is selective for Pb^{2+} (◆) over Zn^{2+} (○), Ca^{2+} (◇), and Cd^{2+} (∇). Addition of buffer (+) to the probe or addition of Pb^{2+} (●) to a control peptide that cannot bind metal does not shift the emission maximum. [Reprinted with permission from S. Deo and H. A. Godwin, *J. Am. Chem. Soc.*, 122, 174–175 (2000). Copyright © 2000 American Chemical Society.]

Bradshaw and co-workers (449) recently reported a series of bis (hydroxyquinoline) derivatives of diazatrithia-15-crown-5 [(e.g., 3-hydroxy-8,14-bis(8-hydroxyquinolin-2-ylmethyl)-8,14-diaza-1,5,11-trithiacyclohexadecane, Fig. 25(c)] that exhibit shifted emission spectra upon binding of Pb(II) and look extremely promising (Table XVII) (449).

In addition, two recent biologically based sensors for lead have been reported that hold particular promise for the development of ratiometric lead sensors that will be useful in vivo: a peptide-based system (38) and a catalytic DNA biosensor (452). In the first case, a fluorescent dye (dansyl) was conjugated to a tetrapeptide (glutamate–cysteine–glutamate–glutamate) to yield a sensor (dns-ECEE) that bound Pb(II) selectively over Ca(II), Zn(II), and Cd(II). In addition, the fluorescence emission shifted to shorter wavelength [yielding a highly desirable ratiometric response to Pb(II)] and increased in intensity upon binding of lead (Fig. 26) (38). Furthermore, the response to lead was reversible. The primary limitations of this sensor are that the affinity for lead ($EC_{50} \approx 120\ \mu M$) is too weak to be useful for most biological applications and the excitation wavelength (337 nm) is too low to be practically useful in cells. However, these are parameters that could be optimized in future generations of sensors, given the modular nature of the system (38).

The catalytic DNA-based sensor reported by Li and Lu (451) also holds extreme promise as a first generation, proof of concept sensor. The DNA-based sensor consists of an in vitro selected deoxyribozyme (17E) that forms a duplex with a DNA substrate (17DS); cleavage of the substrate is catalyzed in the presence of lead (Fig. 27) (451). The enzyme and substrate strands are labeled with fluorescent dyes [4-(4'-dimethylaminophenylazo)benzoic acid (Dabcyl) and 6-carboxytetramethylrhodamine, (TAMRA), respectively] so that cleavage of the substrate can be measured by monitoring the level of energy transfer between the two strands. In the absence of Pb(II), the two strands form a duplex and the emission of TAMRA is quenched by the Dabcyl on the enzyme strand. When Pb(II) is added, the substrate is cleaved and diffuses away from the enzyme strand, resulting in an increased emission intensity (less quenching) of the TAMRA dye. Good affinity and selectivity for Pb(II) were achieved through three rounds of in vitro selection. Although higher affinity and selectivity are probably needed for the probe to be useful in vivo, it should be possible to optimize these parameters via additional rounds of selection (451). Although none of the new generations of sensors have been tested in vivo, the variety of new approaches that are available and their amenity to selection and optimization bodes well for the future of fluorescent sensors for lead.

4. Toxicokinetics

Given the limitations in fluorescent sensors for lead to date, it is not surprising that relatively little is known about the mechanism by which lead is transported throughout the body. However, those studies that have been conducted to date suggest that Pb(II) can be transported by ion channels, can interact with complex metal transport machineries, and may be passively transported across cell membranes.

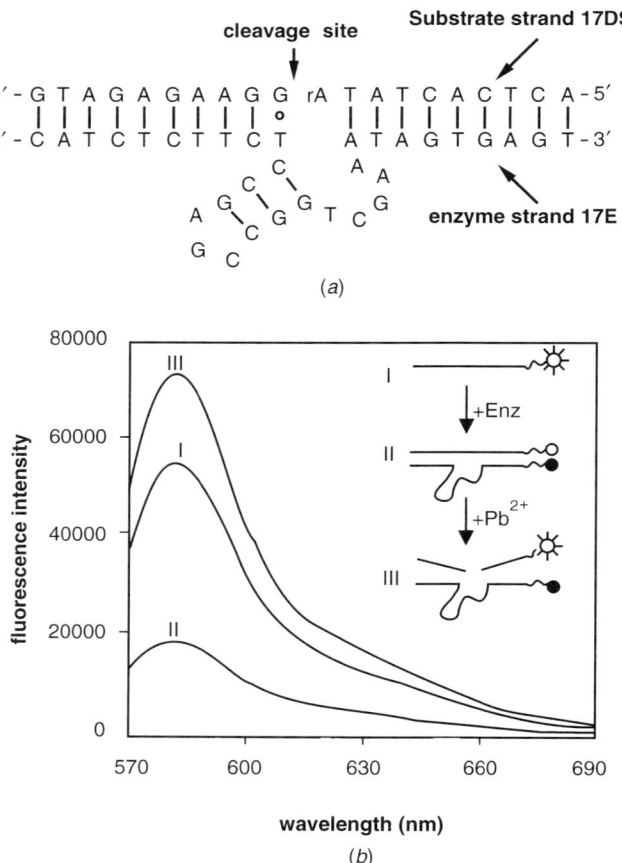

Figure 27. (a) Sequence of the deoxyribozyme (17E)/substrate (17DS) complex and proposed secondary structure. Fluorescence emission spectra (b) of the substrate (**I**), annealed complex (**II**), and cleaved product (**III**) that is formed upon addition of 500 nM Pb(OAc)$_2$. [Reprinted with permission from J. Li and Y. Lu, *J. Am. Chem. Soc.*, 122, 10466–10467 (2000). Copyright © 2000 American Chemical Society.]

Due to the report by Manalis and Cooper in 1973 that lead inhibits calcium-triggered release of neurotransmitters (452), much attention has focused on the possibility that lead interacts with calcium proteins and/or affects calcium homeostasis (40, 173, 338–340, 351, 355, 443, 453–464). Indeed, patch clamp studies reveal that lead blocks voltage-dependent calcium channels (459, 464, 465), but the effects are highly dependent on the channel and cell type (460). In addition, lead has been shown to stimulate efflux from nerve cells via Ca^{II}–ATPase in a calmodulin-dependent fashion, resulting in a net decrease in intracellular calcium levels in the presence of added lead (466). Pioneering studies by Simons and Pocock (467) using atomic absorption spectroscopy and Tomsig and Suszkiw (468) using the fluorescent dye Fura-2 and recent studies by Kerper and Hinkle (443) using the intracellular dye Indo-1 (see Section VI.C.3) suggest that Pb(II) is transported by voltage-sensitive calcium channels and that uptake of Pb(II) is activated by calcium depletion.

In addition, recent developments in our understanding of iron uptake and transport provide strong evidence that these systems may also facilitate uptake and transport of lead. Recent studies on the intestinal iron transporter, Nramp, suggest that this protein may also be responsible for uptake of lead from the intestines into the bloodstream (469).

A critical question is which, if any, of these mechanisms allows lead to cross the blood–brain barrier. Likewise, lead is able to cross the placenta (470), and hence is a major risk to developing fetuses (10). Deane and Bradbury (472) used radioactive ^{203}Pb to look at transport of lead across the blood brain barrier in rats; they found that the blood–brain barrier is very permeable to lead and that uptake into the brain is extremely rapid (occurs in minutes). Lead uptake *increases* when compounds are added that are known to inhibit ATP-dependent Ca(II) pumps (e.g., vanadate and stannic chloride), suggesting that these pumps normally act to *efflux* lead. In addition, influx of ^{203}Pb decreases at lower pH, suggesting that the form of lead that is transported may be a hydroxide complex (e.g., PbOH$^+$) (471). Recent studies by Erdahl et al. (445) suggest that Pb(II) can also diffuse across membranes when it is bound to ionophores such as ionomycin (Fig. 28). In these experiments, mammalian cells in culture are loaded with a fluorescent dye (Indo-1 or Quin-2) that serves as an indicator for Pb(II) and/or calcium uptake (see Section VI.C.3 for a detailed description of how Indo-1 responds to lead and calcium). These studies suggest that ionomycin may be a useful tool for delivering lead to cells in a reliable manner and should prove generally useful for conducting studies on the effects of lead intoxication in cells (445).

D. Molecular Targets for Lead

In addition to interfering with the homeostasis of natural (i.e., "good") metal ions, lead can also substitute for native metal-binding sites in proteins (37, 40, 50, 173, 338, 339, 351, 453). Most research has focused on the interactions between lead and proteins that naturally bind either calcium or zinc (Fig. 17), with the goals of explaining the neurological and developmental toxicity of lead. In addition, there have been some suggestions that lead might target RNA or DNA (Fig. 16). The evidence for and against each of these classes of interactions—both in vitro and in vivo—and the way in which these interactions correlate with known symptoms of lead poisoning are discussed below.

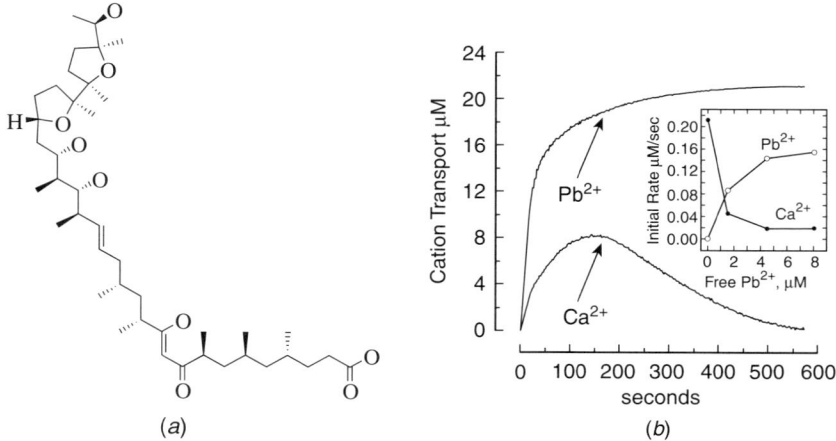

Figure 28. Ionomycin (a) can be used to transport lead across cell membranes. Uptake of lead into cells is rapid (b) and is monitored by quenching of the intracellular dye Indo-1. (See Section VI.C.3 and Fig. 24 for spectral properties of Indo-1.) Competition studies with Ca^{2+} in the presence of Indo-1 (b, inset) reveal that ionomycin is highly selective for Pb^{2+} (445). [Part (b) reprinted from W. L. Erdahl, C. J. Chapman, R. W. Taylor, and D. R. Pfeiffer, *J. Biol. Chem*, 275, 7071–7079, with permission from the American Society for Biochemistry and Molecular Biology.]

1. Interactions with Calcium Proteins

Interactions between lead- and calcium-signaling proteins have been invoked to explain both the neurological and developmental toxicity of lead (339, 340, 434). As early as 1973, Manalis and Cooper observed that lead interferes with the ability of calcium to evoke release of neurotransmitters in frog neuromuscular junctions and first suggested that lead exerts its neurotoxic effects by binding to the "calcium receptor" at synapses (452). This model was widely embraced—despite the fact that the molecule(s) that acts as a neuronal calcium receptor was not known at the time—because it offered a possible molecular explanation for the cognitive and behavioral problems associated with lead poisoning. As the field of calcium signaling matured and specific proteins involved in the signaling pathways were identified, studies were conducted to determine whether lead could replace calcium in binding sites in proteins. Early studies on EF-Hand containing proteins revealed lead can substitute for calcium in calmodulin (350). However, micromolar concentrations of lead are required to activate calmodulin [$EC_{50} \approx 0.5$–1.0 μM for both Pb(II) and Ca(II)] (350) and it seems unlikely that the concentration of *bioavailable* Pb(II) would ever reach micromolar levels within cells. Whereas the intracellular concentration of bioavailable Ca(II) is estimated to be 10^{-8}–$10^{-6} M$ (352, 472, 473) (out of total calcium concentrations of 10^{-6}–10^{-3} M), the concentration of bioavailable lead is estimated to be $\sim 10^{-12}$–10^{-9} M (352) (out of a total lead pool of 10^{-8}–10^{-5} M, depending on the cell type and subcellular location) (10).

The first example of a calcium protein that was found to bind lead tightly enough to actually be a realistic target in vivo is protein kinase C (PKC) (352). Protein kinase C isoforms that are calcium dependent and activated by lead in vitro contain a calcium-binding motif called a C2 domain. This domain is entirely different from the EF-Hand proteins: Whereas EF-hand proteins are largely α-helical and contain mononuclear binding sites (474–476), the C2 domain is primarily composed of β-sheets and has a highly exposed multi-nuclear metal-binding site on one end of the domain (242). Although originally identified in PKC, C2 domains have been identified subsequently in a wide range of signaling pathways (477, 478). The phospholipid-binding activity of calcium-dependent PKC is stimulated by *picomolar* concentrations of lead (352) and recent studies have revealed that lead alters expression of immediate-early genes in a PKC-dependent fashion (340, 353, 463). However, the response observed in vivo occurs even in pathways that involve isoforms of PKC that are calcium-independent (355), suggesting that the simple explanation that lead binds to the calcium-binding domain of PKC cannot explain all of lead's effects.

More recently, studies have also been conducted to test the hypothesis that lead targets the calcium sensor (or receptor) in neurotransmission. Interestingly, this protein—identified by cloning and mutagenesis studies to be the neuronal protein synaptotagmin (syt)—contains two tandem C2 domains (479–483). Studies in PC-12 cells reveal that lead interferes with synaptotagmin's ability to bind to its protein partner, syntaxin (40). In addition, studies on the recombinant protein using radiolabeled phospholipids and the fluorescent metal ion Tb^{3+} reveal that lead potentiates phospholipid binding to syt and that lead binds more tightly than calcium to the protein (Fig. 29) (40). These data, combined with the observation that lead occupies the high affinity calcium-binding site in both known crystal structures of synaptotagmin (240–242) (see Section III.D.2), provide compelling evidence that Pb(II) can compete effectively with Ca(II) for binding site(s) in syt but that the lead-bound form is not a perfect mimic of the native holo-protein. The details of lead binding to this protein and what exactly makes lead-syt different from the calcium-bound form of the protein still remain to be determined. In addition, whether lead targets C2 domain proteins more generally—in addition to PKC and syt—has yet to be investigated.

2. Interactions with Zinc Proteins

Lead has also been shown to bind to and inactivate proteins that naturally bind zinc both in vitro and in vivo. One of the best-documented targets for lead

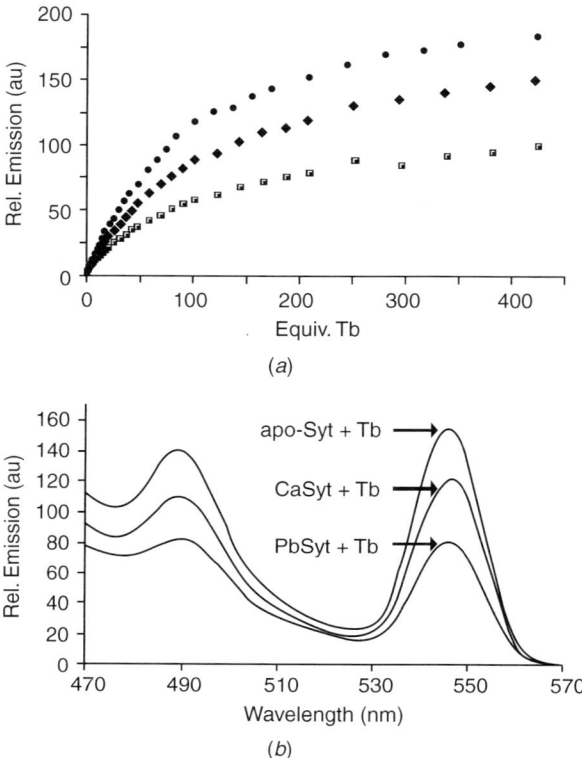

Figure 29. Lead binding to synaptotagmin (syt) was probed by fluorescence spectroscopy by conducting competition experiments with the fluorescent ion Tb^{3+}. (a) Tb^{3+} binding to syt (●) is blocked more effectively by Pb^{2+} (■) than Ca^{2+} (◆). (b) Fluorescence spectra of synaptotagmin I 1–5 with ∼210 equiv Tb^{3+} and with no competing metal ion (apo-Syt + Tb), with 400 equiv Ca^{2+} (CaSyt + Tb), and with 404 equiv Pb^{2+} (PbSyt + Tb). These data suggest that all three metals compete for the same site(s) and that lead binds more tightly than Ca^{2+}. [Figure reprinted from C. M. L. S. Bouton, L. P. Frelin, C. E. Forde, H. A. Godwin, and J. Pevsner, *J. Neurochem.*, 76, 1724–1735 (2001), by permission of Blackwell Science Ltd., publishers.]

is the zinc enzyme ALAD (or porphobilinogen synthase, PBS) (243, 484), which has been used extensively as a biomarker for lead poisoning (see Fig. 30 and Section VI.E) (244, 485–487). δ-aminolevulinic acid dehydratase catalyzes the second reaction in the heme-biosynthetic pathway and has been shown to be inhibited by *femtomolar* concentrations of lead in vitro ($K_i = 0.07$ pM, vs $K_M^{Zn} = 1.6$ pM) (487). Interestingly, inhibition of ALAD by lead in vivo can be reversed by consumption of sufficient quantities of zinc (299), adding support

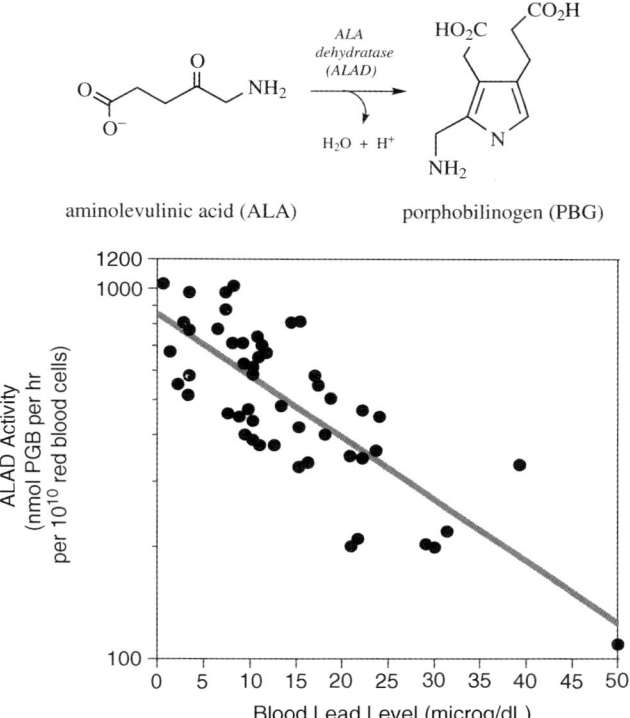

Figure 30. One of the best biomarkers for lead poisoning is the second enzyme in the heme biosynthetic pathway, ALAD, which is inhibited by femtomolar concentrations of lead (467). Activity of ALAD in red blood cells correlates inversely with blood lead level. [Figure reprinted with permission from Elsevier Science (J. A. Millar, V. Battistini, R. L. Cumming, F. Carswell, and A. Goldberg, *The Lancet*, 1970, Vol. 2, 695–698).]

for the call for improved nutrition among children suffering from lead poisoning (10). Lead's ability to inhibit ALAD undoubtedly contributes to much of the anemia associated with lead poisoning.

In the last couple of years, several advances in our understanding on a molecular level of how lead inhibits this important enzyme have been made. The structure of the zinc- and lead-bound forms of yeast ALAD have been determined by X-ray crystallography. These structures reveal that lead substitutes for zinc in the catalytic site of the enzyme (Fig. 17) (245). The crystal structure also resolved an on-going debate about the nature of the metal-binding residues in the catalytic site in ALAD: The active site contains three cysteine

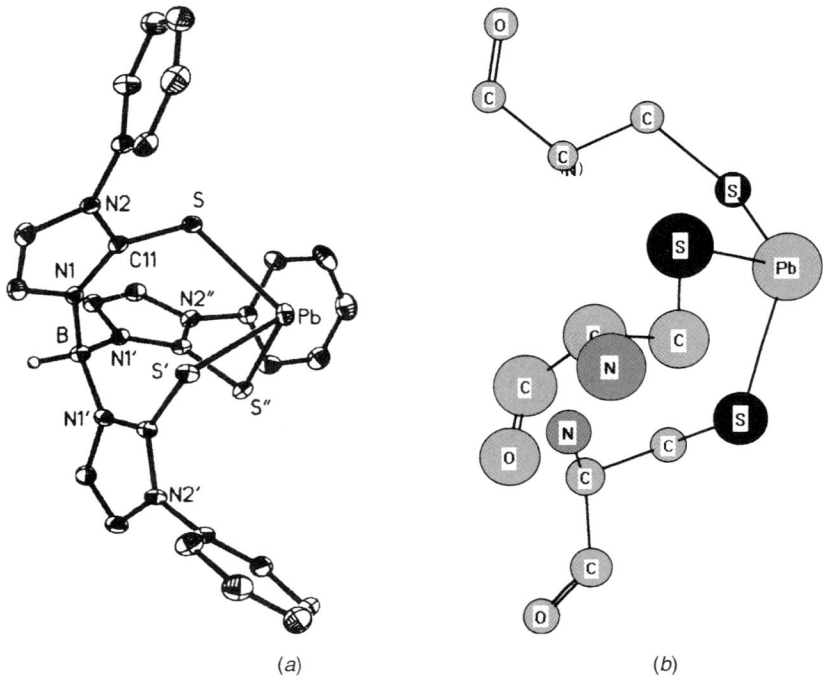

Figure 31. A model system developed by Parkin and co-workers (a), Tris(2-mercapto-1-phenylimidazolyl)hydroborato lead complex, {[Tm^{Ph}]Pb}[ClO$_4$] (197), is strikingly similar in structure to the lead bound in the active site of the target protein ALAD (b), (only lead and coordinated cysteines are shown) (245). In both systems, lead is bound in a trigonal pyramidal geometry by three sulfurs. Metal-binding studies on the model system (198) and inhibition studies on the enzyme reveal that lead has a higher affinity for this type of site than the native zinc ion. [Part (a) reprinted with permission from B. M. Bridgewater and G. Parkin, J. Am. Chem. Soc., 122, 7140–7141 (2000). Copyright © 1999 American Chemical Society. Coordinates for figure part (b) were downloaded from the Protein Databank (pdb file id 1QNV).]

residues that bind either zinc or lead in a trigonal-pyramidal geometry (Fig. 31) (244, 245). This type of active site has not been observed in any other catalytic zinc proteins to date and offers a possible explanation for why ALAD is so much more sensitive to lead than other zinc enzymes. To test whether this environment provides a particularly good binding site for lead, Parkin and co-worker (198) prepared a tris-thiol ligand [tris(2-mercapto-1-phenylimidazolyl)hydroborate] and looked at its ability to chelate both lead and zinc. Lead binds to this ligand in a trigonal geometry that is very similar to that found in ALAD (Fig. 31) and with an affinity that is ∼500 times greater than that of the same ligand for zinc (198).

Studies on model peptide systems also support the notion that lead specifically targets cysteine-rich sites in proteins, which are typically found in structural zinc-binding sites (e.g., zinc finger proteins and steroid receptors). The affinity of lead for a series of zinc finger peptides has been studied using lead-thiolate CT bands (37) (see Section II.D). In this technique, lead is titrated into a solution of the apo-peptide, and the formation of the lead-peptide species is monitored using UV–vis spectroscopy [Fig. 32(a)] (37). From the lead-peptide titration, the stoichiometry of the lead-peptide complex can be determined, in addition to an upper limit of the dissociation constant (lower limit of the binding constant). Because the affinity of lead for these sites is typically too great to be measured by direct titration, the absolute magnitude of the affinity of the K_d(Pb) is determined by conducting a competition experiment with zinc [Fig. 32(b)]. [The affinity of zinc for the binding site must be determined independently in a separate experiment (309, 310)]. By using this approach, we determined the affinity of lead for a series of zinc finger peptides in which the metal-binding sites were systematically varied (cysteine$_2$histidine$_2$, cysteine$_2$histidinecysteine, and cysteine$_4$) (37, 309). These studies revealed that although lead binds tightly to each of the peptides [K_d(Pb) = 10^{-9}–10^{-14} M), that lead has the highest

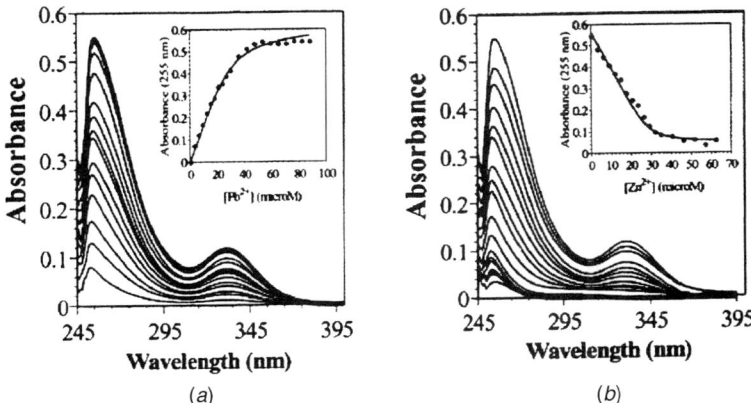

Figure 32. The affinity of lead for structural zinc-binding sites in proteins can be monitored directly using lead-thiolate CT bands. (a) As lead is titrated into a solution of the apo-peptide (here, an 18-amino acid peptide from HIV-nucleocapsid protein), intense CT bands appear in the range of 250–400 nm. This "forward titration" reveals that lead binds tightly to the peptide [K_b(Pb) $\geq 10^{10}$] in a 1:1 complex. (b) Addition of zinc to the lead-peptide complex results in displacement of lead from the peptide and the disappearance of the lead-thiolate CT bands. This "back-titration" allows the relative affinities of lead and zinc for the peptide to be determined [for this example, K_b(Zn)/K_b(Pb) = 12]. [Reprinted with permission from J. C. Payne, M. A. ter Horst, and H. A. Godwin, *J. Am. Chem. Soc.*, 121, 6850–6855 (1999). Copyright © 1999 American Chemical Society.]

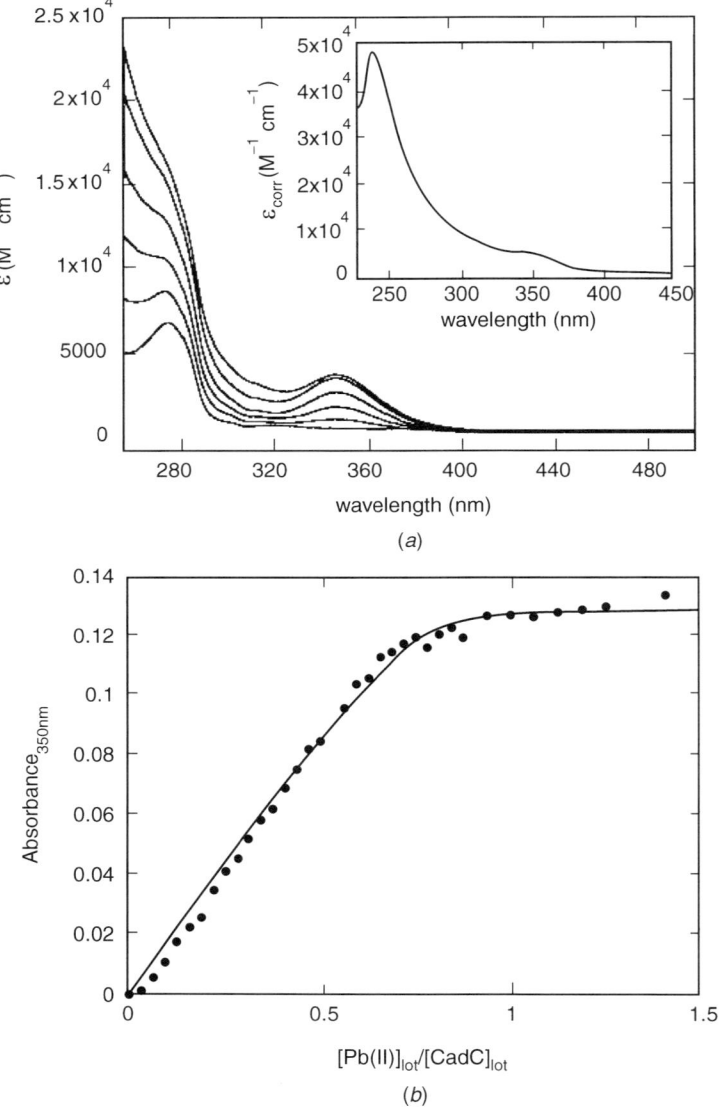

Figure 33. (a) Lead binds to the naturally occurring cysteine$_4$ site in the regulatory protein CadC, as is witnessed from the appearance of intense charge-transfer bands when lead is added to the apo-protein. (b) The titration reveals that lead binds tightly to the peptide $[K_b(\text{Pb}) \geq 10^7]$ in a 1:1 complex. (c) Fluorescence anisotropy experiments with fluorescently labeled DNA reveal that lead binding to CadC disrupts the DNA-binding activity of the protein. [Reprinted with permission L. S. Busenlehner, N. J. Cosper, R. A. Scott, B. P. Rosen, M. D. Wong, and D. P. Giedroc, *Biochemistry*, 40, 4426–4436 (2001). Copyright © 2001 American Chemical Society.]

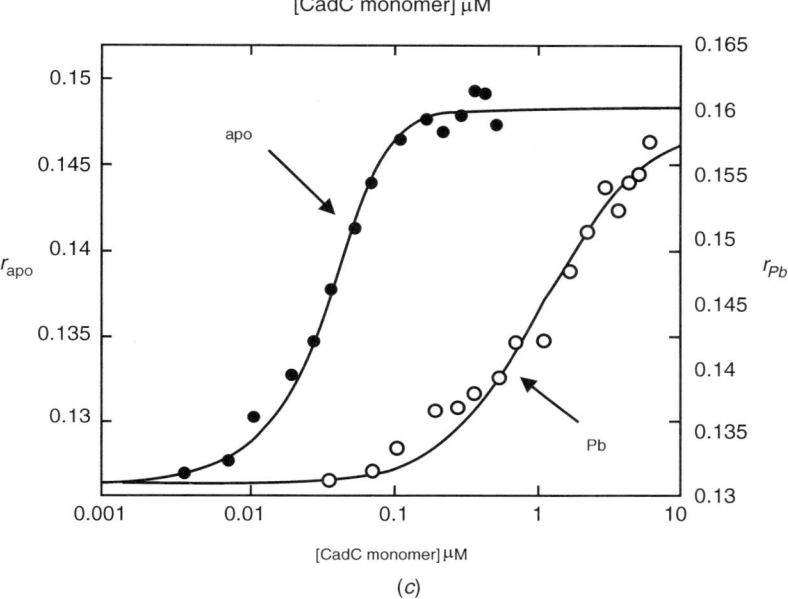

Figure 33 (Continued)

affinity for the cysteine rich site: lead binds even more tightly than zinc to the cysteine$_4$ site [K_b(Zn)/K_b(Pb) = 0.046] (37). In addition, even though lead binds in a 1:1 complex with the peptides, it does not stabilize the native fold of the domain (37). These data suggest that lead may target other cysteine-rich structural zinc-binding sites in proteins and that lead binding should result in inactivation of the protein. Recent studies by Giedroc and co-workers (53) lend credence to this hypothesis: Lead binds to a naturally occurring cysteine$_3$ zinc-binding site in the regulatory protein CadC with a 1:1 stoichiometry (Fig. 33) and causes CadC to dissociate from DNA. Lead has also been shown to bind in vitro to cysteine residues in metallothionein (489) and phytochelatins (51), and to cysteine residues in the chromatin-binding protein HP2 (89), which contains a zinc-binding site that has yet to be defined. In addition, Hanas and co-workers and Zawia and co-workers have shown that lead binds in vitro to two transcription factors: TFIII-A (489) and a synthetic peptide from the transcription factor Sp-1 (90, 490), both of which contain cysteine$_2$histidine$_2$ structural zinc-binding sites. In the case of HP2 and TFIIIA, lead interferes with the ability of the protein to bind to its target sequence of DNA (89, 489). By contrast, lead *promotes* Sp-1 binding to DNA (90). Addition of lead to the transcription factor AP2, which contains *no* structural zinc-binding sites, has no effect on DNA-binding activity (489).

The idea that lead targets structural zinc-binding sites in transcription factors and alters their activity has also been supported by a wide variety of studies in vivo, which suggest that these interactions may account for some of the developmental problems associated with lead poisoning (337). Lead is highly concentrated in the nucleus of cells (491), where it has been shown to alter patterns of gene expression in a variety of pathways (340, 353–355, 492). More specific studies suggest that some of these effects may be due to direct interactions between lead and structural zinc-binding proteins. For example, Zawia and co-workers (493, 494) demonstrated that the DNA binding activity of the transcription factor Sp-1 was altered in PC-12 cells that were exposed to lead and that this resulted in an alteration of the response of the PC-12 cells to nerve growth factor. Similar results were observed in rat pups exposed to lead, which exhibited changes in expression patterns of genes regulated by Sp-1 in the rat pups. By contrast, no alteration was seen in the DNA binding activity and regulatory activity of transcription factors lacking structural zinc-binding sites (e.g., NFκB) (493). These results are especially significant because Sp-1 plays an important role in regulating gene expression in differentiating cells. Combined with the in vitro studies that Zawia and co-workers (90, 490) have conducted on lead interactions with Sp-1 (see above), these data provide compelling evidence that lead targets zinc-sites in transcription factors in cells.

3. *Interactions with RNA and DNA*

Because lead localizes to the nucleus (491) and alters gene expression, there have been multiple suggestions in the literature that lead might directly target RNA and/or DNA (495, 496). In the 1970s, several groups reported that micromolar to millimolar concentrations of Pb(II) ions could catalyze cleavage of mRNA and tRNAs (495, 496). When the structure of yeast tRNA^{Phe} was solved by Klug and co-workers (227, 497) in the 1980s, they noted that lead added to prepare a heavy atom derivative catalyzed site-specific cleavage of the sugar–phosphate backbone, and they hypothesized that this cleavage might be related to the self-splicing activity observed for RNA. Subsequent experiments have borne out the prediction that lead accelerates RNA hydrolysis and can serve as the required metal cofactor in many ribozyme (leadzyme) systems (228–238). The mechanism by which Pb(II) and other divalent cations [e.g., Mg(II)] promote hydrolytic cleavage of RNA is still the topic of ongoing debate, but likely involves stabilization of a leaving group, stabilization of a negatively charged transition state, or activation of an attacking nucleophile by the metal (498). Lead has also been shown to bind to DNA analogues of tRNA^{Phe} (499). However, it is important to note that all of these reactions require micromolar to millimolar concentrations of Pb(II), which are much higher than the concentrations required to alter the activity of calcium and zinc proteins (femtomolar to nanomolar) and are probably too high to be physiologically relevant (see below).

4. Other Possible Targets for Lead

Several other classes of proteins have also been implicated as possible targets for lead, including other proteins in the heme biosynthetic pathway, "lead-binding" proteins in the kidney and brain, and heat shock proteins (342, 500–502). Lead is known to affect several steps in the heme biosynthetic pathway other than that catalyzed by ALAD: Other profound effects include stimulation of δ-aminolevulinic acid synthase (ALAS) and decreased levels of iron incorporation into protoporphyrin by ferrochelatase (see Section VI.E.2 and Fig. 34) (10, 503–505). However, not all of these effects are due to direct interactions between lead and enzymes in the heme biosynthetic pathway. For instance, the widespread assertion that lead "inhibits" ferrochelatase is not supported by studies on the isolated enzyme (506, 507). Furthermore, increased levels of both erythrocyte protoporphyrin IX (EP) and zinc protoporphyrin (ZPP) are observed at high BLLs, suggesting that ferrochelatase is still competent to insert zinc into EP and that the increased levels of EP and ZPP associated with lead poisoning are most likely caused by lead interfering with iron uptake or transport (see Sections VI.C.4 and VI.E) (10, 506, 507).

Other reports of possible molecular targets for lead appear more promising but await detailed characterization. Fowler and co-workers reported "lead-binding proteins" that are localized in the kidney (500–502) and brain (342). These proteins have yet to be completely identified, but preliminary characterization indicates that they are cysteine rich (501). Tiffany-Castiglioni and co-workers (508) recently reported that lead targets the molecular chaperone GRP78: lead both increases expression of the protein and binds to it directly. They also observe that the Menkes protein can chelate Pb(II) away from GRP78, offering an intriguing possibility that lead toxicity and copper homeostasis may be linked (508, 509). Clearly, the conclusion to be reached from these studies is that there are a multitude of targets for lead in vivo, some of which are undoubtably more relevant than others under specific cellular conditions. More comprehensive studies are needed on the effects of lead on a broad range of signaling pathways; recent innovations in genomics (510) and proteomics (511) and the advent of the new field of toxicogenomics make such studies both tractable and appealing.

E. Biomarkers for Lead Poisoning

The holy grail of lead poisoning study is not only to be able to determine an individual's exposure to lead but also to be able to predict his/her susceptibility

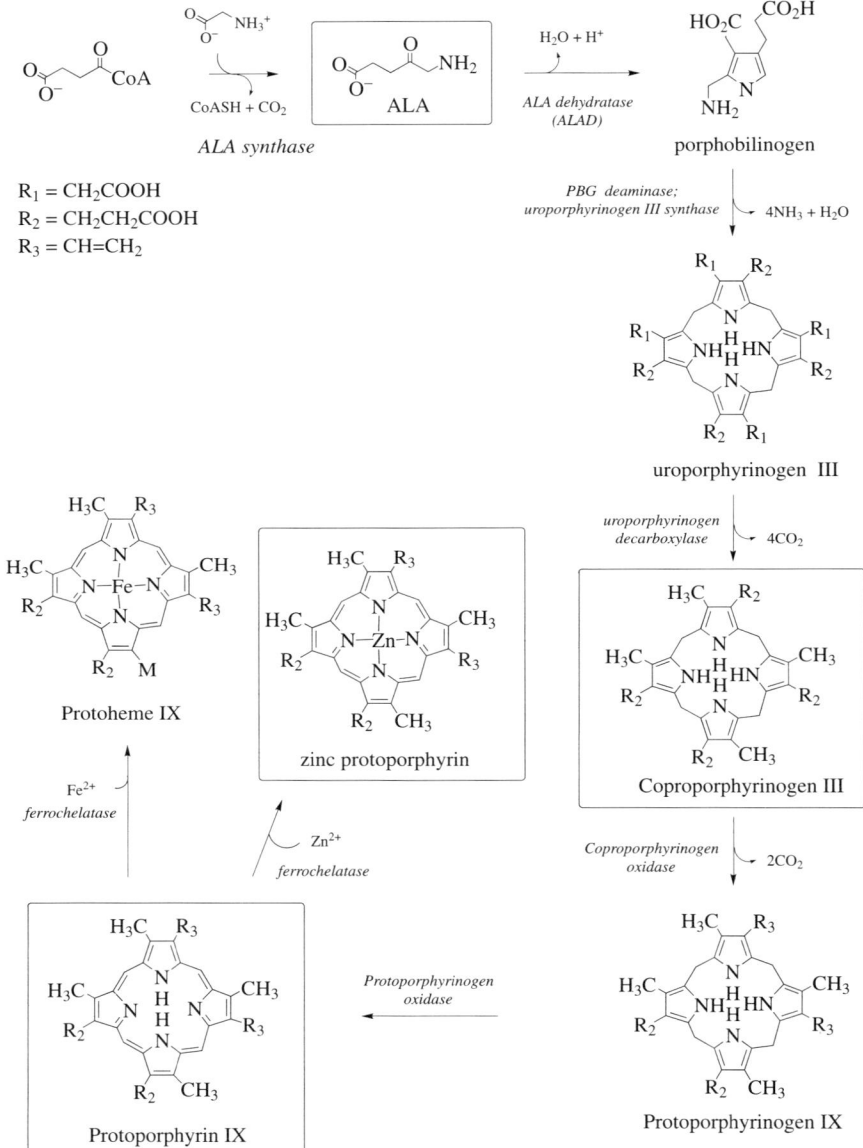

Figure 34. Lead affects several steps in the heme biosynthetic pathway in addition to inhibiting ALAD: in the presence of lead, elevated levels of ALA, coproporphyrinogen, protoporphyrin IX, and ZPP are observed. Species that accumulate in the presence of lead are shown in boxes. Increased levels of ALA result both from inhibition of ALAD and stimulation of ALAS (502–504). Coenzyme A = CoA.

to lead poisoning. To be practically useful, however, such a biomarker for lead poisoning must exhibit the following properties (10):

- The marker must be found in most members of the target population.
- The marker should provide an "early warning" of lead poisoning, rather than being the result of prolonged exposure or widespread destruction of organ systems.
- The marker should be responsive to and correlate quantitatively with changes in blood lead level below 10 µg dL^{-1}.
- The marker should not be significantly dependent on other environmental and physiological parameters or have baseline levels that are highly variable from individual to individual.
- The marker should be easy for clinicians to measure or use technologies that are common to health service laboratories.
- The marker itself, the sampling procedure, and the measurement technique should be relatively immune to contamination.

To this end, a variety of analyses have been developed to determine both lead body burden and the effects of lead on critical cellular processes (Table XVIII) (436).

1. Blood Lead Level

Blood lead levels are the primary criteria by which health professionals determine whether a child is currently suffering from lead poisoning (10). However, it is important to note that BLLs only provide a marker of very recent exposure, because the half-life for lead in blood is 2–3 weeks (5). To determine long-term exposure to lead, effects on hemoglobin biosynthesis (see below) or total bone lead must be used (10). The Centers for Disease Control recommend that all children should have their BLL screened at ages 1 and 2 if they meet one or more of the following criteria (17):

- They reside in a zip code in which ≥27% of the housing was built before 1950.
- They receive public assistance for the poor (e.g., welfare, Medicaid, or food stamps).
- They live in or regularly visit a house built before 1950.
- They live in or regularly visit a house built before 1978 in which there are recent or ongoing renovations.
- They have a sibling or playmate who has (or had) lead poisoning.

TABLE XVIII

Examples of Biological Markers for Monitoring Lead and Their Characteristics[a]

Biological Markers	Half-Life	Comments Regarding Use as Indicator of Accumulated Lead Exposure
Biological Markers of Internal Dose		
Blood lead	Weeks to months	Dependent on urine flow
Urine lead	Weeks	Large intraindividual variability; prone to external contamination or leaching
Hair lead	Months	Prone to external contamination; kinetics uncertain
Nail lead	Months	
Bone lead	Years to decades	Cortical bone has long half-life; trabecular bone has shorter half life
Chelatable lead	Years	Involves injection and timed collection of urine; represents "chelatable" compartment of lead found mostly in soft tissues; partly in bone
Biologically Effective Dose Markers		
Bone lead	Years to decades	May be related to abnormal skeletal development and/or hematopoesis (untested)
Biological Response Marker		
Erythrocytic protoporphyrin	Weeks to months	Level increases due to lead's inhibition of hematopoiesis; integrates exposure over several months; sensitivity poor for sustained mild-to-moderate elevations of blood lead (25–40 μg dL^{-1}); increased levels also seen in iron deficiency
Erythrocytic ALAD	Weeks	Activity inhibited by lead but also by other metals, (e.g., methylmercury and by ethanol intoxication)
Urinary ALAD, coproporphyrins	Weeks	Levels also increased in certain hepatic disorders

[a] Reference 428. [Reproduced with permission from H. Hu, M. Rabinowitz, and D. Smith, *Environ. Health Perspect.*, *106*, 1–8 (1998)].

Blood lead levels are typically determined using whole venous blood that has been chemically degraded (e.g., with HNO_3) (10). The concentration of total lead is usually measured using atomic absorption spectroscopy (AAS), in which the sample is vaporized and the concentration of atomic lead is determined using absorption spectroscopy with a cathode lamp (10, 512, 513). Electrothermal and graphite-furnace AASs are particularly useful because they require relatively little sample (5–20 µL). More recently, ICP–MS has also been validated for use in determining BLLs (514). Although ICP–MS instruments tend to be more expensive and difficult to use, ICP–MS has much lower detection limits than do conventional AAS instruments (514). In addition, ICP–MS can be used to simultaneously quantitate the concentrations of several elements at once, which can be useful for studies in basic metal ion homeostasis (514, 515). For analysis of clinical samples, however, the greatest limitation is typically not sensitivity, but quality control: laboratory standards vary and there has been a general call for improved "clean techniques" and standardization in testing of lead samples (10, 516). Finger stick measurements, although much easier to acquire than venous blood samples, are particularly susceptible to contamination. If a high BLL is observed, regardless of sampling procedure, additional testing to determine whether the result is reproducible is usually recommended (10, 17, 22). Lead levels in blood can also be measured using electrochemical techniques [e.g., differential pulse polarography (DPP) and anodic stripping voltammetry (ASV)]. Electrochemical detection of lead provides a detection limit (≤ 1 pg) that is at least as good as those of AAS methods, it is extremely cost effective, and portable devices are available that can be used by primary health providers to provide immediate feedback to their patients (10). Despite these advantages, electrochemical techniques have not been widely accepted by the medical community.

2. *Alterations in Heme Biosynthesis*

Lead interacts with multiple proteins and pathways involved in heme biosynthesis (see above), and hence alterations in heme biosynthesis can be used as biomarkers for lead poisoning (503–505). However, it is important to note that these are markers for the *effects* of lead poisoning rather than the extent of exposure and that they can be complicated by other physiological and environmental factors (10). For example, hemoglobin levels are frequently used to assess whether an individual has been exposed to lead over a period of months, primarily because hemoglobin levels are easy to assay, lead poisoning is highly correlated with anemia, and the lifetime for erythrocytes in humans is 120 days, providing a long-term marker of exposure (517). However, hemoglobin levels are also highly variable from individual to individual and are highly dependent upon other factors (e.g., gender and diet), so this test cannot be used as an exclusive indicator of lead poisoning. In the past, levels of EP and ZPP were also commonly used as biomarkers for lead poisoning (see Section VI.D.4) because EP/ZPP levels correlate well with BLL and can be screened rapidly and cheaply using fluorescence spectroscopy (Fig. 35) (518, 519). However, EP levels also increase in response to iron

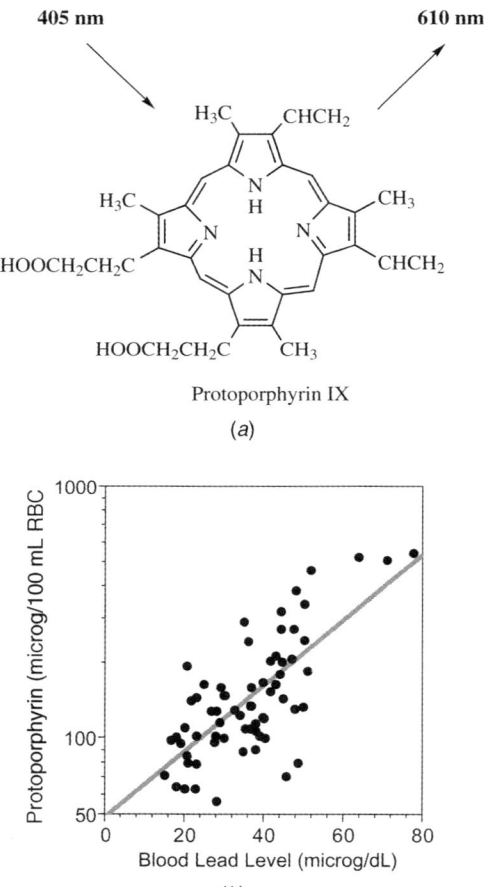

Figure 35. (*a*) Protoporphyrin levels in whole blood can be measured rapidly and cheaply using fluorescence spectroscopy ($\lambda_{ex} = 405$ nm; $\lambda_{em} = 610$ nm) without interference from heme, because the iron quenches fluorescence in heme (10, 518, 519). Although protoporphyrin levels correlate well with BLL (*b*), they are also highly dependent on other environmental and physiological factors (e.g., iron deficiency) and hence are not currently recommended for use as a biomarker for lead poisoning by the CDC (10). [Part (*b*) adapted from reference (518).]

deficiency (the most common nutritional disorder in the United States) and in response to genetic porphyrin disorders. As a result, the CDC no longer recommends that EP levels be used as a marker for lead poisoning (10).

By contrast, ALAD activity levels are less prone to interference from other disorders and seem to correlate well with blood lead levels, even at levels as low as 5 µg dL^{-1} (see Fig. 30) (10). The disadvantages to the assay for ALAD activity are that it is more susceptible to zinc and lead contamination than EP levels and that the assay involves a specialized colorimetric assay for porphobilinogen (520) and hence is not something that would typically be performed by a standard health testing laboratory. Epidemiological studies have suggested that different isoforms of ALAD that are common in the U.S. population might exhibit different susceptibilities to inactivation by lead (521)—and hence might account for different susceptibilities towards lead poisoning exhibited by different individuals. However, detailed studies on recombinant human ALAD have not supported this hypothesis (522). As a group, these heme biosynthesis-based biomarkers for lead poisoning all have fundamental flaws as primary assays for lead poisoning. However, they provide useful insights into the mechanism of lead's toxicity and are useful as biomarkers, if considered in concert with other measures of exposure (e.g., blood lead level).

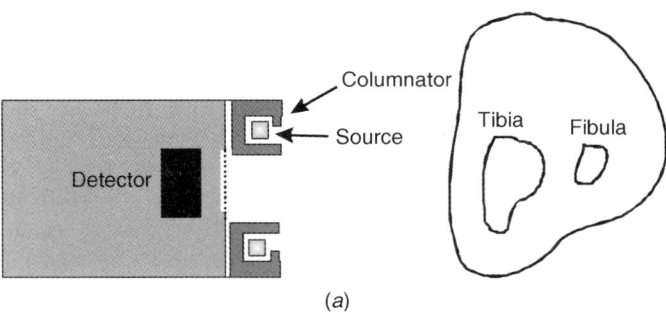

(a)

Figure 36. Lead levels in bone can be measured in vivo using XRF spectroscopy. (a) γ-rays or X-rays are used (source) to eject either L-shell electrons (L-XRF) or K-shell electrons (K-XRF) from lead in bone; when outer-shell electrons fill this vacancy, photons are released (fluorescence) and are monitored by the detector (10, 523). A typical X-ray fluorescence spectrum [(b), e.g., of a 112 µg Pb g^{-1} "phantom") provides the number of counts observed as a function of photon energy. Emissions characteristic of lead occur at 72.8 keV (PbKα_2), 75.0 keV (PbKα_1), and 84.9 keV (Pb Kβ_1) (436, 523). Measurements on actual samples are correlated with those obtained from standard "phantoms" made of plaster-of-paris and doped with a known amount of lead to obtain bone lead concentrations in micrograms of Pb per gram (µg Pb g^{-1} bone). The bone lead levels obtained by this method correlate extremely well with independent measurements of BLL (c). [Parts (a) and (c) adapted from (524). Part (b) adapted from (436).]

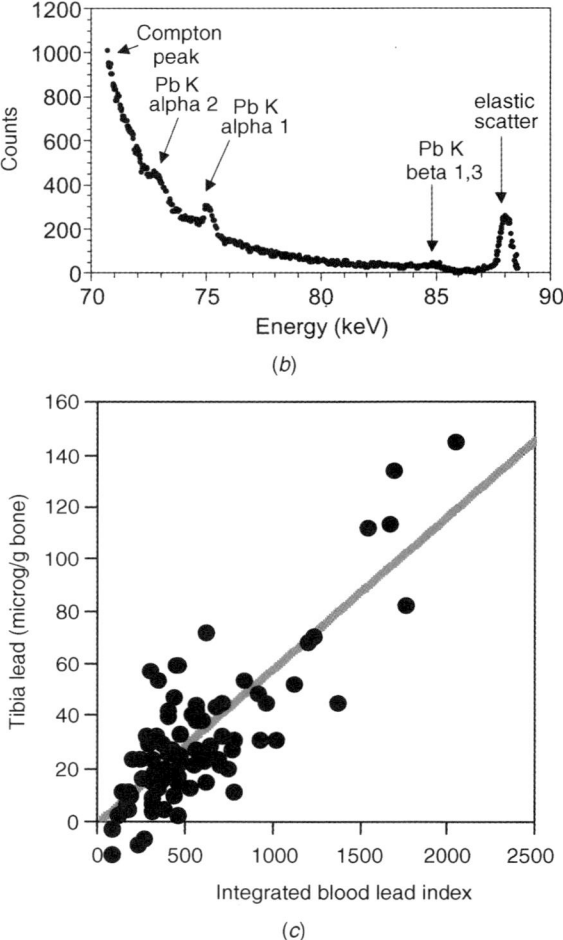

Figure 36 (Continued)

3. Bone Lead Levels

Lead is believed to directly incorporate into the calcium sites in hydroxyapatite in bone. Bone lead levels provide a useful biomarker of long term, cumulative exposure to lead, because the half-life of lead in bone is ≥10 years (5). Lead levels in bone can be measured *in situ* accurately and quantitatively using XRF spectroscopy (Fig. 36) (10, 436, 523). X-ray fluorescence spectrometers [Fig. 36(a)] use γ-rays or X-rays to eject either L-shell electrons

(L-XRF) or K-shell electrons (K-XRF) from lead in bone, both of which are inner shell electrons. When the resulting electronic vacancy is filled by a less tightly bound electron, photons (X-rays) are emitted that can be detected. The energy of the emitted photons is characteristic of the source (here, lead) and XRF spectra are recorded as counts versus energy of emitted radiation [Fig. 36(b)]. The precision of XRF for lead in bone is ± 2–10 µg g^{-1} (10). Bone lead levels measured by XRF correlate extremely well with BLLs [Fig. 36(c)] (523) and are considered to be a good marker for total lead body burden, since a high percentage of the total lead in the body is deposited into bone (see above) (436). The primary disadvantage of XRF is that it requires the use of ionizing radiation, although problems also arise from poor signal to noise ratio, variability between instruments, and physiological properties of bone (e.g., bone remodeling and turnover), which may complicate analysis of XRF data (10). Goals and challenges in this field include correlating bone lead levels with symptoms associated with lead poisoning and developing methods to reduce radiation exposure, to better define dosimetry, and to improve instrumentation calibration (10).

4. Other Potential Markers for Lead Poisoning

Other assays that have been used as biomarkers for lead poisoning include lead levels in urine, teeth, hair, and nails (Table XVIII), but none are particularly useful for clinical studies (5, 10, 436). Urine lead levels can be quantitated easily by AAS but vary widely from individual to individual and are dependent on other parameters (e.g., kidney function) that are difficult to normalize. Urine lead levels are used extensively, however, to monitor the progress and effectiveness of chelation therapy (see below) (10). Lead levels in teeth are a good measure of cumulative lead exposure and can be measured using either AAS, ICP–MS, or ASV, but are not very useful on a practical level because the analysis requires shedding of teeth (10, 524). Hair and nails would appear the ideal source of materials for lead analysis because they are easily procured and rapidly regenerate, but lead levels in hair and nails vary considerably from individual to individual, and these samples are highly susceptible to contamination.

What does the future hold? It is likely that studies on the toxicogenomics of lead will not only provide insights into the mechanisms of lead's toxicity, but also identify genes that are specifically up- or down-regulated in response to lead. These genes can serve as new useful biomarkers for lead poisoning or predictors for what an individual's susceptibility will be to lead poisoning. Until that time, the Centers for Disease Control recommends that screening for lead poisoning be conducted using venous blood samples (10).

F. Chelation Therapy

Although a child is considered to be have an elevated blood lead level if he/she has a BLL $\geq 10\,\mu g\ dL^{-1}$, invasive treatment (i.e., chelation therapy) is usually not recommended unless a child has a BLL $\geq 45\,\mu g\ dL^{-1}$. For children with moderate levels of lead poisoning (BLL = $10\text{--}45\,\mu g\ dL^{-1}$), the CDC recommends taking actions to eliminate sources of exposure (e.g., cleaning household dust, removing children from the sources of exposure, etc.) and frequent retesting (to make sure that the lead poisoning has not worsened) (17, 22). Why the hesitancy to chelate children? Chelation therapy tends to be accompanied by serious side effects: Agents frequently remove essential nutrients (e.g., iron and zinc) and can cause redistribution and mobilization of lead within the body. Furthermore, there are no long-term, well-controlled studies showing whether chelation therapy actually reverses or prevents symptoms associated with childhood lead poisoning (207).

1. Chelation Therapy Agents Currently in Use

The most commonly used chelation therapy agents in the United States today are EDTA and DMSA (or succimer). In addition, penicillamine (PCA) and BAL are used to chelate lead. Each of these agents has numerous disadvantages, ranging from undesirable methods of delivery (intramuscular injection of BAL and intravenous delivery of EDTA), to unpleasant side effects (typically nausea and vomiting), to chelation and increased excretion of necessary metals (e.g., iron and zinc) (Table XIX) (17, 207, 525).

2. Challenges in Developing New Chelation Therapy Agents

Given these limitations, why have "better" treatments not been developed? The reality is that this is a formidable task: The ideal chelation therapy agent should satisfy all of the following criteria (442):

- Bind lead tightly ($K_b \geq 10^{12}$).
- Bind lead preferentially over other bioavailable metal ions (notably calcium, zinc, and iron) by a factor of $\geq 10^3$.
- Exhibit good biodistribution, bioavailability, and solubility.
- Should not mobilize lead from bone.
- Exhibit low toxicity and side effects.
- Be orally available.

As we have discussed in preceding sections (see Section IV.D), it is extremely difficult to design a ligand from first principles that binds lead both

TABLE XIX
Properties of Chelation Therapy Agents Currently in Use (18, 207, 525)

Agent	Advantages	Disadvantages	Recommended Use
Ethylenediamine-N,N,N',N'-tetraacetic acid (EDTA)	More effective than penicillamine	High affinity for zinc and iron; zinc and iron during treatment decrease efficacy; replacement iron therapy may be necessary. Not recommended for oral use; should be delivered intravenously, typically over a course of 5–7 days; requires hospitalization. May cause increased lead absorption from GI tract. Side effects include nephrotoxicity	Administered as CaNa$_2$EDTA Treatment recommended for BLLs ≥ 45 µg dL^{-1}
meso-2,3-Dimercaptosuccinic acid (DMSA)	Administered orally. Effective at mobilizing lead from bone and soft tissue. Minimal toxicity	Drawbacks include: unpleasant taste and "rotten egg" smell. Side effects include: nausea, vomiting, diarrhea, appetite loss, and foul smelling urine, flatus, and stools	Treatment recommended for BLLs > 45 µg dL^{-1}

(*continues*)

TABLE XIX (Continued)

Agent	Advantages	Disadvantages	Recommended Use
D-Penicillamine	Administered orally Can be used on an outpatient basis	Causes increased excretion of other metals Can cause a reaction in individuals allergic to penicillin Side effects include nausea, vomiting, hemotologic reactions, and eosinophilia	Can be used for outpatient treatment for BLLs ≤ 50 µg dL^{-1}
2,3-Dimercaptopropanol (BAL)	Rat studies suggest BAL is more effective than CaNa$_2$EDTA at chelating lead from bone and soft tissue	Delivered as intramuscular injection (can be very painful) May cause reaction in children allergic to peanuts Side effects include fever, nausea, watery eyes, sweating, and unpleasant breath	Recommended in conjunction with CaNa$_2$EDTA for BLLs ≥ 70 µg dL^{-1}

tightly and selectively. However, new combinatorial approaches to ligand design (526) and methods for screening the ability of ligands to coordinate to lead (38) provide much needed methodologies for a more empirical approach to exploring lead coordination chemistry. In addition, advances in biotechnology offer rapid methods for screening new compounds for properties that have traditionally been hard to rationally design, such as toxicity and bioavailability. These advances, coupled with our recently improved understanding of the biological chemistry of lead, suggest that we are on the cusp of a new era in lead coordination chemistry and therapeutics.

VII. CONCLUSIONS

Although relatively understudied compared to transition metals, Pb(II) exhibits a rich coordination chemistry. Because lead is subject to relativistic effects, the electronic properties of Pb(II) complexes are diverse and interesting. Fortunately, a wide variety of spectroscopic techniques are available to probe the properties of these compounds. In addition, detailed studies on the structures, kinetics, and thermodynamics of lead compounds provide important insights into the environmental and biological chemistry of lead(II). Through this chapter, we hope not only to have provided an overview to the chemistry of Pb(II), but also to have corrected several misconceptions about this element:

- The $6s$ lone pair in lead is *not* "inert". Rather, the $6s$ orbital can mix with the $6p$ orbitals on lead, which has a profound effect upon the spectroscopic properties and our understanding of the structures exhibited by Pb(II) compounds.
- Lead is *not* spectroscopically silent. The absorption spectroscopy, photoelectron spectroscopy, and NMR spectroscopy of Pb(II) compounds have all been extensively studied. These techniques not only provide useful insights into the electronic structure and structures of Pb(II) compounds, but have also proven useful in the characterization of lead coordination environments in complex samples (e.g., samples of biological or environmental origin).

In addition, we hope to have highlighted the challenges that remain for chemists in this field:

- To develop a more thorough and systematic understanding of what makes a "good" coordination environment for lead, especially under biologically relevant conditions.

- To use these insights to develop improved chelating agents and fluorescent sensors that bind lead both tightly and selectively.
- To determine the cellular and subcellular distribution of lead and the speciation of lead in vivo.
- To determine the rates and mechanisms of lead binding to proteins and other biomolecules.
- To use the new tools of genomics and proteomics to characterize how lead affects signaling in biological systems.

ACKNOWLEDGMENTS

The original research described herein that was conducted by H.A.G. and co-workers was supported by the National Science Foundation (CHE-9875341, CHE-9810378) and the National Institutes of Health (R01 GM58183), and through a Camille and Henry Dreyfus New Faculty Award, a Burroughs Wellcome Fund New Investigator Award in Toxicology, a Camille Dreyfus Teacher–Scholar Award, and a Sloan Research Fellowship to H.A.G. One of us, J.S.M., acknowledges support from an NIH Institutional NRSA Training Grant in Molecular Biophysics (GM08382) and from the Institute for Environmental Catalysis, funded by the U. S. National Science Foundation and the Department of Energy (CHE-9810378). E.S.C. acknowledges support from the MRSEC program of the National Science Foundation (DMR-0076097). In addition, we thank Scott Barry, Joseph Bressler, Sandhya Deo, Donald Ellis, Ricardo Garcia, Tryg Jensen, Hongbo Li, Elaine Magyar, James Magyar, Donald McClure, John Payne, Mark Ratner, Job Rijssenbeek, Brian Rous, and Samuel Webb for helpful discussions and comments on the manuscript. E.S.C. thanks Thomas O'Halloran and his students for the use of their copy of the *Critical Stability Constants Database*.

ABBREVIATIONS

AAS	Atomic absorption spectroscopy
ALAD	δ-Aminolevulinic acid dehydratase, also known as porphobilinogen synthase, PBS
ALAS	δ-Aminolevulinic acid synthase
ASV	Anodic stripping voltammetry
ATP	Adenosine triphosphate
BAL	British anti-Lewisite, 2,3-dimercaptopropanol
BE	Binding energy
BLL	Blood lead level, usually given in $\mu g\ dL^{-1}$
BSA	Bovine serum albumin
C2	Calcium-binding domain originally identified in protein kinase C, PKC

Ca-ATPase	Transport enzyme that uses ATP to drive transport of calcium
CadC	Cadmium efflux system accessory protein
CDC	Centers for Disease Control
CDS	Cambridge Structural Database
CoA	Coenzyme A
CP–CCCC	Consensus zinc finger peptide with four cysteine residues in the metal-binding site
CP–CCHC	Consensus zinc finger peptide with three cysteine residues and one histidine residue in the metal-binding site
CP–CCHH	Consensus zinc finger peptide with two cysteine residues and two histidine residues in the metal-binding site
CT	Charge transfer
CyDTA	*trans*-1,2-Cyclohexanediaminetetraacetic acid
Cys, C	Cysteine
Dabcyl	4-(4'-Dimethylaminophenylazo)benzoic acid
DBTA	2,3-Diaminobutane-N,N,N',N'-tetraacetatic acid
DMSA	*meso*-2,3-Dimercaptosuccinic acid
DNA	Deoxyribonucleic acid
dns–ECEE	Dansyl–glutamate–cysteine–glutamate–glutamate
DOTA	(12-ane N_4 with 4 pendant acetic acid)
DOTAM	1,4,7,10-Tetrakis(acetamido)-1,4,7,10 tetraazacyclododecane
DPA	D-Penicillamine
DTPA	Diethylenetriamine-N,N,N',N'',N''-pentaacetic acid
EC	N,N'-Ethylene-L-cysteine
EDDDA	Ethylenediamine-N,N'-diacetic-N,N'-dipropionic acid
EDDHA	Ethylenediamine-N-N'-bis(2-hydroxyphenylacetic acid)
DPP	Differential pulse polarography
EDTA	Ethylenediamine-N,N,N',N'-tetraacetic acid
EDTA-N_4	Ethylenediaminetetraacetamide
EF-hand	Calcium-binding motif originally identified in the protein calmodulin
EGTA	1,2-Bis(2-aminoethoxyethane)-N,N,N',N'-tetraacetic acid
E_k	Kinetic energy
EP	Erythrocyte protoporphyrin
EXAFS	Extended X-ray absorption fine structure
Glu, G	Glutamate
GRP78	78-kDa Glucose-regulated protein (heat shock protein/chaperone)
HEDTA	N-(2-Hydroxyethyl)ethylenediamine-N,N',N'-triacetic acid
HEIDA	N-(2-Hydroxyethyl)iminodiacetic acid

His, H	Histidine
HIV	Human immunodeficiency virus
HIV–CCHC	A structural zinc-binding domain from HIV-nucleocapsid protein that contains three cysteine residues and one histidine residue in the metal-binding site
HMQC	Heteronuclear multiple quantum correlation
HP2	Human protamine 2 (chromatin-binding protein)
ICP-MS	Inductively coupled plasma–mass spectrometry
IDMS	High-resolution isotope dilution mass spectrometry
IDA	Iminodiacetic acid
Indo-1	2-[4-[Bis(carboxymethyl)amino]-3-[2-[2-[bis(carboxymethyl)amino]-5- methylphenoxy]ethoxy]phenyl]-1H-indole-6-carboxylic acid
IR	Infrared
K_b	Stepwise stability or binding constant
K_d	Dissociation constant (inverse of binding constant)
LCAO	Linear combination of atomic orbitals
LMCT	Ligand-to-metal charge transfer
MCD	Magnetic circular dichroism
MLCT	Metal-to-ligand charge transfer
MO	Molecular orbital
MS	Mass spectrometry
NFκB	Nuclear factor protein (transcription factor)
NMR	Nuclear magnetic resonance
Nramp	Natural resistance-associated macrophage protein (iron transporter)
NTA	Nitrilo-2,2′,2″-triacetic acid
Pb EF	Lead enrichment factor; Pb EF = $(Pb/Sc)_{sample}/(Pb/Sc)_{background}$
PBS	Porphobilinogen synthase, also known as δ-aminolevulinic acid dehydratase, ALAD
PCA	Penicillamine
PC-12	Rat cell line that can differentiate in the presence of nerve growth factor
1,2-PDTA	1,2-Propylenediaminetetraacetic acid
1,3-PDTA	1,3-Propylenediaminetetraacetic acid
PES	Photoelectron spectroscopy
PKC	Protein kinase C
ppb	Parts per billion (1 ng g^{-1})
ppm	Parts per million (1 μg g^{-1})
ppt	Parts per trillion (1 pg g^{-1})
RNA	Ribonucleic acid

salen	Bis(salicylidene)ethylenediamine
SHE	Standard hydrogen electrode
Sp-1	Transcription factor (Cys_2His_2 zinc-binding site)
STM	Scanning tunneling microscopy
syn	Syntaxin
syt	Synaptotagmin
TAMRA	6-Carboxytetramethylrhodamine
TFIII-A	Transcription factor IIIA (Cys_2His_2 zinc-binding site)
$tRNA^{Phe}$	The transfer RNA for phenylalanine
TPP	Tetraphenylporphyrin
UV	Ultraviolet
vis	Visible
VSEPR	Valence shell electron-pair repulsion
XANES	X-Ray absorption near edge spectroscopy
XRF	X-Ray fluorescence spectroscopy
ZPP	Zinc protoporphyrin
χ_{opt}	Optical electronegativity

REFERENCES

1. J. O'Hara, in *Bartlett's Familiar Quotations*, J. Kaplan, Ed., Little, Brown and Company, Boston, 1992, p. 717.
2. C. Warren, *Brush with Death: A Social History of Lead Poisoning*, Johns Hopkins University Press, Baltimore, 2001.
3. J. O. Nriagu, *Lead and Lead Poisoning in Antiquity*, John Wiley & Sons, Inc., New York, 1983.
4. H. A. Waldron, *Medical History*, *17*, 391 (1973).
5. J. M. Christensen and J. Kristiansen, in *Handbook on Metals in Clinical and Analytical Chemistry*, H. G. Seiler, A. Sigel, and H. Sigel, Eds., Marcel Dekker, New York, 1994, pp. 425–440.
6. D. M. Settle and C. C. Patterson, *Science*, *207*, 1167 (1980).
7. S. Hong, J.-P. Candelone, C. C. Patterson, and C. F. Boutron, *Science*, *265*, 1841 (1994).
8. J. O. Nriagu, *Science*, *281*, 1622 (1998).
9. H. W. Salzberg, *From Caveman to Chemist*, American Chemical Society, Washington, DC, 1991.
10. *Measuring lead exposure in infants, children, and other sensitive populations*, Board Environmental Studies and Toxicology, National Research Council, Washington, DC, (1993).
11. S. C. Gilfillan, *J. Occup. Med.*, *7*, 53 (1965).
12. J. O. Nriagu, *N. Engl. J. Med.*, *308*, 660 (1983).
13. J. Scarborough, *J. Med. Hist.*, *39*, 469 (1984).
14. J. Rich, Ed., *The City in Late Antiquity*, Routledge, New York, 1992.
15. C. F. Boutron, U. Gorlach, J. P. Candelone, M. A. Bol'shov, and R. J. Delmas, *Nature (London)*, *353*, 153 (1991).

16. J. Schwartz and H. Pitcher, *J. Official Statistics*, *5*, 421 (1989).
17. *Screening Young Children for Lead Poisoning: Guidance for State and Local Public Health Officials*, Centers for Disease Control and Prevention, U.S. Department of Health and Human Services, Public Health Service, Atlanta, 1997.
18. *National Report on Human Exposure to Environmental Chemicals*, Centers for Disease Control and Prevention, U.S. Department of Health and Human Services, Public Health Service, Atlanta, 2001.
19. M. Lovei, *Phasing out lead from gasoline: worldwide experience and policy implications*, The World Bank, Washington, DC, *World Bank Technical Paper No. 397*, 1998.
20. V. M. Thomas, R. H. Socolow, J. J. Fanelli, and T. G. Spiro, *Environ. Sci. Technol.*, *33*, 3942 (2000).
21. R. M. Stapleton, *Lead Is a Silent Hazard*, Walker and Company, New York, 1994.
22. I. Kessel and J. T. O'Connor, *Getting the Lead Out: The Complete Resource on How to Prevent and Cope with Lead Poisoning*, Plenum Publishing, New York, 1997.
22. (a) Agency for Toxic Substances and Disease Registry. *Case Studies in Environmental Medicine: Lead Toxicity*, Public Health Service, U.S. Department of Health and Human Services, 1995.
23. A. R. Flegal and D. R. Smith, *Environ. Res.*, *58*, 125 (1992).
24. *Superfund Hazardous Waste Site Basic Query Form*, http://www.epa.gov/superfund/sites/query/basic.htm (accessed April 2001). A search on the Environmental Protection Agency's Superfund website revealed that 844 out of 11,116 sites contain lead.
25. N. J. Shinn, J. Bing-Canar, M. Cailas, N. Peneff, and H. J. Binns, *Environ. Res.*, *82*, 46 (2000).
26. H. W. Mielke, *Am. Sci.*, *87*, 62 (1999).
27. C. Patterson, J. Ericson, M. Manea-Krichten, and H. Shirahata, *Sci. Total Environ.*, *107*, 205 (1991).
28. *Top Ten Hazardous Substances—Agency for Toxic Substances and Disease Registry (ATSDR)*, 1999, http://www.atsdr.cdc.gov/cxcx3.html (accessed July 2001).
29. A. C. Todd, J. G. Wetmur, J. M. Moline, J. H. Godbold, S. M. Levin, and P. J. Landrigan, *Environ. Health Perspect.*, *104, Suppl. 1*, 141 (1996).
30. P. J. Landrigan and A. C. Todd, *West. J. Med.*, *161*, 153 (1994).
31. G. Faure, *Principles of Isotope Geology*, John Wiley & Sons, Ins., New York, 1986.
32. N. N. Greenwood and A. Earnshaw, *Chemistry of the Elements*, Pergamon Press, Oxford, UK, 1984.
33. W. Shotyk, D. Weiss, P. G. Appleby, A. K. Cheburkin, R. Frei, M. Gloor, J. D. Kramers, S. Reese, and W. O. Van Der Knaap, *Science*, *281*, 1635 (1998).
34. R. C. Weast, Ed., *CRC Handbook of Chemistry and Physics*, CRC Press, Boca Raton, FL, 1985.
35. F. A. Cotton, G. Wilkinson, C. A. Murillo, and M. Bochmann, *Advanced Inorganic Chemistry*, John Wiley & Sons, Inc., New York, 1999.
36. T. G. Spiro and W. M. Stigliani, *Chemistry of the Environment*, Prentice Hall, Upper Saddle River, New Jersey, 1996.
37. J. C. Payne, M. A. ter Horst, and H. A. Godwin, *J. Am. Chem. Soc.*, *121*, 6850 (1999).
38. S. Deo and H. A. Godwin, *J. Am. Chem. Soc.*, *122*, 174 (2000).
39. E. S. Claudio, M. A. ter Horst, C. E. Forde, C. L. Stern, M. K. Zart, and H. A. Godwin, *Inorg. Chem.*, *39*, 1391 (2000).
40. C. M. L. S. Bouton, L. P. Frelin, C. E. Forde, H. A. Godwin, and J. Pevsner, *J. Neurochem.*, *76*, 1724 (2001).

41. H. A. Godwin, *Curr. Opin. Chem. Biol.*, **5**, 223 (2001).
42. S. Budavari, Ed., *The Merck Index*, Merck & Co., Rahway, NJ, 1989.
43. S. S. Zumdahl, *Chemistry*, Houghton Mifflin, Boston, 1997.
44. D. R. Lide, Ed., *CRC Handbook of Chemistry and Physics*, CRC Press, Boca Raton, FL, 1995–1996.
45. D. R. Lide, Ed., *CRC Handbook of Chemistry and Physics*, CRC Press, Boca Raton, FL, 2000.
46. T. L. Brown, H. E. Lemay, and B. E. Bursten, *Chemistry: The Central Science*, 8th ed., Prentice Hall, Upper Saddle River, NJ, 2000.
47. J. O. Nriagu, Ed., *The Biogeochemistry of Lead in the Environment. Part A. Ecological Cycles*, Elsevier/North-Holland Biomedical, New York, 1978, Vol. 1.
48. C. Baird, *Environmental Chemistry*, W. H. Freeman, New York, 1999.
49. W. Stumm and J. J. Morgan, *Aquatic Chemistry: Chemical Equilibria and Rates in Natural Waters*, Wiley-Interscience, New York, 1996.
50. I. M. Klotz, J. M. Urquhart, and H. A. Fiess, *J. Am. Chem. Soc.*, **74**, 5537 (1952).
51. R. K. Mehra, V. R. Kodati, and R. Abdullah, *Biochem. Biophys. Res. Commun.*, **215**, 730 (1995).
52. X. Wei and B. Ru, *Chin. J. Biochem. Mol. Biol.*, **15**, 289 (1999).
53. L. S. Busenlehner, N. J. Cosper, R. A. Scott, B. P. Rosen, M. D. Wong, and D. P. Giedroc, *Biochemistry*, **40**, 4426 (2001).
54. H. Fromherz and K.-H. Lih, *Z. Phys. Chem., Abt. A153*, 321 (1931).
55. H. Fromherz, *Z. Elektrochem.*, **37**, 553 (1931).
56. P. Pringsheim, *Rev. Mod. Phys.*, **14**, 132 (1942).
57. S. Khorana, P. B. Merkel, and W. H. Hamill, *J. Chem. Phys.*, **56**, 2807 (1972).
58. M. Bacci, A. Ranfagni, and G. Viliani, *J. Phys. Chem. Solids*, **42**, 1021 (1981).
59. H. Bendiab, J. Meullemeestre, M.-J. Schwing, and F. Vierling, *J. Chem. Res., Synop.*, 280 (1982).
60. H. Bendiab, J. Meullemeestre, M.-J. Schwing, and F. Vierling, *J. Chem. Res., Miniprint*, 2718 (1982).
61. M. Forro, *Z. Physik*, **56**, 534 (1929).
62. R. Hilsch, *Proc. Phys. Soc., London*, **49**, 40 (1937).
63. A. Fukuda, *J. Phys. Soc. Jpn.*, **27**, 96 (1969).
64. R. E. Chaney, P. W. M. Jacobs, and T. Tsuboi, *Can. J. Phys.*, **51**, 2242 (1973).
65. T. Tsuboi, R. E. Chaney, and P. W. M. Jacobs, *Can. J. Phys.*, **53**, 200 (1975).
66. T. Tsuboi and P. W. M. Jacobs, *Kyoto Sangyo Daigaku Ronshu*, **5**, 48 (1976).
67. T. Tsuboi, *Phys. Status Solidi B*, **96**, 321 (1979).
68. T. Tsuboi, *Physica, B* **96**, 341 (1979).
69. J. Ewles, *Proc. R. Soc. London, Ser. A* **167**, 34 (1938).
70. J. Ewles and C. Curry, *Proc. Phys. Soc., London, Ser. A* **63**, 708 (1950).
71. J. Ewles and N. Lee, *J. Electrochem. Soc.*, **100**, 392 (1953).
72. H. F. Folkerts, A. van Dijken, and G. Blasse, *J. Phys.: Condens. Matter*, **7**, 1049 (1995).
73. H. F. Folkerts and G. Blasse, *J. Phys. Chem. Solids*, **57**, 303 (1996).
74. P. A. M. Dirac, *Proc. R. Soc. London, Ser. A* **118**, 351 (1928).
75. P. A. M. Dirac, *Proc. R. Soc. London, Ser. A* **117**, 610 (1927).

76. P. Pyykkö and J.-P. Desclaux, *Acc. Chem. Res.*, *12*, 276 (1979).
77. R. E. Powell, *J. Chem. Educ.*, *45*, 558 (1968).
78. A. Szabo, *J. Chem. Educ.*, *46*, 678 (1969).
79. K. S. Pitzer, *Acc. Chem. Res.*, *12*, 271 (1979).
80. J. E. Huheey, E. A. Keiter, and R. L. Keiter, *Inorganic Chemistry: Principles of Structure and Reactivity*, Harper Collins, New York, 1993.
81. D. R. McKelvey, *J. Chem. Educ.*, *60*, 112 (1983).
82. P. Pyykkö, *Adv. Quantum Chem.*, *11*, 353 (1978).
83. P. Pyykkö, *Chem. Rev.*, *88*, 563 (1988).
84. N. V. Sidgwick, *The Electronic Theory of Valency*, Oxford University Press, Oxford, UK, 1927.
85. C. K. Jørgensen, *Modern Aspects of Ligand Field Theory*, American Elsevier, New York, 1971.
86. J. C. Phillips, *Bonds and Bands in Semiconductors*, Academic, New York, 1973.
87. B. Fricke, *Struct. Bonding*, *21*, 89 (1975).
88. P. Pyykkö and J.-P. Desclaux, *Chem. Phys. Lett.*, *42*, 545 (1976).
89. B. Quintanilla-Vega, D. J. Hoover, W. Bal, E. K. Silbergeld, M. P. Waalkes, and L. D. Anderson, *Chem. Res. Toxicol.*, *13*, 594 (2000).
90. M.'Razmiafshari and N. H. Zawia, *Toxicol. Appl. Pharmacol.*, *166*, 1 (2000).
91. A. B. P. Lever, *J. Chem. Educ.*, *51*, 612 (1974).
92. C. J. Ballhausen and H. B. Gray, in *Coordination Chemistry*, A. E. Martell, Ed. Van Nostrand Reinhold, New York, 1971, Vol. 1, pp. 3–83.
93. A. B. P. Lever, *Inorganic Electronic Spectroscopy*, Elsevier, New York, 1984.
94. E. I. Solomon and A. B. P. Lever, Eds., *Inorganic Electronic Structure and Spectroscopy*, John Wiley & Sons, Inc., New York, 1999.
95. D. S. McClure, in *Solid State Physics: Advances in Research and Applications*, F. Seitz and D. Turnbull, Eds., Academic, New York, 1959, Vol. 9, pp. 399–525.
96. C. K. Jørgensen, in *Solid State Physics: Advances in Research and Applications*, F. Seitz and D. Turnbull, Eds., Academic, New York, 1962, Vol. 13, pp. 375–462.
97. C. K. Jørgensen, *Absorption Spectra and Chemical Bonding in Complexes*, Addison-Wesley, Reading, MA, 1962.
98. C. K. Jørgensen, *Orbitals in Atoms and Molecules*, Academic, New York, 1962.
99. C. K. Jørgensen, *Inorganic Complexes*, Academic, New York, 1963.
100. C. K. Jørgensen, *Adv. Chem. Phys.*, *5*, 33 (1963).
101. C. K. Jørgensen, *Prog. Inorg. Chem.*, *12*, 101 (1970).
102. L. Pauling, *J. Am. Chem. Soc.*, *54*, 3570 (1932).
103. L. Pauling, *Nature of the Chemical Bond*, Cornell University Press, Ithaca, NY, 1939.
104. The optical electronegativity scale is defined based on the shift of metal–halogen charge-transfer bands to lower energy down the group $F > Cl > Br > I$. The corrected wavenumber v_{CT} of the first LMCT band is then defined by $v_{CT} = (30,000 \text{ cm}^{-1})[\chi_{opt}(L) - \chi_{opt}(M)]$, where the constant makes the optical electronegativities similar in magnitude to Pauling's electronegativities and $\chi_{opt}(L)$ and $\chi_{opt}(M)$ are the optical electronegativities of the ligand and the metal, respectively (99, 101). {For MLCT, the relation is similar: $v_{CT} = (30,000 \text{ cm}^{-1})[\chi_{opt}(M) - \chi_{opt}(L)]$ (93)}. Obviously, the optical electronegativity varies with the oxidation state of the metal or ligand but also with the symmetry of the complex since the symmetry affects the spacing of the molecular orbitals and the transition energies.

105. D. R. McMillin, *Bioinorg. Chem.*, *8*, 179 (1978).
106. J. Rawlings, O. Siiman, and H. B. Gray, *Proc. Natl. Acad. Sci. USA*, *71*, 125 (1974).
107. B. P. Kennedy and A. B. P. Lever, *Can. J. Chem.*, *50*, 3488 (1972).
108. M. Vasák, J. H. R. Kägi, and H. A. O. Hill, *Biochemistry*, *20*, 2852 (1981).
109. A. L. Allred, *J. Inorg. Nucl. Chem.*, *17*, 215 (1961).
110. D. W. Fitzgerald and J. E. Coleman, *Biochemistry*, *30*, 5195 (1991).
111. M. Vargek, X. Zhao, Z. Lai, G. L. McLendon, and T. G. Spiro, *Inorg. Chem.*, *38*, 1372 (1999).
112. By contrast, if the opposite assumption is made—that the absorption bands are MLCT rather than LMCT—then the value calculated for $\chi_{opt}(M) = 3.9$, which is the optical electronegativity of fluoride on Jørgensen's scale (101). This result is completely untenable, lending further credence to the assertion that the lead-thiolate charge-transfer bands are LMCT in origin.
113. A. Fukuda, *Sci. Light*, *13*, 64 (1964).
114. A. Ranfagni, D. Mugnai, M. Bacci, G. Viliani, and M. P. Fontana, *Adv. Phys.*, *32*, 823 (1983).
115. F. Seitz, *J. Chem. Phys.*, *6*, 150 (1938).
116. T. Tsuboi, *Phys. Status Solidi B*, *101*, K103 (1980).
117. R. S. Knox, *Phys. Rev.*, *115*, 1095 (1959).
118. D. Bramanti, M. Mancini, and A. Ranfagni, *Phys. Rev. B*, *3*, 3670 (1971).
119. F. E. Williams, *J. Chem. Phys.*, *19*, 457 (1951).
120. P. H. Yuster and C. J. Delbecq, *J. Chem. Phys.*, *21*, 892 (1953).
121. R. S. Knox and D. L. Dexter, *Phys. Rev.*, *104*, 1245 (1956).
122. F. E. Williams, B. Segall, and P. D. Johnson, *Phys. Rev.*, *108*, 46 (1957).
123. S. Sugano, *J. Chem. Phys.*, *36*, 122 (1962).
124. T. Tsuboi, *Physica*, *106*, 97 (1981).
125. E. Rabinowitch, *Rev. Mod. Phys.*, *14*, 112 (1942).
126. L. E. Orgel, *Quart. Rev. (London)*, *8*, 422 (1954).
127. F. R. McFeely, S. Kowalczyk, L. Ley, R. A. Pollak, and D. A. Shirley, *Phys. Rev. B*, *7*, 5228 (1973).
128. G. M. Bancroft and Y. F. Hu, in *Inorganic Electronic Structure and Spectroscopy*, E. I. Solomon and A. B. P. Lever, Eds., John Wiley & Sons, Inc., New York, 1999, Vol. I, pp. 443–512.
129. J. R. Anderson and A. V. Gold, *Phys. Rev. A*, *139*, A1459 (1965).
130. T. L. Loucks, *Phys. Rev. Lett.*, *14*, 1072 (1965).
131. L. Ley, R. Pollak, S. Kowalczyk, and D. A. Shirley, *Phys. Lett. A*, *41*, 429 (1972).
132. A. G. Mathewson, H. P. Myers, and P. O. Nilsson, *Phys. Status Solidi B*, *57*, K31 (1973).
133. E. Doni, G. Grosso, and G. Spavieri, *Solid State Commun.*, *11*, 493 (1972).
134. S. E. Kohn, P. Y. Yu, Y. Petroff, Y. R. Shen, Y. Tsang, and M. L. Cohen, *Phys. Rev. B*, *8*, 1477 (1973).
135. I. C. Schlüter and M. Schlüter, *Phys. Rev. B*, *9*, 1652 (1974).
136. T. Matsukawa and T. Ishii, *J. Phys. Soc. Jpn.*, *41*, 1285 (1976).
137. T. Grandke, L. Ley, and M. Cardona, *Phys. Rev. B*, *18*, 3847 (1978).
138. M. Nizam, M. Allavena, Y. Bouteiller, B. H. Suits, and D. White, *J. Magn. Reson.*, *82*, 441 (1989).
139. M. Itoh, T. Shiokawa, K. Sawada, and M. Kamada, *J. Phys. Soc. Jpn.*, *67*, 2140 (1998).

140. C. Dybowski, S. P. Gabuda, S. G. Kozlova, G. Neue, D. L. Perry, and V. V. Terskikh, *J. Solid State Chem.*, *157*, 220 (2001).
141. A. R. West, *Basic Solid State Chemistry*, John Wiley & Sons, Inc., New York, 1984.
142. A. R. H. F. Ettema, C. Haas, P. Moriarty, and G. Hughes, *Surf. Sci.*, *287/288*, 1106 (1993).
143. H. J. Terpstra, R. A. de Groot, and C. Haas, *Phys. Rev. B*, *52*, 11690 (1995).
144. A. Breeze, *Solid State Commun.*, *14*, 395 (1974).
145. G. W. Watson, S. C. Parker, and G. Kresse, *Phys. Rev. B*, *59*, 8481 (1999).
146. J. Kanbe, H. Onuki, and R. Onaka, *J. Phys. Soc. Jpn.*, *43*, 1280 (1977).
147. M. Scrocco, *Phys. Rev. B*, *25*, 1535 (1982).
148. M. Scrocco, *J. Electron Spectrosc. Relat. Phenom.*, *48*, 363 (1989).
149. P. Llopiz and J. C. Maire, *Bull. Soc. Chim. Fr.*, 457 (1979).
150. K. Fukui, *J. Phys. Soc. Jpn.*, *61*, 1084 (1992).
151. D. T. Sawyer and P. J. Paulsen, *J. Am. Chem. Soc.*, *81*, 816 (1959).
152. Y. Mido, I. Fujiwara, and E. Sekido, *J. Inorg. Nucl. Chem.*, *36*, 1003 (1974).
153. P. G. Harrison, M. A. Healy, and M. Aslam, *Heavy Met. Environ., Int. Conf., 3rd*, 628 (1981).
154. H. A. Tajmir-Riahi, *Bull. Chem. Soc. Jpn.*, *62*, 1281 (1989).
155. H. A. Tajmir-Riahi, *J. Inorg. Biochem.*, *44*, 39 (1991).
156. G. L. J. Trettenhahn, G. E. Nauer, and A. Neckel, *Vib. Spectrosc.*, *5*, 85 (1993).
157. M. Valli, P. Persson, and I. Persson, *Acta Chem. Scand.*, *48*, 810 (1994).
158. U. Kolb, D. Gutwerk, R. Beudert, and H. Bertagnolli, *J. Non-Cryst. Solids*, *217*, 162 (1997).
159. S. Dragan and A. Fitch, *J. Chem. Educ.*, *75*, 1018 (1998).
160. R. L. Davidovich, V. B. Logvinova, and T. A. Kaidalova, *Russ. J. Coord. Chem.*, *24*, 467 (1998).
161. R. L. Davidovich, V. B. Logvinova, and T. A. Kaidalova, *Russ. J. Coord. Chem.*, *25*, 638 (1999).
162. S. Eiden-Assmann, A. M. Schneider, P. Behrens, M. Wiebcke, G. Engelhardt, and J. Felsche, *Chem.–Eur. J.*, *6*, 292 (2000).
163. K. Krishnan and R. A. Plane, *J. Am. Chem. Soc.*, *90*, 3195 (1968).
164. A. A. McConnell and R. H. Nuttall, *Spectrochim. Acta, Part A.*, *33*, 459 (1977).
165. A. Hadrich, A. Lautié, and T. Mhiri, *J. Raman Spectrosc.*, *32*, 33 (2001).
166. W. N. Perera, G. Hefter, and P. M. Sipos, *Inorg. Chem.*, *40*, 3974 (2001).
167. D. T. Sawyer and P. J. Paulsen, *J. Am. Chem. Soc.*, *80*, 1597 (1958).
168. R. E. Sievers and J. C. Bailar, Jr., *Inorg. Chem.*, *1*, 174 (1962).
169. H. G. Langer, *J. Inorg. Nucl. Chem.*, *26*, 767 (1964).
170. P. G. Harrison and A. T. Steel, *J. Organomet. Chem.*, *239*, 105 (1982).
171. T. T. Nakashima and D. L. Rabenstein, *J. Magn. Reson.*, *51*, 223 (1983).
172. B. Wrackmeyer and K. Horchler, *Ann. Rep. NMR Spectrosc.*, *22*, 249 (1989).
173. J. M. Aramini, T. Hiraoki, M. Yazawa, T. Yuan, M. Zhang, and H. J. Vogel, *J. Biol. Inorg. Chem.*, *1*, 39 (1996).
174. R. J. Day and C. N. Reilly, *Anal. Chem.*, *36*, 1073 (1964).
175. K. J. Mordecai, Ph.D. Thesis, "A study of Metal Ions in Aqueous Solution Using NMR Spectroscopy" Bedford College, University of London, London, 1983.
176. K. Abu-Dari, F. E. Hahn, and K. Raymond, *J. Am. Chem. Soc.*, *112*, 1519 (1990).
177. D. L. Reger, *Syn. Lett.*, 469 (1992).

178. D. L. Reger, M. F. Huff, A. L. Rheingold, and B. S. Haggerty, *J. Am. Chem. Soc.*, *114*, 579 (1992).
179. K. Abu-Dari, T. B. Karpishin, and K. N. Raymond, *Inorg. Chem.*, *32*, 3052 (1993).
180. D. L. Reger, Y. Ding, A. L. Rheingold, and R. L. Ostrander, *Inorg. Chem.*, *33*, 4226 (1994).
181. S. Rupprecht, S. J. Franklin, and K. N. Raymond, *Inorg. Chim. Acta*, *235*, 185 (1995).
182. S. Rupprecht, K. Langemann, T. Lugger, J. M. McCormick, and K. N. Raymond, *Inorg. Chim. Acta*, *243*, 79 (1996).
183. D. L. Reger, J. E. Collins, A. L. Rheingold, L. M. Liable Sands, and G. P. A. Yap, *Inorg. Chem.*, *36*, 345 (1997).
184. J. E. Coleman, *Methods Enzymol.*, *227*, 16 (1993).
185. L. M. Utschig, J. W. Bryson, and T. V. O'Halloran, *Science*, *268*, 380 (1995).
186. F. Fayon, I. Farnan, C. Bessada, J. Coutures, D. Massiot, and J. P. Coutures, *J. Am. Chem. Soc.*, *119*, 6837 (1997).
187. J. B. Grutzner, in *Recent Advances in Organic NMR Spectroscopy*, J. B. Lambert and R. Rittner, Eds., Norell Press, Landisville, NJ, 1987, pp. 17–42.
188. A. M. Bond, R. Colton, and A. F. Hollenkamp, *Inorg. Chem.*, *29*, 1991 (1990).
189. J. Kim, C. J. Yoon, H. J. Yoo, G. Kim, and S. J. Kim, *J. Korean Chem. Soc.*, *38*, 41 (1994).
190. L. Pu, B. Twamley, and P. P. Power, *J. Am. Chem. Soc.*, *122*, 3524 (2000).
191. M. Kuchta, J. M. Hahn, and G. Parkin, *J. Chem. Soc. Dalton Trans.*, 3559 (1999).
192. L. Pu, B. Twamley, and P. P. Power, *Organometallics*, *19*, 2874 (2000).
193. W. McFarlane, *Mol. Phys.*, *13*, 587 (1967).
194. A. Bax, R. H. Griffey, and B. L. Hawkins, *J. Magn. Reson.*, *55*, 301 (1983).
195. H. C. Freeman, G. N. Stevens, and I. F. Taylor, Jr., *J. Chem. Soc., Chem. Commun.*, 366 (1974).
196. T. J. McMurry, K. N. Raymond, and P. H. Smith, *Science*, *244*, 938 (1989).
197. L. Shimoni-Livny, J. P. Glusker, and C. W. Bock, *Inorg. Chem.*, *37*, 1853 (1998).
198. B. M. Bridgewater and G. Parkin, *J. Am. Chem. Soc.*, *122*, 7140 (2000).
199. F. Allen and O. Kennard, *Chemical Design Automation News*, *8*, 31 (1993).
200. The April 2001 version of the CCSD database was searched using the Windows software Conquest 1.2. The database was searched for nonionic compounds containing Pb with an R-factor ≤ 0.05 and no disorder. Lead-to-donor distance criterion was set for 0–3.50 Å. A subsearch was done to select for compounds with Pb(II). A total of 295 compounds were found.
201. C. E. Holloway and M. Melnik, *Main Group Metal Chem.*, *20*, 399 (1997).
202. R. G. Pearson, *J. Am. Chem. Soc.*, *85*, 3533 (1963).
203. A. L. Poznyak, G. N. Kupriyanova, I. F. Burshtein, and A. B. Illyuchin, *Koord. Khim.*, *24*, 825 (1998).
204. S. L. Lawton and G. T. Kokotailo, *Inorg. Chem.*, *11*, 363 (1972).
205. R. J. Gillespie, *J. Chem. Educ.*, *40*, 295 (1963).
206. D. F. Shriver, P. Atkins, and C. H. Langford, *Inorganic Chemistry*, W. H. Freeman, New York, 1994.
207. F. A. Cotton, *Chemical Applications of Group Theory*, John Wiley & Sons, Inc., New York, 1990.
208. D. L. Reger, T. D. Wright, C. A. Little, J. J. S. Lambda, and M. D. Smith, *Inorg. Chem.*, *40*, 3810 (2001).
209. D. E. Glotzer and H. Bauchner, *Pediatrics*, *89*, 614 (1992).

210. P. A. W. Dean, J. J. Vittal, and N. C. Payne, *Inorg. Chem.*, *24*, 3594 (1985).
211. A. Bondi, *J. Phys. Chem.*, *68*, 441 (1964).
212. K. N. Raymond, *Coord. Chem. Rev.*, *105*, 135 (1990).
213. J. R. Telford and K. N. Raymond, *Compr. Supramol. Chem.*, *1*, 245 (1996).
214. R. Dietzel and P. Thomas, *Z. Anorg. Allg. Chem.*, *381*, 214 (1971).
215. G. Battistuzzi, M. Borsari, L. Menabue, and M. Saladini, *Inorg. Chem.*, *35*, 4239 (1996).
216. C. F. G. C. Geraldes, A. M. Urbano, M. C. Alpoim, A. D. Sherry, K.-T. Kuan, R. Rajagopalam, F. Maton, and R. N. Muller, *Magn. Reson. Imaging*, *13*, 401 (1995).
217. R. Mathur-de Vre and M. Lemort, *Br. J. Radiol.*, *68*, 225 (1995).
218. G. Sosnovsky, S. W. Li, and N. U. M. Rao, *Z. Naturforsch., B: Chem. Sci.*, *40*, 1558 (1985).
219. K. W. Klinkhammer, T. F. Fässler, and H. Grützmacher, *Angew. Chem. Int. Ed. Engl.*, *37*, 124 (1998).
220. M. Sturmann, W. Saak, H. Marsmann, and M. Weidenbruch, *Angew. Chem. Int. Ed. Engl.*, *38*, 187 (1999).
221. M. Kourgiantakis, M. Matzapetakis, C. P. Raptopoulou, A. Terzis, and A. Salifoglou, *Inorg. Chim. Acta*, *297*, 134 (2000).
222. T. Lis, *Acta Crystallogr., Sect. C*, *C40*, 374 (1984).
223. M. W. Yu and C. J. Frichie, *J. Chem. Soc., Dalton Trans.*, 377 (1975).
224. T. L. Blundell and L. N. Johnson, *Protein Crystallography*, Academic, New York, 1976.
225. H. M. Holden and I. Rayment, *Arch. Biochem. Biophys.*, *291*, 187 (1991).
226. J. D. Robertus, J. E. Ladner, J. T. Finch, D. Rhodes, R. S. Brown, B. F. C. Clark, and A. Klug, *Nature (London)*, *250*, 546 (1974).
227. R. S. Brown, B. E. Hingerty, J. C. Dewan, and A. Klug, *Nature (London)*, *303*, 543 (1983).
228. J. R. Rubin and M. Sundaralingam, *J. Biomol. Struct. Dyn.*, *1*, 639 (1983).
229. D. L. K. Nicoghosian and R. Cedergren, *J. Biol. Chem.*, *260*, 1103 (1985).
230. W. J. Kryzosiak, T. Marciniec, M. Wiewiorowski, P. Romby, J. P. Ebel, and R. Giege, *Biochemistry*, *27*, 5771 (1988).
231. T. Marciniec, J. Ciesolka, J. Wrzesinksi, and W. J. Krzyzosiak, *FEBS Lett.*, *243*, 293 (1989).
232. T. Pan and O. Uhlenbeck, *Biochemistry*, *31*, 3887 (1992).
233. D. E. Otzen, J. Barciszewski, and B. F. C. Clark, *Biochem. Mol. Biol. Int.*, *31*, 95 (1993).
234. D. E. Otzen, *Biochimie*, *76*, 15 (1994).
235. R. Khan, H.-O. Chang, K. Maluarachchi, and D. Giedroc, *Nucleic Acids Res.*, *24*, 3568 (1996).
236. J. Ciesiolka, W.-D. Hardt, J. Schlegl, V. A. Erdmann, and R. K. Hartmann, *Eur. J. Biochem.*, *219*, 49 (1994).
237. D. Winter, N. Polacek, I. Halama, B. Streicher, and A. Barta, *Nucleic Acids Res.*, *25*, 1817 (1997).
238. S. Lemieuz, P. Chartrand, R. Cedergren, and F. Major, *RNA*, *4*, 739 (1998).
239. Y. S. Babu, C. E. Bugg, and W. J. Cook, *J. Mol. Biol.*, *204*, 191 (1988).
240. R. B. Sutton, B. A. Davletov, A. M. Berghuis, T. C. Sudhof, and S. R. Sprang, *Cell*, *80*, 929 (1995).
241. R. B. Sutton, J. A. Ernst, and A. T. Brunger, *J. Cell Biol.*, *147*, 589 (1999).
242. R. B. Sutton, personal communication, 2000.
243. E. K. Jaffe, *Bioenergetics and Biomembranes*, *27*, 169 (1995).

244. M. J. Warren, J. B. Cooper, S. P. Wood, and P. M. Shoolingin-Jordan, *Trends Biochem. Sci.*, 23, 217 (1998).
245. P. T. Erskine, N. Senior, S. Awan, R. Lambert, G. Lewis, I. J. Tickle, M. Sarwar, P. Spencer, P. Thomas, M. J. Warren, P. M. Shoolingin-Jordan, S. P. Wood, and J. B. Cooper, *Nat. Struct. Biol.*, 4, 1025 (1997).
246. E. K. Jaffe, J. Martins, J. Li, J. Kervinen, and J. Roland L. Dunbrack, *J. Biol. Chem.*, 276, 1531 (2001).
247. W. N. Lipscomb, J. G. N. Reeke, J. A. Hatsuck, F. A. Quiocho, and P. H. Bethge, *Philos. Trans. R. Soc. London, Ser. B*, 257, 177 (1970).
248. E. A. V. Ebsworth, D. W. H. Rankin, and S. Cradock, *Structural Methods in Inorganic Chemistry*, CRC Press, Boston, 1991.
249. H. H. Zhang, B. Hedman, and K. O. Hodgson, in *Inorganic Electronic Structure and Spectroscopy*, E. I. Solomon and A. B. P. Lever, Eds., John Wiley & Sons, Inc., New York, 1999, Vol. I, pp. 513–554.
250. K. Clark-Baldwin, D. L. Tierney, N. Govindaswamy, E. S. Gruff, C. Kim, J. M. Berg, S. A. Koch, and J. E. Penner-Hahn, *J. Am. Chem. Soc.*, 120, 8401 (1998).
251. S. S. Hasnain, G. P. Diakun, I. Abrahams, I. Ross, C. D. Garner, I. Bremner, and M. Vasak, *Experientia Suppl.*, 52, 227 (1987).
252. A. J. Dent, C. Beyersmann, C. Block, and S. S. Hasnain, *Biochemistry*, 29, 7822 (1990).
253. A. Manceau, M.-C. Boisset, G. Sarret, J.-L. Hazemann, M. Mench, P. Cambier, and R. Prost, *Environ. Sci. Technol.*, 30, 1540 (1996).
254. A. Manceau, J. C. Harge, G. Sarret, J. L. Hazemann, M. C. Boisset, M. Mench, P. Cambier, and R. Prost, *Colloq. - Inst. Natl. Rech. Agron.*, 85, 99 (1997).
255. D. Hesterberg, P. D. Hansen, W. Zhou, and D. E. Sayers, *Mineral. Mag.*, 62A, 612 (1998).
256. G. Sarret, A. Manceau, L. Spadini, J.-C. Roux, J.-L. Hazemann, Y. Soldo, L. Eybert-Berard, and J.-J. Menthonnex, *Environ. Sci. Technol.*, 32, 1648 (1998).
257. J. D. Ostergren, G. E. Brown, Jr., G. A. Parks, and T. N. Tingle, *Environ. Sci. Technol.*, 33, 1627 (1999).
258. G. Morin, F. Juillot, P. Ildefonse, G. Calas, J.-C. Samama, P. Chevallier, and G. E. Brown, Jr., *Am. Mineral.*, 86, 92 (2001).
259. G. Sarret, J. Vangronsveld, A. Manceau, M. Musso, J. D'Haen, J.-J. Menthonnex, and J.-L. Hazemann, *Environ. Sci. Technol.*, 35, 2854 (2001).
260. D. T. Rickard and J. O. Nriagu, in *The Biogeochemistry of Lead in the Environment*, J. O. Nriagu, Ed., Elsevier, New York, 1978.
261. R. G. Pearson, C. R. Boston, and F. Basolo, *J. Am. Chem. Soc.*, 74, 2943 (1952).
262. F. Basolo, *Chem. Rev.*, 52, 459 (1953).
263. F. Basolo and R. G. Pearson, *Mechanisms of Inorganic Reactions: A Study of Metal Complexes in Solution*, John Wiley & Sons, Inc., New York, 1967.
264. J. H. R. Kägi and H.-J. Hapke, in *Changing Metal Cycles and Human Health*, J. O. Nriagu, Ed., Springer-Verlag, New York, 1984 pp. 237–250.
265. R. D. Hancock and A. E. Martell, *Chem. Rev.*, 89, 1875 (1989).
266. R. D. Hancock, R. Bhavan, P. W. Wade, J. C. A. Boeyens, and S. M. Dobson, *Inorg. Chem.*, 28, 187 (1989).
267. R. D. Hancock, P. W. Wade, M. P. Ngwenya, A. S. de Sousa, and K. V. Damu, *Inorg. Chem.*, 29, 1968 (1990).

268. M. M. Jones, in *Toxicology of Metals: Biochemical Aspects*, R. A. Goyer and M. G. Cherian, Eds., Springer-Verlag, Berlin, 1995, pp. 279–304.
269. O. Andersen, *Chem. Rev.*, *99*, 2683 (1999).
270. A. E. Martell and R. M. Smith, *Critical Stability Constants*, Plenum Publishing, New York, 1975, 1976, 1977, 1982, 1989, Vols. 1–6.
271. W. R. Harris and V. Chen, *J. Coord. Chem.*, *23*, 173 (1991).
272. M. M. Jones, *J. Coord. Chem.*, *23*, 187 (1991).
273. D. W. Margerum, G. R. Cayley, D. C. Weatherburn, and G. K. Pagenkopf, in *Coordination Chemistry*, A. E. Martell, Ed., American Chemical Society, Washington, DC, 1978, Vol. 2, pp. 1–220.
274. M. Kodama, K. Namekawa, and T. Horiuchi, *Bull. Chem. Soc. Jpn.*, *47*, 2011 (1974).
275. K. Bril, S. Bril, and P. Krumholz, *J. Phys. Chem.*, *59*, 596 (1955).
276. K. Bril, S. Bril, and P. Krumholz, *J. Phys. Chem.*, *60*, 251 (1956).
277. N. Tanaka, K. Kato, and R. Tamamushi, *Bull. Chem. Soc. Jpn.*, *31*, 283 (1958).
278. N. Tanaka and K. Kato, *Bull. Chem. Soc. Jpn.*, *32*, 1376 (1959).
279. N. Tanaka and K. Kato, *Bull. Chem. Soc. Jpn.*, *33*, 1236 (1960).
280. N. Tanaka and H. Ogino, *Bull. Chem. Soc. Jpn.*, *36*, 175 (1963).
281. N. Tanaka, H. Osawa, and M. Kamada, *Bull. Chem. Soc. Jpn.*, *36*, 67 (1963).
282. T. Fujisawa and N. Tanaka, *Nippon Kagaku Zasshi*, *87*, A54; 965 (1966).
283. H. Ogino and N. Tanaka, *Bull. Chem. Soc. Jpn.*, *40*, 857 (1967).
284. J. D. Carr, K. Torrance, C. J. Cruz, and C. N. Reilley, *Anal. Chem.*, *39*, 1358 (1967).
285. D. L. Rabenstein and R. J. Kula, *J. Amer. Chem. Soc.*, *91*, 2492 (1969).
286. P. E. Reinbold and K. H. Pearson, *Inorg. Chem.*, *9*, 2325 (1970).
287. J. D. Carr and D. R. Baker, *Inorg. Chem.*, *10*, 2249 (1971).
288. B. J. Fuhr and D. L. Rabenstein, *Inorg. Chem.*, *12*, 1868 (1973).
289. D. L. Rabenstein, *J. Am. Chem. Soc.*, *93*, 2869 (1971).
290. T. P. Radhakrishnan, *Proc. Indian Acad. Sci., Sect. A*, *87A*, 285 (1978).
291. E. Mentasti, *J. Chem. Soc., Dalton Trans.*, 721 (1982).
292. E. Brucher and G. Laurenczy, *Inorg. Chem.*, *22*, 338 (1983).
293. G. Laurenczy, E. Brucher, and V. Novak, *Inorg. Chim. Acta*, *133*, 147 (1987).
294. M. Kodama and E. Kimura, *J. Chem. Soc., Dalton Trans.*, 2269 (1977).
295. J. L. Laing, R. W. Taylor, and C. A. Chang, *J. Chem. Soc., Dalton Trans.*, 1195 (1997).
296. R. M. Izatt, J. S. Bradshaw, S. A. Nielsen, J. D. Lamb, J. J. Christensen, and D. Sen, *Chem. Rev.*, *85*, 271 (1985).
297. L. F. Lindoy, *Prog. Macrocyclic Chem.*, *3*, 53 (1987).
298. C. Grant, Jr., and P. Hambright, *J. Am. Chem. Soc.*, *91*, 4195 (1969).
299. B. Haeger-Aronsen and A. Schutz, *Arch. Environ. Health*, *31*, 215 (1976).
300. F. Basolo and R. C. Johnson, *Coordination Chemistry*, W. A. Benjamin, New York, 1964.
301. A. E. Martell and R. D. Hancock, *Metal Complexes in Aqueous Solutions*, Plenum Publishing, New York, 1996.
302. D. D. Perrin and B. Dempsey, *Buffers for pH and Metal Ion Control*, Chapman & Hall, London, 1974.

303. G. Anderegg, *Pure Appl. Chem.*, *54*, 2693 (1982).
304. A. E. Martell and M. Calvin, *Chemistry of the Metal Chelate Compounds*, Prentice Hall, New York, 1952.
305. S. Chaberek, Jr., and A. E. Martell, *Organic Sequestering Agents*, John Wiley & Sons, Inc., New York, 1959.
306. A. E. Martell and R. J. Motekaitis, *Determination and Use of Stability Constants*, VCH Publishers, New York, 1988.
307. W. R. Harris and A. E. Martell, *Inorg. Chem.*, *15*, 713 (1976).
308. B. A. Krizek, B. T. Amann, V. J. Kilfoil, D. L. Merkle, and J. M. Berg, *J. Am. Chem. Soc.*, *113*, 4518 (1991).
309. B. A. Krizek, D. L. Merkle, and J. M. Berg, *Inorg. Chem.*, *32*, 937 (1993).
310. J. M. Berg and D. L. Merkle, *J. Am. Chem. Soc.*, *111*, 3759 (1989).
311. G. Anderegg, *Critical Survey of Stability Constants of EDTA Complexes*, Pergamon Press, Oxford, UK, 1977.
312. R. M. Town and M. Filella, *Electroanal. Chem.*, *488*, 1 (2000).
313. M. Fillela and R. M. Town, *J. Electroanal. Chem.*, *485*, 21 (2000).
314. R. Luckay, R. D. Hancock, I. Cukrowski, and J. H. Reibenspies, *Inorg. Chim. Acta*, *246*, 159 (1996).
315. L. G. Sillen and A. E. Martell, *1964 Stability Constants of Metal Ion Complexes, Special Publication No. 17*, The Chemical Society, London, 1964.
316. H. Stetter and W. Frank, *Angew. Chem. Int. Ed. Engl.*, *88*, 760 (1976).
317. K. H. Scheller, T. H. J. Abel, P. E. Polanyi, P. K. Wenk, B. E. Fischer, and H. Sigel, *Eur. J. Biochem.*, *107*, 455 (1980).
318. Z.-X. Huang, P. M. May, K. M. Quinlan, D. R. Williams, and A. M. Creighton, *Agents and Actions*, *12*, 536 (1982).
319. P. M. May, M. J. Willes, D. R. Williams, and A. M. Creighton, *Agents and Actions*, *15*, 448 (1984).
320. F. Mulla, F. Marsicano, B. S. Nakano, and R. D. Hancock, *Inorg. Chem.*, *24*, 3076 (1985).
321. V. J. Thom, M. S. Shaikjee, and R. D. Hancock, *Inorg. Chem.*, *25*, 2992 (1986).
322. K. Kumar, M. Magerstaedt, and O. A. Gansow, *J. Chem. Soc., Chem. Commun.*, 145 (1989).
323. R. D. Hancock, A. E. Martell, and R. J. Motekaitis, *Critical Stability Constants Database*, NIST: Gathersburg, MD (1993).
324. H. Sigel, S. S. Massoud, and N. A. Corfu, *J. Am. Chem. Soc.*, *116*, 2958 (1994).
325. Y. Li, A. E. Martell, R. D. Hancock, J. H. Reibenspies, C. J. Anderson, and M. J. Welch, *Inorg. Chem.*, *35*, 404 (1996).
326. A. E. Martell, R. M. Smith, and R. J. Motekaitis, *Critically Selected Stability Constants of Metal Complexes Database*, NIST: College Station, TX (1998).
327. C. P. Da Costa and H. Sigel, *J. Biol. Inorg. Chem.*, *4*, 508 (1999).
328. C. P. Da Costa and H. Sigel, *Inorg. Chem.*, *39*, 5985 (2000).
329. H. Sigel, C. P. Da Costa, and R. B. Martin, *Coord. Chem. Rev.*, *219–221*, 435 (2001).
330. R. D. Hancock, M. S. Shaikjee, S. M. Dobson, and J. C. Boeyens, *Inorg. Chim. Acta*, *154*, 229 (1988).
331. R. D. Hancock and A. E. Martell, *Comments Inorg. Chem.*, *6*, 237 (1988).

332. H. Maumela, R. D. Hancock, L. Carlton, J. H. Reibenspies, and K. P. Wainright, *J. Am. Chem. Soc.*, *117*, 6698 (1995).
333. G. Schwarzenbach, *Helv. Chim. Acta*, *35*, 2344 (1952).
334. R. D. Hancock, *Pure Appl. Chem.*, *58*, 1445 (1986).
335. R. D. Shannon, *Acta Crystallogr., Sect. A.*, *A32*, 751 (1976).
336. D. J. Cram, T. Kaneda, R. C. Helgeson, and S. B. Brown, *J. Am. Chem. Soc.*, *107*, 3645 (1985).
337. F. J. Sunderman and A. Barber, *Ann. Clin. Lab. Sci.*, *18*, 267 (1988).
338. G. W. Goldstein, *Neurotoxicology*, *14*, 97 (1993).
339. T. J. B. Simons, *Neurotoxicology*, *14*, 77 (1994).
340. J. Bressler, K. A. Kim, T. Chakraborti, and G. Goldstein, *Neurochem. Res.*, *24*, 595 (1999).
341. P. Mistry, G. W. Lucier, and B. A. Fowler, *J. Pharmacol. Exp. Ther.*, *232*, 462 (1984).
342. B. Quintanilla-Vega, D. R. Smith, M. W. Kahng, J. M. Hernandez, A. Albores, and B. A. Fowler, *Chem.-Biol. Interact.*, *98*, 193 (1995).
343. D. R. Smith, M. W. Kahng, B. Quintanilla-Vega, and B. A. Fowler, *Chem.-Biol. Interact.*, *115*, 39 (1998).
344. K. S. Larsen and D. S. Auld, *Biochemistry*, *30*, 2613 (1991).
345. D. B. Veprintsev, E. A. Permyakov, L. P. Kalinichenko, and L. J. Berliner, *Biochem. Mol. Biol. Int.*, *39*, 1255 (1996).
346. I. M. Klotz, *Ligand-Receptor Energetics: A Guide for the Perplexed*, John Wiley & Sons, Inc., New York, 1997.
347. J. E. Coleman and B. L. Vallee, *J. Biol. Chem.*, *236*, 2244 (1961).
348. M. P. Waalkes, M. J. Harvey, and C. D. Klaassen, *Toxicol. Lett.*, *20*, 33 (1984).
349. W. Maret, K. S. Larsen, and B. L. Vallee, *Proc. Natl. Acad. Sci. USA*, *94*, 2233 (1997).
350. E. Habermann, K. Crowell, and P. Janicki, *Arch. Toxicol.*, *54*, 61 (1983).
351. G. W. Goldstein and D. Ar, *Life Sci.*, *33*, 1001 (1983).
352. J. Markovac and G. W. Goldstein, *Nature (London)*, *334*, 71 (1988).
353. K. Kim, M. Annadata, G. W. Goldstein, and J. P. Bressler, *Int. J. Devl. Neuroscience*, *15*, 175 (1997).
354. T. Chakraborti, K. A. Kim, G. G. Goldstein, and J. P. Bressler, *J. Neurochem.*, *73*, 187 (1999).
355. F. A. Kim, T. Chakraborti, G. W. Goldstein, and J. P. Bressler, *J. Neurochem.*, *74*, 1140 (2000).
356. A. H. Phillips, *Mineralogy*, Macmillan, New York, 1912.
357. K. Tsuchia, in *Handbook on the Toxicology of Metals*, L. Friberg, G. F. Nordberg, and V. B. Vouk, Eds., Elsevier, Amsterdam, The Netherlands, 1986, Vol. 2.
358. H. G. Seiler, A. Sigel, and H. Sigel, *Handbook on Metals in Clinical and Analytical Chemistry*, Marcel Dekker, New York, 1994.
359. M. Branica and Z. Konrad, Eds., *Lead in the Marine Environment*, Pergamon Press, New York, 1977.
360. J. O. Nriagu, Ed., *The biogeochemistry of lead in the environment. Part B. Biological effects*, Elsevier/North-Holland Biomedical, New York, 1978, Vol. 2.
361. G. R. Smith, in *1999 Minerals Yearbook*, U. S. Geological Survey, Reston, VA, 1999, Vol. I.
362. G. R. Smith, *Mineral Commodity Summaries 2001*, U. S. Geological Survey, Reston, VA, 2001.
363. C. A. DiFrancesco and G. R. Smith, *U.S. Geological Survey Open-File Report 01-006*, 2001.
364. G. R. Smith, *U.S. Geological Survey Open-File Report 01-170*, 2001.

365. D. Greninger, V. Kollonitsch, and C. H. Kline, *Lead Chemicals*, International Lead Zinc Research Organization, New York, 1975.

366. A. F. Wells, *Structural Inorganic Chemistry*, Clarendon Press, Oxford, UK, 1984.

367. M. V. Ruby, A. Davis, and A. Nicholson, *Environ. Sci. Technol.*, *28*, 646 (1994).

368. H. V. Warren, R. E. Delavault, and K. W. Fletcher, *Trace Substances in Environmental Health*, *4*, 94 (1971).

369. H. W. Mielke and P. L. Reagan, *Environ. Health Perspect. Suppl.*, *106*, 217 (1998).

370. E. Callender and K. C. Rice, *Environ. Sci. Technol.*, *34*, 232 (2000).

371. J. T. Kinard, J. Tisdale, and E. Alexander, *J. Environ. Sci. Health*, *A11*, 153 (1976).

372. W. T. Piver, *Environ. Health Perspect.*, *19*, 247 (1977).

373. S. J. LaBelle, P. C. Lindahl, R. R. Hinchman, J. Ruskamp, and K. McHugh, *Pilot Study of the Relationship of Regional Road Traffic to Surface–Soil Lead Levels in Illinois*, Argonne National Laboratory, Energy and Environmental Systems Division, Center for Transportation Research, Publication ANL/ES-1541987.

374. A. Tessier, P. G. C. Campbell, and M. Bisson, *Anal. Chem.*, *51*, 844 (1979).

375. P. Quevauviller, G. Rauret, H. Muntau, A. M. Ure, R. Rubio, J. F. Lopezsanchez, H. D. Fiedler, and B. Griepink, *Fresenius J. Anal. Chem.*, *349*, 808 (1994).

376. A. M. Ure, C. M. Davidson, and R. P. Thomas, *Tech. Instrum. Anal. Chem.*, *17*, 505 (1995).

377. A. Kot and J. Namiesnik, *Trends Anal. Chem.*, *19*, 69 (2000).

378. E. K. Miller and A. J. Friedland, *Environ. Sci. Technol.*, *28*, 662 (1994).

379. P. Michopoulos, *J. Environ. Qual.*, *28*, 1705 (1999).

380. E. X. Wang and G. Benoit, *Environ. Sci. Technol.*, *30*, 2211 (1996).

381. Z. Li and L. M. Shuman, *Environ. Pollut.*, *95*, 219 (1997).

382. Z. Li and L. M. Shuman, *Environ. Pollut.*, *95*, 227 (1997).

383. V. M. Vulava and J. C. Seaman, *Environ. Sci. Technol.*, *34*, 4828 (2000).

384. T. Qiang, S. Xiao-Quan, Q. Jin, and N. Zhe-Ming, *Anal. Chem.*, *66*, 3562 (1994).

385. C. E. Martinez, S. Sauvé, A. Jacobson, and M. B. McBride, *Environ. Sci. Technol.*, *33*, 2016 (1999).

386. J. P. Pinheiro, A. M. Mota, and M. F. Benedetti, *Environ. Sci. Technol.*, *33*, 3398 (1999).

387. G. M. Hettiarachchi, G. M. Pierzynski, and M. D. Ransom, *Environ. Sci. Technol.*, *34*, 4614 (2000).

388. M. E. Hodson, É. Valsami-Jones, and J. D. Cotter-Howells, *Environ. Sci. Technol.*, *34*, 3501 (2000).

389. R. Turpeinen, J. Salminen, and T. Kairesalo, *Environ. Sci. Technol.*, *34*, 5152 (2000).

390. Q. Y. Ma, S. J. Traina, and T. J. Logan, *Environ. Sci. Technol.*, *27*, 1803 (1993).

391. V. Laperche, S. J. Traina, P. Gaddam, and T. J. Logan, *Environ. Sci. Technol.*, *30*, 3321 (1996).

392. X. Chen, J. V. Wright, J. L. Conca, and L. M. Peurring, *Environ. Sci. Technol.*, *31*, 624 (1997).

393. W. R. Berti and S. D. Cunningham, *Environ. Sci. Technol.*, *31*, 1359 (1997).

394. P. Zhang and J. A. Ryan, *Environ. Sci. Technol.*, *33*, 618 (1999).

395. P. Zhang and J. A. Ryan, *Environ. Sci. Technol.*, *33*, 625 (1999).

396. S. Sauvé, M. B. McBride, and W. Hendershot, *Environ. Sci. Technol.*, *32*, 388 (1998).

397. J. O. Nriagu, *Geochim. Cosmochim. Acta*, *38*, 887 (1974).

398. M. V. Ruby, A. Davis, R. Schoof, S. Eberle, and C. M. Sellstone, *Environ. Sci. Technol.*, *30*, 422 (1996).
399. M. V. Ruby, R. Schoof, W. Brattin, M. Goldade, G. Post, M. Harnois, D. E. Mosby, S. W. Casteel, W. Berti, M. Carpenter, D. Edwards, D. Cragin, and W. Chappell, *Environ. Sci. Technol.*, *33*, 3697 (1999).
400. V. Laperche, T. J. Logan, P. Gaddam, and S. J. Traina, *Environ. Sci. Technol.*, *31*, 2745 (1997).
401. J. W. Huang, J. Chen, W. R. Berti, and S. D. Cunningham, *Environ. Sci. Technol.*, *31*, 800 (1997).
402. M. J. Blaylock, D. E. Salt, S. Dushenkov, O. Zakharova, C. Gussman, Y. Kapulnik, B. D. Ensley, and I. Raskin, *Environ. Sci. Technol.*, *31*, 860 (1997).
403. E. M. Cooper, J. T. Sims, S. D. Cunningham, J. W. Huang, and W. R. Berti, *J. Environ. Qual.*, *28*, 1709 (1999).
404. M. P. Bernal, S. P. McGrath, A. J. Miller, and A. J. M. Baker, *Plant Soil*, *164*, 251 (1994).
405. S. L. Brown, R. L. Chaney, J. S. Angle, and A. J. M. Baker, *Soil Sci. Soc. Am. J.*, *59*, 125 (1995).
406. A. Kayser, K. Wenger, A. Keller, W. Attinger, H. R. Felix, S. K. Gupta, and R. Schulin, *Environ. Sci. Technol.*, *34*, 1778 (2000).
407. T. Dunne and L. B. Leopold, *Water in Environmental Planning*, W. H. Freeman, New York, 1978.
408. S. D. Faust and O. M. Aly, *Chemistry of Water Treatment*, Butterworth, Boston, 1983.
409. M. Pourbaix, *Atlas of Electrochemical Equilibria in Aqueous Solutions*, National Association of Corrosion Engineers and CEBELCOR, Houston, TX, 1974.
410. M. P. Taillefert and J. F. Gaillard, *Mineral. Mag.*, *62A*, 1490 (1998).
411. M. Taillefert, C. P. Lienemann, J. F. Gaillard, and D. Perret, *Geochim. Cosmochim. Acta*, *64*, 169 (2000).
412. I. Christl, C. J. Milne, D. G. Kinniburgh, and R. Kretzschmar, *Environ. Sci. Technol.*, *35*, 2512 (2001).
413. A. L. R. Sekaly, R. Mandal, N. M. Hassan, J. Murimboh, C. L. Chakrabarti, M. H. Back, D. C. Gregoire, and W. H. Schroeder, *Anal. Chim. Acta*, *402*, 211 (1999).
414. I. Christl and R. Kretzschmar, *Environ. Sci. Technol.*, *35*, 2505 (2001).
415. F. Berbel, J. M. Díaz-Cruz, C. Ariño, M. Esteban, F. Mas, J. L. Garcés, and J. Puy, *Environ. Sci. Technol.*, *35*, 1097 (2001).
416. D. Weiss, W. Shotyk, and O. Kempf, *Naturwissenschaften*, *86*, 262 (1999).
417. I. Renberg, M. W. Persson, and O. Emteryd, *Nature (London)*, *368*, 323 (1994).
418. R. Bindler, M.-L. Brännvall, I. Renberg, O. Emteryd, and H. Grip, *Environ. Sci. Technol.*, *33*, 3362 (1999).
419. S. N. Chillrud, R. F. Bopp, H. J. Simpson, J. M. Ross, E. L. Shuster, D. A. Chaky, D. C. Walsh, C. C. Choy, L.-R. Tolley, and A. Yarme, *Environ. Sci. Technol.*, *33*, 657 (1999).
420. F. Monna, J. Dominik, J.-L. Loizeau, M. Pardos, and P. Arpagaus, *Environ. Sci. Technol.*, *33*, 2850 (1999).
421. A. C. Heyvaert, J. E. Reuter, D. G. Slotton, and C. R. Goldman, *Environ. Sci. Technol.*, *34*, 3588 (2000).
422. M. A. Vile, R. K. Wieder, and M. Novák, *Environ. Sci. Technol.*, *34*, 12 (2000).
423. Y. Erel, Y. Dubowski, L. Halicz, J. Erez, and A. Kaufman, *Environ. Sci. Technol.*, *35*, 292 (2000).

424. P. Horn, S. Holzl, W. Todt, and D. Matthies, *Isot. Environ. Health Stud.*, *34*, 31 (1998).
425. D. J. Brabander, N. Keon, R. H. R. Stanley, and H. F. Hemond, *Proc. Natl. Acad. Sci. USA*, *96*, 14635 (1999).
426. F. Marcantonio, G. Flowers, L. Thien, and E. Ellgaard, *Environ. Sci. Technol.*, *32*, 2371 (1998).
427. S. A. Watmough, R. J. Hughes, and T. C. Hutchinson, *Environ. Sci. Technol.*, *33*, 670 (1999).
428. S. Tommasini, G. R. Davies, and T. Elliott, *Appl. Geochem.*, *15*, 891 (2000).
429. R. Lobinski, C. Witte, F. C. Adams, P. L. Teissedre, J. C. Cabanis, and C. F. Boutron, *Nature (London)*, *370*, 24 (1994).
430. K. J. Rosman, W. Chisholm, S. Jimi, J.-P. Candelone, C. F. Boutron, P.-L. Teissedre, and F. C. Adams, *Environ. Res.*, *78*, 161 (1998).
431. J. S. Lin-Fu, in *Human Lead Exposure*, H. L. Needleman, Ed., CRC Press, Boca Raton, FL, 1992.
432. Centers for Disease Control and Prevention, *Morbidity and Mortality Weekly Report*, *46*, 141 (1997).
433. J. L. Pirkle, R. B. Kaufmann, D. J. Brody, T. Hickman, E. W. Gunter, and D. C. Paschal, *Environ. Health Perspect.*, *106*, 745 (1998).
434. Y. Finkelstein, M. E. Markowitz, and J. F. Rosen, *Brain Res. Rev.*, *27*, 168 (1998).
435. H. A. Schroeder and I. H. Tipton, *Arch. Environ. Health*, *17*, 965 (1969).
436. H. Hu, M. Rabinowitz, and D. Smith, *Environ. Health Perspect.*, *106*, 1 (1998).
437. M. Manea-Krichten, C. Patterson, G. Miller, D. Settle, and Y. Erel, *Sci. Total Environ.*, *107*, 179 (1991).
438. F. Schweinsberg and L. von Karsa, *Comp. Biochem. Physiol. C*, *95*, 117 (1990).
439. G. Grynkiewicz, M. Poenie, and R. Y. Tsien, *J. Biol. Chem.*, *260*, 3440 (1985).
440. R. Y. Tsien, *Trends Biochem. Sci.*, *11*, 450 (1986).
441. A. Miyawaki, J. Llopis, R. Heim, J. M. McCaffery, J. A. Adams, M. Ikura, and R. Y. Tsien, *Nature (London)*, *388*, 882 (1997).
442. A. W. Czarnik, *Chem. Biol.*, *2*, 423 (1995).
443. L. E. Kerper and P. M. Hinkle, *J. Biol. Chem.*, *272*, 8346 (1997).
444. M. E. Legare, R. Barhoumi, E. Hebert, G. R. Bratton, R. C. Burghardt, and E. Tiffany-Castiglioni, *Toxicol. Sci.*, *46*, 90 (1998).
445. W. L. Erdahl, C. J. Chapman, R. W. Taylor, and D. R. Pfeiffer, *J. Biol. Chem.*, *275*, 7071 (2000).
446. R. Y. Tsien, *Soc. Gen. Physiol. Ser.*, *40*, 327 (1986).
447. U. Resch, K. Rurack, J. L. Bricks, and J. L. Slominski, *J. Fluoresc.*, *7*, 231S (1997).
448. M.-Y. Chae, J. Yoon, and A. W. Czarnik, *J. Mol. Recognit.*, *9*, 297 (1996).
449. R. T. Bronson, J. S. Bradshaw, P. B. Savage, S. Fuangswasdi, S. C. Lee, K. E. Krakowiak, and R. M. Izatt, *J. Org. Chem.*, *66*, 4752 (2001).
450. R. T. Bronson and J. S. Bradshaw, personal communication.
451. J. Li and Y. Lu, *J. Am. Chem. Soc.*, *122*, 10466 (2000).
452. R. S. Manalis and G. P. Cooper, *Nature (London)*, *243*, 354 (1973).
453. G. J. Long, J. F. Rosen, and F. A. X. Schanne, *J. Biol. Chem.*, *269*, 834 (1994).
454. R. A. Rius, S. Govoni, and M. Trabucchi, *Toxicology*, *40*, 191 (1986).
455. R. A. Rius, S. Govoni, S. Bergamaschi, L. Lucci, and M. Trabucchi, *Sci. Total Environ.*, *71*, 441 (1988).
456. E. K. Silbergeld, J. T. Fales, and A. M. Golberg, *Nature (London)*, *247*, 49 (1974).

457. J. G. Pounds and D. A. Cory-Slechta, *Neurotoxicology*, *14*, 4 (1993).
458. G. Audesirk, *Prog. Neurobiol.*, *24*, 199 (1985).
459. G. Audesirk, *Neurotoxicology*, *14*, 137 (1993).
460. G. A. Audesirk, Teresa., *Neurotoxicology*, *14*, 259 (1993).
461. M. J. Gutnick, H. D. Lux, D. Swandulla, and H. Zucker, *J. Physiol. (London)*, *412*, 197 (1989).
462. J. J. Chisholm, in *Lead Toxicity*, R. L. Singhal and J. A. Thomas, Eds., Urban & Schwartzenberg, Baltimore-Munich, 1980, pp. 461–483.
463. J. P. Bressler, L. Belloni-Olivi, S. Forman, and G. W. Goldstein, *J. Neurosci. Res.*, *46*, 678 (1996).
464. T. Narahashi, in *Toxicology of Metals*, L. W. Chang, Ed., CRC Lewis Publishers, Boca Raton, FL, 1996, pp. 677–698.
465. D. Buesselberg, M. L. Evans, H. L. Haas, and D. O. Carpenter, *Neurotoxicology*, *14*, 249 (1993).
466. C. Ferguson, M. Kern, and G. Audesirk, *Neurotoxicology*, *21*, 365 (2000).
467. T. J. B. Simons and G. Pocock, *J. Neurochem.*, *48*, 383 (1987).
468. J. L. Tomsig and J. B. Suszkiw, *Biochim. Biophys. Acta*, *1069*, 197 (1991).
469. H. Gunshin, B. Mackenzie, U. V. Berger, Y. Gunshin, M. F. Romero, W. F. Boron, S. Nussberger, J. L. Gollan, and M. A. Hediger, *Nature (London)*, *388*, 482 (1997).
470. R. A. Goyer, in *Toxicology of Metals: Biochemical Aspects*, R. A. Goyer and M. G. Cherian, Eds., Springer-Verlag, Berlin, 1995, pp. 1–17.
471. R. Deane and M. W. B. Bradbury, *J. Neurochem.*, *54*, 905 (1990).
472. J. J. R. Frausto da Silva and R. J. P. Williams, *The Biological Chemistry of the Elements: The Inorganic Chemistry of Life*, Clarendon Press, Oxford, UK, 1991.
473. E. C. Thiel and K. N. Raymond, in *Bioinorganic Chemistry*, I. Bertini, H. B. Gray, S. J. Lippard, and J. S. Valentine, Eds., University Science Books, Mill Valley, CA, 1994, pp. 1–36.
474. M. Ikura, *Trends Biochem. Sci.*, *21*, 14 (1996).
475. M. R. Nelson and W. J. Chazin, *Biometals*, *11*, 297 (1998).
476. H. Kawasaki, S. Nakayama, and R. H. Kretsinger, *Biometals*, *11*, 277 (1998).
477. E. A. Nalefski and J. J. Falke, *Protein Sci.*, *5*, 2375 (1996).
478. J. Rizo and T. C. Sudhof, *J. Biol. Chem.*, *273*, 15879 (1998).
479. M. S. Perin, N. Brose, R. Jahn, and T. C. Sudhof, *J. Biol. Chem.*, *266*, 623 (1991).
480. N. Brose, A. G. Petrenko, T. C. Sudhof, and R. Jahn, *Science*, *256*, 1021 (1992).
481. J. T. Littleton and H. J. Bellen, *Trends Neurosci*, *18*, 177 (1995).
482. N. E. Reist, J. Buchanan, J. Li, A. DiAntonio, E. M. Buxton, and T. L. Schwarz, *J. Neurosci.*, *18*, 7662 (1998).
483. R. Fernandez-Chacon, A. Konigstorfer, S. H. Gerber, J. Garcia, M. F. Matos, C. F. Stevens, N. Brose, J. Rizo, C. Rosenmund, and T. C. Sudhof, *Nature (London)*, *410*, 41 (2001).
484. E. K. Jaffe, *Acta Crystallogr., Sect. D*, *56*, 115 (2000).
485. J. A. Millar, V. Battistini, R. L. Cumming, F. Carswell, and A. Goldberg, *The Lancet*, *2*, 695 (1970).
486. E. K. Jaffe, S. Bagla, and P. A. Michini, *Biol. Trace Element Res.*, *28*, 223 (1991).
487. T. J. B. Simons, *Eur. J. Biochem.*, *234*, 178 (1995).
488. K. B. Nielson, C. L. Atkin, and D. R. Winge, *J. Biol. Chem.*, *260*, 5342 (1985).

489. J. S. Hanas, J. S. Rodgers, J. A. Bantle, and Y. G. Cheng, *Mol. Pharmacol.*, *56*, 982 (1999).
490. M. Razmiafshari, J. Kao, A. d'Avignon, and N. H. Zawia, *Toxicol. Appl. Pharmacol.*, *172*, 1 (2001).
491. B. Hitzfield and D. M. Taylor, *Mol. Toxicol.*, *2*, 151 (1989).
492. P. L. Goering and B. R. Fisher, in *Toxicology of Metals: Biochemical Aspects*, R. A. Goyer and H. G. Cherian, Eds., Springer-Verlag, Berlin, 1995, Vol. 115, pp. 229–266.
493. N. H. Zawia, R. Sharan, M. Brydie, T. Oyama, and T. Crumpton, *Dev. Brain Res.*, *107*, 291 (1999).
494. T. Crumpton, D. S. Atkins, N. H. Zawia, and S. Barone, Jr., *Neurotoxicology*, *22*, 49 (2001).
495. W. R. Farkas, *Chem.-Biol. Interact.*, *11*, 253 (1975).
496. C. Werner, B. Krebs, G. Keith, and D. Dirheimer, *Biochim. Biophys. Acta*, *432*, 161 (1976).
497. R. S. Brown, J. C. Dewan, and A. Klug, *Biochemistry*, *24*, 4785 (1985).
498. J. K. Bashkin and L. A. Jenkins, *Comments Inorg. Chem.*, *17*, 77 (1994).
499. A. T. Lam, R. Guenther, and P. F. Agris, *BioMetals*, *8*, 8 (1995).
500. A. Oskarsson, K. S. Squibb, and B. A. Fowler, *Biochem. Biophys. Res. Commun.*, *104*, 290 (1982).
501. G. DuVal and B. A. Fowler, *Biochem. Biophys. Res. Commun.*, *159*, 177 (1989).
502. B. A. Fowler and G. DuVal, *Environ. Health Perspect.*, *91*, 77 (1991).
503. M. R. Moore, P. A. Meredith, and A. Goldberg, in *Lead Toxicity*, R. L. Singhal and J. A. Thomas, Eds., Urban and Schwarzenberg, Baltimore-Munich, 1980, pp. 79–117.
504. J. S. Woods, in *Toxicology of Metals: Biochemical Aspects*, R. A. Goyer and M. G. Cherian, Eds., Springer-Verlag, Berlin, 1995, pp. 19–52.
505. H. A. Dailey, *J. Biol. Inorg. Chem.*, *2*, 411 (1997).
506. E. Rossi, K. A. Costin, and P. Garcia-Webb, *Clin. Chem.*, *36*, 1980 (1990).
507. E. Rossi, S. Taketani, and P. Garcia-Webb, *Biomed. Chromatogr.*, *7*, 1 (1993).
508. Y. Qian, E. D. Harris, Y. Zheng, and E. Tiffany-Castiglioni, *Toxicol. Appl. Pharmacol.*, *163*, 260 (2000).
509. Y. Qian, G. Mikeska, E. D. Harris, G. R. Bratton, and E. Tiffany-Castiglioni, *Toxicol. Appl. Pharmacol.*, *158*, 41 (1999).
510. P. O. Brown and D. Botstein, *Nat. Genet.*, *21*, 33 (1999).
511. A. Pandley and M. Mann, *Nature (London)*, *405*, 837 (2000).
512. P. Del Castilho and R. F. M. Herber, in *Clinical Chemistry and Chemical Toxicology of Metals*, S. S. Brown, Ed., Elsevier, Amsterdam, The Netherlands, 1977, pp. 361–365.
513. M. Stoeppler, in *Clinical Chemistry and Chemical Toxicology of Metals*, S. S. Brown, Ed., Elsevier, Amsterdam, The Netherlands, 1977, pp. 307–318.
514. M. A. White, *J. Trace Elements Med. Biol.*, *13*, 93 (1999).
515. M. Krachler and K. J. Irgolic, *J. Trace Elements Med. Biol.*, *13*, 157 (1999).
516. S. S. Brown, in *Clinical Chemistry and Chemical Toxicology of Metals*, S. S. Brown, Ed., Elsevier, Amsterdam, The Netherlands, 1977, pp. 381–392.
517. B. Alberts, D. Bray, J. Lewis, M. Raff, K. Roberts, and J. D. Watson, *Molecular Biology of the Cell*, Garland Publishing, New York, 1994.
518. K. Tomokuni, I. Osaka, and M. Ogata, *Arch. Environ. Health*, *30*, 588 (1975).
519. B. Bush, D. R. Doran, and K. W. Jackson, *Ann. Clin. Biochem.*, *19*, 71 (1982).
520. S. Sassa, P. M. Jordan, and P. N. B. Gibbs, *Enzyme*, *28*, 133 (1982).

521. J. G. Wetmur, G. Lehnert, and R. J. Desnick, *Environ. Res.*, *56*, 109 (1991).
522. E. K. Jaffe, M. Volin, C. R. Bronson-Mullins, R. L. Dunbrack, J. Kervinen, J. Martins, J. F. Quinlan, M. H. Sazinsky, E. M. Steinhouse, and A. T. Yeung, *J. Biol. Chem.*, *275*, 2619 (2000).
523. L. J. Somervaille, D. R. Chettle, M. C. Scott, D. R. Tennant, M. J. McKiernan, A. Skilbeck, and W. N. Trethowan, *Br. J. Ind. Med.*, *45*, 174 (1988).
524. E. Reitznerova, D. Amarasiriwardena, M. Kopcakova, and R. M. Barnes, *Fresenius J. Anal. Chem.*, *367*, 748 (2000).
525. M. E. Mortensen and P. D. Walson, *Clinical Pediatrics*, 284 (1993).
526. M. B. Francis, N. S. Finney, and E. N. Jacobsen, *J. Am. Chem. Soc.*, *118*, 8983 (1996).

Chromium in Biology: Toxicology and Nutritional Aspects

AVIVA LEVINA, RACHEL CODD, CAROLYN T. DILLON, and PETER A. LAY

Centre for Heavy Metals Research
School of Chemistry
University of Sydney
NSW, 2006
Australia

CONTENTS

I. INTRODUCTION

    A. Chromium, the Biological Chameleon
    B. Scope of the Chapter

II. METHODS OF CHARACTERIZATION OF BIOLOGICALLY ACTIVE CHROMIUM COMPLEXES OR COMPLEXES WITH BIOMOLECULES

    A. Applications of EPR Spectroscopy in Chromium Bioinorganic Chemistry
        1. Introduction
        2. Chromium(V)
        3. Chromium(III) and Chromium(IV)
        4. Organic Radicals
        5. Oxygen Radicals and Singlet Oxygen
        6. In Vivo Studies
    B. Applications of XRD and XAS to Structural Characterizations
        1. X-Ray Structures
        2. X-Ray Absorption Spectroscopy
    C. Applications of Electrospray Mass Spectrometry
    D. Other Techniques

III. TOXICOLOGY OF CHROMIUM

    A. Epidemiological and Experimental Evidence for Chromium Toxicity
    B. Mechanisms of Chromium Toxicity at Cellular and Subcellular Levels

Progress in Inorganic Chemistry, Vol. 51, Edited by Kenneth D. Karlin.
ISBN 0-471-26534-9 © 2003 John Wiley & Sons, Inc.

1. Uptake, Distribution, and Metabolism in Cells
2. Cellular Genotoxicity Assays
3. Interactions with Cell Membranes and Cell Walls
4. Damage of DNA and Transcription Factors: Influence on Gene Expression
5. Oxidative Stress and Antioxidants
6. Cell Signaling and Apoptosis
7. Direct and Indirect Pathways in Carcinogenicity
C. Mechanisms of Chromium Toxicity at the Molecular Level
1. Chromium(VI) + Reductant Systems
2. Chromium(III) + Oxidant Systems
3. Interactions of Chromium(III) Complexes with DNA
4. Mechanisms of DNA Damage Induced by a Model Chromium(V) Complex

IV. CHROMIUM AS A NUTRIENT

A. Biological Activity of Chromium(III)
B. Absorption, Metabolism, and Speciation of Chromium(III) In Vivo
C. Search for Natural Biologically Active Forms of Chromium(III)
1. Glucose Tolerance Factor
2. Low Molecular Weight Chromium-Binding Substance
3. Chromodulin
D. Use of Chromium(III) in Food Supplements: Efficacy and Safety
E. Molecular Mechanisms of Chromium(III) Biological Activity

V. CONCLUDING REMARKS

ACKNOWLEDGMENTS

ABBREVIATIONS

REFERENCES

I. INTRODUCTION

A. Chromium, the Biological Chameleon

Chromium is probably the most controversial of the transition metal ions in terms of its toxicity and nutritional value (1). While Cr(VI) complexes were one of the first classes of chemicals to be implicated as human carcinogens (2), Cr(III) has generally been considered as one of the least toxic of the transition metals (3). Moreover, the importance of Cr(III) in nutrition and diabetes and hence, its value as a dietary supplement, are a matter of often heated debate (3–6). The latter is fueled by the fact that Cr nutritional supplements for humans and animals forms the basis of a multimillion dollar industry (6); hence scientific debate has been muddied by competing health and commercial interests. The controversy over Cr(VI) is also driven by these competing interests, since Cr(VI) is the most widely used carcinogen in the workplace and, for many applications, it is still the best or only reagent that is suitable for important technological processes (1, 7). The controversy has escalated as a result of recent massive lawsuits in the United States involving environmental pollution that has adversely affected local populations (8–10).

Overlaying these major commercial and health interests is a complexity of aqueous redox and substitution chemistry that is unprecedented by the other transition metal ions. This complexity is amplified in the cellular and in vivo experiments designed to provide fundamental insights into a range of health issues. Combined with this is the ubiquitous presence of Cr (not unrelated to its commercial importance) in the environments used to perform biological and human tests, which can cause inadvertent contamination of Cr in drinking water or food, and hence invalidate the results (11). These factors combine to yield a complexity of experimental design and chemical ambiguities in biological tests and human trials that readily lends itself to rational misinterpretation of experimental findings.

The chameleon-like nature of Cr in biology is highlighted by the fact that it can be extremely toxic or nontoxic, depending on its chemical composition (2). Its chameleon-like nature in terms of its biological activity has been well camouflaged by its chemical complexity, which has led to conflicting reports of its mode of action and, indeed, whether it is active at all as a dietary supplement (3–6).

B. Scope of the Chapter

As indicated above, there are enormous issues of public interest, health, and commerce associated with the use (and misuse) of Cr. Hence, it is essential that we remove the camouflage that is inherent in the literature in order to rationalize the current state of knowledge. This chapter is aimed at making an important contribution to this debate from the point of view of understanding the fundamental chemical processes that underlie the biological activities. While there are many recent reviews of aspects of Cr chemistry in relation to its biological activities (listed in the corresponding parts of the text), this chapter is aimed at making a critical assessment of the large body of recent research into Cr chemistry and biochemistry in relation to its biological activities. In turn, this will be used to provide insights to rationalize the current state of our knowledge with regard to implications for health. In doing so, it becomes obvious that certain areas demand further study. These will be highlighted where appropriate in the chapter.

II. METHODS OF CHARACTERIZATION OF BIOLOGICALLY ACTIVE CHROMIUM COMPLEXES OR COMPLEXES WITH BIOMOLECULES

The complexity of the chemistry and biochemistry of Cr complexes is such that the characterization of the intermediates in the reactions is a challenge even for the most simple of in vitro studies. These problems are increased many-fold with the biological soup that is applicable to health effects in vivo. The Cr(VI), Cr(V), and Cr(IV) oxidation states are all labile and highly reactive, which makes them difficult to characterize. On the other hand, complexes of the inert Cr(III) oxidation state are normally not amenable to study by nuclear magnetic resonance (NMR) spectroscopy, so most information has been obtained from electronic absorption spectroscopy in solution, or where suitable crystals have been obtained, X-ray crystallography. In addition, a variety of reactive organic and inorganic radicals are produced. Since understanding the health effects of Cr relies on a detailed knowledge of molecular events, it is important to first discuss the methods of characterization of the many intermediates and products formed during biotransformation processes and the reliability and pitfalls of various methods used in their characterization. This section deals with these aspects. More detailed descriptions of some of the methods are given in a recent review (12).

A. Applications of EPR Spectroscopy in Chromium Bioinorganic Chemistry

1. Introduction

Electron paramagnetic resonance (EPR) spectroscopy has been a crucial technique in the development of our understanding of the bioinorganic chemistry and biology of Cr. This importance is based on the fact that it is sensitive to the detection and speciation of Cr(V) intermediates, organic radicals, oxygen-based inorganic radicals, and Cr(III) products. While Cr(IV) species are also paramagnetic, the EPR signals are not observed at frequencies and temperatures normally used for EPR experiments (13, 14).

EPR techniques can be used for fundamental chemical studies, in addition to studies of the reactivity of Cr complexes with cultured cells, or studies of the biotransformations of Cr carcinogens in vivo. While the technique is very powerful in these studies, it is not without its controversies, particularly with its purported use in "establishing" the presence of hydroxyl radicals, which has been discussed elsewhere (12) and in Section II.A.5.

2. Chromium(V)

The presence of Cr(V) species in the reduction of Cr(VI) was first established by EPR spectroscopy by Kon (15, 16) and Garifianov and Usacheva (17) in 1962. Since that time, EPR spectroscopy has been the method of choice to determine the presence of Cr(V) and the structures of Cr(V) complexes in such redox chemistry. The sharp line width (typically 1–10 G) in X-band (~ 9.5 GHz)

EPR spectroscopy characteristic of Cr(V) species enables quantitative speciation of Cr(V) complexes in solution, even when complex equilibria are involved (18, 19). A further feature is that low concentrations of Cr(V) species (≥ 0.1 μM) can be detected selectively in aqueous solutions at room temperature, in the presence of large excesses of Cr species in other oxidation states. An important contribution to the structural characterization of Cr(V) complexes came with the development of empirical parameters that related the values of the isotropic g tensor (g_{iso}) to the ligand donor groups and the coordination number (18). In the same paper, it was shown that there were linear correlations between these EPR parameters and those of analogous isoelectronic V(IV) complexes. This correlation also assisted in the characterization of new Cr(V) complexes, since many V(IV) complexes are sufficiently stable to be crystallized, and their structures have been determined by X-ray diffraction (XRD). This, together with the determination of a number of structures of Cr(V) species by XRD (12, 20) has put the assignment of the structures of Cr(V) complexes by EPR spectroscopy on a firm footing. Since the initial introduction of these empirical parameters, the types of donor groups have been extended to enediolato ligands (21), amines (22, 23), and peroxo ligands (24, 25). As a result, extensive studies have been performed on the chemistry of biologically relevant Cr(V) complexes [(12), and references cited therein]. The large number of Cr(V) complexes that have been characterized by EPR spectroscopy has enabled a reassessment of the generality of the empirical parameters for structural assignments. This finding has resulted in more general parameters that are introduced here, and that have led to the reassessment of some structures. These parameters are an important aid in the characterization of new Cr(V) complexes observed in the bioinorganic chemistry of Cr. The g_{iso} values for new complexes can be calculated by use of Eq. 1, which is based on deviations from the g_{iso} value of the free electron (2.0023) (26), and the empirical parameters contained in Table I. Unlike the previous parameters, which were specific for the coordination number of

TABLE I

Isotropic EPR Δg_{iso} Parameters for Cr(V) Complexes with Different Sets of Donor Groups, Corrections from the Free Electron Value of 2.0023

Donor Groups, L	Δg_{iso}	Donor Groups, L	Δg_{iso}
N^{3-}	0.00	OH^-	0.00505
$RNCOR^-$	0.00030	RCO_2^-	0.00593
RS^-	0.00035	H_2O	0.00604
O^{2-}	0.00210	RNH_2	0.00667
Cl^-	0.00330	$R_2C=NR$	0.0073
RO^- (enediol)	0.00475	O_2^{2-}	0.00750
RO^-	0.00505		

the complex (18), the new parameters are quite general and do not depend on the coordination number.

$$g_{iso} = 2.0023 - \Sigma \Delta g_{iso} \qquad (1)$$

Examples of the application of these parameters to a number of Cr(V) complexes with biologically relevant ligands are summarized in Table II. While by no means comprehensive, this provides examples of how the structures of many Cr(V) species, that are important to understanding the toxicology of Cr(V), have been assigned. Generally, small deviations between calculated and observed values can be accounted for by steric, electronic, and/or solvent effects that modify the donor properties of different functional groups to the Cr(V) center. An example of electronic effects for biologically active diolato (2−) ligands is the electron-withdrawing effects of the oxygen heteroatom in decreasing the donor strengths of diolato (2−) donors next to this endocyclic oxygen atom in sugars (27–29). This electron withdrawal causes a higher Δg_{iso} value than for a simple diolato donor. In general, for donor groups in the same row of the periodic table, the value of Δg_{iso} increases as the strength of the Cr(V) donor bond decreases. The presence of weak donors is often accompanied by an increase in coordination number from five to six and the tendency to form dioxo Cr(V) species [similar to those normally found in V(V) chemistry] rather than the five-coordinate mono-oxo or mono-nitrido species more often observed for complexes with stronger ligand donors. Similar behavior is observed for sulfur donors that are readily displaced by oxygen donors in ligand-exchange reactions (27, 28) and have a strong tendency to form six-coordinate complexes.

As indicated above, the solvent also contributes to the g_{iso} values by affecting the strength of the Cr–L bonds, as demonstrated by the solvent dependence of the EPR spectra of the $[CrO(ehba)_2]^-$ complex (30). Similarly, steric effects are illustrated by the distinct EPR signals that are observed for the two geometric isomers of $[CrO(qa)_2]^-$ (qa = quinato) in which the ligand binds via the 2-hydroxycarboxylate group (31). However, the electronic, steric, and solvent effects on the g_{iso} values are small perturbations on the empirical relationship given by Eq. 1 and as long as they are recognized, this equation holds very well for a large range of mixed-ligand complexes.

Hyperfine coupling involving the ^{53}Cr nucleus ($I = \frac{3}{2}$, natural abundance = 9.55%) also provides useful information on the structures of Cr(V) species. For N and O donor ligands, the magnitude of the A_{iso} increases in going from five- to six-coordinate species, typically by $(2-4) \times 10^{-4} \, cm^{-1}$ (18). While there are some correlations between the values of A_{iso} and the donor atoms of the complexes (18), it is clear that the linear relationships involving g_{iso} values are not as applicable to the A_{iso} values. The reason for this is that compared to the g_{iso} values, the A_{iso} values are much less sensitive to the donor atoms and

TABLE II

Assignment of Solution Structures from Calculated and Experimental g_{iso} Parameters for Some Biologically Relevant Cr(V) Complexes

Complex[a]	g_{iso}(obsd)	g_{iso}(calcd)	References
[CrO(ox)$_2$]$^-$	1.9765	1.9765	19
[CrO(lac)$_2$]$^-$	1.9780	1.9782	18
[CrO(cit)$_2$]$^-$	1.9781	1.9782	18
[CrO(ehba)(ed)]$^-$	1.9791	1.9791	40
[CrO(ed)$_2$]$^-$	1.9801	1.9800	40
[CrO(glyc)$_2$]$^{-b}$	1.980	1.9800	27
[Cr(O)$_2$(OH$_2$)(OH)(ox)]$^-$	1.9765	1.9773	19
[CrO(cit)(phen)]$^+$	1.9744	1.9746	39
[CrO(salen)]$^+$	1.9755	1.9755	38
[CrO(OH$_2$)(ox)$_2$]$^-$	1.9712	1.9704	19
[CrO(OH)(ox)$_2$]$^{2-}$	1.9718	1.9714	19
[CrO(oxH)(ox)$_2$]$^{2-}$	1.9715	1.9706	19
[CrO(OMe)$_3$(ehba)]$^{2-}$	1.9752	1.9741	30
[CrO(OEt)$_3$(ehba)]$^{2-}$	1.9755	1.9741	30
[Cr(O)$_2$(Ala)(OR)$_2$]$^{2-c}$	1.9754	1.9754	23
[CrO(Ala$_3$)(OR)]$^{-c}$	1.9827	1.9820	22
[CrO(tla)$_2$]$^{-d}$	1.9873	1.9876	194
[Cr(O)$_2$(tla)$_2$]$^{3-d}$	1.9848	1.9855	194
[Cr(O)$_2$(SCH$_2$CH$_2$O)]$^{3-d,e}$	1.9879	1.9868	194, 195
[CrO(SCH$_2$CH$_2$S)$_2$]$^{-d,e}$	1.9988	1.9988	194, 195
[Cr(O)$_2$(cit)(bpy)]$^-$	1.9729	1.9725	39
[CrO(cit(3−))(pic)]$^-$	1.971	1.9701	39
[CrO(asc)(OH)(OH$_2$)]^{0f}	1.9791	1.9796	21
[CrO(asc*)(OH)(OH$_2$)]^{0f}	1.9785	1.9790	21
[Cr(O)$_2$(O$_2$)(asc)]$^{3-f}$	1.9824	1.9811	21
[Cr(O)$_2$(O$_2$)(asc*)]$^{3-f}$	1.9818	1.9805	21
[CrO(O$_2$)$_2$(OH$_2$)]$^-$	1.9798	1.9792	24
[Cr(O$_2$)$_3$(OH)]$^{2-}$	1.9764	1.9748	24
[Cr(O$_2$)$_4$]$^{3-}$	1.9939	1.9939	24

[a] See the list of abbreviations for designations of the ligands.
[b] These are models for the binding of Cr(V) to the glycerol tail of sialic acid terminating glycoproteins (42).
[c] R = H, CH$_3$.
[d] These complexes have been reassigned.
[e] The complexes include many biologically relevant thiolato ligands.
[f] These complexes have been reassigned.

much more sensitive to the coordination geometry, which is illustrated by the geometric isomers of [CrO(ehba)$_2$]$^-$ and [CrO(hmba)$_2$]$^-$ [hmba = 2-hydroxy-2-methylbutanato(2−)], where the g_{iso} values are virtually the same, whereas the A_{iso} values differ substantially (30, 31). The origin of this difference has been revealed using density functional calculations (32), which indicate different distortions in the five-coordinate geometries for the two geometric isomers. This finding leads to a change in spin density at the Cr and hence, significantly different A_{iso} values.

The A_{iso} values of the Cr(V) complexes, however, are particularly valuable for determining the structures from well-characterized isoelectronic V(IV) complexes (18), where it has been shown that there are linear correlations involving both g_{iso} and A_{iso} values with well-characterized Cr(V) and V(IV) complexes of the same structure. This tool is also useful in characterizing new Cr(V) complexes.

Where observed, superhyperfine coupling involving ligand atoms is also a very valuable aid in the determination of the structures of Cr(V) complexes, which has been particularly useful in determining the number of amide donors to Cr(V) in peptide and polydentate amide complexes (22, 33–36). The smaller superhyperfine coupling for amine and imine donors, however, is not normally resolved at X-band frequencies (22, 34, 35, 37), with the exception of [CrO(salen)]$^+$ (38) and mixed-ligand citrate complexes with picolinate, bpy, and phen (39).

An important observation with regard to ^{1}H superhyperfine coupling in Cr^V–diolato complexes [of which there are many examples that are biologically relevant (12)] is that the multiplicity of the signal in the EPR spectrum is dependent on whether the ligand is cyclically strained (e.g., sugar) or not (e.g., linear diol). In Cr^V–diolato complexes in which the ligands are not conformationally constrained, such as [CrO(ed)$_2$]$^-$, dynamic processes impart magnetic equivalence to the axial and equatorial protons. This results in superhyperfine coupling to equivalent protons on the EPR timescale (30). Thus the products from ligand-exchange reactions between [CrO(ehba)$_2$]$^-$ and 1,2-ethanediol, [CrO(ehba)(ed)]$^-$, and [CrO(ed)$_2$]$^-$, give rise to five- or nine-line spectra, respectively, due to the ^{1}H-superhyperfine coupling associated with the coordination of either one or two ed ligands (40). Isotropic and anisotropic ^{1}H electron nuclear double resonance (ENDOR) spectra of a series of Cr^V–diolato complexes can be used to distinguish the magnetic environment of the protons in bis-Cr^V–diolato (diol = linear) complexes at subambient temperatures. The superhyperfine coupling constants ^{1}H$_{eq}$ a_{iso} and ^{1}H$_{ax}$ a_{iso} are 0.81×10^{-4} cm^{-1} and 0.37×10^{-4} cm^{-1}, respectively (41). Similar observations were made for systems with cyclic diols, in particular the Cr(V) complexes of cis- or trans-1,2-cyclohexanediol, where the EPR spectra of [CrO(cis-1,2-cyclohexanediolato)$_2$]$^-$ or [CrO(trans-1,2-cyclohexanediolato)$_2$]$^-$ exhibit a triplet and a singlet, respectively. In the latter complex, all of the protons of the carbon atoms bound to the coordinated oxygens are axial and, therefore, there is minimal overlap between the proton orbital and the Cr(V) orbital containing the unpaired electron density (d_{xy}). For [CrO(cis-1,2-cyclohexanediolato)$_2$]$^-$, however, two protons are axial (one from each chelate ring) and two are situated in the Cr(V)-ligand plane, which gives rise to a triplet in the EPR spectrum (27). The interpretation of the ^{1}H superhyperfine coupling has been crucial in the assignment of structures to linkage isomers of Cr(V) complexes of sugars, sialic acid, quinic acid, ascorbate, and many other biologically relevant ligands (12, 18, 21, 27, 28, 31, 42).

Q-band ($\sim$ 34–35 GHz) measurements can be performed for relatively concentrated Cr(V) solutions in vitro and show significant improvements in resolution of overlapping signals from different Cr(V) species compared to those obtained from X-band measurements (18). For example, the complexity of the EPR signals obtained from Cr^V–glucose solutions has been addressed by using Q-, X-, and S-band ($\sim$ 4 GHz) EPR spectroscopies and partially or fully deuterated glucose ligands (18, 27, 28). The Q-band EPR spectroscopy has also been used as an aid in the separation of overlapping peaks of Cr(V) peptide complexes (34, 35) and in the ligand-exchange reactions of the bis-citrato complex [CrO(cit)$_2$]$^-$ [cit = citrato(2−)] with various ligands (39). In general, when complex mixtures of Cr(V) species are obtained, simulation of spectra at more than one frequency gives much more reliable speciation (in terms of structural assignments and relative concentrations) than the use of a single frequency.

Studies on frozen solutions are of limited value in structural assignments because very few Cr(V) complexes exist in solution as a single species, which means that it is difficult to deconvolute the g and A tensors because of the overlap.

3. *Chromium(III) and Chromium(IV)*

Unlike the sharp signals that are observed for Cr(V), those of Cr(III) are typically very broad and difficult to observe at room temperature. The signals are typically centered around g values of 1.98–2.00 (13, 43). While EPR spectroscopy is very useful for the detection of Cr(III) products from redox reactions of Cr in higher oxidation states (44), it is of limited value in determining the nature of the donor atoms bound to the metal ion. Nonetheless, it is still an important technique for study of the bioinorganic chemistry of Cr(III). One application includes studies on frozen solutions (77 K) of the active site for binding of Cr(III) in transferrin (45), which is believed to be the major transport protein for Cr(III) in mammals (Section IV.B).

Signals due to Cr(IV) (d^2 ions) can only be observed in some systems at liquid He temperatures due to the spin-lattice relaxation (13, 14) while most Cr(IV) signals are not observed at all due to the large zero-field splitting; therefore, EPR spectroscopy is not applicable for studies of Cr(IV) species in biological systems. Despite this, Shi and co-workers (46–49) supposedly used EPR spectroscopy to confirm the formation of Cr(IV) species in the reactions of

Cr(VI) with GSH or 2,4-dimethyl-2,4-pentanediol. Isolated solids exhibited broad EPR signals with $g_{iso} = 1.96–1.98$, which were assigned to Cr(IV), while similar signals with $g_{iso} \sim 2.00$ were assigned to Cr(III) species (46). In fact, broad EPR signals with $g_{iso} = 1.96–2.00$ are characteristic of Cr(III) (d^3 ions), but as mentioned above, signals due to Cr(IV) (d^2 ions) cannot be observed at room temperature. Consequently, EPR spectroscopy is unsuitable for observation of Cr(IV) intermediates formed in Cr(VI) reductions and all reports in the literature on the direct observation of Cr(IV) are erroneous (12).

4. Organic Radicals

Organic radicals in the reactions of Cr(VI) with substrates can be detected directly, such as in the case of ascorbate and derivatives (21, 50), vitamin E (α-tocopherol) or its water-soluble analogue, Trolox (51, 52), and catechols (25, 53, 54). For species such as thiyl radicals and carbon-based radicals that are much less stable, spin traps are used to trap the radicals and identify their structures.

Spin traps (precursors of stable nitroxyl radicals) are widely used to detect free radical intermediates formed in biological systems (55). Kadiiska et al. (56, 57) detected adducts of C-based radicals (probably originating from bile acids) in the reactions of bile from Cr(VI) treated rats with a spin trap, α-(4-pyrdyl-1-oxide)-N-tert-butylnitrone (POBN). However, spin trap studies are most often used to invoke involvement of the reactive oxygen species (ROS), particularly •OH, in Cr(VI) induced genotoxicities, and 5,5-dimethylpyrroline-N-oxide (DMPO) is the most widely applied spin trap in such studies. Characteristic EPR signals of DMPO—OH adducts were observed during the reaction of DMPO with blood of Cr(VI) treated mice (58) or in the reactions of Cr(VI) with cell cultures in the presence of DMPO and the reductant, nicotinamide adenine dinucleotide phosphate reduced form (NADPH) (59). Numerous studies of Shi and co-workers [(60–63), and references cited therein], as well as those of several other groups (64–67) have shown the formation of DMPO—OH adducts in the reactions of Cr(VI) + Red + DMPO (Red = GSH, ascorbate, or NADH); Cr(VI) + H_2O_2 + DMPO; or Cr^{III} + H_2O_2 + DMPO in vitro.

5. Oxygen Radicals and Singlet Oxygen

Oxygen-based radicals are often invoked as being important in Cr induced cancers, particularly the hydroxyl radical. This observation has arisen because of the misinterpretation of the results of spin trap experiments, since Cr(V) reacts with spin traps to produce the same products as those produced in the reactions with hydroxyl radicals (68, 69). This result has been discussed in a previous review (12), but is outlined in more detail here.

The spin trap, DMPO, is most often used to determine the presence of reactive oxygen species and, in particular, it is erroneously used to "establish"

$$Cr^{n+} + RH \longrightarrow Cr^{(n-1)+} + R^{\bullet} + H^+ \quad (n = 4\text{--}6) \quad (2)$$

$$Cr^{m+} + H_2O_2 \longrightarrow Cr^{(m+1)+} + {}^{\bullet}OH + H^+ \quad (m = 2\text{--}5) \quad (3)$$

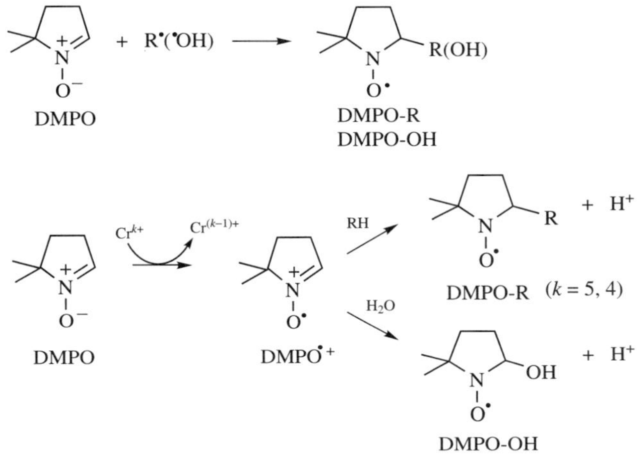

(4) DMPO + R•(•OH) → DMPO-R / DMPO-OH R(OH)

(5) DMPO → DMPO•+ → DMPO-R ($k = 5, 4$) + H+ ; DMPO-OH + H+

Scheme 1. Pathways for the formation of spin-trap adducts in the presence of Cr species.

the presence of •OH. A traditional explanation of DMPO—OH formation in the above-mentioned systems (60, 61) is presented in Scheme 1 (Eqs. 2–4). According to this scheme, organic radicals are formed in the reactions of Cr(VI) with reductants (RH in Eq. 2), while •OH is formed in Fenton-like reactions (Eq. 3, Section III.C.1); either radical is able to react with DMPO (Eq. 4). However, spin traps can react with any strong oxidants, including Cr(V/IV) species; thus, a reaction of DMPO with a model Cr(V) complex, [CrO(ehba)$_2$]$^-$, led to the formation of DMPO—OH in the absence of ROS sources (such as O_2 or H_2O_2) (68, 69). A simplified mechanism for direct DMPO oxidation by Cr(V/IV) species, leading to DMPO—OH or organic radical adducts, is presented in Scheme 1 (Eq. 5). In summary, the formation of nitroxyl radicals, such as DMPO—OH, during the reactions of spin traps with Cr(VI) treated biological materials, points to the presence of strongly oxidizing species in such systems, but it does not distinguish between ROS and Cr derived species. Patterns of DNA damage in Cr(VI) treated cells (Section III.B.2) and in the Cr(VI) + Red + DNA (DNA = deoxyribonucleic acid) reaction in vitro (Section III.C.1), are inconsistent with the involvement of •OH. Taken together, all of the above data provide strong evidence that •OH is not involved in the mechanisms of Cr(VI) induced genotoxicity.

While the evidence points to the lack of involvement of •OH, superoxide radicals ($O_2^{\bullet-}$) are generated in Cr(VI) reactions with catechols (25), thiols (70), and ascorbate (21, 50) as intermediates in the production of hydrogen peroxide. These radicals are not trapped by spin traps such as DMPO since the rate constant [10 M^{-1} s^{-1} (71)] is too low for this reaction to compete with dismutation of superoxide [7.6–8.5 × 10^7 M^{-1} s^{-1} (72)] or subsequent rapid reactions with other species in solution. The presence of superoxide radicals as intermediates is deduced from the presence of hydrogen peroxide products, since the Cr intermediates in the reactions do not react with dioxygen and the organic radicals react with dioxygen in one-electron steps (70).

Singlet oxygen ($^1O_2$) may also be another source of the oxidized spin traps, since it reacts with spin traps to give the same products as •OH (73–75). The potential for the production of singlet oxygen arises because both Cr(VI) and Cr(V) complexes are photochemically active (22) and it is conceivable that they, or Cr(IV) intermediates, could produce $^1O_2$. Evidence for the presence of $^1O_2$ has been gained using the spin trap 2,2,6,6-tetramethyl-4-piperidone (TMPD), which is believed to be specific for $^1O_2$. Experiments with this spin trap indicated the presence of $^1O_2$ in the blood of Cr(VI) treated mice (76). The possible involvement of $^1O_2$ needs to be explored further in more simple in vitro systems.

6. *In Vivo Studies*

Both Cr(V) and Cr(III) EPR signals have been observed when mammalian cells and animals are exposed to Cr(VI). Measurements at L-band (~ 1.2 GHz) are used for in vivo systems, owing to their higher sensitivity (77–79) and sharper signals observed at lower frequencies in these heterogeneous systems. Numerous studies have been reported on the formation of Cr(V) intermediates following the treatment of cells or multicellular organisms with Cr(VI) (Section III.B.1).

B. Applications of XRD and XAS to Structural Characterizations

1. *X-Ray Structures*

There are numerous crystal structures of substitutionally inert Cr(III) complexes [e.g., (80)] and, if suitable crystals can be grown, X-ray crystallography is the method of choice for characterizing Cr(III) complexes. However, the structural characterization of Cr(III) complexes with biomolecules has been rather elusive and there have not been many advances in this area since the early work by Freeman and co-workers [(81–84) and references cited therein] on Cr(III) amino acid and peptide complexes. Some of the Cr(III) dietary supplements have also been characterized by X-ray crystallography (Section IV.D), but none of the species generated in vivo or present in the diet have been structurally characterized.

Complexes of Cr(VI), Cr(V), and Cr(IV) are even more difficult to characterize by X-ray crystallography because they are much more reactive than Cr(III). While the number of Cr(V) complexes that have been characterized by XRD is growing [for a recent review see (12)] and some of these are relevant as models for the study of the genotoxicity, none have been structurally characterized with ligands that are endogenous to humans. While there are numerous structures of chromate and dichromate carcinogens, little XRD information is available on esters with organic substrates that may be relevant to Cr induced cancers. There is only one Cr(VI)–thiolate ester complex that has had its structure determined by XRD (85) and this is not with a biologically relevant ligand. While this may be a model for the Cr(VI)–thiolate ester intermediates that are generated in vivo, it is clearly desirable to develop methods that are suitable for structural determinations of Cr complexes in all oxidation states, including reactive intermediates. Hence, the application of X-ray absorption spectroscopy has been a very valuable addition to structural techniques that are able to further our understanding of the bioinorganic chemistry of Cr.

2. X-Ray Absorption Spectroscopy

K-edge X-ray absorption spectroscopy provides important information on the oxidation state and the coordination geometry of the absorbing Cr atom. The electronic absorption spectrum is derived from exciting a 1 s electron to empty higher energy orbitals giving rise to X-ray absorption near-edge structure (XANES) spectra. Higher energy photons that are of sufficient energy to eject the electron to the continuum cause an interference pattern from the interactions of photoelectrons with atoms in the surrounding environment. The latter gives rise to X-ray absorption fine structure (XAFS). The XAFS spectrum can provide three-dimensional (3D) information about the atoms surrounding the absorbing Cr atom (typically up to ~ 5 Å away) (86, 87). The importance of these techniques to biological work is that they can be applied to any medium, including frozen solutions of reactive Cr intermediates (88). The integration of the resultant fluorescent X-rays from the XAS technique can also be used for synchrotron radiation-induced X-ray emission (SRIXE) elemental mapping of biological samples, such as individual mammalian cells, or XANES can be used to study the oxidation state of Cr in the cells (89–91).

The Cr(III), Cr(IV), Cr(V), and Cr(VI) complexes exhibit distinct XANES spectra (12, 54, 88, 89). For example, Cr(VI) spectra characteristically show relatively large pre-edge peaks due to the symmetry-forbidden $1s \rightarrow 3d$ transition, which achieves considerable intensity from mixing of the p and d orbitals due to the tetrahedral symmetry of the complexes and the strong π bonding involved in the Cr=O bonds. While Cr(V) XANES spectra are similar to those of Cr(VI), their pre-edge peaks are smaller. The XANES spectra of Cr(IV) and

Cr(III) complexes commonly show small and very small pre-edge peaks, respectively, and are ~4 and 5 eV lower in energy than those of Cr(VI) complexes (54, 88). A growing number of Cr(VI), Cr(V), Cr(IV), and Cr(III) complexes have now had their structures determined by XAFS, including many with biologically relevant ligands. The technique also enables the determination of the structures of two different Cr complexes in a single sample, as has been achieved recently for a solid sample containing a mixture of Cr(V) and Cr(III) citrate complexes (39).

C. Applications of Electrospray Mass Spectrometry

In conjunction with XAS and other techniques, such as EPR spectroscopy for the characterization of Cr(V) complexes, electrospray mass spectrometry (ESMS) is an extremely useful tool. Often the parent peak for even reactive Cr(V) and Cr(VI) complexes dominates the spectrum and allows important structural information to be elucidated (92, 93). In water, where complex mixtures are often involved, it is a useful technique to aid in the speciation of complexes, although the redox chemistry and the extensive fragmentation that tends to occur (94) means that the results need to be interpreted with care (usually in conjunction with other independent information on the speciation) (22, 23, 34, 95). Recently, ESMS has been applied to studies of biologically relevant Cr(III) complexes (95) and Cr^{III}–DNA interactions (96). Thus, ESMS is applicable to all oxidation states and will no doubt contribute to further the characterization of reactive Cr species of relevance to cancer, dietary supplements, and naturally occurring Cr complexes with biomolecules.

D. Other Techniques

Electronic absorption spectroscopy is also a useful technique for the characterization of Cr intermediates. When coupled with global kinetic analysis, ligand stoichiometries, and spectra of transient species in the redox chemistry of Cr can be determined. The first applications of this global analysis to biologically relevant systems involved studies of the Cr(VI) reactions with the main biological reductants; Cys, GSH, and ascorbate (70, 97). The distinct electronic absorption features of the Cr(VI), Cr(V), Cr(IV), and Cr(III) oxidation states (12) aid in the characterization of the intermediates.

In addition to the characterization of the intermediates in such reactions, electronic absorption spectroscopy can suggest the nature of donor atoms and often the geometric isomer from the energies and splitting of the d–d transitions in the visible (vis) absorption spectrum (80). However, sometimes these transitions are masked by charge-transfer transitions.

The paramagnetism of all oxidation states except Cr(VI) and the highly reactive nature of Cr(VI) esters with biomolecules, make NMR spectroscopy of limited use for the characterization of Cr complexes; however, multinuclear NMR spectroscopy has been used to characterize Cr(VI) thioester complexes (98, 99) and ^{13}C NMR spectroscopy has enabled Cr(VI)-esters with the oxalate ligand to be characterized (100). While these are challenging experiments and often require the use of labeled ligands and low temperatures to enable spectra to be recorded before the intermediates decompose, it is a valuable technique for some applications.

Despite the paramagnetism of Cr(III) complexes, ^{1}H, ^{2}H, and ^{13}C NMR spectroscopy has been applied to studies of Cr(III) complexes used as nutritional supplements (101, 102), as well as in attempts to determine the structure of Cr(III) complexes isolated from biological sources (103).

Electrochemical techniques can also be used for specific monitoring of Cr(V) in complex reaction mixtures (104) and for studying ligand-exchange reactions of Cr(V) (105), but the techniques have not been exploited in any depth. Potentiometric techniques are useful for determining the redox potentials involving various oxidation states of Cr (106), but are only applicable to relatively stable systems, which rules out most biologically relevant redox reactions.

III. TOXICOLOGY OF CHROMIUM

Medical, environmental, and regulatory aspects of Cr toxicology have been described in detail in several books (1, 2, 7, 107–110) and recent reviews (9, 11, 111–121). This chapter concentrates on the chemical basis of Cr toxicity, primarily using the data published during the period 1996–2001.

A. Epidemiological and Experimental Evidence for Chromium Toxicity

The association between Cr(VI) exposure and cancers of the respiratory tract has been acknowledged for over a century following the observation of nasal cancer in a chrome pigment worker in 1890 (2, 109, 122). Currently, Cr(VI) compounds are among 87 recognized (class I) human carcinogens (only 33 of those are discrete chemicals or groups of chemicals), according to the classification of the International Agency for Research on Cancer (IARC; data of April, 2001) (123). The majority of reported Cr induced cancers have been industry related. These occur over a wide range of chrome manufacturing and handling industries, including chrome electroplating, stainless steel welding, anticorrosives, leather tanning, refractories, timber treatment, and the pigment industry (2, 110, 122, 124–126). The diversity of workplace exposures is such that a document produced by the Institute of Health in Finland rated Cr as the most commonly encountered carcinogen (25%), followed by Ni (20%), asbestos (15%), and benzene (4%) (127). The importance of Cr is further emphasized by the fact that >50,000 tons of Cr are used in the United States each year (128).

The PEL (permissible exposure limit) of Cr(VI) in air for an 8-h workday, 40-h week is 100 µg m^{-3}, while the REL (recommended exposure limit) is much lower, 1 µg m^{-3} (9). In chrome plating workshops with local exhaust, the concentration ranges from 10 to 30 µg m^{-3} and 120 µg m^{-3} with no exhaust (9). Arc stainless steel welding can produce even higher concentrations of up to 1500 µg m^{-3} (9). Inhalation of such high levels of Cr(VI) can cause sneezing, irritation of the nasal mucosa, coughing, and nosebleeds (9). Symptoms of prolonged chronic exposure include dermatitis, asthma, perforations of the nasal septum, and inflammation of the larynx and liver (2, 109, 110, 117, 122, 129). A detailed epidemiological summary of respiratory cancers and the effects of chronic exposure to Cr(VI) can be obtained from the United States EPA in the *Toxicological Review of Hexavalent Chromium* (107).

Despite the implementation of workplace hazard reduction procedures (9) and extensive research toward replacement of Cr(VI) in some industrial applications (130, 131), cancers arising from Cr(VI) exposure still remain a concern, as indicated by the results of several recent epidemiological studies. The most representative study, conducted in the United States, involved 2357 workers employed in Cr(VI) related industries in 1950–1985 and showed a strong dose-response relationship between cumulative Cr(VI) exposure and lung cancer (107, 132). Similar results were obtained in smaller scale studies in the United Kingdom (133) and Argentina (134); however, a study in Sweden failed to reveal a clear connection between exposure to welding fumes and lung cancer (135). In addition to lung cancer, links between Cr(VI) exposure and head and neck or nasal cancers have been established (136–138), and a possible link with pancreatic cancer has been suggested (139). The study of Katabami on ex-chromate workers showed a predominance (16 of 19 cancer cases) of central-type squamous-cell carcinomas, which suggests that there is a direct effect of Cr(VI) on the highly exposed proximal bronchial epithelia (136). This finding is consistent with the accumulation of insoluble chromates in the lung that occurs and results in slow solubilization and uptake of Cr and/or phagocytosis of the particles (Section III.B.1). In the same study, the immunohistochemical markers associated with the onset of cancer were assessed and showed that the frequency of the oncogene cyclin D1 was higher in cancer patients that had been exposed to Cr compared with other cancer patients. The effect on oncogenes is discussed in more detail in Section III.B.4. Recent studies also point to possible links between occupational exposure to Cr(VI) and development of parkinsonism (140) or reproductive disorders (141).

A large number of animal experiments have been performed to investigate the carcinogenicity of Cr(VI), of which only 43% were positive (112, 113, 117,

118). Such tests have employed a large variety of animals (mice, rats, cats, guinea pigs, and rabbits) and a number of exposure routes (intraosseus, intramuscular, subcutaneous, intrapleural, and intraperitoneal injections); however, most of these administration routes, except for inhalation, have little relevance to human occupational exposure (112, 113). In several studies, cancers of the respiratory tract in experimental animals were induced by prolonged inhalation of Cr(VI) compounds [(2, 113), and references cited therein]. For example, inhalation of Cr(VI) containing dust from preserved wood has recently been shown to cause respiratory cancers in rats (142), but As could also present a carcinogenic hazard in such dusts (118). Comprehensive reviews of animal experiments prior to 1995 (2, 113, 117) demonstrated a clear correlation between carcinogenicity and solubility of Cr(VI) compounds, with chromates of medium or poor solubility (Ca^{2+}, Sr^{2+}, Zn^{2+}, or Pb^{2+} salts) being far more potent carcinogens than soluble (Na^+ or K^+) or insoluble (Ba^{2+}) chromates. This finding is consistent with human epidemiological data (136, 143).

There is no established evidence (epidemiological or experimental) for carcinogenicity caused by exposure to Cr(III) or Cr(0) compounds (113, 117). However, allergic contact dermatitis due to occupational exposure to Cr(III) (primarily in the leather tanning industry) is relatively common (144–146). The safety of oral intake of Cr(III) in food supplements has also been questioned (Section IV.D) (5). A concern about the safety of the use of Cr(0) in stainless steel body implants (e.g., artificial joints or fracture fixation plates) was raised following at least one reported case of bone cancer caused by corrosion of the implant (147). In vitro studies have shown that the corrosion of Cr alloys in human serum leads to Cr(III) binding to serum proteins (148). Such corrosion can also release another possible carcinogen, Ni(II) (149). This problem should be overcome by the use of Ti alloys (free of Cr or Ni), which were introduced as an alternative 10–15 years ago (150).

A current debate (9) between De Flora et al. (112, 113, 151, 152) and Costa et al. (117, 153) has centered around the issues of possible systemic toxicity of Cr(VI) compounds (especially via oral ingestion) and the threshold mechanisms in Cr(VI) carcinogenicity. Costa and co-workers (154) have shown that rats given Cr(VI) (3–10 ppm) in drinking water for 44 weeks accumulated increased levels of Cr in bones, kidney, and liver, but not in brain, ovaries, or whole blood. An independent study detected signs of oxidative stress (such as decreased levels of cellular antioxidants, increased activity of antioxidant enzymes, and appearance of lipid peroxidation products) in the brains of mice given Cr(VI) in drinking water (25 mg kg^{-1} day^{-1}) for 1–3 days (155). In other studies, no signs of Cr(VI) induced genotoxicity were observed in animals (156) or human volunteers (157) after ingestion of Cr(VI) (10 ppm) in drinking water, even though blood and urine Cr measurements indicated its significant systemic uptake (157).

De Flora and co-workers (112, 151, 152) argued, based on the estimates of a high capacity in humans to reduce Cr(VI) into noncarcinogenic Cr(III), that Cr(VI) is not a systemic toxicant except at very high doses. According to this group, Cr(VI) induced cancers can only occur at a portal of Cr(VI) entry into organisms (e.g., in the respiratory tract of occupationally exposed humans), and only when the reducing capacity of the airways (e.g., epithelial lining fluid or pulmonary alveolar macrophages) is overwhelmed (112, 151, 152). It is unlikely that Cr(VI) can be carried between organs in blood due to the high reducing capacity of blood plasma and erythrocytes. Orally ingested Cr(VI) is likely to be reduced completely to Cr(III) by saliva and gastric juice (112, 152). On the basis of epidemiological and animal studies, De Flora et al. concluded that there is a minimal risk of Cr(VI) induced cancers, apart from exposure to very high doses of Cr(VI) via inhalation (2, 112, 132). However, their opinion about the exclusively protective nature of Cr(VI) reduction in extracellular body fluids needs to be reconsidered in light of the recent finding that such fluids (e.g., saliva or mucosa) can efficiently stabilize potentially genotoxic Cr(V) intermediates (Sections III.B.3 and III.C.1) (42).

Whether or not prolonged exposure to low concentrations of Cr(VI) in drinking water can lead to cancer (112, 117) is directly related to the current concern about the dangers of environmental exposure to Cr(VI) (8, 9, 107, 158–160). In the United States, 4500 kg of Cr are released into the environment every day (128). Subsequent poor disposal practices, such as dumping of Cr(VI) into unlined ponds, have led to Cr seepage into waterways and consequent contamination of irrigation and drinking waters. A study conducted in China in 1987 reported the effects on 155 villagers subjected to contaminated drinking water (20 mg L^{-1}, with an estimated dose of 0.57 mg kg^{-1} day^{-1}) from a well adjacent to a chromium alloy plant (107). Health problems included oral ulcers, diarrhoea, abdominal pain, indigestion, vomiting, leukocytosis, and the presence of immature neutrophils (107). The most publicized example of the health problems arising from environmental contamination was highlighted in the lawsuit against Pacific Gas & Electric, Hinkley, CA, in 1993 (8–10, 117), where Cr(VI) concentrations of 580 µg L^{-1} in the ground water were reported (cf. state limit of 50 µg L^{-1}). There were reports of health problems associated with the liver, heart, and respiratory tract, as well as reproductive failure and cancers of the brain, kidney, breast, uterus, and gastrointestinal system (9). Despite these links, there has been no unambiguous evidence for cancers induced by environmental exposure to Cr(VI) (9, 119, 161, 162), although improved understanding of this problem, particularly of the effects of possible long-term accumulation of Cr(III) in the body, is required (5, 161).

As described above, although carcinogenicity is not the only health concern associated with Cr(VI) exposure, it has dominated the Cr related medical and biochemical research for several decades. Therefore, in the following sections

(Sections III.B and III.C), cellular and molecular processes that are important in understanding Cr(VI) induced genotoxicity and carcinogenicity (including a possible involvement of lower oxidation states of Cr) will be considered.

B. Mechanisms of Chromium Toxicity at Cellular and Subcellular Levels

The following aspects will be addressed in this part of the chapter: (1) changes caused by treatment of prokaryotic or eukaryotic cells with Cr compounds; (2) changes in cells of multicellular organisms, caused by exposure to Cr compounds; and (3) reactions involving more than one biological macromolecule (e.g., DNA and an enzyme) in a cell-free environment in the presence of Cr compounds.

1. Uptake, Distribution, and Metabolism in Cells

Despite the fact that first links between Cr and cancer date back to 1890, and that the carcinogenicity of Cr(VI) was first officially recognized in 1936 (2, 122, 124), the mechanism(s) of Cr action remain elusive and continue to be an issue of intrigue, interest and debate. This uncertainty may stem from the fact that Cr is diverse in its oxidation state, coordination geometry, and solubility properties, all of which affect its chemical reactivity and, consequently, its biological interactions.

Indisputably, Cr(VI) (as chromate) readily enters cells, which was established as early as 1950 by Gray and Sterling following radioactive tracer analyses with $[^{51}CrO_4]^{2-}$ and erythrocytes (163), and has since been confirmed by gas–liquid chromatography (GLC) (164), atomic absorption spectroscopy (AAS) (165, 166), and more recently by micro-proton-induced X-ray emission (PIXE) and micro SRIXE analyses (90, 91, 167, 168). Chromate ($[^{51}CrO_4]^{2-}$) uptake by erythrocytes increases asymptotically, reaching a maximum value by 2 h (163), while for cultured human foreskin cells, an increase in uptake was observed up to 24–30 h (169). The negative charge and tetrahedral structure of $[CrO_4]^{2-}$ makes it isostructural with HPO_4^{2-} and SO_4^{2-} ions and these properties lead to cellular uptake of Cr(VI) via the anion channels (170, 171). Cellular uptake of Cr(VI) is inhibited significantly in the presence of reagents known to block anion-transport channels (171, 172), which supports the cell uptake mechanism. Mutant cell lines, selected by resistance toward Cr(VI), showed significantly reduced SO_4^{2-} and Cl^- uptake, which indicates that such resistance can be acquired through blocking general anion channels (173). However, another study found that blocking Cl^- channels did not prevent Cr(VI) induced damage to cultured epithelial cells (174); therefore, such channels (unlike for SO_4^{2-} or HPO_4^{2-} channels), are probably not relevant to Cr(VI) transport.

In contrast with Cr(VI), Cr(III) complexes are generally excluded from cells, as demonstrated by radioactive tracer studies (163, 169), GLC (164), AAS (165,

166), and X-ray imaging techniques [(175), and references cited therein] where no appreciable Cr levels were obtained above the concentrations observed in untreated cells. Indeed, a large number of studies have been performed to find exceptions to this "rule" (2, 122). It has been hypothesized that the octahedral symmetry and cationic nature of many Cr(III) complexes is the underlying factor that results in the exclusion of Cr(III) from cells (163). This suggestion has led to the study of Cr(III) uptake (167, 175) as a function of charge and lipophilicity (166, 176). There has been no evidence to suggest that the negatively charged Cr(III) complex, $[Cr(GlyGly)_2]^-$ (GlyGly = glycylglycine) (0.4 mM, 4 h) was permeable to V79 Chinese hamster lung cells above the detection limits of graphite furnace AAS (GFAAS) or micro-PIXE analyses of whole cells (167, 175). Kortenkamp et al. (166) showed some increase in cellular uptake of Cr(III) following 1-h exposure of 2×10^8 human erythrocytes to 1-mM solutions of Cr(III) complexes with lipophilic ligands, cis-$[Cr(bpy)_2Cl_2]^+$ and cis-$[Cr(phen)_2Cl_2]^+$. The results were only approximately twofold greater than those of the other Cr(III) complexes tested, namely, $[Cr(2,4\text{-pentanedione})_3]^0$, $[Cr(Gly)_3]^0$ (gly = glycine) and $[Cr(GSH)_2]^-$ [GSH = glutathione reduced form (γ-Glu-Cys-Gly)], and it was concluded that there was only a weak correlation between uptake and the charge and lipophilicity of the complexes (166). By contrast no Cr uptake above the detection limits was noted when 10^6 V79 cells were exposed for 4 h to subtoxic doses of 0.4 mM $[Cr(phen)_2(OH_2)_2]^{3+}$, then analyzed by GFAAS. Similar results were obtained for $[Cr(en)_3]^{3+}$ (en = 1,2-ethanediamine) and trans-$[Cr(salen)(OH_2)_2]^+$ (175). Neutral Cr(III) amino acid complexes, $[Cr(Aib)_3]^0$ (Aib = α-aminoisobutyrato) and $[Cr(Ala)_3]^0$ (Ala = L-alanine) were not taken up by V79 cells at detectable levels, although increases in Cr uptake were observed for a series of Cr(III) complexes with Gly oligopeptides (triGly < tetraGly < pentaGly) (34).

The majority of Cr permeability studies have focused on the common Cr oxidation states, Cr(III) and Cr(VI). The uptake of Cr(V) complexes has been studied only recently (168), despite findings that relatively long-lived Cr(V) complexes are formed transiently in nature due to the reduction of Cr(VI) to Cr(III) by fulvic and humic acids in soils (177, 178), as well as by extracellular body fluids, such as saliva or other mucosal secretions (42). A stable Cr(V) complex, $[CrO(mampa)]^-$ [mampa = 5,6-(4,5-dichlorobenzo)-3,8,11,13-tetra-oxo-2,2,9,9-tetramethyl-12,12-ethyl-1,4,7,10-tetraazacyclotridecanate(4−)], enters V79 cells at similar concentrations (0.1 mM Cr) as Cr(VI) following 4-h exposure of V79 cells (168). Other less stable Cr(V) complexes, such as $[CrO(ehba)_2]^-$ and $[Cr(O)_2(phen)_2]^+$ (179), were also significantly permeable to cells, since appreciable amounts of Cr were detected in cells following 4-h exposure to 0.1 and 0.4 mM Cr, respectively (168, 175). The Cr(V) complexes produced by PbO_2 oxidation of the Cr^{III}–pentaGly complex were slightly more permeable than the parent Cr(III) complex. Exposure of the cells to a dinuclear Cr^V–Ala complex (0.2 mM; resulting intracellular concentration 208 ± 11.8 ppb) resulted in similar Cr permeability as was observed for Cr(VI) (167.5 ± 11.8 ppb) following exposure under the same conditions. When the Cr(V) complex of triAla (1150 ± 150 ppb) was incubated with the cells, the intracellular Cr concentration increased four to five times that resulting from exposure to the same concentration of Cr(VI). This high permeability for Cr(V) complexes strongly implicates a potential damaging role for Cr(V) species formed in vivo (34). Appreciable concentrations of Cr were also observed in cells following exposure to Cr(V) complexes prepared by PbO$_2$ oxidation of the Cr^{III}–Gly complexes (4.0 mM). The difference in the permeabilities of the Cr(III) complex, cis-[Cr(phen)$_2$(OH$_2$)$_2$]$^{3+}$ (which would exist as cis-[Cr(phen)$_2$(OH)$_2$]$^+$ at the physiological pH 7.4 of the cell medium), and the analogous Cr(V) complex, cis-[Cr(O)$_2$(phen)$_2$]$^+$, is noteworthy (175). It would appear that as both complexes possess similar charge, lipophilicity and coordination geometry, the increased uptake of the Cr(V) complex must be due to its increased reactivity (175).

In all instances, Cr uptake was essentially unidirectional, which is consistent with the absence of a mechanism by which the cell can rid itself of the ultimate Cr(III) product. Consequently, it has been shown that Cr accumulates in cells, resulting in concentrations in excess of 20 times those present extracellularly (166, 168). A recent study of cellular uptake of Cr(VI) has shown that human foreskin cells (keratinocytes) can concentrate [CrO$_4$]$^{2-}$, present in the media, up to 167 times, which is important for understanding skin reactions to Cr(VI) (180). If extracellular Cr concentrations are so high that damage to the cell membrane occurs, or the cell wall has been damaged by another reagent (e.g., in necrosis) (168, 176, 181), release of Cr(III) can occur as shown by experiments with ^{51}Cr(VI).

Before potential mechanisms of the initial molecular steps in Cr(VI) carcinogenicity can be appreciated, it is necessary to understand the intracellular biotransformation of Cr, namely, its ultimate chemical and physical fate. It has been shown by EPR spectroscopy that reactions of Cr(VI) with whole cells or mitochondria produce Cr(V) intermediates (59, 182–184). Recent studies have shown that Cr(V) intermediates can be stabilized by sialoglycoproteins on the outer surfaces of cell membranes and cell walls (Section III.B.3) (12, 42). Relatively long-lived Cr(V) species are formed following Cr(VI) uptake by plants, for example, garlic roots (185) or duckweed (79). Chromium(V) species (g_{iso} = 1.9798, showing unresolved superhyperfine splitting) were observed by L-band EPR spectroscopy in live mice following intravenous injection of Cr(VI) (77). Similar Cr(V) signals (g_{iso} = 1.979) were detected in the skin of live rats after dermal application of Cr(VI) solution (78). Two types of Cr(V) species (g_{iso} = 1.979 and 1.986) were observed by X-band EPR spectroscopy in circulating blood of rats, following intravenous administration of Cr(VI) (186), or in liver and kidney homogenates from animals following Cr(VI) administration (187, 188). Shi and co-workers (77, 78), who observed Cr(V) signals with $g_{iso} = 1.979$ in whole Cr(VI) treated mice or in the skin of Cr(VI) treated rats, assigned the signals to oxo Cr(V) complexes with NAD(P)H ligands with Cr(V) coordinated to *cis*-1,2-diolato moieties of the ribose residues of NAD(P)H, based on the detection of similar EPR signals in vitro for the Cr(VI) + NAD(P)H reactions (60, 61, 77, 78). However, very similar EPR signals are observed for the reactions of Cr(VI) with biological reductants (such as GSH) in the presence of various carbohydrates that possess 1,2-diolato moieties (e.g., glucose or mannose) (18, 27, 189–193). Given very high (~ 5 mM) concentrations of glucose in animal tissues, this is a likely candidate for the ligand responsible for Cr(V) coordination in the $g_{iso} = 1.979$ species (12). The EPR signals with $g_{iso} = 1.986$ have been assigned to a Cr(V) complex possessing both S- and O-donor ligands, by comparison with the spectra of model Cr(V) complexes (186–188). Recent detailed EPR spectroscopic studies of the Cr(V) species, formed in the Cr(VI) + GSH reactions in vitro, suggest that the $g_{iso} = 1.986$ signal is most likely due to $[Cr(O)_2(GS)_2]^-$ species (194, 195). Reduced glutathione, like glucose, is present in cells in millimolar concentrations (196), and is directly involved in cellular reduction of Cr(VI) (197).

Formation of Cr(III) as an ultimate product of Cr(VI) intracellular reduction was initially determined by EPR spectroscopy, typically by treating V79 cells with Na_2CrO_4, and observing the appearance and disappearance of a narrow Cr(V) signal with the concomitant growth of a broad Cr(III) signal (44). Similar EPR signals of Cr(III) species were observed in the livers of Cr(VI) treated mice, following the disappearance of the initially formed Cr(V) intermediates (198). Determination of the oxidation state of the ultimate Cr product has also been confirmed by micro-XAS analysis of Cr(VI) and Cr(V) treated V79 cells (89, 90). Point analysis of several regions of cells using a 200-µm X-ray beam revealed that a Cr(III) product was the dominant species (>90%) (89). The importance of this investigation is that all oxidation states are detectable by XANES, unlike EPR spectroscopy where relevant oxidation states other than Cr(III) and Cr(V) are EPR silent.

One type of biotransformation of Cr(VI) yields DNA bound Cr(III) products (170). Neutron activation analysis of the Cr distribution between different cellular fractions following the uptake of $[^{50}CrO_4]^{2-}$ by yeast cells revealed the highest Cr concentrations in the DNA fraction (48%), followed by RNA (34%) and protein (6%) fractions (199). Recently, this was supported by scanning micro-SRIXE analyses (employing a 0.3 µm-diameter focused X-ray beam) of thin-sectioned chromate-treated V79 cells that showed localization of Cr in P- and Zn-rich regions (which is characteristic of the cell nucleus) (91,

200). Furthermore, it was clearly evident that there was no accumulation of Cr at the cell membrane, either externally or internally, showing that there is reasonable mobilization of Cr within the cell (91, 200).

Unlike soluble chromates, insoluble Cr(VI) particles, such as $PbCrO_4$, are ingested by cells by phagocytosis (143, 201). This uptake mechanism may be more damaging for the cells than the uptake of soluble Cr(VI) through anion channels for several reasons: (1) slow dissolution of phagocytosed particles engulfed in acidic vacuoles creates very high local concentrations of Cr(VI) (143) or more likely Cr(V) intermediates; (2) the acidic environment of the vacuoles (pH 4–5) increases the stability of potentially genotoxic Cr(V/IV) intermediates of Cr(VI) reduction (31, 202); (3) phagocytosis promotes the production of cell-damaging reactive oxygen species (ROS) (181); and (4) Pb(II) can contribute to the cytotoxicity of $PbCrO_4$ (143). Scanning electron microscopy (SEM) of mouse embryo cells treated with $PbCrO_4$ particles showed that the particles were engulfed by the cytoplasm within 24 h of exposure (203). Studies of the phagocytic uptake of insoluble $PbCrO_4$ by human lung small airway epithelial cells showed the predominance of $PbCrO_4$ particles in cytoplasmic vacuoles. Energy-dispersive X-ray analysis performed at point locations within the nucleus revealed that the Cr and Pb coexisted suggesting that the compound had not dissociated into ions (143).

The currently known pathways of cellular uptake and metabolism of Cr in different oxidation states are summarized in a modified uptake-reduction model (Fig. 1) (12), initially proposed [for soluble Cr(VI) compounds] by Wetterhahn and co-workers (170).

The ability of both prokaryotic and eukaryotic cells to uptake Cr(VI), reduce it to Cr(III) and retain the Cr(III) products also has important practical applications. In haematology and immunology, cells are often tagged with small nontoxic doses of $^{51}Cr(VI)$, which is reduced intracellularly to $^{51}Cr(III)$. Release of $^{51}Cr(III)$ by the cell is used as an indicator of membrane damage in tagged cells (e.g., by action of natural killer cells or during blood transfusion) (204, 205). In another application, increasing concern about Cr(VI) as an environmental pollutant (Section III.A) has prompted extensive research on bioremediation, which is based on uptake of Cr(VI) by bacterial or plant cells and its reduction to relatively harmless Cr(III) [(108, 115, 120, 128, 206, 207) and references cited therein].

2. Cellular Genotoxicity Assays

Short-term genotoxicity tests, including bacterial and mammalian cell assays, are widely used to assess potential carcinogenicity of chemical and physical agents, owing to the prominent role of genetic effects in the initiation and promotion of cancer; however, the predictive value of such tests is far from absolute (114). Of the carcinogens listed as class I substances, Cr(VI)

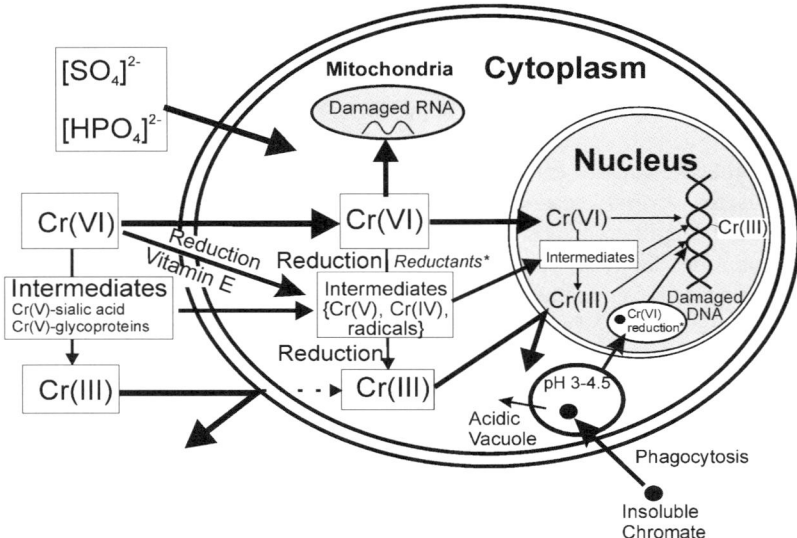

Figure 1. The uptake-reduction model for Cr induced carcinogenesis. [Adapted from (12).] The intracellular reductants include ascorbate, GSH, catecholamines, organic peroxides, NADPH, and other species.

compounds (primarily water-soluble chromates) are those that have shown the most positive responses in genotoxicity assays (2, 119). Hundreds of positive results of such assays with Cr(VI) have been reported (mostly before 1995) and comprehensive reviews of these data have been published (2, 114, 208).

Mutagenicity of Cr(VI) in prokaryotic cells is exhibited across a number of bacterial species (such as *Escherichia coli*, *Bacillus subtilis*, or *Salmonella typhimurium*) and strains, and includes frameshift mutations and base pair substitutions both at GC and AT base pairs (118, 119, 208). Evidence for genotoxicity in eukaryotic cells (predominantly yeasts and mammalian cell cultures) includes gene mutations in several loci, chromosomal aberrations (detected by breaks and exchanges in chromosomes), sister chromatid exchanges, and increased incidence of micronuclei (208). There is a complex dependence of the response of mutagenicity assays on solubility and concentration of Cr(VI) compounds, as well as on the genetic marker and cell lines employed (118, 119). Patierno and co-workers (181) emphasized the importance of using cell lines that are relevant to human respiratory cancers in mechanistic studies of Cr(VI) induced genotoxicity. These authors recently used a combination of two types of cultured human lung cells; small airway epithelial cells, and normal lung fibroblasts, to take into account the likely interactions between various types of cells during the initiation and promotion of cancer in vivo (181). Numerous studies have shown that the observed spectrum of gene mutations and chromosome damage induced by Cr(VI) in vivo [e.g., following intraperitoneal injection of mice with Cr(VI)]; was similar to that observed in cellular assays [(114, 208) and references cited therein].

The mutagenic and genotoxic properties of Cr(V) complexes have been demonstrated by assays with complexes of ehba, mampa, Ala, trisAla, and phen ligands (34, 37, 168, 175, 209, 210). These studies have emphasized the importance of determining the stability of the Cr complexes in the associated cell media using EPR and UV–vis analyses. The Cr(V) complex, [CrO(mampa)]$^-$, is substitutionally inert [no decay of the Cr(V) EPR signal] in the medium employed for the bacterial mutagenicity assay, as well as the cell medium (Eagles MEM, where MEM = minimum essential medium) employed for the genotoxicity assay. The consequent, high mutagenic response of [CrO(mampa)]$^-$ in *S. typhimurium* TA100 and the increased incidence of micronuclei in V79 cells, unambiguously highlight the genotoxic properties of Cr(V) (168, 210). In contrast, [CrO(ehba)$_2$]$^-$ undergoes two types of reactions: disproportionation to Cr(VI) and Cr(III) species, and ligand-exchange reactions with glucose in both the media employed in mutagenicity assays and cell growth media (27, 168). A study of the permeability of molar equivalent doses of [CrO$_4$]$^{2-}$ and [CrO(ehba)$_2$]$^-$, and the associated genotoxicity in V79 cells, showed that the genotoxicity of the Cr(V) complex was somewhat greater than that of Cr(VI). Thus the genotoxicity of the Cr(V) complex was significantly higher than that which could be explained from the highest concentration of Cr(VI) that could be generated from Cr(V) disproportionation (168). Genotoxicities of Cr(V) complexes, generated *in situ* by PbO$_2$ oxidation of Cr(III) complexes, *cis*-[Cr(phen)$_2$(OH$_2$)$_2$]$^{3+}$ and *trans*-[Cr(salen)(OH$_2$)$_2$]$^+$, have also been studied in both bacterial and mammalian cell assays (37, 175). In vitro micronucleus assays showed that a Cr^V–phen complex was significantly more genotoxic than its Cr(III) analogue, due to both its higher stability and higher cellular uptake than the Cr^V–salen complex, which had a similar genotoxicity as its Cr^{III}–salen analogue. The Cr^V–salen complex possessed a very short lifetime (seconds) in the cell medium and showed low levels of genotoxicity (37, 175).

Most Cr(III) compounds are considered nongenotoxic (114, 208). Exceptions include redox-active Cr(III) complexes with imine ligands, such as *cis*-[Cr(bpy)$_2$Cl$_2$]$^+$, *cis*-[Cr(phen)$_2$(OH$_2$)$_2$]$^{3+}$, or *trans*-[Cr(salen)(OH$_2$)$_2$]$^+$, which are genotoxic in both bacterial and mammalian cell assays (37, 175, 211, 212). The levels of genotoxicity of the Cr(III) complexes in V79 cells are relatively low compared with those for Cr(VI) or Cr(V) complexes, as much higher Cr(III) concentrations are required to produce similar genotoxic responses. The extent of the Cr(III) genotoxicity (which may be due to intracellular oxidation) becomes very significant when the intracellular Cr concentrations are considered (i.e., the Cr(III) complexes do not enter cells as readily as Cr(V) and Cr(VI) (175)).

Over a decade ago, Sugden et al. (211) showed that the Cr(III) complex, $[Cr(bpy)_2Cl_2]^+$, was mutagenic in several *S. typhimurium* strains only under aerobic conditions. Chromium(VI) induced mutagenicity was O_2 dependent in *S. typhimurium* strain TA102, which detects oxidative mutagens following base pair substitution of AT to GC (211). These findings triggered extensive research activity into the possible roles of O_2 and ROS (in particular •OH radicals) in the mechanisms of Cr induced genotoxicity (Sections III.B.5–III.B.7, III.C.1, and III.C.2). As part of this research, mutational spectra in the hprt (hypoxanthine phosphoribosyl transferase) locus of human lymphoblasts, caused by Cr(VI) treatment, were compared with those caused by known sources of ROS, such as H_2O_2 or X-rays (213). There was a distinct difference between mutations induced by Cr(VI) and the other two reagents; for example, Cr(VI) produced G:C → A:T at hotspot 243, while H_2O_2 produced G:C → C:G at the same site; and at hotspot 247, Cr(VI) generated A:T → T:A, while X-rays caused deletion of this A:T base pair (119, 213). This finding contradicts the suggestion (60, 61) that •OH is the ultimate species responsible for Cr(VI) or Cr(III) induced genotoxicity. Furthermore, •OH is unlikely to be the DNA damaging species unless it is produced in very close vicinity to the DNA molecule, as •OH will travel no further than 3 nm from its site of generation in the cell, due to its extremely high reactivity (214).

Significant insight can be gained into the mechanisms of Cr genotoxicity based on its mutagenic character in genetically altered bacteria. The mutational spectrum obtained for a number of Cr complexes with various strains of *S. typhimurium* show that all of the Cr(III), Cr(V), and Cr(VI) complexes exhibit greatest mutagenicity for TA102 [Fig. 2 (37)]. As described by Bennicelli et al., TA102 is the most sensitive of nine strains of *S. typhimurium* to Cr(VI) induced mutations (215), whereby the activity increased in the following manner: TA1535 < TA1537 < TA1538 < TA98 < TA1978 < TA92 < TA97a < TA100 < TA102. The mutagenic profiles of a number of Cr complexes are shown in Figure 2. Differences between classes of compounds were observed; for example, the profile for $[CrO_4]^{2-}$ and $[CrO(ehba)_2]^-$ increased in the order TA98 < TA100 < TA97a < TA102. In comparison, the Cr^{III}–salen complex exhibited lowest mutagenicity for TA97a followed by TA100, TA98, and TA102, respectively, and the phen complexes exhibited lowest mutagenicity for TA98 followed by TA97a, TA100, and TA102 (37). The dominant mutagenic activity exhibited by the bacterial strain sensitive to oxidative mutations, TA102, implies that oxidative mechanisms play a role in Cr induced mutagenicity, although the differences in the profiles suggest that the mutagenic properties differ between the complexes. This further suggests that Cr(VI) induced mutagenesis is not based solely on a simple indiscriminant mechanism as might

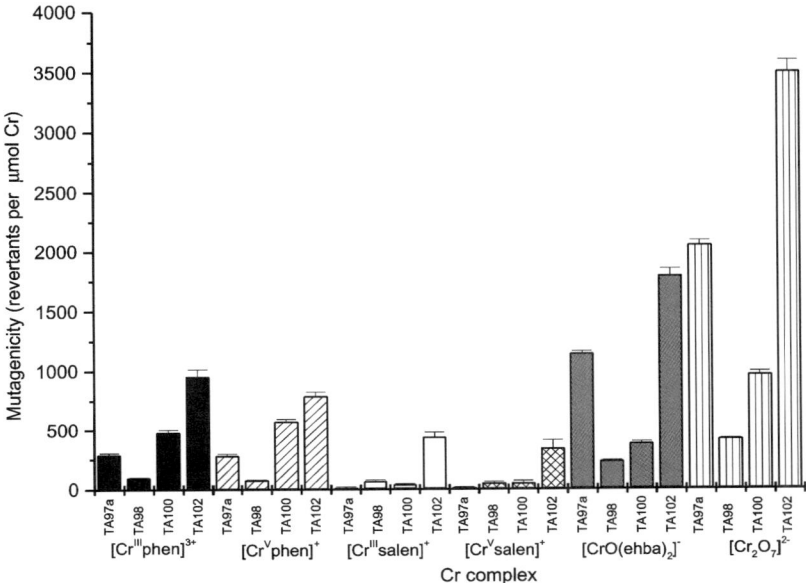

Figure 2. Mutagenicity rates from the linear regions of the dose-response curves of different strains of *S. typhimurium* after exposure to various Cr compounds. [Reproduced with permission from (37), copyright © CSIRO, Australia.]

occur by •OH. Therefore, reactions of Cr species with biological macromolecules are likely to be important in Cr(VI) induced genotoxicity (37).

3. Interactions with Cell Membranes and Cell Walls

There have been few studies that have examined the reaction between Cr(VI) and the isolated or whole components of cell membranes and/or cell walls. A better understanding of the interactions between Cr(VI) and prokaryotic-cell-wall components has been gained by studying the ability of certain bacteria to become tolerant to Cr(VI) uptake (120). Bacterial strains that were tolerant to Cr(VI) showed changes in the composition of cell walls (216) and decreased SO_4^{2-} uptake (which is probably linked to a decreased $[CrO_4]^{2-}$ uptake, Section III.B.1) (173). Currently, an induced tolerance to Cr(VI) and other heavy metals by some bacteria is thought to be due to a coexpression of efflux proteins (217).

A recent study has examined the reaction between Cr(VI) and plasma cell membranes in parental (ergosterol-producing, *33 erg+*) and mutant (ergosterol-free, *erg-2*) strains of an eukaryotic yeast, *C. albicans* (218). In the absence of Cr(VI), the plasma membrane from the *erg-2* strain showed greater rigidity than that of the parental counterpart (*33 erg*$^+$), which was reversed in the presence of Cr(VI) (0.6 mM), whereby the membrane of the *erg-2* strain showed increased fluidity, relative to *33 erg*$^+$ (218). This indicates that the regulation of the fluidity of the plasma cell membrane in yeast is modulated by the concentrations of membrane-bound polyunsaturated fatty acids, and is significantly affected by Cr(VI). The interaction of Cr(VI) with the membrane components also gave rise to an EPR detectable Cr(V) signal ($g = 1.9554$), although the nature of the Cr(V) stabilizing ligand(s) was not described. The cell media used in this and other studies contained ligands, such as glucose, which may stabilize Cr(V) if present in addition to Cr(VI) reductants (such as would be contained in yeast extract, which was also used in this study) (12, 27). These reaction matrix effects need to be considered in the interpretation of results. A decrease of fluidity in rat microsomal membranes has been also observed during exposure to the $CrCl_3 + H_2O_2$ system (219).

Recent work has shown that Cr(VI) affects the expression of the urokinase-type plasminogen activator (uPA) in human type II pneumocytes (220). This study was aimed at an improved understanding of the mechanisms of Cr induced pulmonary fibrosis, which, in addition to respiratory cancers, is another occupational disease resulting from Cr exposure (117, 221). In cells that have suffered damage or an inflammatory event, healing takes place during the deposition of cellular fibrin to stabilize the extracellular matrix. Plasminogen activators ensure that optimal fibrin levels are not exceeded, since damage to this finely tuned process can lead to excess fibrin deposition, which may lead to pulmonary fibrosis. This study found that, in the presence of Cr(VI) (where $[Cr(VI)] > 1$ μM), the concentration and specific activity of uPA in alveolar type II cells significantly decreased compared with control cells (220). This decrease in uPA was coupled with an increase in the concentration of the receptor, uPAR (which upon binding uPA, activates fibrinolysis). While expression of uPA decreased in the presence of Cr(VI), messenger RNA (mRNA) levels were not affected, which suggests that the decreased levels of uPA resulted from translational errors (220). Chromium(VI) increases the stability of the mRNA of uPAR, which leads to increased expression of uPAR. These results suggest that the differential effects of Cr(VI) upon the expression of uPA and uPAR may be important in the mechanism of Cr(VI) induced pulmonary fibrosis.

Treatment of experimental animals or cell cultures with Cr(VI) is known to cause peroxidation of polyunsaturated lipids, which are abundant in cell membranes; this effect is thought to result in membrane damage and to play an important role in tissue injuries induced by Cr(VI) (56, 57, 155, 222–225). Increased levels of malondialdehyde, a product of lipid peroxidation, have been detected in blood and urine of Cr(VI) exposed workers and have been proposed as one of the possible biomarkers of Cr(VI) exposure (226, 227).

In addition to polyunsaturated fatty acids, cell walls and membranes also have significant concentrations of glycoproteins, with the glycosylated moieties predominantly comprising of glucose, galactose, and fucose residues. In many cases, the nonreducing ends of the glycoproteins are terminated by sialic (N-acetylneuraminic) acid groups, which are linked to the penultimate sugar residue by the 2'-hydroxyl group. Sialoglycoproteins are ubiquitous in animals and humans and occur at the cell surface of the majority of cell classes and play crucial roles in cell regulation (228). Sialic acid has special importance in glycoproteins, since it is the only sugar-derived residue that has a negative charge and, therefore, plays an important role in modulating cellular coagulation events (228).

Reactions of free sialic acid or sialoglycoproteins of human saliva with Cr(VI) lead to the formation of stable Cr(V) species (Section III.C.1). The 6' glycerol-like tail is exposed to the extracellular environment in cell-bound sialic acid and is most likely to be responsible for Cr(V) stabilization by sialoglycoproteins (42). The Cr^V–saliva species are formed in the absence of an exogenous reductant and are stable for 2 days at ambient conditions (pH 6–7) (42). Due to the abundance of sialoglycoproteins in the respiratory tract (e.g., saliva, mucins), complexation between this ligand type and Cr(V) has major implications with respect to the occupational uptake of Cr(VI) particles via inhalation. Electroplaters, for example, who are exposed to mists of chromic acid, suffer from nasal septum lesions (229). Chromium(VI) causes significant DNA damage in human nasal or gastric mucosa cells, which are rich with glycoproteins (230, 231). Also, human orosomucoid (α_1-acid glycoprotein) is cleaved by a model Cr(V) complex (232), or by the Cr^{III} + oxidant systems, which are capable of producing Cr(V) intermediates (233). Scanning electron microscopy has been used to study the effect of Cr(VI) on the skin of the catfish, *Saccobranchus fossilis* (234). When kept in test water containing 0.1 mM Cr(VI) for 7 days [the Cr(VI) solution was changed every 24 h], the mucous goblet cells of the fish skin were changed from normal flat, hexagonal, or polygonal disks with microvilli-like structures to cylindrical shapes with needle-shaped forms at their tips (234). The epidermal mucus of fish contain glycoproteins for structural and protective purposes and also contain sialoglycoproteins, such as protein-bound deaminated sialic acid derivative, 3-deoxy-D-glycero-D-galacto-2-nonulosonic acid (235). Human skin in its normal and diseased (e.g., basal cell carcinoma) states also contains significant amounts of sialic acids (236), which may be of significance with respect to dermal contact with chromates. Further in vivo experiments using sialic acid bound biological matrices are currently being conducted to assess the nature of the Cr(VI) induced damage to cell-bound sialic acid residues. Since sialic acid residues are involved in the control of fundamental cell processes (228), any oxidative or alternative damage to these residues may have deleterious consequences on cell behavior.

4. Damage of DNA and Transcription Factors: Influence on Gene Expression

a. Oxidative DNA Damage. Three types of oxidative DNA damage detected in cell cultures and living animals treated with Cr(VI) are single-strand breaks, alkali-labile sites, and oxidized DNA bases [(119, 153) and references cited therein]; the latter (primarily 8-hydroxydeoxyguanidine, 8-OHdG) are most often detected in red blood cells (237). A relatively old method, alkaline filter elution (238) is still used for detection of both single-strand breaks and alkali-labile sites in DNA of Cr(VI) treated cells (239, 240). The alkaline single-cell gel electrophoresis (comet assay) (241) has been used in several recent studies (188, 230, 242, 243). However, the use of the comet assay in the assessment of Cr induced genotoxicity has to be considered with caution, as a recent study (242) showed a similar extent of oxidative DNA damage caused by K_2CrO_7 or $CrCl_3$ when incubated with intact human lymphocytes; a result that totally disagrees with the present knowledge about the effects of different Cr compounds on cells (12, 119, 208). Another study (244) suggested that modifications of the comet assay method (such as a decrease in pH value) are required for the studies of Cr induced genotoxicity.

b. Chromium(III)–DNA Adducts. Cross-linking of proteins to DNA was among the first described genotoxic effects caused by treatment of intact cells with Cr(VI) [(238, 245), and references therein]. More recently, the techniques for detection of such lesions have been significantly improved (244, 246, 247). Treatment of the isolated DNA–protein adducts with ethylenediaminetetraacetic acid (EDTA) has been proposed as a method to distinguish between DNA–Cr^{III}–protein complexes and those involving direct DNA–protein binding (248). However, other studies (249–251) showed that DNA–Cr^{III}–protein adducts are likely to be nonsensitive to EDTA treatment. Determination of the types of proteins, cross-linked to DNA by Cr(III) following the treatment of cells with Cr(VI), is proposed as a useful tool for the studies of protein–DNA interactions in the cells, because stable DNA–Cr^{III}–protein complexes "conserve" the structures of transient DNA–protein adducts (252). The above example shows, however, that much is still to be learned about the composition and reactivity of model nucleotide–Cr^{III}–amino acid complexes before reliable methods for isolation and characterization of DNA–Cr^{III}–protein complexes from cells can be developed (253). Apart from large proteins, free amino acids (mainly Cys and His), and GSH are also extensively crosslinked to DNA by Cr(III) following the exposure of intact mammalian cells to Cr(VI) (254, 255); such DNA–Cr^{III}–amino acid and GSH lesions are mutagenic in human cell cultures (251, 256).

DNA–Cr^{III}–DNA cross-links in cells treated with Cr(VI) were discovered much later than DNA–Cr^{III}–protein lesions, due to the higher lability of the former cross-links to EDTA (treatment with EDTA was used in earlier experimental procedures for isolation of DNA adducts) (249). Patierno and co-workers (181, 257) suggested that DNA–Cr^{III}–DNA cross-links are the most damaging among Cr(VI) induced DNA lesions, leading to DNA polymerase arrest and S-phase blockage of the cell cycle. They also speculated that DNA–protein cross-links are likely to be responsible for disruption of DNA interactions with transcription factors, but the formation of DNA–amino acid (or GSH) adducts can play a protective role by preventing the involvement of Cr(III) in more damaging DNA–DNA or DNA–protein lesions (181, 258). The latter suggestion contradicts the results of Zhitkovich and co-workers (251, 256) on the mutagenicity of DNA–Cr^{III}–amino acid adducts.

Increases in intracellular concentrations of a major cellular reductant, GSH, led to increases in the levels of DNA–protein cross-links (259), suggesting that these lesions are formed during the reduction of Cr(VI) in the vicinity of DNA and proteins, rather than by reactions of pre-formed Cr(III) complexes with the macromolecules (260). Nevertheless, many researchers used the reactions of nonbiologically relevant Cr(III) salts (such as CrCl$_3$) with DNA in vitro (Section III.C.3) to isolate Cr^{III}–DNA adducts, which were then used to study the biological consequences of Cr^{III}–DNA binding (see below). Since structures and reactivities of such Cr^{III}–DNA complexes are likely to differ from those of the adducts formed during the intracellular reduction of Cr(VI) (251, 261), caution is required in interpreting the results of these studies.

c. **DNA Damage as a Biomarker of Exposure to Cr(VI).** Significant recent attention has been paid to development of reliable methods for early detection of genetic damage in humans due to the occupational or environmental exposure to Cr(VI) [(262), and references cited therein]. The use of DNA–protein adducts as biomarkers of Cr(VI) exposure has been studied most extensively, and the method has been validated in several laboratories [(247) and references cited therein]. Among other types of Cr(VI) induced DNA or chromosome damage, the frequencies of both single-strand breaks and sister chromatid exchanges in human lymphocytes correlated with occupational exposure to Cr(VI) (240, 263), while the levels of 8-OHdG in blood cells could not be used as a reliable marker of such exposure (264, 265).

d. **Changes in DNA-Related Enzyme Activity.** It was found almost 20 years ago that the use of CrCl$_3$ treated DNA as a template for in vitro RNA synthesis (catalyzed by RNA polymerase) significantly enhanced the rate of RNA formation (266), but no follow-up studies of this interesting effect were reported. Formation of Cr^{III}–DNA adducts during the reduction of Cr(VI) with GSH or model thiols altered the patterns of DNA cleavage by restriction endonucleases (267). Binding of Cr(III) (as CrCl$_3$) to oligonucleotide primers used in DNA synthesis stimulated the human polymerase β activity and DNA binding affinity, but decreased the fidelity of the enzyme, which led to the formation of potentially mutagenic DNA lesions (268). On the contrary, Cr^{III}–DNA adducts (formed by treatment of plasmid DNA with $CrCl_3$ or with the Cr^{VI} + ascorbate system) caused arrest of DNA replication by mammalian polymerases α and β (269). Notably, some Cr(III) nucleotide complexes are used as inhibitors of polymerase β for structural studies (270, 271).

e. **Influence on Transcription Factors and Inducible Gene Expression.** It has been known for more than a decade that treatment of cultured cells with genotoxic chemicals, including Cr(VI) and As(III) compounds, preferentially alters expression of several inducible genes but has little or no effect on constitutive gene expression (158, 272). One of the potentially adverse consequences of this effect is the inhibition of cellular responses to other toxic substances. Thus, treatment of mammalian cells with 2,3,7,8-tetrachlorodi-benzo-p-dioxin in the absence of Cr(VI) leads to activation of aromatic hydrocarbon receptors, triggering gene expression for several enzymes, which catalyse oxidative degradation of the toxin (273). However, levels of expression of these protective genes were significantly lower in the presence of Cr(VI) (273), which led to synergistic toxicity of Cr(VI) and dioxin. Notably, an antagonistic effect of Cr(VI) on the action of another toxin has also been reported: thus, low levels of Cr(VI) reduced the mutagenicity of benzo[a]pyrene diolepoxide (BPDE) in normal human fibroblasts (274). This effect may be related either to direct destruction of BPDE by reactive intermediates of Cr(VI) intracellular reduction, or to activation of cell protection against BPDE by low doses of Cr(VI) (274).

Reasons for the selective action of Cr(VI) on inducible gene expression are not yet clear, but this effect is likely to be related to disruptions in the functions of transcription factors (proteins that bind to DNA and regulate its transcription) (158, 275). Treatment of cell cultures with subtoxic doses of Cr(VI) significantly altered the levels of DNA binding for several transcription factors, but patterns of these changes were dependent on the Cr(VI) concentration and the cell line (275). Recent experimental evidence points to a possibility of direct oxidative damage of transcription factors during the intracellular reduction of Cr(VI) (Section III.C.1) (95, 276–278). Direct interactions with transcription factors are considered among the most general mechanisms of carcinogenicity induced by transition metal ions (279, 280). However, a possibility of indirect Cr(VI) effects on transcription factors (through oxidative stress and alterations in cell signaling pathways, Sections III.B.5–III.B.7) cannot be discounted (275, 281). The ability of Cr(VI) to damage transcriptional processes is considered to be of particular concern, since it may lead to the expression of aberrant proteins from undamaged reading frames (275).

f. **Mutations in Shuttle Vectors.** Mutagenicity of Cr^{III}–DNA adducts can be studied by generation of such lesions in vitro in plasmid or bacteriophage DNA, followed by transfection of the treated DNA into cells, its replication, isolation from the cells, and analysis of the induced mutations. An early study (282) showed an increase in the mutation frequency in $CrCl_3$ treated bacteriophage DNA following its replication in bacterial cells. More recently, plasmid DNA was modified by treatment with $CrCl_3$ (283) or with the Cr^{VI} + GSH system (284, 285) and replicated in mammalian cells. Liu et al. (284, 285) found that the mutational spectrum induced by treatment of plasmid DNA with the Cr^{VI} + GSH system was similar to that induced by oxidative DNA damage with H_2O_2 or ionizing radiation. This finding was used in support of the hypothesis that generation of ROS (and particularly of •OH radicals) is the primary cause of Cr(VI) induced mutagenicity (60, 61). On the contrary, Zhitkovich et al. (250) achieved selective formation of Cr^{III}–DNA adducts in the absence of oxidative damage by treatment of plasmid DNA with the Cr^{VI} + Cys system, and showed the mutagenic potential of such adducts by replication of the treated DNA in human fibroblasts (251). This work provided the first direct evidence that Cr^{III}–DNA adducts play a dominant role in the mutagenicity caused by the reaction of Cr(VI) with a biological reducing agent (251).

g. **Influence on Cancer-Related Genes.** Mutations in tumor suppressor gene *p53* (coding for a p53 transcription factor, or tumor suppressor protein) have been related to a number of human malignancies, including lung cancer (286). A comparative study of *p53* mutations in tumor cells from patients with Cr(VI) related and unrelated lung cancers has been performed (287). Point mutations in the *p53* gene were found only in 20% of the cases of Cr(VI) related cancers, but such mutations were even less frequent in the Cr(VI) unrelated cancer group; furthermore, the patterns of mutations in these two groups were different (287). These results indicate that *p53* mutations can be caused by Cr(VI) exposure, but they are unlikely to be a major cause of Cr(VI) induced lung cancers (287). The absence of a significant correlation between point mutations in *ras* oncogenes and Cr(VI) induced lung cancer has also been shown (288).

The ability of Cr(VI) to inhibit the expression of several inducible genes (158, 272) raised a question as to whether Cr(VI) inhibits the expression of the *p53* gene, thus triggering neoplastic transformation of the cells (119). Recently, it was found that low levels of Cr(VI) ($< 2 \mu M$) inhibited the expression of *p53*, as well as another tumor suppressor gene, *WAF1*, in human embryonic lung cells, while higher Cr(VI) concentrations promoted the expression of both genes (289). Another study found no significant differences in *p53* expression in Cr(VI) related and unrelated lung cancers; instead, a correlation was found between Cr(VI) exposure and expression of oncogene cyclin D1 in lung cancer tissues (136). Possible roles of the p53 protein (rather than the gene) in Cr(VI) induced carcinogenicity are discussed in Section III.B.6.

A large-scale gene expression analysis has been recently conducted, using Cr(VI) exposed (0.6 mM Cr for 2 h) human epithelial lung cells (290). A complex pattern of Cr(VI) influence to cell functions has been revealed: thus, from 2400 studied genes, the expression of 150 genes was up-regulated and that of 70 genes was down-regulated by Cr(VI) treatment (290). Further studies in this direction (including concentration and time dependencies) can provide important clues for the understanding of biochemical mechanisms of Cr(VI) toxicity.

5. Oxidative Stress and Antioxidants

Reduction of Cr(VI) following its uptake by cells depletes the pool of intracellular reductants, leading to a shift in the redox balance of the cell, known as oxidative stress (181). Changes in the levels of nonenzymatic cellular reductants following Cr(VI) exposure have been used as indicators of oxidative stress. For example, levels of intracellular GSH decreased about sixfold upon treatment of isolated rat hepatocytes with Cr(VI) (0.5 mM for 3 h) (291). Levels of α-tocopherols and sulfhydryl groups in the brain tissues of Cr(VI) treated mice decreased by ∼30%, but the levels of ascorbate did not change significantly (155). Several studies used direct measurements of oxidation products derived from cellular macromolecules to quantify Cr(VI) induced oxidative stress; this includes determinations of malondialdehyde as a product of lipid oxidation (226, 292), or met-hemoglobin [Fe(III) form] as a product of hemoglobin [Fe(II) form] oxidation (293, 294). Another group of experiments involved the reactions of Cr(VI) treated biological materials with redox-sensitive organic compounds such as fluorescent dyes; whereby detection of the oxidized form of the dye is interpreted as a sign of formation of ROS, such as H_2O_2 (61, 295). However, recent studies of the reactions of fluorescent dyes with a model Cr(V) complex, $[Cr^VO(ehba)_2]^-$ (184), or with the Cr(VI) + Cys system (250) showed that reactive intermediates of Cr(VI) reductions, including Cr(V/IV) species, can mimic the action of ROS by directly oxidizing the dyes. Therefore, the formation of oxidized forms of redox-sensitive dyes can indicate the presence of excess oxidant species in cells and tissues, but it cannot distinguish between O_2 derived and Cr derived oxidants (184). The same applies to the use of oxidation products of spin traps as indicators of oxidative stress (Sections II.A.4 and II.A.5) (56, 57, 68). Enhanced O_2 consumption by human lung epithelial cells in the presence of Cr(VI) provides a direct evidence for involvement of O_2 in the mechanisms of Cr(VI) induced cell damage (296).

Changes in the activities of redox enzymes following the exposure of cells to Cr(VI) were studied among the possible indicators of oxidative stress, leading to controversial results. A recent study (155) showed increased activities of catalase and superoxide dismutase (SOD) but unchanged activity of GSH peroxidase, while another (292) found increased activity of GSH peroxidase but unchanged activity of SOD in tissues of Cr(VI) treated animals. In a detailed study by Dubrovskaya and Wetterhahn (297), gene expression for a number of redox enzymes was measured (at mRNA level) in normal and adenocarcinoma human lung cells treated with subtoxic doses of Cr(VI) (5–200 µM). The expression of heme oxygenase, known as a general stress-inducible gene, was increased only in normal lung cells. The expression of several ROS inducible genes, including catalase, SOD, GSH reductase, and GSH peroxidase, was insensitive to Cr(VI) treatment under the experimental conditions, indicating that the formation of ROS is unlikely to be the primary cause of Cr(VI) induced genotoxicity (297, 298).

Many researchers attempted to modulate toxic effects of Cr(VI) by changing the concentrations of cellular antioxidants; however, the results of such studies are often confusing [(119, 181) and references cited therein]. This is likely to be related to the dual role of cellular reductants such as GSH, ascorbate, or catecholamines, because reactions of relatively low concentrations of these reductants with Cr(VI) maximize the formation of reactive intermediates, while higher reductant concentrations suppress the formation of such intermediates (21, 25, 53, 299, 300). Consequently, depletion of intracellular GSH levels (by treatment with buthionine sulfoximine or dimethyl maleate) could either increase (301) or decrease (197, 291) Cr(VI) induced cytotoxicity, depending on experimental conditions. Pretreatment of rat osteoblasts with ascorbate partially prevented Cr(VI) induced cytotoxicity, while α-tocopherol had no effect (301). Increased cellular levels of either ascorbate or α-tocopherol did not affect cytotoxicity of Cr(VI) in human lung fibroblasts (302) or in V79 cells (303). Two independent studies found no significant effects of cellular ascorbate levels on the formation of Cr^{III}–DNA adducts (187, 302). Some markers of Cr(VI) exposure (such as hemoglobin oxidation in human erythrocytes) increased when the cells were pretreated with ascorbate (293). On the other hand, extracellular reduction of Cr(VI) with ascorbate effectively prevents Cr(VI) induced cytotoxicity and genotoxicity by converting Cr(VI) to much less bioavailable Cr(III) (187, 298, 303). There is also a study on humans, which suggests that their levels of ascorbate (based on diet) influences the level of Cr in urine versus blood, that is, the retention and excretion of Cr (304).

Susa et al. (305, 306) used cotreatment of cells with Cr(VI) and an antioxidant (such as melatonin or desferoxamin) to demonstrate a protective effect of these substances (thought to act as free radical scavengers) against Cr(VI) induced cytotoxicity and genotoxicity in rat hepatocytes. Apparently, extracellular reduction of Cr(VI) was not the primary cause of the protective effect of these substances, as Cr uptake by cells was not affected (305, 306).

The lack of information on the chemical nature of Cr(VI) interactions with these antioxidants, however, makes interpretation of the results difficult. In a similar study, an antihypertensive drug, todralazine, prevents Cr(VI) induced mutagenicity in bacterial cells by extracellular reduction of Cr(VI) to Cr(III) (307).

Due to the ability of most biological antioxidants to act as prooxidants at low concentrations (308), a beneficial role of dietary antioxidants as protective measures against Cr(VI) induced carcinogenesis is being debated; thus, recent data suggest that N-acetylcysteine may offer better protection than ascorbate (309, 310).

6. Cell Signaling and Apoptosis

a. Modulation of Activity of Transcription Factors. Treatment of cells with Cr(VI) leads to changes in functions of several groups of regulatory proteins. Recent attention has focused upon a role for NFκB (nuclear factor κB), a redox-sensitive transcription factor, in Cr(VI) induced genotoxicity (311, 312). Unlike other transcription factors, which are expressed in response to certain stimuli, NFκB is stored in the cytosol in an inactive form. In response to oxidative stress, NFκB is activated and rapidly moved to the cell nucleus (translocated) where it binds to DNA and promotes gene expression for autoimmune and inflammatory factors (312). Consequently, activation of NFκB and its binding to DNA is a sign of a protective reaction by the cell; however, this causes the increased production of ROS, which, in turn, amplifies the NFκB activation and can lead to cell death (312). Alternatively, oxidative damage of NFκB and other transcription factors can cause disturbances in cell signaling and gene expression systems and may potentially lead to cancer (281).

The described dualism (i.e., either activation or damage) in the action of oxidants towards NFκB is reflected in the current controversy over the role of this protein in Cr(VI) induced carcinogenesis. Shi and co-workers (313–316) observed the formation of the active form of NFκB in response to treatment of mammalian cells with Cr(VI). The authors attributed this activation to the increased production of ROS induced by Cr(VI). By contrast, Shumilla et al. (317) observed the inhibition of transcriptional activity of NFκB as a result of treatment of human lung carcinoma cells with nontoxic levels of Cr(VI). In vitro studies indicated that this effect is likely to be caused by Cr(III) binding to the active sites of NFκB during the intracellular reduction of Cr(VI) (318).

Another transcription factor, p53, has been extensively studied in relation to its roles in carcinogenicity, apoptosis, and cell responses to toxic substances (286, 311, 319, 320). Several lines of evidence implicate a significant role of p53 in cellular responses to Cr(VI). Thus, increased serum levels of p53 group proteins have been found in 90% of the studied Cr(VI) exposed workers (321). Increased susceptibility of p53 deficient mice to a number of xenobiotics, including Cr(VI), has been shown (322, 323). Finally, several independent studies showed significant increases in p53 levels in cell cultures exposed to Cr(VI) (302, 324). Recent studies suggested that the activation of p53 in response to treatment of cells with Cr(VI) occurs at the protein level, rather than at the transcriptional level (320).

b. Promotion of Apoptosis. Work conducted over the last decade has shown that Cr(VI) induces apoptosis in mammalian cells (325–333). Model Cr(V) complexes, $[Cr^VO(ehba)_2]^-$ and $[Cr^VO(hmba)_2]^-$, induce apoptosis in human peripheral blood lymphocytes (334). Some Cr(III) complexes with Schiff base ligands also induced apoptosis (327); this is likely to be related to the ability of such complexes to form highly mutagenic Cr(V) species during oxidation under biologically relevant conditions (Sections III.B.2 and III.C.2) (175). The overexpression of p53, an apoptotic initiator, has been used as a marker for Cr(VI) induced apoptosis (181, 326, 328, 335). Extensive mechanistic studies of Cr(VI) induced apoptosis have been performed by Patierno and co-workers [(181, 335, 336) and references cited therein]. Recent work by this group studied Cr(VI) treated cells that were either pre- or cotreated with cyclosporin A (CsA), which inhibits the release of mitochondrial cytochrome c into the cytosol, in order to better understand the mechanisms of Cr(VI) induced apoptosis (337). The study showed that in the absence of CsA, cells treated with Cr(VI) (3 or 6 μM) underwent apoptosis and were unable to regain replicative competency (clonogenicity). However, when co- or pretreated with CsA, the Cr(VI) treated cells showed lower levels of internucleosomal DNA fragmentation (also, an indicator of apoptosis) and also showed significant clonogenicity (337).

The effects of Cr(VI) on human cell lines have been tested using human lung fibroblasts (HLF) (302, 335). The growth of Cr treated (9 μM) HFL cells was arrested for 3 days and microscopic analysis revealed cell apoptosis shortly thereafter (observed as cell shrinkage and chromatin condensation) (302). Pre- or cotreatment of Cr(VI) treated HLF cells with either vitamin C or E did not affect levels of apoptosis or clonogenicity and also did not affect the levels of Cr–DNA adducts that were formed under the same conditions. While vitamins C or E do not affect the clonogenicity of Cr(VI) treated cells, these vitamins do decrease the level of clastogenicity (chromosome damage) exhibited by Cr(VI) treated cells (201). This finding indicates that Cr(VI) treated cells in the presence of vitamins may exhibit clonogenic and/or clastogenic behaviors and that it is most likely, therefore, that the mechanisms of these cellular events are differentially tuned by Cr(VI). Normal human lung epithelial cells that were treated with lead chromate, phagocytose lead particles and form intracellular lead-inclusion bodies (143). Both cell apoptosis and the formation of Cr–DNA adducts was dependent on Cr(VI) concentrations (143).

Whether apoptosis processes initiate or inhibit Cr(VI) induced carcinogenesis has been raised as a matter of debate by Patierno and co-workers (181, 338). Prevention of cell death by CsA or any other mitochondrial permeability transition inhibitor, for example, may promote the survival of Cr(VI) damaged cells that become resistant to apoptosis and may lead to replication of apoptotic-resistant cells that have additionally compromised cell machinery (181, 337, 338).

c. **Activation of Protein Kinases.** Mitogen-activated protein (MAP) kinases mediate cellular response to a variety of stimuli, including oxidative stress, through phosphorylation of protein Ser or Thr residues (339). Treatment of mammalian cells with Cr(VI) (micromolar levels) activates several types of MAP kinases, as well as protein kinase C (295, 340–343). Such kinase activation has been related to Cr(VI) induced production of ROS, based on the observed effects of antioxidants and radical scavengers (295, 340, 341); however, as indicated in Section III.B.5, such effects can be similar for ROS and Cr(V/IV) species. Activation of MAP kinases is considered to be an early step of cellular response to Cr(VI) induced oxidative stress, followed by activation of transcription factors and apoptosis (275, 320, 341, 343). However, such activation may also be due to the action of Cr(III) complexes [products of intracellular Cr(VI) reduction], since some Cr(III) complexes are known to activate phosphorylation–dephosphorylation enzymes (Section IV.E).

d. **Other Protective Responses to Cr(VI).** Several groups of DNA binding proteins, such as the elongation factor complex and high mobility group proteins 1 and 2, bind specifically to Cr(III) modified DNA (344, 345). This effect may be important in repair of the damaged DNA sites. Low levels of Cr(VI) ($\sim 0.5~\mu M$) induce gene expression of stress proteins, known as heat shock proteins, in human cell lines (346). Such proteins are likely to participate in cell protection against xenobiotics; expression of these proteins may also be used as a biomarker for exposure to low doses of Cr(VI) in the environment (346). Exposure of rats to Cr(VI) compounds (by inhalation) induced immune responses in pulmonary macrophages, including increased production of NO, interleukins, and tumor necrosis factor α (347). New mechanisms of cellular responses to toxic substances, including Cr compounds, are likely to emerge with the refinement of biochemical techniques (348).

7. *Direct and Indirect Pathways in Carcinogenicity*

Free radicals and ROS have been implicated as primary causes of degenerative diseases such as atherosclerosis, diabetes, or cancer (349). As a consequence, much recent attention has been directed toward oxidative-damage hypotheses related to metal-induced genotoxicity and carcinogenesis (60, 61, 281, 350–353). This hypothesis states that uncontrolled reactions of biomolecules with O_2 (catalyzed by transition metal ions) lead to excessive formation of ROS, resulting in inflammation, lipid peroxidation, oxidative DNA damage, derangement of cell signaling system, or impairment of DNA repair (281, 351).

This theory most directly applies to carcinogenesis induced by Ni(II), since Ni(II) binds preferentially to histones (proteins of the cell nucleus) rather than to DNA (354–373). The Ni^{II}–protein complexes can then catalyze oxidative DNA damage by ROS (149). Catalytic rather than stoichiometric action of Ni(II) complexes in oxidative DNA damage is consistent with significant genotoxic effects caused by small concentrations of the metal ion (350–368). Indirect (i.e., ROS mediated) oxidative mechanisms are also considered among the major causes of Cd(II) and Cr(VI) induced carcinogenicity (351, 353).

Another recently proposed hypothesis relates the mutagenic and carcinogenic actions of metal ions to their direct interference with mechanisms of DNA repair, including the replacement of Zn(II) in zinc finger proteins (transcription factors) (279, 280) or inhibition of enzymes involved in repair of oxidized DNA sites (355). This mechanism has been considered for Cd(II), As(III), Co(II), and Ni(II) (280, 355); however, the recent results on Cr(VI/V/IV) induced damage in zinc finger proteins and their models (Section III.C.1) (95, 276, 277) suggest that Cr may be similarly involved in the potential carcinogenic pathway.

Among transition metal ions, Cr(VI) possesses the highest potency and widest spectrum of genotoxic action in cells (119, 181, 208), which is probably related to high permeability of Cr(VI) and its ability to form stable DNA–Cr^{III}–DNA and DNA–Cr^{III}–protein cross-links during intracellular reduction (249, 250, 356). Possible reasons for misfunctioning of DNA, related to changes in its electronic structure caused by binding of transition metal ions, have been recently reviewed (357).

There has been a long-standing debate on the relative importance of direct interactions of Cr species with biological macromolecules, versus ROS mediated pathways, in Cr(VI) induced genotoxicity and carcinogenicity (12, 61, 298, 358–360). While much of the evidence in support of Cr(VI) induced formation of ROS has been based on misinterpretation of experimental data (Sections II.A.5 and III.B.5), it is indisputable that oxidative stress can be caused by depletion of intracellular reductants during the reduction of Cr(VI) (181, 291), as well as by the formation of highly oxidizing Cr(V/IV) complexes, stabilized by intracellular ligands (12, 298). Such oxidative stress can lead to activation of redox-sensitive transcription factors, such as NFκB, that mimic the cellular response to ROS. The current dominant opinion is that both direct and indirect pathways can play a particular role in Cr(VI) induced carcinogenicity (12, 181, 298, 299, 361), as summarized in Scheme 2. Separate pathways, included in Scheme 2, have been described in detail in Sections III.B.1–III.B.6.

C. Mechanisms of Chromium Toxicity at the Molecular Level

This part of the chapter focuses on the interactions of Cr complexes in various oxidation states with biological macromolecules (nucleic acids, proteins, and

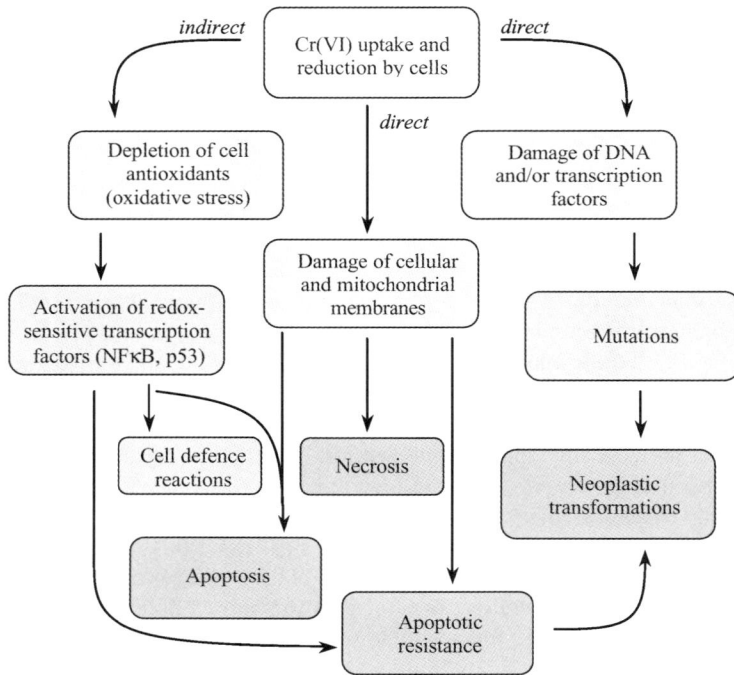

Scheme 2. Primary pathways for the cellular metabolism of Cr(VI). [Adapted from (181, 361)]. The data indicate that the initiation of cell defense reactions or necrosis are responses unique to indirect or direct pathways, respectively.

lipids) and models in cell-free systems. The formation and structures of some reactive Cr species have been reviewed in detail elsewhere (12) and the methods of characterization are outlined in Section II.

1. *Chromium(VI) + Reductant Systems*

It is well established that Cr(VI) (as $[CrO_4]^{2-}$) does not react with isolated DNA to a significant extent (362, 363). However, in the presence of low molecular weight biological reductants, Cr(VI) is capable of inducing a variety of DNA lesions, including single-strand breaks (364–366), abasic sites (367), oxidized nucleobases (365), and Cr^{III}–DNA adducts (365); the latter include DNA–Cr^{III}–amino acid and DNA–Cr^{III}–DNA cross-links (368, 369). This spectrum of DNA damage is similar to that observed in cell cultures and living organisms exposed to Cr(VI) (153). No significant damage occurs if DNA is added to the reaction mixture after the complete reduction of Cr(VI) to Cr(III) (260, 261). These data support the hypothesis of Wetterhahn and co-worker

(170) on the roles of Cr(VI) reduction intermediates [such as Cr(V) and Cr(IV) complexes, as well as the reductant and O_2 derived radicals] in Cr(VI) induced genotoxicity and carcinogenicity. Reactions involving the Cr^{VI} + Red + DNA system have been the subject of numerous studies, summarized in several recent reviews (60, 61, 298–300). Ascorbate and GSH are believed to be the main intracellular reductants of Cr(VI) (291, 298, 370–372). However, the possible roles of other low molecular weight compounds (such as cysteine or catechol derivatives) (25, 53, 54, 250, 255, 370) and enzymes [such as NAD(P)H dependent reductases or heme proteins] (291, 360, 373) cannot be disregarded. Detailed studies of DNA damage in vitro as a result of Cr(VI) reductions with GSH (368, 374–379), Cys (250, 251, 368, 374), ascorbate (369, 375, 380–382), dehydroascorbate (383), or catecholamines (25, 53, 54) have been performed. Little is known about the DNA damaging potential of Cr(VI) reactions with several important classes of biomolecules, such as tocopherols (Vitamin E and its analogues) (12, 51) or oxidized functional groups of proteins (25). While numerous in vitro studies of DNA cleavage have been performed using the Cr^{VI} + H_2O_2 or Cr^{VI} + Red + H_2O_2 systems [(25, 60, 61, 64, 384–386), and references cited therein], the biological relevance of these studies is ambiguous (298, 300). In most cellular compartments, very low (nanomolar) concentrations of H_2O_2 are maintained by scavenging enzymes (387, 388), but up to millimolar levels may be achieved in mitochondria (389) as well as in macrophages (390). Significant amounts of H_2O_2 can be formed in cells under the conditions of oxidative stress, and at the sites of acute inflammation, and this may be of particular relevance to smokers exposed to Cr(VI) (25).

Mechanistic studies of the Cr^{VI} + Red + DNA reactions are complicated by generally low steady-state concentrations of reactive intermediates and by the lack of direct methods for their detection (12, 97, 298). Even though Cr(V) intermediates formed in the Cr(VI) + Red reactions can be detected by EPR spectroscopy (Section II.A.2), the concentrations of the reagents required for the observation of Cr(V) species are usually higher than those used in DNA cleavage experiments. Therefore, EPR spectroscopic data obtained in concentrated solutions have to be extrapolated to dilute solutions used in DNA cleavage studies (25, 383). No direct methods exist for the detection of low concentrations of Cr(IV) and most free radical species in the Cr(VI) + Red systems (12). Possible pathways for the formation of intermediates, capable of causing oxidative DNA damage [based on the results of mechanistic studies of Cr(VI) reductions; (12) and references cited therein], are summarized in Schemes 3–6. As oxygen is required to induce significant oxidative damage to DNA during the reductions of Cr(VI) with GSH, ascorbate, or catecholamines (25, 376–378, 380, 381), possible reactions of O_2 are included in Schemes 3–6.

Unlike other biological reductants, GSH forms a relatively stable Cr(VI) complex (thioester; **A** in Scheme 3) at biologically relevant pH values (12, 98,

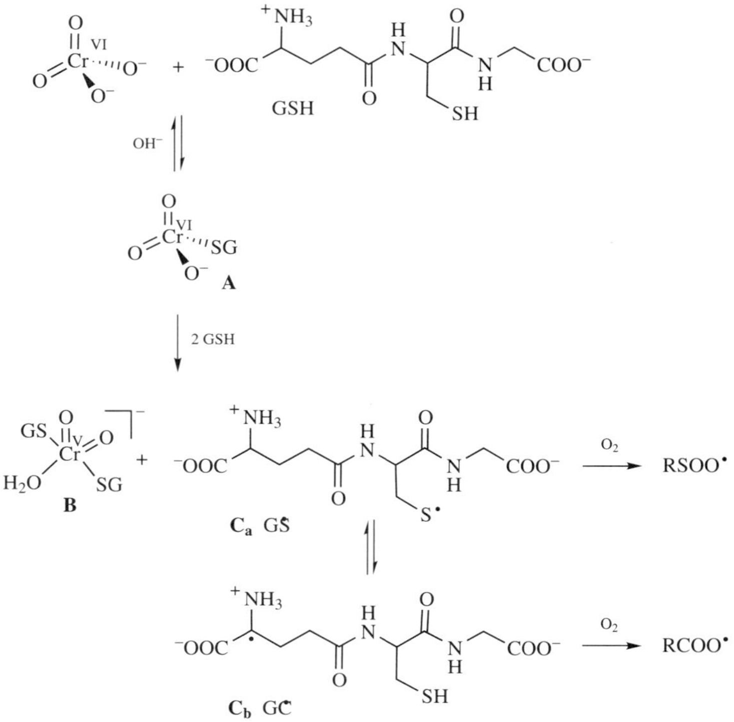

Scheme 3. Formation of potential DNA cleaving species in the $Cr^{VI} + GSH + O_2$ system in neutral aqueous solutions.

391). At least five different Cr(V) species have been detected by EPR spectroscopy during the $Cr^{VI} + GSH$ reactions at different reagent concentrations and pH values (194, 392–394). However, only one species [a narrow EPR signal with $g_{iso} = 1.9858$ and $A_{iso}(^{53}Cr) = 7.8 \times 10^{-4}$ cm^{-1}] with the proposed structure **B** (Scheme 3) has been detected at biologically relevant conditions ([GSH]$_0 \leq 10$ mM (196); [Cr^{VI}]$_0 \leq 1.0$ mM; pH 6–8) (195). Reduction of Cr(VI) with GSH is accompanied by the formation of thiyl radicals (**Ca** in Scheme 3), which isomerize in neutral aqueous solutions into carbon-based radicals (**Cb** in Scheme 3) (395). Both types of radicals can react with O$_2$ with the formation of peroxyl radicals [Scheme 3; (70) and references cited therein]. These radical species can either react with DNA directly (299) or participate in secondary reactions with the formation of damaging Cr(V/IV) species (175).

Reactions of Cr(VI) with ascorbate (Scheme 4) at pH 3–10 resulted in at least nine EPR detectable Cr(V) species (21, 396). The most abundant species at

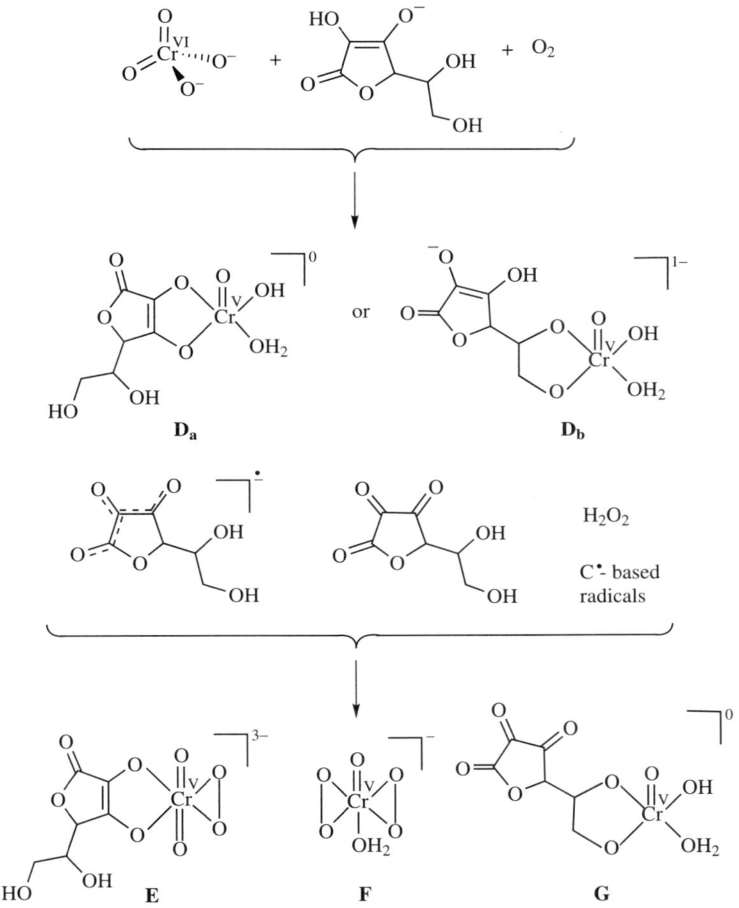

CHROMIUM IN BIOLOGY: TOXICOLOGY AND NUTRITIONAL ASPECTS 187

Scheme 4. Formation of potential DNA cleaving species in the Cr^{VI} + ascorbate + O_2 system in neutral aqueous solutions.

pH ~ 7 and [ascorbate]$_0$/[Cr^{VI}]$_0$ = 1–2 [g_{iso} = 1.9785–1.9791; A_{iso}(^{53}Cr) = 16.4 × 10^{-4} cm^{-1}] were proposed to be monoligated complexes with Cr(V) binding by either the enediolato moiety or the aliphatic side chain of the ligand (**Da** or **Db** in Scheme 4, respectively; Table II) (21, 50). Significant amounts of H$_2$O$_2$ are formed during the autooxidation of ascorbate either in the presence or absence of Cr(VI) (50, 70, 397), leading to the appearance of detectable amounts of Cr^V–ascorbate–H$_2$O$_2$ and Cr^V–H$_2$O$_2$ complexes (such as **E** and **F** in Scheme 4; Table II) in the Cr(VI) + ascorbate + O$_2$ reaction mixture (21, 50).

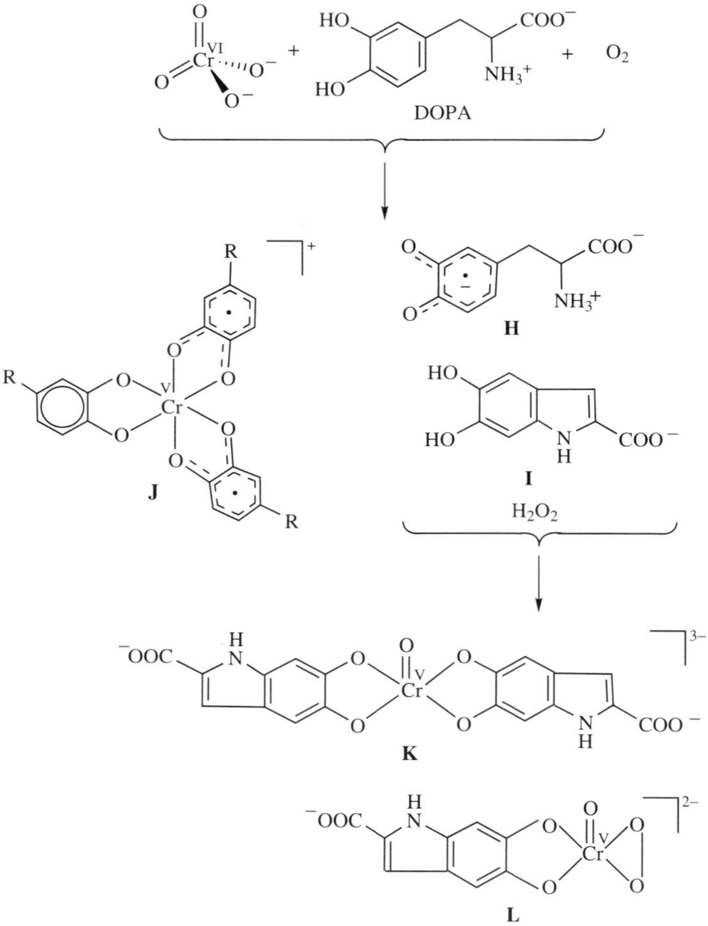

Scheme 5. Formation of potential DNA cleaving species in the $Cr^{VI} + DOPA + O_2$ system in neutral aqueous solutions.

Oxidation of ascorbate produces a relatively stable anion radical (directly detectable by EPR spectroscopy), and dehydroascorbic acid (398); the latter is capable of complexation with Cr(V) (**G** in Scheme 4) (21, 383). Carbon-based radicals have been detected in Cr(VI) + ascorbate reaction mixtures using spin-trapping techniques (399) but the structures of these radicals are still unclear (298).

A method for indirect detection of Cr(IV) intermediates, based on the decrease in the intensity of EPR signals of Mn(II) during the Cr(VI) + Red

$$O_2 + Red \rightarrow O_2^{-\bullet} \quad (6)$$

$$2\,O_2^{-\bullet} + 2\,H^+ \rightarrow H_2O_2 + O_2 \quad (7)$$

$$Cr^{n+} + O_2^{-\bullet} \rightarrow Cr^{(n-1)+} + O_2 \quad (8)$$

$$Cr^{(n-1)+} + H_2O_2 \rightarrow Cr^{n+} + {}^{\bullet}OH + OH^- \quad (9)$$

Red = GSH, Cys, ascorbate, NAD(P)H

$n = 3–6$

Scheme 6. Proposed mechanism [based on the data of (60, 61)] for the formation of ROS in the reactions of Cr species with intracellular reductants and O_2 (or H_2O_2).

reactions, has been proposed on the assumption that Mn(II) reacts with Cr(IV), but not with Cr(V) (378, 383, 399). However, it was subsequently shown that Cr(V) complexes with GSH, ascorbate, or H_2O_2 are able to react with Mn(II) under the same conditions (50, 194). Reports on the isolation of water-stable (pH ~ 7) Cr(IV) complexes, capable of damaging DNA in the presence of H_2O_2, (46, 47, 314) were based on erroneous assignment of the isolated solids as Cr(IV) complexes [in fact, these were mixtures of Cr(VI) and Cr(III) complexes] (12). At present, there is no reliable evidence for the formation of significant amounts of Cr(IV) complexes under conditions of the Cr^{VI} + Red + DNA reactions (12). However, well-characterized Cr^{IV} 2-hydroxycarboxylato complexes are known to cause single-strand breaks in DNA in vitro (400).

Reactions of catecholamines [such as 3,4-dihydroxyphenylalanine (DOPA), Scheme 5] with Cr(VI) produce intermediates such as semiquinone **H** and leucodopachrome **I** (Scheme 5) as precursors of polymeric oxidation products, such as melanin [(53) and references cited therein]. These reactions are accompanied by significant O_2 consumption and H_2O_2 production (25). Both reduced and oxidized catecholamines, as well as H_2O_2, can form Cr(V) complexes, such as **J–L** in Scheme 5 [detected by EPR spectroscopy (25, 53, 54)].

Reactions of Cr(VI) with biological reductants in the presence of O_2 or H_2O_2 have been proposed to trigger a redox cycle (Scheme 6), involving the Cr species in Cr(II) to Cr(VI) oxidation states, and leading to the formation of hydroxyl radicals ($^{\bullet}OH$; Eq. 9 in Scheme 6) as the ultimate DNA damaging species (60, 61). The key steps in the proposed mechanism (Scheme 6), including the formation of $^{\bullet}OH$ and the involvement of Cr(IV) species, are based on misinterpretation of EPR spectroscopic data (Sections II.A.3 and II.A.5). Furthermore, the patterns of DNA damage induced by Cr(VI) reactions with major biological reductants such as GSH or ascorbate are inconsistent with the action of $^{\bullet}OH$ (compared with genuine $^{\bullet}OH$ sources such as γ-radiation) (298, 300, 379).

At present, it is generally not established which of the active species are responsible for various kinds of DNA damage induced by the Cr^{VI} + Red + O_2 systems (298, 300). However, growing experimental evidence points to a significant role for Cr(V) complexes with the reductant and O_2 derived ligands in oxidative DNA damage (12, 25, 377, 379, 380, 401). The conditions for maximal formation of Cr^V–Red–H_2O_2 complexes (Red = ascorbate or DOPA), determined by EPR spectroscopy, corresponded to the conditions of maximal DNA cleavage by the Cr^{VI} + Red + O_2 systems (21, 25, 50, 382). A Cr^V–ehba–H_2O_2 intermediate is formed during the reduction of a well-characterized complex, $[Cr^VO(ehba)_2]^-$, by H_2O_2 (68, 402), and DNA cleavage induced by this Cr(V) complex is significantly enhanced in the presence of H_2O_2 (403, 404).

There are few mechanistic details on the formation and structures of the DNA oxidation products in the Cr^{VI} + Red + O_2 + DNA systems. Reduction of Cr(VI) with GSH or ascorbate induced simultaneous formation of comparable amounts of single-strand breaks and abasic sites in isolated DNA (376, 377, 381). Malondialdehyde was detected (by reaction with thiobarbituric acid) as an oxidation product of DNA or 2-deoxyribose in the Cr^{VI} + Red + O_2 systems (Red = GSH or ascorbate) (379). These features correspond to DNA oxidation by hydrogen abstraction at C4' atom of the deoxyribose moiety (405, 406). More detailed studies of DNA oxidation mechanisms were performed using a model Cr(V) complex, $[Cr^VO(ehba)_2]^-$ (Section III.C.4).

Mapping of oxidative DNA damage at the nucleotide level has been performed using the M + ascorbate + H_2O_2 systems [M = Cr(VI), Fe(III), or Cu(II)] (386, 407). For all metal ions used, guanine bases were predominantly damaged; however, the use of Cr(VI) led to damage of several specific thymine positions. Oxidative damage in the *p53* gene of human DNA (believed to play a key role in cancer suppression; Section III.B.4) (408) by the Cr^{VI} + ascorbate + H_2O_2 system has been detected (386); however, these results have limited implications for the in vivo systems due to the high concentrations of H_2O_2 used (5–50 mM) (298, 300).

Unlike GSH, ascorbate, or DOPA, the reaction of Cr(VI) with Cys at pH ~ 7 does not produce detectable amounts of Cr(V) species (97, 194, 394), and leads to the formation of Cr^{III}–DNA adducts in the absence of significant oxidative DNA damage (250, 356, 374). Formation of transient positively charged Cr(III) complexes, which efficiently bind to the negatively charged phosphate backbone of DNA (or, in parallel reactions, to OH^- and the anionic forms of Cys and the buffer), has been proposed as the reason for Cr^{III}–DNA binding during the Cr^{VI} + Cys reaction (250, 251).

Relatively few studies have been performed on the Cr(VI) induced oxidations of biological macromolecules, other than DNA, and their models. Consequently, there are no studies on the types of lesions or products that result from reactions between Cr(VI) and RNA. The levels and stability of mRNA in cells in response to Cr(VI) exposure have been examined (220, 297), but no attempts have been made to see whether Cr–RNA complexes are also produced in these reaction systems. An almost two-decade-old study showed that Cr(V) is stabilized by ribonucleotides but not by the analogous deoxyribonucleotides (409), which has prompted extensive studies directed at understanding the types of species formed between Cr(V) and 1,2-diolato donors (18, 27, 31, 190–193). Chromium(VI) induced damage to RNA is likely to represent a minor biological assault compared with Cr induced DNA damage, since damage to a single nucleotide is more likely to be repaired or have minimal biological consequences relative to the proliferation of faulty biomolecules replicated from high-copy genetic material (410).

The reaction of Cr(VI) with bovine serum albumin (BSA) at pH 7.4 results in the formation of Cr(V) species (detected by EPR spectroscopy, $g_{iso} = 1.9790$) (411). The reported g_{iso} value is characteristic of five-coordinate Cr^V–oxo complexes with 1,2-diolato coordination (**I** in Chart 1; Section II.A.2) (12, 18, 27). This complex, however, may be an artifact due to the presence of trace carbohydrates in BSA, as no significant EPR signals were observed during the reactions of high purity BSA with Cr(VI) (412). Photocatalyzed reactions of Cr(VI) with amino acids (Gly or Ala) or with small peptides (Gly_n or Ala_n, $n = 3$–5) in methanol solutions led to the formation of relatively stable Cr^V–amino acid(peptide) complexes (such as **II** in Chart 1), characterized by EPR spectroscopy and electrospray mass spectrometry (22, 23, 34). An EPR silent Cr^V–Ala dimer (**III** in Chart 1) has been isolated and characterized by XAFS spectroscopy; this is the first structurally characterized Cr^V–amino acid complex (23).

The Cr^{VI} + ascorbate system or the well-characterized Cr(V) and Cr(IV) complexes ($[Cr^VO(ehba)_2]^-$ and $[Cr^{IV}O(qaH)_2]^0$, **IV** and **V** in Chart 1, respectively), cause oxidation of $[Zn(CysOEt)_2]^0$ (Scheme 7), a mimic of a tetrahedral 2S,2N Zn(II) binding site in zinc finger proteins (413), with the formation of S—S bonds (Scheme 7) (95). This reaction may serve as a model for Cr(VI/V/IV) induced oxidative damage of Zn(II) dependent transcription factors, including NFκB and p53 [(95), and references cited therein]. Oxidation of zinc finger proteins by the Cr^{VI} + GSH system (278) or by a model Cr(V) complex, **IV** (276, 277), prevented their binding to DNA.

The reaction of Cr(VI) with whole human saliva (in the absence of added reductants) leads to the formation of relatively stable Cr(V) species ($g_{iso} = 1.9799$) (42). The Cr^V–saliva species were very similar to Cr(V) species formed from the reaction of Cr(VI) with GSH in the presence of excess 2,3-dehydro-2-deoxy-N-acetylneuraminic acid. This provided strong evidence for the formation of Cr(V) species with 2′-bound sialic acid terminated salivary glycoproteins (**VI** in Chart 1) (42). Further in vitro studies on the chemical mechanisms of Cr(VI) interactions with sialoglycoproteins have been performed

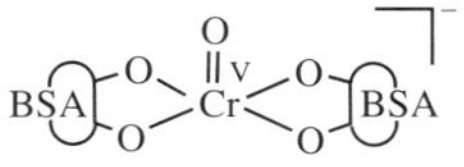

I

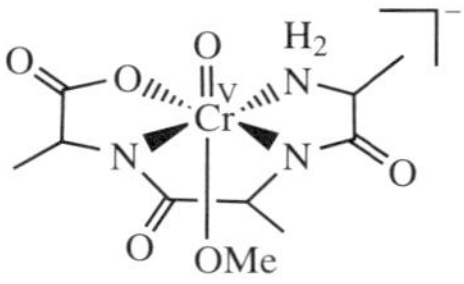

II

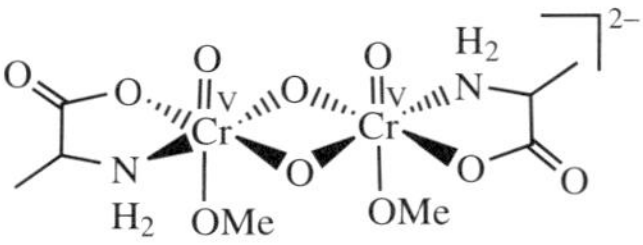

III

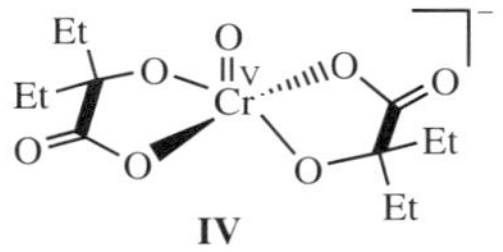

IV

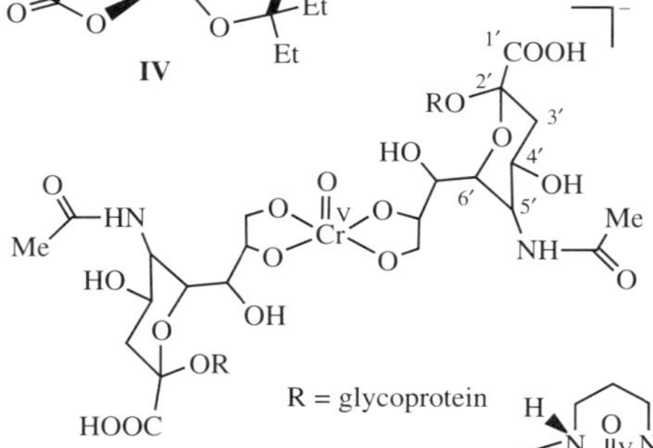

VI

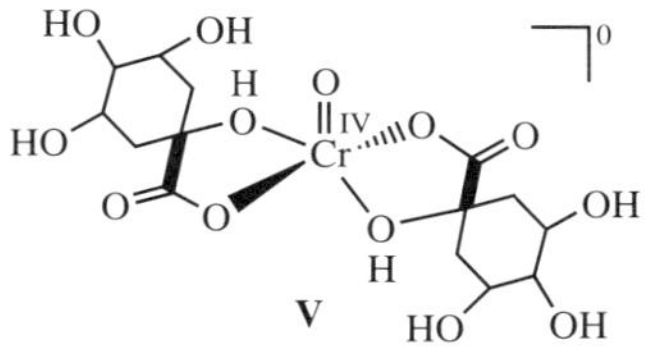

V

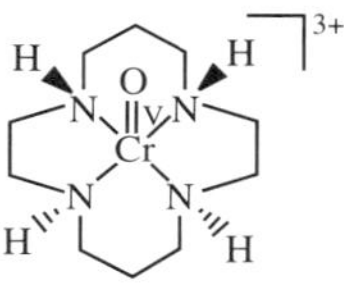

VII

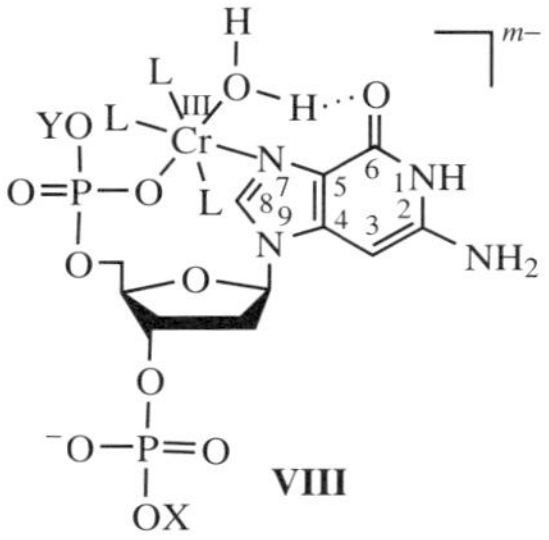

VIII

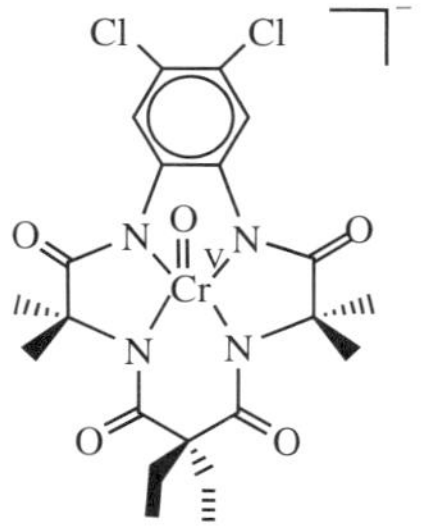

IX

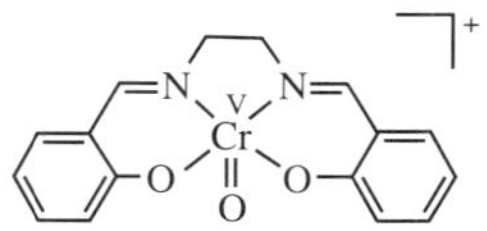

X

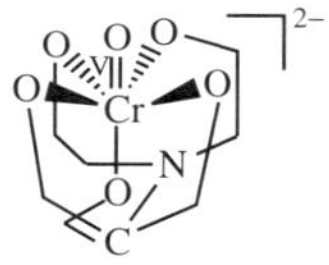

XI

Chart 1

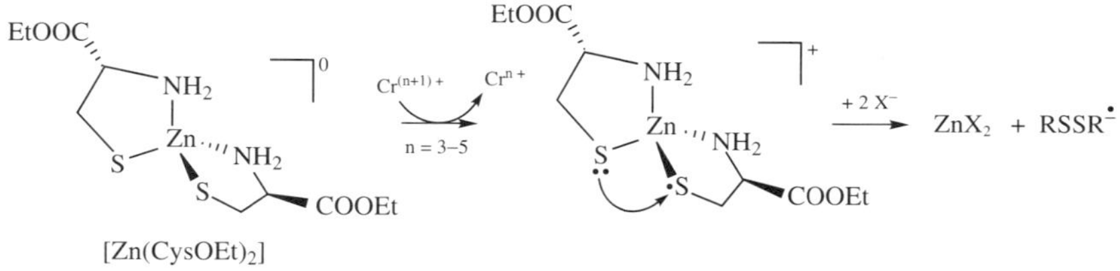

Scheme 7. Proposed mechanism for the oxidation of $[Zn(CysOEt)_2]^0$ with Cr(VI/V/IV) complexes. [Adapted from (12, 95).]

using a model reaction of Cr(VI) with GSH in the presence of excess sialic (N-acetylneuraminic) acid (414). Strong EPR signals of Cr^V–sialic acid complexes are observed in these solutions and the signals are pH dependent, with 2-hydroxy acid or diolato coordination favored under acidic or alkaline pH conditions, respectively (414). Sialoglycoproteins may be responsible for the formation of relatively stable and potentially genotoxic Cr(V) species during the interactions of Cr(VI) with cell membranes in relation to Cr(VI) uptake via the respiratory tract (Section III.B.3).

Reactions of Cr(VI), as well as Cr(V) and Cr(IV) complexes ($[Cr^VO(ehba)_2]^-$; $[Cr^VO(hmba)_2]^-$; $[Cr^V(O_2)_4]^{3-}$; or $[Cr^{IV}(O_2)_2(NH_3)_3]^0$) with mouse liver homogenates in the presence of O_2, lead to lipid peroxidation (224, 225). Notably, lipid peroxides (generated in Cr independent reactions) have been reported to cause oxidative DNA damage in the presence of peroxidases (415). Compared to Cr(VI) and Cr(V), Cr(IV) induced the greatest amount of lipid peroxidation. Based on this trend, the authors postulated hydroxyl radicals as the most likely candidates for lipid peroxidation, since the reaction between Cr(IV) and H_2O_2 can produce two times the concentrations of radicals compared to the Cr(V)/H_2O_2 system (225). The authors did not, however, consider the fate of the Cr species following incubation with the liver homogenates (37°C, 10 min) and also that the oxidizing ability of a metal is dependent on the types of coordinated ligands and upon the conditions of the biological matrix. While Cr(VI) is considered a powerful oxidant, chemical studies have illustrated the superior oxidizing ability of Cr(IV) and Cr(V) by potentiometric titrations of $[Fe(CN)_6]^{4-}$ with $[Cr^VO(ehba)_2]^-$ (106). The reduction potential values [vs normal hydrogen electrods (NHE)] for the $[Cr^VO(ehba)_2]^-/[Cr^{IV}O(ehbaH)_2]^0$ and the $[Cr^{IV}O(ehba)_2]^{2-}/[Cr^{III}(ehba)_2(OH_2)_2]^-$ couples are 0.44 and 1.24 V, respectively (106). Therefore, when coordinated with the ehba ligand (which is a good model for hydroxy acids found in vivo), the order of increasing oxidizing power is $[Cr^{VI}O_4]^{2-} < [Cr^VO(ehba)_2]^- < [Cr^{IV}O(ehbaH)_2]^0$ (106). This work highlights the importance of interpreting Cr biological data in the context of detailed speciation studies and redox chemistry.

2. *Chromium(III) + Oxidant Systems*

The uptake of Cr(VI) by cells leads to accumulation of Cr(III) complexes in cellular compartments (Section III.B.1) (89, 91, 167). Formation of DNA damaging species during the intracellular reoxidation of these complexes may be important for Cr induced genotoxicity (175). While it was previously considered that oxidation of Cr(III) to Cr(VI) in neutral media can only be accomplished with strong nonbiologically relevant oxidants such as $KMnO_4$ (416), more recent studies have shown that Cr(III) complexes can be oxidized to highly reactive Cr(V/IV) species under relatively mild conditions (37, 175, 417).

Several studies on oxidative DNA damage by the $CrCl_3 + H_2O_2$ systems have been performed (65, 418–422). Maximal damage of isolated DNA is achieved at pH 7–8 (65, 418). In addition to single-strand breaks (418), significant formation of double-strand breaks has been observed (419). The latter lesions are more difficult for cell defense systems to repair, and hence are more damaging; thus, generation of DNA double-strand breaks is proposed as the main reason for the anticancer activity of Fe^{II}–bleomycin (423). The $CrCl_3 + H_2O_2$ + DNA reactions also led to oxidation of DNA guanine bases, as 8-OHdG was detected in the reaction mixtures after enzymatic digestion of DNA (65, 419–422). In contrast, no significant amounts of 8-OHdG are detected in the Cr^{VI} + Red + DNA systems in the absence of added H_2O_2 (298, 300). The relevance of the $CrCl_3 + H_2O_2$ + DNA reactions to biological systems is limited by the following: (1) millimolar concentrations of H_2O_2 are used, which are several orders of magnitude higher than the expected H_2O_2 levels in cells, except for some special conditions (Section III.C.1); and (2) $CrCl_3$ is rapidly hydrolyzed at physiological pH values, making its existence in cells unlikely (80, 424).

There are few reports on the Cr(III) + H_2O_2 + DNA reactions with different Cr(III) compounds. The Cr(III) products of Cr(VI) oxidations of ascorbate or glucose induced single-strand breaks in plasmid DNA in the presence of H_2O_2 (425). A trinuclear Cr(III) acetate complex, $[Cr_3O(OAc)_6(OH_2)_3]^+$, causes single strand breaks in DNA and catalyzes the hydrolysis of phosphate mono- and diesters in the presence of H_2O_2 at pH 8.0 (426). In addition, some Cr(III) complexes, such as $[Cr(salen)(OH_2)_2]^+$ and $[Cr(edta)(OH_2)]^-$, induce oxidative degradation of proteins (BSA or human orosomucoid) in the presence of H_2O_2, while other complexes, such as $[Cr(en)_3]^{3+}$, do not cause this effect (233, 427).

Formation of •OH as the ultimate DNA damaging species in the Cr^{III} + H_2O_2 + DNA systems (by a Fenton-like mechanism; Eq. 9 in Scheme 6, with $n = 4$) has been proposed (65, 418–421, 426) on the basis of the following observations: (1) formation of DMPO—OH adducts detected by EPR spectroscopy (418); (2) inhibition of DNA damage by commonly used radical scavengers, such as NaN_3, mannitol, vitamin E, or melatonin (65, 418, 420–422); and (3) the absence of EPR spectroscopic evidence for the formation of Cr(V) complexes (426). However, none of these observations presents unambiguous evidence for the involvement of •OH and indeed are readily explained by known chemistry that does not involve such radicals (Sections II.A.5 and III.C.4).

An alternative mechanism of DNA damage in the Cr^{III} + oxidant systems involves direct interactions of DNA with the Cr(V/IV) complexes, formed by Cr(III) oxidation (12, 175). Recently, direct evidence has been obtained for the DNA damaging potential of Cr(V) complexes generated by oxidation of Cr(III) complexes under biologically relevant conditions (37, 428). A macrocyclic Cr(V) complex, $[CrOL]^{3+}$ (**VII** in Chart 1; L = 1,4,8,11-tetraazacyclotetradecane), capable of inducing single-strand breaks in plasmid DNA in the absence of H_2O_2, has been generated from a corresponding Cr(III) superoxo complex, $[CrL(OO)(OH_2)]^{2+}$ (which, in turn, is generated in the reaction of a Cr(II) complex, $[CrL(OH_2)]^{2+}$, with O_2) (417, 428). Imine complexes of Cr(V), $[CrO(salen)]^+$, and $[Cr(O)_2(phen)_2]^+$, generated in aqueous buffer solutions (pH 3.3 or 7.4) by oxidation of the corresponding Cr(III) complexes ($[Cr(phen)_2(OH_2)_2]^{3+}$ or $[Cr(salen)(OH_2)_2]^+$) with PbO_2 or PhIO, induce strand breaks in plasmid DNA (37). Oxidation of $[Cr(pic)_3]^0$ (widely used as a food supplement) with the formation of Cr(V/IV) species is likely to be responsible for the mutagenicity of this Cr(III) complex (Section IV.D) (429). More extensive studies on the reactions of different types of Cr(III) complexes with biological oxidative systems (both enzymatic and nonenzymatic) are required to determine the mutagenic potential of intracellular reoxidations of Cr(III), in relevance to the long-term health effects caused by exposure to Cr(VI) and to some types of Cr(III) complexes (5, 181, 430).

3. Interactions of Chromium(III) Complexes with DNA

It is well known that protein–Cr^{III}–DNA, amino acid–Cr^{III}–DNA, and DNA–Cr^{III}–DNA cross-links form in cells and organisms exposed to Cr(VI) (Section III.B.4). The potential importance of Cr^{III}–DNA lesions in Cr(VI) induced genotoxicity is determined by the following: (1) kinetically inert (80) Cr(III) complexes with DNA and other bioligands are less likely to be recognized and repaired by enzymatic systems than oxidative DNA damage; and (2) Cr(III) complexes can accumulate in cells of Cr(VI) exposed individuals over many years (431). Interactions of Cr(III) complexes with isolated DNA and nucleotides have been studied in order to elucidate the structures and biological roles of such crosslinks (250, 356).

Numerous studies in the last 20 years have established that additions of highly acidic Cr(III) complexes, such as $Cr(NO_3)_3$ (i.e., $[Cr(OH_2)_6]^{3+}$) or $CrCl_3$ (i.e., $[CrCl_2(OH_2)_4]^+$) (80), to solutions of DNA in neutral aqueous media, cause extensive Cr^{III}–DNA binding and changes in DNA conformation (363, 431–435). As established by IR spectroscopy, the most likely Cr(III) binding sites in the DNA molecule are the phosphate backbone and the N7 atoms of guanine bases (435); one of the possible structures is represented by **VIII** in Chart 1 (251). A complex of Cr(III) with a dinucleotide (dGpG) and a tripeptide (Gly-Cys-Gly) has been synthesized as a possible model for protein–Cr^{III}–DNA crosslinks (253, 436).

Based on the abilities of Cr(III) salts to bind to DNA, as well as the lack of Cr(VI) interactions with isolated DNA, Beyersmann and Koster (432) suggested that Cr(III) is the ultimate carcinogenic form of Cr. This approach has been criticized by Kortenkamp et al. (261), and Wetterhahn and co-workers (260), who showed that the neutral or anionic Cr(III) complexes of amino acids, ascorbate, or GSH, which are likely to form during the intracellular reduction of Cr(VI), do not bind to DNA significantly at pH ≥ 7. These observations led to the assumption that Cr^{III}–DNA binding in vivo occurs during the intracellular reduction of Cr(VI), and that Cr(V/IV) intermediates are likely to be responsible for such lesions [(299), and references cited therein]. On the other hand, significant levels of binding of Cr^{III}–amino acid complexes (amino acid = Cys or His) to DNA have been detected in slightly acidic media (pH 5.6–6.1) (356, 437), which is likely to be relevant to cellular uptake and reduction of insoluble Cr(VI) particles (Section III.B.1) (12, 31). Such binding is less favorable at higher pH values or in the presence of inorganic phosphate due to the concurrent hydrolysis of Cr(III) (250). Polynuclear hydroxo complexes, formed by the hydrolysis of Cr(III) salts at pH ≥ 7 (80, 424), do not bind to DNA (261); however, insoluble hydrolysis products can precipitate DNA by adsorption (438). Complexes of Cr(III) with hexose ligands have also been reported to bind to DNA (439). Finally, stable cationic Cr(III) complexes, such as $[Cr(phen)_3]^{3+}$ or trans-$[Cr(salen)(OH_2)_2]^+$, bind electrostatically to the negatively charged phosphate backbone of DNA (37, 440–442), but such interactions are unlikely to be biologically relevant.

4. Mechanisms of DNA Damage Induced by a Model Chromium(V) Complex

Difficulties with mechanistic studies of the damage in biological macromolecules induced by the Cr^{VI} + Red systems (Section III.C.1), as well as the proposed role of reactive Cr(V) species in Cr(VI) induced genotoxicity (170), have prompted many researchers to study the interactions of isolated DNA or proteins with model Cr(V) complexes. Crystallographically characterized Cr(V) complexes, $[CrO(ehba)_2]^-$ (443, 444), $[CrO(mampa)]^-$ (33), and $[CrO(salen)]^+$ (38) (**IV, IX**, and **X** in Chart 1, respectively) have been shown to induce single-strand breaks in plasmid DNA in the absence of added oxidants or reductants (37, 209, 210). A cationic Cr(V) complex, **X**, has recently been shown to act as a specific oxidant of dG sites in DNA (445), in contrast with an anionic complex, **IV**, which predominantly oxidizes deoxyribose residues (446). Complex **X** also induces oxidative degradation of a glycoprotein from human blood plasma (232), and may be considered as a model of intracellular Cr^V–peptide complexes (22, 175, 445). The reactions of DNA or mono- and oligonucleotides with **IV** have been studied in detail (209, 400, 403, 446–452), and the proposed mechanism of DNA damage, based on all of these results, is shown in Schemes 8–10.

Scheme 8 presents the proposed mechanism of DNA interactions with $Cr^{V/IV}$–ehba complexes (phosphate buffers, pH 6–8) (400, 451). Kinetic studies have shown that **IV** is responsible for the Cr^V–DNA binding at pH 6–8 (**M**, Eq. 10 in Scheme 8) (451) while, at lower pH values, monoligated Cr^V–ehba complexes

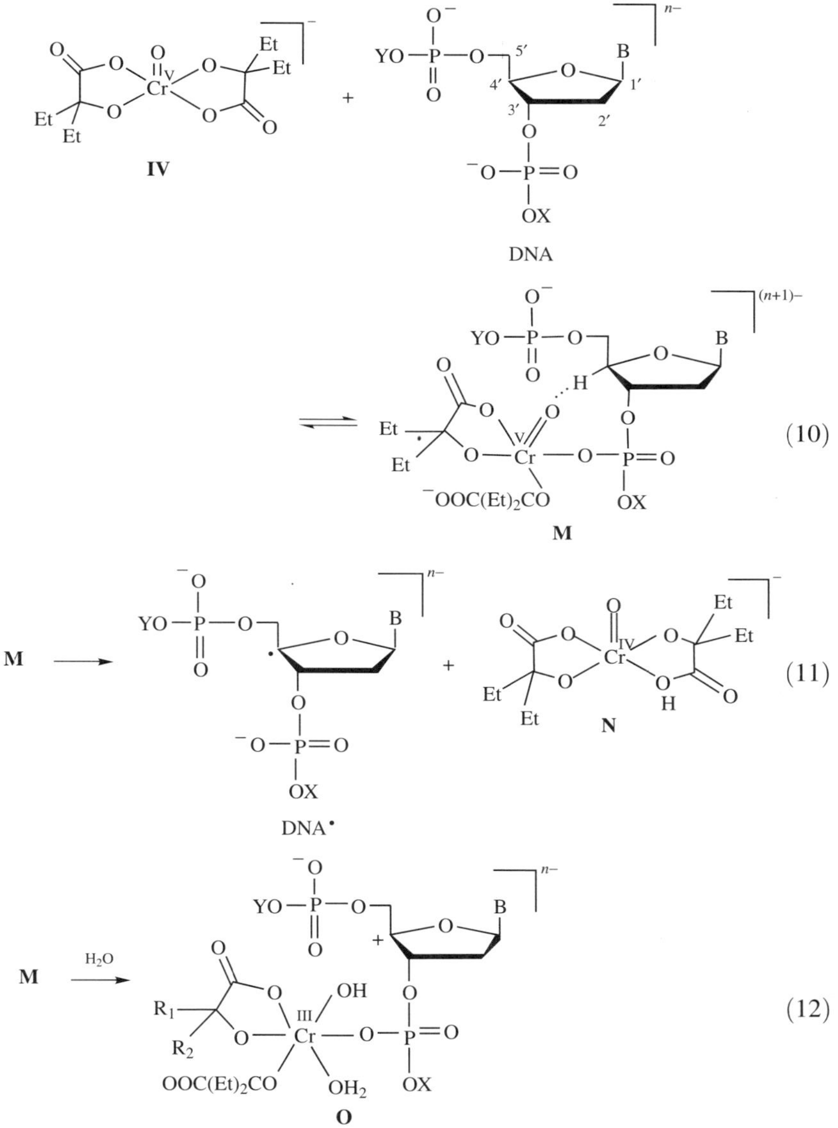

Scheme 8. Proposed mechanism for the reaction of $Cr^{V/IV}$–ehba complexes with DNA in neutral aqueous solutions. [Adapted from (400, 451).]

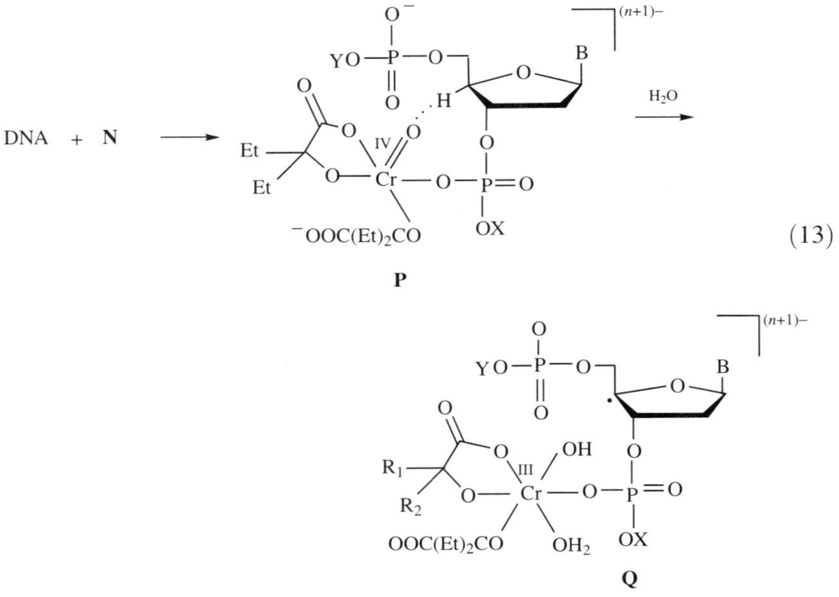

$$\text{DNA} + \mathbf{N} \longrightarrow \mathbf{P} \xrightarrow{\text{H}_2\text{O}} \mathbf{Q} \quad (13)$$

Scheme 8 (Continued)

are likely to be involved in DNA binding (209, 400, 451). Binding of Cr(V) to the phosphate backbone of DNA in **IV** is supported by EPR spectroscopic data on Cr(V)–phosphate interactions (448, 453). Complex **M** undergoes intramolecular one- or two-electron transfers with the formation of a DNA radical (DNA•) and a Cr(IV) complex **N** (Eq. 11 in Scheme 8), or of a Cr(III) complex of a DNA cation **O** (Eq. 12 in Scheme 8), respectively. Oxidations at the C4′ position of the deoxyribose ring (Eqs. 11 and 12) have been suggested on the basis of product studies for the oxidations of DNA and nucleotides with **IV** (403, 446, 447); these are also consistent with the results of product studies for the $\text{Cr}^{\text{VI}} + \text{Red} + \text{DNA}$ reactions (Red = GSH or ascorbate) (379). The Cr(IV) complex **N** is capable of reacting with DNA (as shown in studies with the well-characterized Cr(IV) 2-hydroxycarboxylates) (400) to produce a Cr^{IV}–DNA intermediate **P** and subsequently a Cr(III) complex of a DNA radical **Q** (Eq. 13 in Scheme 8). Dissociation of Cr(III) complexes from the oxidized DNA strands in **O** and **Q** is likely to be slow in the timescale of DNA oxidation (400, 452), due to the kinetic inertness of Cr(III) (80).

Possible pathways of further reactions of **Q** are depicted in Scheme 9 (Eqs. 14–16; similar reactions are expected for DNA•). Reactions of 4′-DNA radicals with O_2 (Eq. 14 in Scheme 9) are well characterized (405, 406) and lead to peroxo radicals **R**, which are converted after a series of rearrangements to

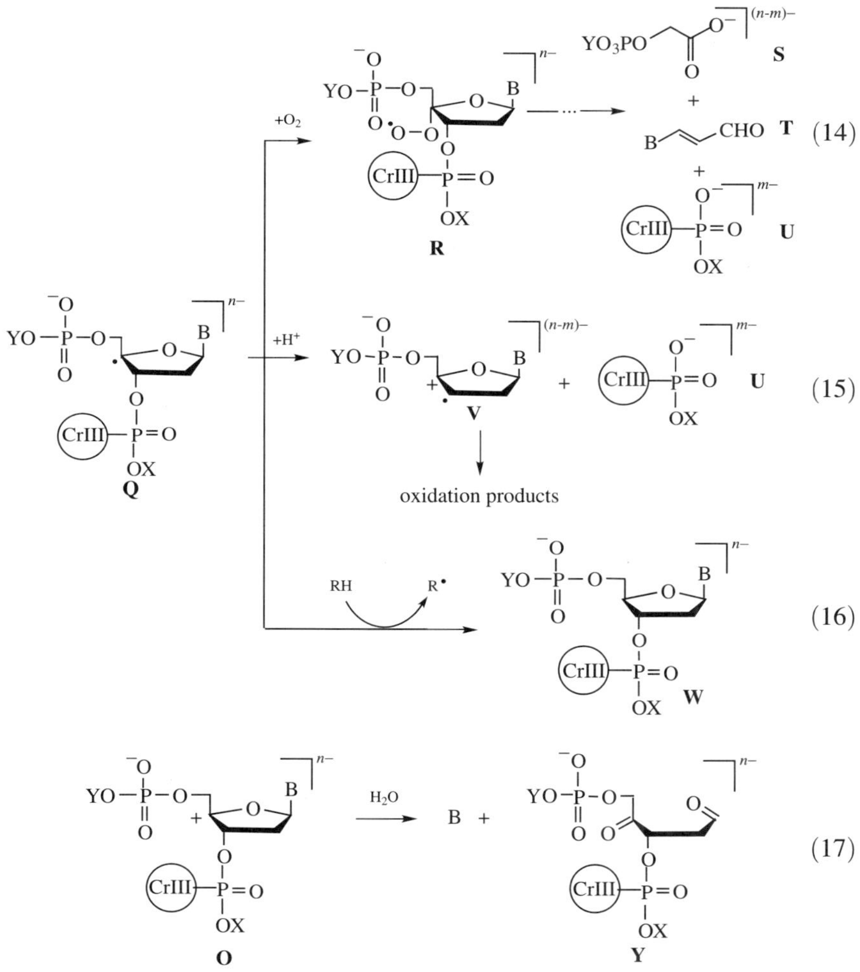

Scheme 9. Proposed mechanism for the further reactions of oxidized DNA intermediates. [Adapted from (400).]

3′-phosphoglycolate **S** and base propenal **T**, leaving the Cr(III) complex of the 5′-phosphorylated end of the DNA fragment **U**. An alternative oxygen-independent pathway of **Q** decomposition (454), which is more characteristic for the reactions in slightly acidic media (400), involves the formation of cation radical **V**, as well as **U** (Eq. 15 in Scheme 9). The DNA cleavage by **IV** at pH ∼ 7 is efficiently inhibited by organic molecules (H donors), such as

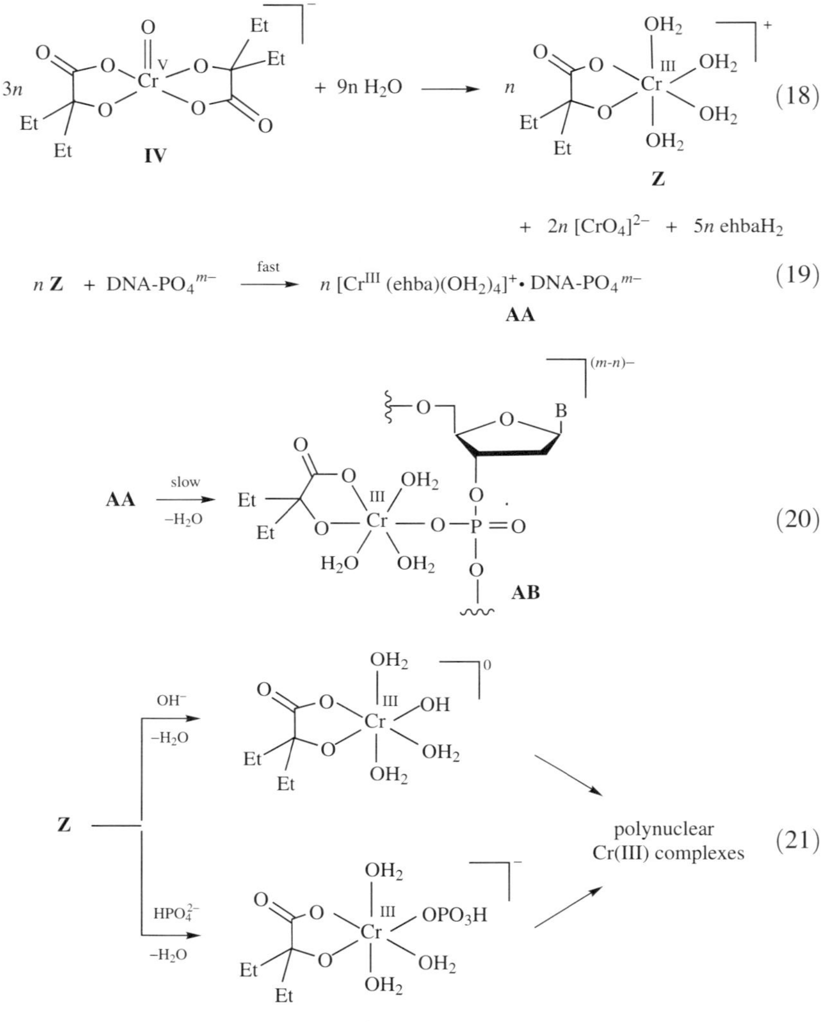

Scheme 10. Proposed mechanism for Cr^{III}–DNA binding, induced by disproportionation of $[Cr^VO(ehba)_2]^-$. [Adapted from (452).]

alcohols, carboxylic acids, dimethyl sulfoxide (DMSO), or common buffers [tris(hydroxymethyl)aminomethane (Tris) and 4-(2-hydroxyethyl)-1-piperazeneethanesulfuric acid (HEPES)], presumably through the thermodynamically favorable repair reactions (Eq. 16 in Scheme 9) (454), as none of the used inhibitors reacts significantly with **IV** (400). Notably, the repaired DNA molecule retains the bound Cr(III) complex (**R** in Scheme 9), which implies that the formation of Cr^{III}–DNA complexes during the reactions of DNA with Cr(V) in vivo may be more significant than the generation of single-strand breaks, as the latter are easily repaired by H-donors, which are abundant in cellular media (400). Finally, a DNA cation **O** is likely to be converted, through O_2 independent base release, into the oxidation product **Y** (Eq. 17 in Scheme 9) (405, 406). This reaction is supported by the observation of O_2 independent free base release during the reactions of DNA and nucleotides with **IV** (403, 446, 447). Formation of abasic sites (revealed by reaction with putrescine) (250, 376, 377, 381) has been observed during the DNA + **IV** reactions in phosphate buffers (pH 7.4). In agreement with the proposed mechanism (Schemes 8–9), these lesions were much less sensitive to the action of organic inhibitors than single-strand breaks (455).

Complex **IV** disproportionates to Cr(III) and Cr(VI) in neutral aqueous solutions at a timescale of minutes at room temperature (456). The stoichiometry of the disproportionation of **IV** at pH 6–8 is shown in Eq. 18 in Scheme 10 and formation of cationic Cr(III) product **Z** has been proposed on the basis of pH changes during the reaction (451). Incubation of DNA (fish sperm or calf thymus) with **IV** in HEPES buffers at pH 7.4 leads to nearly quantitative binding of the formed Cr(III) products to DNA (detected by AAS after removal of unbound Cr by gel filtration) (452), while no oxidative damage in plasmid DNA was observed under these conditions (400). The extent of Cr^{III}–DNA binding significantly exceeds that caused, under similar conditions, by Cr(III) salts, such as $Cr(NO_3)_3$ or $CrCl_3$, as well as by the Cr(VI) + Red reactions (Sections III.C.1 and III.C.3) (452). The mechanism of Cr^{III}–DNA binding caused by **IV** disproportionation (Eqs. 19 and 20 in Scheme 10), proposed on the basis of stoichiometric and kinetic studies (452), involves fast formation of electrostatic Cr^{III}–DNA ion pair **AA**, followed by slower ($\sim$ 1 h at 37°C) formation of covalent Cr(III)–DNA bonds in the complex **AB**. This does not exclude the formation of Cr^{III}–DNA complexes during the oxidative cleavage of DNA by Cr(V) (Eqs. 12 and 13 in Scheme 8), but the extent of such lesions ($\sim$ 1/5000 nucleotides) was below the detection limits of AAS (400, 452). Similar levels of Cr(III) binding were observed during the reactions of **IV** with synthetic polynucleotides, poly(dG-dC) or poly(dA-dT). This finding implies the absence of significant Cr(III) interactions with guanine bases in **AB** (452). The Cr(III)–DNA binding induced by the disproportionation of **IV** is almost completely prevented by prolonged reaction of the Cr(V) complex with the buffer before the addition of DNA, or by the presence of inorganic phosphate (P_i) (452). In these cases, the initial Cr(III) product **Z** is likely to bind OH^- or HPO_4^{2-} ions with the formation of neutral or anionic Cr(III) species that are unreactive toward DNA (Eq. 21 in Scheme 10). The initial Cr^{III}–DNA ion pair adducts **AB** were stable toward a strong complexant EDTA, but released Cr(III) in the presence of excess Na^+ or Mg^{2+} (452). Similar properties were observed for the Cr^{III}–DNA adducts formed during the Cr^{VI} + Cys + DNA reaction (250). The proposed mechanism (Schemes 6–8) implies that stabilized Cr(V) species formed in Cr(VI) treated cells (Section III.B.1) (12) can induce both oxidative (Schemes 8 and 9) and nonoxidative (Scheme 10) DNA damage. In both cases, potentially mutagenic (251) Cr^{III}–DNA adducts can be formed.

The proposed mechanism of DNA damage by a model Cr(V) complex (Schemes 8–10) solves several uncertainties in the mechanistic studies of the Cr(VI) + Red + DNA reactions (299, 300). Thus, inhibitions of DNA cleavage by radical scavengers, which were often interpreted in favor of •OH promoted DNA damage (60, 61, 64, 66, 67, 364, 457), are in fact in agreement with the direct DNA oxidation by Cr(V/IV) species. One group of inhibitors (such as formate or DMPO) (375, 382) act by reducing Cr(V) to Cr(III) (68, 69). Polyol compounds, such as mannitol (66), glucose, or sucrose (375) form relatively stable Cr(V) complexes that are less reactive towards DNA than the transient Cr(V) species formed in Cr(VI) reductions (12). Finally, some substances, such as HEPES buffer, inhibit DNA cleavage without affecting the amount and speciation of Cr(V) formed in the Cr(VI) + Red reactions. This finding was claimed to rule out the role of Cr(V) in DNA cleavage (380). However, a variety of organic substances, including HEPES, inhibit DNA cleavage with **IV** by repairing DNA radicals through a hydrogen-abstraction mechanism (Eq. 16 in Scheme 9); these reactions are driven by high thermodynamic stability of double-stranded DNA (454). The observed O_2 dependence of oxidative DNA damage induced by the Cr^{VI} + Red systems (25, 376–378, 380, 381) is at least partially explained by the reactions of DNA radicals with O_2 (Eq. 14 in Scheme 9). Simultaneous formation of DNA strand breaks, abasic sites, and Cr^{III}–DNA complexes in such reactions (376, 377, 379, 381) is in agreement with the modes of DNA damage induced by **IV** (Schemes 8–10).

Detailed mechanistic studies of DNA damage induced by a model Cr(V) complex **IV** have also revealed some significant differences compared with the mechanisms of Cr^{VI} + Red + DNA reactions. Oxidative damage of DNA induced by Cr(VI) reductions with GSH, ascorbate, or DOPA is strongly inhibited by catalase, which implicates the involvement of peroxo species [possibly Cr(V) peroxo complexes] in DNA oxidation (25, 376, 377, 380, 381); but DNA cleavage by **IV** is unaffected by catalase (400). Oxidative DNA cleavage by the Cr(VI) + Red systems is observed in the presence of an excess of strong reductants, such as GSH, ascorbate, or DOPA (25, 299, 300), which would be expected to act as potent inhibitors of DNA cleavage according to Eq. 16 in Scheme 9. Therefore, further studies are required to elucidate the mechanisms of DNA damage by the Cr(VI) + Red systems.

Mechanistic studies of DNA damage by a model Cr(V) complex, as well as the Cr(VI) + Red systems, revealed the potential importance of intracellular environment for the extent and types of DNA damage in Cr(VI) exposed cells and organisms. The most likely Cr(V) intermediates in the intracellular Cr(VI) reductions are Cr(V) complexes of carbohydrates or their derivatives (12, 27, 31, 42). Preliminary studies (51, 66, 375, 458) have shown that complexation with carbohydrates significantly affects the reactivity of Cr(V) toward DNA, but detailed research on these effects is yet to be performed. The influence of the structures of Cr(V) complexes on the mechanisms of their interactions with DNA is illustrated by observations (448–450) that a polyol complex [CrO(BisTris)]$^{2-}$ (**XI** in Chart 1; formed *in situ* by the reaction of **IV** with excess BisTris ligand) attacks primarily C1′ and C5′ positions of the deoxyribose ring (as determined by the studies of DNA oxidation products). Thus, the nature of the Cr(V) complex may determine the geometry of Cr^V–DNA adduct (such as **M** in Scheme 8), and hence the mechanism of DNA cleavage (298). The presence of hydrogen-donors (i.e., most organic compounds) is likely to reduce the extent of DNA strand breaks, but not that of abasic sites or Cr^{III}–DNA binding (Scheme 9). In addition, ligands with a high affinity for Cr(III) (such as phosphate) will significantly reduce the amount of Cr(III) bound to DNA by the nonoxidative route (Scheme 10). Taking these effects into account is likely to be the next step in the in vitro studies aimed toward better understanding of the mechanisms of Cr induced toxicity.

IV. CHROMIUM AS A NUTRIENT

A number of recent reviews (3, 4, 6, 459–469) have focused on medical and nutritional issues related to the proposed role of Cr(III) in glucose and fat metabolism. This chapter concentrates on the origins of proposed Cr(III) dependent biomolecules, possible chemical mechanisms of Cr(III) biological activity, and safety concerns over the use of Cr(III) in food supplements. Controversies in Cr(III) biochemistry are discussed, with the emphasis on distinguishing between the experimental evidence on the one hand, and hypotheses and unproved claims on the other.

A. Biological Activity of Chromium(III)

Metal ions have been loosely divided into groups, according to whether they have biological activity, as being essential, beneficial, abiological, or detrimental (470). Curiously, Cr(III) has been placed into each of these categories by different authors. Most nutritionists regard Cr(III) as an essential micronutrient for humans (3, 4, 6, 459–465). The first evidence suggesting the involvement of Cr(III) in glucose metabolism was obtained in animal experiments by Mertz and co-worker in the 1950s (471, 472). The main evidence for the essential nature of

Cr(III) in humans comes from several independent cases of patients on total parenteral nutrition (TPN), when the absence of Cr(III) in TPN solutions led to severe symptoms of glucose intolerance that were reversed by Cr(III) supplementation (461, 473). However, the elevated glucose levels in most of the studied TPN patients were unaffected by Cr(III) supplementation (5, 474). Most studies failed to detect any significant Cr(III) related metabolic changes in healthy humans on normal diets [(4, 6) and references cited therein].

Attempts to induce Cr deficiency in experimental animals have also led to controversial results [(6) and references cited therein]. Recently, Anderson and co-workers (475, 476) formulated the conditions for inducing Cr deficiency in rats: (1) careful removal of Cr sources from food and environment (e.g., stainless steel cages cannot be used); and (2) sucrose- or fat-based diet. These conditions led to abnormal insulin levels, characteristic of the first stage of type II diabetes, that could be prevented by addition of $CrCl_3$ (5 ppm) to drinking water. In these experiments, Cr(III) apparently acts as an insulin potentiator under the conditions of dietary stress, which corresponds to the beneficial rather than essential metal ion category (470, 477).

From the classical bioinorganic chemistry point of view, Cr(III) can still be regarded as a nonbiologically relevant (or abiological) metal ion, as no Cr(III) dependent enzyme or cofactor has been definitively described as yet (478) (Section IV.C). Finally, in a recent review by Stearns (5), Cr(III) was considered as a detrimental ion based on its genotoxicity under certain conditions (Sections III.B.2 and IV.D). Controversies related to the proposed beneficial roles of Cr(III) were also discussed (5).

The majority of the reported beneficial actions of Cr(III) in vivo have been related to the improved insulin function, which included decreasing the levels of blood glucose (459, 479–481), cholesterol (482), and triglycerides (480, 482); lowering blood pressure (468, 483); and curing insulin-dependent neurological disorders (473, 484). In addition, improvements in immune responses and increased litter sizes in farm animals receiving Cr(III) supplementation, have been described (485, 486). Also, Cr(III) induced enhancement of RNA synthesis in mice (with uncertain physiological consequences) has been reported (487, 488).

Several types of tests have been employed to demonstrate the insulin-potentiating activity of Cr(III) in vitro. In the early studies of Mertz and co-workers (489, 490), glucose uptake or CO_2 production by isolated epididymal fat tissue (from Cr-deficient rats) in the presence or absence of Cr(III) complexes were compared. This method required a large number of animals per small number of samples and did not allow for data comparison among different laboratories (491). To overcome this limitation, Anderson et al. (491) proposed the use of isolated adipocytes from rats raised on low-Cr diets. Significantly enhanced CO_2 production due to glucose oxidation by adipocytes was observed in the presence of micromolar concentrations of some Cr(III) complexes, thought to be models of a natural Cr(III) containing factor (Section IV.C.1), while simple Cr(III) salts such as $CrCl_3$ had no effect (491). However, the positive results could not be reproduced in subsequent studies (82). The main limitations of the method were (1) unknown initial concentrations of Cr(III) in isolated adipocytes, that is, it was uncertain whether the cells were truly Cr deficient (82, 476); (2) unknown residual concentrations of insulin in the isolated cells, which complicated interpretation of the results in the presence or absence of added insulin (492, 493); and (3) changes in Cr(III) speciation during the experiments (82), which, in our opinion, was the main factor invalidating the experiment. In particular, incubations of Cr(III) complexes with adipocytes were performed for 2 h at 37°C in phosphate buffers (pH 7.4) containing mineral salts and albumin (491). These conditions would lead to Cr(III) hydrolysis with the formation of insoluble Cr^{III}–hydroxo–phosphato complexes, especially in the absence of organic ligands (80); indeed, precipitates were formed in most assays (82). Binding of Cr(III) complexes to albumin is also possible (494). Thus, the different abilities of various Cr(III) complexes to promote glucose metabolism in adipocytes (491) are likely to correlate with the stabilities of these complexes toward hydrolysis in neutral aqueous solutions, rather than with their specific biological activity.

Morris et al. (495) observed a Cr(III) dependent increase in 2-deoxyglucose uptake by cultured mouse myotubes in the presence of insulin. The effect was caused by the addition of nanomolar concentrations of $CrCl_3$ to the reaction medium [a standard minimal essential cell culture medium, containing amino acids, mineral salts, and glucose, which was purified with Chelex resin to remove trace heavy metal ions before the addition of Cr(III)]. The same Cr free medium was used for cell growth [growing the cells in the presence of a trace amount of Cr(III) reduced their sensitivity in the assays]. Thus, unlike the previous studies (491), an apparent insulin-potentiating effect of $CrCl_3$ was demonstrated. However, Cr(III) most likely formed complexes with amino acids of the medium (80), which prevented its hydrolysis in a neutral aqueous solution and made it more available to the cells.

Vincent and co-workers (496, 497) studied the insulin-potentiating action of Cr(III) complexes at a subcellular level. The activity of tyrosine kinase (in the isolated fragment of a cell membrane insulin receptor) increased threefold to eightfold in the presence of some types of Cr(III) complexes (50–500 nM Cr) and insulin, compared with insulin alone (Section IV.C.3). The active Cr(III) species included one isolated from the reaction of Cr(VI) with bovine liver homogenate (Section IV.C.3) (496) and a well-characterized trinuclear Cr(III) propionate complex (Section IV.D) (497). Integrity of the latter complex in the reaction medium (50 mM Tris buffer, pH 7.4) was verified by NMR spectroscopy (497).

The above assays (glucose uptake and oxidation by isolated rat adipocytes; hexose transport by muscle cells; and activation of the insulin receptor tyrosine kinase) are among the standard methods for the in vitro screening for potential antidiabetic drugs (498, 499). The previous examples show that interpretation of the results of such tests for metal complexes should commence with verification of the stability of the complex (or speciation of the metal ion) in the reaction medium.

B. Absorption, Metabolism and Speciation of Chromium(III) In Vivo

It is believed that Cr(III) is ingested mainly in the form of carboxylato (e.g., oxalato or citrato) or amino acid complexes with food of plant or animal origin, respectively, but the characteristics of Cr(III) speciation in food sources are incomplete (11). Gastrointestinal absorbance of Cr(III) in humans and experimental animals is low (0.4–2%). It is apparently stimulated by organic acid ligands (such as nicotinate or ascorbate), although no significant dependence of the absorption levels on the type of Cr(III) complexation has been observed in some studies (4, 500). Most biologically relevant Cr(III) complexes are expected to convert to $CrCl_3$ in gastric media (HCl + NaCl, pH $\sim$ 2), and significant formation of insoluble hydroxo complexes probably occurs in the small intestine ($NaHCO_3$ + NaCl, pH $\sim$ 7.5) (501). However, decomposition of Cr(III) complexes is relatively slow due to their kinetic inertness (80), so partial absorption of Cr(III) complexes can occur prior to their decomposition (500).

While several models of Cr(III) pharmacokinetics in mammals have been developed (111, 430, 502), little is known about Cr(III) speciation in organisms. Transferrin, a major transport protein for many metal ions (503), is likely to be the main Cr(III) carrier in blood (504). Although iron transport is a primary function of transferrin, the strength of its binding to Cr(III) is comparable to that for Fe(III) (45, 505); and Cr^{III}–transferrin complexes [unlike for the complexes of Mn(II), Cu(II), or Zn(II)] mimic those of Fe(III) in interactions with cell transferrin receptors (506). Preliminary results of X-ray absorption spectroscopic studies suggest that Cr(III) may occupy the same sites in the transferrin protein as Fe(III) (507).

A detailed study of Cr(III) speciation in human blood plasma was performed on continuous peritoneal dialysis patients (494). Total amounts and molecular mass distribution (by gel filtration) of ^{51}Cr containing species in plasma samples were measured for 2 days after the addition of a single dose of $^{51}CrCl_3$ into the dialysis solution. In the first 1–6 h, Cr(III) was mainly bound to transferrin ($\sim$ 80%; in agreement with earlier results of animal studies) [(508) and references cited therein], and to a lesser extent to albumin. At later times, a shift of Cr(III) to an unidentified compound with molecular mass of $\sim$ 5 kDa (possibly a small peptide) was observed. Binding of metal ions to small peptides in biofluids is probably very important in physiology, but little is known about such peptide complexes due to the isolation and purification problems (509, 510).

Generally, analysis of Cr(III) in biofluids is a demanding task, due to low concentrations (~ 0.1–1 ppb; values reported before 1980 were largely overestimated because of sample contamination) (511–513). Suitable methods are the following: (1) neutron activation analysis with ^{50}Cr enriched samples (514); (2) radioactive tracing with ^{51}Cr enriched samples (494); (3) inductively coupled plasma–mass spectrometry (ICP–MS) (513); and (4) graphite-furnace atomic absorption spectrometry (512). The two last methods, though most readily available, require special care in sample preparation (515).

Problems with chemical analysis are reflected in the lack of reliable indicators of Cr(III) deficiency in patients (516). Thus, different studies found that Cr levels in blood plasma were increased (517), decreased (462, 518), or unchanged (519, 520) in diabetic patients compared with control groups. This inconsistency could be a factor in the diet of the people used in the studies, for example, vitamin C was alluded to as an important factor in a paper published in 1997 (304). Significantly increased Cr losses in urine (462) and lower Cr content in hair (521) in diabetic patients were recently reported, but more studies are required before these methods can be validated. At present, the only indicator of Cr deficiency is the improvement of the patient's condition after Cr supplementation (516).

C. Search for Natural Biologically Active Forms of Chromium(III)

1. Glucose Tolerance Factor

The history of glucose tolerance factor (GTF) is described in detail in several reviews (6, 464, 522). Briefly, Mertz and co-worker (471, 472) were the first to isolate a fraction containing Cr(III), amino acids and nicotinate, from brewer's yeast. This fraction was shown to improve glucose tolerance in Cr deficient rats and was accordingly called GTF. Further studies on GTF revealed a number of inconsistencies. The structure of GTF has not been established as yet; furthermore, a substance originally described as GTF (472) could be separated by chromatography into several Cr(III) containing fractions, from which only one ($\sim 6\%$ of total Cr) was biologically active (523). Fermentation of yeast in the presence of Cr(III) or Cr(VI) compounds was used in most studies to increase the yield of a Cr(III) containing fraction (524–527), although a significant amount of Cr(III) is present even in a commercial yeast extract (528). The relationship between the compounds isolated from yeast grown in the presence of Cr(III), Cr(VI), or in the absence of added Cr, is not clear. Vincent and co-worker (529) suggested that GTF was an artifact of the hydrolysis of a natural Cr(III) containing compound under harsh isolation conditions (refluxing for 18 h in 5 M HCl) (472, 523). For example, nicotinate could have originated from a Cr^{III}–NAD complex (530). Subsequent isolations of GTF used milder conditions, but the composition of the isolated products was not studied (528).

Biological testing of GTF brought further uncertainties. Several research groups failed to detect a significant improvement in glucose tolerance in Cr deficient rats by GTF (531, 532), which was later attributed to the presence of a trace of Cr(III) in animal diets (475, 476). More definitive results were obtained for diabetic rats (streptozotocin-induced), when injections of a Cr(III) containing substance from yeast significantly reduced blood glucose and free fatty acid levels (480). However, some researchers were able to separate non-Cr(III) containing biologically active fractions from the yeast extract (533, 534).

The current consensus (6, 464) is that a Cr(III) containing fraction, isolated from brewer's yeast [as well as some synthetic Cr(III) complexes], is able to affect glucose metabolism in mammals under carefully controlled conditions. Although it is likely that this ability is related to Cr(III) [given the other evidence on Cr(III) biological activity], the yeast extracts may also contain other insulin-potentiating substances. It is unlikely that a well-defined Cr(III) containing dietary factor can be isolated from plants or fungi, including yeast. Indeed, no such factors are known for metal ions, except for the Co(II) containing vitamin B_{12} (535). Many plants are known to accumulate Cr(III) (536, 537), and a correlation between Cr contents and the insulin-potentiating activity has been established (522). Interestingly, a Cr(III) containing fraction of yeast extract, added to Cr depleted yeast fermentation medium, significantly enhanced the rate of glucose uptake and oxidation by yeast cells whereas inorganic Cr(III) salts did not cause this effect (528, 538). Further studies on Cr(III) biological activity in yeast may shed some light on the mechanisms of Cr(III) action in higher organisms (6).

2. *Low Molecular Weight Chromium-Binding Substance*

There is a single report on the isolation of a Cr containing protein ($\sim$ 70 kDa, 4–6 Cr atoms) from the liver of mice injected with $CrCl_3$ (488). This, however, may be an artifact of the Cr^{III}–transferrin complex ($\sim$ 80 kDa), which binds most of the administered Cr(III) (Section IV.B) (494, 508). The majority of attempts to isolate a natural Cr(III) containing factor from mammals focused on low molecular weight substances (<5 kDa). In the early 1980s, Yamamoto and co-workers intended to study the mechanisms of Cr(VI) detoxification by injecting mice (539, 540), dogs (541), or rabbits (542) with Cr(VI) ($K_2Cr_2O_7$, 100–280 μmol Cr kg^{-1}). An anionic fraction ($\sim$ 1.5 kDa) containing Cr(III) and amino acid residues, was subsequently isolated from animal liver homogenates by ethanol precipitation, followed by anion-exchange and gel-filtration chromatographies. It consisted mainly of Asp, Glu, Gly, and Cys (in $\sim$ 1:2:2:1 ratio) with a Cr/amino-terminal residue ratio of 4:1 (542). A low molecular weight Cr-containing fraction was also detected in plasma and urine of Cr(VI) treated animals (539). While initially regarded as a form in which toxic Cr is excreted from the body (539), this species (called low molecular weight chromium-binding substance or LMCr) was found to enhance glucose metabolism in isolated rat adipocytes, that is, it showed similar activity to that of GTF (Sections IV.A and IV.C.1) (542). Notably, both GTF and LMCr preparations contained high proportions of Glu, Cys, and Gly (which are the components of an ubiquitous cellular tripeptide, GSH) (196) but, unlike for LMCr, the originally described GTF was cationic (523). Anionic Cr(III) containing fractions have been also isolated from yeast, but their compositions and biological activities were not studied in detail (543). A Cr(III) compound similar to LMCr in amino acid composition, but with much lower Cr/amino-terminal residue ratio (0.25:1), was isolated from bovine colostrum (with no added Cr); it also exhibited a GTF-like activity and was regarded as a Cr unsaturated form of LMCr (544). The Cr depleted form of LMCr (apo-LMCr) was prepared by treating LMCr (from rabbit liver) with EDTA (3.5 mM, pH 3.0) at 110°C for 1 h (545). Apo-LMCr did not show significant activity in the glucose uptake assay, but the activity could be restored (50–90%) after the addition of 4 mol equiv of $CrCl_3$ (1 h at 37°C, pH $\sim$ 7) (545).

More recently, Davis and Vincent (103, 546) claimed to have isolated LMCr from the reaction of Cr(VI) ($K_2Cr_2O_7$, 6.8 mmol Cr kg^{-1}) with bovine liver homogenate in vitro; the isolation procedure and composition of the obtained product were similar to those reported for rabbit liver (542). This substance [subsequently called chromodulin by analogy with a Ca(II) binding protein, calmodulin] (547) could be obtained in relatively large quantities and was studied in more detail (Section IV.C.3).

3. Chromodulin

The ability of chromodulin (from bovine liver) to enhance glucose uptake by isolated rat adipocytes (103), similar to that of the rabbit liver LMCr (542), was used to confirm its biological activity. However, the assays were tedious and led to large experimental errors (103). The use of commercially available fragments of insulin receptors (InR) and tyrosine kinase assay kits allowed simpler and more direct assessment of the potential role of chromodulin in glucose metabolism (496). The following types of activities were found for Cr^{III}–chromodulin (496, 497, 548): (1) activation ($\sim$2-fold) of phosphotyrosine phosphatase in rat adipocitic membranes; (2) insulin-dependent activation (up to $\sim$8-fold) of tyrosine protein kinase in rat adipocytic membranes or in isolated InR from rat liver; and (3) activation ($\sim$3.5-fold) of protein tyrosine kinase in the isolated active site fragment of the β-subunit of human InR. In agreement with the earlier results with LMCr (542), the activity was dramatically decreased for apo-chromodulin (Cr depleted form) and was restored by addition of 4 mol equiv of Cr(III) (but not other biologically relevant metal ions) (496, 497). There is an apparent difference in the role of chromodulin in the regulation of enzyme activities compared with the recently discovered fungal metabolite (a p-benzoquinone derivative, no metal ions involved), which binds to the β-subunit of InR and activates InR tyrosine kinase independently from insulin (499), as well as with that for V(IV/V) compounds that activate InR independent kinases (498).

The proposed mode of action of chromodulin in regulation of glucose metabolism in vivo (6, 547, 549–552) includes the following steps: (1) binding of insulin to the α-subunit of the cell InR (on the cell membrane surface); (2) phosphorylation of signal proteins by tyrosine kinase of activated InR (on the β-subunit of InR, within the cell), which triggers the cell response to insulin; (3) transfer of Cr(III) from the Cr^{III}–transferrin complex in blood plasma to apo-chromodulin within the cell; (4) binding of holo-chromodulin [Cr(III) loaded] to the β-subunit of InR (within the cell), which presumably assists to maintain the receptor in its active conformation and amplifies its kinase activity; and (5) excretion of holo-chromodulin from the cell. This mechanism is similar to that for a small regulatory protein calmodulin, which binds four Ca(II) ions in response to Ca(II) flux; the Ca(II) loaded form of calmodulin binds to various kinases and phosphatases, stimulating their activity (553). The proposed mechanism of chromodulin action is supported by the following results of biological studies: (1) the decrease in Cr(III) concentrations in blood plasma and increase of that in urine following oral glucose load (554); and (2) the glucose-induced Cr(III) binding to insulin-sensitive tissues (555). However, the proposed fast Cr(III) binding to apo-chromodulin in response to glucose stress is inconsistent with the well-known kinetic inertness of Cr(III) (Section IV.E) (80).

A mechanism for the Cr(III) promoted activation of InR, proposed by Vincent et al. (6, 547, 549–551), was preceded by several earlier hypotheses. Mertz et al. (556) suggested the formation of a ternary complex between insulin, Cr^{III}–GTF (Section IV.C.1.) and the insulin receptor. Ramasami and co-workers (557–559) used an analogy with Cr(III) induced stabilization of collagen in leather tanning to suggest that Cr(III) binding to insulin may provide a required conformation of the insulin molecule and prevent its destruction by proteases. However, further studies have shown that Cr(III) complexes interact with β-subunits of cell InR rather than with insulin itself, as selective blockages of these subunits prevented the activity of Cr(III) complexes (3, 496). The proposed similarity in the mechanisms of action of chromodulin and calmodulin (496, 547) may have arisen from the observation (560) that Cr(III) (added as $CrCl_3$, the speciation in the reaction media was not studied) can activate calmodulin in vitro in the absence of Ca(II). The proposed interactions of biological Cr(III) complexes with InR (a protein–lipid complex situated within the cell membrane) may be related to the reported changes in the properties of biological membranes (such as the fluidity or the intermembrane potential) associated with the binding of Cr(III) complexes (561, 562); the molecular mechanisms of Cr(III) membrane interactions are yet to be studied.

Many questions remain unanswered about the origin and structure of chromodulin (as well as LMCr, since both substances were isolated by similar methods and had similar compositions) (103, 542). The purified chromodulin [from the reaction of Cr(VI) with bovine liver homogenate] (103) was anionic in neutral aqueous solutions and included $\sim$7% of Cr found in the crude reaction mixture, or $\sim$2% of the added Cr(VI) [it is not clear what proportion of Cr(VI) was reduced to Cr(III) in the reaction]. The compositions and biological activities of other Cr containing fractions were not determined (103). The results of chemical analyses and gel-filtration studies (103) suggested that a chromodulin molecule consists of 2 Asp, 4 Glu, 2 Cys and 2 Gly residues, a maximum of 4 Cr atoms, and some oxo or hydroxo bridging ligands and has a mass of $\sim$1.48 kDa, but no elemental analysis data were available. Because addition of protease inhibitors to the reaction mixture did not change the yield and composition of chromodulin, it was probably not a fragment of a larger peptide (103). The ultraviolet–visible (UV–vis) and EPR spectroscopic studies of chromodulin (103) suggested that all Cr is present in the Cr(III) oxidation state and a spin coupling exists between several (probably four) Cr(III) centers within the molecule; the calculated parameters from the ligand-field splitting indicated predominantly oxygen coordination of Cr(III). The results of NMR spectroscopic and magnetic susceptibility studies (103) were consistent with the existence of a polynuclear Cr(III) assembly, but did not directly confirm it. These authors (103), as well as earlier researchers (542), were unable to assign a shoulder at 250–300 nm in the UV spectrum of Cr^{III}–chromodulin. The absorbance with $\lambda_{max} \sim 260$ nm for the GTF preparations (Section IV.C.1) was attributed to nicotinate (523), but no nicotinate was detected in the LMCr or chromodulin preparations (103, 542). Such absorbance could also be due to Cr^{III}—S bonds (97, 563), but this was not the case, since the Cr(III) depletion of chromodulin did not significantly decrease its absorbance at 260 nm (103, 545). However, the absorbance at 250–300 nm is readily assigned to S—S bonds; Figure 3 shows a comparison of UV spectra for the reduced and oxidized forms of glutathione (GSH and GSSG, respectively; GSH = γ-Glu-Cys-Gly) (196) under similar conditions to those of the chromodulin studies (103). Qualitatively, the UV spectrum of GSSG (Fig. 3) is similar to that of Cr^{III}–chromodulin (assuming that 1 mol of GSSG corresponds to 3.5 mol of Cr) (103, 529), but the absorbance at 250–300 nm for the latter was $\sim$1.5 times higher, possibly due to the contributions of Cr^{III}–ligand bonds. The presence of S—S bonds in chromodulin is in agreement with the following observations (103): (1) both

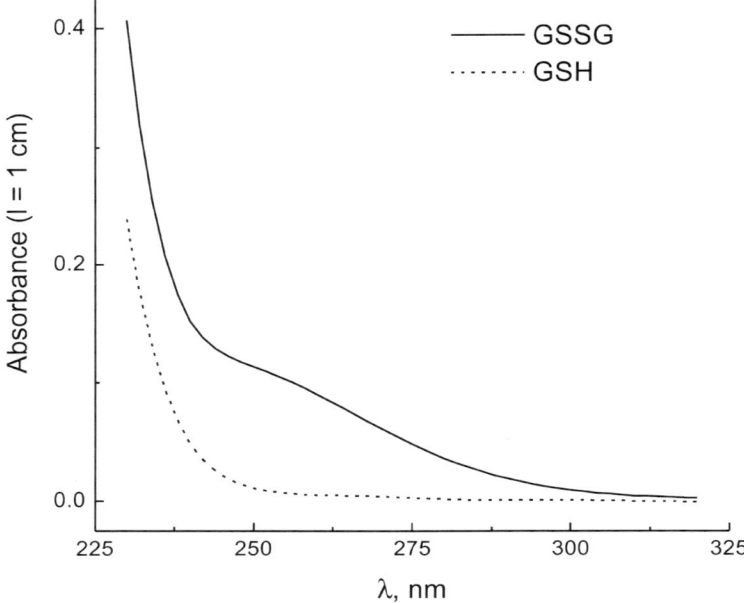

Figure 3. The UV spectra of oxidized (GSSG; 0.31 mM) and reduced (GSH; 0.62 mM) glutathione solutions in 50 mM NH$_4$OAc (Ar saturated; pH 6.5; 22°C). All reagents were from Sigma.

apo and holo forms of chromodulin were apparently stable toward aerial oxidation during the isolation and biological assays (no inert atmosphere was used), despite a high proportion of Cys residues (which would be rapidly oxidized by O$_2$ in neutral aqueous solutions with the formation of S—S bonds) (564, 565); (2) reductive alkylation of Cys residues during the attempted sequencing of the polypeptide led to its cleavage; and (3) the even numbers of all the amino acid residues in the chromodulin molecule indicate that it may consist of two equal subunits, bound by a S—S bond.

Given the composition of chromodulin (which is clearly related to GSSG), and its isolation method [based on the reduction of Cr(VI) to Cr(III) by liver homogenate] (103), the most likely product is a Cr(III) complex of GSSG, stabilized by additional Glu and Asp residues, which act as carboxylato donors for Cr(III). Liver cells contain GSH in millimolar concentrations (much higher than for any other small peptide) (196), and it is well known that GSH reduces Cr(VI) with the formation of Cr^{III}–GSH–GSSG complexes in neutral aqueous solutions (566–568). Polynuclear Cr(III) complexes could be formed during the ethanol precipitation, which was the initial step of chromodulin isolation (103).

Complexes of Cr(III) with GSH or GSSG are usually isolated from the reaction mixtures by methanol or ethanol precipitation (82, 566–568), probably resulting in the formation of polynuclear Cr(III) species (82). Recently, preliminary results have been reported on the isolation and characterization of polynuclear Cr(III) complexes from the reactions of Cr(VI) with GSH and amino acids under similar conditions to those used for the isolation of chromodulin (but in the absence of liver homogenate) (569, 570).

The possibility that chromodulin is a complex of Cr(III) with GSSG, Glu, and Asp, rather than with an integral polypeptide, is seemingly ruled out by those experiments where most of Cr(III) was removed by EDTA treatment, and the residue (apo-chromodulin) still had a molecular mass of $\sim$1.5 kDa, as measured by gel filtration (103, 545). However, there are large errors associated with determining molecular masses by gel filtration (571). This finding can be illustrated by the observation that the removal of four Cr atoms plus the bridging ligands from chromodulin (as a result of the EDTA treatment) did not significantly change its retention time in gel filtration, while the molecular mass was expected to decrease by $\sim$20% (103, 545). Mass spectrometric methods [ESMS and matrix-assisted laser desorption-ionization, time of flight (MALDI–TOF)] are now routinely used for precise determination of molecular masses of peptides (572). Unpublished data on MALDI–TOF mass spectrometry of chromodulin ($m/z = 1439$ for $[M + nH^+]$) have been cited in a recent work (573), but it is unclear whether this number is related to holo- or apo-form of chromodulin (only the latter would provide a definitive evidence for the existence of chromodulin as an integral peptide). The results of equilibrium studies of Cr(III) binding to apo-chromodulin (574) are also questionable. According to the published procedure (574), $CrCl_3$ was added to a solution of apo-chromodulin at pH 7.4, and after equilibrium was achieved (12 h at 4°C), the Cr^{III}–peptide complex was separated by ultrafiltration and the concentration of free Cr(III) in the eluent was measured. First, the use of a highly acidic complex, $CrCl_3$ (i.e., $[CrCl_2(OH_2)_4]^+$) (80), is not relevant to the possible interactions of chromodulin with stable anionic or neutral Cr(III) complexes in vivo (the same applies to the studies of $CrCl_3$ interactions with DNA, Section III.C.3) (261). Second, insoluble Cr(III) hydroxo complexes, formed during the hydrolysis of $CrCl_3$ (80), could be retained in the ultrafiltration compartment; therefore, the measurements of [Cr(III)] in the eluent would overestimate the binding constants (no data from control experiments in the absence of chromodulin were presented) (574). This possibility is supported by the observation that the UV–vis spectral changes during the addition of $CrCl_3$ to chromodulin corresponded to those due to the formation of Cr(III) hydroxo complexes (574). The reported high binding constants of Cr(III) to chromodulin (574) are inconsistent with the notion that holo-chromodulin loses Cr with time (103, 496, 497, 529). The results of equilibrium studies were explained in terms of highly cooperative binding of four Cr(III) ions by apo-chromodulin (574). However, these results also correspond with the binding of positively charged, polynuclear Cr(III) complexes (including tetramers), formed at early stages of $CrCl_3$ hydrolysis, to the negatively charged residues of GSSG, Asp and Glu (the likely constituents of chromodulin) (424, 575).

A great majority of reports by Vincent and co-workers on the observation, isolation and characterization of a purported natural Cr(III) containing factor, including a patented method of chromodulin isolation (546) and the recent pharmacokinetic studies (573), used a nonbiological source of Cr ($[CrO_4]^{2-}$). This, as admitted by the authors (549, 573), is a major limitation of their studies. Isolation of chromodulin from bovine liver without added Cr(VI) has been mentioned (103), but this material was isolated in trace quantities and not studied in any detail.

In summary, the existence of chromodulin (or LMCr) as a natural Cr(III) dependent biologically active substance (6, 103, 496, 497, 542, 544, 545, 547, 549–551, 573) is questionable. More direct (mass spectrometric or, ideally, X-ray crystallographic) evidence is required to confirm the integrity of chromodulin. If proven to be an integral polypeptide, chromodulin may belong to a family of small GSH related peptides that are involved in binding and detoxification of heavy metal ions (5, 576). Finally, the question about the relation of chromodulin to a Cr(III) binding substance ($\sim$ 5 kDa), found in blood plasma (494), remains unresolved. Despite all of the contradictions in the studies of Vincent and co-workers (496, 497, 548), related to chromodulin, their finding that some Cr(III) complexes are able to activate phosphorylation and dephosphorylation enzymes, particularly those of insulin receptors, is an important step towards the understanding of the possible biological roles of Cr(III).

D. Use of Chromium(III) in Food Supplements: Efficacy and Safety

Despite the contradictions surrounding the biological roles of Cr(III) (Section IV.A), this ion is currently officially recognized as an essential micronutrient for humans (577, 578). The recently published values of Dietary Reference Intakes of Cr(III) in the United States and Canada (ranging from 0.2 to 5.5 µg day^{-1} for infants or from 20 to 45 µg day^{-1} for adults) (578) are significantly lower than the previously accepted values of Recommended Dietary Allowances (RDA) (from 10 µg day^{-1} for infants to 200 µg day^{-1} for adults) (577). Measurements of Cr(III) contents in food sources suggested that up to 90% of the population in industrialized countries receive less than adequate amounts of dietary Cr(III). These amounts can be further decreased by carbohydrate- and fat-based diets, which promote Cr(III) excretion (3). As the biological activity of Cr(III) has been related to improvements in glucose and fat metabolism (579), the use of

CHROMIUM IN BIOLOGY: TOXICOLOGY AND NUTRITIONAL ASPECTS 215

supplemental Cr(III) was readily seen as one of the weapons in the fight against some of the most important health problems of the modern world: type II diabetes, obesity, and related cardiovascular diseases. More efficient glucose and fat metabolism in athletes could contribute to improved performance and body composition. In farm animals, Cr(III) supplementation was expected to lead to more efficient food consumption and reduced fat content in the meat. Not surprisingly, the production of Cr(III) containing food supplements for humans and animals has quickly become a multimillion dollar industry (6).

Chromium(III) tris-picolinate ([Cr(pic)$_3$], **XIIa** in Chart 2), first tested as a dietary supplement by Evans and co-workers (580–582), is currently the most widely used Cr(III) containing food supplement (6). The main reason for the choice of **XIIa** was its expected high bioavailability compared with other Cr(III) complexes, due to its neutrality, stability, and high lipophilicity (583, 584); although the latter has been disputed (102). Gastrointestinal absorption of **XIIa** in humans was reported to be significantly higher than that for CrCl$_3$ (1.5–5.2% vs 0.4–2.0%) (585, 586); **XIIa** was stable for several hours in artificial gastric juice (500). Unlike Cr(III) chloride or nicotinate, **XIIa** increased glucose uptake by skeletal muscle cultures, which was attributed to the ability of **XIIa** to increase the fluidity of biomembranes (584, 587). The proposed high bioavailability

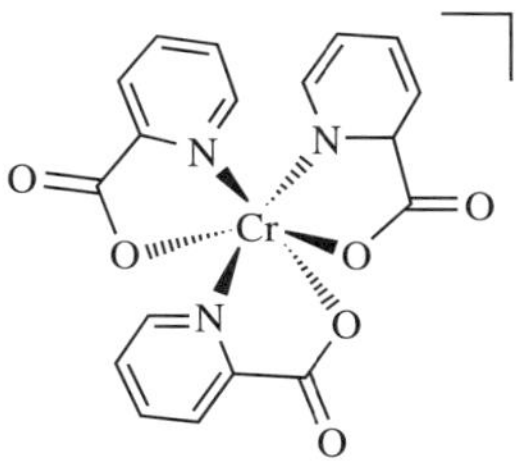

XIIa

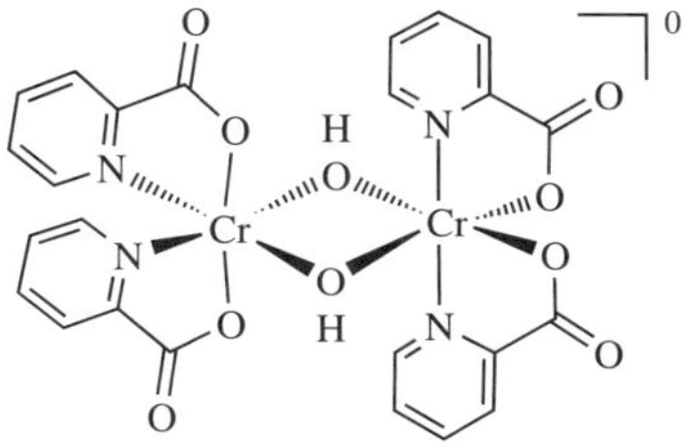

XIIb

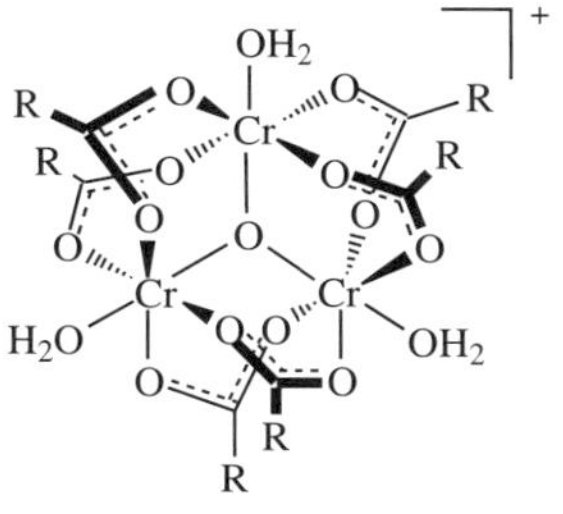

XIIIa R = Et
XIIIb R = Me

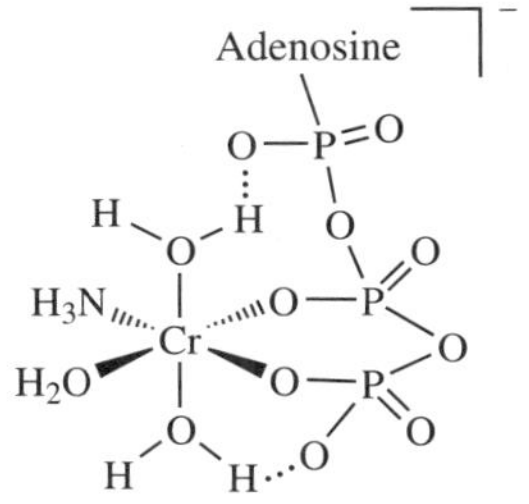

XIV

Chart 2

of **XIIa** is in agreement with that of [V^{IV}O(pic)$_2$], which is among the most potent V(IV) based insulin mimetics (588). An analogue of **XIIa**, [Cr(dipic)$_2$]$^-$ [dipic = 2,6-pyridinedicarboxylato(2−)] has been recently reported to cause an acute lowering of blood glucose levels in diabetic rats, an effect that is not observed for **XIIa** (589). The observed insulin-dependent potentiation of phagocytosis in cultured macrophages by **XIIa** (590) may be related to the improved immune responses in animals, supplemented with **XIIa** (486). The additions of **XIIa** have also been shown to modulate insulin-dependent Ca(II) transport in cultured vascular smooth muscle cells (591). This action may be related to the beneficial effects of Cr(III) supplementation in hypertension, caused by insulin resistance (468, 483).

Crystallographically characterized **XIIa** has been synthesized by the reaction of Cr(NO$_3$)$_3$ with excess ligand in dilute HNO$_3$ (pH 1.0) (592). An increase in the pH value led to significant formation of a dimer **XIIb** (Chart 2) (592, 593). Commercial preparations of **XIIa** are likely to contain significant amounts of **XIIb** (6).

The second most widely used Cr(III) food supplement is nicotinate (3-pyridinecarboxylate), a poorly characterized substance, probably a mixture of oligomeric Cr(III) complexes (6, 584). The use of nicotinate as a ligand for Cr(III) arose from its hypothetical role in GTF (Section IV.C.1) (6). Complexes of Cr(III) with other carboxylates (such as propionate) (594) and amino acids (such as L-methionine) (595) are used in food supplements for farm animals. Chromium-enriched brewer's yeast, probably containing Cr(III) as a mixture of amino acid complexes (Section IV.C.1), is used as a more "natural" alternative to synthetic Cr(III) food supplements (480, 524, 526, 596).

A critical assessment of the effects of Cr(III) supplementation in humans and animals has been reported in several recent reviews (4, 6, 460, 466, 467, 597). Although the results of independent studies were often highly contradictory, some conclusions could be drawn with a reasonable degree of confidence. For non-diabetic humans, Cr(III) supplementation is unlikely to improve glucose tolerance, body composition, or to assist in weight loss (4, 6, 460, 597, 598). This finding is reflected in two recent regulatory acts: (1) the conclusion of the United States Federal Trade Commission that "there is no basis for claims that chromium picolinate promotes weight loss and fat loss in humans" (599); and (2) an interim final rule of the United States Food and Drug Administration (FDA) to "prohibit the use on foods of a claim relating to the relationship between chromium and the risk in adults of hyperglycemia and the effects of glucose intolerance," as there is no unambiguous scientific basis for such a claim (600). On the other hand, significant positive effects (although not for all patients) of Cr(III) supplementation (200–1000 μg Cr day^{-1} as **XIIa**) were observed in type II and steroid-onset diabetes (3, 459, 479, 481, 601). Despite a large number of studies, the positive effects and economic soundness of Cr(III) supplementation in farm animals are still ambiguous, possibly because of the difficulties with the control of Cr(III) contents in non supplemented animal diets (4). No systematic studies were performed on the Cr(III) contents in meat or milk of supplemented animals and subsequent health effects for the consumers. Chromium(III) supplements showed some promise in the treatment of diabetes in cats, but not in dogs (602–604).

A trinuclear Cr(III) propionate ($[Cr_3O(OCOPr)_6(OH_2)_3]^+$; **XIIIa** in Chart 2) has been recently patented by Vincent and Davis for treatment of disorders related to Cr deficiency, or as a nutritional supplement (605). The structure of **XIIIa**, which is a member of a large family of trinuclear oxo-bridged transition metal ion carboxylates, has been determined by X-ray crystallography (606). Biological activity of **XIIIa** was first suggested on the basis of the existing structural and mechanistic information for a hypothetical Cr(III) dependent biomolecule, chromodulin [a complex of a carboxylato-rich polypeptide with a tetranuclear Cr(III) core; Section IV.C.3] (497). The observed abilities of **XIIIa** to activate phosphorylation–dephosphorylation enzymes (involved in insulin signaling) in vitro were similar, but somewhat lower (typically 1.2–2-fold), than for chromodulin under similar conditions (Section IV.C.3); thus, **XIIIa** was considered as a functional biomimetic of chromodulin (497). Surprisingly, **XIIIb** (Chart 2), the acetate analogue of **XIIIa**, exhibited no biological activity (497), suggesting that a higher degree of lipophilicity in **XIIIa** compared with **XIIIb** was required for the interactions with cell membrane enzymes (496, 548). Unlike chromodulin, **XIIIa** was expected to be relatively stable in gastric juices, as **XIIIa** can be recrystallized from dilute acidic solutions (497); however, no studies of stability and speciation of **XIIIa** in biological media have been performed as yet. The most prominent biological action of **XIIIa** was its ability to cause significant decreases in blood cholesterol and triglycerides (in rats, 20 μg Cr kg^{-1} day^{-1} administered intravenously for 12 weeks); no such effect was observed for chromodulin, presumably due to its easy degradation or excretion in vivo (482). In contrast to the results of Anderson and co-workers (475, 476) (Section IV.A), no measures to induce Cr deficiency in the rats were required to observe the positive results of supplementation with **XIIIa** (482). Recent pharmacokinetic studies in rats, using ^{51}Cr- or ^{14}C-labeled **XIIIa**, suggested that the trimeric complex enters cells intact, but degrades within 24 h after injection (552).

Unlike Cr(VI), a recognized human carcinogen (Section III.A) (2), Cr(III) is often considered as one of the least toxic transition metal ions on the basis of its poor absorptivity and kinetic inertness, as well as from the lack of acute toxicity of large doses of Cr(III) compounds in rats (3). A concern about the safety of the use of Cr(III) in food supplements firstly arose after the finding of Stearns et al. (607) that **XIIa** (0.050–1.0 mM Cr) caused chromosome damage in cultured mammalian cells. In addition, mutations in the *hprt* locus, ultrastructural changes in mitochondia, and apoptosis, caused by **XIIa** in CHO cells, have been recently reported (608). These effects were due, at least in part, to the picolinato ligand, rather than Cr(III) ion (607, 608). Nonetheless, it begs the question as to the safety of ingesting the chemical. Recent studies have shown that **XIIa** is efficiently metabolized in hepatic cell cultures, leading to the release of Cr(III) and the formation of N-1-methylpicotinamide as a primary organic metabolite (609). McCarty (610, 611) (the exclusive patent rights holder for the manufacturing of **XIIa** in the United States) (612) disputed the significance of these results, as the concentrations of **XIIa** used (607) were $\sim$1000-fold higher than those of nutritionally relevant doses. Another study (613) showed low levels of oxidative DNA damage and lipid peroxidation in cell cultures by both **XIIa** and Cr(III) nicotinate. No appreciable oxidative DNA damage was found in humans receiving recommended doses of **XIIa** (400-μg Cr day^{-1} for 8 weeks), although only one of the possible markers of DNA damage (the formation of 5-hydroxymethyl-2'-deoxyuridine) was determined (614). However, there may be an increased risk of genotoxicity resulting from accumulation of **XIIa** or the products of its metabolism in cells during prolonged use (430, 612).

In search of the possible mechanisms of genotoxicity due to **XIIa**, Vincent and co-workers (429) found that oxidative cleavage of plasmid DNA in neutral aerated aqueous solutions can be caused by physiologically relevant concentrations of **XIIa** (120 μM) in the presence of a strong reductant (Red = ascorbate or dithiothreitol; 5.0 mM); no DNA cleavage by **XIIIa** was observed under similar conditions (615). This difference was explained by the higher Cr(III)/Cr(II) redox potential of **XIIa**, which facilitates the formation of Cr(II) in the presence of a reductant, leading to O$_2$ activation (Scheme 11) (212, 429, 574).

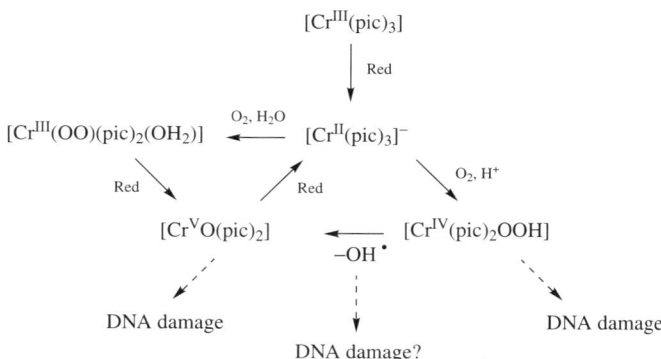

Scheme 11. Possible pathways in DNA damage induced by the **XIIa** + Red + O$_2$ (Red = ascorbate or dithiothreitol) systems in neutral aqueous solutions. [Based on the data of (417, 428, 429).]

Participation of ROS in DNA cleavage induced by the **XIIa** + Red + O_2 systems is supported by the following observations (429): (1) inhibition of DNA cleavage by the absence of O_2 [although this does not always mean that ROS are the primary oxidants of DNA (Section III.C.4) (400)], or by the presence of catalase; and (2) enhancement of DNA cleavage by the presence of H_2O_2. Hydroxyl radicals (•OH) were originally proposed as the main DNA damaging species in the **XIIa** + Red + O_2 systems (Scheme 11) (429), however, participation of •OH in the redox reactions involving transition metal ions is unlikely (Section II.A.5).

Thus, the proposed risk factor for Cr(III) containing food supplements, according to Vincent and co-workers (429, 615), was the ability of Cr(III) to be reduced to Cr(II) by biological reductants such as ascorbate, with the subsequent formation of ROS (Scheme 2). However, the described type of reactions (Fenton chemistry) is also characteristic for Fe(III) and Cu(II), and DNA cleavage by the ascorbate + O_2 system in the presence of trace amounts of Fe(III) or Cu(II) complexes is well known (380). As both Fe(III) and Cu(II) are essential bioelements, mechanisms have evolved to prevent the oxidative damage caused by these ions, which includes the ubiquitous presence of ROS scavenging enzymes such as catalase and SOD (390). The observation that DNA cleavage by the **XIIa** + Red + O_2 system could not be completely prevented by catalase (429) suggests the presence of an alternative mechanism of DNA damage, most likely involving Cr(IV) or Cr(V) species (Scheme 11), particularly those with O_2 derived ligands (21, 25). This suggestion is supported by a number of recent findings (1) the in vitro DNA cleavage by a model Cr(V) complex does not require the presence of ROS and it is not affected by catalase (400); (2) autooxidation of ascorbate produces ROS, which may oxidize Cr(III) complexes with the formation of stabilized Cr(V) species (21, 70); (3) the picolinate ligand stabilizes Cr(IV) and Cr(V) oxidation states at physiological pH values (39, 202); (4) Cr(V) oxo complexes are known to form [with the intermediate formation of Cr(III) or Cr(IV) peroxo complexes] during the oxidation of Cr(II) by air oxygen (417, 428); and (5) there is a clear correlation between the ability of Cr(III) complexes to be oxidized to relatively stable Cr(V) species and their mutagenicity in the cell assays (37, 175). The latter led to the suggestion (12) that Cr(III) complexes that accumulate in cells can be reoxidized (enzymatically or by organic peroxides) to Cr(IV) or Cr(V) complexes, which, in turn, can cause DNA damage. Reoxidation of accumulated Cr(III) could lead to adverse health effects following the prolonged use of Cr(III) containing supplements (5, 430, 612). The potential importance of long-term effects of Cr(III) toxicity has been emphasized by the recent finding (616) that preconceptional exposure of male mice to a single large dose of Cr(III) (intraperitoneal injection of 1.0 mmol Cr kg^{-1} as CrCl$_3$, pH 4.1) was carcinogenic in the offspring. The abilities of different types of Cr(III) complexes to be oxidized under biological conditions by enzymatic or nonenzymatic systems with the formation of DNA damaging Cr(V/IV) species are yet to be determined.

In summary, the acquired evidence on the use of Cr(III) containing food supplements in humans (4, 6, 460, 597) does not support the opinion (3) that there is a widespread Cr(III) deficiency in the general population. This suggestion was based on the previous values for safe and adequate dietary intake of Cr(III) (577), which were likely to be overestimated (5) from outdated analytical results on Cr(III) contents in food sources (511). A recent study in Germany found that the adequate Cr(III) intake for adults was < 20 µg day^{-1}, and higher intakes could lead to Cr(III) related allergies (617). This is close to the currently accepted values in the United States (578). Evidence is growing for the positive effects of Cr(III) supplementation in type II diabetes and other disorders related to insulin resistance (4, 479, 481, 484). It is likely that the use of Cr(III) compounds will shift from uncontrolled intake in food supplements to medically supervised applications. However, much still needs to be learned about the activity, selectivity (Section IV.E), and safety (including long-term effects) (5, 6) of various types of Cr(III) complexes before their clinical use can be recommended.

E. Molecular Mechanisms of Chromium(III) Biological Activity

An overview of the published data (Sections IV.A–IV.D) suggests that at least some Cr(III) complexes are able to affect glucose and fat metabolism both in vitro and in vivo, most likely through the interactions with cellular insulin receptors (6, 547, 549–551). It is not yet clear whether this ability represents an intrinsic biological role of Cr(III) (3, 6, 465), or whether it acts as an exogenous pharmacological agent (5). Currently, there is a similar state of knowledge about the insulin-mimetic action of V(IV/V) compounds (498, 618). At present, one can only speculate on possible chemical reactions involved in Cr(III) biological activity; however, a review of published hypotheses in relation to the known chemical properties of Cr(III) may provide useful suggestions for future mechanistic studies.

General properties of Cr(III) coordination compounds include the following (80): (1) they have six-coordinate octahedral structures; (2) they are kinetically inert; (3) there is a high affinity for negatively charged oxygen donors, such as phosphate or carboxylate; and (4) base-catalyzed hydrolysis occurs with the formation of polynuclear hydroxo- and oxo-bridged complexes. The inertness of Cr(III) complexes implies their structural, rather than catalytic, role in biological systems, because catalysis requires fast ligand-exchange processes (4). On the contrary, the currently best developed hypothesis on the mechanisms of Cr(III) action (Section IV.C.3) (6, 547, 549–551) includes a fast Cr(III) binding to a specific peptide, chromodulin, in response to increased blood insulin levels.

Several methods are known in chemistry to increase (up to several orders of magnitude) the rates of ligand-exchange reactions in Cr(III) complexes: (1) to use multidentate ligands such as porphyrins or Schiff bases, or polycarboxylates such as EDTA (619); (2) to perform the ligand-exchange reactions at an aqueous/organic interface, that is, to use micellar catalysis (620); and (3) to convert Cr(III) into a kinetically labile oxidation state [Cr(II), Cr(IV), or Cr(V)] (80, 97, 574). The first two ways are directly relevant to biological systems, and may be involved in the enzyme activation by Cr(III) complexes (Section IV.C.3) (496, 497, 548). The third mechanism is likely to be related to the detrimental action of excessive Cr(III) used as a food supplement (Section IV.D) (12, 175, 429, 574, 615).

The proposed biologically active forms of Cr(III) (such as chromodulin or **XIIIa**, Sections IV.C.3 and IV.D) are hydroxo- or oxo-bridged polynuclear assemblies, stabilized by carboxylato-rich ligands (103, 496, 497). Formation of such structures may have analogies in the mechanism of collagen stabilization (leather tanning) by Cr(III), where polynuclear Cr(III) complexes, formed during the hydrolysis of $CrCl_3$ or $Cr_2(SO_4)_3$ in tanning solutions (at pH ~ 4), bind to carboxylato groups of the collagen macromolecule (103, 496, 497, 559, 621, 622). Given very low Cr(III) concentrations in biological systems (511–513), it is highly unlikely that a polynuclear Cr^{III}–peptide complex of a certain structure (such as chromodulin) (103) can form spontaneously in the cell. If chromodulin is a natural biologically active form of Cr(III) (6, 547, 549–551), then a yet undiscovered mechanism for its synthesis should exist. Possible analogies for such a mechanism may be found in the formation of polynuclear Fe^{III}–oxo–hydroxo assemblies in the Fe^{III} storage protein, ferritin (623).

A possible link between the effects of Cr(III) and Cr(VI) on carbohydrate metabolism and enzyme activation has been overlooked in the studies of Cr(III) as a nutrient (5). Treatment of rats with large doses of Cr(VI) ($Na_2Cr_2O_7$; 20–40 mg kg^{-1} subcutaneously) induced a severe, but short term, decrease in blood insulin levels (624). A significant insulin-independent stimulation of 3-O-methylglucose uptake by isolated rat adipocytes was achieved in the presence of $Na_2Cr_2O_7$ [50–300 μM Cr(VI)]; this effect was strictly ATP dependent, which implies a relationship to phosphorylation reactions (625). A detailed study of this effect by Yurkow and Kim (295, 626) showed that treatment of intact rat hepatoma cells with Cr(VI) (100 μM) induced the insulin-independent activation of certain types of protein kinases, which led to increased phosphorylation levels in various proteins; these phosphorylated proteins could then participate in insulin signaling pathways. Treatment with Cr(VI) did not affect the insulin-dependent phosphorylation on β-subunits of insulin receptors. Thus, Cr(VI), similar to V(V) (618) but unlike the biologically active Cr(III) complexes (496, 497, 618), acts as an insulin mimetic rather than potentiator. Unlike V(V), however, the action of Cr(VI) resulted from kinase activation rather than phosphatase inhibition. Alterations in protein phosphorylation patterns in Cr(VI) treated cells were related to an increase in cellular ROS levels (determined with a redox-sensitive dye or with radical scavengers) (295, 296). However, the positive results using these assays could be due to the formation of reactive Cr(V/IV) species during the intracellular reduction of Cr(VI) to Cr(III) (Section III.B.5) (184). The rates of Cr(VI) reduction in the cells are probably determined by Cr(VI) diffusion through the cell membranes; thus, Cr(III) was the dominant (>90%) oxidation state in the Cr(VI) treated mammalian cells (as shown by microprobe XANES spectroscopy) (89). Therefore, it is unlikely that Cr(VI) as such is responsible for the kinase activation seen in intact Cr(VI) treated cells (295, 626). The role of Cr(VI), which permeates the cell membranes much more easily than Cr(III) (Section III.B.1), is probably its delivery of high intracellular Cr(III) concentrations. Such high intracellular Cr(III) concentrations can then act as an insulin mimetic [while insulin potentiation is observed at lower Cr(III) concentrations] (495).

The disturbances in protein phosphorylation patterns in Cr(VI) treated cells are considered among the possible reasons for Cr(VI) toxicity and carcinogenicity (295, 626). The question then arises, as to whether the proposed beneficial action of Cr(III) in activation of insulin receptor tyrosine kinase (496, 497) is, in fact, a sign of Cr(III) toxicity (5). Unpredictable changes in the concentrations of phosphorylated proteins in the presence of excess Cr(III) may lead to abnormalities in the cell signaling pathways and ultimately to cancer (5). An answer to this dilemma may lie in selectivity studies (which are yet to be performed) of different types of Cr(III) complexes toward various kinases or phosphatases. Clearly, the Cr(III) complexes of potential use as anti-diabetics should be highly selective in the activation of protein tyrosine kinase of the β-subunit of the insulin receptor (496, 497). On the other hand, the potential ability of some Cr(III) complexes to selectively activate non-insulin dependent protein kinases may lead to beneficial effects, such as stimulation of immune responses or antitumor activity (627, 628).

The ability of some Cr(III) complexes to activate ATP dependent kinases (496, 497) may seem surprising, as stable Cr^{III}–ATP complexes (such as **XIV** in Chart 2; one of the many possible stereoisomers is shown) (629) are well known to act as kinase inhibitors. Chromium(III) in these complexes is a kinetically inert replacement of a natural ATP binding ion, Mg(II) (629–631). However, a Cr(III) complex, $[Cr(NH_3)_5(OH)]^{2+}$, has recently been shown to promote the in vitro phosphorylation by ATP of the hydroxo groups of Ser and Thr residues in bovine serum albumin at pH 7.4, while several other Cr(III) complexes with N- and O-donor ligands did not possess such activity (632). The proposed phosphorylation mechanism (Scheme 12) (632) involves the formation of a ternary complex of Cr(III), ATP, and the amino acid residue. This mechanism may be related to the kinase-catalyzed phosphorylation of protein Tyr residues,

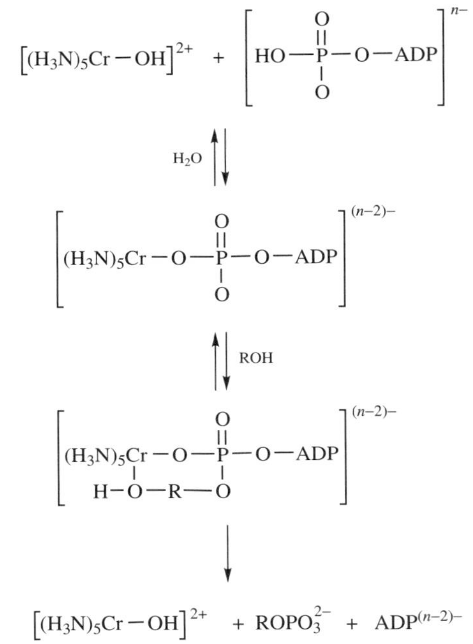

$$[(H_3N)_5Cr-OH]^{2+} + \left[HO-\underset{\underset{O}{|}}{\overset{\overset{O}{\|}}{P}}-O-ADP\right]^{n-}$$

$$\Updownarrow H_2O$$

$$\left[(H_3N)_5Cr-O-\underset{\underset{O}{|}}{\overset{\overset{O}{\|}}{P}}-O-ADP\right]^{(n-2)-}$$

$$\Updownarrow ROH$$

$$\left[\begin{array}{c}(H_3N)_5Cr-O-\underset{\underset{H-O-R-O}{|}}{\overset{\overset{O}{\|}}{P}}-O-ADP\end{array}\right]^{(n-2)-}$$

$$\downarrow$$

$$[(H_3N)_5Cr-OH]^{2+} + ROPO_3^{2-} + ADP^{(n-2)-}$$

Scheme 12. Proposed mechanism of phosphorylation of Ser and Thr residues in bovine serum albumin by ATP, promoted by $[Cr(NH_3)_5(OH)]^{2+}$. [Adapted from (632).]

promoted by chromodulin or by **XIIIa** (496, 497, 632). The phosphorylation in the presence of $[Cr(NH_3)_5(OH)]^{2+}$ was relatively slow (hours timescale) (632); however, in biological systems, similar reactions may be accelerated by the formation of Cr^{III}–kinase complexes (496, 497). Alternatively, the phosphorylation mechanism may not directly involve the Cr(III) atoms in chromodulin or **XIIIa**. Instead, the formation of a supramolecular assembly (633) between the Cr(III) complex and the kinase molecule may provide a required conformation of the kinase active center (547, 553). In this case, it seems unlikely that similar biological activities (497) can be achieved by molecules so different in mass ($\sim$ 1480 vs 664 Da), charge (anionic vs cationic), and composition, as chromodulin and **XIIIa** (Sections IV.C.1 and IV.D).

If kinetically inert Cr(III) complexes (80) have some biological functions, these are likely to involve chemical changes in Cr(III) bound ligands (95). Several types of ligand-centered reactions in biologically relevant Cr(III) complexes have been reported: (1) intramolecular lactonization (634) in a Cr^{III}–Cys complex (Scheme 13a) (97); (2) hydrolysis of the Cr(III) bound Cys ethyl ester (Scheme 13b) (95); and (3) Cr(III) catalyzed hydrogen isotope

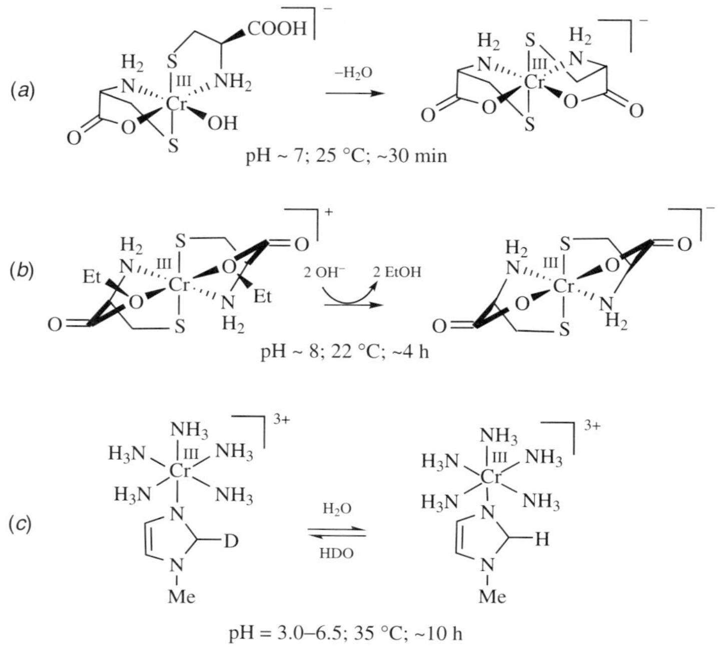

Scheme 13. Some types of intramolecular reactions in Cr(III) complexes: lactonization [(a) adapted from (97)]; ester hydrolysis [(b) adapted from (95)] and hydrogen isotope exchange [(c) adapted from (635).]

exchange in methylimidazoles (Scheme 13c) (635). The first two reactions are probably driven by the high affinity of Cr(III) for carboxylato ligands (80), and may be involved in conformational changes of Cr(III) complexes with Cys-rich peptides in relation to possible interactions with insulin receptors (95). The third reaction demonstrates the pronounced ability of Cr(III) to increase the acidity of the C—H bond in the 2-position of 1-methylimidazole; for example, the isotope exchange rates for the Cr(III) bound ligand are $\sim 10^5$ times higher than those for the Co(III) bound ligand and ~ 20 times higher than those for the protonated ligand (635). The validity of the latter results, however, has been recently disputed (636). Biological implications of these reactions, as well as of other possible intramolecular rearrangements in Cr(III) complexes, are yet to be explored.

Stearns (5) proposed two alternatives to the mechanism of Cr(III) induced promotion of glucose metabolism, developed by Vincent et al. (6, 547, 549–551). According to the first, the enhanced glucose uptake by intact cells represents a nonspecific cell reaction toward a toxic substance [in this case Cr(III)] (637). However, this possibility conflicts with the observations of specific interactions of Cr(III) complexes with insulin receptors (496, 497). The second alternative, which was also suggested by Vincent and co-workers (573), considers the participation of Cr(III) in the Fe(III) metabolic pathways. The two metal ions have similar sizes and coordination modes (638), and share the same transport protein (transferrin; Section IV.B) (494, 504, 573). Previously, the decrease in Fe(III) uptake as a result of Cr(III) supplementation was considered among the possible detrimental effects of Cr(III), but this decrease was found to be too small to cause a significant Fe(III) deficiency (4). On the contrary, the reduced Fe(III) uptake was proposed as the basis of a beneficial effect of Cr(III) supplementation (5), as excessive levels of Fe(III) have been related to oxidative stress, which may trigger many diseases, including diabetes (312, 390, 639). While the reduction of oxidative stress may prevent the development of diabetes, it is unclear how it can cause the decrease in blood glucose levels in existing diabetic conditions (479, 481). Furthermore, since Cr(III) is a redox-active ion (Sections III.C.2 and IV.D), the replacement of Fe(III) with Cr(III) is unlikely to cause a significant reduction of the oxidative stress. Nevertheless, the possibilities of Cr(III) sharing the metabolic pathways with Fe(III), including the formation of relatively kinetically labile Cr^{III}–porphyrin complexes (619), and the consequences of such sharing for the biological activity of Cr(III), are worthy of further investigation.

In summary, the many controversies in Cr(III) biochemistry over the last 40 years (6), provide a rich ground for further studies. One of the most promising directions seems to be the investigation of interactions of different types of Cr(III) complexes with various kinases and phosphatases, particularly those related to insulin signaling (547, 549–551). The aims of such studies should be (1) the clarification of the roles of Cr(III) in glucose metabolism; and (2) the development of selective Cr(III) based anti-diabetic drugs. One of the major limitations in Cr(III) biochemistry is the lack of knowledge about Cr(III) speciation in biological media, which complicates the interpretation of the results of biological assays (Section IV.A). Methods for characterization of metal complexes developed in the last ten years, such as ESMS (95, 640–642) and XAS (86, 87), may assist in solving these problems of speciation (Section II).

V. CONCLUDING REMARKS

This part of the chapter presents a brief summary of the authors' opinion on the current position of Cr biochemistry. While genotoxicity and carcinogenicity of Cr(VI) are well established, most Cr(III) compounds are considered nontoxic due to their poor bioavailability (2, 112, 208). However, Cr(III) complexes of biological macromolecules formed within the cells following the uptake and reduction of Cr(VI) are highly genotoxic (181, 251). The essential nature of Cr(III) for glucose and fat metabolism in humans (3) is questionable (5); however, some Cr(III) complexes have been reported to promote insulin action through interactions with cellular insulin receptors (6).

Progress in understanding the mechanisms of Cr biological activity (both beneficial and detrimental) at cellular and subcellular levels parallels the developments in cell and molecular biology. While a number of cellular responses to Cr compounds have been revealed, the mechanisms of Cr action in cells are far from fully understood. Current interest in the mechanisms of Cr(VI) induced genotoxicity and carcinogenicity spreads beyond the possible adverse health effects of exposure to Cr(VI) at the workplace or through environmental pollution, notwithstanding the importance of these problems (9). With Cr(VI) being a paradigm of carcinogenicity for transition metals (181), insight into the general mechanisms of carcinogenicity caused by environmental factors can be gained through the studies of Cr(VI) interactions with cells and biomolecules. It can be expected that new approaches [such as gene expression microarray data analysis (290, 643) or quantitative proteome analysis (644)] will be widely applied in the near future to the studies of Cr genotoxicity.

Understanding the mechanisms of Cr biological activity is complicated by the richness of its oxidation–reduction and ligand-exchange chemistry in biological media (12). Studies on Cr biochemistry often suffer from the lack of consideration of the fundamental chemical properties of this element. Some of the most common mistakes are the following: (1) the detection of DMPO—OH adducts in EPR spectra of the Cr^{VI} + Red + O_2 + DMPO systems is interpreted as the evidence for the formation of discrete •OH radicals (such adducts are more likely to form from the direct oxidation of DMPO by Cr(V/IV) species; Section II.A.5); and (2) $CrCl_3$ or $Cr(NO_3)_3$ are used as "representative" Cr(III) compounds, without considering their rapid hydrolysis in neutral aqueous media (Section III.C.3). There is limited knowledge of the solution speciation of Cr in various oxidation states [particularly Cr(III)] in biological media. Future in vitro studies of the insulin-potentiating activity of Cr(III) are likely to concentrate on the interactions of different types of Cr(III) complexes with phosphorylation–dephosphorylation enzymes, and on solution structures of the active Cr(III) species. Further insight into the mechanisms of Cr(VI) induced DNA damage is likely to be provided through the in vitro studies of the Cr^{VI} + Red + O_2 + DNA reactions in the presence of important intracellular ligands, such as carbohydrates, taking into account the existing mechanistic information on DNA reactions with model Cr(V/IV) complexes (Section III.C.4).

ACKNOWLEDGMENTS

We thank our many collaborators (listed in the references) for their part in this research, and Dr. Antonio Bonin (University of Sydney) for valuable comments on the manuscript. Research on Cr biochemistry in the group was supported by Australian Research Council (ARC) Large Grants and RIEFP grants for the EPR spectrometers, ESMS equipment, and 10-element Ge detector, the National Health and Medical Research Council, and the University of Sydney Cancer Research Fund. X-ray absorption spectroscopy was performed at the Australian National Beamline Facility (ANBF) at the Photon Factory and SRIXE experiments at the Advanced Photon Source (APS) at Argonne National Laboratories. Travel to both were supported by the Australian Synchrotron Research Program, which is funded by the Commonwealth of Australia under the Major National Research Facilities Program. Use of the APS was supported by the U.S. Department of Energy, Basic Energy Sciences, Office of Science, under Contract No. W-31-109-Eng-38. The XAS research at Brookhaven National Laboratories and SSRL was also supported by the Access to Major Facilities Program Access funded by the Department of Industry, Science and Resources and managed by the Australian Nuclear Science and Technology Organisation. Work done (partially) at SSRL, which is operated by the Department of Energy, Office of Basic Energy Sciences. The SSRL Biotechnology Program is supported by the National Institutes of Health, National Center for Research Resources, Biomedical Technology Program, and by the Department of Energy, Office of Biological and Environmental Research.

ABBREVIATIONS

A or dA	Deoxyadenosine phosphate
AAS	Atomic absorption spectroscopy
Aib	α-Aminoisobutyrato = 2-amino-2-methylpropanoato(−)
Ala	L-Alanine
Ala$_3$	Tri-L-alanine
Asp	L-Aspartic acid
asc	L-Ascorbate(2−) bound via the enediolato group
asc*	L-Ascorbato(2−) bound via the ethanediolato group
ATP	Adenosine 5′-triphosphate
bpy	2,2′-Bipyridine
bisTris	Bis(hydroxyethyl)amino tris(hydroxymethyl)methane
BPDE	Benzo[a]pyrene diolepoxide
BSA	Bovine serum albumin
C or dC	Deoxycytidine phosphate
CHO	Chinese hamster ovarian
cit	Citrato(2−)
CsA	Cyclosporin A

Cys	L-Cysteine
CysOEt	L-Cysteine ethyl ester
3D	Three dimensional
dipic	2,6-Pyridinedicarboxylato(2−)
DNA	Deoxyribonucleic acid
DMPO	5,5-Dimethylpyrroline-N-oxide
DMSO	Dimethylsulfoxide
DOPA	3,4-Dihydroxyphenylalanine
ed	1,2-Ethanediolato(2−)
EDTA	edta(ligand) = Ethylenediaminetetraacetate
ehba	2-Ethyl-2-hydroxybutanoato(2−)
en	Ethylenediamine = 1,2-ethanediamine
ENDOR	Electron nuclear double resonance
EPA	Environment Protection Agency
EPR	Electron paramagnetic resonance
ESMS	Electrospray mass spectrometry
FDA	Food and drug administration
G or dG	Deoxyguanosine phosphate
GFAAS	Graphite-furnace AAS
GLC	Gas–liquid chromatography
Glu	L-Glutamic acid
Gly	Glycine
GlyGly	Glycylglycine
glyc	Glycerolato(2−)
GSH	Glutathione reduced form (γ-Glu-Cys-Gly)
GSSG	Glutathione oxidized form
GTF	Glucose tolerance factor
HEPES	4-(2-Hydroxyethyl)-1-piperazineethanesulfonic acid
His	L-Histidine
HLF	Human lung fibroblast
hmba	2-Hydroxy-2-methylbutanoato(2−)
hprt	Hypoxanthine phosphoribosyl transferase
IARC	International Agency for Research on Cancer
ICP–MS	Inductively coupled plasma–mass spectrometry
IR	Infrared
InR	Insulin receptor
lac	Lactato(2−)
LMCr	Low molecular weight chromium-binding substance
MALDI-TOF	Matrix-assisted laser desorption-ionization, time of flight
mampa	5,6-(4,5-Dichlorobenzo)-3,8,11,13-tetraoxo-2,2,9,9-tetramethyl-12,12-diethyl-1,4,7,10-tetraazacyclotridecanate(4−)
MAP kinases	Mitogen-activated protein kinases

MEM	Minimum essential medium
Met	L-Methionine
mRNA	Messenger RNA
NAD(P)H	Nicotinamide adenine dinucleotide (phosphate), reduced form
NAD	Nicotinamide adenine dinucleotide, oxidized form
NFκB	Nuclear factor κB (transcription factor)
NHE	Normal hydrogen electrode
NMR	Nuclear magnetic resonance
8-OHdG	8-Hydroxydeoxyguanidine
ox	Oxalato(2−)
PEL	Permissible exposure limit
phen	1,10-Phenathroline
P_i	Inorganic phosphate
pic	Picolinato = 2-pyridinecarboxylato(−)
PIXE	Proton-induced X-ray emission
POBN	α-(4-Pyridyl-1-oxide)-*N-tert*-butylnitrone
qa	Quinato, (1*R*,3*R*,4*R*,5*R*)-1,3,4,5-tetra hydroxycyclohexanecarboxylato(2−)
RDA	Recommended dietary allowance
Red	Reductant
REL	Recommended exposure limit
RNA	Ribonucleic acid
ROS	Reactive oxygen species
salen	*N,N*-Ethylenebis(salicylideneiminato)(2−)
SEM	Scanning electron microscopy
Ser	L-serine
SOD	Superoxide dismutase
SRIXE	Synchrotron-radiation-induced X-ray emission
uPA	Urokinase-type plasminogen activator
uPAR	Urokinase-type plasminogen activator receptor
Red	Reductant
T or dT	Deoxythimidine phosphate
Thr	L-Threonine
tla	Thiolactato(2−)
TMPD	2,2,6,6-Tetramethyl-4-piperidone
TPN	Total parenteral nutrition
Tris	Tris(hydroxymethyl)aminomethane
Tyr	L-Tyrosine
UV–vis	Ultraviolet–visible
XAS	X-Ray absorption spectroscopy
XAFS	X-Ray absorption fine structure

| XANES | X-Ray absorption near-edge structure |
| XRD | X-Ray diffraction |

REFERENCES

1. *The Biological and Environmental Chemistry of Chromium*, S. A. Katz and H. Salem, Eds.,VCH Publishers, New York, 1994.
2. IARC, *IARC Monographs on the Evaluation of Carcinogenic Risk to Humans: Chromium, Nickel and Welding.*, Vol. 49, International Agency for Research on Cancer, Lyon, France, 1990.
3. R. A. Anderson, *J. Am. Coll. Nutr.*, *17*, 548 (1998).
4. H. C. Lukaski, *Ann. Rev. Nutr.*, *19*, 279 (1999).
5. D. M. Stearns, *BioFactors*, *11*, 149 (2000).
6. J. B. Vincent, *Polyhedron*, *20*, 1 (2001).
7. *Chromium in the Natural and Human Environments*, J. O. E. Nriagu and E. E. Nieboer, Eds., John Wiley & Sons, Inc., New York, 1988.
8. M. Cone, *Los Angeles Times*, November 13, 2000.
9. C. Pellerin and S. M. Booker, *Environ. Health Perspect.*, *108*, A402 (2000).
10. M. G. Steinpress and A. C. Ward, *Ground Water*, *39*, 321 (2001).
11. D. E. Kimbrough, Y. Cohen, A. M. Winer, L. Creelman, and C. Mabuni, *Crit. Rev. Environ. Sci. Techn.*, *29*, 1 (1999).
12. R. Codd, C. T. Dillon, A. Levina, and P. A. Lay, *Coord. Chem. Rev.*, *216–217*, 533 (2001).
13. E. König, in *Physical Methods in Advanced Inorganic Chemistry*, H. A. O. Hill and P. Day, Eds., Wiley-Interscience, London, 1968, pp. 329–339.
14. B. M. Weckhuysen, R. A. Schoonheydt, F. E. Mabbs, and D. Collison, *J. Chem. Soc., Faraday Trans.*, *92*, 2431 (1996).
15. H. Kon, *Bull. Chem. Soc. Jpn.*, *35*, 2054 (1962).
16. H. Kon, *J. Inorg. Nucl. Chem.*, *25*, 933 (1963).
17. N. S. Garifianov and N. F. Usacheva, *Dokl. Akad. Nauk SSSR*, *145*, 565 (1962).
18. G. Barr-David, M. Charara, R. Codd, R. P. Farrell, J. A. Irwin, P. A. Lay, R. Bramley, S. Brumby, J.-Y. Ji, and G. R. Hanson, *J. Chem. Soc., Faraday Trans.*, *91*, 1207 (1995).
19. R. P. Farrell, P. A. Lay, A. Levina, I. A. Maxwell, R. Bramley, S. Brumby, and J.-Y. Ji, *Inorg. Chem.*, *37*, 3159 (1998).
20. R. P. Farrell and P. A. Lay, *Comments Inorg. Chem.*, *13*, 133 (1992).
21. L. Zhang and P. A. Lay, *J. Am. Chem. Soc.*, *118*, 12624 (1996).
22. H. A. Headlam and P. A. Lay, *Inorg. Chem.*, *40*, 78 (2001).
23. H. A. Headlam, C. L. Weeks, P. Turner, T. W. Hambley, and P. A. Lay, *Inorg. Chem.* *40*, 5097 (2001).
24. L. Zhang and P. A. Lay, *Inorg. Chem.*, *37*, 1729 (1998).
25. D. I. Pattison, M. J. Davies, A. Levina, N. E. Dixon, and P. A. Lay, *Chem. Res. Toxicol.*, *14*, 500 (2001).
26. A. Bencini and D. Gatteschi, in *Inorganic Electronic Structure and Spectroscopy*, Vol. 1, E. I. Solomon and A. B. P. Lever, Eds., John Wiley & Sons, Inc., New York, 1999, pp. 93–159.

27. J. A. Irwin, Ph.D. Thesis, Cr(V)-Sugar Complexes: Possible Intracellular Intermediates of Importance to Chromate-Induced Carcinogenesis, University of Sydney, Australia, 1998.
28. J. A. Irwin, G. R. Hanson, and P. A. Lay, to be submitted.
29. J. A. Irwin, G. R. Hanson, and P. A. Lay, to be submitted.
30. R. Bramley, J.-Y. Ji, R. J. Judd, and P. A. Lay, *Inorg. Chem.*, 29, 3089 (1990).
31. R. Codd and P. A. Lay, *J. Am. Chem. Soc.*, 121, 7864 (1999).
32. P. Jackson, J. A. Irwin, J. Reimers, and P. A. Lay, to be submitted.
33. T. J. Collins, C. Slebodnick, and E. S. Uffelman, *Inorg. Chem.*, 29, 3433 (1990).
34. H. A. Headlam, Ph.D. Thesis, The Role of Cr(III) and Cr(V) Peptide and Amino Acid Complexes in Cr-induced Carcinogenesis, University of Sydney, Australia, 1998.
35. H. A. Headlam and P. A. Lay, to be submitted.
36. C. L. Weeks, Ph.D. Thesis, Chromium and Nickel Complexes with Tetradentate Diamide Ligands, University of Sydney, Australia, 2001.
37. C. T. Dillon, P. A. Lay, A. M. Bonin, N. E. Dixon, and Y. Sulfab, *Aust. J. Chem.*, 53, 411 (2000).
38. K. Srinivasan and J. K. Kochi, *Inorg. Chem.*, 24, 4671 (1985).
39. C. M. Cawich, A. Ibrahim, K. L. Link, A. Bumgartner, M. D. Patro, S. N. Mahapatro, P. A. Lay, A. Levina, S. S. Eaton, and G. R. Eaton, to be submitted.
40. R. P. Farrell, R. J. Judd, P. A. Lay, R. Bramley, and J.-Y. Ji, *Inorg. Chem.*, 28, 3401 (1989).
41. M. Branca, G. Micera, U. Segre, and A. Dessi, *Inorg. Chem.*, 31, 2404 (1992).
42. R. Codd and P. A. Lay, *J. Am. Chem. Soc.*, 123, 11799, (2001).
43. R. P. Bonomo, A. J. Di Bilio, and F. Riggi, *Chem. Phys.*, 151, 323 (1991).
44. M. Sugiyama, *Environ. Health Perspect.*, 102, 31 (1994).
45. P. Aisen, R. Aasa, and A. G. Redfield, *J. Biol. Chem.*, 244, 4628 (1969).
46. A. Chiu, N. Chiu, X. Shi, J. Beaubier, and N. S. Dalal, *Environ. Carcin. Ecotox. Rev.*, C16, 135 (1998).
47. K. J. Liu, X. Shi, and N. S. Dalal, *Biochem. Biophys. Res. Commun.*, 235, 54 (1997).
48. H. Luo, Y. Lu, X. Shi, Y. Mao, and N. S. Dalal, *Ann. Clin. Lab. Sci.*, 26, 185 (1996).
49. H. Luo, Y. Lu, Y. Mao, X. Shi, and N. S. Dalal, *J. Inorg. Biochem.*, 64, 25 (1996).
50. L. Zhang and P. A. Lay, *Aust. J. Chem.*, 53, 7 (2000).
51. A.-M. Dalla-Pozza, B.Sc. Honours Thesis, Formation of DNA-Damaging Cr(V) Complexes during the Reactions of Cr(VI) with a Vitamin E Analogue and D-Glucose, The University of Sydney, Australia, 1996.
52. A.-M. Dalla-Pozza, J. A. Irwin, and P. A. Lay, to be submitted.
53. D. I. Pattison, P. A. Lay, and M. J. Davies, *Inorg. Chem.*, 39, 2729 (2000).
54. D. I. Pattison, A. Levina, M. J. Davies, and P. A. Lay, *Inorg. Chem.*, 40, 214 (2001).
55. M. J. Davies and G. S. Timmins, in *Biomedical Applications of Spectroscopy*, R. J. H. Clark and R. E. Hester, Eds., John Wiley, & Sons, Inc., Chichester, UK, 1996, pp. 217–266.
56. M. B. Kadiiska, Q.-H. Xiang, and R. P. Mason, *Chem. Res. Toxicol.*, 7, 800 (1994).
57. M. B. Kadiiska, J. D. Morrow, J. A. Awad, L. J. Roberts, and R. P. Mason, *Chem. Res. Toxicol.*, 11, 1516 (1998).
58. Y. Hojo, A. Okado, S. Kawazoe, and T. Mizutani, *Biol. Trace Elem. Res.*, 76, 75 (2000).
59. X. Shi, N. S. Dalal, and V. Vallyathan, *Arch. Biochem. Biophys.*, 290, 381 (1991).

60. X. Shi, A. Chiu, T. C. Chen, B. Halliwell, V. Castranova, and V. Vallyathan, *J. Toxicol. Environ. Health*, *B2*, 87 (1999).
61. S. Leonard, S. Wang, L. Zang, V. Castranova, V. Vallyathan, and X. Shi, *J. Environ. Pathol., Toxicol. Oncol.*, *19*, 49 (2000).
62. K. J. Liu and X. L. Shi, *Molec. Cell. Biochem.*, *222*, 41 (2001).
63. K. J. Liu, J. Husler, J. P. Ye, S. S. Leonard, D. Cutler, F. Chen, S. W. Wang, Z. Zhang, M. Ding, L. Y. Wang, and X. L. Shi, *Molec. Cell. Biochem.*, *222*, 221 (2001).
64. M. J. Molyneux and M. J. Davies, *Carcinogenesis*, *16*, 875 (1995).
65. T.-C. Tsou, C.-L. Chen, T.-Y. Liu, and J.-L. Yang, *Carcinogenesis*, *17*, 103 (1996).
66. T.-C. Tsou, H.-J. Lai, and J.-L. Yang, *Chem. Res. Toxicol.*, *12*, 1002 (1999).
67. X. Liu, J. Lu, and S. Liu, *Mutat. Res.*, *440*, 109 (1999).
68. R. J. Judd, Ph.D. Thesis, The Chemistry of Cr(V), V(V) and V(IV) 2-Hydroxybutanoato Complexes, University of Sydney, Australia, 1992.
69. K. D. Sugden and K. E. Wetterhahn, *Inorg. Chem.*, *35*, 651 (1996).
70. P. A. Lay and A. Levina, *J. Am. Chem. Soc.*, *120*, 6704 (1998).
71. E. Finkelstein, G. M. Rosen, and E. J. Rauckman, *J. Am. Chem. Soc.*, *102*, 4994 (1980).
72. D. Behar, G. Czapski, J. Rabani, L. M. Dorfman, and H. A. Schearz, *J. Phys. Chem.*, *74*, 3209 (1970).
73. J. R. Harbour, S. L. Issler, and M. L. Hair, *J. Am. Chem. Soc.*, *102*, 7778 (1980).
74. J. B. Feix and B. Kalyanaraman, *Arch. Biochem. Biophys.*, *291*, 43 (1991).
75. P. Bilski, K. Reszka, M. Bilska, and C. F. Chignell, *J. Am. Chem. Soc.*, *118*, 1330 (1996).
76. Y. Hojo, A. Okado, S. Kawazoe, and T. Mizutani, *Biol. Trace Elem. Res.*, *76*, 85 (2000).
77. K. J. Liu, X. Shi, J. Jiang, F. Goda, N. Dalal, and H. M. Swartz, *Ann. Clin. Lab. Sci.*, *26*, 176 (1996).
78. K. J. Liu, K. Mäder, X. Shi, and H. M. Swartz, *Magn. Reson. Med.*, *38*, 524 (1997).
79. K. J. Appenroth, M. Bischoff, H. Gabrys, J. Stoeckel, H. M. Swartz, T. Walczak, and K. Winnefeld, *J. Inorg. Biochem.*, *78*, 235 (2000).
80. L. F. Larkworthy, K. B. Nolan, and P. O'Brien, in *Comprehensive Coordination Chemistry*, Vol. 3, G. Wilkinson et al., Eds., Pergamon Press, Oxford, UK, 1987, pp. 699–969.
81. P. de Meester, D. J. Hodgson, H. C. Freeman, and C. J. Moore, *Inorg. Chem.*, *16*, 1494 (1977).
82. C. M. Murdoch, Ph.D. Thesis, Chromium(III) and the Glucose Tolerance Factor, University of Sydney, Australia, 1989.
83. C. M. Murdoch, M. K. Cooper, T. W. Hambley, W. H. Hunter, and H. C. Freeman, *J. Chem. Soc., Chem. Commun.*, 1329 (1986).
84. K. Madafiglio, T. M. Manning, C. M. Murdoch, W. R. Tulip, M. K. Cooper, T. W. Hambley, and H. C. Freeman, *Acta Crystallogr., Ser. C46*, 554 (1990).
85. W. Mazurek, G. D. Fallon, P. J. Nichols, and B. O. West, *Polyhedron*, *9*, 777 (1990).
86. H. H. Zhang, B. Hedman, and K. O. Hodgson, in *Inorganic Electronic Structure and Specroscopy*, Vol. 1, E. I. Solomon and A. B. P. Lever, Eds., John Wiley & Sons, Inc., New York, 1999, pp. 513–554.
87. J. E. Penner-Hahn, *Coord. Chem. Rev.*, *190–192*, 1101 (1999).
88. A. Levina, P. A. Lay, and G. J. Foran, *J. Chem. Soc., Chem. Commun.*, 2339 (1999).
89. C. T. Dillon, P. A. Lay, M. Cholewa, G. J. F. Legge, A. M. Bonin, T. J. Collins, K. L. Kostka, and G. Shea-McCarthy, *Chem. Res. Toxicol.*, *10*, 533 (1997).

90. D. X. Balaic, Z. Barnea, J. N. Varghese, M. Cholewa, C. T. Dillon, P. A. Lay, and G. Shea-McCarthy, *Proc. SPIE*, *3449*, 157 (1998).
91. C. T. Dillon, P. A. Lay, B. J. Kennedy, A. P. J. Stampfl, Z. Cai, P. Ilinski, W. Rodrigues, D. G. Legnini, B. Lai, and J. Masor, *J. Biol. Inorg. Chem.*, *7*, 640 (2002).
92. R. Codd, A. Levina, L. Zhang, T. W. Hambley, and P. A. Lay, *Inorg. Chem.*, *39*, 990 (2000).
93. A. Levina and P. A. Lay, unpublished results.
94. R. B. Cole, *J. Mass. Spectrom.*, *35*, 763 (2000).
95. A. Levina, A. M. Bailey, G. Champion, and P. A. Lay, *J. Am. Chem. Soc.*, *122*, 6208 (2000).
96. K. P. Madhusudanan, S. B. Katti, R. Vijayalakshmi, and B. U. Nair, *J. Mass Spectrom.*, *34*, 880 (1999).
97. P. A. Lay and A. Levina, *Inorg. Chem.*, *35*, 7709 (1996).
98. S. L. Brauer and K. E. Wetterhahn, *J. Am. Chem. Soc.*, *113*, 3001 (1991).
99. S. L. Brauer, A. S. Hneihen, J. S. McBride, and K. E. Wetterhahn, *Inorg. Chem.*, *35*, 373 (1996).
100. R. P. Farrell, Ph.D. Thesis, Thermodynamic and Kinetic Studies on the Reactions of Cr(VI) and Cr(V) with Oxalic Acid, University of Sydney, Australia, 1993.
101. C. L. Broadhurst, W. F. Schmidt, J. B. Reeves, M. M. Polansky, K. Gautschi, and R. A. Anderson, *J. Inorg. Biochem.*, *66*, 119 (1997).
102. K. F. Kingry, A. C. Royer, and J. B. Vincent, *J. Inorg. Biochem.*, *72*, 79 (1998).
103. C. M. Davis and J. B. Vincent, *Arch. Biochem. Biophys.*, *339*, 335 (1997).
104. J. M. Eckert, R. J. Judd, and P. A. Lay, *Inorg. Chem.*, *26*, 2189 (1987).
105. R. Bramley, R. P. Farrell, J.-Y. Ji, and P. A. Lay, *Aust. J. Chem.*, *43*, 263 (1990).
106. R. N. Bose, B. Fonkeng, G. Barr-David, R. P. Farrell, R. J. Judd, P. A. Lay, and D. F. Sangster, *J. Am. Chem. Soc.*, *118*, 7139 (1996).
107. P. C. Grevatt, *Toxicological Review of Hexavalent Chromium*. United States Environmental Protection Agency, Washington, DC, 1998.
108. *Chromium in Soil: Perspectives in Chemistry, Health, and Environmental Regulation*, D. M. Proctor, B. L. Finley, M. A. Harris, D. J. Paustenbach, and D. Rabbe, Eds., Lewis Publishers, Boca Raton, FL, 1997.
109. World Health Organization, *Environmental Health Criteria 61. Chromium*, WHO, Geneva, Switzerland, 1988.
110. D. Burrows, *Chromium: Metabolism and Toxicity*, CRC Press, Boca Raton, FL, 1983.
111. E. J. O'Flaherty, B. D. Kerger, S. M. Hays, and D. J. Paustenbach, *Toxicol. Sci.*, *60*, 196 (2001).
112. S. De Flora, *Carcinogenesis*, *21*, 533 (2000).
113. S. De Flora, P. Zanacci, and C. Bennicelli, in *Handbook of Metal-Ligand Interactions in Biological Fluids. Bioinorganic Medicine*, Vol. 2, G. Berthon, Ed., Marcel Dekker, New York, 1995, pp. 716–725.
114. S. De Flora, A. Camoirano, M. Bagnasco, and P. Zanacchi, in *Handbook of Metal–Ligand Interactions in Biological Fluids. Bioinorganic Medicine*, Vol. 2, G. Berthon, Ed., Marcel Dekker, New York, 1995, pp. 1020–1036.
115. S. Fendorf, B. W. Wielinga, and C. M. Hansel, *Int. Geol. Rev.*, *42*, 691 (2000).
116. D. G. Barceloux, *Clin. Toxicol.*, *37*, 173 (1999).
117. M. Costa, *Crit. Rev. Toxicol.*, *27*, 431 (1997).
118. R. P. Farrell and M. Costa, in *Comprehensive Toxicology*, Vol. 12, I. G. Sipes et al., Eds., Pergamon Press, New York, 1997, pp. 225–254.

119. C. B. Klein, in *Toxicology of Metals*, L. W. Chang, Ed., Lewis Publishers, Boca Raton, FL, 1996, pp. 205–219.
120. C. Cervantes, J. Campos-Garcia, S. Devars, F. Gutierrez-Corona, H. Loza-Tavera, J. C. Torres-Guzman, and R. Moreno-Sanchez, *FEMS Microbiol. Rev.*, 25, 335 (2001).
121. L. A. Burns-Naas, *Pulm. Immunotoxicol.*, 241 (2000).
122. IARC, *IARC Monographs on the Evaluation of Carcinogenic Risk to Humans: Chromium and Chromium Compounds*, Vol. 23, International Agency for Research on Cancer, Lyon, France, 1980.
123. IARC, http:// www.iarc.fr (2001).
124. A. Yassi and E. Nieboer, in *Chromium in the Natural and Human Environments*, J. O. Nriagu and E. Nieboer, Eds., Wiley-Interscience, New York, 1988, pp. 443–489.
125. National Occupational Health and Safety Commission (Australia), *National Code of Practice and Guidance Note for the Safe Handling of Timber Preservatives and Treated Timber.* Australian Government Publishing Service, Canberra, Australia, 1989.
126. E. W. Merry, *The Chrome Tanning Process. Its Theory, Practical Application and Chemical Control.* A. Harvey Publisher, London, 1936.
127. T. Kauppin, *ASA 1990 Institute of Occupational Health*, Institute of Occupational Health (Finland), Helsinki, 1990.
128. J. M. Chen and O. J. Hao, *Crit. Rev. Environ. Sci. Techn.*, 28, 219 (1998).
129. H. J. Gibb, P. S. J. Lees, P. F. Pinsky, and B. C. Rooney, *Am. J. Ind. Med.*, 38, 127 (2000).
130. P. C. Wynn and C. Bishop, *Plat. Surf. Finish.*, 88, 12 (2001).
131. D. Duclos, *Adv. Mater. Proc.*, 159, 49 (2001).
132. H. J. Gibb, P. S. J. Lees, P. F. Pinsky, and B. C. Rooney, *Am. J. Ind. Med.*, 38, 115 (2000).
133. T. Sorahan and J. M. Harrington, *Occupat. Environ. Med.*, 57, 385 (2000).
134. E. L. Matos, M. Vilensky, and D. Mirabelli, *J. Occupat. Environ. Med.*, 42, 653 (2000).
135. T. E. Danielsen, S. Langard, and A. Andersen, *J. Occupat. Environ. Med.*, 42, 101 (2000).
136. M. Katabami, H. Dosaka-Akita, T. Mishina, K. Honma, K. Kimura, Y. Uchida, K. Morikawa, H. Mikami, S. Fukuda, Y. Inuyama, Y. Ohsaki, and Y. Kawakami, *Human Pathol.*, 31, 973 (2000).
137. F. W. Sunderman, *Ann. Clin. Lab. Sci.*, 31, 3 (2001).
138. V. J. Feron, J. H. E. Arts, C. F. Kuper, P. J. Slootweg, and R. A. Woutersen, *Crit. Rev. Toxicol.*, 31, 313 (2001).
139. I. A. Ojajarvi, T. J. Partanen, A. Ahlbom, P. Boffetta, T. Hakulinen, N. Jourenkova, T. P. Kauppinen, M. Kogevinas, M. Porta, H. U. Vainio, E. Weiderpass, and C. H. Wesseling, *Occupat. Environ. Med.*, 57, 316 (2000).
140. B. A. Racette, L. McGee-Minnich, S. M. Moerlein, J. W. Mink, T. O. Videen, and J. S. Perlmutter, *Neurology*, 56, 8 (2001).
141. N. H. I. Hjollund, J. P. E. Bonde, T. K. Jensen, T. B. Henriksen, A. M. Andersson, H. A. Kolstad, E. Ernst, A. Giwercman, N. E. Skakkebaek, and J. Olsen, *Scand. J. Work Environ. Health*, 26, 187 (2000).
142. R. G. Klein, P. Schmezer, F. Amelung, H. G. Schroeder, W. Woeste, and J. Wolf, *Int. Arch. Occupat. Environ. Health*, 74, 109 (2001).
143. J. Singh, D. E. Pritchard, D. L. Carlisle, J. A. Mclean, A. Montaser, J. M. Orenstein, and S. R. Patierno, *Toxicol. Appl. Pharmacol.*, 161, 240 (1999).
144. A. T. Haines and E. Nieborer, in *Chromium in the Natural and Human Environment*, J. O. Nriagu and E. Nieborer, Eds., John Wiley & Sons, Inc., New York, 1988, pp. 497–532.

145. D. Basketter, L. Horev, D. Slodovnik, S. Merimes, A. Trattner, and A. Ingber, *Contact Dermatitis*, *44*, 70 (2001).
146. T. Estlander, R. Jolanki, and L. Kanerva, *Contact Dermatitis*, *43*, 114 (2000).
147. P. Laffargue, H. F. Hildebrand, M. Lecomte-Houcke, V. Biehl, J. Breme, and J. Decoulx, *Rev. Chirurg. Orthopediq. Reparat. Appar. Moteur*, *87*, 84 (2001).
148. N. J. Hallab, K. Mikecz, C. Vermes, A. Skipor, and J. J. Jacobs, *Molec. Cell. Biochem.*, *222*, 127 (2001).
149. M. Costa, in *Toxicology of Metals*, L. W. Chang, Ed., Lewis Publishers, Boca Raton, FL, 1996, pp. 245–251.
150. M. F. Swiontkowski, J. Agel, J. Schwappach, P. McNair, and M. Welch, *J. Orthopaed. Trauma*, *15*, 86 (2001).
151. R. M. Balansky, F. D'Agostini, A. Izzotti, and S. De Flora, *Carcinogenesis*, *21*, 1677 (2000).
152. S. De Flora, A. Camoirano, M. Bagnasco, C. Bennicelli, G. Corbett, and B. D. Kerger, *Carcinogenesis*, *18*, 531 (1997).
153. M. D. Cohen, B. Kargacin, C. B. Klein, and M. Costa, *Crit. Rev. Toxicol.*, *23*, 255 (1993).
154. J. E. Sutherland, A. Zhitkovich, T. Kluz, and M. Costa, *Biol. Trace Elem. Res.*, *74*, 41 (2000).
155. M. Travacio, J. M. Polo, and S. Llesuy, *Toxicology*, *150*, 137 (2000).
156. J. C. Mirsalis, C. M. Hamilton, K. G. O'Loughlin, D. J. Paustenbach, B. D. Kerger, and S. Patierno, *Environ. Mol. Mutagen.*, *28*, 60 (1996).
157. J. R. Kuykendall, B. D. Kerger, E. J. Jarvi, G. E. Corbett, and D. J. Paustenbach, *Carcinogenesis*, *17*, 1971 (1996).
158. J. W. Hamilton, R. C. Kaltreider, O. V. Bajenova, M. A. Ihnat, J. McCaffrey, B. W. Turpie, E. E. Rowell, J. Oh, M. J. Nemeth, C. A. Pesce, and J. P. Lariviere, *Environ. Health Perspect. Suppl. 4*, *106*, 1005 (1998).
159. C. S. Dubey, B. K. Sahoo, and N. R. Nayak, *Bull. Environ. Contamin. Toxicol.*, *67*, 541 (2001).
160. B. Boxer, *Clean Water Rep.*, *39*, 1 (2001).
161. A. L. Rowbotham, L. S. Levy, and L. K. Shuker, *J. Toxicol. Environ. Health, Part B*, 145 (2000).
162. J. P. Fryzek, M. T. Mumma, J. K. McLaughlin, B. E. Henderson, and W. J. Blot, *J. Occupat. Environ. Med.*, *43*, 635 (2001).
163. S. J. Gray and K. Sterling, *J. Clin. Invest.*, *29*, 1604 (1950).
164. P. Debetto, A. Lazzarini, A. Tomasi, M. Beltrame, and P. Arslan, *Cell Biol. Int. Rep.*, *10*, 214 (1986).
165. H. H. Popper, E. Grygar, E. Ingolic, and O. Wawschinek, *Inhal. Toxicol.*, *5*, 345 (1993).
166. A. Kortenkamp, D. Beyersmann, and P. O'Brien, *Toxicol. Environ. Chem.*, *14*, 23 (1987).
167. M. Cholewa, I. F. Turnbull, G. J. F. Legge, H. Weigold, S. M. Marcuccio, G. Holan, E. Tomlinson, P. J. Wright, C. T. Dillon, P. A. Lay, and A. M. Bonin, *Nucl. Instr. Met. Phys. Res.*, *B104*, 317 (1995).
168. C. T. Dillon, P. A. Lay, A. M. Bonin, M. Cholewa, G. J. F. Legge, T. J. Collins, and K. L. Kostka, *Chem. Res. Toxicol.*, *11*, 119 (1998).
169. K. A. Biedermann and J. R. Landolph, *Cancer Res.*, *50*, 7835 (1990).
170. P. H. Connett and K. E. Wetterhahn, *Struct. Bonding (Berlin)*, *54*, 93 (1983).
171. P. Arslan, M. Beltrame, and A. Tomasi, *Biochim. Biophys. Acta*, *931*, 10 (1987).
172. D. Beyersmann, A. Koster, and B. Buttner, in *Carcinogenic and Mutagenic Metal Compounds*, E. Merian et al., Eds., Gordon and Breach Science Publishers, London, 1985, pp. 303–310.

173. Y.-Y. Lu and J.-L. Yang, *J. Cell. Biochem.*, *57*, 655 (1995).
174. P. C. Dartsch, S. Hildenbrand, R. Kimmel, and F. W. Schmahl, *Int. Arch. Occup. Environ. Health*, *71*, S40 (1998).
175. C. T. Dillon, P. A. Lay, A. M. Bonin, M. Cholewa, and G. J. F. Legge, *Chem. Res. Toxicol.*, *13*, 742 (2000).
176. B. Buttner and D. Beyersmann, *Xenobiotica*, *15*, 735 (1985).
177. D. M. L. Goodgame, P. B. Hayman, and D. E. Hathway, *Inorg. Chim. Acta*, *91*, 113 (1984).
178. S. L. Boyko and D. M. L. Goodgame, *Inorg. Chim. Acta*, *123*, 189 (1986).
179. Y. Sulfab and M. Nasreldin, *Transit. Met. Chem.*, *26*, 147 (2001).
180. M. Ermolli, C. Menne, G. Pozzi, M. A. Serra, and L. A. Clerici, *Toxicology*, *159*, 23 (2001).
181. J. Singh, D. L. Carlisle, D. E. Pritchard, and S. R. Patierno, *Oncol. Rep.*, *5*, 1307 (1998).
182. K. Wetterhahn Jennette, *J. Am. Chem. Soc.*, *104*, 874 (1982).
183. S. C. Rossi, N. Gorman, and K. E. Wetterhahn, *Chem. Res. Toxicol.*, *1*, 101 (1988).
184. B. D. Martin, J. A. Schoenhard, and K. D. Sugden, *Chem. Res. Toxicol.*, *11*, 1402 (1998).
185. G. Micera and A. Dessi, *J. Inorg. Biochem.*, *34*, 157 (1988).
186. H. Sakurai, K. Takechi, H. Tsuboi, and H. Yasui, *J. Inorg. Biochem.*, *76*, 71 (1999).
187. J.-M. P. Yuann, K. J. Liu, J. W. Hamilton, and K. E. Wetterhahn, *Carcinogenesis*, *20*, 1267 (1999).
188. S. Ueno, T. Kashimoto, N. Susa, Y. Furukawa, M. Ishii, K. Yokoi, M. Yasuno, Y. F. Sasaki, J.-i. Ueda, Y. Nishimura, and M. Sugiyama, *Toxicol. Appl. Pharmacol.*, *170*, 56 (2001).
189. M. Branca, A. Dessi, H. Kozlowski, G. Micera, and J. Swiatek, *J. Inorg. Biochem.*, *39*, 217 (1990).
190. S. Signorella, V. Daier, S. Garcia, R. Cargnello, J. C. Gonzalez, M. Rizzotto, and L. F. Sala, *Carbohydr. Res.*, *316*, 14 (1999).
191. S. Signorella, M. I. Frascaroli, S. Garcia, M. Santoro, J. C. Gonzalez, C. Palopoli, V. Daier, N. Casado, and L. F. Sala, *J. Chem. Soc., Dalton Trans.*, 1617 (2000).
192. V. P. Roldán, V. A. Daier, B. Goodman, M. I. Santoro, J. C. González, N. Calisto, S. R. Signorella, and L. F. Sala, *Helv. Chim. Acta*, *83*, 3211 (2000).
193. M. Rizzotto, V. Moreno, S. Signorella, V. Daier, and L. F. Sala, *Polyhedron*, *19*, 417 (2000).
194. L. Zhang, Ph.D. Thesis, EPR and XAFS Studies of Biologically Relevant Chromium(V) Complexes and Manganese(II)-Activated Aminopeptidase P, University of Sydney, 1998.
195. L. Zhang, A. Levina, and P. A. Lay, to be submitted.
196. E. M. Kosower, in *Glutathione: Metabolism and Function*, I. M. Arias and W. B. Jacoby, Eds., Raven Press, New York, 1976, pp. 1–15.
197. J. Ning and M. H. Grant, *Toxicol. in Vitro*, *14*, 329 (2000).
198. S. Ueno, N. Susa, Y. Furukawa, and M. Sugiyama, *Toxicol. Appl. Pharmacol.*, *135*, 165 (1995).
199. W. J. Ding, Q. F. Qian, X. L. Hou, W. Y. Feng, and Z. F. Chai, *J. Radioanalyt. Nucl. Chem.*, *244*, 259 (2000).
200. M. Cholewa, C. Dillon, P. Lay, D. Phillips, T. Talarico, and B. Lai, *Nucl. Instr. Meth. Phys. Res. B*, *181*, 715 (2001).
201. L. J. Blankenship, D. L. Carlisle, J. P. Wise, J. M. Orenstein, L. E. I. Dye, and S. R. Patierno, *Toxicol. Appl. Pharmacol.*, *146*, 270 (1997).
202. R. Codd, P. A. Lay, and A. Levina, *Inorg. Chem.*, *36*, 5440 (1997).
203. S. R. Patierno, D. Banh, and J. R. Landolph, *Cancer Res.*, *48*, 5280 (1988).

204. A. E. Munson and K. E. Phillips, *Methods Mol. Biol. (Totowa, N. J.)*, *121*, 359 (2000).
205. C. R. Valeri, L. E. Pivacek, G. P. Cassidy, and G. Ragno, *Vox Sanguinis*, *80*, 48 (2001).
206. C. Michel, M. Brugna, C. Aubert, A. Bernadac, and M. Bruschi, *Appl. Microbiol. Biotechnol.*, *55*, 95 (2001).
207. L. J. DeFilippi, *Environ. Sci. Pollut. Control Ser.*, *22*, 177 (2000).
208. S. De Flora, M. Bagnasco, D. Serra, and P. Zanacchi, *Mutat. Res.*, *238*, 99 (1990).
209. R. P. Farrell, R. J. Judd, P. A. Lay, N. E. Dixon, R. S. U. Baker, and A. M. Bonin, *Chem. Res. Toxicol.*, *2*, 227 (1989).
210. C. T. Dillon, P. A. Lay, A. M. Bonin, N. E. Dixon, T. J. Collins, and K. L. Kostka, *Carcinogenesis*, *14*, 1875 (1993).
211. K. D. Sugden, R. B. Burris, and S. J. Rogers, *Mutat. Res.*, *244*, 239 (1990).
212. K. D. Sugden, R. D. Geer, and S. J. Rogers, *Biochemistry*, *31*, 11626 (1992).
213. J. Chen and W. G. Thilly, *Mutat. Res.*, *323*, 21 (1994).
214. A. E. Aust and J. F. Eveleigh, *Exp. Biol. Med.*, *222*, 246 (1999).
215. C. Bennicelli, A. Camoirano, S. Petruzzelli, P. Zanacchi, and S. D. Flora, *Mutat. Res.*, *122*, 1 (1983).
216. A. M. Saad and M. A. Abdel-Hadi, *Bull. Natl. Res. Cent. (Egypt)*, *25*, 297 (2000).
217. G. Vincze, J. Vallner, Á. Balogh, and F. Kiss, *Bull. Environ. Contam. Toxicol.*, *65*, 772 (2000).
218. J. Belagyi, M. Pas, P. Raspor, M. Pesti, and T. Páli, *Biochim. Biophys. Acta*, *1421*, 175 (1999).
219. M. Karbownik, J. J. Garcia, A. Lewinski, and R. J. Reiter, *J. Bioenerg. Biomembr.*, *33*, 73 (2001).
220. J. A. Shumilla and A. Barchowsky, *Toxicol. Appl. Pharmacol.*, *158*, 288 (1999).
221. I. J. Yu, K. S. Song, H. K. Chang, J. H. Han, K. J. Kim, Y. H. Chung, S. H. Maeng, S. H. Park, K. T. Han, K. H. Chung, and H. K. Chung, *Toxicol. Sci.*, *63*, 99 (2001).
222. S. Ueno, N. Susa, Y. Furukawa, K. Aikawa, I. Itagaki, T. Komiyama, and Y. Takashima, *Jpn. J. Vet. Sci.*, *50*, 45 (1988).
223. N. Susa, S. Ueno, Y. Furukawa, N. Michiba, and S. Minoura, *Jpn. J. Vet. Sci.*, *51*, 1103 (1989).
224. Y. Hojo, K. Nishiguchi, S. Kawazoe, and T. Mizutani, *J. Health Sci.*, *45*, 329 (1999).
225. Y. Hojo, K. Nishiguchi, S. Kawazoe, and T. Mizutani, *J. Health Sci.*, *46*, 75 (2000).
226. Y.-L. Huang, C.-Y. Chen, J.-Y. Sheu, I. C. Chuang, J.-H. Pan, and T.-H. Lin, *J. Toxicol. Environ. Health A*, *56*, 235 (1999).
227. A. Elis, P. Froom, A. Ninio, L. Cahara, and M. Lishner, *Int. J. Occup. Environ. Health*, *7*, 206 (2001).
228. R. Schauer, S. Kelm, G. Reuter, P. Roggentin, and L. Shaw, in *Biology of the Sialic Acids*, A. Rosenberg, Ed., Plenum Press, New York, 1995, pp. 7–67.
229. H.-W. Kuo, J.-S. Lai, and T.-I. Lin, *Int. Arch. Occup. Environ. Health*, *70*, 272 (1997).
230. J. Blasiak, A. Trzeciak, E. Malecka-Panas, J. Drzewoski, T. Iwanienko, I. Szumiel, and M. Wojewodzka, *Teratog., Carcinog., Mutagen.*, *19*, 19 (1999).
231. N. H. Kleinsasser, F. Gamarra, A. Bergner, B. C. Wallner, U. A. Harreus, J. Juchhoff, E. R. Kastenbauer, and R. M. Huber, *ORL - J. Oto-Rhino-Laryngol. Rel. Spec. 63*, 141 (2001).
232. H. Y. Shrivastava and B. U. Nair, *Biochem. Biophys. Res. Commun.*, *279*, 980 (2000).
233. H. Y. Shrivastava and B. U. Nair, *Biochem. Biophys. Res. Commun.*, *285*, 915 (2001).
234. B. S. Khangarot and D. M. Tripathi, *J. Environ. Sci. Health*, *A27*, 1141 (1992).
235. M. Kimura, Y. Hama, T. Sumi, M. Asakawa, R. B. N. Narasinga, A. P. Horne, S.-C. Li, Y.-T. Li, and H. Nakagawa, *J. Biol. Chem.*, *269*, 32138 (1994).

236. C. Fahr and R. Schauer, *J. Invest. Dermatol.*, *116*, 254 (2001).
237. M. Misra, J. A. Alcedo, and K. E. Wetterhahn, *Carcinogenesis*, *15*, 2911 (1994).
238. J. W. Hamilton and K. E. Wetterhahn, *Carcinogenesis*, *7*, 2085 (1986).
239. C. Bjoerge, G. Brunborg, R. Wiger, J. A. Holme, T. Scholz, E. Dybing, and E. J. Soederlund, *Reprod. Toxicol.*, *10*, 509 (1996).
240. U. Werfel, V. Langen, I. Eickhoff, J. Schoonbrood, C. Vahrenholz, A. Brauksiepe, W. Popp, and K. Norpoth, *Carcinogenesis*, *19*, 413 (1998).
241. J. A. Ashby, H. Tinwell, P. A. Lefevre, and M. A. Browne, *Mutagenesis*, *10*, 85 (1995).
242. J. Blasiak and J. Kowalik, *Mutat. Res.*, *469*, 135 (2000).
243. K. D. Devi, R. Rozati, B. S. Banu, K. Jamil, and P. Grover, *Food Chem. Toxicol.*, *39*, 859 (2001).
244. O. Merk, K. Reiser, and G. Speit, *Mutat. Res.*, *471*, 71 (2000).
245. M. Costa, *Environ. Health Perspec.*, *92*, 45 (1991).
246. M. Costa and A. V. Zhitkovich, *U.S. Patent No.* 5,545,529, (1996), *Chem. Abstr.*, Vol. *125*, 188200 (1996).
247. A. Zhitkovich, V. Voitkun, T. Kluz, and M. Costa, *Environ. Health Perspect. Suppl.*, *106*, 969 (1998).
248. S. N. Mattagajasingh and H. P. Misra, *Mol. Cell. Biochem.*, *199*, 149 (1999).
249. J. Singh, L. C. Bridgewater, and S. Patierno, *Toxicol. Sci.*, *45*, 72 (1998).
250. A. Zhitkovich, S. Shrager, and J. Messer, *Chem. Res. Toxicol.*, *13*, 1114 (2000).
251. A. Zhitkovich, Y. Song, G. Quievryn, and V. Voitkun, *Biochemistry*, *40*, 549 (2001).
252. S. N. Mattagajasingh and H. P. Misra, *J. Biol. Chem.*, *271*, 33550 (1996).
253. E. R. Civitello, K. A. Hossler, and D. M. Stearns, *Abstr. Pap. - Am. Chem. Soc.*, *216*, TOXI-021 (1998).
254. A. Zhitkovich, V. Voitkun, and C. Costa, *Carcinogenesis*, *16*, 907 (1995).
255. G. Quievryn, M. Goulart, J. Messer, and A. Zhitkovich, *Molec. Cell. Biochem.*, *222*, 107 (2001).
256. V. Voitkun, A. Zhitkovich, and M. Costa, *Nucl. Acids Res.*, *26*, 2024 (1998).
257. J. Xu, G. J. Bubley, B. Detrick, L. J. Blankenship, and S. R. Patierno, *Carcinogenesis*, *17*, 1511 (1996).
258. T. O'Brien, J. Xu, and S. R. Patierno, *Molec. Cell. Biochem.*, *222*, 173 (2001).
259. M. Capellmann, A. Mikalsen, M. Hindrum, and J. Alexander, *Carcinogenesis*, *16*, 1135 (1995).
260. A. S. Hneihen, A. M. Standeven, and K. E. Wetterhahn, *Carcinogenesis*, *14*, 1795 (1993).
261. A. Kortenkamp, B. Curran, and P. O'Brien, *Carcinogenesis*, *13*, 307 (1992).
262. A. Kortenkamp, *NATO ASI Ser., Ser. 2*, *26*, 35 (1997).
263. F. Y. Wu, F. J. Tsai, H. W. Kuo, C. H. Tsai, W. Y. Wu, R. Y. Wang, and J. S. Lai, *Mutat. Res.*, *464*, 289 (2000).
264. M. Toraason, *Biomarkers*, *4*, 3 (1999).
265. H. Kim, S.-H. Cho, and M.-H. Chung, *Ind. Health*, *37*, 335 (1999).
266. S. Okada, M. Taniyama, and H. Ohba, *J. Inorg. Biochem.*, *17*, 41 (1982).
267. K. M. Borges and K. E. Wetterhahn, *Chem. Res. Toxicol.*, *4*, 638 (1991).
268. J. Singh and E. T. Snow, *Biochemistry*, *37*, 9371 (1998).
269. L. C. Bridgewater, F. C. R. Manning, and S. R. Patierno, *Mol. Carcinog.*, *23*, 201 (1998).
270. X. Zhong, S. S. Patel, and M.-D. Tsai, *J. Am. Chem. Soc.*, *120*, 235 (1998).

271. J. W. Arndt, W. Gong, X. Zhong, A. K. Showalter, J. Liu, C. A. Dunlap, Z. Lin, C. Paxson, M.-D. Tsai, and M. K. Chan, *Biochemistry*, *40*, 5368 (2001).
272. K. E. Wetterhahn and J. W. Hamilton, *Sci. Total Environ.*, *86*, 113 (1989).
273. A. Maier, T. P. Dalton, and A. Puga, *Mol. Carcinog.*, *28*, 225 (2000).
274. Y. Tesfai, D. Davis, and D. Reinhold, *Mutat. Res.*, *416*, 159 (1998).
275. R. C. Kaltreider, C. A. Pesce, M. A. Ihnat, J. P. Lariviere, and J. W. Hamilton, *Mol. Carcinog.*, *25*, 219 (1999).
276. A. M. O'Connell, B.Sc. Honours Thesis, *The Effect of Cr(VI/V) on the DNA Binding Protein GATA-1*, University of Sydney, Australia, 1997.
277. A. M. O'Connell, M. Crossley, and P. A. Lay unpublished results.
278. D. Krepkiy, W. E. Antholine, C. Myers, and D. H. Petering, *Molec. Cell. Biochem.*, *222*, 213 (2001).
279. A. Hartwig, *Toxicol. Lett.*, *102*, 235 (1998).
280. A. Hartwig, *Pure Appl. Chem.*, *72*, 1007 (2000).
281. G. S. Buzard and K. S. Kasprzak, *J. Environ. Pathol., Toxicol. Oncol.*, *19*, 179 (2000).
282. E. T. Snow, *Environ. Health Perspec.*, *92*, 77 (1991).
283. T.-C. Tsou, R.-J. Lin, and J.-L. Yang, *Chem. Res. Toxicol.*, *10*, 962 (1997).
284. S. Liu and K. Dixon, *Environ. Mol. Mutagen.*, *28*, 71 (1996).
285. S. Liu, M. Medvedovic, and K. Dixon, *Environ. Mol. Mutagen.*, *33*, 313 (1999).
286. S. P. Hussain, M. H. Hollstein, and C. S. Harris, *Ann. New York Acad. Sci.*, *919*, 79 (2000).
287. K. Kondo, N. Hino, M. Sasa, Y. Kamamura, S. Sakiyama, M. Tsuyuguchi, M. Hashimoto, T. Uyama, and Y. Monden, *Biochem. Biophys. Res. Commun.*, *239*, 95 (1997).
288. A. A. Ewis, K. Kondo, J. Lee, M. Tsuyughuchi, M. Hashimoto, T. Yokose, K. Mukai, T. Kodama, T. Shinka, Y. Monden, and Y. Nakahori, *Am. J. Ind. Med.*, *40*, 92 (2001).
289. G. Jia, S. Liu, Y. Lu, W. Li, and S. Zhou, *Zhonghua Laodong Weisheng Zhiyebing Zazhi*, *16*, 201 (1998). *Chem. Abstr.*, *130*, 192961 (1998).
290. J. P. Ye and X. L. Shi, *Molec. Cell. Biochem.*, *222*, 189 (2001).
291. M. Gunaratnam and M. H. Grant, *Chem.-Biol. Interact.*, *134*, 191 (2001).
292. M. Sai Ram, B. Anju, T. Pauline, D. Prasad, A. K. Kain, S. S. Mongia, S. K. Sharma, B. Singh, R. Singh, G. Ilavazhagan, D. Kumar, and W. Selvamurhty, *J. Ethnopharmacol.*, *71*, 235 (2000).
293. M. A. S. Fernandes, C. Geraldes, C. R. Oliveira, and M. C. Alpoim, *Toxicol. Lett.*, *114*, 237 (2000).
294. M. A. S. Fernandes, C. Geraldes, C. R. Oliveira, and M. C. Alpoim, *Ecotoxicol. Environ. Safety*, *47*, 39 (2000).
295. G. Kim and E. J. Yurkow, *Cancer Res.*, *56*, 2045 (1996).
296. Y. Qian, B. Jiang, D. C. Flynn, S. S. Leonard, S. W. Wang, Z. Zhang, J. P. Ye, F. Chen, L. Y. Wang, and X. L. Shi, *Molec. Cell. Biochem.*, *222*, 199 (2001).
297. V. A. Dubrovskaya and K. E. Wetterhahn, *Carcinogenesis*, *19*, 1401 (1998).
298. K. D. Sugden and D. M. Stearns, *J. Environ. Pathol., Toxicol. Oncol.*, *19*, 215 (2000).
299. D. M. Stearns and K. E. Wetterhahn, *NATO ASI Ser., Ser. 2*, *26*, 55 (1997).
300. A. Kortenkamp, M. Casadevall, P. Da Cruz Fresco, and R. O. J. Shayer, *NATO ASI Ser., Ser. 2*, *26*, 15 (1997).
301. J. Ning and M. H. Grant, *Toxicol. in Vitro*, *13*, 879 (1999).

302. D. L. Carlisle, D. E. Pritchard, J. Singh, B. M. Owens, L. J. Blankenship, J. M. Orenstein, and S. R. Patierno, *Toxicol. Sci.*, 55, 60 (2000).
303. C. T. Dillon and P. A. Lay, unpublished data.
304. L. W. Miksche and J. Lewalter, *Biomarkers Occup. Health Progr. Persp.*, 313 (1995).
305. N. Susa, S. Ueno, Y. Furukawa, and M. Sugiyama, *Arch. Toxicol.*, 71, 345 (1997).
306. N. Susa, S. Ueno, Y. Furukawa, J. Ueda, and M. Sugiyama, *Toxicol. Appl. Pharmacol.*, 144, 377 (1997).
307. K. Gasiorowski, K. Szyba, D. Wozniak, and B. Gulanowski, *Mutagenesis*, 12, 411 (1997).
308. B. Halliwell, *Adv. Pharmacol.*, 38, 3 (1997).
309. S. M. Bradberry and J. A. Vale, *J. Toxicol. Clin. Toxicol.*, 37, 195 (1999).
310. F. D'Agostini, R. M. Balansky, A. Camoirano, and S. De Flora, *Int. J. Canc.*, 88, 702 (2000).
311. G. Powis, J. R. Gasdaska, and A. Baker, *Adv. Pharmacol.*, 38, 329 (1997).
312. E. Ho and T. M. Bray, *Proc. Soc. Exp. Biol. Med.*, 222, 205 (1999).
313. J. Ye, X. Zhang, H. A. Young, Y. Mao, and X. Shi, *Carcinogenesis*, 16, 2401 (1995).
314. X. Shi, M. Ding, J. Ye, S. Wang, S. S. Leonard, L. Zang, V. Castranova, V. Vallyathan, A. Chiu, N. Dalal, and K. Liu, *J. Inorg. Biochem.*, 75, 37 (1999).
315. X. Shi, Z. Dong, C. Huang, W. Ma, K. Liu, J. Ye, F. Chen, S. S. Leonard, M. Ding, V. Castranova, and V. Vallyathan, *Molec. Cell. Biochem.*, 194, 63 (1999).
316. F. Chen, M. Ding, V. Castranova, and X. L. Shi, *Molec. Cell. Biochem.*, 222, 159 (2001).
317. J. A. Shumilla, R. J. Broderick, Y. Wang, and A. Barchowsky, *J. Biol. Chem.*, 274, 36207 (1999).
318. J. A. Shumilla, K. E. Wetterhahn, and A. Barchowsky, *Arch. Biochem. Biophys.*, 349, 356 (1998).
319. C. E. Canman and M. B. Kastan, *Adv. Pharmacol.*, 41, 429 (1997).
320. S. W. Wang and X. L. Shi, *Carcinogenesis*, 22, 757 (2001).
321. T. Hanaoka, Y. Yamano, N. Katsuno, J. Kagawa, and S. Ishizu, *Scand. J. Work, Environ. Health*, 23, 37 (1997).
322. D. Bagchi, J. Balmoori, M. Bagchi, X. Ye, C. B. Williams, and S. J. Stohs, *Free Radical Biol. Med.*, 28, 895 (2000).
323. D. Bagchi, M. Bagchi, and S. J. Stohs, *Molec. Cell. Biochem.*, 222, 149 (2001).
324. S. W. Wang, S. S. Leonard, J. P. Ye, M. Ding, and X. L. Shi, *Am. J. Physiol., Cell Physiol.*, 279, C868 (2000).
325. J. P. Wise, J. C. Leonard, and S. R. Patierno, *Mutat. Res.*, 278, 69 (1992).
326. L. J. Blankenship, F. C. R. Manning, J. M. Orenstein, and S. R. Patierno, *Toxicol. Appl. Pharmacol.*, 126, 75 (1994).
327. R. Rajaram, B. U. Nair, and T. Ramasami, *Biochem. Biophys. Res. Commun.*, 210, 434 (1995).
328. J. Ye, S. Wang, S. S. Leonard, Y. Sun, L. Butterworth, J. Antonini, M. Ding, Y. Rojanasakul, V. Vallyathan, V. Castranova, and X. Shi, *J. Biol. Chem.*, 274, 34974 (1999).
329. A. Flores and J. M. Pérez, *Toxicol. Appl. Pharm.*, 161, 75 (1999).
330. D. Bagchi, S. S. Joshi, M. Bagchi, J. Balmoori, E. J. Benner, C. A. Kuszynski, and S. J. Stohs, *J. Biochem. Mol. Toxicol.*, 14, 33 (1999).
331. E. Rudolf, J. Peychl, and M. Cervinka, *Acta Med. (Hradec Kralove, Czech Rep.)*, 43, 83 (2000).
332. Z. Zhang, S. S. Leonard, S. W. Wang, V. Vallyathan, V. Castranova, and X. L. Shi, *Molec. Cell. Biochem.*, 222, 77 (2001).
333. F. Chen, V. Vallyathan, V. Castranova, and L. Shi, *Molec. Cell. Biochem.*, 222, 183 (2001).

334. C. Vasant, K. Balamurugan, R. Rajaram, and T. Ramasami, *Biochem. Biophys. Res. Commun.*, 285, 1354 (2001).
335. D. L. Carlisle, D. E. Pritchard, J. Singh, and S. R. Patierno, *Mol. Carcinog.*, 28, 111 (2000).
336. M. P. Waalkes, D. A. Fox, J. C. States, S. R. Patierno, and M. J. McCabe, *Toxicol. Sci.*, 56, 255 (2000).
337. D. E. Pritchard, J. Singh, D. L. Carlisle, and S. R. Patierno, *Carcinogenesis*, 21, 2027 (2000).
338. F. C. R. Manning and S. R. Patierno, *Cancer Invest.*, 14, 455 (1996).
339. T. Moriguchi, Y. Gotoh, and E. Nishida, *Adv. Pharmacol.*, 36, 121 (1996).
340. S. M. Chuang, G. Y. Liou, and J. L. Yang, *Carcinogenesis*, 21, 1491 (2000).
341. F. Chen, M. Ding, Y. Lu, S. S. Leonard, V. Vallyathan, V. Castranova, and X. Shi, *J. Environ. Pathol., Toxicol. Oncol.*, 19, 231 (2000).
342. D. Bagchi, M. Bagchi, L. Tang, and S. J. Stohs, *Toxicol. Lett.*, 91, 31 (1997).
343. S. M. Chuang and J. L. Yang, *Molec. Cell. Biochem.*, 222, 85 (2001).
344. J. F. Wang, M. Bashir, B. N. Engelsberg, C. Witmer, H. Rozmiarek, and P. C. Billings, *Carcinogenesis*, 18, 371 (1997).
345. J. F. Wang, B. N. Engelsberg, S. W. Johnson, C. Witmer, W. C. Merrick, H. Rozmiarek, and P. C. Billings, *Arch. Toxicol.*, 71, 450 (1997).
346. F. Delmas, S. Schaak, Y. Gaubin, F. Croute, C. Arrabit, and J. C. Murat, *Cell Biol. Toxicol.*, 14, 39 (1998).
347. M. D. Cohen, J. T. Zelikoff, L.-C. Chen, and R. B. Schlesinger, *Toxicol. Appl. Pharmacol.*, 152, 30 (1998).
348. R. J. Isfort, *Ann. New York Acad. Sci.*, 919, ix (2000).
349. T. M. Bray, *Proc. Soc. Exp. Biol. Med.*, 222, 195 (1999).
350. K. S. Kasprzak, in *Toxicology of Metals*, L. W. Chang, Ed., Lewis Publishers, Boca Raton, FL, 1996, pp. 299–320.
351. K. S. Kasprzak, *NATO ASI Ser., Ser. 2*, 26, 73 (1997).
352. K. S. Kasprzak, W. Bal, D. W. Porter, and K. Bialkowski, *NATO ASI Ser., Ser. A*, 302, 193 (1999).
353. S. J. Stohs, D. Bagchi, E. Hassoun, and M. Bagchi, *J. Environ. Pathol., Toxicol. Oncol.*, 19, 201 (2000).
354. W. Bal, H. Kozlowski, and K. S. Kasprzak, *J. Inorg. Biochem.*, 79, 213 (2000).
355. K. S. Kasprzak and K. Bialkowski, *J. Inorg. Biochem.*, 79, 231 (2000).
356. A. Zhitkovich, V. Voitkun, and M. Costa, *Biochemistry*, 35, 7275 (1996).
357. J. Müller, R. K. O. Sigel, and B. Lippert, *J. Inorg. Biochem.*, 79, 261 (2000).
358. K. S. Kasprzak, *Chem. Res. Toxicol.*, 4, 604 (1991).
359. C. B. Klein, K. Frenkel, and M. Costa, *Chem. Res. Toxicol.*, 4, 592 (1991).
360. A. M. Standeven and K. E. Wetterhahn, *Chem. Res. Toxicol.*, 4, 616 (1991).
361. K. E. Wetterhahn and E. J. Dudek, *New J. Chem.*, 20, 199 (1996).
362. M. J. Tsapakos and K. E. Wetterhahn, *Chem.-Biol. Interact.*, 46, 265 (1983).
363. A. Köster and D. Beyersmann, *Toxicol. Environ. Chem.*, 10, 307 (1985).
364. S. Kawanishi, S. Inoue, and S. Sano, *J. Biol. Chem.*, 261, 5952 (1986).
365. J. Aiyar, K. M. Borges, R. A. Floyd, and K. E. Wetterhahn, *Toxicol. Environ. Chem.*, 22, 135 (1989).
366. A. Kortenkamp, Z. Ozolins, D. Beyersmann, and P. O'Brien, *Mutat. Res.*, 216, 19 (1989).

367. M. Casadevall and A. Kortenkamp, *Carcinogenesis*, *15*, 407 (1994).
368. K. M. Borges and K. E. Wetterhahn, *Carcinogenesis*, *10*, 2165 (1989).
369. L. C. Bridgewater, F. C. R. Manning, and S. R. Patierno, *Carcinogenesis*, *15*, 2421 (1994).
370. P. H. Connett and K. E. Wetterhahn, *J. Am. Chem. Soc.*, *107*, 4282 (1985).
371. A. M. Standeven and K. E. Wetterhahn, *Carcinogenesis*, *12*, 1733 (1991).
372. A. M. Standeven and K. E. Wetterhahn, *Carcinogenesis*, *13*, 1319 (1992).
373. P. J. Jannetto, W. E. Antholine, and C. R. Myers, *Toxicology*, *159*, 119 (2001).
374. K. M. Borges, J. S. Boswell, R. H. Liebross, and K. E. Wetterhahn, *Carcinogenesis*, *12*, 551 (1991).
375. A. Kortenkamp and P. O'Brien, *Environ. Health Perspect.*, *102*, 237 (1994).
376. M. Casadevall and A. Kortenkamp, *Carcinogenesis*, *16*, 805 (1995).
377. A. Kortenkamp, M. Casadevall, and P. Da Cruz Fresco, *Ann. Clin. Lab. Sci.*, *26*, 160 (1996).
378. A. Kortenkamp, M. Casadevall, S. P. Faux, A. Jenner, R. O. J. Shayer, N. Woodbridge, and P. O'Brien, *Arch. Biochem. Biophys.*, *329*, 199 (1996).
379. M. Casadevall, P. D. Fresco, and A. Kortenkamp, *Chem.-Biol. Interact.*, *123*, 117 (1999).
380. P. Da Cruz Fresco and A. Kortenkamp, *Carcinogenesis*, *15*, 1773 (1994).
381. P. Da Cruz Fresco, F. Shacker, and A. Kortenkamp, *Chem. Res. Toxicol.*, *8*, 884 (1995).
382. D. M. Stearns, L. J. Kennedy, K. D. Courtney, P. H. Giangrande, L. S. Phieffer, and K. E. Wetterhahn, *Biochemistry*, *34*, 910 (1995).
383. D. M. Stearns and K. E. Wetterhahn, *Chem. Res. Toxicol.*, *10*, 271 (1997).
384. J. Aiyar, H. J. Berkovits, R. A. Floyd, and K. E. Wetterhahn, *Chem. Res. Toxicol.*, *3*, 595 (1990).
385. S. P. Faux, M. Gao, J. K. Chipman, and L. S. Levy, *Carcinogenesis*, *13*, 1667 (1992).
386. H. Rodriguez and S. A. Akman, *Free Radical Res.*, *29*, 499 (1998).
387. K. E. Tomaszewski, D. K. Agarwal, and R. L. Melnick, *Carcinogenesis*, *7*, 1871 (1986).
388. S. I. Liochev, *Metal Ions Biol. Syst.*, *36*, 1 (1999).
389. B. Chance, H. Sies, and A. Boveris, *Physiol. Rev.*, *59*, 527 (1979).
390. B. Halliwell and J. M. C. Gutteridge, *Methods Enzymol.*, *186*, 1 (1990).
391. P. H. Connett and K. E. Wetterhahn, *J. Am. Chem. Soc.*, *108*, 1842 (1986).
392. P. O'Brien, J. Barrett, and F. Swanson, *Inorg. Chim. Acta*, *108*, L19 (1985).
393. D. M. L. Goodgame and A. M. Joy, *J. Inorg. Biochem.*, *26*, 219 (1986).
394. S. Kitagawa, H. Seki, F. Kametani, and H. Sakurai, *Inorg. Chim. Acta*, *152*, 251 (1988).
395. R. Zhao, J. Lind, G. Merényi, and T. E. Eriksen, *J. Chem. Soc., Perkin Trans. 2*, 569 (1997).
396. D. M. L. Goodgame and A. M. Joy, *Inorg. Chim. Acta*, *135*, 115 (1987).
397. G. R. Buettner, *Free Rad. Res. Commun.*, *1*, 349 (1986).
398. B. H. J. Bielski, in *Ascorbic Acid: Chemistry, Metabolism, and Uses*, P. A. Seib and B. M. Tolbert, Eds., American Chemical Society, Washington DC, 1982, pp. 81–99.
399. D. M. Stearns and K. E. Wetterhahn, *Chem. Res. Toxicol.*, *7*, 219 (1994).
400. A. Levina, R. Barr-David, R. Codd, P. A. Lay, N. E. Dixon, A. Hammershøi, and P. Hendry, *Chem. Res. Toxicol.*, *12*, 371 (1999).
401. Y. Lefebvre and H. Pezerat, *Chem. Res. Toxicol.*, *5*, 461 (1992).
402. S. K. Ghosh and E. S. Gould, *Inorg. Chem.*, *25*, 3357 (1986).
403. K. D. Sugden and K. E. Wetterhahn, *Chem. Res. Toxicol.*, *10*, 1397 (1997).

404. G. Barr-David, Ph.D. Thesis, In Vitro Cr(V)-DNA Reactions in the Context of Cr(VI)-Induced Cancers, University of Sydney, Australia, 1998.
405. G. Pratviel, J. Bernadou, and B. Meunier, *Angew. Chem., Int. Ed. Engl.*, *34*, 746 (1995).
406. W. K. Pogozelski and T. D. Tullius, *Chem. Rev.*, *98*, 1089 (1998).
407. H. Rodriguez, G. P. Holmquist, R. D'Agostino, Jr., J. Keller, and S. A. Akman, *Cancer Res.*, *57*, 2394 (1997).
408. C. Méplan, K. Mann, and P. Hainaut, *J. Biol. Chem.*, *274*, 31663 (1999).
409. D. M. L. Goodgame, P. B. Hayman, and D. E. Hathway, *Polyhedron*, *1*, 497 (1982).
410. C. J. Burrows and J. G. Muller, *Chem. Rev.*, *98*, 1109 (1998).
411. K. M. Ananth, R. Rajaram, and T. Ramasami, *Biochem. Biophys. Res. Commun.*, *273*, 1138 (2000).
412. D. I. Pattison, M. J. Davies, and P. A. Lay, unpublished data.
413. A. Klug, *J. Mol. Biol.*, *293*, 215 (1999).
414. R. Codd and P. A. Lay, to be submitted.
415. W. Adam, A. Kurz, and C. R. Saha-Möller, *Chem. Res. Toxicol.*, *13*, 1199 (2000).
416. F. L. Petrilli and S. DeFlora, *Mutat. Res.*, *58*, 167 (1978).
417. A. Bakac and W.-D. Wang, *J. Am. Chem. Soc.*, *118*, 10325 (1996).
418. T.-C. Tsou and J.-L. Yang, *Chem.-Biol. Interact.*, *102*, 133 (1996).
419. D. R. Lloyd, P. L. Carmichael, and D. H. Phillips, *Chem. Res. Toxicol.*, *11*, 420 (1998).
420. W. B. Qi, R. J. Reiter, D. X. Tan, J. J. Garcia, L. C. Manchester, M. Karbownik, and J. R. Calvo, *Environ. Health Persp.*, *108*, 399 (2000).
421. W. B. Qi, R. J. Reiter, D. X. Tan, L. C. Manchester, A. W. Siu, and J. J. Garcia, *J. Pineal Res.*, *29*, 54 (2000).
422. S. Burkhardt, R. J. Reiter, D. X. Tan, R. Hardeland, J. Cabrera, and M. Karbownik, *Int. J. Biochem. Cell Biol.*, *33*, 775 (2001).
423. J. Stubbe, J. W. Kozarich, W. Wu, and D. E. Vanderwall, *Acc. Chem. Res.*, *29*, 322 (1996).
424. H. Stünzi and W. Marty, *Inorg. Chem.*, *22*, 2145 (1983).
425. A. Sreedhara, N. Susa, and C. P. Rao, *Inorg. Chim. Acta*, *263*, 189 (1997).
426. A. Parand, A. C. Royer, T. L. Cantrell, M. Weitzel, N. Memon, J. B. Vincent, and M. W. Crowder, *Inorg. Chim. Acta*, *268*, 211 (1998).
427. H. Y. Shrivastava and B. U. Nair, *Biochem. Biophys. Res. Commun.*, *270*, 749 (2000).
428. A. Bakac and W.-D. Wang, *Inorg. Chim. Acta*, *297*, 27 (2000).
429. J. K. Speetjens, R. A. Collins, J. B. Vincent, and S. A. Woski, *Chem. Res. Toxicol.*, *12*, 483 (1999).
430. D. M. Stearns, J. J. Belbruno, and K. E. Wetterhahn, *FASEB J.*, *9*, 1650 (1995).
431. B. T. Picard, M. E. Ketterer, and D. M. Stearns, *Abstr. Pap. - Am. Chem. Soc.*, *221*, INOR-527 (2001).
432. D. Beyersmann and A. Koster, *Toxicol. Environ. Chem.*, *14*, 11 (1987).
433. T. Wolf, R. Kasemann, and H. Ottenwalder, *Carcinogenesis*, *10*, 655 (1989).
434. B. Gulanowski, J. Swiatek, and H. Kozlowski, *J. Inorg. Biochem.*, *48*, 289 (1992).
435. H. Arakawa, R. Ahmad, M. Naoui, and H. A. Tajmir-Riahi, *J. Biol. Chem.*, *275*, 10150 (2000).
436. E. R. Civitello, R. G. Leniek, K. A. Hossler, K. Haebe, and D. M. Stearns, *Bioconjugate Chem.*, *12*, 459 (2001).

437. B. Gulanowski, M. Cieslak-Golonka, K. Szyba, and J. Urban, *BioMetals*, 7, 177 (1994).
438. E. Kejnovsky and J. Kypr, *Nucleic Acids Res.*, 26, 5295 (1998).
439. C. P. Rao, S. P. Kaiwar, and M. S. S. Raghavan, *Polyhedron*, 13, 1895 (1994).
440. R. T. Watson, N. Desai, J. Wildsmith, J. F. Wheeler, and N. A. P. Kane-Maguire, *Inorg. Chem.*, 38, 2683 (1999).
441. R. Vijayalakshmi, M. Kanthimathi, V. Subramanian, and B. U. Nair, *Biochim. Biophys. Acta*, 1475, 157 (2000).
442. R. Vijayalakshmi, M. Kanthimathi, V. Subramanian, and B. U. Nair, *Biochem. Biophys. Res. Commun.*, 271, 731 (2000).
443. M. Krumpolc and J. Roček, *J. Am. Chem. Soc.*, 101, 3206 (1979).
444. R. J. Judd, T. W. Hambley, and P. A. Lay, *J. Chem. Soc., Dalton Trans.*, 2205 (1989).
445. K. D. Sugden, C. K. Campo, and B. D. Martin, *Chem. Res. Toxicol.*, 14, 1315 (2001).
446. K. D. Sugden and K. E. Wetterhahn, *J. Am. Chem. Soc.*, 118, 10811 (1996).
447. K. D. Sugden, *J. Inorg. Biochem.*, 77, 177 (1999).
448. R. N. Bose and B. S. Fonkeng, *J. Chem. Soc., Chem. Commun.*, 2211 (1996).
449. R. N. Bose, B. S. Fonkeng, S. Moghaddas, and D. Stroup, *Nucleic Acids Res.*, 26, 1588 (1998).
450. R. N. Bose, S. Moghaddas, P. A. Mazzer, L. P. Dudones, L. Joudah, and D. Stroup, *Nucleic Acid Res.*, 27, 2219 (1999).
451. A. Levina, P. A. Lay, and N. E. Dixon, *Inorg. Chem.*, 39, 385 (2000).
452. A. Levina, P. A. Lay, and N. E. Dixon, *Chem. Res. Toxicol.*, 14, 946 (2001).
453. K. D. Sugden and K. E. Wetterhahn, *Inorg. Chem.*, 35, 3727 (1996).
454. B. Giese, A. Dussy, E. Meggers, M. Petretta, and U. Schwitter, *J. Am. Chem. Soc.*, 119, 11130 (1997).
455. A. Levina and P. A. Lay, unpublished data.
456. M. Krumpolc and J. Roček, *Inorg. Chem.*, 24, 617 (1985).
457. A. Kortenkamp, G. Oetken, and D. Beyersmann, *Mutat. Res.*, 232, 155 (1990).
458. L. Shi, A. Levina, and P. A. Lay, unpublished data.
459. R. A. Anderson, *Diabetes Metab.*, 26, 22 (2000).
460. H. C. Lukaski, *Am. J. Clin. Nutr.*, 72, 585S (2000).
461. K. N. Jeejeebhoy, *Nutr. Rev.*, 57, 329 (1999).
462. B. W. Morris, *J. Trace Elem. Exp. Med.*, 12, 61 (1999).
463. M. K. Hellerstein, *Nutr. Rev.*, 56, 302 (1998).
464. W. Mertz, *J. Am. Coll. Nutr.*, 17, 544 (1998).
465. W. Mertz, *Nutr. Rev.*, 56, 174 (1998).
466. E. B. Kegley and J. W. Spears, *J. Trace Elem. Exp. Med.*, 12, 141 (1999).
467. M. D. Lindemann, *J. Trace Elem. Exp. Med.*, 12, 149 (1999).
468. R. A. Anderson, A. M. Roussel, and J. Nève, in *Handbook of Hypertension. Vol. 20: Epidemiology of Hypertension.*, C. J. Bulpitt, Ed., Elsevier Science, Amsterdam, The Netherlands, 2000, pp. 313–336.
469. D. W. Laight, *Expert Opin. Ther. Patents*, 10, 1703 (2000).
470. G. L. Christie and D. R. Williams, in *Handbook of Metal–Ligand Interactions in Biological Fluids. Bioinorganic Medicine*, Vol.1, G. Berthon, Ed., Marcel Dekker, New York, 1995, pp. 29–37.

471. K. Schwartz and W. Mertz, *Arch. Biochem. Biophys.*, *72*, 515 (1957).
472. K. Schwarz and W. Mertz, *Arch. Biochem. Biophys.*, *85*, 292 (1959).
473. K. N. Jeejeebhoy, *J. Trace Elem. Exp. Med.*, *12*, 85 (1999).
474. R. A. Anderson, J. S. Borel, M. M. Polansky, N. A. Bryden, T. C. Majerus, and M. P. B., *J. Trace Elem. Exper. Med.*, *1*, 9 (1988).
475. J. S. Striffler, M. M. Polansky, and R. A. Anderson, *Metab.-Clin. Exp.*, *47*, 396 (1998).
476. J. S. Striffler, M. M. Polansky, and R. A. Anderson, *Metab.-Clin. Exp.*, *48*, 1063 (1999).
477. R. A. Anderson, in *Handbook of Metal–Ligand Interactions in Biological Fluids. Bioinorganic Medicine*, Vol.1, G. Berthon, Ed., Marcel Dekker, New York, 1995, pp. 261–265.
478. R. H. Holm, P. Kennepohl, and E. I. Solomon, *Chem. Rev.*, *96*, 2239 (1996).
479. N. Z. Cheng, X. X. Zhu, H. L. Shi, W. L. Wu, J. M. Chi, J. Y. Cheng, and R. A. Anderson, *J. Trace Elem. Exp. Med.*, *12*, 55 (1999).
480. N. Mirsky, *J. Inorg. Biochem.*, *49*, 123 (1993).
481. A. Ravina, L. Slezak, A. Rubal, and N. Mirsky, *J. Trace. Elem. Exp. Med.*, *8*, 183 (1995).
482. Y. J. Sun, K. Mallya, J. Ramirez, and J. B. Vincent, *J. Biol. Inorg. Chem.*, *4*, 838 (1999).
483. H. G. Preuss, N. Talpur, V. Manohar, N. Venkataramiah, and R. A. Anderson, *J. Trace Elem. Exp. Med.*, *12*, 125 (1999).
484. M. N. McLeod and R. N. Golden, *Int. J. Neuropsychopharmacol.*, *3*, 311 (2000).
485. M. D. Lindemann, C. M. Wood, A. F. Harper, E. T. Kornegay, and R. A. Anderson, *J. Anim. Sci.*, *73*, 457 (1995).
486. B. A. Mallard, P. Borgs, M. J. Ireland, B. W. McBride, B. D. Brown, and J. A. Irwin, *J. Trace Elem. Exp. Med.*, *12*, 131 (1999).
487. S. Okada, M. Suzuki, and H. Ohba, *J. Inorg. Biochem.*, *19*, 95 (1983).
488. S. Okada, H. Tsukada, and M. Tezuka, *Biol. Trace Elem. Res.*, *21*, 35 (1989).
489. W. Mertz, E. E. Roginsky, and K. Schwarz, *J. Biol. Chem.*, *236*, 318 (1961).
490. W. Mertz and D. E. Thurman, *Fed. Proc.*, *27*, 482 (1968).
491. R. A. Anderson, J. H. Brantner, and M. M. Polansky, *J. Agric. Food. Chem.*, *26*, 1219 (1978).
492. J. B. Vincent, *J. Nutr.*, *124*, 117 (1994).
493. W. Mertz, *J. Nutr.*, *124*, 119 (1994).
494. F. Borguet, R. Cornelis, J. Delanghe, M.-C. Lambert, and N. Lameire, *Clin. Chim. Acta*, *238*, 71 (1995).
495. B. Morris, T. Gray, and S. MacNeil, *J. Endocrinol.*, *144*, 135 (1995).
496. C. M. Davis and J. B. Vincent, *Biochemistry*, *36*, 4382 (1997).
497. C. M. Davis, A. C. Royer, and J. B. Vincent, *Inorg. Chem.*, *36*, 5316 (1997).
498. I. Goldwaser, D. Gefel, E. Gershonov, M. Fridkin, and Y. Shechter, *J. Inorg. Biochem.*, *80*, 21 (2000).
499. B. Zhang, G. Salituro, D. Szalkowski, Z. Li, Y. Zhang, I. Royo, D. Vilella, M. T. Díez, F. Pelaez, C. Ruby, R. L. Kendall, X. Mao, P. Griffin, J. Calaycay, J. R. Zierath, J. V. Heck, R. G. Smith, and D. E. Moller, *Science (Washington, DC)*, *284*, 974 (1999).
500. B. Gammelgaard, K. Jensen, and B. Steffansen, *J. Trace Elements Med. Biol.*, *13*, 82 (1999).
501. A. F. Hofmann, in *Handbook of Metal-Ligand Interactions in Biological Fluids*, Vol. 1, G. Berthon, Ed., Marcel Dekker, New York, 1995, pp. 38–47.
502. T. H. Lim, T. Sargent, and N. Kusubov, *Am. J. Physiol.*, *244*, R445 (1983).
503. H. Sun, H. Li, and P. J. Sadler, *Chem. Rev.*, *99*, 2817 (1999).

504. R. A. Anderson, in *Handbook of Metal–Ligand Interactions in Biological Fluids. Bioinorganic Medicine*, Vol. 1, G. Berthon, Ed., Marcel Dekker, New York, 1995, pp. 418–421.
505. H. Li, P. J. Sadler, and H. Sun, *Eur. J. Biochem.*, 242, 387 (1996).
506. P. Aisen, in *Inorganic Biochemistry*, Vol. 1, G. L. Eichhorn, Ed., Elsevier, Amsterdam, 1973, pp. 280–305.
507. A. Levina and P. A. Lay, unpublished data.
508. E. Nieboer and A. A. Jusys, in *Chromium in the Natural and Human Environments*, J. O. Nriagu and E. Nieboer, Eds., John Wiley & Sons, Inc., New York, 1988, pp. 21–79.
509. C. Harford and B. Sarkar, in *Handbook of Metal–Ligand Interactions in Biological Fluids. Bioinorganic Medicine*, Vol.1, G. Berthon, Ed., Marcel Dekker, New York, 1995, pp. 62–70.
510. S. Stojkovski, W. Goumakos, and B. Sarkar, *Biochim. Biophys. Acta*, 1137, 155 (1992).
511. F. Salem, in *The Biological and Environmental Chemistry of Chromium*, S. A. Katz and H. Salem, Eds., VCH, New York, 1994, pp. 121–153.
512. C. Veillon, *Methods Enzymol.*, 158, 334 (1988).
513. C. Veillon and K. Y. Patterson, *J. Trace Elem. Exp. Med.*, 12, 99 (1999).
514. W. Y. Feng, Q. F. Qian, W. J. Ding, and Z. F. Chai, *J. Radioanalyt. Nucl. Chem.*, 244, 321 (2000).
515. D. Behne, P. Braetter, H. Gessner, G. Hube, W. Mertz, and U. Roesick, *Fresenius' Z. Anal. Chem.*, 278, 269 (1976).
516. R. A. Anderson, *J. Trace Elem. Exp. Med.*, 11, 241 (1998).
517. J. D. Kruse-Jarres and M. Rukgauer, *J. Trace Elem. Med. Biol.*, 14, 21 (2000).
518. C. Ekmekcioglu, C. Prohaska, K. Pomazal, I. Steffan, G. Schernthaner, and W. Marktl, *Biol. Trace Elem. Res.*, 79, 205 (2001).
519. J. E. Gunton, G. Hams, R. Hitchman, and A. McElduff, *Am. J. Clin. Nutr.*, 73, 99 (2001).
520. T. Zima, O. Mestek, V. Tesar, P. Tesarova, K. Nemecek, A. Zak, and M. Zeman, *Biochem. Molec. Biol. Int.*, 46, 365 (1998).
521. S. Kazi, T. G. Kazil, S. S. Ali, and G. H. Kazi, *Abstr. Pap. - Am. Chem. Soc.*, 219, ANYL-124 (2000).
522. J. Barrett, P. O'Brien, and J. Pedrosa De Jesus, *Polyhedron*, 4, 1 (1985).
523. E. W. Toepfer, W. Mertz, M. M. Polansky, E. E. Roginski, and W. R. Wolf, *J. Agric. Food Chem.*, 25, 162 (1977).
524. A. Demirci and A. L. Pometto, *J. Agric. Food Chem.*, 48, 531 (2000).
525. R. A. Anderson, M. M. Polansky, E. E. Roginski, and W. Mertz, *J. Agric. Food Chem.*, 26, 858 (1978).
526. P. Raspor, M. Batic, P. Jamnic, D. Josic, R. Milacic, M. Pas, M. Recek, V. Rezic-Dereani, and M. Skrt, *Acta Microbiol. Immunol. Hung.*, 47, 143 (2000).
527. V. G. Zetic, V. Stehlik-Tomas, S. Grba, L. Lutilsky, and D. Kozlek, *J. Biosci.*, 26, 217 (2001).
528. N. Mirsky, A. Weiss, and Z. Dori, *J. Inorg. Biochem.*, 13, 11 (1980).
529. K. H. Sumrall and J. B. Vincent, *Polyhedron*, 16, 4171 (1997).
530. M. Beran and R. Stahl, *Analyst*, 120, 979 (1995).
531. J. Woolliscroft and J. Barbosa, *J. Nutr.*, 107, 1702 (1977).
532. P. R. Shepherd, C. Elwood, P. D. Buckley, and L. F. Blackwell, *Biol. Trace Elem. Res.*, 32, 109 (1996).
533. S. J. Haylock, P. D. Buckley, and L. F. Blackwell, *J. Inorg. Biochem.*, 19, 105 (1983).
534. E. S. Holdsworth, D. V. Kaufman, and E. Neville, *Br. J. Nutr.*, 65, 285 (1991).

535. M. J. Kendrick, M. T. May, M. J. Plishka, and K. D. Robinson, *Metals in Biological Systems*, Ellis Horwood, New York, 1992, pp. 70–74.
536. E. Garcia, C. Cabrera, M. L. Lorenzo, and M. C. Lopez, *Sci. Total Environ.*, *247*, 51 (2000).
537. G. H. Starich and C. Blinkoe, *J. Agric. Food Chem.*, *30*, 458 (1982).
538. N. Mirsky, A. Weiss, and Z. Dori, *J. Inorg. Biochem.*, *15*, 275 (1981).
539. A. Yamamoto, O. Wada, and T. Ono, *Toxicol. Appl. Pharmacol.*, *59*, 515 (1981).
540. A. Yamamoto, O. Wada, and T. Ono, *J. Inorg. Biochem.*, *22*, 91 (1984).
541. O. Wada, G. Y. Wu, A. Yamamoto, S. Manabe, and T. Ono, *Environ. Res.*, *32*, 228 (1983).
542. A. Yamamoto, O. Wada, and T. Ono, *Eur. J. Biochem.*, *165*, 627 (1987).
543. A. Knoechel and G. Weseloh, *Fresenius' J. Anal. Chem.*, *363*, 533 (1999).
544. A. Yamamoto, O. Wada, and H. Suzuki, *J. Nutr.*, *118*, 39 (1988).
545. A. Yamamoto, O. Wada, and S. Manabe, *Biochem. Biophys. Res. Comm.*, *163*, 189 (1989).
546. J. B. Vincent and M. Davis, *U.S. Patent No.* 5,872,102 (1999), *Chem. Abstr.*, *130*, 150644 (1999).
547. J. B. Vincent, *Acc. Chem. Res.*, *33*, 503 (2000).
548. C. M. Davis, K. H. Sumrall, and J. B. Vincent, *Biochemistry*, *35*, 12963 (1996).
549. J. B. Vincent, *J. Am. Coll. Nutr.*, *18*, 6 (1999).
550. J. B. Vincent, *J. Nutr.*, *130*, 715 (2000).
551. J. B. Vincent, *Nutr. Rev.*, *58*, 67 (2000).
552. A. A. Shute, N. E. Chakov, and J. B. Vincent, *Polyhedron*, *20*, 2241 (2001).
553. M. R. Nelson and W. J. Chazin, in *Calmodulin and Signal Transduction*, L. J. Van Eldik and D. M. Watterson, Eds., Academic, San Diego, 1998, pp. 17–65.
554. B. W. Morris, S. MacNeil, K. Stanley, T. A. Gray, and R. Fraser, *J. Endocrinol.*, *139*, 339 (1993).
555. B. W. Morris, T. A. Gray, and S. MacNeil, *Clin. Sci.*, *84*, 477 (1993).
556. W. Mertz, E. W. Toepfer, E. E. Roginsky, and M. M. Polansky, *Fed. Proc., Fed. Am. Soc. Exp. Biol.*, *33*, 2275 (1974).
557. K. Govindaraju, T. Ramasami, and D. Ramaswamy, *J. Inorg. Biochem.*, *35*, 137 (1989).
558. R. Gayatri, A. Rajaram, R. Rajaram, K. Govindaraju, J. R. Rao, B. U. Nair, and T. Ramasami, *Proc. - Indian Acad. Sci., Chem. Sci.*, *109*, 307 (1997).
559. R. Gayatri, R. Rajaram, B. U. Nair, F. Chandrasekaran, and T. Ramasami, *Proc. Ind. Acad. Sci. - Chem. Sci.*, *111*, 133 (1999).
560. S. MacNeil, R. Dawson, T. Lakey, and B. Morris, *Cell Calcium*, *8*, 207 (1987).
561. B. A. Bovykin, K. E. Kilivnik, and G. D. Zegzhda, *Biofizika*, *44*, 461 (1999).
562. M. Pesti, Z. Gazdag, and J. Belágyi, *FEMS Microbiol. Lett.*, *182*, 375 (2000).
563. T. M. Santos, J. Pedrosa De Jesus, and P. O'Brien, *Polyhedron*, *11*, 1687 (1992).
564. A. Hanaki, *Bull. Chem. Soc. Jpn.*, *68*, 831 (1995).
565. M. Scarpa, F. Momo, P. Viglino, F. Vianello, and A. Rigo, *Biophys. Chem.*, *60*, 53 (1996).
566. M. Cieslak-Golonka, A. Adach, and K. R. Zurowski, *J. Therm. Anal.*, *63*, 91 (2000).
567. M. Cieslak-Golonka, M. Raczko, and Z. Staszak, *Polyhedron*, *11*, 2549 (1992).
568. P. O'Brien, G. Wang, and P. B. Wyatt, *Polyhedron*, *11*, 3211 (1992).
569. E. Gaggelli, F. Berti, D. Valensin, and G. Valensin, in *34th International Conference on Coordination Chemistry*, Royal Society of Chemistry, Edinburgh, UK, 2000, poster P0861.
570. G. Valensin, *J. Inorg. Biochem.*, *86*, 106 (2001).
571. E. A. Dawes, *Quantitative Problems in Biochemistry*, 6 ed., Longman, London, 1980, pp. 20–24.

572. S. J. Gaskell, *J. Mass Spectrosc.*, *32*, 677 (1997).
573. B. J. Clodfelder, J. Emamaullee, D. D. D. Hepburn, N. E. Chakov, H. S. Nettles, and J. B. Vincent, *J. Biol. Inorg. Chem.*, *6*, 608 (2001).
574. Y. J. Sun, J. Ramirez, S. A. Woski, and J. B. Vincent, *J. Biol. Inorg. Chem.*, *5*, 129 (2000).
575. A. Drljaca, M. J. Hardie, C. L. Raston, and L. Spiccia, *Chem. Eur. J.*, *5*, 2295 (1999).
576. R. Österberg, in *Handbook of Metal–Ligand Interactions in Biological Fluids*, Vol.1, G. Berthon, Ed., Marcel Dekker, New York, 1995, pp. 10–28.
577. National Research Council (USA), in *Recommended Dietary Allowances, Food and Nutrition Board*, National Academy, Washington, DC, 1989, pp. 241–243.
578. Institute of Medicine (USA), *Dietary Reference Intakes for Vitamin A, Vitamin K, Arsenic, Boron, Chromium, Copper, Iodine, Iron, Molybdenum, Nickel, Silicon, Vanadium and Zinc. Food and Nutrition Board*. National Academy Press, Washington, DC, 2001.
579. W. Mertz, *Physiol. Rev.*, *49*, 163 (1969).
580. G. W. Evans, *U.S. Patent No.* 4,315,927 (1981), *Chem. Abstr.*, *95*, 23384 (1981).
581. R. I. Press, J. Geller, and G. W. Evans, *West. J. Med.*, *152*, 41 (1990).
582. H. Boynton and G. W. Evans, *U.S. Patent No.* 5,087,623 (1992), *Chem. Abstr.*, *116*, 207833 (1992).
583. M. F. McCarty, *J. Appl. Nutr.*, *43*, 58 (1991).
584. G. W. Evans and D. J. Pouchnik, *J. Inorg. Biochem.*, *49*, 177 (1993).
585. M. L. Gargas, R. L. Norton, D. J. Paustenbach, and B. L. Finley, *Drug Metab. Dispos.*, *22*, 522 (1994).
586. R. A. Anderson, M. M. Polansky, N. A. Bryden, K. Y. Patterson, C. Veillon, and W. H. Glinsmann, *J. Nutr.*, *113*, 276 (1983).
587. G. W. Evans and T. D. Bowman, *J. Inorg. Biochem.*, *46*, 243 (1992).
588. K. Fukui, Y. Fujisawa, H. Ohya-Nishiguchi, H. Kamada, and H. Sakurai, *J. Inorg. Biochem.*, *77*, 215 (1999).
589. L. Yang, D. C. Crans, V. G. Yuen, J. McNeill, and G. R. Willsky, *Abstr. Pap. - Am. Chem. Soc.*, *220*, INOR-047 (2000).
590. D. N. Lee, H. T. Yen, T. F. Shen, and B. J. Chen, *Biol. Trace Elem. Res.*, *77*, 53 (2000).
591. J. W. Moore, M. A. Maher, W. J. Banz, and M. B. Zemel, *J. Nutr.*, *128*, 180 (1998).
592. D. M. Stearns and W. H. Armstrong, *Inorg. Chem.*, *31*, 5178 (1992).
593. N. E. Chakov, R. A. Collins, and J. B. Vincent, *Polyhedron*, *18*, 2891 (1999).
594. L. D. Bunting, T. A. Tarifa, B. T. Crochet, J. M. Fernandez, C. L. Depew, and J. C. Lovejoy, *J. Dairy Sci.*, *83*, 2491 (2000).
595. E. B. Kegley, D. L. Galloway, and T. M. Fakler, *J. Animal Sci.*, *78*, 3177 (2000).
596. N. Mirsky, A. Aharoni, C. Rabinowitz, and I. Izhaki, *J. Trace Elem. Exp. Med.*, *12*, 111 (1999).
597. P. M. Clarkson and E. S. Rawson, *Crit. Rev. Food Sci. Nutr.*, *39*, 317 (1999).
598. S. L. Volpe, H.-W. Huang, K. Larpadisorn, and I. I. Lesser, *J. Am. Coll. Nutr.*, *20*, 293 (2001).
599. Federal Trade Commision (USA), Docket No. C-3758, 1997.
600. The Food and Drug Administration (USA), *Fed. Regist.*, *63*, 34104 (1998).
601. R. A. Anderson, *Nutrition*, *15*, 720 (1999).
602. E. N. Behrend and D. S. Greco, *Compend. Contin. Educ. Pract. Vet.*, *22*, 423 (2000).
603. R. W. Nelson, *J. Small Anim. Prac.*, *41*, 486 (2000).
604. S. Schachter, R. W. Nelson, and C. A. Kirk, *J. Vet. Intern. Med.*, *15*, 379 (2001).

605. J. B. Vincent and C. M. Davis, U.S. Patent No. 6,197,816 (2001), Chem. Abstr., 134, 207223 (2001).
606. A. Earnshaw, B. N. Figgis, and J. Lewis, J. Chem. Soc. A, 1656 (1966).
607. D. M. Stearns, J. P. Wise, Sr., S. R. Patierno, and K. E. Wetterhahn, FASEB J., 9, 1643 (1995).
608. D. M. Stearns, J. A. Hager, A. M. Luke, K. R. Manygoats, B. T. Picard, K. K. Wolf, and M. Yazzie, Abstr. Pap. - Am. Chem. Soc., 222, TOXI-092 (2001).
609. S. A. Kareus, C. Kelley, H. S. Walton, and P. R. Sinclair, J. Haz. Mat., 84, 163 (2001).
610. M. F. McCarty, FASEB J., 10, 365 (1996).
611. M. F. McCarty, Med. Hypotheses, 48, 263 (1997).
612. D. M. Stearns and K. E. Wetterhahn, FASEB J., 10, 367 (1996).
613. D. Bagchi, M. Bagchi, J. Balmoori, X. Ye, and S. J. Stohs, Res. Comm. Mol. Pathol. Pharmacol., 97, 335 (1997).
614. I. Kato, J. H. Vogelman, V. Dilman, J. Karkoszka, K. Frenkel, N. P. Durr, N. Orentreich, and P. Toniolo, Eur. J. Epidemiol., 14, 621 (1998).
615. J. K. Speetjens, A. Parand, M. W. Crowder, J. B. Vincent, and S. A. Woski, Polyhedron, 18, 2617 (1999).
616. W. Yu, M. A. Sipowicz, D. C. Haines, L. Birely, B. A. Diwan, C. W. Riggs, K. S. Kasprzak, and L. M. Anderson, Toxicol. Appl. Pharmacol., 158, 161 (1999).
617. M. Anke, R. Muller, A. Trupschuch, M. Seifert, M. Jaritz, S. Holzinger, and S. Anke, J. Trace Microprobe Tech., 18, 541 (2000).
618. K. H. Thompson and C. Orvig, J. Chem. Soc., Dalton Trans., 2885 (2000).
619. C. L. Beswick, R. D. Shalders, and T. W. Swaddle, Inorg. Chem., 35, 991 (1996).
620. C. Wang, D. F. Martin, and B. B. Martin, J. Environ. Sci. Health, A33, 1631 (1998).
621. K. Takenouchi, Hikaku Kagaku, 43, 137 (1997). Chem. Abstr., 128, 49758, (1997).
622. A. D. Covington, G. S. Lampard, O. Menderes, A. V. Chadwick, G. Rafeletos, and P. O'Brien, Polyhedron, 20, 461 (2001).
623. M. J. Kendrick, M. T. May, M. J. Plishka, and K. D. Robinson, in Metals in Biological Systems, Ellis Horwood, New York, 1992, pp. 90–91.
624. E. Kim and K. J. Na, Toxicol. Appl. Pharmacol., 110, 251 (1991).
625. Y. Goto and K. Kida, Jap. J. Pharmacol., 67, 365 (1995).
626. E. J. Yurkow and G. Kim, Molec. Pharmacol., 47, 686 (1995).
627. G. L. Schieven, U.S. Patent No. 5,877,210 (1999), Chem. Abstr., 130, 218273 (1999).
628. J. Puente, M. A. Salas, C. Canon, D. Miranda, M. E. Wolf, and A. D. Mosnaim, Int. J. Clin. Pharmacol. Ther., 34, 212 (1996).
629. L. M. Lester, L. A. Rusch, G. J. Robinson, and D. C. Speckhard, Biochemistry, 37, 5349 (1998).
630. W. W. Cleland and A. S. Mildvan, in Advances in Inorganic Biochemistry, Vol. 1, G. L. Eichhorn and L. G. Marzilli, Eds., Elsevier, New York, 1979, pp. 163–191.
631. H. E. Van Wart, Methods Enzymol., 158, 95 (1988).
632. K. Balamurugan, C. Vasant, R. Rajaram, and T. Ramasami, Biochim. Biophys. Acta, 1427, 357 (1999).
633. M. W. Hosseini, J.-M. Lehn, K. C. Jones, K. E. Plute, K. B. Mertes, and M. P. Mertes, J. Am. Chem. Soc., 111, 6330 (1989).
634. C. J. Boreham, D. A. Buckingham, D. J. Francis, A. M. Sargeson, and L. G. Warner, J. Am. Chem. Soc., 103, 1975 (1981).

635. O. Clement, I. Onyido, and E. Buncel, *Can. J. Chem.*, *78*, 474 (2000).
636. C. R. Clark, A.G. Blackman, and A. J. Clarkson, *J. Am. Chem. Soc.*, *123*, 8131 (2001).
637. C. A. Pasternak, *Ind. J. Biochem. Biophys.*, *27*, 363 (1990).
638. F. A. Cotton, G. Wilkinson, C. A. Murillo, and M. Bochmann, *Advanced Inorganic Chemistry*, 6 Edn., John Wiley & Sons, Inc., New York, 1999.
639. R. D. Crawford, *Biochem. Mol. Med.*, *54*, 1 (1995).
640. A. Van den Bergen, R. Colton, M. Percy, and B. O. West, *Inorg. Chem.*, *32*, 3408 (1993).
641. I. I. Stewart and G. Horlick, *J. Analyt. Atomic Spectrosc.*, *11*, 1203 (1996).
642. W. Henderson, B. K. Nicholson, and L. J. McCaffrey, *Polyhedron*, *17*, 4291 (1998).
643. M. J. Cunningham, S. Liang, S. Fuhrman, J. J. Seilhamer, and R. Somogyi, *Ann. New York Acad. Sci.*, *919*, 52 (2000).
644. R. Aebersold, B. Rist, and S. P. Gygi, *Ann. New York Acad. Sci.*, *919*, 33 (2000).

Laterally Nonsymmetric Aza-Cryptands

PARIMAL K. BHARADWAJ

Department of Chemistry
Indian Institute of Technology Kanpur
208016, India

CONTENTS

I. INTRODUCTION AND SCOPE

II. SYNTHESIS OF LATERALLY NONSYMMETRIC CRYPTANDS

    A. Methodology
    B. Strategy
        1. Tripodal Coupling
        2. Tripodal Capping

III. METAL COMPLEXATION

    A. Cryptands with N and S as Donors
    B. Cryptands with N and O as Donors

IV. FLUORESCENT SIGNALING

    A. General Design Principles
    B. Metal Ion Sensing
    C. Chemical Logics
    D. Transition Metal Ions in Fluorescent Signaling
    E. Cryptand-Based Fluorescent Sensors

V. CRYPTANDS AS NONLINEAR OPTICAL MATERIALS

    A. Nonlinear Optical Effect
    B. Cryptand Derivatives as D-π-A Nonlinear Systems

VI. CRYPTAND-BASED AMPHIPHILES

    A. Vesicular Aggregation
    B. Amphiphiles at Air–Water Interface
        1. Langmuir–Blodgett Film
        2. Molecular Recognition

VII. CONCLUSIONS

ACKNOWLEDGMENTS

ABBREVIATIONS

REFERENCES

I. INTRODUCTION AND SCOPE

Macropolycyclic cryptands occupy an important place in supramolecular chemistry. These molecules have found a wide range of applications (1–18) that cover most areas of chemistry and several fields of biological and material sciences as well. Cryptands provide a number of donor atoms in space, whose topology can be varied via ligand design to accommodate ions, neutral molecules or both forming inclusion complexes known as cryptates. The nature and selectivity of a cryptand can be tailored by the selection of appropriate donor atoms at strategic positions. However, the desired selectivity of a cryptand toward a single metal ion in presence of several others remains an elusive goal although several studies have been carried out (19–22) toward this end. This chapter deals with bicyclic cryptands where the two bridgehead atoms like N or C, or two benzene units, are connected by three bridges. There are many examples starting from the N-bridgehead polyether cryptands that fit this description. A number of books and review articles have been devoted (1, 23–30) to their chemistry. However, the discussions here will be focused on bicyclic cryptands where three secondary amino nitrogens are located about one bridgehead while the other end contains donor atoms such as S or O making them laterally nonsymmetric (Fig. 1). These cryptands constitute an important class of molecules with several potentially useful applications. For the cryptands with a small cavity, a single metal ion can occupy the entire space inside. In such situations, placing donor atoms at strategic positions can impose a low symmetry of coordination around the metal ion. Such cryptands are, therefore, suitable to mimic the *intrinsic active sites* (31) of metalloproteins. The metal ion being trapped inside can exhibit a rapid electron-transfer capability, since it is prevented from relaxing to its preferred geometry in either redox state. Such systems can, therefore, act as electron carriers in H_2

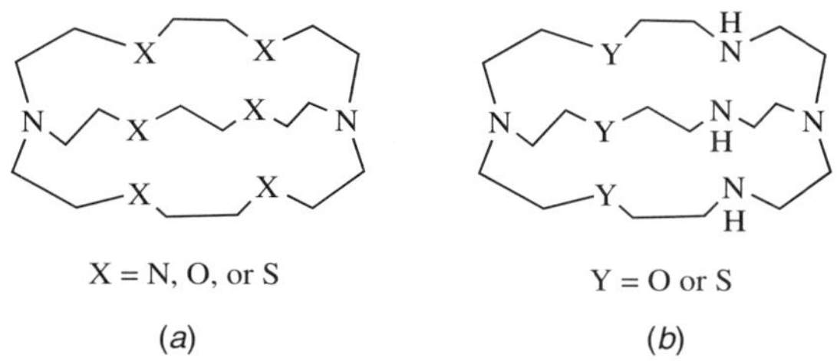

Figure 1. Examples of laterally symmetric (a) and nonsymmetric (b) cryptands.

production via photolysis of water. On the other hand, when the cavity is large a single metal ion can be led via ligand design to a predetermined site leaving a vacant space inside. Molecules or ions can enter the cavity and bond to the metal ion. The amino nitrogens in these cryptands can be derivatized to impart a variety of specific properties to the cryptands making them useful in several contemporary areas of research.

After a brief discussion of the synthetic methods adopted for these molecules, their complexing abilities toward different metal ions are covered. Coordination chemistry of these cage molecules is not discussed in detail, as an excellent review on this topic is available (25). A comprehensive discussion on the derivatization of the secondary amino nitrogens with different groups leading to the exhibition of specific molecular–supramolecular properties is presented.

II. SYNTHESIS OF LATERALLY NONSYMMETRIC CRYPTANDS

Since the early years of its discovery (32), there have been a number of synthetic approaches made and a large number of cryptands are now reported in the literature. The synthesis of these molecules has two aspects: (1) methodology and (2) strategy, which are briefly discussed next.

A. Methodology

The commonly adopted methods for cryptand synthesis are (1) high dilution, (2) use of templates, (3) rigid-group principle, and (4) low temperature.

Allowing reactions to proceed under high dilution is the most extensively used technique and is accomplished by simultaneous addition of the reactants in a flask having a large amount of the solvent. The appropriate concentration level for the cyclization step is determined empirically. However, under high-dilution conditions, the reactions should be facile to reduce the scope of forming undesired side products and also to maintain a very low stationary concentration of the reactants. In addition, the desired product must be stable under the prevailing reaction conditions. An apparatus for controlling the reaction conditions in the high-dilution synthesis has been described (33). This apparatus consists of motorized burets, which can add measured quantities of the reactants to the reaction vessel at a very slow rate. Although the high-dilution technique has been known (34) for a long time, it was Ziegler who developed (35) and widely used this method for macrocyclization reactions. Detailed theoretical studies (36) regarding the factors that influence a macrocyclization process and practical hints (37) are available. As high-dilution syntheses take a long time to complete, other methods like use of a template or allowing the reaction to proceed at low temperature have been used wherever possible. A template organizes (38) the reactants with respect to each other to achieve a particular linking to form the desired compound. The use of a template in enhancing the yield of a desired product is known as the *template effect*. The template effect of a metal ion can be of kinetic or of thermodynamic origin or a combination of both (39). When the kinetic template effect is operative, the geometrical arrangement of ligands within the coordination sphere of the templating ion provides constraints that can be used to control the structure of a product formed from the reactions of coordinated ligands. The thermodynamic template effect, on the other hand, acts on stabilizing a ligand structure that might otherwise be disfavored in the pure organic chemical system at equilibrium. Pioneering research by Curtis in the macrocyclization of isomeric Ni(II) complexes (40) paved the way for template synthesis. Both transition as well as nontransition metal ions have been used as templates in cryptand synthesis (41, 42). When a kinetically inert metal ion is used, the final product is isolated as a metal cryptate; the removal of the metal ion can sometimes be difficult (42). Kinetically labile alkali, alkaline earth, and some main group metal ions can also be efficient templates in directing cyclization reactions and in the event of any encapsulation can be trans-metalated easily.

According to the rigid-group principle (43), the synthetic difficulties associated with cyclization reactions from acyclic components owe mainly to the possibility of rotation about the bonds in the molecule. Unfavorable rotation effectively lowers the possibility of reactive groups coming in proximity within the reactive zone to yield the desired macrocycle. To lower the conformational mobility of an acyclic compound, rigidity may be increased in the structure by incorporating aromatic groups or unsaturation to aid in preorganizing the reactive fragments to react in a directed fashion thereby reducing the possibility of oligomerization.

Low-temperature synthesis is a new technique (44). At low temperature, movement of the reactive groups slow down, which minimizes polymerization to a great extent even if there is no templating metal ion used. Of course, the functional groups should react readily at low temperature to make the synthesis a success.

B. Strategy

Some major strategies (45) developed for the synthesis of cryptands are illustrated in Scheme 1 and are termed (1) stepwise, (2) tripodal coupling, (3) tripodal or single capping, and (4) double capping. In case of laterally nonsymmetric cryptands as defined above (hereafter only cryptands), two synthetic strategies used are tripodal coupling, and tripodal (or single) capping.

1. Tripodal Coupling

In tripodal coupling (Scheme 1b), [1 + 1] condensation of two tripodal units is carried out in a one-pot synthesis (46). A number of cryptands have been synthesized (46–50) by Schiff base condensation of tripodal amines with tripodal trialdehydes at 40°C under a dinitrogen blanket followed by reduction of the imine groups formed with NaBH$_4$. An alkali metal ion such as Rb(I) or Cs(I) is used as a template to organize the reactants with respect to each other. The entire operation requires <12 h and the yields are usually very high (range, ~40–55%). The cryptands are isolated as crystalline solids through fractional crystallization. The Schiff base condensation involves initial attack on the carbonyl carbon by the lone pair of nitrogen; hence, availability of this lone pair is important. When a transition metal ion is used as the template, it coordinates strongly to the lone pair of amino nitrogen attenuating its nucleophilicity. So, the reaction does not proceed in the desired direction even when the temperature is raised to a higher value using a different solvent.

Comparable yields of the products are also obtained (48–50) when these reactions are carried out at a low temperature (~5°C) without employing any templating ion (Fig. 2). However, choice of solvent although empirically determined is crucial for the success of this method (44). For the condensation of the podand with three terminal aldehydes with a podand amine (Fig. 2), methanol has been found to be the solvent of choice. The aldehyde is taken in methanol in a round-bottom flask and the solution of the amine in methanol is added dropwise at a slow rate. The podand aldehyde is slightly soluble in methanol at low temperature and with the progress of the reaction, more and more of the aldehyde comes into the solution phase. This condition keeps the amount of the reactants low, mimicking a high-dilution reaction condition. Besides, at low temperature movement of the tripodal arms slow down considerably leading to higher yields of the desired product. In this method, only a minute amount of the [2 + 2] product is formed. Concentration of the

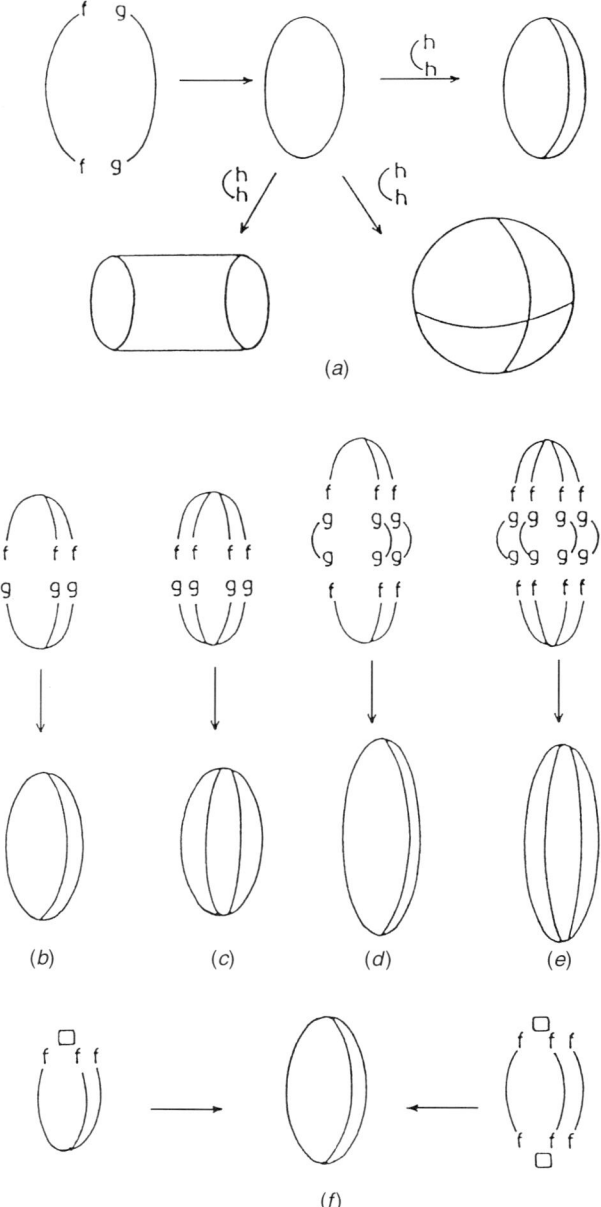

Scheme 1

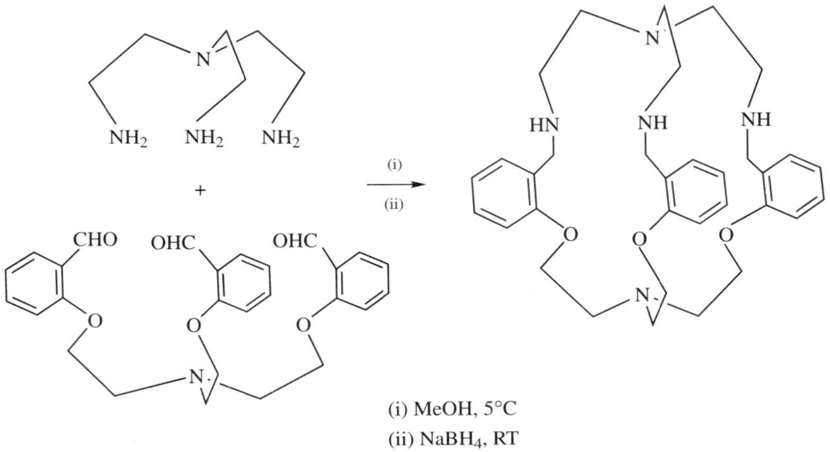

Figure 2. Synthesis of a laterally nonsymmetric aza-cryptand via tripodal coupling strategy.

(i) MeOH, 5°C
(ii) NaBH$_4$, RT

reactants in solution can be raised to ~20 mM by suitable choice of solvent [tetrahydrofuran, (THF)], which affords the [2 + 2] condensation product (Fig. 3) as the major component (51, 52) along with a mixture of side products of indefinite composition. Simple modification of the two podand partners can lead to the synthesis of a large number of cryptands by this method. However, if the two tripodal units are mis-matched in size and stereochemistry, the [1 + 1] product is not formed. Instead, the [2 + 2] product is formed (53) along with a number of side products.

2. Tripodal Capping

In the tripodal capping strategy (Scheme 1f) the Co(III) ion is used (42, 54) as a template in bringing about the reaction to the desired cage compound. Sargeson and co-workers contributed significantly in this area (42, 54). The idea behind using Co(III) as a template is that octahedral Co(III) is kinetically inert. So, upon formation of the complex with a podand, it can hold the reactive ends in proper orientation to facilitate capping. By using this method, macrobicycles with six nitrogen donors can be obtained in high yields. However, in case of laterally nonsymmetric cryptands like the one shown in Fig. 4, only moderate yields are achieved (55–57). The presence of thioethers in the podand leads to reduction of Co(III) to kinetically labile Co(II). In addition, a thioether is a poorer donor compared to amine. Both these factors lead to the loss of the cobalt ion from the podand under the capping conditions and results in a poorer yield

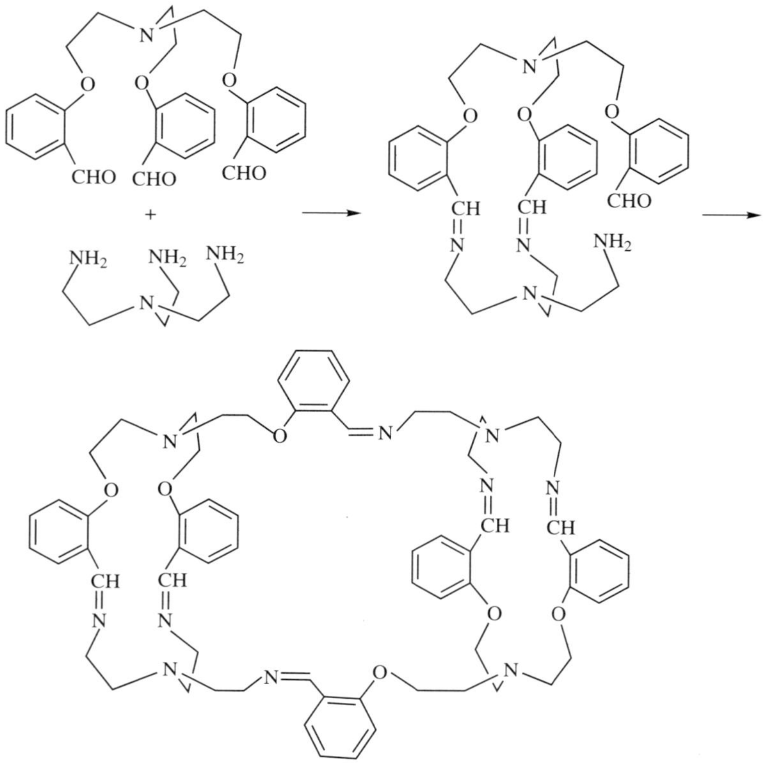

Figure 3. The [2 + 2] condensation of podands under high concentration.

of the desired product. Once the podand is capped, then the Co(III) ion inside is both kinetically as well as thermodynamically inert, which makes removal of the metal to isolate the product quite a difficult task. The encapsulated Co(III) is first reduced to Co(II) by means of a strong reducing agent like Zn dust. In the form

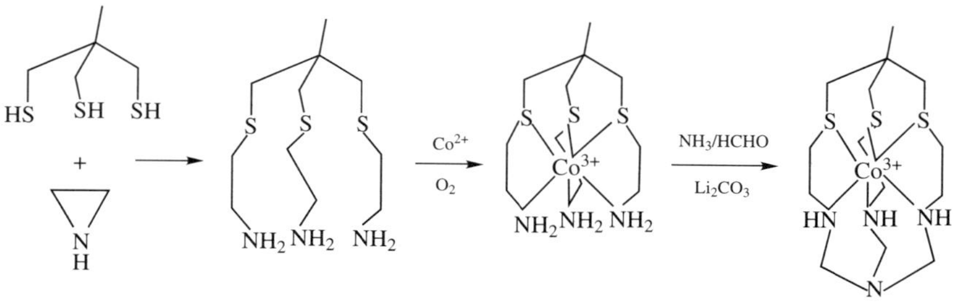

Figure 4. Synthesis of a laterally nonsymmetric aza-cryptand via tripodal capping strategy.

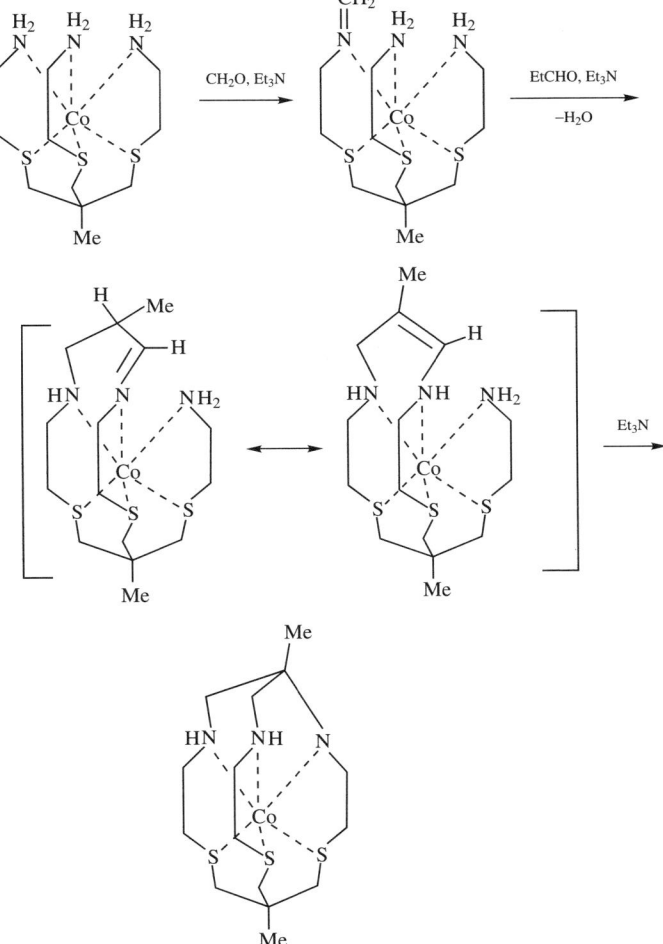

Figure 5. Synthesis of a laterally nonsymmetric aza-cryptand with a contracted cavity.

of Co(II) ion, the cobalt can be removed using cyanide ion or concentrated boiling HBr. The capping strategy can be used to form a macrobicyclic cage with a contracted cavity (58). The proposed reaction sequence for its formation is shown in Fig. 5.

This strategy provides an easy access (59) to functionalized cage molecules (Fig. 6) as well.

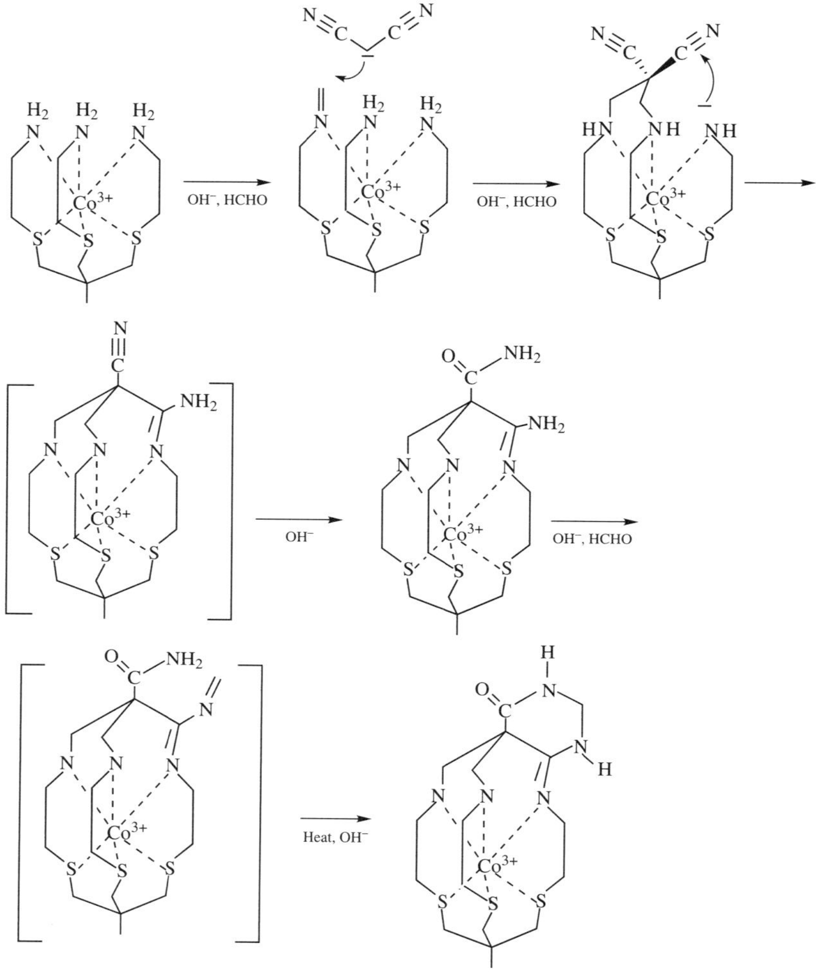

Figure 6. Synthesis of an amidine-functionalized aza-cryptand.

III. METAL COMPLEXATION

A. Cryptands with N and S as Donors

The cryptands shown in Fig. 7 were synthesized by the tripodal capping strategy. They readily form mononuclear inclusion complexes with a number of transition metal ions. Replacement of secondary amines in the original N_6 cage

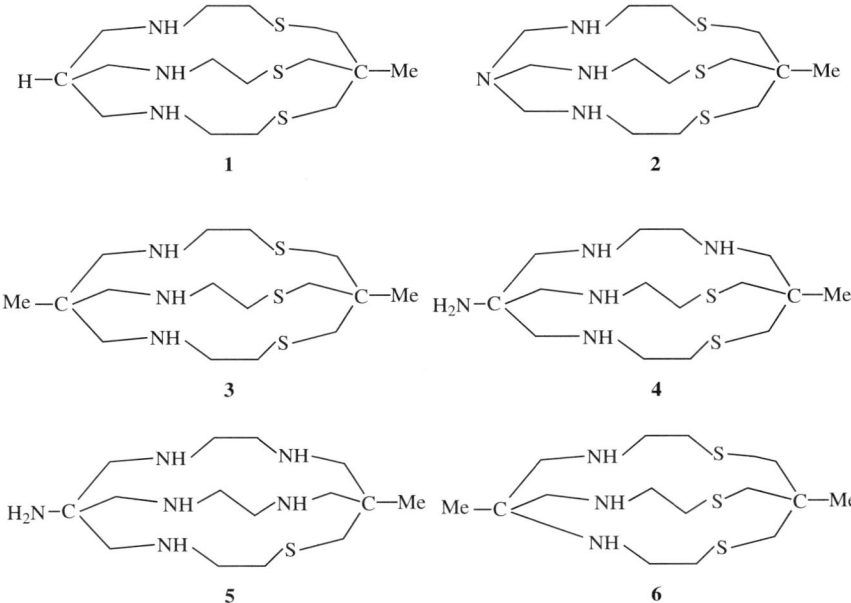

Figure 7. Laterally nonsymmetric cryptands with N and S as donors.

molecules (42, 54) by softer thioether S has a profound effect on the spectroscopic as well as electrochemical properties of the metal complexes. The Co(III) complexes of all these cage ligands are known making structure–property correlation possible for these complexes. The electronic spectra of these complexes show (55–57) two spin-allowed ($^1A_{1g} \rightarrow {}^1T_{1g}$ and $^1A_{1g} \rightarrow {}^1T_{2g}$) ligand-field transitions characteristic of low-spin octahedral Co(III). However, when the coordination symmetry is low (i.e., with 5) two spin-forbidden transitions ($^1A_{1g} \rightarrow {}^3T_{1g}$ and $^1A_{1g} \rightarrow {}^3T_{2g}$) are also observed (45) at low (11 K) temperature. For the Co(III) complex of the contracted cage (6), the corresponding spin-allowed ligand-field transitions are shifted (58) to higher energy by $\sim 750\,\text{cm}^{-1}$ each. The band positions are insensitive to the nature of the apical atom (i.e., whether C or N) as they do not take part in coordination. Assuming an approximate octahedral coordination symmetry around Co(III) for the complexes with N_3S_3, N_4S_2, and N_5S donors (Figs. 8–10), the ligand-field parameters have been calculated based on the two spin-allowed transitions, to find a possible correlation between the number of thioether sulfur donors and the parameters $10Dq$ and Racah B. The data reveal that on going from N_6 to S_6 donor, the value of $10Dq$ reduces by $\sim 400\,\text{cm}^{-1}$ initially, that is, from the N_6 to

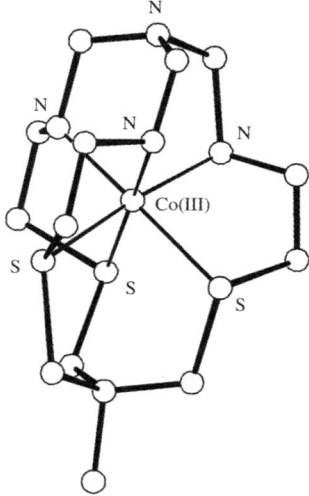

Figure 8. Structure of Co(III) inclusion complex of an N_3S_3-cryptand.

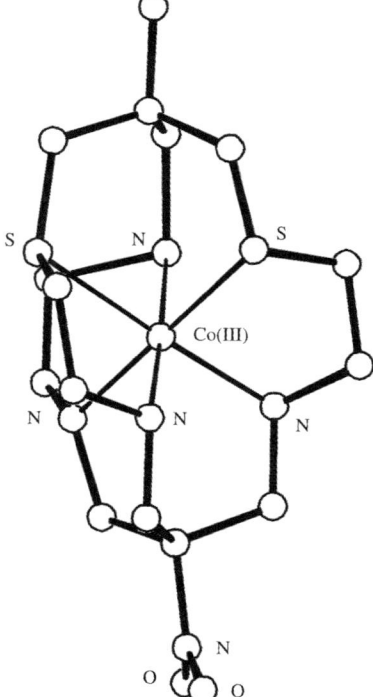

Figure 9. Structure of Co(III) inclusion complex of an N_4S_2-cryptand.

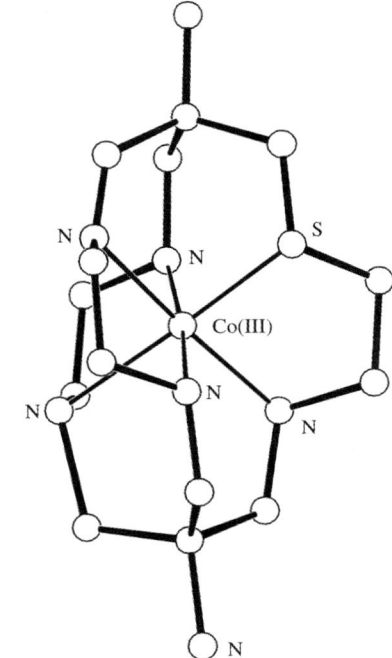

Figure 10. Structure of Co(III) inclusion complex of an N_5S-cryptand.

N_5S complex. Thereafter, the reduction is much less and is only $\sim 300 \text{ cm}^{-1}$ from the N_5S to S_6 complex. These data indicate that the ligand-field strength of thioether sulfur is only slightly weaker compared to that of amino nitrogen in these complexes. The *Racah B* parameter, on the other hand, is reduced by $\sim 30 \text{ cm}^{-1}$ on introduction of a thioether sulfur in place of an amino nitrogen.

The ligands **1–6** do not allow the normal substitution chemistry of kinetically labile Co(II) ion. Besides, the metal ion is prevented from relaxing to its preferred geometry upon change in its oxidation state (60). Both these factors contribute to the chemical as well as electrochemical reversibility of the Co(III)/Co(II) couple. However, the potential values are sensitive to (1) the number of thioether donors, (2) the nature of the apical atom and its substituents, and (3) the size of the cavity (58, 60). The $E_{1/2}$ values for the Co(III)/Co(II) potential of **1–6** and their derivatives are collected in Table I. The complexes with N_6 and S_6 donors are also included for comparison. An examination of Table I will show that the redox potential moves to the positive direction upon replacement of an amino N by a thioether S. This replacement causes (1) an increase of the cavity

TABLE I

Redox Potential Values for the Co(III)/Co(II) couple versus SHE at 293 K

Complex[a]	$E_{1/2}(V)^b$
$[Co(Me_2-N_6sar)]^{3+}$	−0.484
$[Co(H,Me-N_4S_2sar)]^{3+}$	−0.156[c]
$[Co(Cl,Me-N_4S_2sar)]^{3+}$	−0.037[c]
$[Co(Me_2-N_3S_3sar)]^{3+}$	−0.104
$[Co(H,Me-N_3S_3sar)]^{3+}$	−0.092
$[Co(HO,Me-N_3S_3sar)]^{3+}$	−0.031
$[Co(Cl,Me-N_3S_3sar)]^{3+}$	+0.010
$[Co(Me_2-N_3S_3absar)]^{3+}$	−0.181
$[Co(HOCH_2Me-N_3S_3absar)]^{3+}$	−0.160
$[Co(ClCH_2Me-N_3S_3absar)]^{3+}$	−0.121
$[Co(N_3S_3azacapten)]^{3+}$	+0.09[d]
$[Co(Me_2-S_6sar)]^{3+}$	+0.244

[a] See Abbreviation section for definition of ligands used.
[b] Potential value in water with ionic strength = 0.2 M NaCl. Values in 0.1 M NaCl are found to be 10–20 mV more positive.
[c] 0.1 M NaClO$_4$.
[d] In acetonitrile with 0.1 M Me$_4$NCF$_3$SO$_3$ versus Ag/AgCl electrode.

size due to longer C—S bond distance and (2) a greater shift of electron density from the metal to the more covalent M—S bond. Both of these make the Co(II) state more stable inside a N_xS_{6-x} cage compared to that in the N_6 cage. Keeping the donor set as well as the cavity size unchanged, the potential depends on the nature of the substituents when the bridgehead is carbon. If the substitutent has a negative inductive effect (like a bound chloride), the potential shifts to the positive direction. On the other hand, a substituent like a methyl group with a positive inductive effect causes the potential to shift to the negative direction. Any attempt to reduce the Co(II) to Co(I) inside the N_6 cage invariably leads to rupture of the cage, as its cavity is not large enough to accommodate a Co(I) ion. In contrast, metastable Co(I) species can be generated electrochemically (60) with **1** and **2** without any bond rupture. The Co(I) complex is characterized by an eight-line pattern centered at $g = 2.17$. However, none of the Co(III) cage complexes of **1–6** exhibit a Co(IV)/Co(III) couple consistent with the known chemistry of the metal.

The high chemical stability of the cobalt cages allows the Co(III)/Co(II) redox states to be cycled repeatedly without decomposition. Besides, these compounds are soluble in water. Thus, Co(III) complexes with N_3S_3 donors can be used as electron relay compounds for the photoinduced H_2 production (Scheme 2).

$$S + h\nu \longrightarrow S^* \qquad (1)$$

$$S^* + Co^{III}L \longrightarrow S^+ + Co^{II}L \qquad (2)$$

$$Co^{II}L + H^+ \xrightarrow{Pt/pva^a} \tfrac{1}{2} H_2 + Co^{III}L \qquad (3)$$

$$Co^{II}L + S^+ \longrightarrow S + Co^{III}L \qquad (4)$$

^{a}Colloidal platinum dispersion on poly(vinyl alcohol)

Scheme 2

As the Eqs. 2 and 3 indicate, electron transfer from the excited sensitizer (S^*) to Co(III) will be favored if the Co(III)/Co(II) potential is less negative, whereas a more negative potential will favor H_2 production. These two opposing influences are optimized in N_3S_3 donors with different apical substituents and they have been used for dihydrogen production from water at modest rates (61, 62).

These aza-macrobicycles have been used to coordinate other metal ions as well. With **4**, Cu(II) forms a distorted tetragonal complex (Fig. 11) with one thioether and one nitrogen occupying axial positions (63) forcing a significant distortion of the pseudo-threefold symmetry of the molecule. Low-temperature electronic and EPR spectral data of this complex in solution is consistent with its solid-state structure. Ligand **1** forms a stable Ru(II) inclusion complex (64) in which the three thioether sulfurs appear to stabilize the Ru(II) state by ~0.9 V relative to the corresponding N_6 ligand. The increase in the reduction potential is believed to be mainly a function of donor (and acceptor) properties of the sulfur atom rather than differences in the cavity size from N_6 to N_3S_3. On oxidation of the metal center, the cage ligand undergoes stepwise dehydrogenation of the (Scheme 3) coordinated amine to form coordinated imine. Such behavior is, however, not unprecedented and also occurs (65) with N_6 macrobicycles. Since ligand **5**, is very unsymmetrical, it can enforce a distorted octahedral geometry onto the Cr(III) ion both in the solid state and in solution (66).

B. Cryptands with N and O as Donors

Metal complexation of cryptands incorporating N and O donors (Fig. 12) has been studied in solution and in the solid state. Each of the cryptands shown in

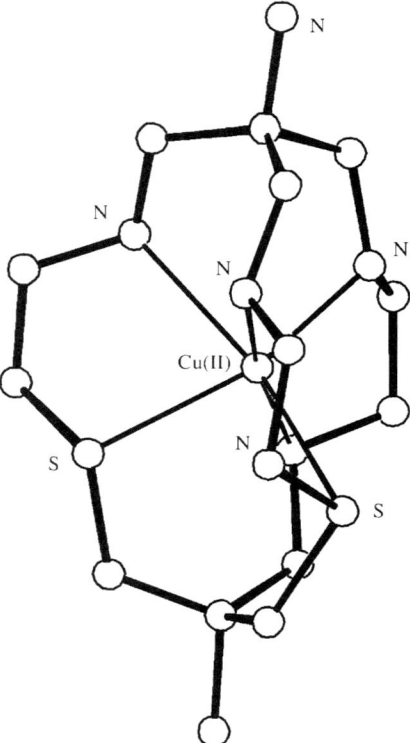

Figure 11. Structure of Cu(II) inclusion complex of an N_4S_2-cryptand.

Fig. 12 has a large cavity with two distinct binding sites separated by a hydrophobic spacer. The N_4 moiety represents a suitable binding site for transition metal ions while the ethereal moiety can be expected to be effective in the binding of harder cations. The cryptands **7–9** are sparingly soluble in water, which allow studies (67) of their complexation properties toward Cu(II), Zn(II), Ni(II), Cd(II), and Pb(II), as well as of alkali and alkaline earth metal ions in aqueous media. Other metal ions could not be studied due to extensive precipitation of the corresponding cryptates. Such complexation studies became necessary for the eventual uses of the metal cryptates in several other fields of research as discussed later. It became clear from complexation studies that these cryptands do not show any tendency to bind alkali or alkaline earth cations. Although the set of stability constants is not complete, some interesting conclusions regarding the coordination properties of these ligands can be drawn. In the case of the metal ions Zn(II), Cd(II), and Pb(II) for which no ligand-field

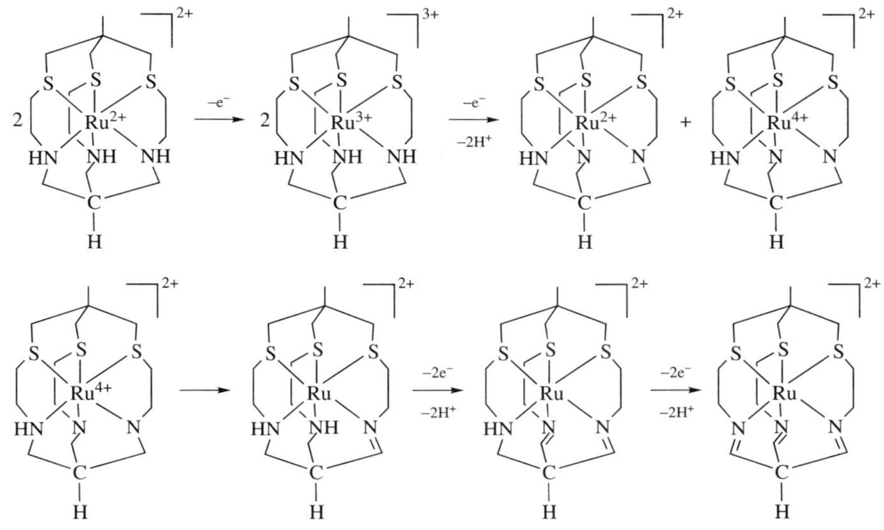

Scheme 3

stabilization is possible, the complex stability increases with increasing metal ion size. Such coordination tendency is evidently connected with the macrobicyclic nature of the ligands since an opposite stability trend is observed (68) for Zn(II) and Cd(II) complexes with tris(2-aminoethyl)amine (tren). Cryptands are more rigid compared to acyclic molecules and, consequently, are less prone to modification of the donor atoms' topology to satisfy the metal ion coordination requirements. Cryptands **7–9** are too large to give a good fitting with all the studied metal ions, justifying the stability increase with increasing cation dimensions. Accordingly, the difference in stability between the complexes of tren and those of the cryptands diminishes with increasing cation size. A significant difference in complexation ability is observed depending on the ligand structure. Compared to **7** and **8**, cryptand **9** shows lower tendency to form complexes. Surprisingly, complexes formed by **7** are more stable (Table II) than those with **8**, the difference in stability being greater by more than three orders of magnitude for the Cd(II) and Pb(II) complexes. This difference can be attributed to the more rigid framework in **8** compared to that in **7** such that the entering metal ion cannot adjust the coordination geometry to its liking in **8**.

The inclusion complexes of **7–9** with a number of first-row transition metal ions have been characterized in the solid state. X-ray crystal structures of a few of these complexes are available (47, 48, 69–70) along with those of Ag(I) and Pb(II) (71) of **7**. The N_4 moiety is suitable to bind a transition metal ion although

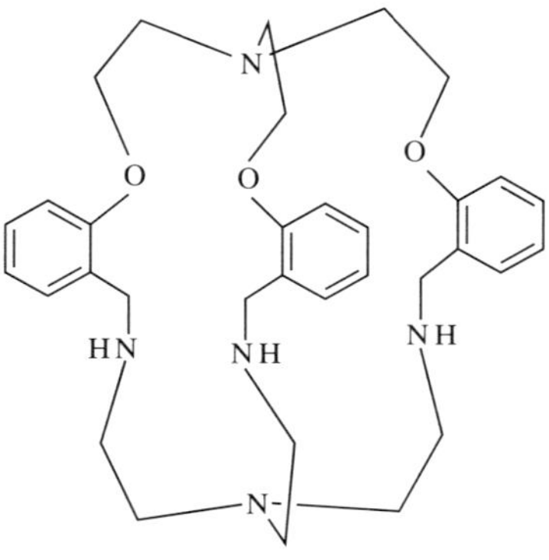

Figure 12. Laterally nonsymmetric cryptands with N and O as donors.

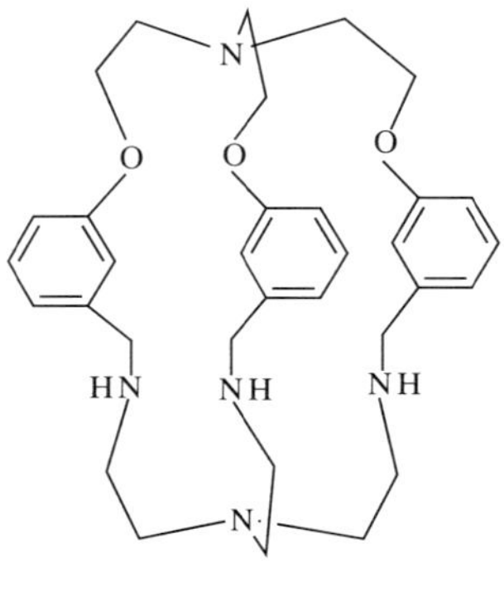

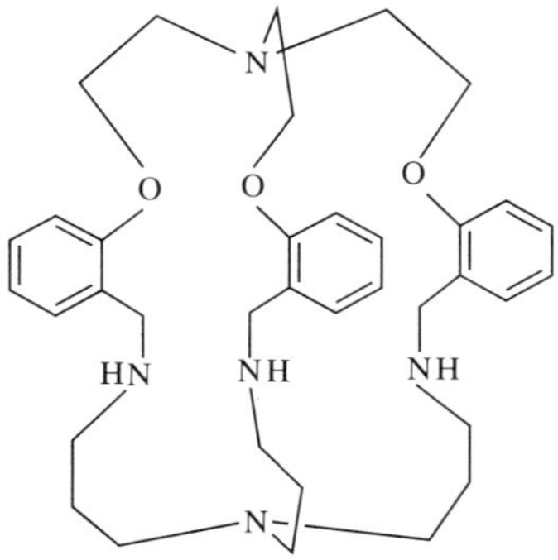

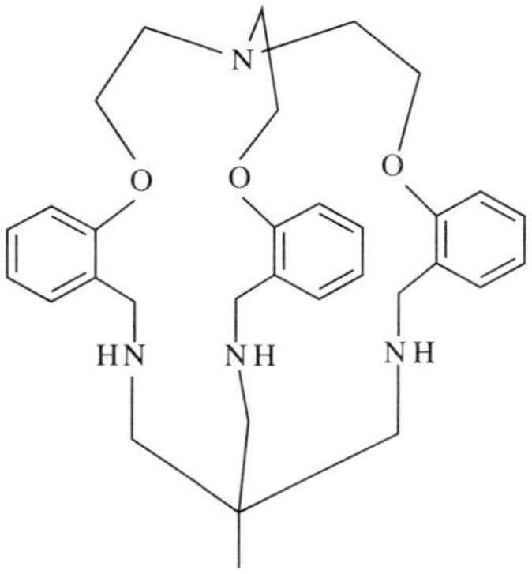

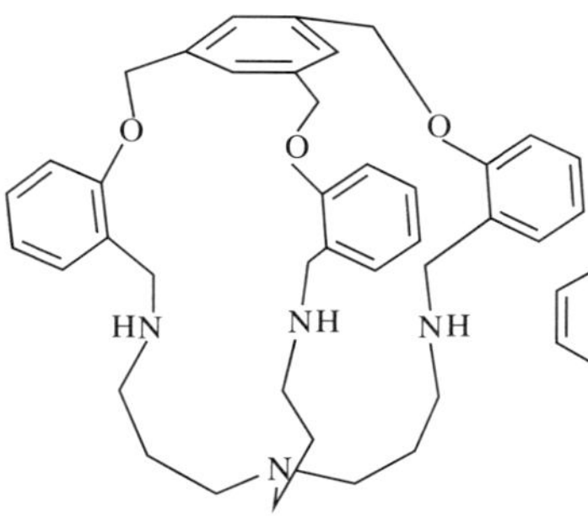

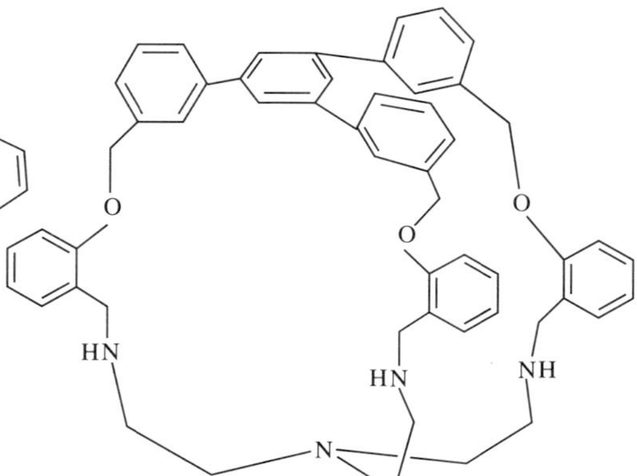

TABLE II

Logarithms of Complexation Constants[a] ($\log K$) in $0.10 M$ $Me_4N^+NO_3^-$ at 298.1 K

Complexation	7	8	9
$Cu^{2+} + L = CuL^{2+}$	15.22(3)	14.36(3)	
$Cu^{2+} + HL^+ = CuLH^{3+}$		11.2(1)	7.44(3)
$CuL^{2+} + H^+ = CuLH^{3+}$		6.6(1)	8.70(8)
$CuL^{2+} + OH^- = CuLOH^+$		5.6(1)	4.38(8)
$Zn^{2+} + L = ZnL^{2+}$	8.83(5)	7.06(6)	
$Zn^{2+} + HL^+ = ZnLH^{3+}$		5.07(6)	4.8(1)
$ZnL^{2+} + H^+ = ZnLH^{3+}$		7.76(6)	7.45(2)
$ZnL^{2+} + OH^- = ZnLOH^+$	3.78(7)		3.66(6)
$Cd^{2+} + L = CdL^{2+}$	11.14(2)	7.35(8)	
$Cd^{2+} + HL^+ = CdLH^{3+}$		6.94(8)	
$CdL^{2+} + H^+ = CdLH^{3+}$		7.94(9)	
$CdL^{2+} + OH^- = CdLOH^+$	3.83(3)		
$Pb^{2+} + L = PbL^{2+}$	13.09(2)	9.52(9)	
$Pb^{2+} + HL^+ = PbLH^{3+}$	6.95(9)		
$PbL^{2+} + H^+ = PbLH^{3+}$	7.16(9)		
$PbL^{2+} + OH^- = PbLOH^+$	3.25(5)	3.8(1)	

[a] Values in parentheses are standard deviation in the last significant figure.

the number of donor atoms is insufficient to saturate the coordination sphere of many metal ions. Besides, in both cases **7** and **8**, the tren moiety pushes the metal ion out of the plane described by the three secondary amines exposing the metal ion to further attack by other ligands. As a result, the metal bound at this site can draw other ions or molecules from the medium to saturate its coordination ability as exemplified by the Ni(II) cryptate (70) of **7** (Fig. 13). To attain octahedral coordination, the metal ion binds a water molecule inside the cavity while it coordinates to a MeCN molecule from outside the cavity. The H atoms of the water molecule are hydrogen bonded to two nearest ethereal oxygen atoms at the upper deck of the cryptand. Although the cavity of **8** is big enough to accommodate both a cation and an anion, its top part tilts away upon binding of an anion to avoid hydrophobic aromatic spacers (Fig. 14) such that a part of the anion is outside the cavity. However, a neutral molecule like water can reside completely inside the cavity as its hydrogen atoms can take part in hydrogen bonding with the aromatic spacer units (Fig. 15). Both Pb(II) and Ag(I) readily form inclusion complexes with **7**, which are structurally characterized. These metals do not show any tendency to induct another molecule/ion inside the cavity to attain higher coordination. The Pb(II) ion makes a weak bonding interaction (71) with one of the ethereal oxygen atoms and distorts the cavity to

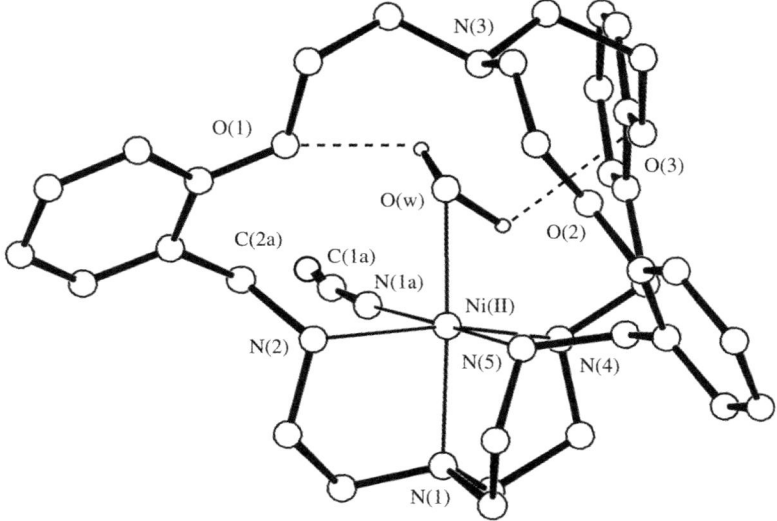

Figure 13. Structure of Ni(II) cryptate of **7**.

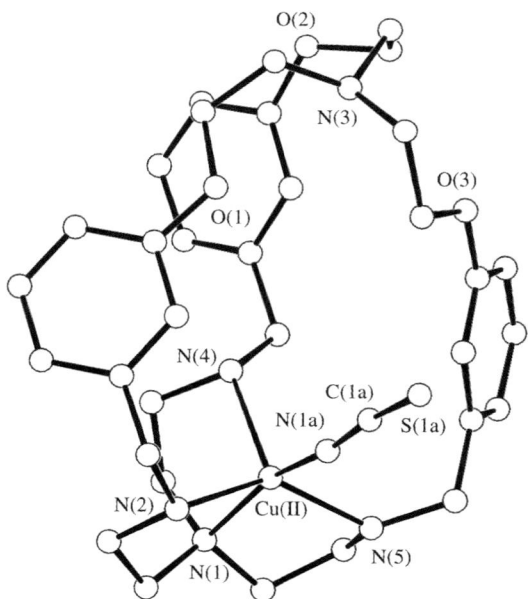

Figure 14. Structure of Cu(II) cryptate of **8** showing the SCN⁻ group bound to the metal ion.

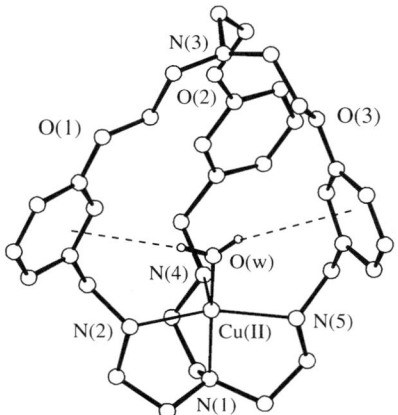

Figure 15. Structure of Cu(II) cryptate of **8** where a H$_2$O molecule is coordinated to the metal inside the cavity.

a significant extent (Fig. 16). Cryptand **9**, with a tris(3-aminopropyl)amine (trpn) unit in place of tren and ortho-substituted benzene groups in the three bridges, has a flexible cavity that can enlarge upon metal binding (Table III). The trpn unit can wrap around an ion like Zn(II) in such a way that the metal lies (Fig. 17) below the plane described by the three secondary nitrogen atoms and

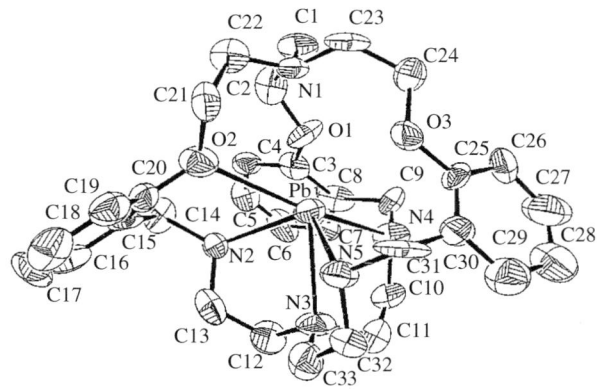

Figure 16. A perspective view of the structure of Pb(II) cryptate of **7**.

TABLE III

Selected Nonbonded Distances (Å) in Free and Metal Cryptates of **7–9**

	7	[**7** ⊂ Ni(MeCN)(H$_2$O)]2ClO$_4$ · MeCN
N(1)...N(3)	6.249	6.410
N(2)...N(4)	4.036	3.082
N(2)...N(5)	4.336	4.295
N(4)...N(5)	3.536	3.354
O(1)...O(2)	4.862	4.603
O(1)...O(3)	5.365	5.450
O(2)...O(3)	3.992	3.602
	8	[**8** ⊂ Cu(H$_2$O)] · 2picrate · H$_2$O
N(1)...N(3)	9.904	9.221
N(2)...N(4)	3.738	3.452
N(2)...N(5)	3.684	4.102
N(4)...N(5)	3.717	3.284
O(1)...O(2)	3.672	4.491
O(1)...O(3)	3.579	5.332
O(2)...O(3)	3.520	3.632
	9	[**9** ⊂ Zn]2ClO$_4$ · 2H$_2$O
N(1)...N(3)	5.291	7.263
N(2)...N(4)	4.960	3.342
N(2)...N(5)	6.526	3.350
N(4)...N(5)	6.410	3.336
O(1)...O(2)	4.974	4.589
O(1)...O(3)	5.321	4.309
O(2)...O(3)	5.805	4.346

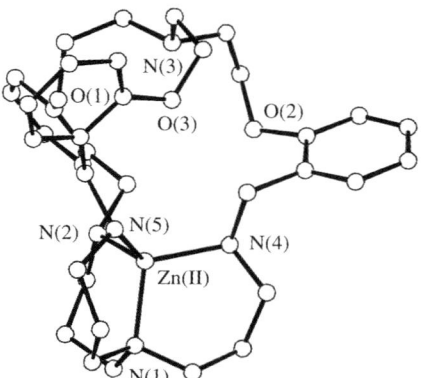

Figure 17. Structure of Zn(II) cryptate of **9**.

toward the bridgehead N atom it is bonded to. Various nonbonding distances in the free ligands and in the complexes (Table III) suggest that the donor atoms do not move from each other to a significant extent upon complexation in the cases of **7** and **8**, while in case of **9** they move significantly. This means, while **7** and **8** can impose a coordination geometry onto the metal, **9** can adjust itself to a certain degree to suit the coordination symmetry requirement of the entering metal ion. However, the X-ray structures of the cryptates of **7–9** show (69, 70) that out of four nitrogens at the tren end, three bind a metal ion with normal metal–nitrogen distances (range, 1.94–2.10 Å) while the fourth metal–nitrogen distance is much longer (~2.40 Å). Thus these cryptands not only imposes a low coordination symmetry onto the metal ion but makes the ligand field around the bound metal weak as well. Consequently, the cryptates exhibit low-energy ligand-field bands in the ultraviolet–visible (UV–vis) spectra and small A_{II} values in the electron paramagnetic resonance (EPR) signals [for Cu(II) cryptates]. The spectroscopic data are collected in Table IV. For the ligands, **10–12** no cryptate could be isolated in the solid state. However, electronic spectral studies indicate that these cryptands form 1:1 complexes with first-row transition metal ions showing low-energy ligand-field bands.

The chemistry of metal cryptates is interesting from the points of view of their novel structure and bonding and the associated spectroscopic properties these systems exhibit. Suitably designed cryptands can impose low coordination symmetry around the metal ion in the cavity, and thus can be excellent candidates to serve as electronic structural analogues of *intrinsic active sites* (31) of several metalloproteins. A metal ion anchored inside a cryptand with low coordination symmetry and being shielded to a considerable extent from the bulk solvent, can be potentially useful in homogeneous catalysis (72). Laterally nonsymmetric cryptands with a large cavity are suitable for this purpose because in such cryptands, donor atoms can be placed at strategic positions to lead a single metal ion into one end of the cavity leaving a vacant space inside to carry out catalytic reactions. Both Cu(II) and Co(II) cryptates of **8** were found (69) to be quite effective in oxidizing a number of organic substrates with dioxygen in the presence of a terminal oxidant like isobutyraldehyde at ambient pressure and temperature. Cryptates can become important homogeneous catalysts because of the increased kinetic and thermodynamic stability. Great possibilities exist with such systems especially if they contain chiral centers for possible stereoselective catalytic transformations.

It is now obvious that metal–cryptate chemistry will continue to draw attention from chemists. The activities mostly centered on utilizing the cavity of the cryptand and its tunability with respect to shape, size, number as well as types of donors atoms. However, when these cryptands are derivatized with specific groups their scope is likely to increase enormously as the outside groups

TABLE IV

Ligand Field Bands (cm^{-1}) and Magnetic and EPR Spectral Data for the Cryptates of **7–9**

Compound	d–d^a(cm^{-1})	$\mu/\mu_B{}^b$	$g_\parallel$	$g_\perp$	$A_\parallel(\times 10^{-4}$ cm$^{-1})$
[**7** ⊂ Cu]$^{2+}$	10,100(sh)	1.89	2.07(g_{av})^c		
	12,350(335)		2.12^d	2.07	63
	14,720(260)		2.18^e	2.06	95
[**7** ⊂ Cu(N$_3$)]$^+$	12,400(175)	2.04	2.09(g_{av})^c		
	15,020(250)		2.03^d	2.13	65
			2.01^e	2.26	64
[**7** ⊂ Ni]$^{2+}$	10,640(20)	2.98			
	17,700(15)				
	27,800(55)				
[**7** ⊂ Ni(N$_3$)]$^+$	10,525(15)	3.07			
	17,540(15)				
[**8** ⊂ Cu(H$_2$O)]$^{2+}$	9,600(95)	1.94	2.07(g_{av})^c		
	11,000(115)		2.14^d	2.06	60
	14,200(210)		2.18^e	2.05	91
[**8** ⊂ Ni (H$_2$O)]$^{2+}$	10,980(30)	3.02			
	15,150(20)				
[**8** ⊂ Cu(CN)]$^+$	11,400(145)	2.01	2.09(g_{av})^c		
	15,100(215)		2.03^d	2.12	65
[**8** ⊂ Cu(NCS)]$^+$	11,300(150)	1.98	2.08(g_{av})^c		
	14,800(180)		2.04^d	2.14	70
			2.03^e	2.24	75
[**8** ⊂ Cu(N$_3$)]$^+$	11,400(170)	2.03	2.06(g_{av})^c		
	15,200(220)		2.03^d	2.13	70
			2.02^e	2.21	85
[**9** ⊂ Cu]$^{2+}$	11,415(375)	1.97	2.07(g_{av})^c		
	13,425(235)				
[**9** ⊂ Ni]$^{2+}$	10,710(30)	3.01			
	18,100(25)				
	27,650(40)				

^{a} At 298 K in MeOH or in MeCN, molar absorption coefficients (ε in dm^3 mol^{-1} cm^{-1}) are in parentheses.
^{b} At 298 K.
^{c} Solid sample at 298 K.
^{d} In MeCN at 298 K.
^{e} In MeCN glass at 77 K.

can be utilized in tandem with the cavity inside for having smart molecules. The cryptands described here are suitable for attachment of various groups by simple substitution reactions for the expression of specific properties, which are discussed under separate headings.

IV. FLUORESCENT SIGNALING

A. General Design Principles

Molecular systems that combine binding ability with photophysical properties are of great interest as they are potentially useful in a variety of applications such as sensors, photonic molecular devices, and so on. In these systems, the binding of a guest species to a host molecule results in a change in the measurable photophysical property of the system. Two different processes occur during sensing–logic operations: (1) metal ion–molecular binding to the receptor, and (2) signal transduction to the fluorophore leading to photophysical changes. The design of such systems essentially comprises a signaling (fluorophore) and a guest-binding moiety (receptor). The two components can either be separated by a spacer [Fig. 18(a)] or can be integrated into one unit [Fig. 18(b)]. The signal transduction can be realized via any one of the processes, namely, (1) electron transfer, (2) charge transfer, (3) energy transfer, and (4) excimer and/or exciplex formation–disappearance. Of course, more than one process can be operative as well. The photophysical changes exhibited by the fluorophoric unit can be in terms of (1) appearance of the spectrum, (2) emission quantum yield, and (3) excited-state lifetime. The emphasis has been on the transduction of discrete and stoichiometric recognition events into fluorescence signals. Knowledge of the concentration of the guest can be gathered from the fluorescence quantum yield. Note, however, that nonstoichiometric interactions such as solvation can complicate the matter although it can provide us with important information as well. Photoinduced electron transfer (PET) is the most commonly used signal transduction mechanism for the design of sensors and chemical logics (73). The configuration shown in Fig. 18(a) is mostly used for this purpose where the fluorophore unit is the site for photonic activities of excitation and emission and the receptor unit is for guest complexation and decomplexation. The spacer besides holding the two units together can also tune

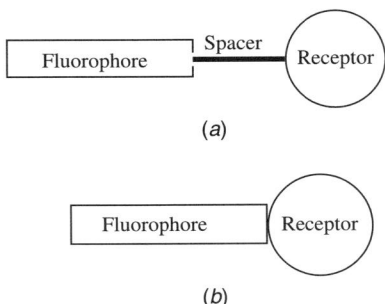

Figure 18. Two different configurations for fluorescent signaling.

the PET process. Besides, the design being modular in nature, both the fluorophore and the receptor units can be varied to get a large number of fluorescence signaling systems. The fluorophore can be varied for higher quantum efficiency, lower energy of excitation, and so on while the receptor can be changed to bind a cation or an anion or a neutral molecule. While Weller's pioneering work (74) provides the thermodynamic basis of PET, the intramolecular PET kinetics are best described following Marcus (75). The simple frontier orbital energy diagram can be invoked to understand (73, 76) the PET process (Scheme 4). In the format of Fig. 18(a), the PET from the receptor (where usually an amino nitrogen is connected via spacer to the fluorophore) to the fluorophore initiates thermal back-electron transfer thereby quenching fluorescence of the system (Scheme 4a). Inclusion of a guest engages the lone pair of the nitrogen atom preventing PET from the nitrogen to the fluorophore, which causes the fluorescence to be recovered (Scheme 4b). In other words, the presence of a guest is signaled by fluorescence enhancement of the system. The frontier orbital energy diagrams are given on the right of each scheme. In case of the arrangement described in Fig. 18(b), some atoms of the fluorophoric unit can participate in complexation of the guest molecule complicating the situation.

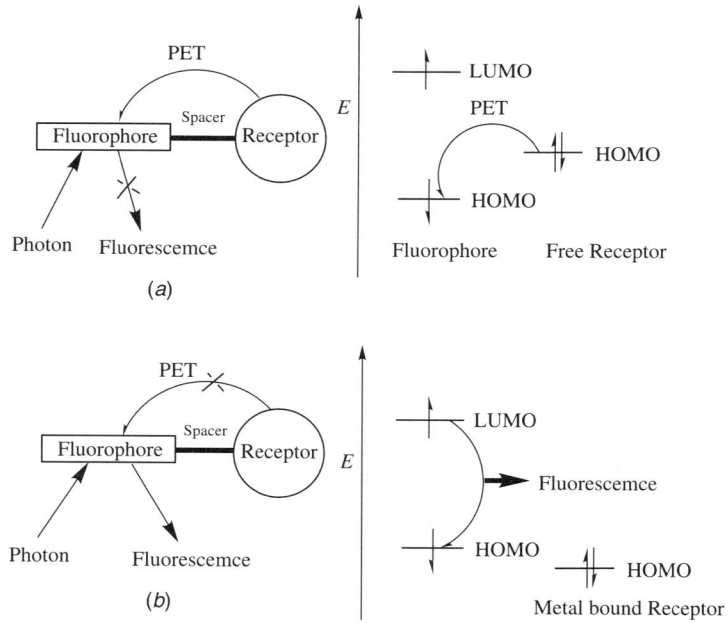

Scheme 4

B. Metal Ion Sensing

Fluorescence-based chemosensors can be designed for signal transduction upon analyte binding in the modular format of "fluorophore–spacer–receptor" offering the possibility to examine the concentration of the analyte in real-time and real-space. Note here that the dissociation constant of a metal complex has to match the expected range of its concentration under the conditions of its detection. The need for such sensors of biological, clinical, as well as environmental interests cannot be overemphasized and has spurred intense research in this area (76–82). Research has also focused utilizing optical fiber technology on developing sensors for remote monitoring of toxic and carcinogenic metals that impose serious human and environmental health hazards (83, 84). Sensors have been developed for the detection of mainly alkali–alkaline earth and main group elements. For this, a number of acyclic, macrocyclic, as well as cryptand hosts have been designed (76–82). While most of the studies have been carried out in nonaqueous media, a few systems are known (81) that are operative in aqueous medium. Sensors for transition metal ions are particularly important as some of them are present in trace quantities in all biological systems (85).

C. Chemical Logics

A chemical logic is defined as an assembly of molecular components that, because of the specific arrangement of the components in the dimensions of space and energy, is able to perform light-induced logic functions. Present day computers use semiconductor technology for performing binary arithmetic and logic operations. With the rapid technological advancement in electronics and semiconductor technology, computers have become smaller and smaller as a result of miniaturization of the components. This "top-down" approach attempts to put an increasing number of electronic components on a computer chip for computation at ever increasing speed. However, at the present rate of miniaturization, the process will reach its physical limit in the very near future. Chemistry offers a "bottom-up" approach where molecules, which are the smallest entities, that can be handled and manipulated under normal circumstances as simple components with useful properties to perform logic operations. However, wholly organic molecules possibly cannot replace silicon-based technology because of their sluggish response (86). A combination of organic molecules and metal ions with fluorescence as the monitoring parameter can respond much faster. The type of chemistry required for such activities is still at the research stage although some systems performing elementary logic operations are available (87–95). Depending on the input–output functionality, chemical logics have been divided (87) as simple YES, NO logic functions or as OR, AND, XOR, and so on logic gates. Some of the logic gates are illustrated in Scheme 5.

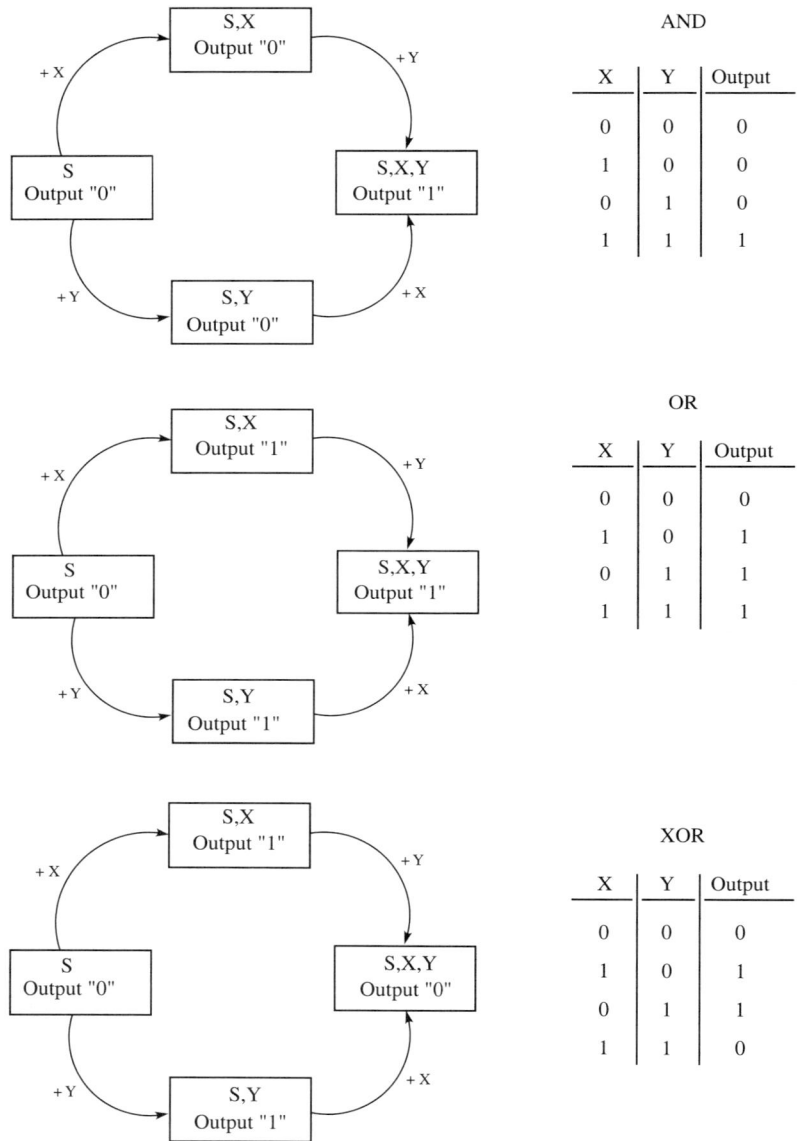

Scheme 5

D. Transition Metal Ions in Fluorescent Signaling

Transition metal ions quench fluorescence (96) very effectively such that the fluorescence enhancement resulting from the suppression of PET (Scheme 4) is nullified by the inherent quenching ability of these ions. A number of mechanisms have been forwarded (96) to account for the quenching ability of transition metal ions. These are (1) conversion of electronic energy to kinetic via collisions, (2) heavy atom effects, (3) redox perturbation, (4) perturbations by paramagnetic metal ions, (5) formation of charge-transfer complexes, (6) electronic energy transfer from/to the metal ion, and so on. Mechanisms like collisional energy transfer and heavy atom effects can be easily discarded by proper choices of the fluorophoric systems and the metal ion. For the other mechanisms to be operative, the metal ion and the fluorophore should have bonding interactions via the spacers or the orbitals of the metal ion and the fluorophore moieties of proper symmetry should come closer in case the spacer is flexible enough to make bonding interactions. In simple terms, blocking of the fluorescence quenching by metal ions requires that metal ion–receptor (M–R) interaction should be greater than the metal ion–fluorophore (M–F) communication.

One possible way to lower the M–F communication is by using an electron-deficient fluorophore by suitable attachment of electron-withdrawing substituents or by simple structural modification of the entire system. Of course, electron-deficient fluorophores have another added advantage; in such systems the PET will be more facile when the receptor is not bound to a metal ion contributing to the higher efficiency of the system. Samanta and co-workers (97) and Mitchel et al. (98) reported such fluorophoric systems where the receptor consists of only one aliphatic amine group (Fig. 19). When the receptor is not bonded to a metal ion or proton, no fluorescence is observed due to an efficient PET process from the nitrogen to the fluorophore. The fluorescence can be recovered upon metal binding by $\sim$30 times that of the metal-free state.

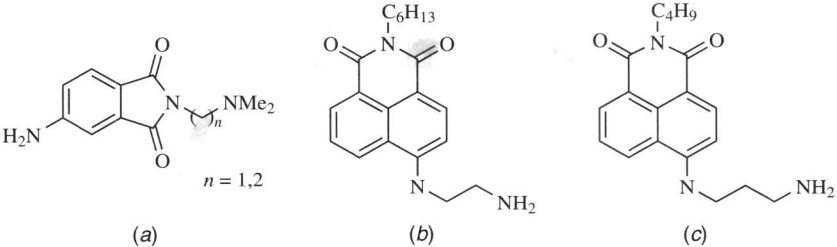

Figure 19. Electron deficient fluorophores for metal ion signaling.

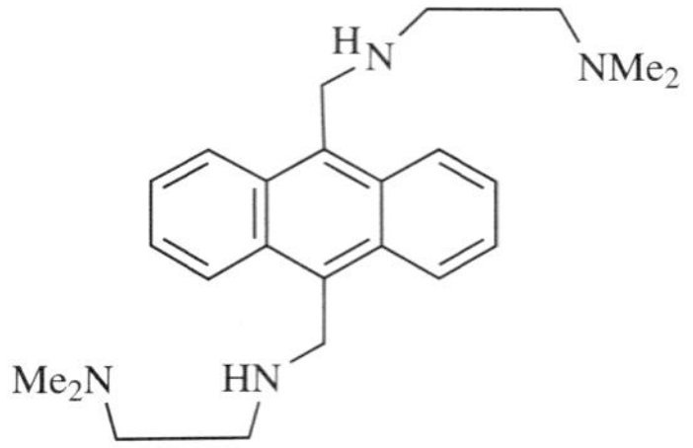

Figure 20. A fluorescent signaling system for Zn(II) ion.

Disadvantages of these systems being unknown complex stoichiometry apart from low complex stability that influence the detection limit. To overcome these problems, receptor units have been constructed (99) to impart higher stability to the metal-binding process through *chelate* effect. However, in most of these cases the enhancement factor seldom reaches >10. When a transition metal salt that is hydrated is dissolved in an organic solvent, it generates protons. These protons can also engage the nitrogen lone pair responsible for the PET process causing fluorescence enhancement. To nullify such a possibility, several control experiments should be carried out. A notable example where the enhancement increases by 1000-fold was provided by Czarnik and co-workers (100) who connected two N,N-dimethylethylenediamine groups to the 9- and 10-positions of anthracene via methylene spacers (Fig. 20). The Zn(II) ion, however, has a filled d orbital and thus fluorescence quenching mechanism(s) of a typical transition metal ion is nonoperative here. Several systems have been reported where fluorescence quenching in the presence of a transition metal ion has been exploited in designing sensors (101). However, fluorescence enhancement rather than quenching is better understood in sensors and photonic devices. A recent review is available (79), which gives an account of the systems known for transition metal ion induced fluorescence enhancement.

E. Cryptand-Based Fluorescent Sensors

Designing of an efficient sensor/device for transition metal ions based on the fluorophore–spacer–receptor configuration [Fig. 18(*a*)] depends significantly on the total architecture of the whole system. The desired architecture of a molecular photonic device/sensor demands the following: (1) the PET causing quenching of the fluorescence should be very fast and efficient in the free molecule, (2) the PET should be more or less completely prevented when the receptor unit accepts a metal ion, and (3) the M–F communication should be negligible while at the same time M–R interactions should be high, and (4) the distance and orientation of the metal ion entering the receptor unit with respect to the fluorophore π system should be such that the spin–orbit coupling that facilitates the S–T intersystem crossing would be minimal. Laterally nonsymmetric aza-cryptands are excellent receptors for transition metal ions forming highly stable inclusion complexes. Once a metal ion enters the cavity, it is isolated from the surroundings to a great extent, which can block most if not all of the quenching pathways.

By keeping the design principles in mind, cryptands **7–9** were derivatized (102) with anthryl groups to afford L_1, L_2, and L_3, respectively (Fig. 21). In the absence of a metal ion, they show well-resolved anthracene monomer emission along with a broad structureless emission centered $\sim$550 nm in THF at 298 K. Intensity of the monomer emission decreases while that of the structureless emission increases (Fig. 22) with concentration. This emission is due to an intramolecular exciplex involving the nitrogen lone pair and the anthracene (103, 104). The position of the emission maximum is found to be dependent (Fig. 23) on the dielectric strength of the solvent (low polarity) showing the charge-transfer character of the emitting complex. The linear relationship observed (Fig. 24) between v_{ex}(max) and the solvent polarity parameter is consistent with the Weller equation (105) for exciplex formation. The quantum yields of monomer emission, Φ_{FM}, and that of the exciplex, Φ_{FE}, are extremely low (Table V) for all three systems, indicative of very efficient PET from the highest occupied molecular orbital (HOMO) of tertiary nitrogen to the corresponding anthracene moiety (Scheme 4a). Anthraceno cryptands (i.e., fluorophore and receptor are integrated as one unit as in Fig. 18b) designed for alkali and alkaline earth metal ion signaling (104) exhibited significantly higher quantum yields for both Φ_{FM} and Φ_{FE}.

With a first-row transition metal ion as input to any of L_1–L_3 in dry THF, the Φ_{FE} vanishes and the Φ_{FM} increases quite significantly (Fig. 25). The extent of enhancement show dependence on the nature of the metal ion as well as the receptor (Table V). The cavity in L_3 is most flexible while that in L_2 is least so, as mentioned in Section III. Both in the case of L_1 and L_3, the donor atoms at the N_4 end of the cavity can adjust more compared to that of L_2 for efficient binding of a metal ion. Therefore, with a metal ion like Co(II), Cu(II), or Zn(II), which prefer tetra coordination, the fluorescence recovery is maximum due to strong donation of the nitrogen lone pairs. Metal ions like Mn(II), Ni(II), and Fe(III), which prefer higher coordination, also exhibit high intensity of fluorescence. Of course, these metal ions can achieve higher coordination by binding solvent molecules (70). With L_2, the fluorescence enhancement is considerably less due to the quite rigid nature of the N_4 donor set that lead to less strong bonding interactions between the nitrogen lone pairs and the metal ion. The binding abilities of L_1, L_2, and L_3 toward a metal ion, for example, Cu(II), are nicely reflected in the Φ_{FM} values (Table V). In all three systems, Zn(II) shows the highest enhancement among the metal ions used as inputs due to its

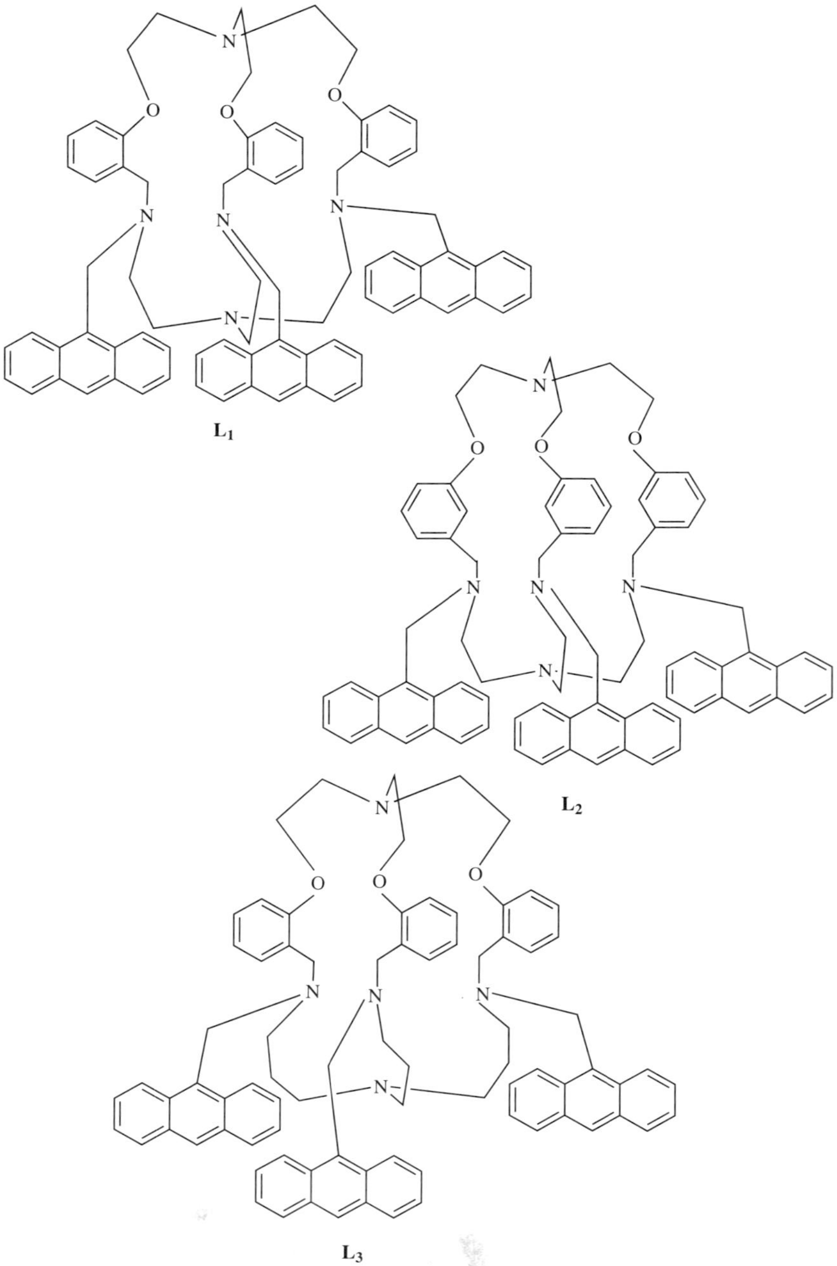

Figure 21. Cryptand-based fluorophores with methylene as spacer.

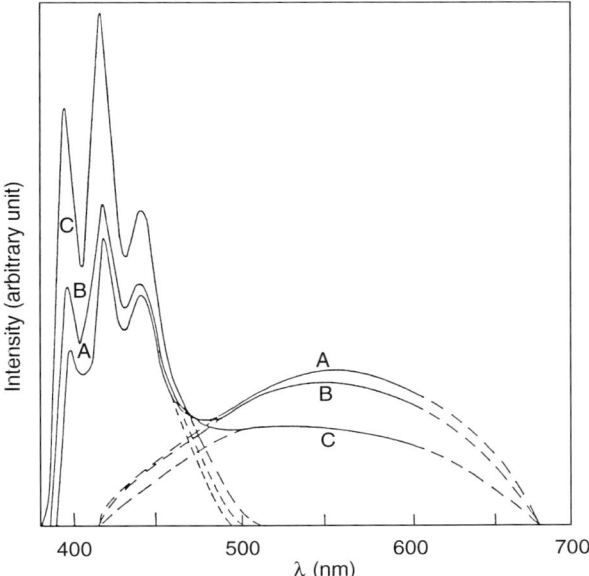

Figure 22. Corrected fluorescence emission of metal-free **L₁** in dry THF showing the effect of concentration on the exciplex: (A) $5 \times 10^{-5} M$, (B) $2 \times 10^{-5} M$, and (C) $1 \times 10^{-5} M$.

nonquenching nature. In all cases, the emission spectra show slight bathochromic shifts (Table V) attributable to the counteranion induced change in polarity around the fluorophore. No exciplex emission is observed with a metal ion as input as the nitrogen lone pairs are engaged to the metal ion. As the data in Table V indicates, these three systems show fluorescence enhancement by >100 times even in the presence of a paramagnetic transition metal ion that is unprecedented and make these systems potentially useful as transition metal sensors. Besides, the detection limit can be low as the metal ions form inclusion complexes that are very stable due to the *cryptate* effect. The emission signal also undergoes drastic enhancement (Table V) when the Pb(II) ion is used as an input, which presents the possibility of using such systems in the management of lead pollution as well.

Pioneering work in using a lanthanide ion for emission was reported by Lehn and co-workers (107). Those studies are based on 2,2'-bipyridine (bpy), 1,10-phenanthroline (phen), or 3,3'-bis(isoquinoline) (biqn) units incorporated into macrobicyclic structures conforming to the design as in Fig. 18(*b*). When a lanthanide ion such as Eu(III) or Tb(III) is added, they show strong emission that is red shifted by interacting with the triplet state of the heterocyclic units.

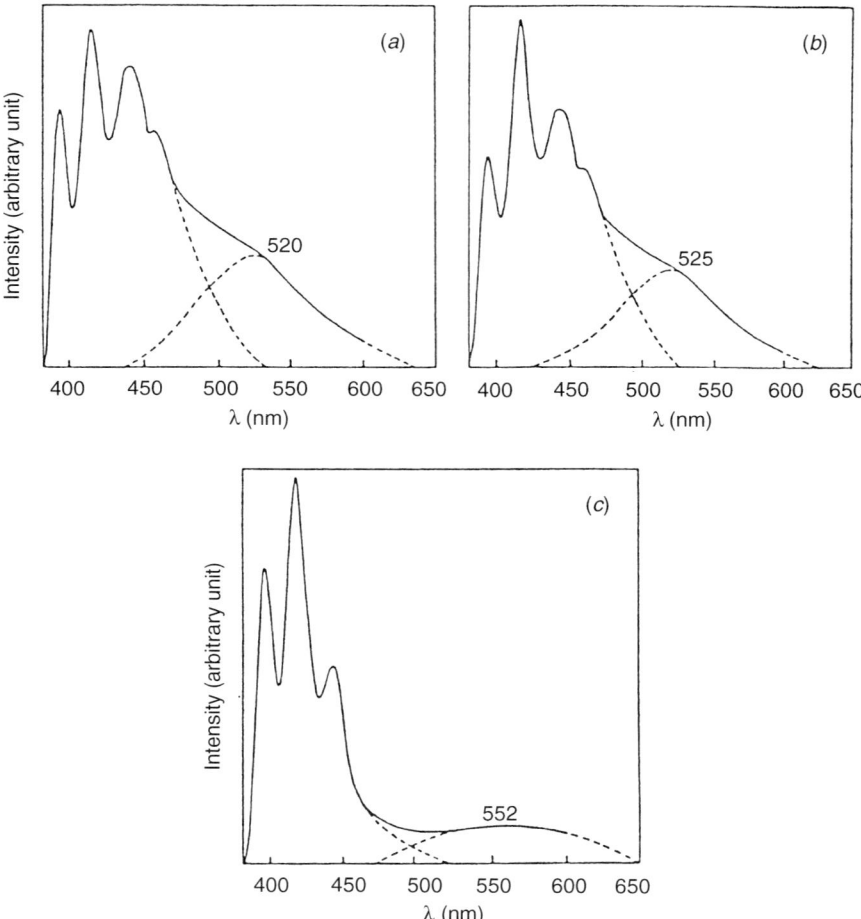

Figure 23. Corrected fluorescence emission of metal-free **L$_1$** in dry (a) n-hexane, (b) cyclohexane, and (c) dichloromethane at $2 \times 10^{-2} M$.

These systems are promising as potential labels due to high emission quantum yields and excited-state lifetimes that can be as long as several tenths of a millisecond (108). A cyclen (12-ane-N$_4$) unit connected to a phenanthridine moiety in fluorophore–spacer–receptor configuration (Fig. 26) exhibit strong Tb(III) based luminescence (109) in the absence of protons and oxygen. Few other luminescent lanthanide complexes are available in the literature (110, 111). In all these systems, there is a substantial energy transfer between the

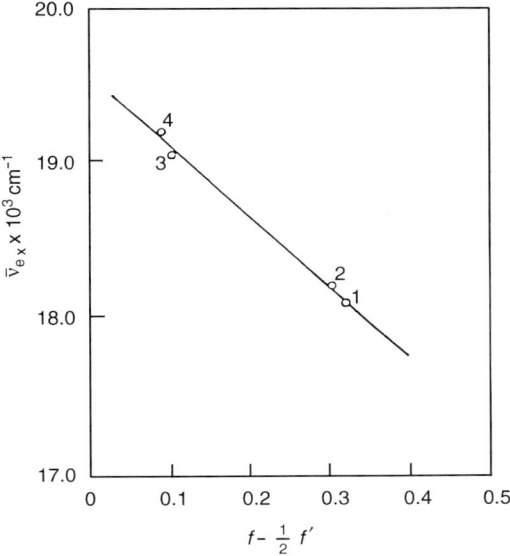

Figure 24. Wavenumber of the exciplex emission maximum at 298 K of L_1 as a function of solvent polarity parameter $f - \frac{1}{2}f'$; (1) dichloromethane, (2) THF, (3) cyclohexane, and (4) n-hexane. [see (102) and (105)].

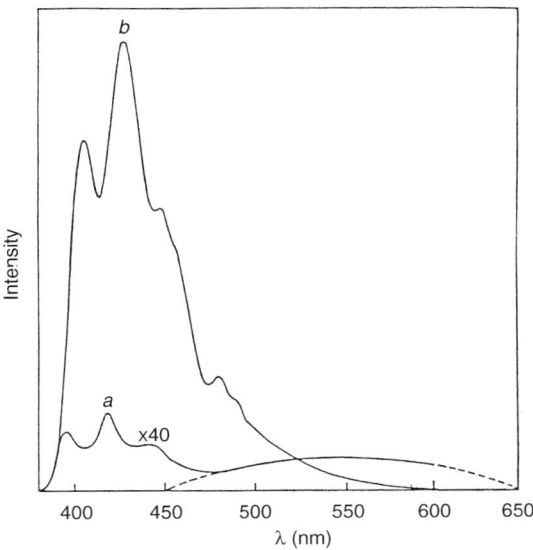

Figure 25. Corrected fluorescence spectra of metal-free L_1 (curve a) and of L_1 in the presence of Cu(II) in dry THF at 298 K.

TABLE V

Fluorescence Output with Different Cation Input in L_1-L_3[a]

Fluorophore	Ionic Input	(0,0) Band Position(nm)	Fluorescence Output Quantum Yield Φ_F		
			Φ_{FM}	Φ_{FE}	Φ_{FT}
L_1	Nil	394.8	0.0005	0.0005	0.001
	Mn(II)	405.0	0.301		0.301
	Fe(III)	404.0	0.226		0.226
	Co(II)	405.6	0.215		0.215
	Ni(II)	403.2	0.210		0.210
	Cu(II)	405.0	0.280		0.280
	Zn(II)	405.8	0.310		0.310
	Pb(II)	405.0	0.180		0.180
	Eu(III)	394.8	0.077		0.077
	Tb(III)	395.0	0.100		0.100
	H^+	404.0	0.245		0.245
L_2	Nil	395.5	0.001	0.0005	0.0015
	Mn(II)	401.0	0.136		0.136
	Fe(III)	403.0	0.156		0.156
	Co(II)	397.0	0.131		0.131
	Ni(II)	395.5	0.080		0.080
	Cu(II)	403.0	0.110		0.110
	Zn(II)	396.0	0.150		0.150
	Pb(II)	395.6	0.148		0.148
	Eu(III)	395.8	0.024		0.024
	Tb(III)	395.8	0.036		0.036
	H^+	399.6	0.230		0.230
L_3	Nil	394.6	0.001	0.001	0.002
	Mn(II)	403.6	0.370		0.370
	Fe(III)	404.2	0.289		0.289
	Co(II)	403.4	0.512		0.512
	Ni(II)	404.6	0.309		0.309
	Cu(II)	403.2	0.266		0.266
	Zn(II)	403.4	0.603		0.603
	Pb(II)	404.4	0.420		0.420
	Eu(III)	394.4	0.005	0.001	0.006
	Tb(III)	394.8			0.007
	H^+	403.6	0.250		0.250

[a] Experimental conditions: medium, dry THF; conc. of L_1 and L_3, 10^{-5} M; conc. of ionic input, 10^{-3} M; conc. of L_2, 5×10^{-6} M; conc. of ionic input, 5×10^{-4} M. Excitation at isobestic point 370 nm with excitation band-pass of 5 nm; emission band-pass, 5 nm; temperature, 298 K; Φ_F calcd. by comparison of corrected spectrum with that of anthracene ($\Phi_F = 0.297$) (106) taking area under the total emission. The error in Φ_F within $\pm 10\%$ in each case except free ligands where error in Φ_F within $\pm 15\%$.

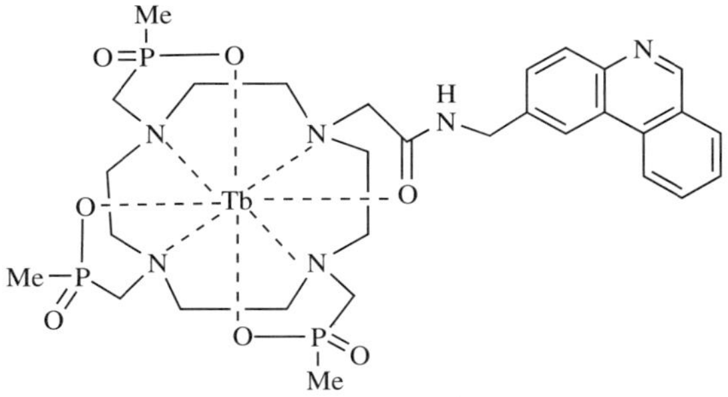

Figure 26. A macrocycle-based luminescent signaling system for Tb(III).

lanthanide and the fluorophore. In the case of L_1 and L_2, however, a lanthanide ion like Eu(III) or Tb(III) can cause (102b) fluorescence enhancement without showing any significant shift of the (0,0) band position (Table V). Interestingly, neither Eu(III) nor Tb(III) exhibits similar enhancement with L_3, which is attributable to significantly lower metal–receptor interactions between L_3 and either lanthanide ion. It is known that lanthanide ions prefer (112) higher than hexa-coordination. The lanthanide ion can be in the middle of the cavity in L_1 and L_2 establishing many bonding interactions with N and O donors sensing the size and shape of the receptor. The cavity in L_3 being quite large cannot effectively bind either lanthanide ion. With L_3, the exciplex emission is observed (Fig. 27) in the presence of the Eu(III)/Tb(III) ion as in metal-free L_3, which further shows that no significant donation occurs (104, 113) from nitrogen lone pairs to the metal ion. However, a definite clue to the binding of lanthanide ions in the cavity can be gathered from X-ray structures that are not yet available. With either lanthanide ion added to L_1 or L_2, no spectral shift is observed and the resolution remains unaffected. This result is due to energy of the lowest triplet state of anthracene being lower than the emitting states of Eu(III) and Tb(III) ions besides the fact that the two do not interact when the metal ion is inside the cavity. The lower Φ_{FT} observed for the lanthanide complexes of L_1 and L_2 compared to that of transition metal complexes could be due to a combination of less binding ability of the receptor (114) and a nonradiative decay of the fluorescent state of anthracene to the lowest triplet state through the rare earth (III) ion states. In the case of L_2, the fluorescence recovery with either lanthanide ion as input is much less compared to that with L_1 due to less metal–receptor interactions. These systems can potentially function as an OR logic gate since no shift of the emission band occurs upon metal binding (Fig. 28).

The large enhancement of fluorescence in L_1–L_3 in the presence of a paramagnetic transition metal ion is unprecedented. To confirm that the

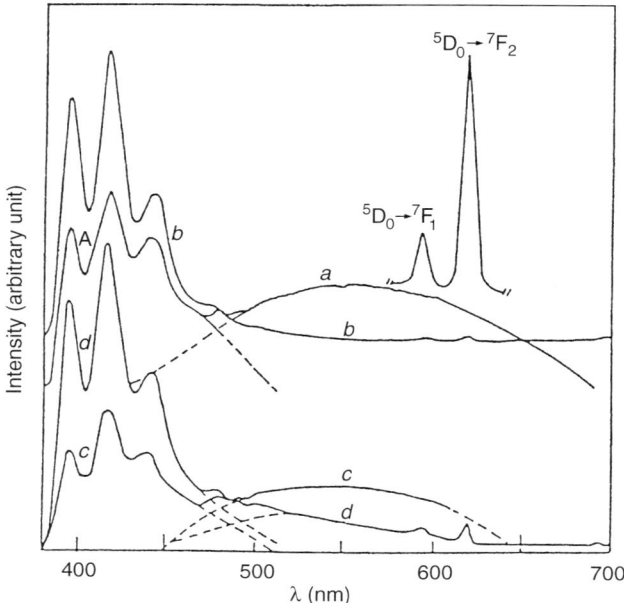

Figure 27. Corrected fluorescence emission spectra of metal-free L_1 (a), of L_1 in the presence of Eu(III) (b), of metal-free L_2 (c), and of L_2 in the presence of Eu(III) (d) in dry THF. Both (c) and (d) are showing exciplex emission. The extra peaks in (b) and (d) are due to Eu(III) as shown.

enhancement is due to the metal ions and not due to protons generated from the hydrated metal salts, several control experiments are performed (102) that clearly rule out the possibility of involvement of the protons.

In the compound L_4, the amino nitrogens are derivatized with the anthroyl group (Fig. 29) to have a carbonyl group in place of methylene as the spacer

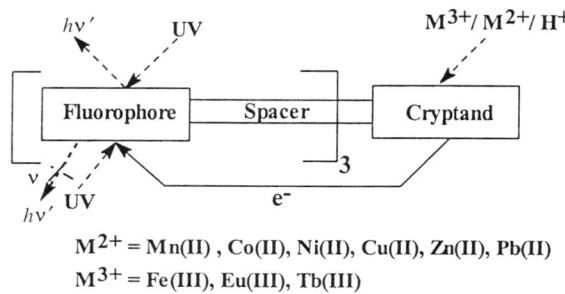

M^{2+} = Mn(II), Co(II), Ni(II), Cu(II), Zn(II), Pb(II)
M^{3+} = Fe(III), Eu(III), Tb(III)

Figure 28. Plan of an OR logic gate with a cryptand receptor.

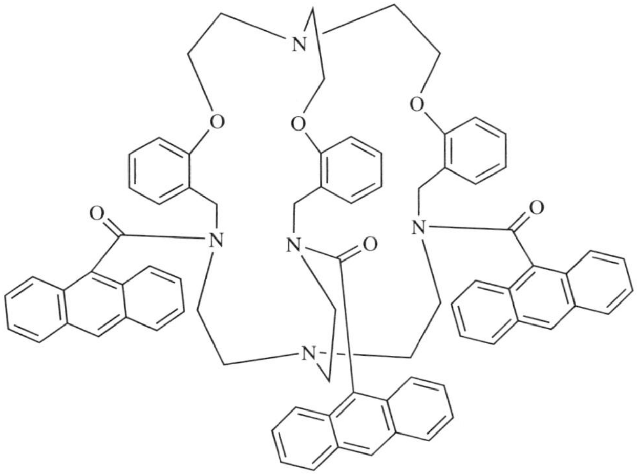

Figure 29. A cryptand-based fluorescent signaling system with a carbonyl group as spacer.

(115), which has altered the photophysical properties of the system drastically compared to those of L_1. Introduction of the carbonyl group as spacer is expected to affect the interaction of the lone pair of nitrogen with the anthracene moiety and flexibility of the fluorophore units with respect to the cryptand influencing the PET process. Free L_4 shows a well-resolved anthracene monomer emission similar to that of L_1 without showing any exciplex emission (Fig. 30). The emission spectra of carbonyl-substituted anthracenes have been found to be greatly influenced by the nature of the carbonyl substituent (116, 117). Thus, although 9-substituted anthryl ketones are virtually nonfluorescent at room temperature in aprotic solvents due to a suitably placed (n, π^*) triplet level that enhances intersystem crossing, the anthracene amide carbonyls have the (n, π^*) too high in energy to affect S_1 decay and are fluorescent in nature. Earlier studies revealed (116, 118, 119) that in the ground state (S_0) of 9-substituted carbonyl anthracenes like 9-anthramide and N,N-diethyl-9-anthramide, steric hindrance between the carbonyl group and the ring hydrogens keep the carbonyl group twisted almost 90° out of plane of the anthracene ring that precludes extensive conjugation between the two. These compounds thus exhibit anthracene-like structured emission. However, the shape as well as position of the emission spectra of 9-anthramide has been found to be quite solvent dependent (116) due to increasing conjugation between the carbonyl group and the anthracene π system. In contrast, the spectra of N,N-diethyl-9-anthramide

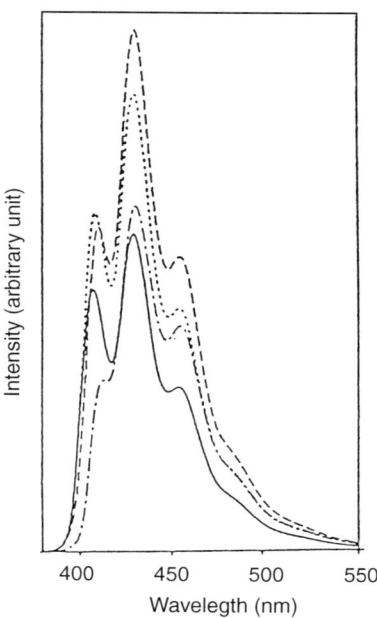

Figure 30. Corrected fluorescence emission spectra of meta-free L_4 at different concentrations: $1 \times 10^{-5} M$ (—), $2 \times 10^{-5} M$ (- - -), $5 \times 10^{-5} M$ (· · · ·), and $1 \times 10^{-4} M$ (—·—). No exciplex emission is seen in any of the spectra.

are solvent independent (120) and show the same structured emission as unsubstituted anthracene due to much higher steric hindrance to rotation of the carbonyl group with respect to the anthracene moiety. The emission spectra of L_4 is very similar compared to N,N-diethyl-9-anthramide indicating restricted rotation of the carbonyl group in the molecule. This restriction has two significant effects: (1) L_4 does not show any exciplex like L_1 and (2) the PET process in L_4 is less efficient compared to that in L_1 (Table VI) due to unfavorable donor–acceptor orientation (121) in the former. Consequently, the quantum yield of monomer emission Φ_{FM} increases only slightly upon addition of a metal ion as input (Table VI). The fluorophoric system L_5 (Fig. 31) was probed (122) to understand more of the photophysical properties of cryptand-based fluorophoric systems. The $-COCH_2-$ linkage in the molecule hinders efficient PET as in the case of L_4. This system also exhibits a broad structureless emission centered ~ 520 nm assignable (123) as intramolecular exciplex emission. Among the metal ions studied as input, the Φ_{FM} increases marginally in the presence of Zn(II), Co(II), and Mn(II) while binding of Ni(II) or Cu(II) does not

TABLE VI
Fluorescence Output with Different Cation Input in L_4 and L_5[a]

Fluorescence	Ionic Input	Fluorescence Output (Φ_F)
L_4	Nil	0.13
	Cu(II)	0.18
	Co(II)	0.26
	Zn(II)	0.19
L_5	Nil	0.42
	Zn(II)	0.92
	Co(II)	0.94
	Mn(II)	0.81
	Ni(II)	0.15
	Cu(II)	0.32

[a] Experimental conditions: medium, dry THF; conc. of L_4 and L_5, $10^{-3}\,M$; concentration of ionic input, $10^{-3}\,M$, excitation band-pass = 5 nm; emission band-pass = 5 nm; temperature, 298 K; Φ_F calculated taking the total area under the respective emission spectrum; error in Φ_F within $\pm 10\%$.

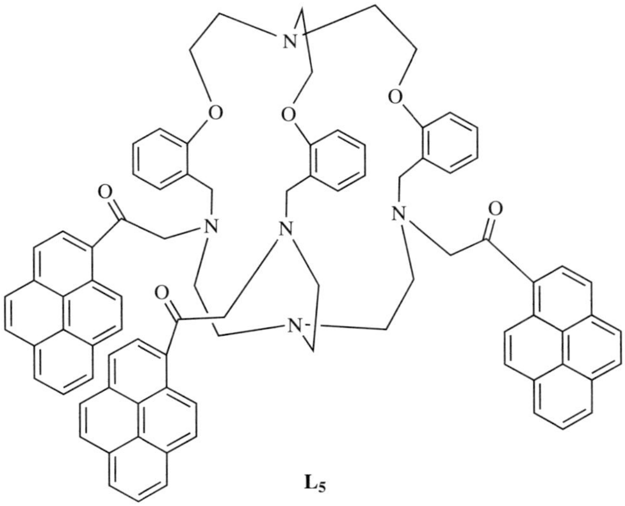

Figure 31. A cryptand-based fluorescent signaling system with a carbonyl group as spacer and pyrene as the fluorophore.

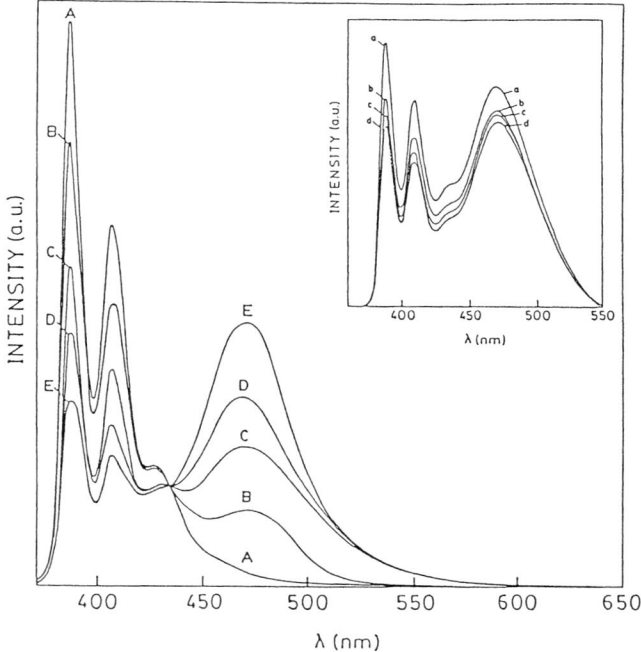

Figure 32. Corrected fluorescence emission spectra of L_5 ($1 \times 10^{-5} M$) in dry THF for various concentration of H^+: (A) 0, (B) $2.0 \times 10^{-3} M$, (C) $3.5 \times 10^{-3} M$, (D) $5.0 \times 10^{-3} M$, and (E) $6.5 \times 10^{-3} M$ showing the built-up of the excimer. Inset: The fluorescence spectra (E) taken at different temperatures: (a) 285 K, (b) 299 K, (c) 305 K, and (d) 313 K.

result in any fluorescence enhancement (Table VI). In the presence of protons, the molecule exhibits an intramolecular excimer (124) emission (Fig. 32) with concomitant decrease in the monomer emission due to close approach of two pyrene moieties in the excited state when the nitrogen lone pairs are engaged to protons. This system is thus one of the rare examples showing monomeric, excimeric, and exciplex emissions depending on the environment. In order to characterize these three emissions, time-resolved fluorescence measurements were carried out in various environments (125). The results are summarized in Table VII. The monomer region show a lifetime of 11 ns, which is shorter compared to that of free pyrene (126) indicative of PET from the cryptand N atom to the pyrene moiety. The presence of a longer lifetime component of ~55 ns (only ~4% contribution) is attributed to a different conformation of the cryptand. The major component of the lifetime of the exciplex emission is 2.9 ns for the metal-free system. This value is of the same order as that of systems

TABLE VII

Lifetime Measurement Data for L_5 in Various Environments in THF at 298 K[a]

System	Lifetime Monomer (ns)	Lifetime Excimer (ns)	Lifetime Exciplex (ns)
L_5	$l_1 = 11.02$ (96.5%) $l_2 = 55.7$ (3.5%)		$l_1 = 2.9$ (55%) $l_2 = 13.1$ (45%)
$L_5 (10^{-5} M)$ + Zn(II) $(10^{-3} M)$	$l_1 = 10.8$ (45.2%) $l_2 = 21.8$ (54.8%)		
$L_5 (10^{-5} M)$ + $H^+ (10^{-3} M)$	$l_1 = 5.2$ (85.7%) $l_2 = 13.8$ (14.4%)	$l = 6.2$ (100%)	

[a] Percentages in the parentheses refer to the amount of contribution of the components.

where tertiary amino groups are semirigidly attached to polycyclic aromatic hydrocarbons (127). However, this value is much shorter compared to those reported for unlinked hydrocarbon–tertiaryamine systems (128), the later being of the order of 10 ns in a polar or semipolar solvent. In the presence of protons, the system switches over to a conformer that favors excimer formation. The lifetime of the excimer as well as that of the monomer in the presence of H^+ are, however, shorter compared to that of other pyrene-based systems (126), which indicates that all the N atoms in the cryptand are not protonated so that the process of PET is not totally blocked. Thus, while L_1–L_3 can potentially function as OR logic gates, L_5 can function as fluorescent "ON–OFF" logical switch in the presence of H^+ ion where the excimer is switched "on" while the monomer emission is reduced simultaneously.

Once the importance of the nature of the spacer unit in the PET process was realized, it was natural to probe a few more systems with different spacer units to understand the effect of M–F and M–R communications on the PET process and their photophysical behavior. Thus, the systems shown in Fig. 33 were designed (129) with the aim to understand the PET process and its manipulation to create new fluorescent signaling systems for transition metal ions. An examination of Table VIII shows that the efficiency of the PET process in the metal-free state depends on the nature of spacer and the amount of fluorescence recovery varies not only with the metal ion but with the spacer as well. In the case of L_8, the fluorescence enhancement is negligible as the fluorophore can come close to the metal center through rotation of the ethylene bonds. In L_{10}, the fluorophore is integrated with the receptor and the enhancement is slightly higher only in case of Zn(II) ion as input.

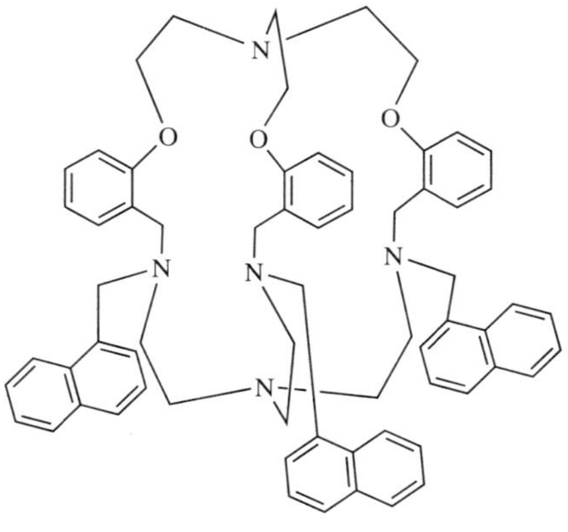

$\mathbf{L_6}$

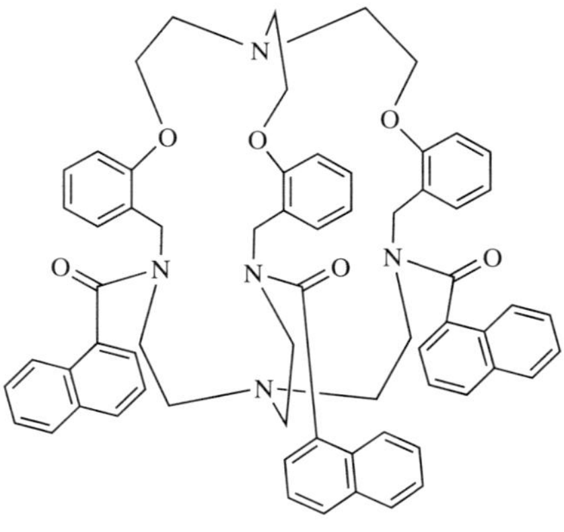

$\mathbf{L_7}$

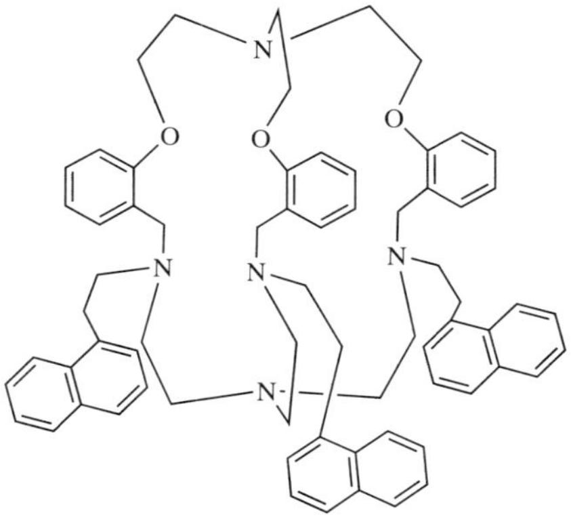

$\mathbf{L_8}$

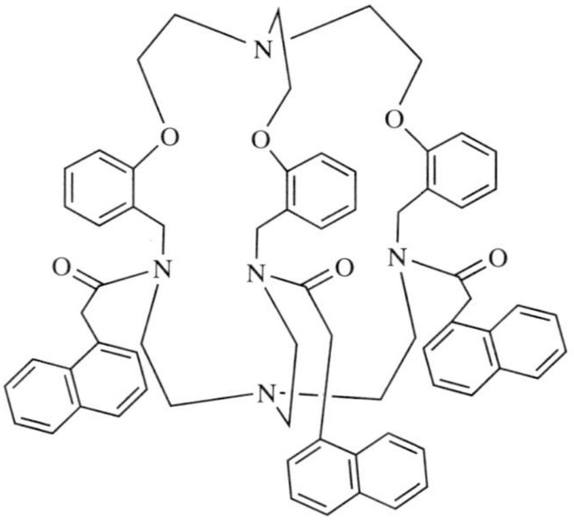

$\mathbf{L_9}$

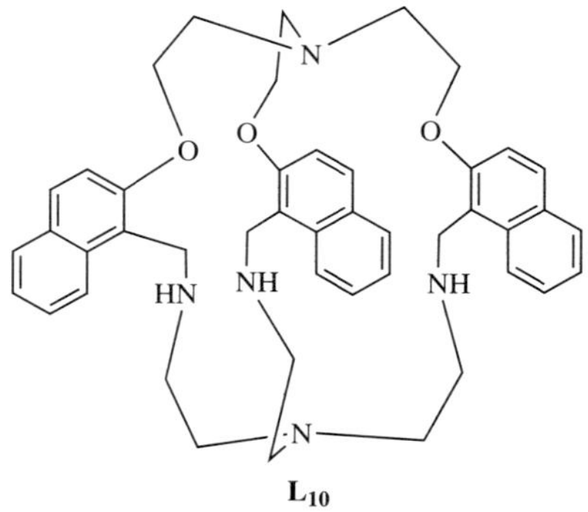

$\mathbf{L_{10}}$

Figure 33. Naphthyl fluorophores connected to Cryptand **7** via different spacers.

TABLE VIII
Fluorescence Output with Different Cation Input in L_6–L_{10}[a]

Fluorophore	Ionic Input	Fluorescence Output (Φ_F)
L_6	Nil	0.04
	Cu(II)	0.14
	Ni(II)	0.22
	Co(II)	0.20
	Mn(II)	0.12
	Zn(II)	0.24
L_7	Nil	0.11
	Cu(II)	0.19
	Ni(II)	0.25
	Co(II)	0.33
	Mn(II)	0.35
	Zn(II)	0.49
L_8	Nil	0.02
	Co(II)	0.02
	Mn(II)	0.02
	Ni(II)	0.02
	Cu(II)	0.02
	Zn(II)	0.04
L_9	Nil	0.26
	Cu(II)	0.09
	Ni(II)	0.11
	Co(II)	0.11
	Mn(II)	0.19
	Zn(II)	0.32
L_{10}	Nil	0.11
	Cu(II)	0.03
	Co(II)	0.08
	Zn(II)	0.16

[a] Experimental conditions: medium, dry THF; conc. of L_n, $10^{-3} M$; conc. of ionic input, $10^{-3} M$; excitation band pass = 5 nm, emission band-pass = 5 nm; temperature, 298 K; Φ_F calculated taking the total area under the respective emission spectrum; error in Φ_F within $\pm 10\%$.

V. CRYPTANDS AS NONLINEAR OPTICAL MATERIALS

A. Nonlinear Optical Effect

Compounds displaying nonlinear optical (NLO) properties are currently drawing considerable attention due to their possible applications in the emerging technologies of optoelectronics and photonic devices (130–135). Before discussing the usefulness of cryptands as NLO materials, it is of relevance to introduce in a very rudimentary way the essentials of NLO effects. Nonlinear optical effects arise from nonlinear polarization of molecules and materials. In the presence of low-intensity light, the induced polarization of a molecule is linearly proportional to the strength of the electric field associated with light. However, in the presence of high-intensity light such as laser radiation, the polarization of the molecule is no longer linear in the field and can be expressed as a nonlinear function of the field strength, E as shown in Eq. (5), where P is the

$$P = \alpha E + \beta E^2 + \gamma E^3 + \cdots \quad (5)$$

induced molecular polarization, α is the linear molecular polarizability, and β and γ, are the first and second molecular hyperpolarizability, respectively. Second-order NLO effects arise via the βE^2 term in Eq. (5) and occurs in molecules that do not have a center of symmetry. At the molecular level, compounds likely to exhibit large values of molecular hyperpolarizability, β, must have (1) excited states close in energy to the ground state for easy access by visible–infrared (vis–IR) light, (2) a large oscillator strength for the transition, and (3) a large difference between the ground- and excited-state dipole moments. At the macroscopic level, the induced polarization can be expressed as in Eq. (6), where the coefficients $\chi^{(2)}$ and $\chi^{(3)}$ are the second- and third-order

$$P = \chi^{(1)} E + \chi^{(2)} E^2 + \chi^{(3)} E^3 + \cdots \quad (6)$$

bulk polarizability, respectively, of the molecular material. The most important manifestation of second-order bulk nonlinearity is frequency doubling also known as second harmonic generation (SHG). Detailed mathematical explanations of the nature and origins of NLO effects are available (136, 137). Second-order NLO effects, including SHG and electrooptic modulation are important parameters for interfacing electronic data to wide-band optical communication.

Although β in a molecule is closely related to bulk nonlinearity χ^2 in the solid state, large values of β does not mean that when the molecule is crystallized it will show a higher value of χ^2. For this, the molecule must crystallize in a noncentrosymmetric space group. It has been a belief for a long time that extended π systems with a considerable molecular dipole character are most promising candidates as second-order NLO materials. In this dipolar approach, β is associated with only one HOMO–LUMO electronic transition of charge-transfer (CT) character (138). These molecules lack significant off-diagonal components; hence β is termed one dimensional (1D). A major problem (137) of traditional dipolar chromophores is the nonlinearity–transparency trade-off, whereby the desirable increase in second-order polarizability is accompanied by a bathochromic shift (139) of the electronic transition, leading to reabsorption of the second harmonic light making them ineffective in frequency doubling applications. Moreover, these molecules are difficult to crystallize in noncentrosymmetric space groups.

At the microscopic level, strategies have been developed during the past decade in order to circumvent these eventual drawbacks by extending the CT dimension from one to two or even to three (140). These two- and three-dimensional (2D and 3D) chromophores (141–143) with C_3, D_3, or T symmetry, have several advantages: (1) they are more transparent as the lack of a permanent dipole moment results in negligible solvatochromism, (2) enhanced nonlinearity due to coupling of the excited states at no cost of transparency, and (3) greater probability of crystallization in a noncentrosymmetric space group due to very low dipole moment. Based on this principle, several molecules have been designed. Noteworthy among them are 1,3,5-triamino-2,4,6-trinitrobenzene (TATB) (144), crystal violet (145), tris(4-methoxyphenyl) cyclopropenylium bromide, and triazene derivatives (146).

B. Cryptand Derivatives as D-π-A Nonlinear Systems

Laterally nonsymmetric cryptands incorporating secondary amino nitrogens can serve as excellent skeletons onto which acceptor groups can be added for having trigonal nonlinear materials. The cryptands **7** and **8** were derivatized (147, 148) with different acceptor units (π-A) to have 3D NLO chromophores with pseudo-threefold symmetry (Scheme 6). The central core or the cryptand has been varied to alter the donating ability of the nitrogen atoms and also to probe any effect rigidity of the cryptand has on the NLO behavior. The nitrogen atoms give the cryptand it's donor (D) character and the π-A units were grafted by simple aromatic nucleophilic substitution (ArSN) reactions (149). The ease of formation of the trigonal chromophores and their sharp and high melting points make them more attractive compared to traditional nitroaniline-based molecules (150) showing bulk NLO properties. The different π-A units incorporated in the cryptand strands are 2,4-dinitrobenzene, 5-nitropyridine, and 4-nitrobenzene as these molecules have been used as efficient π-A units in important NLO molecules like methyl(2,4-dinitrophenyl)aminopropanoate (MAP) (151), prolinolnitropyridine (PNP) (152), 4-nitrophenyl-L-prolinol (NPP) (153), and so on.

The lack of a permanent dipole moment in the cryptand-based molecules makes the classical electric field-induced second harmonic generation (EFISH) experiment (154) unsuitable for the determination of molecular nonlinearity β which, however, can be obtained in solution employing the hyper-Rayleigh scattering (HRS) technique (155). The powder SHG measurements were carried out using the Kurtz–Perry method (156) using the fundamental wavelength (1064 nm) of a Q-switched Nd: YAG laser having a pulse width of 8 ns.

As each cryptand core inherits threefold symmetry, there arises two possibilities by which these trigonal molecules can crystallize (Fig. 34): a planar centrosymmetric hexagonal lattice formed by the interactions between identical groups resulting in SHG inactive molecules or noncentrosymmetric trigonal lattice formed by the interaction between different groups leading to the SHG

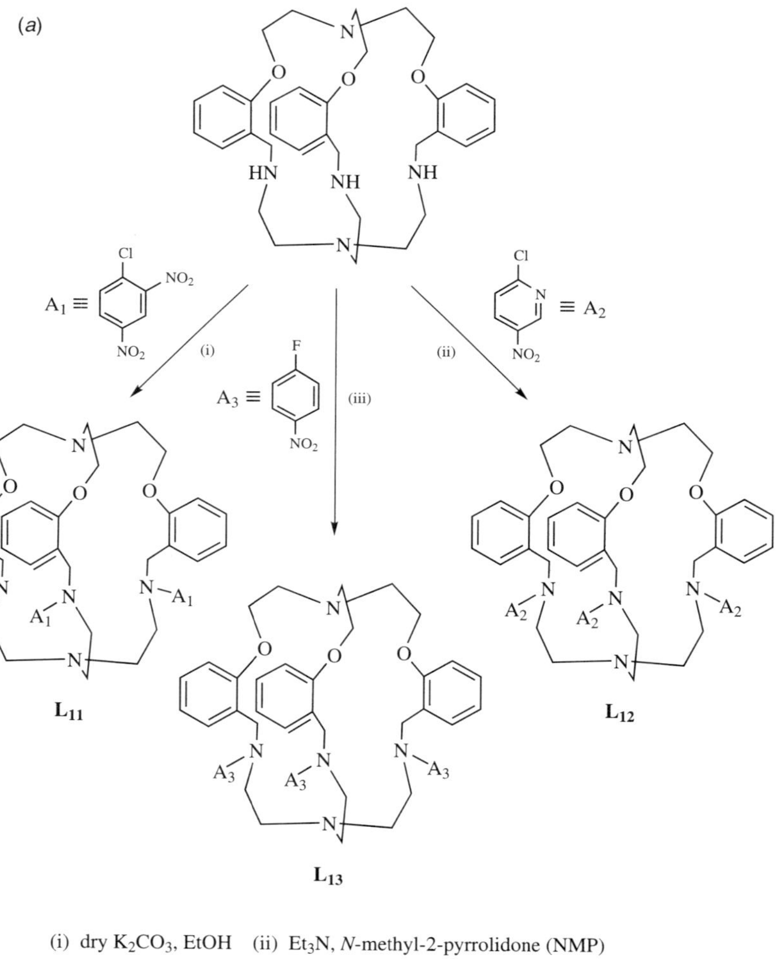

(i) dry K_2CO_3, EtOH   (ii) Et_3N, N-methyl-2-pyrrolidone (NMP)
(iii) dry K_2CO_3, dimethyl sulfoxide (DMSO)

Scheme 6a

(b)

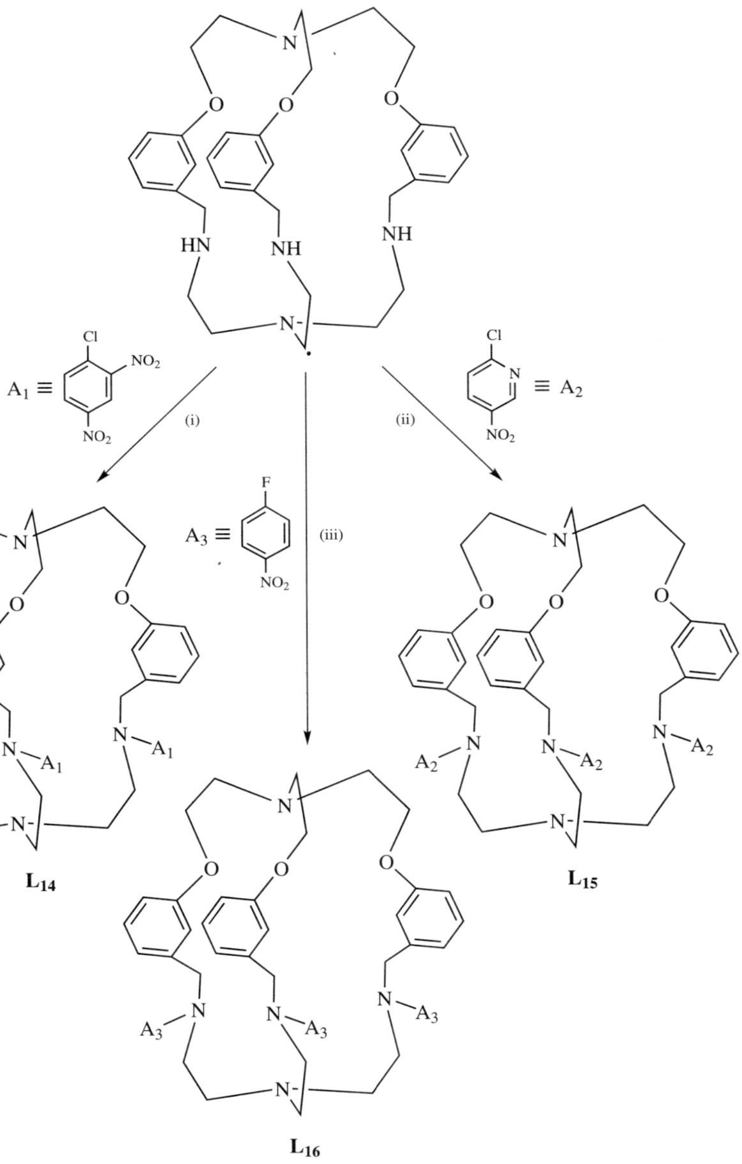

L14 **L15** **L16**

(i) dry K$_2$CO$_3$, EtOH (ii) Et$_3$N, N-methyl-2-pyrrolidone (NMP) (ii) dry K$_2$CO$_3$, dimethyl sulphoxide (DMSO)

Scheme 6b

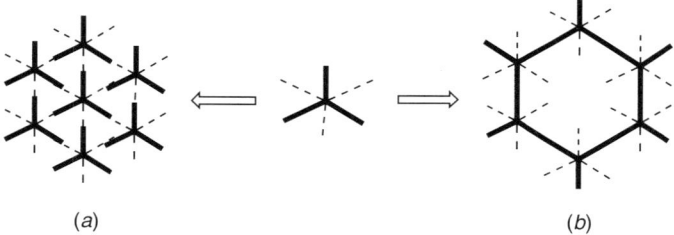

Figure 34. Two possible ways a tripodal molecule can crystallize: trigonal noncentrosymmetry (*a*) and hexagonal centrosymmetry (*b*).

active form. Till now, crystal structures of three compounds, **L$_{11}$**, **L$_{13}$**, and **L$_{14}$** (see Scheme 6), are available (148). While **L$_{11}$** adopt a trigonal space group **L$_{13}$** and **L$_{14}$** crystallizes in orthorhombic and triclinic space groups, respectively. All three crystal structures maintain a pseudo-threefold symmetry along the two bridgehead N atoms (Figs. 35–37). For **L$_{11}$**, two different kinds of π interactions,

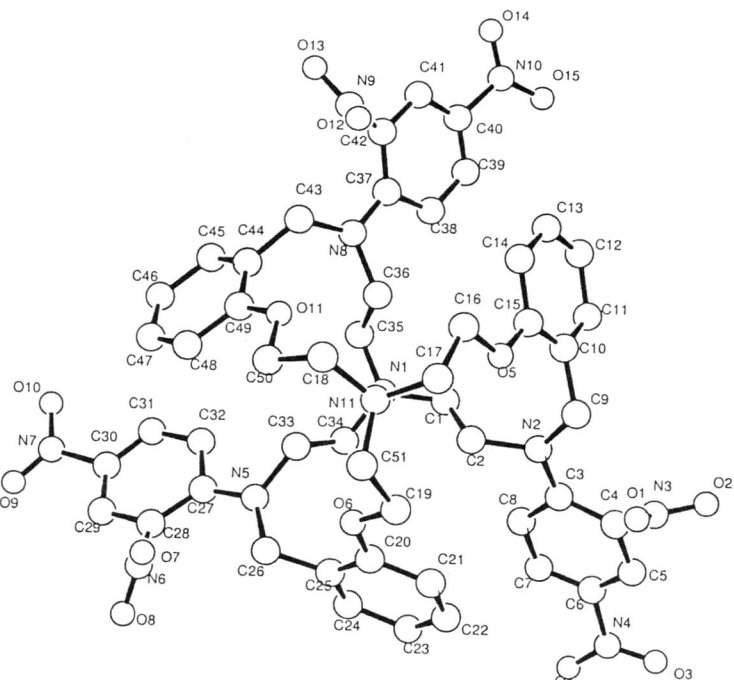

Figure 35. A perspective view of **L$_{11}$** looking down the pseudo-threefold axis passing through the bridgehead nitrogen atoms.

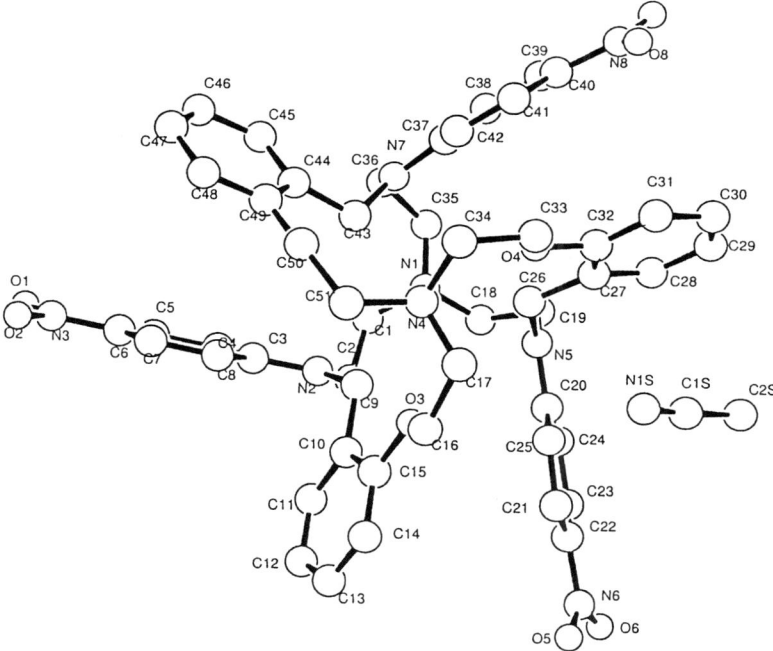

Figure 36. A perspective view of L_{13} looking down the pseudo-threefold axis passing through the bridgehead nitrogen atoms.

namely, face-to-face and edge-to-face are present as intermolecular interactions that assist the cryptand molecule to assemble in a perfectly trigonal network (Fig. 38). In addition, the polar hydrogen bonds steer this molecule to a noncentrosymmetric structure. The molecule L_{13} on the other hand, crystallizes in the orthorhombic centrosymmetric space group *Pbca*. The molecule L_{14} crystallizes in the triclinic centrosymmetric space group $P\bar{1}$. This molecule also exhibits face-to-face and edge-to-face hydrogen-bonding interactions and packs into a pseudo-threefold network. These chromophores (L_{11}–L_{16}) do not fluoresce at room temperature and are essentially transparent at 532 nm for nonresonant HRS measurements at $\lambda = 1064$ nm. Besides, a plot of $I(2\omega)/I(\omega^2)$ (where ω is 532 nm) against the number density of any of the chromophores (L_{11}–L_{16}) show a perfectly linear dependence as expected from a system of independent scatterers. This finding again confirms the suitability of these cryptand derivatives as NLO materials. The nonlinear $\sqrt{\langle\beta^2\rangle}$ coefficients of L_{11}–L_{16} are collected in Table IX. The corresponding $\sqrt{\langle\beta^2\rangle}(0)$ static values derived from a degenerate two-level dispersion model (157) are also presented.

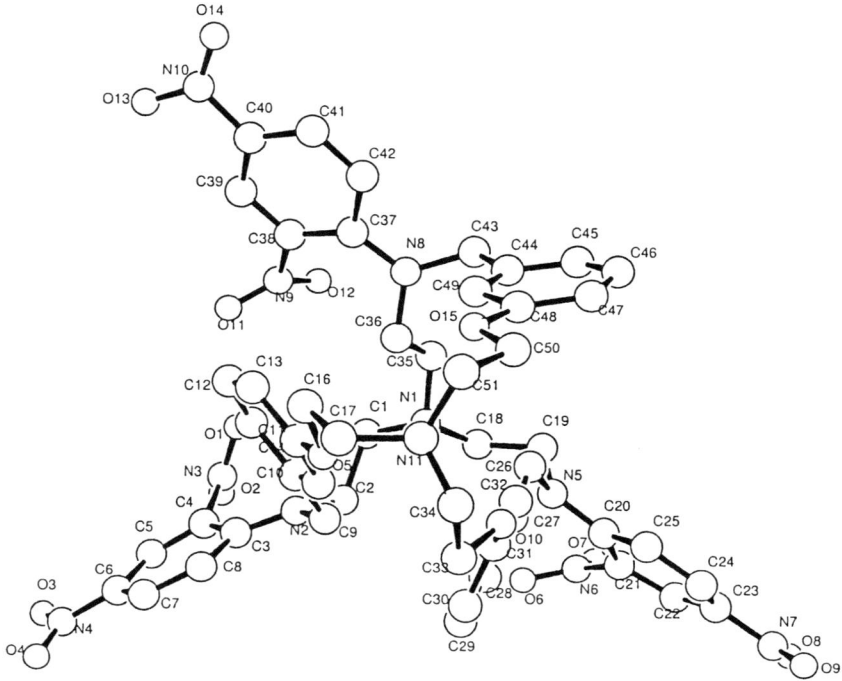

Figure 37. A perspective view of L_{14} looking down the pseudo-threefold axis passing through the bridgehead nitrogen atoms.

The nonlinearity of L_{11}–L_{13} are comparable to the classical *para*-nitroaniline (*p*NA) molecule (158), with $\sqrt{\langle\beta^2\rangle}(0) = 10 \times 10^{-30}$ esu as measured in $CHCL_3$ under similar experimental conditions. In comparison, the nonlinearity of L_{14}–L_{16} show a slightly higher value compared to *p*NA, which has a more rigid framework in **8** compared to that in **7**, both maintaining pseudo-threefold symmetry (13a).

Compound L_{11} gives a SHG powder signal that is 0.6 times that of urea in conformity with its acentric crystallization. Compound L_{13}, in spite of being crystallized in a centrosymmetric space group, shows a very weak SHG powder signal efficiency of 0.05 times that of urea, which points out some defect sites in the crystal. Although the molecules crystallize in a centrosymmetric space group, there may be regions or zones having defects. In fact, such defects are known in organic molecules. The X-ray crystallographic data pertaining to the macroscopic ordered region of the crystal say nothing about the molecular packing in the defect region. The L_{13} structure is stabilized by hydrogen

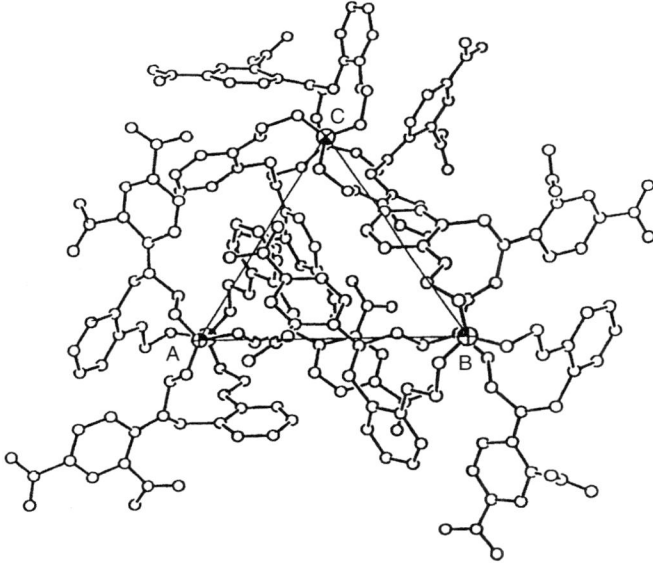

Figure 38. Packing of L_{11} in a trigonal network showing both intra- and intermolecular hydrogen-bonding interactions.

bonding and stacking interactions in a different fashion compared to the other two cases. Perhaps this is the reason why it adopted a centrosymmetric packing. In other words, stacking interactions between the molecules in a unit cell seems necessary for adopting a noncentrosymmetric crystal structure in the solid state and, hence, for efficient macroscopic SHG. Further experiments are necessary to determine the exact nature of the defects in L_{13}. Surprisingly, compound L_{14} in spite of being crystallized in a centrosymmetric space group, exhibits a SHG

TABLE IX

Results of Hyper-Rayleigh Scattering Measurements on Compounds L_{11}–L_{16} at 1064 nm

	pNA^a	L_{11}	L_{12}	L_{13}	L_{14}	L_{15}	L_{16}
λ_{max}	347	370	376	390	364	368	380
$\beta \times 10^{-30}$ esu	18.0	16	16.1	19.0	20	22	27
$\beta_0 \times 10^{-30}$ esu	10	10.6	10.5	12.2	14	15	18

^{a} p-Nitroaniline = pNA.

powder signal of 0.36 times that of urea. This signal could be due to polymorphism of the crystalline structure or to a slight difference in the molecular orientations along one of the axes. However, this result is not unprecedented and was earlier observed in case of TATB, which also shows substantial SHG activity (144).

Using a cryptand as the skeleton has other added advantages. The cryptand can include a redox active metal ion in the cavity that can act as a signal transducer offering switching possibilities (159) or it can include a paramagnetic transition metal ion that might be useful in aligning in strong fields. The ability to switch the NLO response of a molecule "ON" and "OFF" reversibly should add significant utility to NLO molecules. Interestingly, the use of metals in altering the properties of nonlinear materials have not been explored extensively (160–162).

VI. CRYPTAND-BASED AMPHIPHILES

Amphiphilic molecules are crucial to life starting from forming cell walls to translocation of biomolecules in vivo. They are known to form a great variety of supramolecular structures such as micelles, vesicles, monolayers, bilayers, multilayer rods, and so on. Naturally, a large number of synthetic amphiphiles with both acyclic as well as cyclic head groups and one or more hydrophobic tails with different chemical structures have been reported (163–169). Such amphiphiles can be important as model systems for membrane structure (170) and functional (171) studies. They can also be useful in drug delivery and targeting (172), medical imaging (173), catalysis (174), energy conversion (175), and separation (176). Synthetic surfactants with a cryptand headgroup offer several design options—a cryptand has (1) a preformed cavity whose topology can be designed to recognize the ion–molecule of interest, (2) the rigidity of the macrobicyclic structure favors ordered structures, and (3) the possibility to have amphiphilic molecules with different chemical structures by selective derivatization of the amino groups. The first surfactant molecule with a macrobicyclic headgroup was reported by Sargeson and co-workers (177). They were able to attach alkyl chains to the N_6 macrobicycle either by reductive alkylation or via diazotization reaction. A laterally nonsymmetric N_3S_3 macrobicycle had also been derivatized this way. Some of these molecules are shown in Fig. 39. These Co(III) cage surfactants are effective in bursting cell walls of the tapeworm, *Hymenolepis diminuta*. The destabilization of the tegumental membrane is caused by a combination of large size and high charge of the head group, which is able to alter the curvature and charge distribution of the bilayer membrane. Such studies are possible when the amphiphilic molecules are soluble in water. Surfactants with metal ions other than Co(III) inside

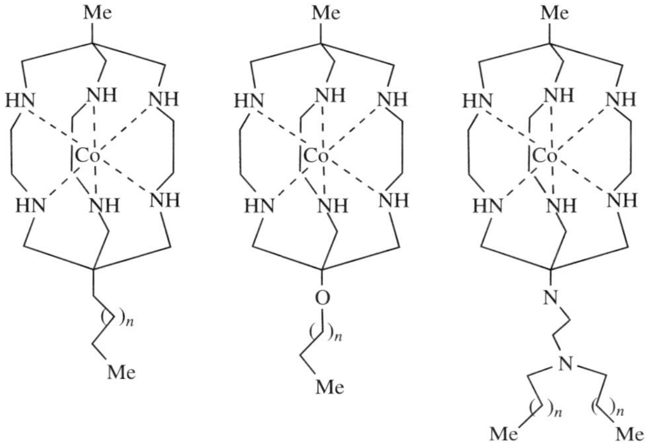

Figure 39. Examples of surfactant molecules with the N_6 cage as the headgroup.

the cage had also been reported (178) by this group. The secondary amino groups in laterally nonsymmetric cryptands can be derivatized with acid chlorides to form a new generation of surfactant molecules. Discussions on these amphiphiles are made here in the context of their ability to form vesicular dispersions and to self-assemble as monolayers at the air–water interface.

A. Vesicular Aggregation

The process of vesicle formation via self-assembly is an important biological event. Natural as well as synthetic amphiphiles can self-organize into vesicular aggregates depending on several factors such as solubility, alignment of hydrophobic tails, intermolecular packing of rigid segments, hydrogen bonding among neighboring amphiphiles, and so on. The secondary amino groups in cryptands **7** and **8** were derivatized with acid chlorides of different chain lengths (179–181) to form triple-tailed amphiphiles (Scheme 7). The cryptand headgroup can accommodate a Cu(II) ion forming a new class of synthetic surfactants. Both these free as well as Cu(II) complexed amphiphiles can self-organize into vesicular structures when an ethanolic solution of an amphiphile is dispersed in water. Vesicles form spontaneously in vivo in the existing biological environment while most synthetic amphiphiles require either considerable mechanical energy-like sonication, extrusion, and so on or chemical treatments like detergent dialysis, reverse-phase evaporation, and so on for vesicle formation (170). Few systems are, however, known (182–184) that spontaneously

Scheme 7

7 + RCOCl ⟶ L_n

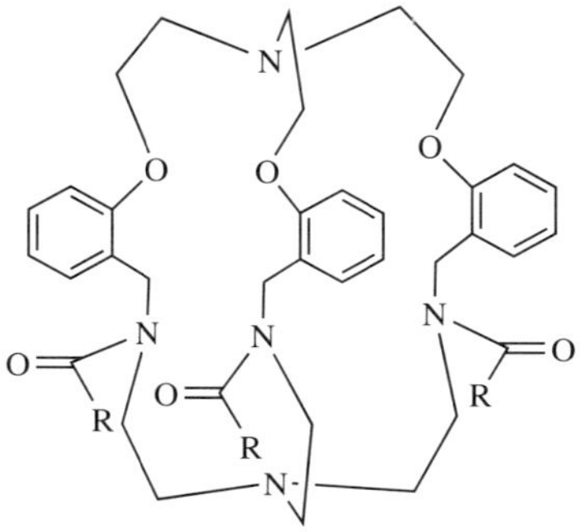

n	R
17	$CH_3-(CH_2)_{14}-$
18	$CH_3-(CH_2)_{10}-$
19	$CH_3-(CH_2)_8-$
20	$CH_3-(CH_2)_6-$
21	$CH_3-(CH_2)_5-$
22	$CH_3-(CH_2)_3-$

8 + RCOCl ⟶ L_n

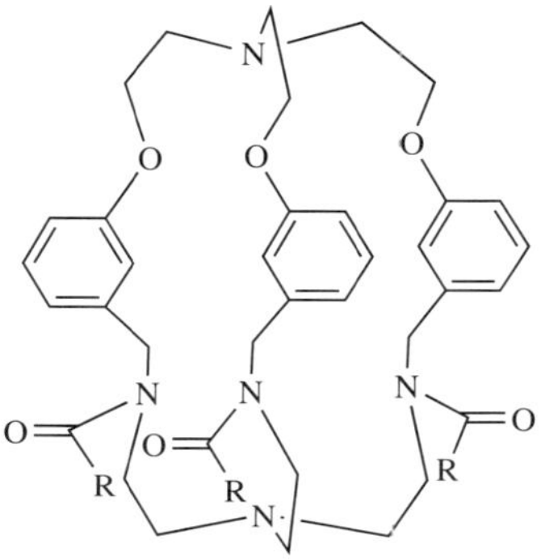

n	R
23	$CH_2-\{(CH_2)_7-CH=\}_2$
24	$CH_3-(CH_2)_{14}-$
25	$CH_3-(CH_2)_{10}-$
26	$CH_3-(CH_2)_8-$
27	$CH_3-(CH_2)_6-$
28	$CH_3-(CH_2)_5-$
29	$CH_3-(CH_2)_3-$

form vesicles in solution and this list will definitely increase at a faster rate in the future. Cryptand-based surfactants can also form stable vesicular dispersions spontaneously due to the special topology of the cryptand architecture and the presence of carbonyl units as spacers between the headgroup and the hydrophobic tails that allow compact chain packing without conformational constraint on the headgroup. Of course, they can also form vesicles by the extrusion method as well (169). Both tunneling electron microscopy (TEM) (Fig. 40) and dynamic light scattering (DLS) studies show that the average diameter of these vesicles lie within the range, 250–350 nm. The stability of vesicular dispersions (185) with respect to their size and shape, increase with the increase of the hydrophobic chain length. Generally, vesicular dispersions made by the extrusion method are found to be more stable compared to the corresponding ones made spontaneously (Table X). Changing the cryptand headgroup from **7** to **8** does not affect the stability of vesicular dispersions to any noticeable extent. With time, the vesicles fuse, they form clusters, and finally coagulate. Lowering the temperature to 278 from 298 K increases the stability of the dispersions due to slowing down of the fusion process. A TEM micrograph of a vesicular dispersion taken after a week clearly shows (Fig. 41) fusion of vesicles prior to coagulation. X-ray crystallographic studies on the structures of some of the

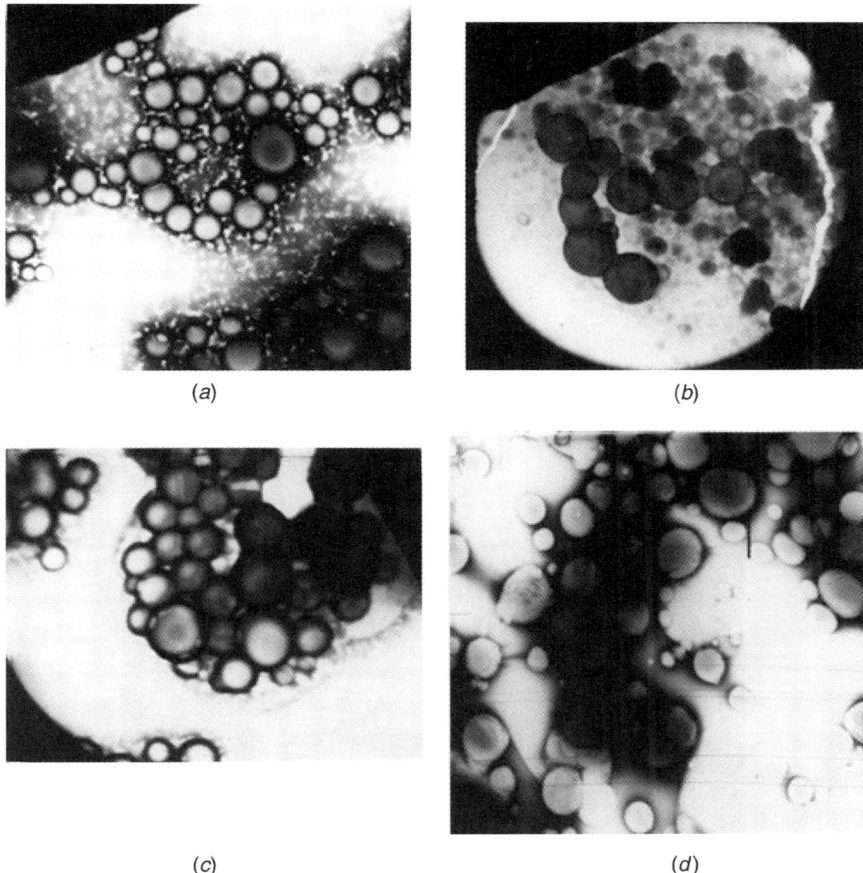

Figure 40. Representative negative-stain transmission electron micrographs of spontaneously made vesicles of **L$_{20}$** [(a), magnification, ×10000]; its Cu(II) inclusion complex [(b), magnification, ×12000]; micrographs of vesicles made by extrusion method of **L$_{20}$** [(c), magnification, ×20000]; its Cu(II) inclusion complex [(d), magnification, ×20000].

TABLE X

Turbidity Measurements of Spontaneously Made Vesicles of L_{17}–L_{29} Monitored at 400 nm at 298 K

Amphiphile	Optical Density				
	fresh	1 day	2 days	4 days	7 days
L_{17}	0.70	0.85	0.75	0.65	0.10
L_{18}	0.65	0.75	0.55	0.40	0.08
L_{19}	0.72	0.60	0.45	0.25	0.08
L_{20}	0.77	0.97	1.00	1.20	1.40
L_{21}	0.82	0.60	0.20	0.04	0.04
L_{22}	0.89	0.07			
L_{23}	0.75	0.65	0.50	0.45	0.08
L_{24}	0.70	0.75	0.60	0.40	0.05
L_{25}	0.72	0.70	0.55	0.35	0.05
L_{26}	0.85	0.65	1.00	0.70	0.08
L_{27}	0.90	0.95	0.60	0.45	0.05
L_{28}	0.90	0.95	0.50	0.30	0.05
L_{29}	0.90	0.05			

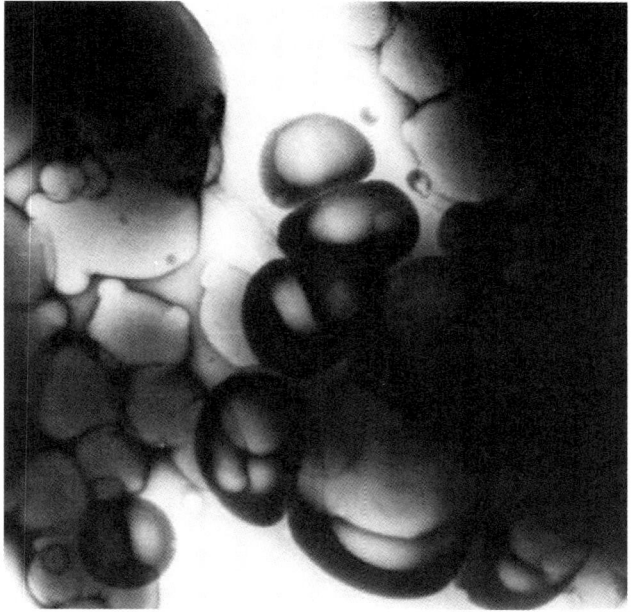

Figure 41. Representative negative-stain transmission electron micrograph of the spontaneously made vesicles of L_{20} ($\times 10000$) after 1 week.

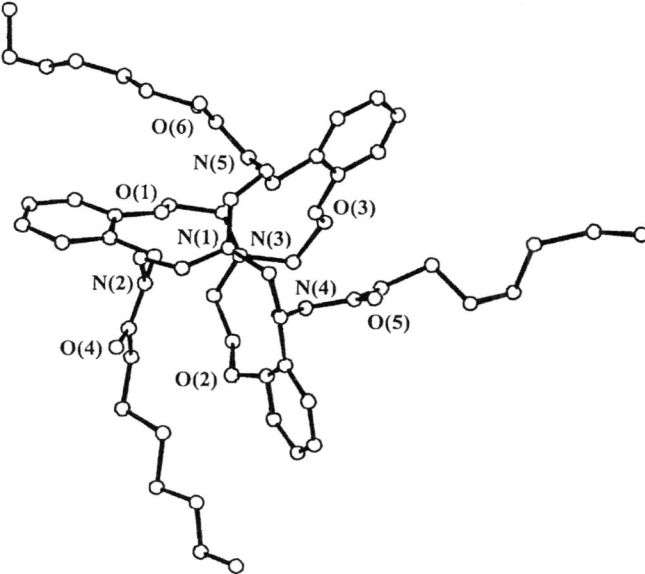

Figure 42. A perspective view of L_{20} along the pseudo-threefold axis passing through the two bridgehead nitrogen atoms.

amphiphiles (180, 181) show that the hydrophobic chains favor curvature (Fig. 42). These surfactants pack in the lattice in an intermolecular digitizing and nondigitizing fashion in the crystal lattice, which is not common with amphiphilic molecules (186–188).

Two out of three amino groups in **7** and **8** can be protected by the *tert*-butoxycarbonyl (Boc) group and the third amino group can be treated with acid chlorides to generate a new class of amphiphiles (Scheme 8). These amphiphiles also form vesicular dispersions (189) spontaneously when an alcoholic solution is dispersed in water. However, unlike the previous cases, the vesicular dispersions are not stable and within hours they coagulate underlining the need for an optimum hydrophilic–hydrophobic balance to make stable vesicles. Upon treatment with diacid chlorides of different chain lengths, **7** gives a new class of amphiphiles where two cryptand headgroups (190) are connected via spacers (Fig. 43) to form what is known as a bola-amphiphile, where the noun "bola" relates to the shape of a South American missile weapon (191). This molecule forms very thin monolayer lipid membranes (MLMs) reproducing the unusual architecture of natural *archaeal* macrocyclic bola-amphiphilic lipids (192). These natural molecules offer several advantages for the construction of liposomes that are characterized by high mechanical and thermal stabilities due

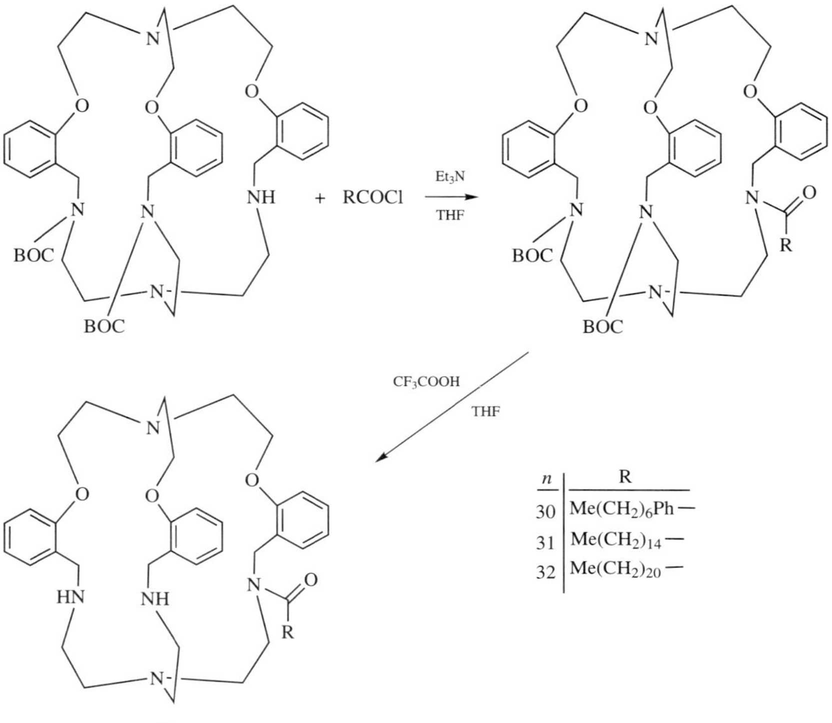

Scheme 8

n	R
30	Me(CH$_2$)$_6$Ph—
31	Me(CH$_2$)$_{14}$—
32	Me(CH$_2$)$_{20}$—

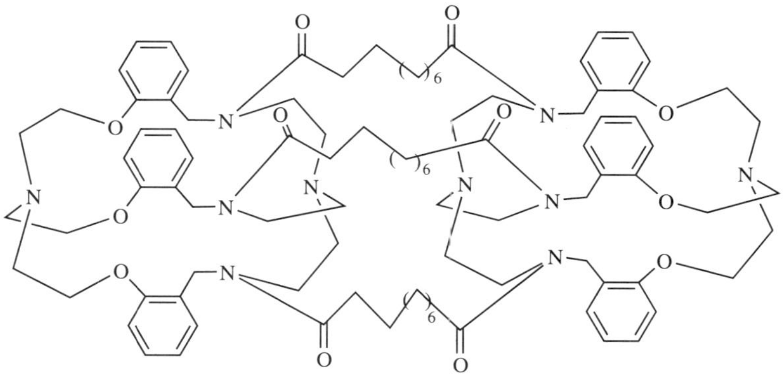

L$_{33}$

Figure 43. Chemical structure of a cryptand based Gemini amphiphile.

to the organization of the membrane (184). The bola-amphiphile from **7** form vesicular structures with an average diameter of ∼500 nm that are stable for almost 1 week at 298 K.

Unsymmetrical derivatization of a cryptand can afford yet another class of compounds known as gemini-amphiphiles (193) as shown in Scheme 9. However, vesicles from these surfactant molecules are not stable for more than 1 h at 298 K.

B. Amphiphiles at the Air–Water Interface

1. Langmuir–Blodgett Film

A wealth of useful information starting from molecular sizes to intermolecular forces can be obtained from studies of monolayers at the air–water interface. Great resurgence of interest in this area of science has been largely due to the fact that films can be transferred from the water surface onto a solid substrate using what has become universally known as the Langmuir–Blodgett (LB) technique. The LB technique (194) is one of the most effective ways of depositing extremely thin films of amphiphiles with precise molecular dimension and high structural order. The LB technique of monolayer transfer has often been used for the construction of highly ordered ultrathin films. This technique has assumed greater importance in recent times (195–202) with the demand for materials with tailored interfacial properties. Besides, organized molecular layers provide unique environments for molecular interactions and consequently for molecular recognition (203). Using monolayer assemblies that may possess characteristics uniquely different from those in homogeneous media can develop new molecular recognition systems. These supramolecular systems should be important in applications such as chemical sensors, in understanding molecular interactions on biological cell surfaces, and in developing novel 2D molecular assemblies composed of multiple chemical species (204). Cryptand-based amphiphiles are attractive because they have a closed cavity with donor atoms whose topology can be tailored via ligand design to recognize a specific metal ion–molecular guest. The amphiphiles shown in Scheme 7 give stable monolayers (181). The patterns of the isotherms are shown in Fig. 44. The Cu(II) complexed amphiphiles exhibit similar isotherms. Several generalization can be made on the nature of the isotherms vis a vis the amphiphiles. The stability of the monolayers decreases monotonically with the decrease in the length of the hydrophobic chains. The monolayers made from L_{17} and L_{24} can sustain a surface pressure of 25 mN m^{-1} for over 30 min without any significant area loss. As the length of the hydrophobic chain increases, the monolayer can sustain higher surface pressure due to better packing of the amphiphiles. Interestingly, the Cu(II) complexed amphiphile can sustain higher pressure compared to the

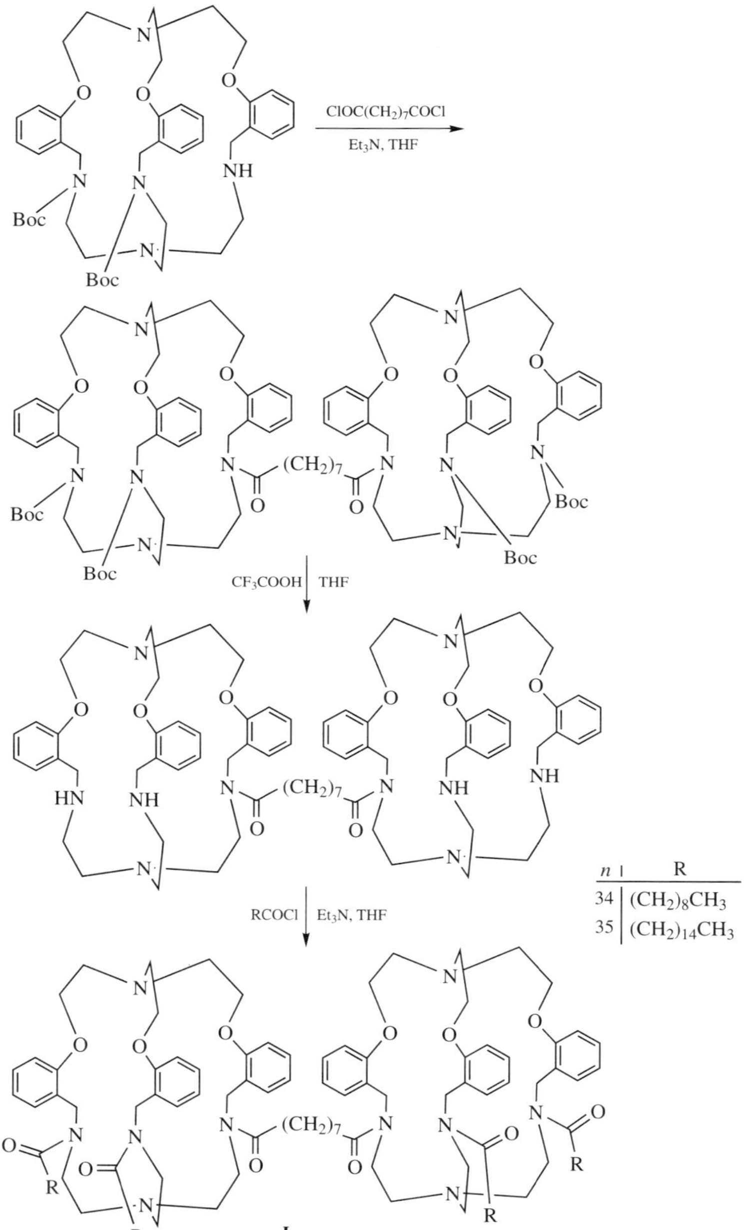

n	R
34	$(CH_2)_8CH_3$
35	$(CH_2)_{14}CH_3$

L$_n$

Scheme 9

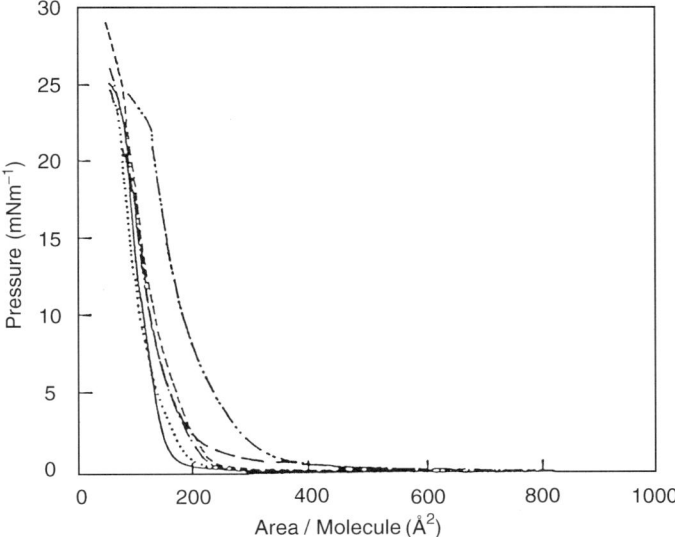

Figure 44. Pressure–area (Π–A) isotherms for L_{17}–L_{22} at 298 K: L_{17}(—), L_{18}(- - -), L_{19}(–·–), L_{20} (···), L_{21} (–··–), L_{22} (– – –).

corresponding metal-free amphiphile (Fig. 45). In all cases, the pressure–area isotherms are reproducible without any hysterisis (Fig. 46) due to the robust nature of the amphiphiles. The molecular radii of the amphiphiles are calculated to be in the range, 6.4–7.0, Å which are slightly larger compared to the average radius of the cryptand headgroup as determined from X-ray crystallography (180, 181). This suggests that the cryptand headgroups are slightly tilted with respect to the water surface reflecting the association tendencies and orientations of the hydrocarbon chains. The set of amphiphiles formed from **8** are able to sustain higher surface pressure compared to those formed with **7** (Fig. 47). Both sets of surfactants form stable mixed monolayers with stearic acid (SA) with the following characteristics: (1) when mixed with SA, maximum pressure sustained by the monolayer increases, (2) the shape of the isotherm becomes steeper, and (3) with increasing SA concentration in the spreading solution, limiting area–molecule decreases monotonically. All these observations can be rationalized by the fact that no attractive interactions exist between the two types of amphiphiles. The packing behavior of amphiphiles with small headgroup depends strongly on the nature of the hydrophobic chain(s). Stearic acid, which has a saturated alkyl chain, condenses to a solid-like incompressible film, whereas the presence of a double bond, as in oleic acid, leads to a relatively expanded compressible liquid-like film. The amphiphile L_{23} (Scheme 7), which

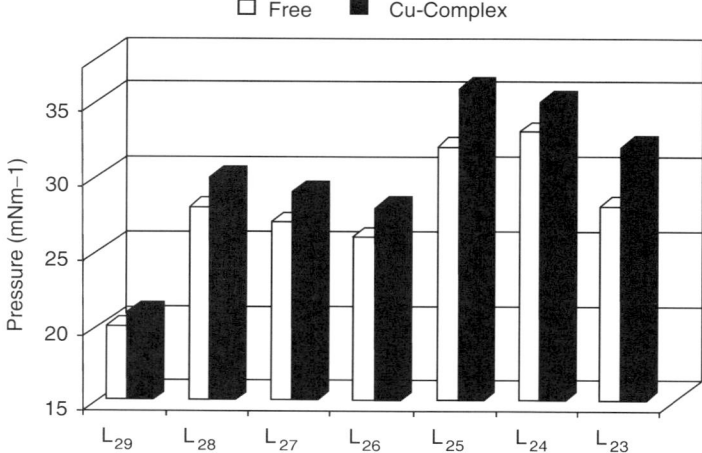

Figure 45. Comparison of maximum surface pressures (mN m^{-1}) sustained by **L$_{17}$–L$_{22}$** and their Cu(II) inclusion complexes.

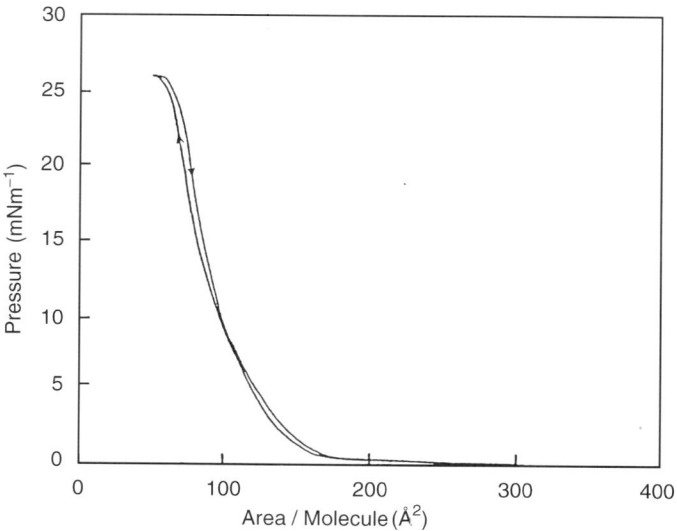

Figure 46. Representative successive compression and expansion cycles (hysteresis) with a monolayer of **L$_{18}$**.

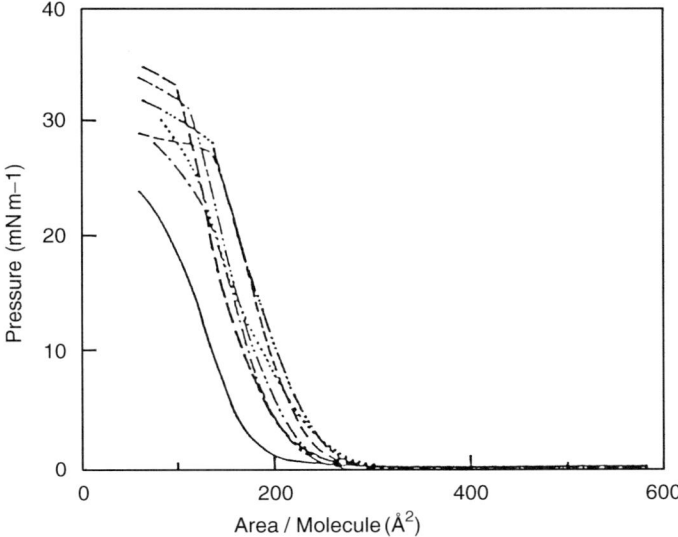

Figure 47. Pressure–area (Π–A) isotherms for L_{23}–L_{29} at 298 K: L_{23} (–···–), L_{24} (- - -), L_{25} (–··–), L_{26} (···), L_{27} (–···–), L_{28} (– – –), L_{29} (—).

has three oleoyl chains, however, does not show any significant differences in the packing behavior due to the cryptand headgroup being large and rigid.

The LB films for a few monolayers can be transferred onto substrates such as glass, quartz, indium–tin oxide (ITO) coated glass and fluorite with a Y-type deposition (194). The transfer ratio varied from 65 to 80% when the glass is untreated. However, when the glass is precoated with SA, the transfer ratio can reach up to 95%. Typically, 50 layers can be transferred without encountering any problem. The scanning electron micrographs of the LB films deposited onto ITO coated glass plates of free amphiphiles appear inhomogeneous [Fig. 48(a)] and remain so after a year exposed to air [Fig. 48(b)]. The Cu(II) complexed amphiphiles, on the other hand, show distinct platelike appearance [Fig. 48(c)] with sharp edges attributable (205) to ordered aggregates in the film that disappears and becomes inhomogeneous after 1 year at RT and exposed to air [Fig. 48(d)]. In triple-tailed amphiphiles (Scheme 7) the three hydrophobic tails cannot pack in a parallel fashion due to which they cannot sustain >25-mN m^{-1} surface pressure. Sustenance of high surface pressure by a monolayer signifies large attractive interactions between the amphiphiles with the formation of rigid films. It also indicates that the individual amphiphilic molecules pack effectively in the monolayer that is important for its eventual use(s) as materials. The

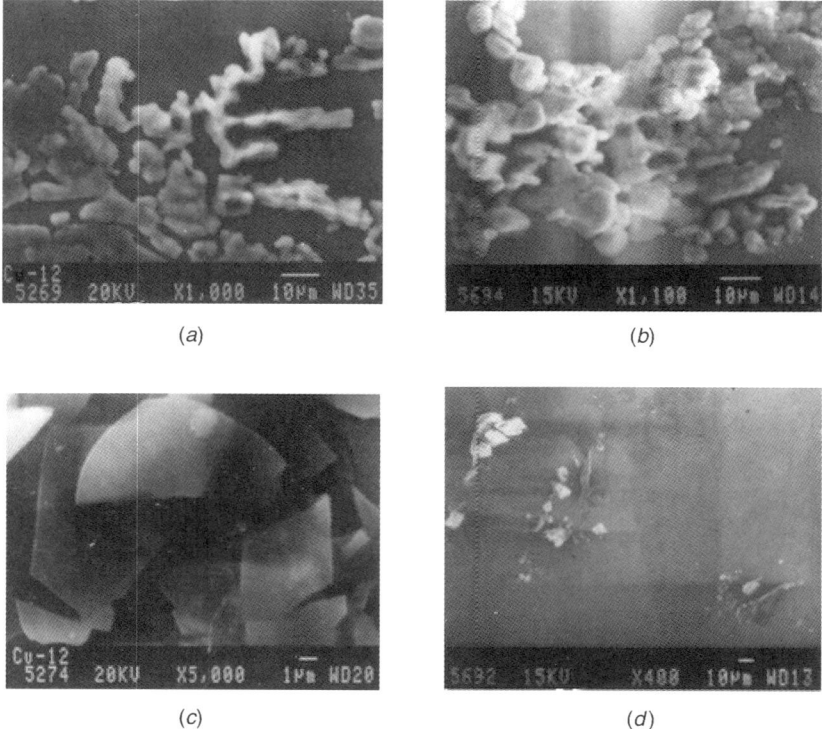

Figure 48. Representative scanning electron micrographs of the LB films of (a) L_{18}, (b) L_{18} after 12 months, (c) Cu(II) inclusion complex of L_{18}, and (d) Cu(II) inclusion complex of L_{18} after 12 months exposed to air.

mono-substituted cryptand-based amphiphiles (Scheme 8) exhibit isotherms shown in Fig. 49 and no hysterisis is observed on repeated expansion and compression experiments. Interestingly, the monolayers can sustain a pressure of $\sim 45\,\text{mN}\,\text{m}^{-1}$, which is significantly higher compared to the triple-tailed amphiphiles due to better packing ability of the former as envisaged. However, the big headgroup with a single hydrophobic tail reduces the stability of the monolayers and they can sustain a pressure of $40\,\text{mN}\,\text{m}^{-1}$ for only ~ 10 min. This instability reduces the transferability of the films to ITO coated glass plates to $\sim 60\%$ in a Y-type deposition. However, a mono-substituted cryptand-based amphiphile offers other possibilities; it can be a starting point for the attachment of fluorophoric or donor–acceptor groups that can lead to thin films with either sensing or NLO capabilities.

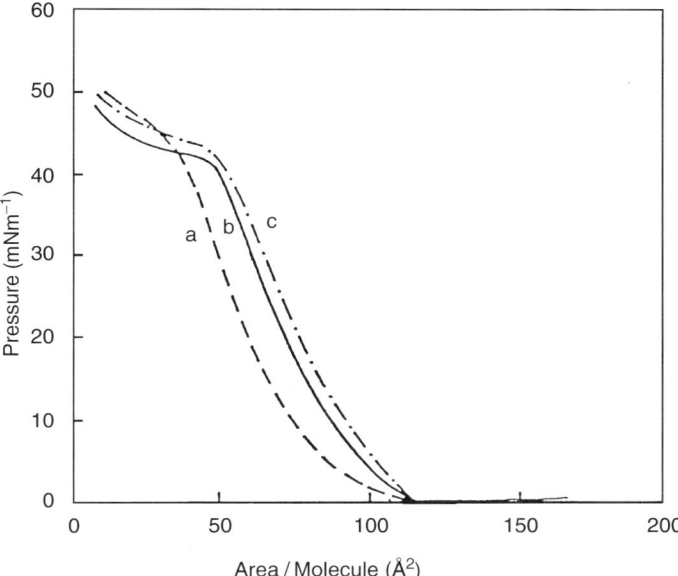

Figure 49. Pressure–area (Π–A) isotherms for **L₃₀**–**L₃₂** at 298 K: **L₃₀** (—), **L₃₁** (–·–), **L₃₂** (– – –).

2. Molecular Recognition

Molecular recognition relies upon the complementarity of size, shape, and intermolecular forces (206). It explores and exploits intermolecular forces, the week attractions that act over short distances between molecules. Hydrogen bonding is a highly directional secondary valence force compared with other noncovalent interactions such as electrostatic, van der Waals, and hydrophobic forces. It plays decisive roles in biological molecular recognition such as replication of nucleic acids, maintenance of the tertiary structure of proteins, and substrate recognition of enzymes. Intensive effort has been made recently to develop organic host molecules that specifically bind substrates by complementary hydrogen bonding. Unlike biological molecular recognition, most of the artificial systems are not effective in aqueous media due to strong hydrogen bonding with water. To circumvent this problem, several strategies have been applied for effective hydrogen bonding in aqueous media (207–210). Shimomura et al. (211) investigated the binding of aqueous nucleosides to a cytosine-functionalized monolayer. Fluorescence microscopy observation revealed that the monolayer produced spiral domains on aqueous guanosine. Hydrogen-bond

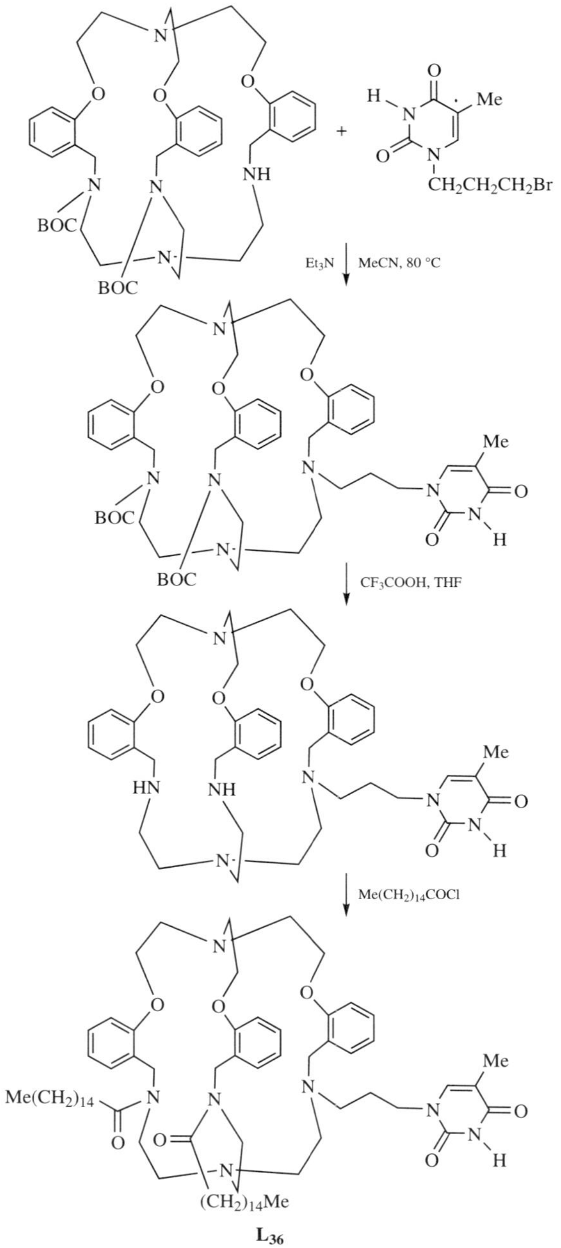

Scheme 10

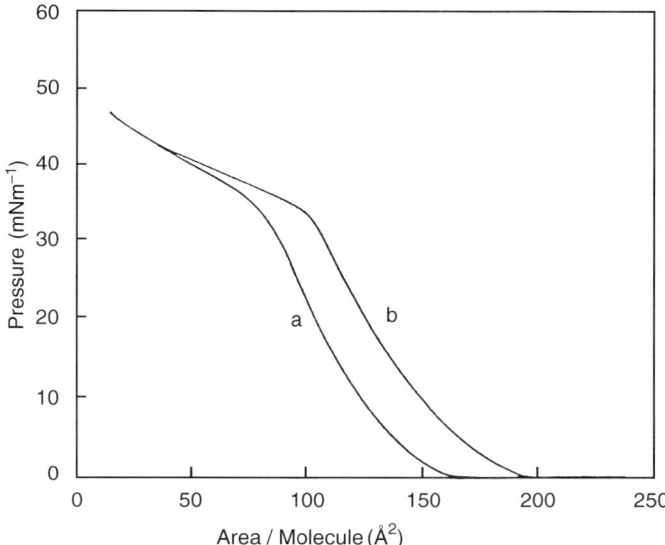

Figure 50. Pressure–area isotherms for L_{36} (a) pure water as subphase, (b) 0.1 M adenine in pure water as subphase at 298 K.

based recognition systems other than the nucleic acid mimics have also been found effective at the air–water interface (211–214).

Since the pioneering work on the base pair mimic at the air–water interface by Kitano and Ringsdorf (215), several studies have been made. They investigated pressure–area isotherms of an adenine-functionalized amphiphile on aqueous nucleosides and proposed (216) that larger expansion of the isotherm on aqueous thymidine is ascribable to formation of the complementary A-T type pair at the interface. Cryptands **6–11** with a derivatizable amino groups offer an opportunity to study molecular recognition at the air–water interface. Thus, a thymine group has been added to an amphiphile made from **7** following the synthetic Scheme 10.

It gives a well-behaved isotherm in pure water as the subphase. The isotherm is expanded (217) by ∼20% when adenine is present in the subphase (Fig. 50) due to the formation of A-T base pair at the interface. Such an expansion does not take place if any other nucleic acid base is present in the subphase.

VI. CONCLUSIONS

This chapter gives a brief account of the chemistry of laterally nonsymmetric aza-cryptands. Since these molecules can now be isolated in large amounts in pure form starting from easily available acyclic precursors, they open-up possibilities for their use in several applications. The secondary amino groups present can be derivatized to impart special properties to these molecules. While it has been possible to have transition metal induced fluorescent signaling systems with these cryptands, major efforts should now be directed to have reversible molecular photonic devices. Such devices are still not reported in the literature and if successful should be regarded as a big leap forward in the practical use of chemical logics in computing. Selective derivatization is possible with these cryptands that can lead to isolation of systems exhibiting complex chemical logics. When a metal ion is included in the cavity of a cryptand, it can be utilized, in principle, for signal transduction in the NLO active compounds for switchable NLO effects. Although a diverse range of organic as well as organometallic compounds are available with quadratic NLO properties, incorporation of switchability in an NLO active compound will definitely increase its potential for applications in the emerging optoelectronic and photonic technology. A few NLO active molecules are known that exhibit switching ability due to isomerization or tautomerism requiring a substantial change in their structures. A redox-active metal mediated switching is definitely a better option as far as reversibility and speed are concerned. Cryptand-based amphiphiles offer possibilities to have ultrathin films with tailored interfacial properties. They can also be useful in studying molecular recognition characteristics at the air–water interface mimicking the biological systems more closely. Moreover, if these cryptands were derivatized with groups incorporating metal donor sites they would provide an easy way to assemble a number of cryptands around a metal ion via exocyclic coordination forming cryptand dendrimers with a metal ion at the core. Besides, the metal ion coordinated outside the cavity can communicate with the one inside the cavity. Chemistry of compounds with exocyclic coordination is still not reported in the literature. Several ideas can be built around such supramolecular systems. It is now clear that this type of molecules would continue to attract a lot of attention in the near future and should prove valuable in many contemporary areas of research.

ACKNOWLEDGMENTS

Partial financial support for this work from the Department of Science and Technology, New Delhi, India is gratefully acknowledged. The author wishes to express his sincere thanks to all his students and co-workers over the years whose names appear in the references. Most of this chapter was written while the author was on a sabbatical leave at the Chemistry Department, Pohang University of Science and Technology, Republic of Korea. The author wishes to thank Professor Kimoon Kim for his kind cooperation and hospitality during the stay.

ABBREVIATIONS

A	Acceptor
azacapten	[structure]
bpy	2,2'-Bipyridine
biqn	3,3'-Bisisoquinoline
Boc	*tert*-Butoxycarbonyl
Cl,Me—N_4S_2sar	[structure]
Cl,Me—N_3S_3sar	[structure]
cylen	12-ane-N_4
D	Donor
1D	One dimensional
DLS	Dynamic light scattering
DMF	Dimethylformamide
DMSO	Dimethyl sulfoxide
EFISHG	Electric field-induced second harmonic generation
EPR	Electron paramagnetic resonance
esu	Electrostatic unit
H,Me—N_4S_2sar	[structure]
H,Me—N_3S_3sar	[structure]

HO,Me—N$_3$S$_3$sar

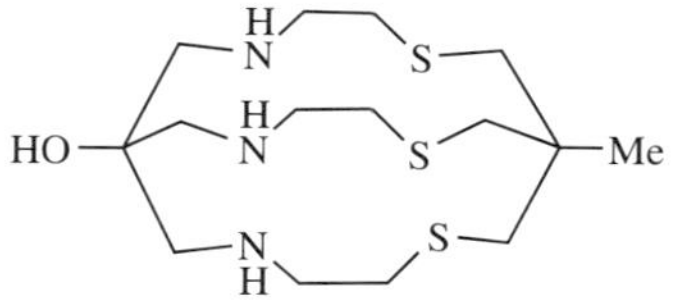

HOMO	Highest occupied molecular orbital
HRS	Hyper-Rayleigh scattering
ITO	Indium–tin oxide
IR	Infrared
LB	Langmuir–Blodgett
LUMO	Lowest unoccupied molecular orbital
MAP	Methyl(2,4-dinitrophenyl)aminopropanoate
MLM	Monolayer lipid membrane

Me,OHCH$_2$—N$_3$S$_3$absar

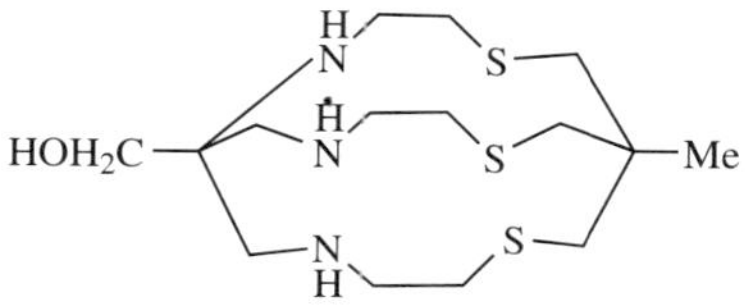

Me,ClCH$_2$-N$_3$S$_3$absar

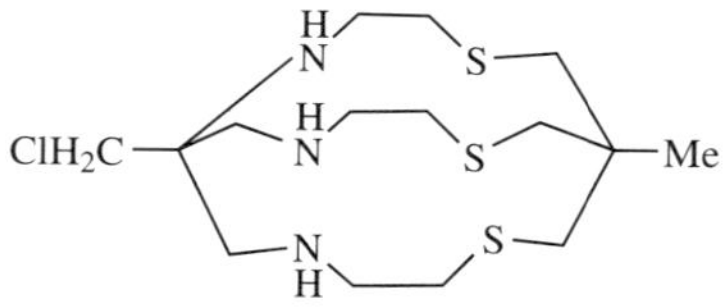

Me$_2$—N$_3$S$_3$absar

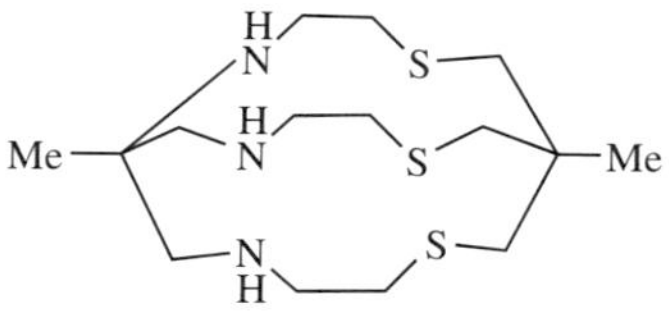

Me$_2$—N$_6$sar

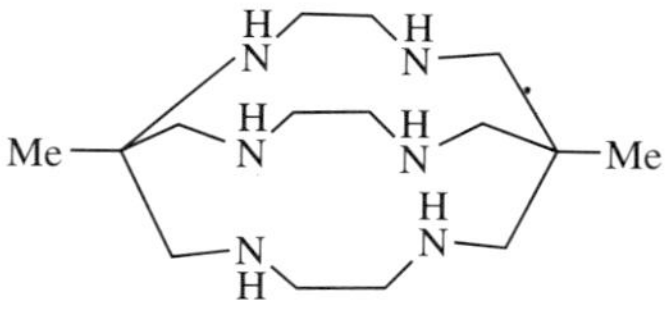

Me$_2$—S$_6$sar

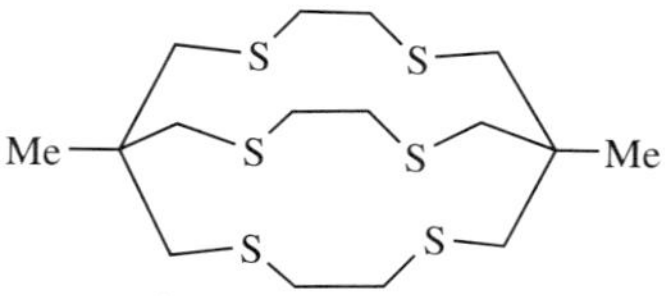

NLO	Nonlinear optics
NMP	N-Methyl-2-pyrrolidone.
NMR	Nuclear magnetic resonance
NO_2—capten	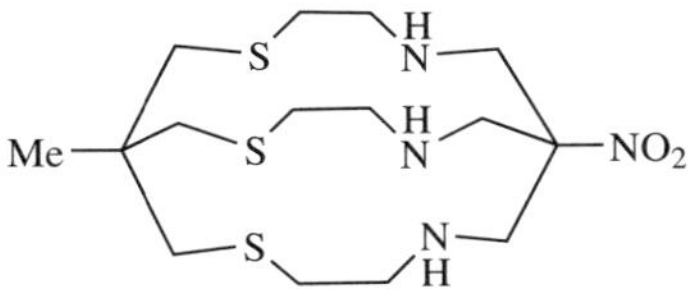 (structure shown)
NPP	4-Nitrophenyl-L-prolinol
PET	Photoinduced electron transfer
phen	1,10-Phenanthroline
pNA	p-Nitroaniline
PNP	Prolinolnitropyridine
pva	Poly(vinyl alcohol)
RT	Room temperature
SA	Stearic acid
SHE	Standard hydrogen electrode
SHG	Second harmonic generation
TATB	1,3,5-Triamino-2,4,6-trinitrobenzene
TEM	Tunneling electron microscopy
THF	Tetrahydrofuran
tren	Tris(2-aminoethyl)amine
trpn	Tris(3-aminopropyl)amine
UV–vis	Ultraviolet–visible

REFERENCES

1. (a) J. L. Atwood, J. E. D. Davies, D. D. MacNicol, F. Vögtle, and J.-M. Lehn, Eds., *Comprehensive Supramolecular Chemistry*, Pergamon, New York, 1996; (b) B. Dietrich, in *Inclusion Compounds*, J. L. Atwood, J. E. Davies, and D. D. MacNicol, Eds., Academic, New York, 1984; (c) Y. Inoue and G. W. Gokel, Eds., *Cation Binding by Macrocycles*, Marcel Dekker, New York, 1990.

2. (a) P. Chang and J. S. Shih, *Anal. Chim. Acta*, *380*, 55 (1999); (b) S. S. Tsai and J. S. Shih, *Separ. Sci. Technol.*, *33*, 1407 (1998).

3. J.-M. Lehn, *Supramolecular Chemistry—Concepts and Perspectives*, VCH Publishers, Weinheim, Germany, 1995.

4. (a) M. J. Wagner, A. S. Ichimura, R. H. Huang, R. C. Phillips, and J. L. Dye, *J. Phys. Chem. B*, *104*, 1078 (2000); (b) J. L. Dye, *Inorg. Chem.*, *36*, 3816 (1997).

5. F. Demol, F. X. Sauvage, A. Devos, and M. G. De Backer, *Synthetic Metals*, *99*, 155 (1999).

6. T. L. Niederhauser, J. Halling, N. A. Polson, and J. D. Lamb, *J. Chromatogr. A*, *804*, 69 (1998).

7. J. D. Lamb and M. D. Christenson, *J. Incl. Phenom. Mol. Recog.*, *32*, 107 (1998).

8. D. Tait, G. Haase, and A. Wiechen, *J. Radioanal. Nucl. Chem.*, **226**, 225 (1997).
9. R. Cacciapaglia, L. Mandolini, and V. V. Castelli, *J. Org. Chem.*, **62**, 3089 (1997).
10. R. Z. Gnann, R. I. Wagner, K. O. Christe, R. Bau, G. A. Olah, and W. W. Wilson, *J. Am. Chem. Soc.*, **119**, 112 (1997).
11. M. P. TeuladeFichou, J. P. Vigneron, and J.-M. Lehn, *J. Chem. Soc. Perkin Trans. 2*, 2169 (1996).
12. B. G. Keevil, S. P. Nichols, and D. P. Tierney, *Ann. Clin. Biochem.*, **33**, 459 (1996).
13. (a) D. K. Chand, H.-J. Schneider, A. Bencini, A. Bianchi, C. Giorgi, S. Ciattini, and B. Valtancoli, *Chem. Eur. J.*, **6**, 4001 (2000); (b) S. Moghaddas, P. Hendry, R. J. Geue, C. Qin, A. M. T. Bygott, A. M. Sargeson, and N. E. Dixon, *J. Chem. Soc. Dalton Trans.*, 2085 (2000).
14. G. D. Christian, *J. Pharmaceut. Biomed.*, **14**, 899 (1996).
15. T. R. Park and S. A. Solin, *Phys. Rev.*, **B53**, 11273 (1996).
16. S. K. Srivastava, V. K. Gupta, and S. Jain, *Anal. Chem.*, **68**, 1272 (1996).
17. M. C. Lonergan, M. A. Ratner, and D. F. Shriver, *J. Am. Chem. Soc.*, **117**, 2344 (1995).
18. P. B. Smith, J. L. Dye, J. Cheney, and J.-M. Lehn, *J. Am. Chem. Soc.*, **103**, 6044 (1981).
19. X. X. Zhang, R. M. Izatt, J. S. Bradshaw, and K. E. Krakowiak, *Coord. Chem. Rev.*, **174**, 179 (1998).
20. P. G. Potvin and J.-M. Lehn, in *Synthesis of Macrocycles: the Design of Selective Complexing Agents*, R. M. Izatt and J. J. Christensen, Eds., John Wiley & Sons, Inc., New York, 1987.
21. R. D. Hancock and A. E. Martell, *Chem. Rev.*, **89**, 1875 (1989).
22. L. F. Lindoy, *The Chemistry of Macrocyclic Ligands*, Cambridge University Press, Cambridge, UK, 1989.
23. (a) E. Weber and F. Vögtle, *Top. Curr. Chem.*, **98**, 1 (1981); (b) D. J. Cram and K. N. Trueblood, *Top. Curr. Chem.*, **98**, 43 (1981).
24. E. Graf and M. W. Hosseini, *Coord. Chem. Rev.*, **178**, 1193 (1998).
25. J. Nelson, V. McKee, and G. Morgan, *Prog. Inorg. Chem.*, **47**, 167 (1998).
26. V. Alexander, *Chem. Rev.*, **95**, 273 (1995).
27. D. H. Busch, *Chem. Rev.*, **93**, 847 (1993).
28. H. An, J. S. Bradshaw, and R. M. Izatt, *Chem. Rev.*, **92**, 543 (1992).
29. A. E. Martell, R. D. Hancock, and R. J. Motekaitis, *Coord. Chem. Rev.*, **392**, 39 (1994).
30. For earlier references see J.-M. Lehn, *Pure Appl. Chem.*, **50**, 871 (1978); F. Montanari, D. Landini, and F. Rolla, *Top. Curr. Chem.*, **101**, 147 (1982); F. Vögtle, H. Sieger, and W. M. Muller, *Top. Curr. Chem.*, **98**, 107 (1981).
31. B. Vallee and R. J. P. Williams, *Proc. Natl. Acad, Sci. USA*, **59**, 498 (1968).
32. (a) C. H. Parks and H. E. Simmons, *J. Am. Chem. Soc.*, **90**, 2428 (1968); (b) B. Dietrich, J.-M. Lehn, and J.-P. Sauvage, *Tetrahedron Lett.*, 2885 (1969).
33. B. Dietrich, J.-M. Lehn, J.-P. Sauvage, and J. Blanzat, *Tetrahedron*, **29**, 1629 (1973); (b) S. Karbech, W. Lohr, and F. Vögtle, *J. Chem. Res. (S)*, 314 (1981).
34. P. Ruggli, *Liebigs Ann. Chem.*, **92**, 392 (1912).
35. K. Ziegler, *Methoden der Organischen Chemie*, Georg Thieme, Stuttgart, Germany, 729 (1955).
36. G. Illuminati and L. Mandolini, *Acc. Chem. Res.*, **14**, 95 (1981).
37. (a) L. Rossa and F. Vögtle, *Top. Curr. Chem.*, 1123 (1983); (b) P. Knops, N. Sendhoff, H.-B. Mekelburger, and F. Vögtle, *Top. Curr. Chem.*, **161**, 1 (1991).
38. D. H. Busch, *J. Incl. Phenom.* **12**, 389 (1992).

39. D. H. Busch and C. Cairns, in *Progress in Macrocyclic Chemistry*, R. M. Izatt and J. J. Christensen, Eds., John Wiley & Sons, Inc., New York, Vol. 3, 1987, p. 1.
40. N. F. Curtis, *Coord. Chem. Rev.*, *3*, 3 (1968).
41. T. J. Hubin and D. H. Busch, *Coord. Chem. Rev.*, *200*, 5 (2000) and references cited therein.
42. A. M. Sargeson, *Pure Appl. Chem.*, *50*, 905 (1978).
43. W. Baker, J. F. W. McOmie, and W. D. Ollis, *J. Chem. Soc.*, 200 (1951).
44. P. H. Smith, M. E. Barr, J. R. Brainard, D. K. Ford, H. Freiser, S. Muralidharan, S. D. Reilly, R. R. Ryan, L. A. Silks III, and W. Yu, *J. Org. Chem.*, *58*, 7939 (1993).
45. J.-M. Lehn, J. Simon, and J. Wagner, *Angew. Chem. Int. Ed. Engl.*, *12*, 578 (1973).
46. K. G. Ragunathan and P. K. Bharadwaj, *Tetrahedron Lett.*, *33*, 7581 (1992).
47. P. Ghosh, S. Sengupta, and P. K. Bharadwaj, *J. Chem. Soc. Dalton Trans.*, 935 (1997).
48. D. K. Chand and P. K. Bharadwaj, *Inorg. Chem.*, *35*, 3580 (1996).
49. D. K. Chand and P. K. Bharadwaj, *Tetrahedron*, *53*, 10517 (1997).
50. P. Ghosh. R. Shukla, D. K. Chand, and P. K. Bharadwaj, *Tetrahedron*, *51*, 3265 (1995).
51. G. Das, P. Tripathi, A. Tripathi, and P. K. Bharadwaj, *Tetrahedron*, *56*, 1501 (2000).
52. P. Lipkowski, D. T. Gryko, J. Jurczak, and J. Lipkowski, *Tetrahedron Lett.*, *39*, 3833 (1998).
53. D. K. Chand and P. K. Bharadwaj, *Tetrahedron Lett.*, *37*, 8443 (1996).
54. R. J. Geue, T. W. Hambley, J. McB. Harrowfield, A. M. Sargeson, and M. R. Snow, *J. Am. Chem. Soc.*, *106*, 5478 (1984).
55. L. R. Gahan, T. W. Hambley, A. M. Sargeson, and M. R. Snow, *Inorg. Chem.*, *21*, 2699 (1982).
56. T. M. Donlevy, L. R. Gahan, T. W. Hambley, and R. Stranger, *Inorg. Chem.*, *31*, 4376 (1992).
57. J. I. Bruce, L. R. Gahan, T. W. Hambley, and R. Stranger, *Inorg. Chem.*, *32*, 5997 (1993).
58. P. Osvath, A. M. Sargeson, A. McAuley, R. E. Mendelez, S. Subramanian, M. J. Zaworotko, and L. Broge, *Inorg. Chem.*, *38*, 3634 (1999).
59. P. M. Angus, A. M. Sargeson, and A. C. Willis, *Chem. Commun.*, 1975 (1999).
60. L. R. Gahan, G. A. Lawrance, and A. M. Sargeson, *Inorg. Chem.*, *23*, 4369 (1984).
61. (a) C. Königstein, A. W. H. Mau, P. Osvath, and A. M. Sargeson, *Chem. Commun.*, 423 (1997); (b) I. I. Creaser, L. R. Gahan, R. J. Geue, A. Launikonis, P. A. Lay, J. D. Lydon, M. G. McCarthy, A. W.-H. Mau, A. M. Sargeson, and W. H. F. Sasse, *Inorg. Chem.*, *24*, 2671 (1985); (c) R. V. Dubs, L. R. Gahan, and A. M. Sargeson, *Inorg. Chem.*, *22*, 2523 (1983).
62. P. A. Lay, J. Lydon, A. W.-H. Mau, P. Osvath, A. M. Sargeson, and W. H. F. Sasse, *Aust. J. Chem.*, *46*, 641 (1993).
63. T. M. Donlevy, L. R. Gahan, T. W. Hambley, G. R. Hanson, K. L. McMahon, and R. Stranger, *Inorg. Chem.*, *33*, 5131 (1994).
64. P. Bernhard, D. J. Bull, W. T. Robinson, and A. M. Sargeson, *Aust. J. Chem.*, *45*, 1241 (1992).
65. P. Bernhard and A. M. Sargeson, *J. Am. Chem. Soc.*, *111*, 597 (1989).
66. J. I. Bruce, L. R. Gahan, T. W. Hambley, and R. Stranger, *Chem. Comm.*, 702 (1993).
67. C. Bazzicalupi, P. Bandyopadhyay, A. Bencini, A. Bianchi, P. K. Bharadwaj, C. Giorgi, D. Bharadwaj, B. Valtancoli, and R. J. Butcher, *Eur. J. Inorg. Chem.*, 2111 (2000).
68. R. M. Smith and A. E. Martell, *NIST Critical Stability Constants Database, version 4.0*, 1997.
69. D. K. Chand and P. K. Bharadwaj, *Inorg. Chem.*, *36*, 5658 (1997).

70. P. Ghosh, S. Sengupta, and P. K. Bharadwaj, *J. Chem. Soc. Dalton Trans.*, 935 (1997).
71. P. Bandyopadhyay and P. K. Bharadwaj, unpublished result.
72. J.-M. Lehn, *Science*, 227, 849 (1985).
73. A. P. de Silva, H. Q. N. Gunaratne, T. Gunnlaugsson, A. J. M. Huxley, C. P. McCoy, J. T. Rademacher, and T. E. Rice, *Chem. Rev.*, 97, 1515 (1997).
74. A. Weller, *Pure Appl. Chem.*, 16, 115 (1968).
75. R. A. Marcus, *Angew. Chem. Int. Ed. Engl.*, 32, 1111 (1993).
76. B. Valeur and I. Leray, *Coord. Chem. Rev.*, 205, 3 (2000).
77. F. Szurdoki, D. Ren, and D. R. Walt, *Anal. Chem.*, 72, 5250 (2000).
78. J.-P. Desvergne and A. W. Czarnik, Eds., *Chemosensors of Ion and Molecule Recognition*, NATO ASI Series, Kluwer, Dordrecht, The Netherlands, 1997; A. W. Czarnik, Ed., *Fluorescent Chemosensors for Ion and Molecule Recognition*, ACS Symposium Series 538, American Chemical Society, Washington DC, 1993.
79. L. Prodi, F. Bolletta, M. Montalli, and N. Zaccheroni, *Coord. Chem. Rev.*, 205, 59 (2000).
80. R. Y. Tsien, *Biochemistry*, 19, 2396 (1980); R. Y. Tsien, *Annu. Rev. Neurosci.*, 12, 227 (1989).
81. A. W. Czarnik, *Acc. Chem. Res.*, 27, 302 (1994).
82. M.-Y. Chae, J. Yoon, and A. W. Czarnik, *J. Mol. Recognit.*, 6, 297 (1996).
83. (a) Z. H. Lin, K. S. Booksh, L. W. Burgess, and B. R. Kowalski, *Anal. Chem.*, 66, 2552 (1994); (b) M. G. Schweyer, J. C. Andle, D. J. McAllister, and J. F. Vetelino, *Sens Actuators B*, 35, 170 (1996); (c) I. Oehme and O. S. Wolfbeis, *Mikrochim. Acta*, 126, 177 (1997); (d) J. Herdan, R. Feeney, S. P. Kounaves, A. F. Flannery, C. W. Storment, G. T. A. Kovacs, and R. B. Darling, *Environ. Sci. Technol.*, 32, 131 (1998).
84. O. S. Wolfbeis, *Fibre Optic Chemical Sensors and Biosensors, Vol. I-II*, CRC Press, Boca Raton, FL, 1991.
85. J. J. R. F. da Silva and R. J. P. Williams, *The Biological Chemistry of the Elements*, Clarendon Press, Oxford, UK, 1993.
86. P. G. Collins, M. S. Arnold, and P. Avouris, *Science*, 292, 706 (2001).
87. A. P. de Silva, H. Q. N. Gunaratne, and C. P. McCoy, *Nature (London)*, 364, 42 (1993).
88. M. Inouye, K. Akamatsu, and H. Nakazumi, *J. Am. Chem. Soc.*, 119, 9160 (1997).
89. H.-F. Ji, R. Dabestani, and G. M. Brown, *J. Am. Chem. Soc.*, 122, 9306 (2000).
90. F. Pina, M. Maestri, and V. Balzani, *Chem. Commun.*, 107 (1999).
91. A. Rouque, F. Pina, S. Alves, R. Ballardini, M. Maestri, and V. Balzani, *J. Mater. Chem.*, 9, 2265 (1999).
92. A. P. de Silva, I. M. Dixon, H. Q. N. Gunaratne, T. Gunnlaugsson, P. R. S. Maxwell, and T. E. Rice, *J. Am. Chem. Soc.*, 121, 1393 (1999).
93. T. Gunnlaugsson, D. A. MacDonail, and D. Parker. *Chem. Commun.*, 93 (2000).
94. (a) A. Credi, V. Balzani, S. J. Langford, and J. F. Stoddart, *J. Am. Chem. Soc.*, 119, 2679 (1997); (b) A. P. de Silva and N. D. McClenaghan, *J. Am. Chem. Soc.*, 122 3965, (2000).
95. F. M. Raymo and S. Giordani, *J. Am. Chem. Soc.*, 123, 4651 (2001).
96. (a) A. W. Varnes, R. B. Dodson, and E. L. Wehry, *J. Am. Chem. Soc.*, 94, 946 (1972); (b) J. A. Kemlo and T. M. Shepard, *Chem. Phys. Lett.*, 47, 158 (1977).
97. (a) B. Ramachandran and A. Samanta, *Chem. Commun.*, 1037 (1997); (b) B. Ramachandran and A. Samanta, *J. Phys. Chem. A*, 102, 10579 (1998); (c) B. Ramachandran, G. Saroja, N. B. Sankaran, and A. Samanta, *J. Phys. Chem. B*, 104, 11824 (2000).

98. K. A. Mitchell, R. G. Brown, D. Yuan, S.-C. Chang, R. E. Utecht, and D. E. Lewis, *J. Photochem. Photobiol. A*, *115*, 157 (1998).
99. (a) L. Fabbrizzi, M. Licchelli, P. Pallavicini, A. Perotti, A. Taglietti, and D. Sacchi, *Chem. Eur. J.*, *2*, 75 (1996); (b) J. A. Sclafani, M. T. Maranto, T. M. Sisk, and S. A. Van Arman, *Tetrahedron Lett.*, *13*, 2193 (1996).
100. M. E. Huston, K. W. Haider, and A. W. Czarnik, *J. Am. Chem. Soc.*, *110*, 4460 (1988).
101. M. Costa, L. Fabbrizzi, M. Licchelli, P. Pallavicini, L. Parodi, L. Prodi, F. Bolletta, M. Montalti, and N. Zaccheroni, *J. Chem. Soc. Dalton Trans.*, 1381 (1999).
102. (a) P. Ghosh, P. K. Bharadwaj, S. Mandal, and S. Ghosh, *J. Am. Chem. Soc.*, *118*, 1553 (1996); (b) P. Ghosh, P. K. Bharadwaj, J. Roy, and S. Ghosh, *J. Am. Chem. Soc.*, *119*, 11903 (1997).
103. M. Gordon and W. R. Ware (Eds.), *The Exciplex*, Academic, New York, 1975.
104. (a) F. Fages, J. P. Desvergne, and H. B. Laurent, *J. Am. Chem. Soc.*, *111*, 96 (1989); (b) F. Fages, J. P. Desvergne, H. B. Laurent, P. Marsau, J.-M. Lehn, F. K. Hibert, A. M. A. Gary, and M. A. Joubbeh, *J. Am. Chem. Soc.*, *111*, 8672 (1989).
105. H. Beens, H. Knibbe, and A. Weller, *J. Chem. Phys.*, *47*, 1183 (1967).
106. J. B. Birks, *Photophysics of Aromatic Molecules*, Wiley-Interscience, New York, 1970.
107. J.-C. Rodriguez-Ubis, B. Alpha, D. Plancherel, and J.-M. Lehn, *Helv. Chim. Acta*, *67*, 2264 (1984).
108. B. Alpha, V. Balzani, J.-M. Lehn, S. Perathoner, and N. Sabbatini, *Angew. Chem. Int. Ed. Engl.*, *26*, 1266 (1987).
109. D. Parker and J. A. G. Williams, *Chem. Commun.*, 245 (1998).
110. D. Parker, K. P. Senanayake, and J. A. G. Williams, *Chem. Commun.*, 1777 (1997).
111. A. P. de Silva, H. Q. N. Gunaratne, and T. E. Rice, *Angew. Chem. Int. Ed. Engl.*, *35*, 2116 (1996).
112. G. Wilkinson, R. D. Gillard, and J. A. McCleverty, Eds., *Comprehensive Coordination Chemistry*, Pergamon Press, Oxford, UK, 1987, p. 83 Vol 1.
113. J. P. Konopelski, F. K. Hibert, J.-M. Lehn, J. P. Desvergne, F. Fages, A. Castellan, and H. B. Laurent, *J. Chem. Soc. Chem. Commun.*, 433 (1985).
114. D. K. Chand, P. K. Bharadwaj, and H.-J. Schneider, *Tetrahedron*, *57*, 6727 (2001).
115. G. Das, P. K. Bharadwaj, M. B. Roy, and S. Ghosh, *J. Photochem. Photobiol. A*, *135*, 7 (2000).
116. T. C. Werner and J. Rodgers, *J. Photochem. 32*, 59 (1986).
117. R. J. Strugeon, S. G. Schulman, *J. Pharma. Sci.*, *65*, 1833 (1976).
118. R. S. Shon, D. O. Cowan, and W. W. Schmiegel, *J. Phys. Chem.*, *79*, 2087 (1975).
119. T. C. Werner in *Modern Fluorescence Spectroscopy*, E. L. Wehry, Ed., Plenum Press, New York, 1976, Vol. 2.
120. T. C. Werner and R. M. Hoffman, *J. Phys. Chem.*, *77*, 1611 (1973).
121. G. J. Kavarnos and N. J. Turro, *Chem. Rev.*, *86*, 401 (1986).
122. P. Bandyopadhyay, P. K. Bharadwaj, M. B. Roy, R. Dutta and S. K. Ghosh, *Chem. Phys.*, *255*, 325 (2000).
123. (a) A. M. Swinnen, M. van der Auweraer, F. C. de Schryver, M. Katikani, T. Okada, and N. Mataga, *J. Am. Chem. Soc.*, *109*, 321 (1987); (b) F. D. Lewis, J. M. Wagner-Brennman, and J. M. Denari, *J. Phys. Chem. A*, *102*, 519 (1998).
124. (a) A. Ueno, I. Suzuki, and T. Osa, *J. Am. Chem. Soc.*, *111*, 6391 (1989); (b) C. Parker and C. G. Hatchard, *Trans. Faraday Soc.*, *59*, 284 (1963); (c) T. Jin, K. Ichikawa, and T. Koyama, *Chem. Commun.*, 499 (1992); (d) K. Kano, H. Matsumoto, Y. Yoshimura, and S. Hashimoto, *J. Am. Chem. Soc.*, *110*, 204 (1988).

125. M. B. Roy, S. Ghosh, P. Bandyopadhyay, and P. K. Bharadwaj, *J. Luminesc.*, 92, 115 (2001).
126. I. R. Gould, P. L. Kuo, and N. J. Turro, *J. Phys. Chem.*, 89, 3030 (1985).
127. A. M. Brun, A. Harriman, Y. Tsuboi, T. Okada, and N. Mataga, *J. Chem. Soc. Faraday Trans.*, 91, 4047 (1995).
128. K. Bhattacharya and M. Chowdhury, *Chem. Rev.*, 93, 507 (1993).
129. G. Das, S. Ghosh, and P. K. Bharadwaj, unpublished result.
130. S. R. Marder, B. Kippelen, A. K.-Y. Jen, and N. Peyghambarian, *Nature (London)*, 388, 845 (1997).
131. T. Verbiest, S. Houbrechts, M. Kauranen, K. Clays, and A. Persoons, *J. Mater. Chem.*, 7, 2175 (1997).
132. S. R. Marder, J. S. Sohn, and G. D. Stucky, *Materials for Nonlinear Optics: Chemical Perspectives*, ACS Symposium Series, 455 (1991).
133. J. Zyss, *Molecular Nonlinear Optics: Materials, Physics and Devices*, Academic, Boston, 1994.
134. D. S. Chemla and J. Zyss, Eds., *Nonlinear Optical Properties of Organic Molecules and Crystals*, Academic, Orlando, FL, 1987.
135. N. Peyghambarian, S. W. Koch, and A. Mysyrowicz, *Introduction to Semiconductor Optics*, Prentice Hall, Englewood Cliffs, NJ, 1993.
136. P. Prasad and D. J. Williams, *Introduction to Nonlinear Optical Effects in Molecules and Polymers*, John Wiley & Sons, New York, 1991.
137. J. Zyss and I. Ledoux, *Chem. Rev.*, 94, 77 (1994).
138. J.-L. Oudar and D. S. Chemla, *J. Chem. Phys.*, 66, 2664 (1977); J. L. Oudar, *J. Chem. Phys.*, 67, 446 (1977); R. Wortmann, P. Kramer, C. Glania, S. Lebus, and N. Detzer, *Chem. Phys.*, 173, 99 (1993).
139. L.-T. Cheng, W. Tam, S. H. Stevenson, G. R. Meredith, G. Rikken, and S. R. Marder, *J. Phys. Chem.*, 95, 10631 (1991).
140. J. Zyss, *Nonlinear Opt.*, 1, 3 (1991).
141. (a) M. Joffre, D. Yaron, R. J. Silbey, and J. Zyss, *J. Chem. Phys.*, 97, 5607 (1992); (b) J. L. Bredas, F. Meyers, B. M. Pierce, and J. Zyss, *J. Am. Chem. Soc.*, 114, 4928 (1992); (c) J. Zyss, T. C. Van, C. Dhenaut, and I. Ledoux, *Chem. Phys.*, 177, 281 (1993); (d) A. Sastre, T. Torres, M. A. Diaz-Garcia, F. Agullo-Lopez, C. Dhenaut, S. Brasselet, I. Ledoux, and J. Zyss, *J. Am. Chem. Soc.*, 118, 2746 (1996).
142. S. Stadler, F. Feiner, C. Brauchle, S. Brandl, and R. Gompper, *Chem. Phys. Lett.*, 245, 292 (1995).
143. R. Wortmann, C. Glania, P. Kramer, R. Matschiner, J. J. Wolff, S. Craft, B. Treptow, E. Barbu, D. Langle, and G. Gorlitz, *Chem. Eur. J.*, 3, 1765 (1997).
144. I. Ledoux, J. Zyss, J. Siegel, J. Brienne, and J.-M. Lehn, *Chem. Phys. Lett.*, 172, 440 (1990).
145. T. Verbiest, K. Clays, C. Samien, J. Wolff, D. N. Reinhoudt, and A. Persoons, *J. Am. Chem. Soc.*, 116, 9320 (1994).
146. (a) V. R. Thalladi, S. Brasselet, H.-C. Weiss, D. Blaser, A. K. Katz, H. L. Carrell, R. Boese, J. Zyss, A. Nangia, and G. R. Desiraju, *J. Am. Chem. Soc.*, 120, 2563 (1998); (b) V. R. Thalladi, R. Boese, S. Brasselet, I. Ledoux, J. Zyss, R. K. R. Jetti, and G. R. Desiraju, *Chem. Commun.*, 1639 (1999).
147. P. Mukhopadhyay, P. K. Bharadwaj, G. Savitha, A. Krishnan, and P. K. Das, *Chem. Commun.*, 1815 (2000).
148. P. Mukhopadhyay, P. K. Bharadwaj, G. Savitha, A. Krishnan, and P. K. Das, in press.

149. (a) R. E. Parker, *Adv. Fluorine Chem.*, *3*, 63 (1963); (b) H. E. Smith, W. I. Cozart, T. de Paulis, and F. M. Chen, *J. Am. Chem. Soc.*, *101*, 5186 (1979).
150. J.-L. Oudar, *J. Chem. Phys.*, *67*, 446 (1977).
151. J.-L. Oudar and R. Hierle, *J. Appl. Phys.*, *48*, 2699 (1997).
152. R. J. Tweig and C. W. Dirk, *J. Chem. Phys.*, *85*, 3537 (1986).
153. J. Zyss, J. F. Nicoud, and M. Coquillay, *J. Chem. Phys.*, *81*, 4160 (1984).
154. (a) B. F. Levine and C. G. Bethea, *J. Chem. Phys.*, *63*, 2666 (1975); (b) J.-L. Oudar, *J. Chem. Phys.*, *67*, 446 (1977).
155. (a) R. W. Terhune, P. D. Maker, and C. M. Savage, *Phys. Rev. Lett.*, *14*, 681 (1965); (b) P. D. Maker, *Phys. Rev. A*, *1*, 923 (1970); (c) K. Clays, A. Persoons, and L. de Maeyer, *Adv. Chem. Phys.*, *85*, 455 (1993).
156. S. K. Kurtz, *J. Appl. Phys.*, *39*, 3798 (1968).
157. S. J. Lalama and A. F. Garito, *Phys. Rev. A*, *20*, 1179 (1979).
158. L.-T. Cheng, W. Tam, S. H. Stevenson, G. R. Meredith, G. Rikken, and S. R. Marder, *J. Phys. Chem.*, *95*, 10631 (1991).
159. J. A. McCleverty, in *Transition Metals in Supramolecular Chemistry*, L. Fabbrizzi and A. Poggi, Eds., NATO ASI Series, Kluwer Academic Publishers, Dordrecht, The Netherlands, 1994.
160. B. Coe, *Chem. Eur. J.*, *5*, 2464 (1999).
161. S. Houbrechts, Y. Kubo, T. Tozawa, S. Tokita, T. Wada, and H. Sasabe, *Angew. Chem. Int. Ed. Engl.*, *39*, 3859 (2000).
162. M. Malaun, Z. R. Reeves, R. L. Paul, J. C. Jeffery, J. A. McCleverty, M. D. Ward, I. Asselberghs, K. Clays, and A. Persoons, *Chem. Commun.*, 49 (2001).
163. T. Kunitake, Y. Okahata, M. Shimomura, S. Yasunami, and K. Takarabe, *J. Am. Chem. Soc.*, *103*, 5401 (1981).
164. H. Ringsdorf, B. Schlarb, J. Venzmer, *Angew. Chem. Int. Ed. Engl.*, *27*, 113 (1988).
165. (a) A. P. H. Schenning, M. C. Freiters, and R. J. M. Nolte, *Tetrahedron Lett.*, 7077 (1993); (b) A. P. H. Schenning, B. de Bruin, M. C. Freiters, and R. J. M. Nolte, *Angew. Chem. Int. Ed. Engl.*, *33*, 1662 (1994).
166. F. M. Menger and Y. Yamasaki, *J. Am. Chem. Soc.*, *115*, 3840 (1993).
167. S. Munoz, J. Mallen, A. Nakano, Z. Chen, I. Gay, L. Echegoyen, G. W. Gokel, *J. Am.Chem. Soc.*, *115*, 1705 (1993).
168. S. Bhattacharya and S. De, *J. Chem. Soc. Chem. Comm.* 651 (1995).
169. A. M. Carmona-Ribero, *Chem. Rev.*, *92*, 209 (1992).
170. J. H. Fendler, *Membrane Mimetic Chemistry*, John Wiley & Sons, Inc., New York, 1982.
171. T. Kunitake, *Angew. Chem. Int. Ed. Engl.*, *31*, 709 (1992); H. Ringsdorf, B. Schlarb, and J. Venzmer, *Angew. Chem. Int. Ed. Engl.*, *27*, 113 (1988).
172. G. Gregoriadis, Ed., *Liposomes as Drug Carriers: Recent Trends and Progress*, John Wiley & Sons, Inc., Chichester, 1988.
173. R. W. Storrs, F. D. Tropper, H. Y. Li, C. K. Song, J. K. Kuniyoshi, D. A. Sipkins, K. C. P. Li, and M. D. Bednarski, *J. Am. Chem. Soc.*, *117*, 7301 (1995).
174. P. Scrimm, P. Tecilla, and U. Tonellato, *J. Am. Chem. Soc.*, *114*, 5086 (1992).
175. A. Kay, M. Gratzel, *J. Phys. Chem.*, *97*, 6272 (1993).
176. J. H. van Zanten, H. G. Monbouquette, *Biotechnol. Prog.*, *8*, 546 (1992).

177. C. A. Behm, I. I. Creaser, B. K. Daszkiewicz, R. J. Geue, A. M. Sargeson, and G. W. Walker, *J. Chem. Soc. Chem. Comm.*, 1844 (1993).
178. A. M. Sargeson, *Coord. Chem. Rev.*, *151*, 89 (1996).
179. P. Ghosh, T. K. Khan, and P. K. Bharadwaj, *J. Chem. Soc. Chem. Comm.*, 189 (1996).
180. P. Ghosh, S. Sengupta, and P. K. Bharadwaj, *Langmuir*, *14*, 5712 (1998).
181. (a) G. Das, P. K. Bharadwaj, U. Singh, R. A. Singh, and R. J. Butcher, *Langmuir*, *16*, 1910 (2000); (b) G. Das, P. Ghosh, P. K. Bharadwaj, U. Singh, and R. A. Singh, *Langmuir*, *13*, 3582 (1997).
182. S. Chiruvolu, H. E. Warriner, E. Naranjo, S. H. J. Idziak, J. O. Radler, R. J. Plano, J. A. Zasadzinski, and C. R. Safinya, *Science*, *266*, 1222 (1994).
183. P. Herve, D. Roux, A.-M. Bellocq, F. Nallet, and T. Gulik-Krzwicki, *J. Phy. II*, *3*, 1255 (1993).
184. J. Guilbot, T. Benvegnu, N. Legros, and D. Plusquellec, *Langmuir*, *17*, 613 (2001).
185. D. D. Lasic, *Angew. Chem. Int. Ed. Engl.*, *33*, 1685 (1994).
186. K. Okuyama, Y. Soboi, N. Iijima, K. Hirabayashi, T. Kunitake, and T. Kajiyama, *Bull. Chem. Soc. Jpn.*, *61*, 1485 (1988).
187. A. M. Fahrnow, W. Saenger, D. Fritsch, P. Schneider, and J.-H. Furhop, *Carbohydr. Res.*, *242*, 11 (1993).
188. Y. Abe, K. Harata, M. Fujiwara, and K. Ohbu, *Langmuir*, *12*, 636 (1996).
189. P. Tripathi, P. K. Bharadwaj, unpublished results.
190. P. Bandyopadhyay, P. K. Bharadwaj, *Langmuir*, *14*, 7537 (1998).
191. J.-H. Furhop and D. Fritsch, *Acc. Chem. Res.*, *19*, 130 (1986).
192. (a) A. Gambacorca, A. Gliozzi, and M. De Rosa, *World J. Microbiol. Biotechnol.*, *11*, 115, (1995); (b) G. D. Sprott, *J. Bioenerg. Biomembr.*, *24*, 555 (1992).
193. F. M. Menger and K. D. Gabrielson, *Angew. Chem. Int. Ed. Engl.*, *34*, 2091 (1995).
194. A. Ulman, *An Introduction to Ultrathin Films: from Langmuir-Blodgett to Self-Assembly*, Academic Press, San Diego, CA, 1991.
195. M. K. Dearmond and G. A. Fried, *Prog. Inorg. Chem.*, *44*, 97 (1997).
196. (a) G. J. Ashwell, A. J. Whittam, M. A. Amiri, R. Hamilton, A. Green, and U. W. Grummt, *J. Mater. Chem.*, *11*, 1345 (2001); (b) G. J. Ashwell, K. Skjonnemand, G. A. N. Paxton, D. W. Allen, J. P. L. Mifflin, and X. Li, *J. Mater. Chem.*, *11*, 1351 (2001).
197. A. N. Shipway, E. Katz, and I. Willner, *Chem. Phys. Chem.*, *1*, 18 (2000).
198. M. Bardosova and R. H. Tredgold, *Mol. Cryst. Liq. Cryst.*, *355*, 289 (2001).
199. R. W. Munn and O. Szczur, *Mol. Cryst. Liq. Cryst.*, *355*, 305 (2001).
200. W. X. Lu, H. B. Wang, W. H. Guo, and P. S. He, *J. Mater. Sci. Lett.*, *20*, 423 (2001).
201. D. Q. Yang, Y. Sun, R. F. Wang, and Y. Guo, *Thin Solid Films*, *385*, 239 (2001).
202. A. Chowdhury and A. J. Pal, *Thin Solid Films*, *385*, 266 (2001).
203. K. Ariga and T. Kunitake, *Acc. Chem. Res.*, *31*, 371 (1998).
204. J. Nagel, U. Oertel, P. Friedel, H. Komber, and D. Mobius, *Langmuir*, *13*, 4693 (1997).
205. A. K. Dutta, T. N. Misra, and A. J. Pal, *J. Phys. Chem.*, *98*, 12844 (1994).
206. H.-J. Schneider, *Angew. Chem. Int. Ed. Engl.*, *30*, 1417 (1991).
207. M. Torneiro and W. C. Still, *J. Am. Chem. Soc.*, *117*, 5887 (1995).
208. V. M. Ritello, E. A. Viani, G. Deslongchamps, B. A. Murray, and J. Rebek, Jr., *J. Am. Chem. Soc.*, *115*, 797 (1993).
209. J. S. Nowick, T. Cao, and G. Noronha, *J. Am. Chem. Soc.*, *116*, 3285 (1994).

210. R. P. Bonar-Law, *J. Am. Chem. Soc.*, *117*, 12397 (1995).
211. M. Shimomura, F. Nakamura, K. Ijiro, H. Taketsuna, M. Tanaka, H. Nakamura, and K. Hasebe, *J. Am. Chem. Soc.*, *119*, 2341 (1997).
212. M. Bohanon, S. Denzinger, R. Fink, W. Paulus, H. Ringsdorf, and M. Weck, *Angew. Chem. Int. Ed. Engl.*, *34*, 58 (1995).
213. Y. Ebara, K. Itakura, and Y. Okahata, *Langmuir*, *12*, 5156 (1996).
214. I. Weissbuch, M. Berfeld, W. Bouwman, K. Kajaer, J. Als-Nielsen, M. Lahav, and L. Leiserowitz, *J. Am. Chem. Soc.*, *119*, 933 (1997).
215. H. Kitano and H. Ringsdorf, *Bull. Chem. Soc. Jpn.*, *58*, 2826 (1985).
216. D. Y. Sasaki, K. Kurihara, and T. Kunitake, *J. Am. Chem. Soc.*, *113*, 9685 (1991); D. Y. Sasaki, M. Yanagi, K. Kurihara, and T. Kunitake, *Thin Solid Films*, *210*, 776 (1992).
217. P. Tripathi, R. A. Singh, and P. K. Bharadwaj, unpublished results.

Coordination Complexes in Sol–Gel Silica Materials

STEPHEN P. WATTON, COLLEEN M. TAYLOR, GRANT M. KLOSTER, and STEPHANIE C. BOWMAN

Department of Chemistry
Virginia Commonwealth University
Richmond, VA

CONTENTS

I. INTRODUCTION: DOPED SOL–GELS IN GENERAL

    A. Background Information
        1. Sol–Gel Materials in General
        2. Sol–Gel Processing
        3. Historical Development
    B. Scope of this Chapter

II. OVERVIEW: SILICA AND SOL–GELS

    A. Silica Gels
        1. Overview of Silica Chemistry
    B. Sol–Gel Processing
        1. Basics
    C. The Sol–Gel Environment

III. ENTRAPPED COMPLEXES

    A. Background
    B. Related Silica Chemistry
        1. Origins
        2. More Recent Advances
    C. From ICC to ISSC (Inner-Sphere Sol–Gel Chemistry)
    D. Noncoordinated Interactions
        1. Background
        2. Luminescent Probes
        3. Redox Active Probes

Progress in Inorganic Chemistry, Vol. 51, Edited by Kenneth D. Karlin.
ISBN 0-471-26534-9 © 2003 John Wiley & Sons, Inc.

IV. COVALENTLY ATTACHED COMPLEXES

    A. Introduction
    B. Preparation of Precursors
        1. Preparation of Silylated Ligands
        2. Preparation of Silylated Metal Complexes
        3. Purification
    C. Preparation of Materials
        1. Sol–Gel Immobilization Techniques
        2. Postmodification Approaches
        3. Issues Concerning Survival of Metal Centers
    D. Characterization of Materials
    E. Environment of Tethered TM Complexes

V. APPLICATIONS

    A. Sensors
    B. Photonics
    C. Catalysts
        1. Entrapment Versus Immobilization
        2. Site Accessibility
        3. Leaching
        4. Site Isolation

VI. SUMMARY TOPICS

    A. Synthesis Issues
    B. Conclusions and Future Prospects

ABBREVIATIONS

ACKNOWLEDGMENTS

REFERENCES

I. INTRODUCTION: DOPED SOL–GELS IN GENERAL

A. Background Information

1. Sol–Gel Materials in General

The sol–gel method has gained immense popularity in recent years as a means for the preparation of oxide materials, which have a multitude of technologically useful properties. Sol–gel processing has been widely used because it permits fabrication of oxide materials under mild conditions and with a wide range of adjustable experimental parameters. These characteristics permit extensive manipulation of the resultant material properties.

New modifications of silica are continually emerging—most of which are prepared by sol–gel processing. These modifications include microstructured silicates [e.g., micelle-templated materials such as the mobil composition of matter (MCM) (1)] and composite materials such as hybrid-organic inorganic materials (ORganically MOdified SILicates, or ORMOSILs). Traditionally, silicates have been regarded as inorganic polymers—classed with minerals such as clays and quartz, and distinct from their "organic" counterparts, such as the silicones. The boundaries between these classes of polymer have become blurred, however, as hybrid materials such as ORMOSILs have become increasingly prominent. The palette of accessible materials has expanded rapidly to include homopolymers derived from organosilicate precursors (2), organosilicate–silicate copolymers (3), as well as polymer composites and blends (4–6). The chemical variations in silicon-based polymers are mirrored in a range of properties that can be modulated: these include polarity, hydrophilicity, surface acidity and charge, porosity, flexibility, solvent-swelling behavior, and optical transparency. Each of these attributes will have a different degree of importance depending on the application for which a material is intended. Significant advances have been made toward rational tailoring of material structures and properties, and the scientific foundation for designing tailored materials is rapidly gaining momentum. In principle, and increasingly in practice, the sol–gel approach permits infinite variation of the composition, structure, and properties of the oxide matrices themselves. Of particular interest to chemists, however, is the fact that sol–gel methods can be used to introduce a seemingly unlimited array of guest species (dopants), including a wide array of metal complexes, into oxide materials.

Sol–gel chemistry is a broad area that may be unfamiliar to the typical coordination chemist seeking to prepare and study sol–gel materials doped with metal complexes, for whatever reason. Pertinent literature can be found in journals from many fields of different interest, spanning various areas of chemistry, materials science, physics, chemical engineering, biochemistry, and others. Because of the diversity of interest that the sol–gel processing of materials has engendered, specific issues that are addressed in two papers that pertain to transition metal complexes in sol–gel materials can be radically different, which makes navigating the literature to find relevant information and comparing the results from two studies a very time-consuming, and often frustrating undertaking. We have made an effort to summarize some of the many issues that are associated with inclusion of metal complexes within sol–gel silica-based media, with both being defined in as broad a sense as was practical. Hopefully, this broad view will help in providing an entry into the area for a classically trained coordination chemist, regardless of the purpose for which he/she is interested in preparing doped sol–gels. Specifics concerning evaluation and understanding of the performance of many of the advanced materials are covered briefly, and the reader is directed to more detailed application-oriented reviews for further information.

2. Sol–Gel Processing

To give a very brief definition, a sol–gel procedure encompasses any process that involves polymerization of soluble precursor molecules to afford a polymeric material, via the intermediate formation of a colloidal sol phase. Sol–gel processes are widely used to prepare oxide-based materials, including silica. When made at the appropriate pH, which varies depending on the silicate concentration, the materials can be obtained as optically transparent gels: A very significant attribute that has led to many technological applications for sol–gel materials. The procedures encountered most frequently in the silica sol–gel literature involve hydrolysis and condensation of alkyl orthosilicates rather than the aqueous-based chemistry that initially gave rise to the field. The basic chemistry associated with formation of a typical silica sol–gel material is summarized in very simplified form in Figure 1. The chemistry associated with formation of aqueous sol–gels is very similar, but does not require the first, alcohol generating, step.

Compared with classical methods for the production of silica glasses (i.e., heating of sand and a few other additives to temperatures of ~ 800–$1000°C$, followed by rapid cooling), the sol–gel process affords a number of distinct advantages for the preparation of hybrid materials. The benchtop procedures associated with sol–gel processing have placed materials science at the fingertips of any chemist. Sol–gel materials are very easy to prepare, and can be made in almost limitless variety. Their utility for the preparation of materials containing temperature-sensitive molecules is fundamental to the research area summarized here. Before proceeding with one small aspect of this vast field, we will give a very brief summary of the development of sol–gel chemistry, emphasizing some aspects of the evolution of our particular focus area.

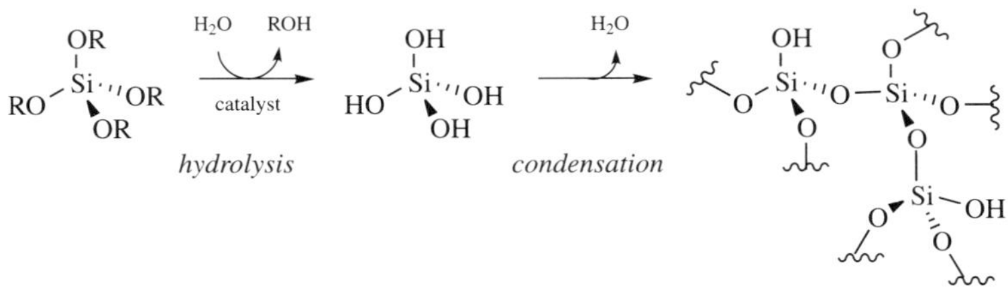

Figure 1. Basic steps of a typical sol–gel process. Hydrolysis of an alkylorthosilicate affords silicic acid derivatives that undergo condensation reactions to form a silicate network.

3. Historical Development

The basic principles underlying sol–gel chemistry have been known since the mid-1800s (7, 8). Silica chemistry has since been widely explored, as silica materials have a vast range of applications (9, 10). The first reports documenting hydrolysis of alkoxide-based precursors, rather than silicate salts, appeared ~1930 (11, 12). Initial studies of alkoxide-based sol–gels were driven primarily by their usefulness in preparing optical components, since stable transparent materials could be obtained using drying periods of days, rather than weeks, months, or even years (13). Interest in the use of sol–gels for optical applications is still thriving, and now attracts the attention of a diverse range of chemists, chemical engineers, and physicists (14). Early interest in sol–gel processing among inorganic chemists arose because the approach is advantageous for the preparation of mixed oxide materials (15). The intimate mixing of component precursors eliminates many technical difficulties associated with the high temperatures used in melt chemistry. Research in solid-state chemistry gradually revealed advantages of sol–gel chemistry to inorganic chemists with a variety of interests. Central to the current discussion is the fact that the ambient conditions associated with sol–gel processing allow for preparation of hybrid materials that contain a wide variety of molecules (dopants) that would not survive the high-temperature conditions for "classical" glass preparation. The recognition that sol–gel materials can accommodate temperature-sensitive compounds has resulted in another rapid expansion of the sol–gel literature, since inclusion of almost any dopant raises possibilities for an immense range of technological applications (14, 16–19).

The genesis of doped sol–gels occurred with Avnir's studies of Rhodamine G in silicate sol–gels (20), which demonstrated that the low temperatures and chemically innocuous conditions of the sol–gel process are appropriate for inclusion of sensitive molecules in the materials. The initial studies with organic dyes paved the way for a rapid expansion of the number and types of species that could be doped into sol–gels. Dopants reported to date include organic dyes (21, 22), metal complexes (19, 23, 24), enzymes (16, 25–29), antibodies (30, 31), whole cells (32), and higher biological structures (33).

In the case of transition metal complexes intended for catalytic purposes, immobilization by sol–gel methods was a natural extension of work that had been vigorously pursued in the 1970s and 1980s to heterogenize metal complexes on inert supports, including silica (34, 35). The earliest reports of sol–gels being used for this purpose appeared in the early 1980s (36, 37). The sol–gel method provided an attractive alternative route compared with immobilization on preformed silica surfaces. The more intimate mixing of metal complexes with the silica during formation of the material allows for improved dispersion of the metal centers through the material and higher loading capacities, as well as improved stability of the materials toward degradation during catalytic turnover. Further, sol–gel silica materials can be used to entrap metal complexes within the material structures. Creating supported catalysts by noncovalent inclusion of metal complexes offers a combination of significant improvements in catalyst recyclability when compared with conventional surface impregnation techniques, and a considerable saving in synthetic effort, when compared to covalent immobilization. Covalently immobilized metal complexes have also been studied extensively, principally for catalytic purposes, but new applications are emerging—particularly in the area of imprinting of materials with receptor sites for the selective binding of molecules or ions (38–44).

Sol–gel materials are typically amorphous, but an adaptation of chemistry originally developed for the preparation of large pore zeolites, which are crystalline aluminosilicates, has shown that surfactant molecules can direct assembly of sol–gel silicates into materials with semiregular porous structures. These micelle-templated materials, which are typically called MCMs or MTSs (micelle templated silicas) are under intensive study because of the size and shape selectivity conferred by the regular pore structure (1). Coverage of the rapidly expanding literature on MCM type materials is largely beyond the scope of the current discussion, although some illustrative examples will be included.

B. Scope of This Chapter

For the purposes of this chapter, dopants are broadly divided into two major categories: those bearing covalent attachments to the silica matrix (tethered) and those that are noncovalently included within the silica matrix (untethered). Noncovalent inclusion is accomplished simply by mixing a solution of the dopant with a sol prior to gelation. In these cases, a number of interactions can contribute to association of the dopant with the material. These include physical entrapment of the dopant molecules by entanglement within the pore structures of the polysilicate network, specific association with the silicate matrices through hydrogen bonding, ion pairing with ionized $Si-O^-$ residues, and coordination at the material–solvent interface. For many applications, particularly those that rely on the optical behavior of dopant metal ions, noncovalent entrapment results in a material that is sufficiently stable for the purpose. Since no special functional groups are required to provide covalent attachment points, little or no synthetic effort needs to be expended in preparation of the dopant: In general, noncovalently doped materials are much more readily prepared than their covalent counterparts. The absence of covalent attachment, however, means that the dopants are relatively mobile within the sol–gel matrix, and may be prone to migration out of the gel matrix when exposed to an external solution. For applications such as the fabrication of sensors or recyclable catalysts, this degradation by leaching can represent a serious limitation (45–47). Despite the additional synthetic efforts involved, covalent attachment of the dopants has become the preferred approach for such applications.

Covalent attachment is normally achieved by adding molecules with pendant —Si(OR)$_3$ to a silicate sol. The silylated side chains copolymerize with the silica sol to afford stable linkages. The most common applications for materials of this type are in catalysis, where leaching of metal centers into solution during use can lead to significant decrease in functional utility. Covalently attached ligands have also been introduced into sol–gel materials to afford metal-specific materials for sensor, separation, and catalytic applications (15, 40, 43, 44). Typically, the orthosilicate is present in large excess with respect to the dopant (at least 100-fold) and complete attachment occurs as the side chains and silicates condense together.

The considerations pertaining to untethered and tethered complexes are quite different, so the two situations will be discussed separately, after an introduction to silica chemistry and sol–gel processing.

II. OVERVIEW: SILICA AND SOL–GELS

A. Silica Gels

1. Overview of Silica Chemistry

a. Chemistry of Silica-Based Materials

i. Range of Structures. The properties of sol–gel polymers are readily modulated by changes in the synthetic parameters. Adjustable parameters include the monomer composition (including the nature and number of types of precursor, as well as their absolute and relative concentrations), conditions for processing, such as the solvents used to homogenize the reaction mixtures, the amount of water available relative to the number of silicon alkoxides available for hydrolysis and condensation reactions ($R_w = $[Si—OR]/[H$_2$O]), pH, temperature, the nature of the active catalyst, and many other factors. Conditions used to treat the gel after initial formation (postprocessing) also have a profound impact on the structure and behavior of the silica material. Even chemically similar materials can be prepared with widely varying structural features and bulk properties by changing the processing parameters. For example, by using various sol–gel procedures the silica structure can range, from dense, nonporous microspheres through the semiregular microporous MCM materials (1) to the extremely low-density aerogel materials (48–50).

It has been pointed out (2) that we do not yet understand the complexities of sol–gel chemistry sufficiently well to accurately predict the microscopic structure, and hence the macroscopic properties of the material that will result from a particular polymerization recipe. Nevertheless, some generalities have been established that are useful in narrowing down the choices that need to be made in formulating a sol–gel recipe, and it is becoming increasingly possible to tailor material properties of silicon-based materials to suit specific applications. A brief description of the mechanism of some of the elementary steps in formation of sol–gel materials can be found below, which emphasizes the relation between some of the synthetic variables and the resultant structures. First, some of the relevant properties of silica gels are described.

ii. Fundamentals of Silica Structure. Typical amorphous silicas are built up from aggregates of small colloidal particles of silica having dimensions somewhere between 1 and 100 nm. Many important parameters that determine the properties of a silica gel—surface area, porosity, and so on are determined by the size of these particles, and the degree to which they are packed together.

When initially formed, the silica structure is very open, with large solvent-filled pores and channels pervading the material. The arrangement of particles at the point of gel formation represents a very early but fragile stage of evolution of the silica material. The initial structure of a sol–gel material is highly tenuous: The open and porous network structure of the original gel is retained in aerogels (51). Normal processing conditions usually result in considerable collapse of this initial structure (Section II.B.1.b.ii), giving rise to gels with much less porous structures. The nature of aerogels is of great importance to the study of metal complexes in sol–gel media. The aerogel structure provides insight into the nature of the starting point for structural evolution of a sol–gel material, and represents the state of the gel structure during events such as sol–gel immobilization of metal complexes.

As silica monomers become integrated into sol–gel materials, condensation occurs to varying degrees, depending on the conditions. A silicate center is capable of making up to four siloxane linkages, although complete condensation happens for only a limited population of the silicate centers. Complete condensation of all silica centers can only occur in crystalline arrangements, so the more common situation in amorphous materials is for the silica sites to make two or three linkages. The degree of condensation is readily measured by a number of techniques, such as solid-state nuclear magnetic resonance (NMR) (52). This parameter affects many properties of the silica material, such as porosity and surface structure. Most important from the point of view of determining the chemistry of sol–gel materials are those silica sites that do not undergo complete condensation. These silica sites afford one or more active hydroxyl groups, present at concentrations of approximately five silanol groups per squre nanometer (nm^2) (46). The hydroxyl groups have a strong influence on the nature of the solvent–surface interface, and are capable of reaction with external reagents, such as metal complexes or tethering side chains.

iii. **ORMOSILs.** The prototypical materials prepared by sol–gel methods are silicates, derived from tetraalkylorthosilicates such as tetramethylorthosilicate (TMOS) or tetraethylorthosilicates (TEOS) and whose centers can make up to four siloxane linkages termed "Q4" centers. Three siloxanes and an uncondensed silanol or alkoxysilyl group give a "Q3" center, and so on. There is increasing interest, however, in materials containing organofunctional silicates based on $R_nSi(OH)_{(4-n)}$, where $n = 1$, 2, or 3 and the designations are "T" (monoalkyl), "D" (dialkyl), and "M" (trialkyl) substituted centers, respectively. The resulting ORMOSILs extend the range of accessible gel structures and properties considerably. Most of the ORMOSIL materials mentioned herein are mixed silicate–organosilicate materials. The most important class for our purposes is the mixed Q–T systems, since the use of an alkylsilicate modified ligand for immobilization of metal complexes creates ORMOSIL materials. Some understanding of the behavior of immobilized transition metal complexes in sol–gels can thus be obtained by considering results from the ORMOSIL literature. This point will be discussed further when immobilized complexes are under consideration (Section IV).

Mixing-and-matching any number of Q, T, D, and M silicate precursors can create a vast array of ORMOSILs (53). Pure homopolymers are well known in all cases where polymer formation is possible: It has already been shown that silica gels are derived from Q species. Homopolymers of T-species (silsesquioxanes), are somewhat less familiar, but are the basis of many ORMOSIL materials. Finally, D-species form silicones: Strictly speaking these materials are ORMOSILs, but they have been of such great industrial importance for so long that they are rarely considered as such. The M species (e.g., Me_3Si-OR) cannot form polymers, but these species are ubiquitous and familiar components that are used to modulate surface properties of glasses of all types.

In addition to building blocks based on a single silicon center, polyfunctional units having organic linkers between two or more silicon atoms have also been used, which further extend the range of materials that can be prepared (53). Given the number of modifications that can be made to the basic alkoxide precursor, a large palette of starting materials is available for synthesis of homo- and copolymers. Coupled with the range of choices that can be made for polymerization conditions, an enormous number of starting conditions can be selected. With so many possibilities available, the structures, and properties of sol–gel silicates can in principle be tailored over a wide range.

b. Material Properties. Material properties that can be modulated by changes in the composition or processing of silica-based materials include porosity, density, surface hydrophilicity, flexibility, and optical transparency. These properties are interrelated, and in some instances can be mutually exclusive, so that it is not always possible to balance all desirable attributes in a single material. The mechanistic basis underlying many of the observed structural properties are quite well established, and some basic choices concerning the best procedure to use can be made depending on which material property is of particular importance for the application at hand. Only a brief summary of some pertinent aspects will be presented at this point. The recent book by Schubert and Husing (51) provides a lucid discussion that covers these issues in considerably more detail.

The surface areas of silica materials have been of great importance for development of surface-heterogenized catalysts, since they represent the areas over which metal complexes can be dispersed or immobilized during impregnation or surface grafting procedures. The advent of sol–gel chemistry meant that accessible surface sites could be increased relative to pre-formed silica gel, especially for dopant species that require silanol sites for covalent inclusion or inner-sphere coordination within the material. Consequently, it is commonly asserted that the use of sol–gel processing allows higher loadings of metal complexes per unit mass of the support (15, 17, 54, 55). This assertion is quite logical given that researchers have made materials with mole ratios of metal/Si as large as 1:1, although it has also been shown that such high loadings typically lead to collapse of the gel structure affording a nonporous material (56). One is hard pressed, however, to find meaningful quantitative comparisons of maximum loading of a given metal complex on preformed silica and in a sol–gel material. We have found that the loading, [metal]/[Si], of [Co(2,9-dimethyl-1,10-phenanthroline)]$^{2+}$ (Section III.C), which forms an inner-sphere complex with surface silanol groups, can be increased using the sol–gel method by at least 30 times that achieved by adsorption on a common commercially available silica gel with surface area of 480 m^2 g^{-1}. In this case, however, high loadings required a high pH to activate surface silanols, at the expense of the transparency of the material. Nevertheless, greater loading, roughly 2-fold that of the commercial silica gel, can still be achieved while still maintaining a transparent monolith (57). Note that the surface-immobilization method virtually guarantees that any immobilized metal complex is, by definition, in an accessible region of the gel surface. The same cannot necessarily be said for a complex that has been immobilized by a sol–gel technique, since pore size may limit access to a significant population of dopant species (17).

The literature is not yet clear on whether sol–gel synthesis methods can be manipulated in such a manner as to allow all dopant sites to be accessible, although surface modification of micelletemplated materials (1, 38–40) is an approach that has particular promise in this regard.

Typical silica gels are composed of small-to-medium sized particles (1–100 nm) that are aggregated into a bulk material. The voids between these particles make a certain fraction of the surface area available, and define the pores. If the average particle size is small, the spheres will have a high ratio of surface area to mass, but will tend to pack together more efficiently and create smaller pores. Conversely, larger particles will have smaller surface areas per unit mass, but will yield a larger average pore volume than smaller particles packed with comparable efficiency (10). For typical silica materials, it should be appreciated that the surface area of accessible pores largely defines the measured surface area—the actual external surface area of a silica granule contributes relatively little to the total.

Measurement of surface area, porosity, and so on is usually accomplished by measurement of gas adsorption isotherms, typically using gases such as N_2, Ar, or Kr. Dinitrogen is most commonly used for routine work, due to it's lower cost. The measured surface areas depend critically on the probes used for measurement, as well as on the material composition, processing conditions and other factors. Measurements on comparable silica gels using different gases, for example, N_2, Ar, or Kr will typically return different results, as larger sized gas molecules do not fit as well into smaller pore openings on surface. For measurements made by N_2 adsorption, surface areas typical range from 5 $m^2 g^{-1}$ to over 1000 $m^2 g^{-1}$ in aerogel materials. For calibration, typical materials used for chromatography have surface areas falling in the 100–500-$m^2 g^{-1}$ range (10).

Porosity is a major contributing factor to the function of most materials—particularly those used in catalysis, sensor, and separations applications, where access to functional sites is a prerequisite for effective operation—and has consequently been examined in great detail (58). Pore size affects mass transport into and out of material, the sizes and shapes of substrates, or analytes that are able to access active sites, and product release following catalysis. For most sol–gel materials, pore sizes fall within the microporous-to-mesoporous range, from <2 nm to between 2 and 50 nm (59).

Pore size is more difficult to determine than specific surface area, particularly where microporous materials with a broad distribution of pore sizes are concerned. A number of data treatments are used to obtain information concerning the pore sizes and their distributions from N_2 absorption–desorption data. The most appropriate method for determining pore volumes is a matter of some debate, and will not be covered further here (58, 60–62). The reader is referred to the paper of Lukens, Jr. et al. (61) for a discussion of the relative merits and shortcomings of the various methods. Newer approaches for measuring porosity, such as the use of variable-temperature NMR, which avoids the need for dedicated instrumentation, are also emerging (63).

B. Sol–Gel Processing

1. Basics

a. Formation of the Typical Sol–Gel Material. The range of materials that are accessible using sol–gel routes is tremendous, due to wide variability of processing parameters that can be used. We have included as much detail as possible without overloading the reader and suggest the works of Brinker and Scherer (9), Iler (10), and Hench and West (13) for further information on sol–gel chemistry.

Starting with the most basic description of the procedure, one can better understand the nuances that affect the state of the final material. The sol–gel process can be broken down into four key stages, which are sol formation, gelation, aging, and drying. The initial mixture generally consists of alkoxide precursors, most commonly tetramethyl- and tetraethylorthosilicates (TMOS and TEOS, respectively), water, acid, or base catalyst and often a cosolvent such as methanol or ethanol to obtain an admixture. Hydrolysis and condensation of the alkoxide precursors occur in a complicated sequence of nonconsecutive pathways that consume water, release alcohol, then produce water again (Fig. 1). The relative rates of the overlapping hydrolysis and condensation processes are dependent on the starting solution and are arguably the single most important factor in determining the final structure of the material. The alteration of any of the starting materials, their relative molar amounts, or addition of a dopant molecule can also have an effect on the final structure.

A colloidal sol, which is comprised of subnanometer sized oxide particles in a liquid solvent, forms from the mixture as a result of the hydrolysis and condensation reactions that initially occur. At this point the sol is very fluid, and can be cast into molds, coated onto the surface of a substrate, such as a glass plate or electrode, or drawn into fibers. Subsequent cross-linking and aggregation of the colloidal particles eventually results in the formation of a polymeric network. At some point, the viscosity of the fluid sol increases sharply, and gelation occurs. The time at which this readily observable phase transition occurs is called the gel point. The materials thus obtained are porous, solvent-filled gels, called alcogels. Evolution of the silica matrix continues after gelation in a process known as aging. The aging process can be aided by slow evaporation of the solvent material at mild temperatures until the release of solvent is essentially over. At this point, the material is called a xerogel. High processing temperatures can be used to drive off any residual solvent producing an aerogel. Further postprocessing procedures of the material are frequently employed: These include surface modification with silylating agents, attachment of metal complexes, or high-temperature processing to remove organic components and to further modify the silica structure.

The host of modified monomers that have been included in sol–gel reactions, both in homopolymerization and copolymerization reactions, have been mentioned previously. In addition to the monomer composition, many other reaction parameters can be varied, including temperature, and the nature and amounts of catalysts, cosolvents, or reaction modifiers (such as surfactants and buffers). Each of the above variables, as well as the postprocessing steps, contribute to the ultimate outcome of the sol–gel reactions and can have a strong influence on the ultimate structure and properties of a material.

b. Stages of the Sol–Gel Process

i. Initial Stages of the Sol–Gel Reaction. Although the conditions used for the sol–gel reaction have a strong influence over the ultimate material structure, the fundamental features of the reactions taking place during the early stages of the sol–gel reaction are essentially the same regardless of conditions. The initial steps involve obligatory hydrolysis of the orthosilicate precursor, and condensation of the hydrolyzed products to form small (3–4 silicon) particles, which aggregate to form the larger colloidal silica particles that actually condense to form the final network structure. A common misconception is that silica gels are made up of a random assortment of particle sizes: In fact, the particles making up any given gel structure tend to have a fairly narrow size distribution, because particle nucleation and growth rates are all controlled by the same kinetic parameters under the conditions of that particular reaction (10). The size of these ultimate particles will determine many of the properties of a sol–gel material, such as optical transparency, mechanical strength, and catalytic activity. Ultimate particle size depends on many influences, most of which act by changing the relative magnitudes of the rate constants for the many elementary reactions that can occur during a sol–gel reaction. Schubert and Husing (51) recently summarized the influences of many processing parameters on material properties. A brief description of some salient aspects will be presented here.

When alkoxysilanes are used as the silicon source, both hydrolysis and condensation reactions will be taking place simultaneously for most of the sol–gel reaction. The picture is considerably simplified when sodium silicate is used as the silicon source, since only condensation reactions need to be considered. The sequence of reactions shown initially in Fig. 1 is a gross oversimplification of the sol–gel mechanism. The reaction actually involves a large number of potential reactant species and intermediates, even in the simplest case of a gel being created from a single Q-type alkoxide precursor. Each silicon center in the reaction mixture can undergo a total of four hydrolyses and condensations, but the sequence in which each of these steps takes place, with the exception of the first hydrolysis, can vary considerably. For a single Q-type silicon center, there

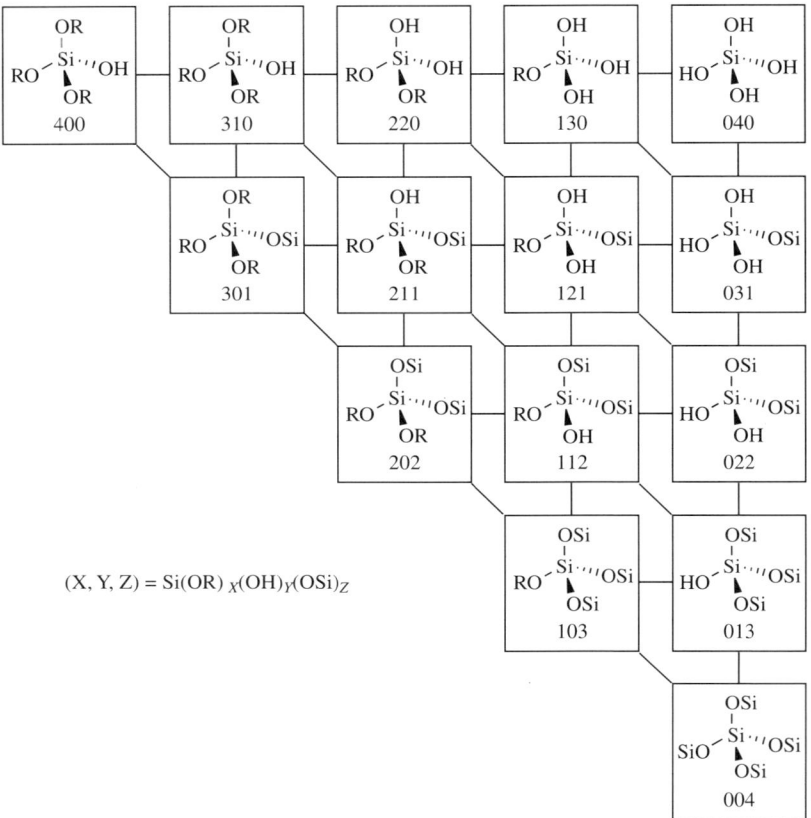

Figure 2. Summary of the various pathways for hydrolysis and condensation reactions at a single silicon center during a sol–gel reaction. [Adapted from (65).]

are a total of 15 distinct species that can be considered as potential reactants and products (Fig. 2). A total of 165 rate constants are required to describe all of the reactions that could potentially occur between these species on the way to forming a silica polymer (64, 65).

With so many different reactions to consider, it can be readily appreciated how small changes in the conditions, which will clearly have a different impact on the rates of some steps in comparison to others, can result in significant changes in the overall material structure. A complete description of any given sol–gel reaction is virtually impossible, but some clear and useful generalizations can be drawn from the kinetic studies. In particular, the influence of acidic or basic catalysis on the sol–gel reaction has been studied in great detail (10). The

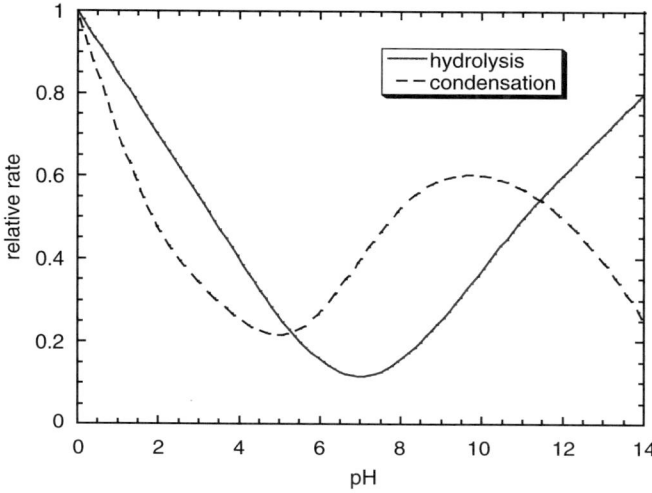

Figure 3. Dependence of hydrolysis and condensation rates on pH. [Redrawn from (51).]

influence of this parameter on ultimate particle size can be readily understood if one considers that the rate dependence on pH for each individual process in turn affects the degree of overlap of both processes (Fig. 3) (51).

The available kinetic information for acid and based catalyzed can be summarized as follows: Under acid-catalyzed conditions, condensation reactions are rate limiting, and the initial sol contains many small oligomers and particles. Under basic catalysis, by contrast, hydrolysis and particle nucleation steps are rate limiting, so the sol will contain fewer, larger particles than in the acid-catalyzed case (Fig. 4) (13).

The gradual aggregation of these particles formed at the early stages of sol evolution eventually gives rise to the gel structure. Acid-catalyzed gels are therefore aggregates of very small ultimate particles, which form into chains and cross-link to generate the final gel network. The very small sizes of the particles that comprise an acid-catalyzed gel minimize light scattering effects, and give rise to the optical transparency. Base-catalyzed gels are composed of larger, denser particles that tend to result in lower surface area gels with high porosities. These properties are preferable if a gel is to be used in an application such as catalysis, where accessibility of substrates to supported metal centers is essential, and optical transparency is expendable.

ii. Gelation, Aging, and Drying. Aggregation of silica particles across the entire sample is the fundamental event at the sol–gel transition. At this point

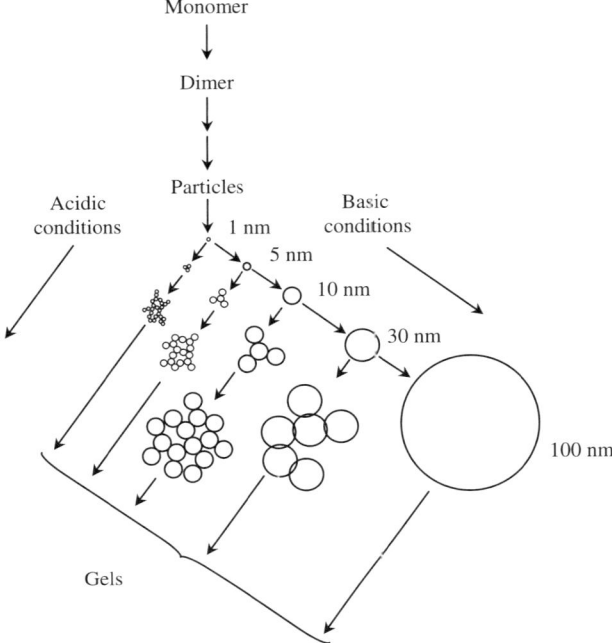

Figure 4. The pH dependence of growth and condensation of silica particles during gel formation. [Adapted from (13).]

the material ceases to flow, normal solution mobility is suddenly lost, and the material becomes segregated into regions of solid and regions of entrapped pore solvent. At the gel point, the silica structure is essentially locked in, and further sol–gel chemistry only tends to modify the existing framework.

Many studies have shown that the sol–gel structure continues to evolve long after gelation. For a period immediately following gelation, the gel is extremely fragile. An optically transparent (most likely acid-catalyzed) gel, when first prepared, cannot be removed from its container without turning into a powdery material. Before such gels can be used as lenses, nonlinear optics (NLO) devices, and so on, they must be allowed an aging period before the monolithic structure has stabilized sufficiently to permit manipulation of the sample. During the aging period, the silica particles undergo continued condensation, which stabilizes the structure. As with gelation time, the duration of the aging period that is necessary to stabilize the gel depends greatly on the conditions used in the sol–gel recipe, but typical aging periods are on the order of days or weeks.

If, following aging, the solvent is allowed to evaporate slowly from the container, the gel will undergo an additional phase of evolution. As solvent molecules evaporate from the exterior of the gel, molecules will migrate to the surface, depleting the interior of solvent. This solvent depletion perturbs the pseudoequilibrium conditions that had been established during the aging period, and generates a driving force (hydrostatic or hydrodynamic force) for further rearrangement of the gel structure. As drying proceeds, the pore structure collapses considerably, resulting in a dramatic shrinkage of the gel monolith. In many cases, forces can be sufficiently strong to tear the network apart, producing substantial cracks and fissures.

Control of the drying process can be attained using drying chemical control agents (DCCAs) to modulate the structural evolution of the materials. The DCCAs such as formamide, are most commonly included in sol–gel processes used for fabrication of optical components in order to minimize shrinkage and cracking during the drying processes. The reagents act principally by modulating the pH during the course of the sol–gel reaction (66). Surfactants may also be included to perform a similar function, by effectively insulating the silica surface from the bulk solvent, lowering the interfacial surface tension, and reducing drying stresses (1, 65–69).

Another common approach to minimizing drying stresses is the preparation of the materials as thin films on appropriate supports, which speeds drying by increasing the surface area of the material relative to the volume of the bulk material. This approach has been particularly useful in sensor development, but has also been applied in the preparation of catalytic membranes, and separations media.

Supercritical drying, that is, removal of the solvent at a temperature and pressure above the critical point, eliminates the liquid–vapor interface and hence surface tension effects. Solvent removal is thus accomplished without pore collapse, affording a very low-density material with very high surface area, called an aerogel (49). These materials can be optimized for their optical properties, and have been widely investigated for use in optical components such as the construction of large lenses for telescopes (51), where their extraordinary light weight simplifies engineering problems. Because of their high porosities, aerogels have also attracted attention for use in catalytic applications (70, 71). Despite their potential utility in many applications, preparation and study of classical aerogel materials suffers from a significant disadvantage in that their preparation requires specialized equipment to generate the conditions for supercritical drying. Dunn and co-worker (72) recently reported a method for creating aerogel-like materials by a sequential solvent-exchange process, which eliminates drying stresses by gradually changing the internal solvent environment, and does not require use of supercritical conditions. This method has not yet been widely applied, but it is expected that the advent of these "ambigels" will lead to a rapid expansion in research on low-density materials, including those based on silica sol–gels.

c. Additional Sol–Gel Parameters that Influence Material Properties

i. Catalysts for the Sol–Gel Process. The effects of acid and base catalysis have already been discussed, but some other catalysts are used sufficiently frequently to merit mention. In addition to acids and bases, other common catalysts are nucleophiles, which act by generating pentacoordinate silicon intermediates, which are more reactive toward substitution by water or silanols than are the four-coordinate precursor molecules (73). These catalysts include fluoride, *N*-methylimidazole, hexamethylphosphoric triamide (HMPA), and *N,N*-dimethylaminopyridine (DMAP) (9). Of these, fluoride is most commonly used for preparation of supported catalysts as it gives rise to porous, high-surface area materials. Another reagent that has found some use as a sol–gel catalyst is $SnBu_2(OAc)_2$, which presumably promotes reactions by acting as a Lewis acid. The influence of this catalyst on final gel structure does not appear to have been studied extensively. The catalyst is particularly useful in cases where acid- or base-sensitive functional groups are present within the dopants. For example, Lindner and co-workers (74a) used it to avoid problems with hydrolysis of ether linkages associated with their phosphine ligands. The catalyst also appears to be useful in syntheses where the presence of basic ligands inactivate HCl as a catalyst (73, 75). Note that an account by Parish et al. (76) showed NMR evidence for contamination of silica materials by this catalyst, which has a relatively bulky structure that can become entrapped within the material structure (76). Lindner et al. (74b) observed no evidence for such contamination, however, probably because the flexibility of their materials allowed the catalyst to be washed out effectively.

Two-catalyst procedures have also been developed, which use both acid and base catalysis. For example, Schumann et al. (77) used a two-stage procedure to speed up entrapment of air-sensitive rhodium complexes. The procedure involved acid hydrolysis to form an initial sol, which was doped with metal complex, and base was added to induce condensation. The effect of this two-step procedure on gel structure was not examined: The rhodium complex was effectively trapped within the silicate matrix, however, and was an active catalyst for hydrogenation of styrene and nitrobenzene.

ii. Catalysis of the Sol–Gel Process by Transition Metal Ions. Catalysis of the sol–gel process by transition metal ions has only been sparsely studied, but it is clear that the presence of certain metal complexes can affect the sol–gel process. Cerium(IV), for example, is a strong oxidant that was doped into a sol in an attempt to probe the site accessibility of a covalently attached metal complex (78). Although present in the gel at only millimolar concentrations, it had a profound effect on the time it took the sol to gel. Lanthanide ions are strong Lewis acids that complex with water to give lanthanide hydroxides and excess protons, thereby significantly altering the pH of a sol, and altering the gelation kinetics. More detailed studies have been conducted by Mayo et al. (79), who showed that a number of metal bis(acetylacetonate) complexes can promote the sol–gel reaction, apparently by facilitating condensation steps. The mechanism shows a half-order dependence on the concentration of metal complex, which was explained in terms of dissociation of one of the acetylacetonato (acac) ligands to form the active catalyst.

iii. Solvent. Sol–gels can be prepared either with or without addition of solvents, besides the water required for the sol–gel reaction itself (80). When water alone is used for the sol–gel reaction, the initial reaction mixture is a two-phase system, which must be vigorously mixed or sonicated in order to effect formation of a homogeneous sol. It has been suggested (71) that the sonication step should be considered as an additional sol–gel parameter, although at present the influence of the ultrasound treatment on gel structure does not appear to have been established. Alternatively, additional cosolvents are frequently added to homogenize the reaction mixture or to assist with dissolution of dopants. The influences of these cosolvents on the outcome of the sol–gel reactions are not fully established, but there is no doubt that they contribute significantly. In one extreme case, for example, processing of a palladium bis(alkene) complex in the presence of different solvents resulted in a complete change in the chemistry, the resultant structures and behaviors of the materials (81). Processing with tetrahydrofuran (THF) as a cosolvent resulted in a material containing (alkene) complexes in which the Pd centers were buried within the material structure and catalytically inactive. By contrast, processing in the presence of methanol resulted in loss of trimethoxysilyl groups from the organic ligands, affording catalytically active, surface-bound Pd–allyl complexes (Fig. 5).

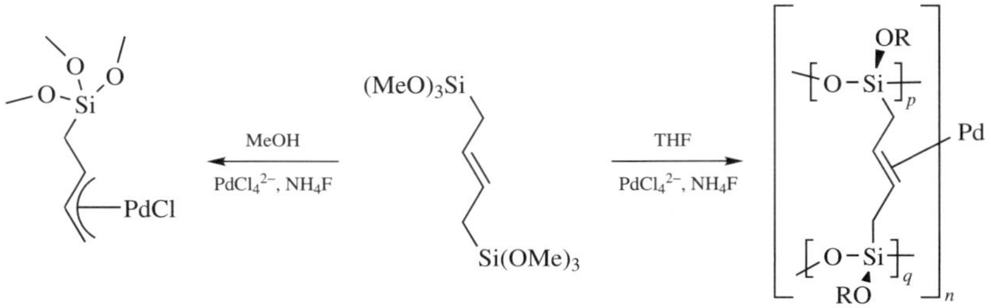

Figure 5. Effect of added cosolvent on the results of immobilization of a palladium alkene complex. [Adapted from (81).]

iv. *Postprocessing Procedures.*

1. In the first procedure, sintering is used to remove solvent and organic components from the silica gel, and when sufficiently high temperatures are used, results in coarsening of the silica structure (10).

   Since organic ligands are destroyed during the sintering process, sintered materials are of only limited interest to the current discussion. However, it has been shown that sintering of immobilized complexes dispersed through silica sol–gels helps with controlling the resultant particle size when sintering is performed with the purpose of generating nanoparticles within the silica matrix. Many immobilized complexes have therefore been prepared for this purpose, and the synthetic procedures can be adaptable for other applications.

2. In the second procedure, surface modification, silica gels can be postmodified by reaction with a variety of alkyl silicon compounds, such as TMS—Cl, TMS—OMe, or alkyltrialkoxysilanes. These procedures are no different from the surface modification of silica gels prepared by conventional methods (37). Capping of uncondensed silanols in this manner serves a number of purposes. The principal reasons are to inactivate uncondensed silanol groups, to change the hydrophilicity of the surface, or to introduce additional functional groups that can be used for further modification of the material. Surface capping also appears to diminish the migration of tethered complexes across silica surface, as demonstrated by Drago (82). In that case, capping of the surface silanols removed sites to which tethering side chains could migrate during catalyst turnover, which increased the long-term stability of a mononuclear rhodium catalyst by preventing deactivation through the formation of inactive binuclear complexes.

C. The Sol–Gel Environment

A considerable amount of work has been directed at understanding the influences of the silica environment on dopant molecules, due the importance of these effects on the behavior of the composite materials. Much of this work has been performed using organic dyes as probes of the sol–gel interior, which continue to provide a wealth of useful information (83). Detailed study of proteins has also given considerable insight (16, 26, 28, 29).

An excellent summary of work reported up to 1997 was presented by Dunn and Zink (22), who put forth a simple model for the interior of a sol–gel, which is very useful in considering interactions of molecules with the sol–gel environment. The pore structure was considered as a collection of small test tubes whose walls are made up of component silica particles, and which contain a bulk solvent region. This model of the silica interior is represented in Figure 6,

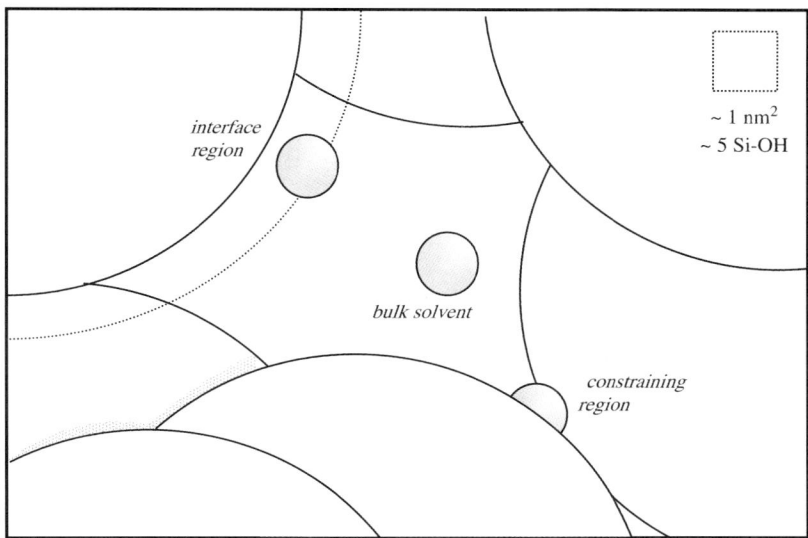

Figure 6. Environments for dopants entrapped in sol–gel pores. Three distinct regions are identified within the pore structure—a bulk solvent phase, the interface between the solvent and the silica surface, and the constraining region, located at the interfaces between silica particles. The structure as represented has a relatively large pore volume, as may be found in a solvent-swollen gel prior to drying. [Adapted from (22)].

which indicates the relative dimensions for a typical metal complex in relation to the pore dimensions of a typical microporous material. Molecules in the bulk solvent region of gel material experience little effect on their behavior, as has been observed for many neutral dopants. However, interactions with the pore walls can become significant when molecules capable of hydrogen bonding (83), electrostatic interaction and ion pairing (84), and even inner-sphere coordination are considered (85).

III. ENTRAPPED COMPLEXES

A. Background

Sol–gel materials have been considered chemically innocuous—a perception brought about by their ability to accommodate and stabilize an enzyme (16, 26, 27, 86), or to host a living cell (32, 33). In reality, however, for many metal complexes, the sol–gel medium can offer a range of environments, from the very harsh and potentially damaging to the mild and solution-like. Regardless of the

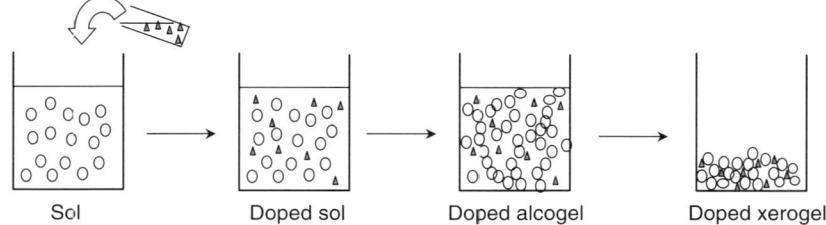

Figure 7. Steps in forming an entrapped metal complex. [Adapted from (19).]

eventual effect on the metal complex in question, the actual introduction into the sol–gel material to achieve "encapsulation" is usually a straightforward process (Fig. 7).

In general, an aqueous or alcohol solution of the desired species is introduced after a short period of hydrolysis of the orthoester precursor. The sol–gel process proceeds quite often as if the dopant molecule were not present, as long as the sol pH is not drastically affected. Treatments following gelation such as aging and drying of the sol material depend entirely on the purpose for which the material is to be used or studied. Aging and drying processes both involve a continuous evolution of the metastable silica gel (Section II.B.1.b.ii), which tends to contract considerably during drying—a process that can unpredictably affect the chemical behavior of metal complexes within the material. During the initial encapsulation process, metal complexes will be subjected to a variety of conditions that may affect their coordination environment depending on the recipe used for the sol–gel synthesis. Acid- or base-promoted sol–gels will require solution speciation of the metal complexes to be considered—an aspect that can be used to advantage, but that requires careful consideration of solution behavior. Substitution-inert complexes suffer less from interactions with the sol–gel medium, because the thermodynamics of metal–matrix interactions tend to be overcome by the kinetic barriers associated with ligand substitution reactions. Consideration of the sol–gel environment is of primary importance when labile transition metal complexes are to be encapsulated. To provide a background of what an inorganic chemist should consider when dealing with sol–gels, we must first delve slightly into the interaction of preformed silica with metal complexes. Many of the concepts found in the silica literature can be applied to sol–gel systems.

B. Related Silica Chemistry

1. Origins

Interest in combining the disciplines of inorganic chemistry and surface science can be traced back at least as far as the 1960s (87, 88). Equilibrium constants and thermodynamic data ($\Delta H°$ and $\Delta S°$) values for the absorption of 20 metal ions on a silica surface can be found in an article published in 1964 by Dugger et al. (88). The interest in silica by inorganic chemists was driven by the observation that metal ions tend to form inner-sphere complexes with silica silanol groups even at pH values where the surface should be positively charged. A great number of the studies that address this phenomenon can by found in a review by Iler (10). The typical experimental procedures of the studies reviewed involved the combination of a readily exchangeable form of a metal salt in solution with silica, followed by measurement of pH changes that accompanied the event of adsorption. The amount of metal uptake could be determined independently by some colorimetric test or by flame ionization techniques. In many cases, color changes that occurred during the process are noted and, in other cases, with the absence of diffuse reflectance technologies, creative solutions for obtaining absorption spectra were employed (87). Some of the important generalities discussed in Iler's review are as follows:

- The tendency of a metal ion to form an inner-sphere complex occurs at a pH value just below that required to precipitate the corresponding hydroxide complex.
- A linear relationship exists between $\log K_s B_2$ and $\log K_1 B_2$, where $K_s B_2$ and $K_1 B_2$ represent stability constants of the inner-sphere complex formed by chelation of two surface silanols and the hydroxylated metals, respectively.
- The tendency to form covalent bonds with surface groups decreases for metal ions in the order $Ga^{3+} > Al^{3+} > Fe^{3+} > Cr^{3+} > Y^{3+} > Co^{2+} > Cu^{2+} > Zn^{2+} > Ba^{2+} > Sr^{2+} > Ca^{2+} > Mg^{2+} > Rb^+ > K^+ > Na^+$.
- The Al^{3+}, Fe^{3+}, and Cr^{3+} ions have a tendency to form polymers on silica surfaces. The driving force is quite strong for polymer formation, as SiO_2 can catalyze the oxidation of ferrous ion to ferric ion.
- Several lines of evidence have led researchers to deduce that SiO^- behaves very similarly to water but with slightly higher ligand field strength.

2. More Recent Advances

More recently, with the advancement of technologies and expanded applications, the interest in oxide materials among inorganic chemists has been rejuvenated (85, 89–111). In particular Che and co-worker (85) brought together the fields of inorganic chemistry and oxide surface research by applying the appropriate name of "interfacial coordination chemistry," or ICC. The basic concept behind ICC is to bring the application of traditional inorganic chemistry techniques to surface chemistry by treating the silica surface in much the same way as any other ligand. The study of the changes observed as a result of metal–oxide interactions or lack thereof can allow one to gain information about the surface in question.

The design of heterogeneous catalysts is a particular application that shows the usefulness of ICC. The knowledge of surface microstructure, factors that effect loading and speciation can aid in the optimization of the desired system. Often a metal complex can act as both a reporter of the surface chemistry and a participant in the catalytic reaction that is to be studied. Familiar characterization tools that have been used in the study of ICC interactions include ultraviolet–visible (UV–vis) absorption, electron paramagnetic resonance (EPR), infrared (IR), thermomagnetic analysis (TGA), elemental analysis, X-ray absorption spectra [X-ray absorption near-edge spectroscopy (XANES) and extended X-ray absorption fine strucutre (EXAFS)], Brunauer–Emmett–Teller (absorption isotherm) (BET) and X-ray photoelectron spectroscopy (XPS). Diffuse reflectance accessories to study powdered materials are available for the spectroscopic range of near-UV to far-IR.

Some of the most extensively studied metal silica systems include those containing the complexes of Ni(II), Co(II), Mo(VI), and combinations thereof (87, 92, 93, 96–100, 102, 104, 106–109, 112–114). Interest in these complexes stems from their use in various catalytic systems including olefin dimerization, olefin metathesis, oxidation of CO, side-chain oxidation of alkyl aromatic compounds, and hydrodesulfurization of thiophene (85, 95, 106, 108). Most often metal complexes are used with mixed-ligand systems with at least one labile ligand that undergoes surface exchange with a silanol group, and one or more chelate rings that complete the coordination sphere. Ethylenediamine (en) is quite often used because its complexes are relatively resistant to substitution and its complexes are stable during moderately high-temperature processing (96, 97). One of the most applicable results for ICC comes out of a study by Lambert et al. (92) using cis-$[Ni(en)_2(H_2O)_2]^{2+}$. In their study, a variety of oxide surfaces including silica, zeolite, alumina, and clays were impregnated with the nickel complex. The techniques used to monitor spectroscopic changes of impregnated materials were UV–vis diffuse reflectance, near-IR, and IR. Three d–d transitions, v_1, v_2, and v_3 are known for octahedral $[Ni(en)_3]^{2+}$. The splitting of the d–d transitions that accompanies a change of the symmetric molecule to one with lower symmetry such as trans-substituted D_{4h} and cis-substituted C_{2v}, were used to interpret the speciation accompanying surface interactions. Through interpretation of the visible spectroscopy they determined a distinct preference of silica and alumina oxide materials to form the inner sphere cis-$[Ni(en)_2(SiO)_2]$, while the other materials showed mixed-speciation and electrostatic interactions with the surface. By using values obtained for v_2 and v_3 from the spectra of the adsorbed complexes, they provided supporting evidence for the placement of the oxide materials in the following spectrochemical order: $<\Delta_0(Cl) < \Delta_0 (AlO) < \Delta_0 (ZrO) < \Delta_0 (SiO) \sim \Delta_0 (H_2O)$.

C. From ICC to ISSC (Inner-Sphere Sol–Gel Chemistry)

Research with cobalt complexes bridges the gap between sol–gel and silica chemistry. In sol–gel, Co(II) can become part of the oxide matrix, forming a tetrahedral species and producing a light blue glass. Reisfeld and Jorgenson (24, 115) describe the spectroscopy of the cobalt material in detail. The tetrahedral cobalt species in glass has a unique spectral characteristic, which manifests itself in significant splitting of the v_3 band in the visible region into three peaks located at 645, 595, and 525 nm. The splitting cannot be solely attributed to spin–orbit coupling, as with tetra-chlorocobaltate, and is most likely the result of a greater distortion in symmetry (116).

In silica and sol–gel materials, octahedral cobalt species containing one or two dichelate ligand(s) in combination with more labile aqua or amino groups have been shown to form inner-sphere complexes with oxide surface groups (57, 87, 94, 105). The exchange occurs with the labile ligands while the bis-(chelate) ligands remain coordinated. As early as 1965, Burwell et al. (87) observed the formation of a rose colored $[Co(en)_2Cl(SiO)]^+$ from $[Co(en)_2Cl_2]^+$ on a silica surface upon treatment with base or after long equilibration times. The formation of a Co—O—Si bond has been observed by at least three different research groups using the systems described below and was confirmed by EXAFS data (57, 105, 106). Beland et al. (105) used postsynthesis ion exchange to diffuse $[Co(en)_2Cl_2]^+$ into an MCM material after drying. Iwasawa, et al. (106) also used $[Co(en)_2Cl_2]^+$ but in their case, impregnated on a γ-alumina surface. Both the Beland and Iwasawa research groups reported a similar bond distance for Co—O in the Co—O—Si unit, 1.92 and 1.97 Å, respectively. Work by our research group (57) has shown that the pink octahedral $[Co(neo)]^{2+}$ ion (neo = 2,9-dimethy-1,10-phenanthroline) forms a tetrahedral cobalt species on silica surfaces and in sol–gels (Fig. 8).

The splitting in the visible region is very similar to that of cobalt silicate (see above) but is far more dramatic with three maxima located at 654, 584, and 510 nm, indicating a larger degree of electronic asymmetry at the Co(II) center. The spectrum of the surface adduct resembles that of a crystallographically characterized carboxylate complex of $[Co(neo)]^{2+}$, which shows similar splitting of the high-energy term into three distinct components (117). The formation of the surface complex on silica simply requires the admixture of an aqueous or methanol 1:1 cobalt nitrate/neocuproine solution with silica. The pH profile for adsorption shows the event releases 1 equiv of H^+ ion per Co^{2+} ion adsorbed suggesting the exchange of one surface ligand in the inner-coordination sphere (Fig. 9).

When diffuse reflectance measurements are taken of the adsorbed tetrahedral species and the transmittance data are transformed into Kubelka–Munk units (Eq. 1), a plot versus loading affords a linear correlation analogous to the

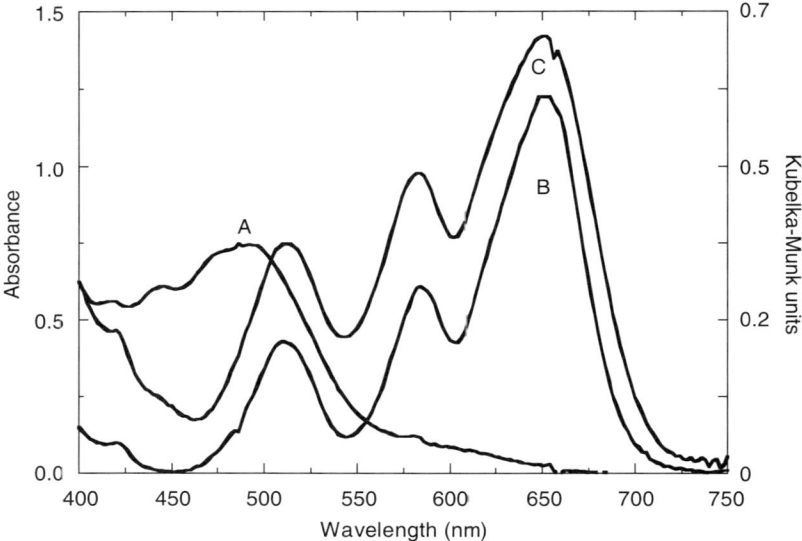

Figure 8. Absorption spectra of (**A**) 0.05 M [Co(neo)]$_{aq}^{2+}$ in aqueous methanol (50 vol.%), (**B**) 0.005 M [Co(neo)]$_{aq}^{2+}$ in a silica sol, and (**C**) diffuse reflectance spectrum of 6.7 μmol of [Co(neo)]$_{aq}^{2+}$ absorbed on 1 g of Merck silica grade 9385. [Adapted from (57).]

Beer–Lambert relationship (Fig. 10), consistent with the formation of one distinct species.

$$\text{Kubelka–Munk function} = \frac{(1-f_T)^2}{2f_T} \qquad (f_T = \text{Fraction of light transmitted}) \tag{1}$$

Synthesis of the tetrahedral complex in sol–gel requires the addition of base to the acid-catalyzed gel with a calculated ratio of between 1 and 2 equiv of hydroxide per Co^{2+} ion. This result suggests that sol–gel may require additional activation in order to form the same species as is formed on dry silica powder (57).

D. Noncoordinated Interactions

1. Background

Overall, ICC is most usefully applied to optimizing catalytic systems by probing a particular coordination site or sites on a silica or sol–gel surface.

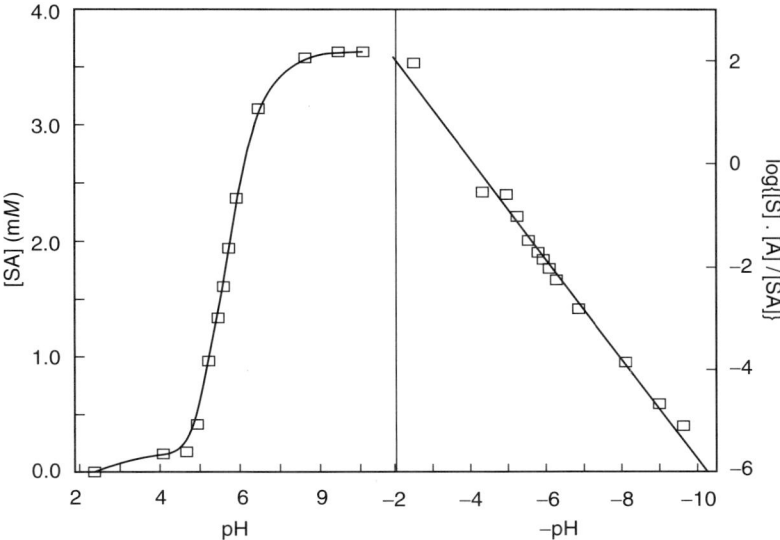

Figure 9. The pH titration for surface binding of $[Co(neo)]_{aq}^{2+}$ to silica. Titrations were performed at a 3.6 mM concentration (subsaturation) of $[Co(neo)]_{aq}^{2+}$, 1 g of silica grade 9385 with a final volume of 22 mL containing 12% volume/volume MeOH/H_2O. The silica sample was pretreated with 2 mL of 0.1 M HCl and titrated with 0.1 M Et_3N. The linear fits are in accord with a simple pH dependence on binding: $SH + A \rightarrow SA + H^+$ where SH = surface sites (SiOH), A = adsorbent ($[Co(neo)]_{aq}^{2+}$) and SA = $[Co(neo)]_{aq}^{2+}$ bound to the surface. The equilibrium concentration of surface sites was predicted using the mass law relationships, $[SA]_{max} = [SH]_{tot}$ and $[SH]_{eq} = [SH]_{tot} - [SA]_{eq}$. The total number of surface sites was predicted by the Langmuir data (not shown). [Adapted from (57).]

However, many transition metal (TM) ions and their complexes never form inner-sphere compounds with oxide surfaces; they instead interact electrostatically with the surface to varying degrees or exist in particular regions inside pores within the material (22). This type of interaction is expected when the coordination sphere consists exclusively of ligands such as en, 1,10-phenanthroline(phen), 2,2'-bipyridyl(bpy), or other heterocyclic ligands that resist exchange with surface oxide groups. Alternatively, metal complexes without a complete array of di- or multidentate ligands may not fully interact with the sol–gel matrix in the early stages of development of the material. By using these complexes or probing early stages of the material development, one can access a much broader range of microenvironments within sol–gels that are unavailable to their ICC counterparts (22). The optical properties of the lanthanides and the flexibility of the polypyridyl complexes of ruthenium make them excellent examples of the information that can be gained by such methods.

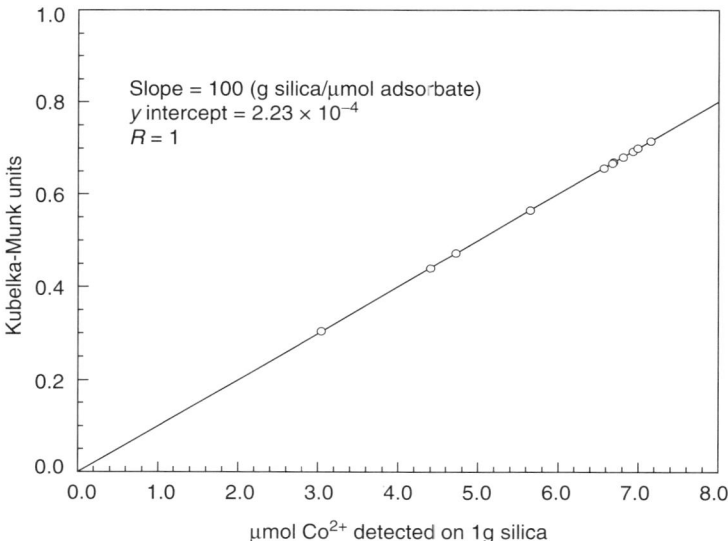

Figure 10. Linearity of Kubelka–Munk function with surface loading of $[Co(neo)]_{aq}^{2+}$. Transmittance values were obtained from diffuse reflectance UV–vis spectra taken at 654 nm. Data are shown for low loadings of $[Co(neo)]^{2+}$ on Merck silica grade 9385 and the values were converted to Kubelka–Munk units. Transmittance values $< 5\%$ were not used. The concentration of Co^{2+} was determined by removal of the adsorbed complex by treatment with $1\,M$ HCl followed by colorimetric determination with ammonium thiocyanate. [Adapted from (57).]

2. Luminescent Probes

a. **Lanthanide Complexes.** Lanthanide ions, in particular Eu^{3+} and Tb^{3+}, and their complexes incorporated within silica matrixes, allow for the development of optical or solar devices that take advantage of the fluorescent characteristics of these dopant molecules (20, 118–132). Most of the published data demonstrate methodologies designed to enhance the lifetimes of fluorescent emissions under the ambient conditions of sol–gel processing. Researchers have directly or indirectly provided insight into the environment of the sol–gel matrix using the unique spectroscopic properties of the rare earth ions. Several emission bands brought about by excitation of Eu^{3+} and Tb^{3+} occur in the visible region due to relaxation of the $^5D_0 \rightarrow {}^7F_J$ transition levels. Many of the emission bands show splitting due to deviations in inversion symmetry, however, the $J=2$ and 5 bands for Eu^{3+} and Tb^{3+}, respectively, are the most sensitive to such inversions (20, 121, 124, 125, 130, 133, 134). Several research groups have observed broadening of fluorescence bands as well as excitation band asymmetry within

TMOS, TEOS, ORMOSIL, and TEOS/polystyrene hybrid gels compared to corresponding crystalline solids (124, 125, 129, 130). The presence of heterogeneous microenvironments that encapsulate the rare earth complexes within xerogel matrices explains this broadening phenomenon (128).

Quenching of fluorescence occurs via two main pathways: (1) concentration effects that are mainly due to cross-relaxation between neighboring rare earth ion centers, and (2) weak vibronic coupling of ion excited states in the presence of O—H oscillators (121, 129). Both the lifetimes and the intensities of the emissions of the rare earth salts in solutions, sol–gels, and crystalline solids are enhanced by the introduction of ligands such as β-diketones, aromatic carboxylic acids and heterocyclic ligands (119). The increase in intensities is due to efficient ligand-to-metal transfer (LMT) when excited in the near-UV, and the increase in emission lifetimes is attributed to decreased contacts between metal ion centers and protection from silanol groups (118, 121, 129). Within the same group of lanthanide complexes, the fluorescence lifetimes increase from solution-to-solid-to-sol–gel matrices (120). Ligated lanthanide ions have shown at least a 40-fold increase in fluorescence intensity in TEOS derived xerogels when compared to the corresponding salts in the same system (118, 121). In their study of lanthanide complexes in sol–gels, Yan et al. (122, 128) attribute increases in fluorescence lifetimes at least in part to damping of the vibrational modes of acac, phen, dibenzoylmethanide (DBM), and p-aminobenzoic (P-ABA) ligands within the rigid silica matrix. The decrease in vibrations of the ligands is thought to eliminate a nonradiative relaxation pathway. Both IR and Raman spectroscopies used by these researchers show a decrease in, or absence of, vibrations of the ligands from complexes within xerogels they examined. These vibrations would appear within the same region where silica strongly absorbs. The IR data from other researchers, however, do not demonstrate the same results. In order to clarify this apparent discrepancy, it would be helpful to see comparisons of IR response as a function of a range of dopant concentrations within a particular sol–gel matrix. Such a study would help to eliminate the possibility that the strongly absorbing matrix masks the bands.

Rare earth salts exhibit only slightly different emission lifetimes in mixed-water–ethanol solutions when compared to TEOS derived sol–gels during the sol, gel, and xerogel stages (130). In general, the lifetimes of the emissions from the dopant molecule decrease as the gelation process unfolds due to an increase in water content and a greater likelihood of contact with pendant silanols. Matthews et al. (121) demonstrated that nonradiative relaxation is due to interactions with matrix silanol groups in the case of certain lanthanide compounds, which showed an increase up to twofold in emission intensity when doped in a hybrid ORMOSIL host versus a traditional TMOS xerogel. The ORMOSIL material contains pendant methyl groups instead of silanols. Further, Lochhead and Bray (130) showed that Eu^{3+} salts (nitrate, chloride, and perchlorate) within xerogels that were heated between 100 and 250°C, which resulted in a loss of surface hydroxyls, showed an increase in fluorescence intensities and lifetimes of the dopants still associated with their counterion.

Studies such as those of Lochhead and Bray (130) and Levy and co-workers (20), are unique in the approach used to examine the silica matrix, in that they use Eu^{3+} salts with differing counterions as opposed to chelated rare earth metals. Their choice of metal systems allows for more informative reporting of matrix effects as a consequence of the developing material. In the absence of additional coordinating ligands, changes in the spectral behavior of the metal ions are more sensitive to the changes in material environment. Luminescence changes were observed earlier in the evolution of the material structure, continued over a longer period of aging and drying, and were more extensive. Reisfeld and Jorgenson (115) report a greater availability of surface silanol groups to quench the fluorescence of guest lanthanide ions in TEOS versus TMOS systems resulting from the increased density of the former. The possibility of forming inner-sphere coordination compounds with the matrix increases when the sol–gel material is heated at temperatures > or $\sim 300°C$. Thus, at high-processing temperatures, the lanthanide ions may be better reporters of local structure on the surface than are ICC compounds on silica.

b. Ruthenium Polypyridyl Complexes. The polypyridyl complexes of ruthenium, especially $[Ru(bpy)_3]^{2+}$, are particularly suited for reporting matrix–dopant interactions when incorporated into sol–gels. The suitability of this complex for such studies stems from its unique optical properties and excited-state dynamics. The photo-redox behavior of ruthenium complexes also holds promise for examining the effects of charge separation within a solid support but, to date, offers little information about the sol–gel environment. Excellent discussions of $[Ru(bpy)_3]^{2+}$ photochemistry within solid supports can be found in the reviews by Gafney (135) and Castellano and Meyer (14).

The study of $[Ru(bpy)_3]^{2+}$ is facilitated by an intense and relatively long-lived emission maximum in the vis region due to relaxation of its triplet metal-to-ligand charge transfer (MLCT) excited to ground state (115, 136, 137). Compared to the behavior in solution, an increase in lifetime and a decrease in nonradiative decay occurs within a variety of matrices, and this behavior can be directly correlated to the density of the material as show in Table I (137).

The most intriguing aspect of the emission response is the blue shift that occurs upon the transition from fluid to rigid media, as in the transformation of a sol to a xerogel (115, 136–140). Formerly attributed to "rigidochromism" this phenomenon is actually a result of inhibition of solvent reordering around the dipole, which is created by an asymmetric excited-state charge distribution (136, 140). The solvent molecules that reorient themselves around the dipole lower

TABLE I

Fluorescence Lifetimes[a] as a Function of Glass Density[b]

Matrix	Refractive Index	Lifetime (μs) Measured	Nonradiative Rate (μs^{-1})
H_2O	1.33	0.62	1.54
Sol–gel	1.34	1.34	0.67
PVA[c]	1.50	1.74	0.48
Boric glass	1.64	2.00	0.39
Heavy glass[d]	1.90	1.98	0.36

[a] Measured at room temperature.
[b] Adapted from (115).
[c] Poly(vinyl alcohol) = PVA.
[d] Heavy glass = lead–tin fluorophosphate glass.

the energy of the Frank–Condon excited state and, following intersystem crossing, a Stokes shifted emission occurs (Fig. 11) (140).

When the rate of solvent reorientation is slow, intersystem crossing occurs from an energy state that is not thermally relaxed and thus is at a higher energy. In turn, the triplet MLCT excited state is at higher energy when emission occurs. The blue shift in emission is sensitive to the changes in structure as a sol–gel develops, as well as the protonation state of the newly formed surface.

Innocenzi et al. (140) performed a particularly interesting study investigating the blue shift phenomenon described above. They observed a change in the emission spectrum of $[Ru(bpy)_3]^{2+}$ from 620 to 603 nm as a result of the sol–gel conversion for a TEOS based thin film. A temporary reversal of this trend, or relative red shift, occurred during drying between 100 and 200°C. The normal blue shift trend returned between 200 and 300°C. The authors hypothesized the following series of events as an explanation for the "blue-to-red-to-blue" shift phenomenon.

During the sol–gel conversion, the typical blue shift occurs as the solvent molecules become more ordered at the developing surface where the ruthenium complex is most likely to reside in gels > pH 2, which is usually quoted as being the point of zero charge for the surface (10).

As shown by the IR data included in this study, the onset of adsorbed surface water loss occurs at 100°C, the same temperature at which the relative red shift in emission begins. Among a choice of explanations given, the authors hypothesized that the pores within the matrix at this point become void of solution, temporarily allowing greater rotational movement of $[Ru(bpy)_3]^{2+}$. As the pores collapse around the molecule, $[Ru(bpy)_3]^{2+}$ molecules are again restricted in their motion by the matrix acting as the solvent.

It is unclear if the argument used by the authors for the intermediate red shift is the most appropriate choice out of the other explanations presented within the

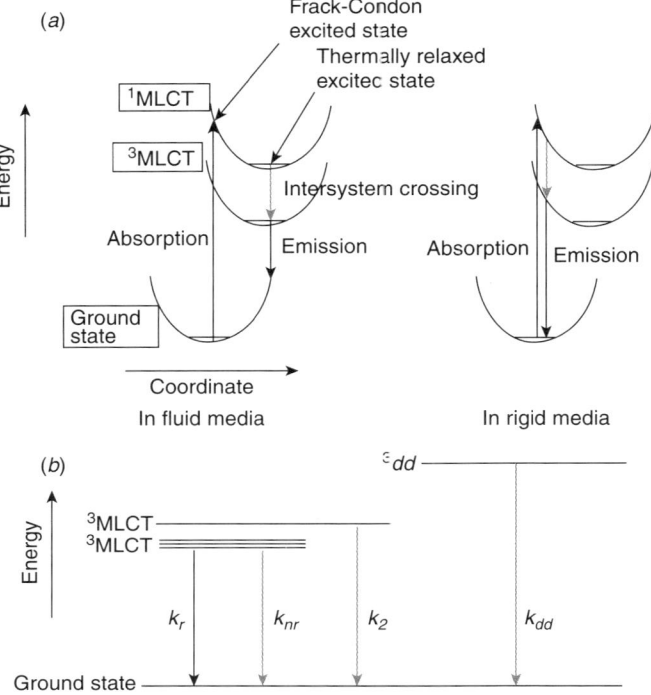

Figure 11. Origins of the blue shifted emission of ruthenium polypyridyl complexes in a rigid environment due to slow solvent reordering around the dipole of the excited-state complex relative to the timescale of the emission. [Adapted from (140).]

article. An alternative explanation given was that a decrease in pore volume resulted in the development of a slight increase in acidity of the pore walls. The positively charged complex may be repelled temporarily to a different region within the pore allowing for greater mobility and thus a "relative red-shift" occurs.

Matsui et al. (141) showed the luminescence changes upon the sol–gel conversion of monolithic gels that are doped with $[Ru(bpy)_3]^{2+}$ at different pH values and various H_2O/TEOS ratios. A similar relative red shift of λ_{max}, as with the study of Innocenzi, is observed after gelation of a sample made at pH 1 (Fig. 12) (141).

The ruthenium molecules do not undergo the same behavior within gels at pH 3. In that case, compared to the thin-film study reported by Innocenzi, the pH effect on the dopant location is more obvious. As the surface is positively charged at pH 1, the positively charged complex will tend to migrate away from

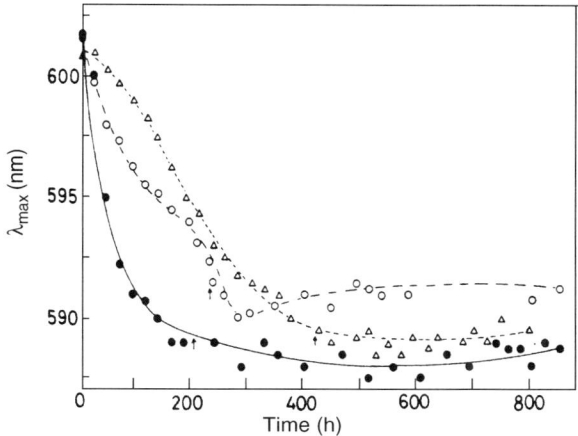

Figure 12. Maximum emission wavelength of $[Ru(bpy)_3]^{2+}$ over time during the sol–gel process under three preparation conditions: darkened circles, $r = 6.2$, pH 3; triangles, $r = 3.1$, pH 3; open circles $r = 6.2$, pH 1, where $r =$ molar ratio of water/TEOS. The gelation point for each is indicated by an arrow. [Adapted from (141).]

the matrix and deeper into the pore (i.e., bulk solvent rather than ordered surface as solvent). In addition, the shift in maximum of the emission of $[Ru(bpy)_3]^{2+}$ from the gel-to-xerogel stage in the pH 1 sample is smaller overall compared to the sample made at pH 3. The same behavior was observed by Mongey et al. (138, 139) for a series of polypyridyl ruthenium complexes that are doped in TEOS based monoliths. For $[Ru(bpy)_3]^{2+}$ xerogels, λ_{max} lies at 597 and 588 nm for gels made at pH 1 and 5, respectively. Additionally, the fluorescence lifetimes of all the complexes follow a similar trend where all samples made in gels of pH 5 exhibit longer lifetimes than those made at pH 1. The maximum difference in lifetimes of emissions between samples of pH 1 and 5 (5.5-fold) was seen in gels doped with the dimeric $[Ru(phen)_2(H_3Mptr)]^{2+}$ ($H_3Mptr = 3$-methyl-5-(pyridin-2-yl)-1,2,4-triazole).

The luminescence decay rate of ruthenium complexes doped in sol–gels does not follow a simple exponential decay. In theory, the interpretation of the excited-state dynamics of $[Ru(bpy)_3]^{2+}$ has the potential to offer a wealth of information about the distribution of environments within sol–gels. Ironically, the same nonexponential behavior in decay rate constants that offers the potential for this information makes attaining detailed information all the more difficult. In general, decay rate constants are fit assuming a bimodal distribution where there is a long component that represents the bulk of the distribution and a short component that represents a different and smaller distribution. The origin and explanation of this treatment can be found in the review of Castellano and Meyer (14).

Oxygen quenching by excited-state ruthenium allows for the development of devices for environmental and medical purposes. The use of the "diffusion-limited" reaction (142) can reveal information about the degree of porosity of different sol–gel matrices. An example of the information gained by such a study can be seen in the work of McDonagh et al. (143). The researchers show an increase in diffusibility of dissolved oxygen into ORMOSIL thin-film matrices containing $[Ru(4,7\text{-diphenyl-phen})_3]^{2+}$ as compared to more polar TEOS thin films. The increase in diffusibility was attributed to the nonpolar environment, which was more hospitable to an O_2 guest. Further, an increase in diffusibility of gas-phase O_2 was observed when the ratio of water to TEOS is decreased from 4 to 2. Decreasing the molar equivalents of water to TEOS from 4 to 2 results in a decrease in the hydrolysis rate and a subsequent increase in pore size (144). It would be interesting if the diffusion effect showed some predictable trend associated with pore size as would be revealed by a broader systematic variation of the ratio of H_2O/TEOS. Alternatively, the same effect can be attained by other methods to effect pore size such as the addition of DCCAs or pH changes.

3. Redox Active Probes

a. Introduction. As with the oxygen $[Ru(bpy)_3]^{2+}$ system, information about the diffusion of molecules into or within sol–gel matrices can be obtained by using electrochemical voltammetric or amperometric methods. The discussion of electrochemistry and sol–gel materials will be limited in this chapter to those that address such mobility issues. The review by Lev et al. (145) presents a much broader discussion of the topic of sol–gel materials in electrochemistry.

There are three primary methods for modifying an electrochemical system with sol–gel materials.

1. The working electrode can be coated with a thin film using the inert silica matrix with or without incorporation of dopant metal complexes (146–151).
2. The entire electrode assembly can be embedded in a monolith to measure solid-state electrochemistry (84, 152, 153).
3. The working electrode can be prepared from a composite of sol–gel material and conductive material (e.g., graphite powder), with incorporation of redox active molecules (154–156).

b. Thin-Film Studies. A thin film is usually adhered to a working electrode such as indium tin oxide (ITO), Pt, glassy carbon, or graphite by spin coating with a prepared sol and aging under controlled atmospheric conditions. The redox active species can be incorporated into the working electrode before spin coating or mixed with the electrolyte solution. Applying a thin film on the electrode can be problematic with respect to precisely controlling the thickness and maintaining the integrity of the material. It is also inherently difficult to determine the concentration of metal complex in the thin film once aged. The film will shrink and may not retain all of the metal complexes initially added to the batch. However, concentration can be estimated by the known diameter of the electrode, estimation of the thickness of the film by scanning electron microscopy (SEM), and removal of the metal complex from a commensurately prepared electrode (145, 146). Once the working electrode is prepared, it can be used in a standard three-electrode assembly, with the working, counter, and reference electrode, usually Ag/AgCl, placed in a supporting electrolyte solution.

Collinson et al. (146) performed one of the most extensive studies to probe metal ion mobility within the thin film on a working electrode. Several species, chosen to represent a range of molecular charges and size were incorporated in acid-catalyzed (pH ~ 3) TMOS thin films. The dopants including $[Mo(CN)_8]^{4-}$, $[FeCN_6]^{3-}$, $[IrCl_6]^{2-}$, $[Ru(NH_3)_6]^{3+}$, $[Ru(bpy)_3]^{2+}$, and $[Fc(CH_2OH)]$. Cyclic voltammograms obtained from the electrochemical experiments with modified working electrodes were compared to the response at a bare electrode in the corresponding metal-doped electrolyte solutions. All of the smaller complexes $[Mo(CN)_8]^{4-}$, $[FeCN_6]^{3-}$, $[IrCl_6]^{2-}$, $[Ru(NH_3)_6]^{3+}$,) showed typical symmetric cyclic voltammograms with Faradaic peak currents proportional to scan rate, v, indicative of finite diffusion at the electrode. By contrast, the larger $[Ru(bpy)_3]^{2+}$, showed a tail-shaped CV and $v^{1/2}$ proportionality indicative of a semiinfinite linear diffusion. For the anionic species, $[Mo(CN)_8]^{4-}$, $[FeCN_6]^{3-}$, $[IrCl_6]^{2-}$, the peak currents increased over time during repeated scans in electrolyte solutions and the redox potentials were similar to those measued with bare electrodes. The cationic and neutral species showed a decrease in current over time. In the system containing $[Ru(bpy)_3]^{2+}$, this decrease in current was correlated to the amount of leaching of the complex into the electrolyte solution determined from UV–vis absorption spectra. By estimation of the surface concentrations in the films containing $[FeCN_6]^{3-}$ and $[Ru(bpy)_3]^{2+}$, it was determined that only ~ 3–9% of the encapsulated molecules were electroactive. Interestingly, the negatively charged complex $[FeCN_6]^{3-}$, was found not to leach from the matrix even though in the experimental conditions used (pH < 2) the matrix should be negatively charged.

A good comparison to Collinson's study can be found in the work of Petite-Dominguez et al. (147), using the metal complexes $[FeCN_6]^{3-}$, and $[Ru(bpy)_3]^{2+}$. In their study, the redox active species were put in the supporting electrolyte solution and the square wave voltammetry peak current was used to compare the responses of bare, sol–gel and sol–gel–polymer hybrid modifed electrodes. Positively and negatively charged polymers, poly(dimethyldiallylammonium chloride) (PDMDAAC), and poly(vinylsulfonic acid, sodium salt) (PVSA), respectively, were used in the hybrid materials. The sensitivity of the electrodes for $[Ru(bpy)_3]^{2+}$, are enhanced in the order bare < sol–gel < modified sol–gel, while for $[FeCN_6]^{3-}$, the observed sensitivity showed an increase according to sol–gel < bare < modified sol–gel. A correlation between leaching of $[Ru(bpy)_3]^{2+}$ and loss in peak current was noted, in accord with the study of Collinson et al. (146). Uptake and repulsion of $[Ru(bpy)_3]^{2+}$ at the sol–gel electrode was almost perfectly reversible. The peak current increased over time as $[Ru(bpy)_3]^{2+}$ in electrolyte supporting solution was put in contact with the electrode. When this equilibrated peak current was reached and the electrode was placed in a new solution with only electrolyte, nearly all the complex leached off (85%) as the peak current returned nearly to its original value. However, when the composite negatively charged PVSA/sol–gel electrode was used, the positively charged ruthenium complex was not leached completely from the material. To further address the issue of electrostatic influence on the uptake and peak current of $[Ru(bpy)_3]^{2+}$, they looked at the effect of changing the sol–gel pH from 1 to 11 for the sol–gel modified electrode. The peak current for the ruthenium complex plotted against pH of the sol used to modify the electrode was remarkably similar to typical pH titration curve (Fig. 13) (147).

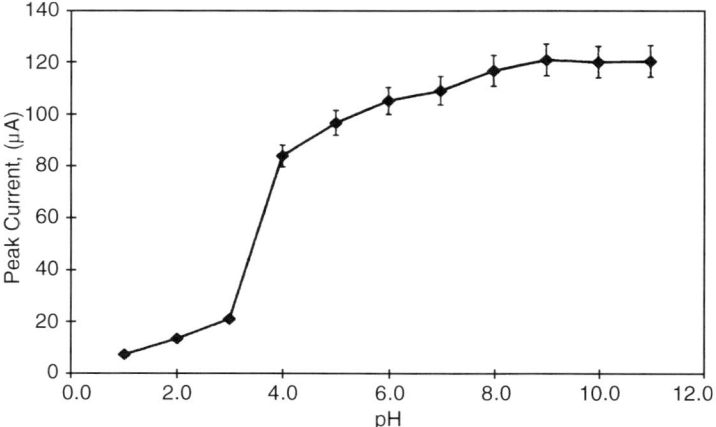

Figure 13. Effect of solution pH on pure sol–gel modified electrode: in 2.8 mM $[Ru(bipy)_3]^{2+}$, 0.1 M KCl. Working electrode: sol–gel modified graphite: film coating at 3500 rpm, 2 min; immersion time 10 min. Error bars show standard deviation for four different electrodes. [Adapted from (147).]

A similar absence of leaching of the negatively charged [FeCN$_6$]$^{3-}$ from sol–gel modified electrodes, as was noted in Collinson's work, was seen by this group. The complex also showed a resistance to diffusing into the sol–gel-modified electrode as deduced from the slow development of a maximum peak current. Apparently, the negatively charged silica surfaces within the confining regions of the sol–gel provide a barrier to diffusion once the complex reaches the interior pores. An alternative explanation given by Collinson's group was that including ferrocyanide in the initial sol mixture might effect the structure of the developing material. However, in the work of Petit-Dominguez the complex was not included in the initial processing steps and thus this explanation for lack of diffusion out of the matrix appears inconsistent with the alternative explanation given by Collinson.

Hu et al. (148, 149) described methods for increasing the affinity of a thin-film coated working electrode toward a particular metal. In their study using Nafion or sodium poly(styrenesulfonate) NaPSS modified sol–gel and PVA matrix supports on a working electrode they found that the formal potential of [Re(DMPE)$_3$]$^+$ (DMPE = 1,2-bi(dimethylphosphino)ethane) is not affected by the identity of support matrix but is affected by the type of modifier due to hydrophobicity differences of the respective resulting materials. The Nafion–silica composite gave the maximum voltammetric response for the redox active metal complex. In a separate study using Nafion modifier and TEOS derived thin films, they examined the systematic variation of the ratio of H$_2$O/TEOS and Nafion/TEOS. The rate at which the peak potential was reached during preconcentration decreased with increasing Nafion/TEOS but the magnitude of the equilibrium peak current increased. The alternative was true for increasing the water content of the initial sol recipe, the response time increased but the equilibrium peak current magnitude of [Re(dmpe)$_3$]$^+$ decreased.

c. Studies in Monoliths. Designing apparati for the measurement of voltammetry in sol–gel monoliths is a particularly clever experiment that has allowed researchers to determine apparent diffusion coefficients of encapsulated redox active metal complexes (84, 152, 153). A typical setup is shown schematically in Fig. 14 where the working electrodes in these experiments are ultramicroectrodes on the order of 13 μm in diameter.

As with the thin-film method it is difficult to define concentration in the final monolith because of the shrinkage that occurs during aging. Potential step amperometric methods or simultaneous equations can be used to overcome these problems, and self-consistent values for the diffusion coefficients can be obtained. An excellent treatment of both methods of solving for diffusion can be found in the paper by Cox et al. (153). As noted by the authors it is important to keep in mind that diffusion coefficients obtained in this manner cannot be solely attributed to the physical exchange of redox molecules, since electron

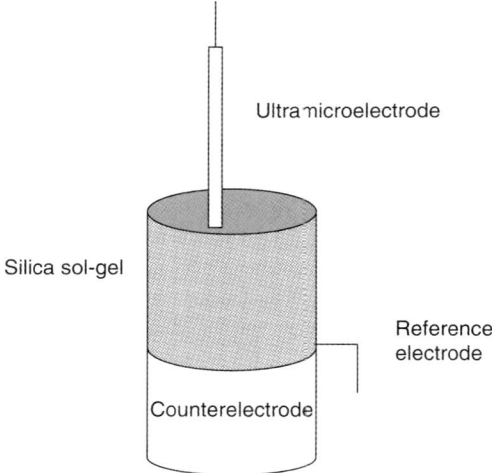

Figure 14. A typical experimental setup for examination of the solid-state electrochemistry of a sol–gel monolith. [Adapted from (152).]

self-exchange and counterion mobility may also contribute to the observed diffusion. The groups of Cox (153) and Holstrom (152) observed similar diffusion behavior for two polyoxoacids of tungsten, $H_4SiW_{12}O_{40}$ (silicotungstic acid, STA) and $H_3PW_{12}O_{40}$ (phosphotungstic acid, PTA), respectively. Cox et al. (153) determined that the diffusion coefficient for the sol–gel encapsulated STA is very similar to that observed in solution - demonstrating that even though the system used is in the solid-state, it still retains some fluid chemical properties. Holstrom et al. (152) demonstrated that addition of surfactants can increase the diffusion of PTA within the matrix by a factor as great as 20 when compared to the diffusion in unmodified gels. This increase in diffusion as a result of sol–gel modification is similar to the increase seen for O_2 diffusion in ORMOSIL materials based on ruthenium luminescence quenching. As shown by Collinson et al. (84), it is not always straightforward to predict the changes in diffusibility as a function of gel aging and drying (Fig. 15). The diffusion of neutral $FcCH_2OH$ changes markedly over time as a monolithic TMOS gel is aged and dried. Conversely, the mobility of $[Fe(CN)_6]^{3-}$ remains unchanged during aging even though there appears to be a size decrease of at least fivefold in the gel dimensions (Fig. 15).

d. Other Techniques. There are many examples of more complex systems using modified electrodes. Ogura et al. (157) demonstrate that a working electrode can be modified with a sol–gel thin film containing an active enzyme

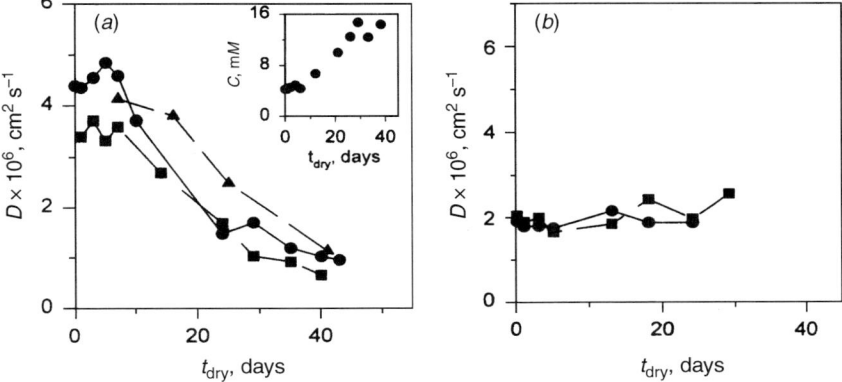

Figure 15. Variation in diffusion coefficients of gel-encapsulated FcCH$_2$OH (a) and [Fe(CN)$_6$]$^{3-}$ (b) determined using solid-state electrochemistry. The gels consists of 1:10:3:2 × 10^{-4} TMOS/H$_2$O/ MeOH/HCl. The dopant concentrations were ~4–5 mM in the initial mixture. Inset of (a), change in concentration of FcCH$_2$OH as a result of drying and aging of the gel [Adapted from (84).]

and that the enzyme still maintains its structure and activity. In their system, they were able to detect urea with encapsulated urease using potentiometric techniques. Gun and Lev (155, 156) show the utility of designing a complex multicomponent sol–gel graphite composite electrode containing the active enzyme glucose oxidase and covalently tethered ferrocene functional groups that provide signal transduction. In their system, they were able to detect glucose using amperometric sensing.

Doping of covalently modified ferrocenes into sols has been used by several other research groups to prepare modified electrodes (157–160). These tethered molecules cannot physically diffuse through the material, so the diffusion coefficients determined by such studies can give insight into the other mechanisms available to transport electrons to the working electrode. For comparison purposes, the diffusion coefficient for ferrocene is reduced by at least three orders of magnitude when covalently attached within a TMOS thin film compared to a similar study mentioned above by Collinson using noncovalently attached ferrocene methanol (FcCH$_2$OH) in a sol–gel monolith (84, 160). Although the two systems used by the individual researchers differ, the implication is that actual physical mobility of the molecules may contribute significantly to the measured diffusion coefficients. It would be interesting to compare both tethered and nontethered ferrocenes within the same sol–gel system. As will be discussed below, the effects of tethering molecules within silica matrixes can have dramatic effects on the chemistry of such molecules, but consistent conditions are essential for meaningful comparisons to be drawn.

IV. COVALENTLY ATTACHED COMPLEXES

A. Introduction

Attachment of metal complexes to silica through covalent tethering has been used in preparing supported catalysts for many years. More recent developments, such as imprinting for selective binding of small molecules and metal ions have expanded the scope and utility of this class of materials considerably (38, 40, 161–164). Covalent attachment of molecules improves the stability of the metal–silica interactions, and can enhance the recyclability of catalytic materials considerably. Use of sol–gel processing also improves the dispersion of transition metal complexes through the support material, and allows for increased loadings to be obtained, relative to classical surface attachment approaches. Sol–gel immobilization may also result in significantly enhanced stability toward degradation of metal complexes during catalytic reactions. Often, such effects arise from site-isolation of the individual metal complexes, which prevents reactions between catalytic sites that form inactive decomposition products (77). Covalent attachment in some fashion is practically essential for applications such as structural imprinting, where a memory of the template structure must be maintained permanently. Preparation and sol–gel immobilization of metal complexes presents significantly greater synthetic challenges than simple entrapment procedures. We will therefore begin our discussion of immobilized complexes with an overview of some of the methods that have been used to prepare silylated ligands and metal complexes for immobilization purposes. Methods used for immobilization are then discussed, followed by a brief summary of issues associated with characterizing the resultant materials.

B. Preparation of Precursors

1. Preparation of Silylated Ligands

A variety of routes have been developed for preparing metal-binding ligands with silylated side chains, and methods compatible with a range of chemical functionality are now available. Schubert et al. (17) reviewed many of these synthetic approaches in some detail. Viable preparative routes to silylated organic molecules initially emerged around the 1940s, largely pioneered by a group at Degussa (37). Development of these commercial-scale procedures was originally driven by interest in the bifunctional reagents as adhesion promoters for inorganic–organic polymer composites such as glass fiber-reinforced materials and mineral-filled elastomers. Further developments arose from the realization that attachment of catalytically active metal complexes to silica gels via silylated ligands afforded supported complexes whose properties combined the positive attributes of both heterogeneous and homogeneous catalysts. The historical development of these so-called "heterogenized homogeneous" catalysts is particularly well documented in the introduction to the book by Yermakov et al. (35).

Examples of reactions used to prepare bifunctional ligands for immobilization on silica are illustrated in Fig. 16. By far the most widely used methods are hydrosilylation [Fig. 16(1)] and nucleophilic substitution reactions [Fig. 16(2)] on silylated alkyl chlorides. Additional routes include organometallic exchange reactions [Fig. 16(3)], isocyanate-based coupling reactions [Fig. 16 (4a and 4b)], and amide formation [Fig. 16(4c)]. Many useful bifunctional silane reagents for use in ligand preparations are commercially available, including those having chloride [Fig. 16(2)], isocyanate [Fig. 16 (4a and b)], and amine [Fig. 16 (4c) functional groups. Metal-binding ligands, including those with thiol, isocyanide, pyridine, and alkylene polyamine (e.g., endiethylenetriamine) are also readily available, as are derivatives of ethylenediaminetetraacetic acid (EDTA) (165).

Hydrosilylation [Fig. 16(a)] is a facile reaction that can be accomplished using a variety of catalysts, or by photochemical means (43). Since catalytic hydrosilylations with trialkoxysilanes are typically sluggish, it is customary to perform the reaction with trichlorosilane ($HSiCl_3$) instead (43, 166). The product alkyltrichlorosilanes, however, are highly reactive compounds that are difficult to purify and handle. The high reactivity of these compounds is desirable for attachment of metal complexes to solid surfaces, but can cause problems with sol–gel reactions. The principal problems are (1) phase segregation during cocondensation with alkyl orthosilicates (Section IV.C.1), and (2) release of HCl—a sol–gel catalyst—during the hydrolysis step, which may significantly influence the final gel structure (166). Consequently, the chlorides are typically converted to the less reactive alkoxy derivatives by reaction with the appropriate alcohol in the presence of triethylamine to neutralize the liberated HCl. Reversion to the chlorides, if desirable, may be accomplished using trichloroacetyl chloride as a chlorinating agent (166). Attachment of TM complexes to silica surfaces may also be carried out by addition of a vinyl-substituted ligand to a silica surface modified with silane functional groups [Fig. 16(b)] (167). Photochemical hydrosilylation has been used to prepare silylated precursors in a number of instances. For example, Krocher et al. (168) used the approach to form silylated ligands from reaction of vinyl triethoxy-silane and dialkyl- or diaryl- phosphines. Yields of up to 97% were obtained for these reactions after purification by fractional distillation.

Silyl tethered amines are used frequently in amide-forming coupling reactions [e.g., Fig. 16(c)] (169, 170). Isocyanate compounds are also commercially available, and have been used in the immobilization of enzymes within the sol–gel pores (171). These isocyanates can also be used to prepare silylated phen ligands that have been used in our own studies (172). Silylated isocyanates have

1. Hydrosilylation

   (a)

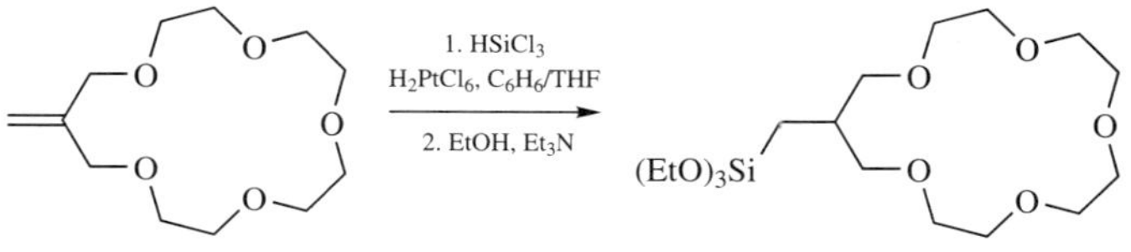

   [reaction scheme with reagents: 1. HSiCl₃, H₂PtCl₆, C₆H₆/THF; 2. EtOH, Et₃N]

   (ref. 43)

   (b)

   [reaction scheme on SiO₂ surface]

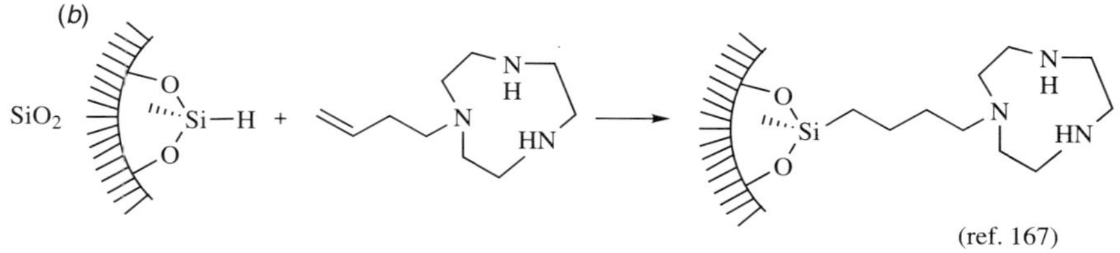

   (ref. 167)

2. Nucleophilic displacement of alkyl halides

   (a)

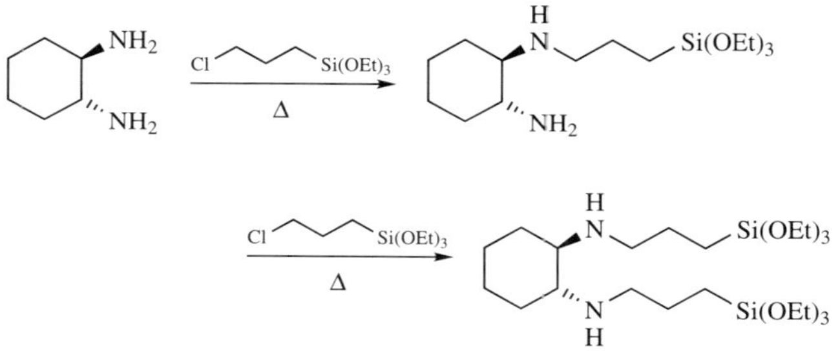

   [reaction scheme with Cl~~~Si(OEt)₃, Δ]

   (ref. 185)

   (b)

   [reaction scheme with Cl~~~Si(OR₃), NaI, DMF or DMSO]

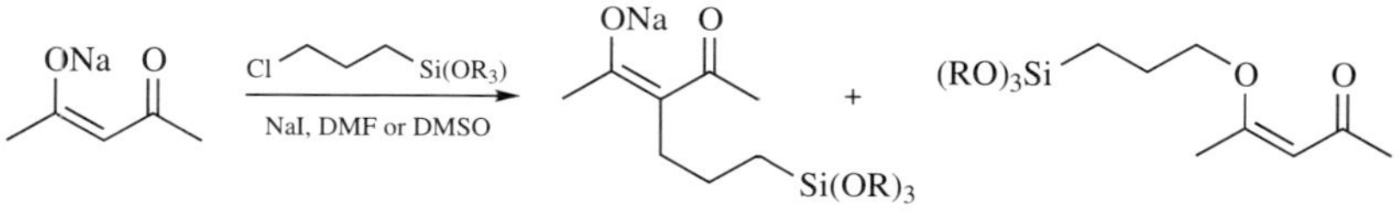

   (DMF = N,N-dimethylformamide)
   (DMSO = dimethyl sulfoxide)

   75%   25%

   (ref. 175)

   (c)

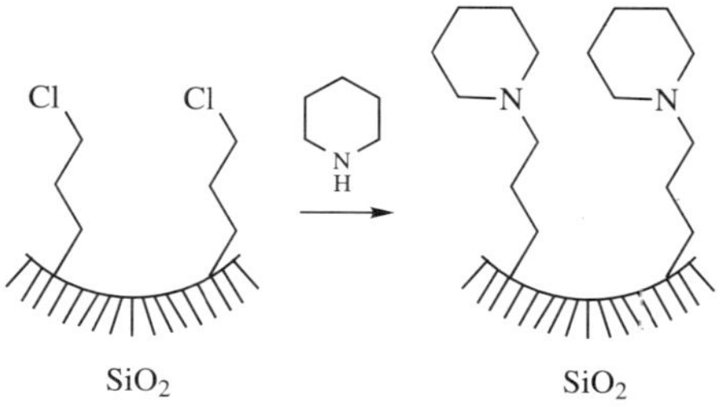

   [reaction scheme on SiO₂ surface with piperidine]

   (ref. 201)

Figure 16. Examples of commonly used methods for preparing silylated ligands.

3. Grignard Reactions

(a) (ref. 178)

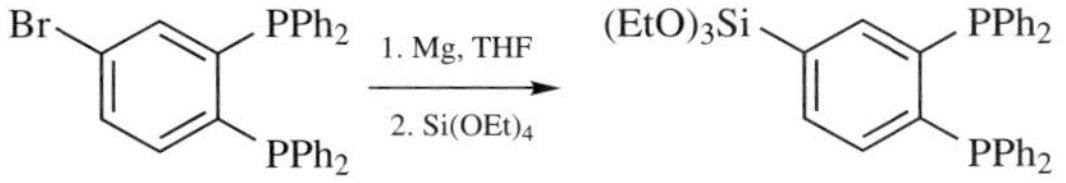

(b) (ref. 177)

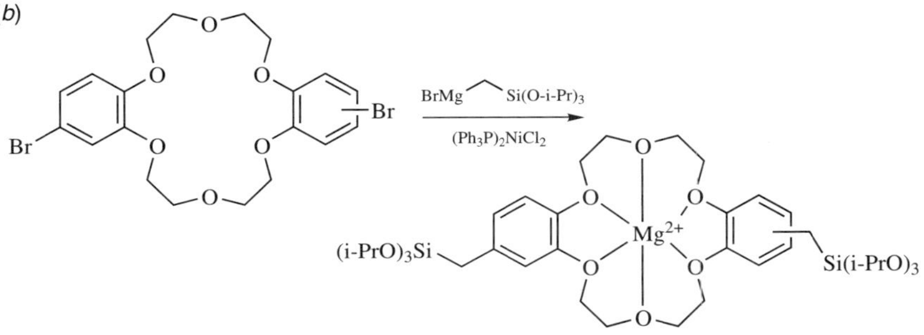

4. Coupling reactions

(a) (ref. 181)

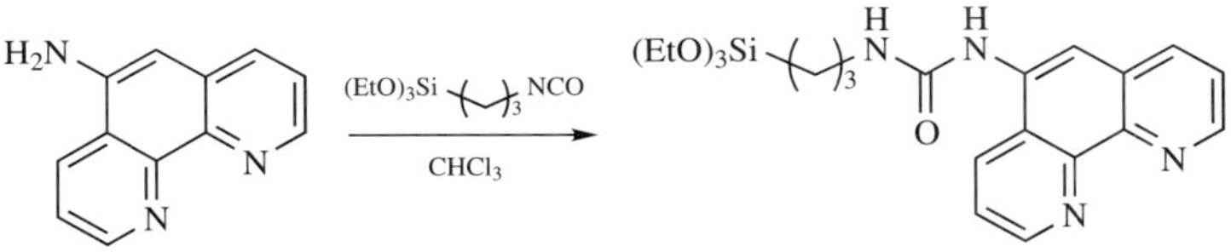

(b) (ref. 173)

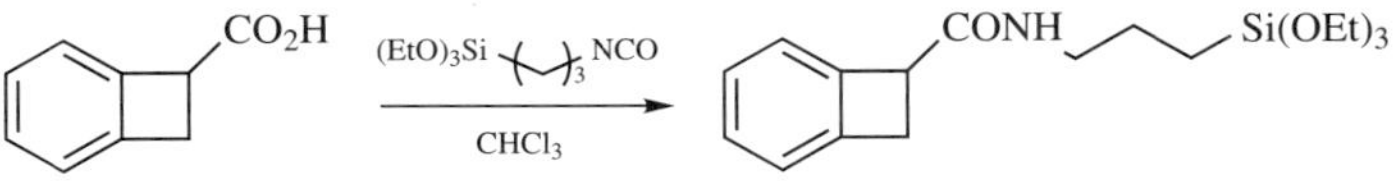

(c) (Boc = *tert*-butoxycarbonyl) (ref. 169)

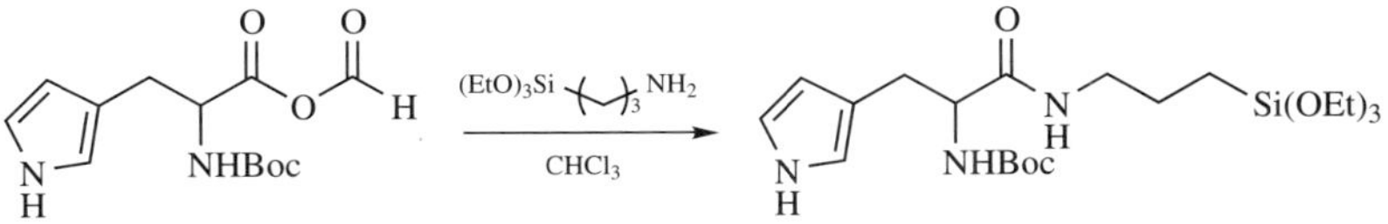

Figure 16 (Continued)

also been used in decarbonylative formation of amide linkages during modification of C_{60} for covalent immobilization. Although potentially useful for the purpose, this reaction apparently has not been applied in the synthesis of a metal-binding ligand or silylated metal complex (173).

Nucleophilic displacement of the chloride in chloropropyltrialkoxysilane derivatives (Fig. 16, reaction 2) has been widely applied for the preparation of tethering ligands (37). Ligands prepared by this route include amines, phosphines, cyclopentadienyls, thiols (via nucleophilic substitution with thiourea) and polysulfides. Similarly, quaternary ammonium salts may be prepared by displacement of chloride using a tertiary amine. Avnir and co-workers (174) recently demonstrated the utility of the latter phase-transfer functional groups for enhancing noncovalent association between anionic dopants and silicate matrices, as a means for improving the stability during catalysts. Inclusion of such surface modifying agents appears to reduce leaching of the metal complexes into solution, and thus improves the long-term stability of catalysts based on entrapped, rather than tethered, metal complexes. Despite their widespread application, the alkyl chlorides do suffer from some drawbacks associated with their low reactivities. For example, Urbaniak and Schubert (175) pointed out that alkylation of acetylactonate salts with trialkoxysilyl chlorides leads to inconvenient levels of O-alkylation of the acetylacetonate ligands. They reported significant improvements in yields of the desired C-alkylated products via conversion to the alkyl iodide using a modified Finkelstein reaction.

Grignard reagents have also been used for preparation of silylated ligands, such as the derivatized crown ether [Fig.16(a)]. Although Grignard reagents typically react readily with alkoxysilanes to form undesirable products, the use of tri-isopropyloxysilyl sidechains results in clean cross-coupling reactions (176). An example of such a reaction is that used to derivatize dibromo-dibenzo-18-crown-6 with silylated sidechains. The resultant ligand was subsequently used to form materials with efficient and reasonably selective alkali metal ion-binding properties (177). Grignard reaction of phosphine-derivatized phenylmagnesium bromide with TEOS has also been applied (178).

2. *Preparation of Silylated Metal Complexes*

Although there are a few scattered examples to the contrary [Fig. 17(a) and (b)] (179, 180), introduction of the silylated sidechains into preformed metal complexes is rarely performed. Instead, it is much more common for the metal complex to be prepared by reaction of an appropriate metal precursor with a silylated ligand.

One of the most prominent synthetic problems associated with preparing metal complexes bearing silylated ligands is that the sidechains are sensitive to water, and therefore susceptible to prematurely undergoing the hydrolysis and

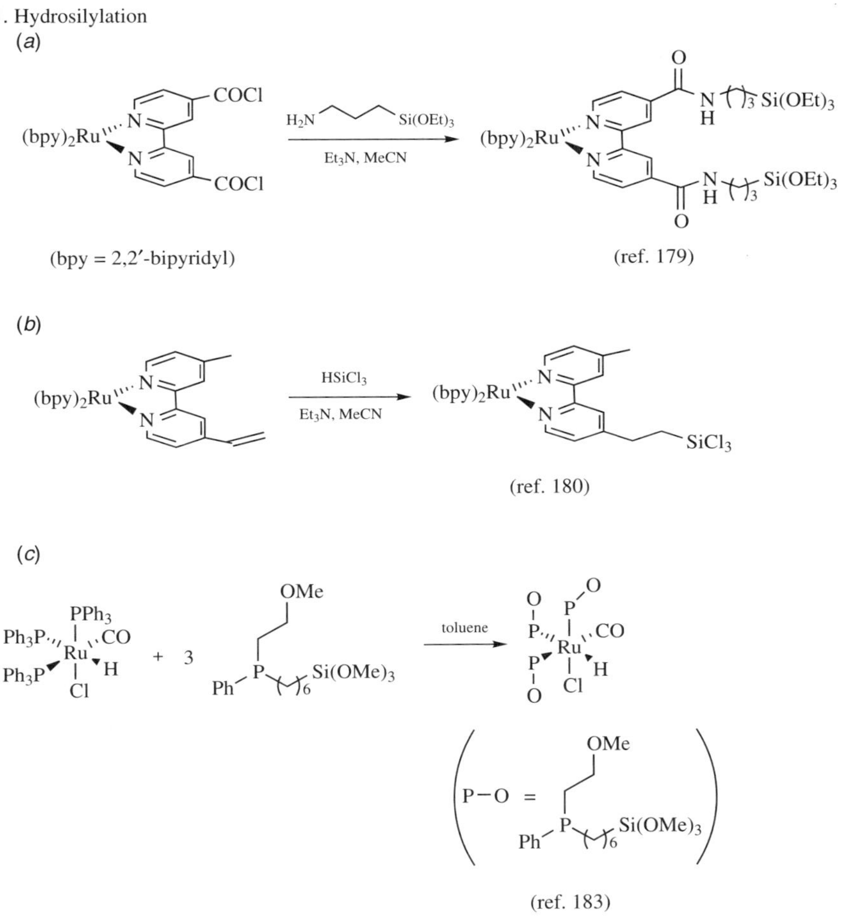

COORDINATION COMPLEXES IN SOL–GEL SILICA MATERIALS 377

1. Hydrosilylation

(a)

(bpy = 2,2'-bipyridyl)

(ref. 179)

(b)

(ref. 180)

(c)

(ref. 183)

Figure 17. Examples of synthetic methods for the preparation of silylated metal complexes.

condensation reactions for which they are intended. Consequently, the side chains are generally incompatible with the conditions used for preparation and isolation of many coordination complexes, and alternative nonaqueous routes for synthesis must be sought unless the desired complex forms quantitatively in solution and can be used directly for doping of a sol–gel. Such is the case with preparation of silylated ferroin derivatives (172, 181), but these complexes are exceptional in that the sequential binding constants of the phenanthroline ligands to the Fe^{2+} centers are anomalous (182), and ensure complete formation of the tris-phenanthroline complex without need for an excess of ligand to drive the coordination equilibria. For the majority of coordination complexes, and particularly those with complex polymetallic architectures, the silylated ligands will rarely survive the conditions required for assembly of the desired metal complex. The problem is compounded by the fact that the majority of TM salts are commercially available in hydrated form. Syntheses of metal complexes intended for immobilization thus require careful consideration of conditions, and novel synthetic procedures must often be developed. For example, Lindner et al. (183) developed a scheme for the preparation of rhodium phosphine hydrides by displacement of triphenylphosphine ligands with more basic alkoxyphosphines [Fig. 17(c)]. This procedure avoided the standard conditions for preparation of the metal complex, which involve use of $RhCl_3 \cdot xH_2O$. In some cases, use of hydrated precursors has afforded good yields of silylated metal complexes without special precautions—as, for example, in the preparation of ruthenium derivatives of trimethoxysilylated phosphines by Krocher et al. (184). In that case, the presence of a significant excess of silylated phosphine and the large volume of methanol used may both have contributed to the successful outcome.

One aspect that should be borne in mind is the fact that anhydrous alcohols corresponding to the alkoxy groups on the trialkoxysilyl side chains are ideal solvents for complexes that are polar or charged (to the extent that ionic complexes are soluble in the appropriate alcohol). Exchange of alkoxy groups at the silicon atom is a nondestructive reaction, which results in formation of precisely the same functionality after solvolysis as was present before. For example, incorporation of Rh centers into a triethoxysilylated chiral cyclohexanediamine ligand in ethanolic solution was reported by Adima et al. (185), although yields and analytical data for the products were not reported. The complexes were used in the preparation of enantioselective catalytic silsesquioxane and Q,T-hybrid materials (Section II.A.1.a.iii). In fact, it has been observed that reactions may be performed in the presence of different alcohols without the occurrence of exchange reactions (possibly due to the absence of appropriate catalysts for the alkoxide exchange reaction). For example, demetalation of the $MgCl_2$ adduct of a triisopropyl-substituted benzo-18-crown-6 derivative by refluxing in anhydrous methanol afforded a clean sample of the isopropyl-substituted compound (177). Note, however, that the compound was isolated in low yield (39%) after chromatography. The conditions for purification were not specified, but it is possible that the demetalation conditions actually led to a significant amount of exchange of some isopropyl groups for methoxy groups, affording derivatives that were reactive with the chromatographic media, which would account for the low yield obtained.

An approach that takes advantage of the noninterfering nature of homoalkoxide exchange is the use of alcohol-ligated complexes as precursors to silyl-containing sol–gel dopants. For example, di- and trinuclear ruthenium clusters can be prepared with alcohol ligands at exchangeable sites (186). These oxo-bridged clusters cannot be prepared in the absence of water, since it is required to form the bridging oxo group, so direct assembly of the clusters in the presence of silylated ligands is precluded. The appropriate alcohol-terminated precursors, however, can be obtained either as solids or as anhydrous solutions in alcohols. Exchange of the alcohol ligands for silylated ligands such as 2-(triethoxysilyl) ethylpyridine is facile, and the approach thus affords convenient routes to intact silylated complexes under conditions that retain the desired cluster structures. These complexes have been immobilized in sol–gel materials, and show catalase (H_2O_2 disproportionation) activities. The immobilized complexes are currently being examined as catalysts for oxidations. We are also examining the utility of the inert ruthenium clusters for imprinting of architecturally defined ligand arrays into the silicate matrix, for generation of more active catalysts based on binuclear iron and copper centers.

A continued challenge to the ingenuity of the synthetic coordination chemist is to devise viable routes that circumvent the practical difficulties associated with generation of suitably modified complexes for immobilization. Very few generally applicable routes are available, although the emergence of such procedures would greatly expand the range of transition metal complexes that can be immobilized on silica and other oxide materials.

3. Purification

Purification of silylated compounds can be an even more challenging task than their synthesis. Due to the sensitivity of the silylated side chains, many silylated ligands and complexes are used in immobilization procedures as obtained from the synthetic reactions without purification (179). It is no accident that most of the TM complexes bearing silylated side chains that have been successfully prepared and purified to date are neutral species, and have mostly been organometallics. These compounds can be prepared reliably from water-free precursors, and nonaqueous preparation, and purification routes are readily available.

Very few reports provide detailed accounts documenting purification of silylated ligands or complexes, unless the compounds are crystalline or amenable to distillation. The presence of the trialkoxysilyl functional group introduces significant problems for purification by chromatography, since the functional groups are, unsurprisingly, reactive with solid phases such as silica or alumina. Stille's group reported chromatographic purification of a triethoxysilyl compound on silica that had been pretreated with ethyl acetate, but reported low yields due to reaction of the compounds with the silica gel (178). Avnir and co-workers (77) purified neutral complexes on Florisil (magnesium silicate) using ether or ether–pentane mixtures for elution. Kraus et al. (173)

purified a silylated C_{60} derivative using a polystyrene gel, eluting with THF. In all cases, the successfully purified complexes were electrically neutral.

Krocher et al. (184) reported preparation of neutral Ru(II), Ir(I), and Pt(II) trimethoxysilyl phosphine complexes that were obtained in high yield and purity by synthesis in methanol followed by evaporation of solvent, extraction into CH_2Cl_2, filtration through Celite and evaporation, which afforded the products as oils. Reaction of appropriate metal–cod (cod = 1,4-cyclooctadiene) precursors resulted in mixtures of isomers, however, which were not separated. In particular, ruthenium complexes proved troublesome—up to five isomeric complexes were discernible by NMR, and purification was not possible. The authors noted that the presence of the silylated sidechains hampered purification by standard methods (see above). This result illustrates the difficulties associated with preparation of pure metal-containing precursors for sol–gel syntheses.

A recent report by Barness et al. (166) indicates that chromatographic separation of trimethoxysilyl compounds is facilitated by inclusion of a small percentage of TMOS in the eluent mixture, which is believed to act by passivating the silica surface. This new development has not yet been widely applied to the purification of silylated ligands or metal complexes, but offers a promising solution to a major difficulty associated with obtaining purified compounds. The purification routes are generally restricted to neutral complexes, however, because of the high affinity of charged complexes for the surfaces of most chromatographic supports, including silica gel (Section III.B.1).

C. Preparation of Materials

1. Sol–Gel Immobilization Techniques

Immobilization of metal complexes within, or on, a silicate material can be accomplished using one of four basic methods, which differ with respect to the order in which the various components are formed or introduced (Fig. 18) (15):

1. In method (a) of Fig. 18, the alkoxysilane-functionalized ligand is first used to functionalize a silica surface. The modified material is then mixed with a metal precursor to generate the desired supported complex.
2. In method (b) of Fig. 18, the silylated metal complex is stirred with the support material to allow for covalent attachment to the silica.
3. In method (c) of Fig. 18, the sol–gel process can be used to produce a material in a method analogous to route (a). First, the substituted alkoxysilane is immobilized by a sol–gel route, typically by stirring with a sol–gel precursor such as TEOS or TMOS, to distribute the ligand throughout the material. A metal precursor is then added to the material, and covalently bonds to the accessible ligand sites.

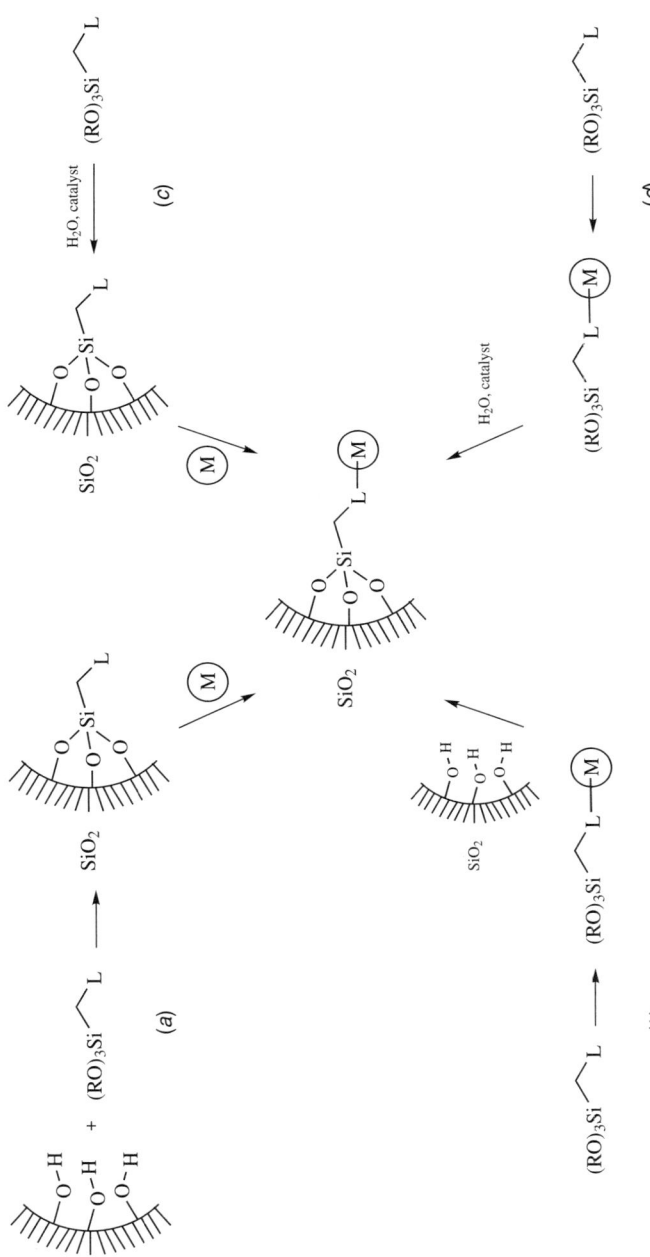

Figure 18. Four possible routes used for preparing silica-immobilized transition metal complexes. [Adapted from (15).]

4. In method (*d*) of Fig. 18, in the sol–gel counterpart of (*b*), the alkoxysilane-substituted metal complex is formed prior to its incorporation into the material by sol–gel processing.

Methods (*a*) and (*b*) represent the classical methods for surface anchoring of metal complexes, whereas (*c*) and (*d*) represent the use of sol–gel chemistry to afford materials containing metal complexes dispersed throughout the silicate material. It should also be noted, however, that method (*c*) does not allow for the complexation of the metal by sites that were "closed-off" during the sol–gel processing with the organic alkoxysilane (76). In early studies that established some of the principal advantages of sol–gel processing for preparing supported catalysts, the four methods were compared for an extensive series of metal complexes. Preparation of materials through these methods has a considerable influence on the catalytic activities and stabilities of the resultant hybrid materials. In general, materials made using method (*d*) were observed to be significantly more active than those materials generated by methods (*a*) and (*b*), and at least as active, if not more so, than those catalysts prepared by method (*c*) (15, 17, 55, 187–189). In addition, method (*d*) afforded the most leachproof catalysts, when compared to the materials obtained from the other three methods. Note, however, that preparation of materials by method (*d*) did not *always* afford the most active catalyst. Capka et al. (188) showed that preparation of a supported rhodium complex by method (*d*) did not result in the most catalytically active material, and in fact the highest activities were observed with materials generated by surface grafting [method (*b*)]. Even though the preformed metal complex/sol–gel material was not the most active, it still exhibited minimal leaching and the researchers proposed that the lack of activity observed was due to the inaccessibility of the metal sites within the silicate matrix. Comparative studies of this type were important in establishing the utility of sol–gel processing for catalyst preparation, but are rarely performed any more. It is important that the reader keeps in mind that although preparation of catalytic materials through the sol–gel process is generally far more effective at generating active catalysts, *this will not always be the case.*

Immobilization of a silylated ligand or metal complex during a sol–gel reaction is typically a straightforward matter of introducing the metal complex into the sol–gel mixture, either at the outset or after the other precursors have been allowed to react to form a sol. After the effort needed to prepare and purify the necessary silylated ligands and metal complexes, the sol–gel reactions represent a relatively easy step in the overall scheme for material preparation. The simplicity of the sol–gel procedures belies the real complexity of the immobilization reactions, however, and it is clear from many studies that choice of the sol–gel conditions must be made carefully. One of the most common reasons cited for the use of sol–gel processing is the intimate molecular-level mixing of components that can be obtained at the outset of the reactions. Despite the fact that homogeneous distributions can be obtained at the outset of the reactions, this homogeneity may not persist in the final material, since the kinetics of hydrolysis and condensation reactions of different components can differ by orders of magnitude. If the differences in polymerization rates for the various components are sufficiently large, the result can be a heterogeneous material comprising separate domains of homopolymers, instead of the desired copolymerized material. Such phase separation will obviously have a major influence on the final properties of the material, and will interfere with the tailoring of important properties such as site isolation (190). It is by no means valid to assume that the material structure is composed of a random copolymer of the two monomers. Electrochemical investigations of hybrid gels created from silylated ferrocenes and TMOS precursors performed by Corriu and co-workers (159) illustrate this point—the electrochemical accessibility of the supported ferrocenes was strongly dependent on the material composition, structure and synthetic conditions. Even the disposition of silylated side chains on otherwise identical metal complexes can have an unpredictable effect on material structure, as illustrated by the observations of Corriu and co-workers (191), who varied the numbers and positions of T-groups on $ArCr(CO)_3$ moieties and noted that the effects on the surface areas of the resulting gels were inherently unpredictable.

Introduction of an alkyl silicate component into a sol–gel mixture creates an extremely complicated kinetic situation (192, 193). Typically, these "network modifiers" are diluted with an orthosilicate species such as TMOS (network formers) to afford a porous material. Condensation of pure alkyltrialkoxysilanes tends to result in nonporous materials, as is apparent from the plot of surface area versus silylated metal complex: TEOS ratio shown below (Fig. 19).

Such nonporous materials can be used to advantage: The flexible structures can be swollen in the presence of organic solvents to generate porous, solvent filled materials, termed interphases, which are useful both for chromatography and for catalysis (53). Typically, however, a rigid gel matrix is sought. When mixed-monomer compositions are considered, the relative kinetic behavior of both silicate components is important to the material structure obtained. As with single-precursor sol–gels, the conditions used for the sol–gel reaction have a strong influence on the results. The behavior of typical mixtures under acid and basic conditions have been studied in detail, notably by Schubert and co-workers (190), who evaluated phase separation phenomena in mixed-monomer systems by preparing and studying aerogel materials, which retain the silica structure present at gelation. Alkyl substituents on the silicon center (as in a T-species) have an electron-donating effect that diminishes the electrophilicity of the silicon center (51). Under acidic conditions, this inductive effect will tend to promote hydrolysis and condensation steps relative to the behavior of an

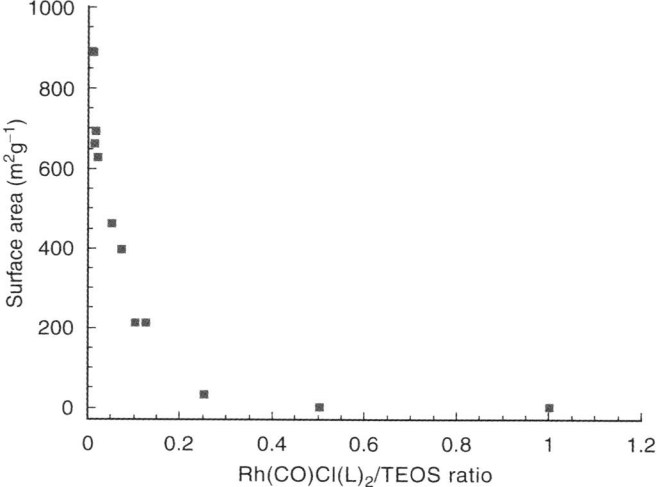

Figure 19. Plot of material surface area as a function of tethered complex, showing collapse of pore structure as the amount of metal complex increases. [Adapted from (56).]

orthosilicate (Q) species, whereas the situation is reversed under basic conditions. Consequently, in acid media, there may be a tendency for a T-species to undergo hydrolysis and condensation before the Q-species, so that polymerization of the T-species may occur before formation of the bulk silica network. Such behavior was observed in gels formed by copolymerization of a phosphine ligand and TEOS (194). There is therefore a possibility that aggregated T-species may be formed initially, and become entrapped in inaccessible regions of the silicate network as the Q-species condense around them in later stages of the reaction (74). Conversely, basic conditions appear to favor formation of the bulk silica network first, with the T-species acting essentially as a cosolvent for the reaction (195). Attachment of the alkylated component after formation of the silica network will tend to place the R groups at the surface of the silica particles, and these groups will thus be most accessible to external species.

The importance of these kinetic phase-separation effects will depend strongly on the relative concentrations of the components, of course, since the kinetics of every elementary step leading to gel formation are concentration dependent. From the point of view of creating functional materials that require site accessibility, the base-catalyzed situation is preferable. Unfortunately for those applications that rely on optical properties as well as site accessibility (such as fabrication of optically based sensors), the use of base catalysis typically affords opaque materials. For catalysis and separations applications, however, the optical properties are of little importance (except for characterization of the material), so the base-catalyzed process is particularly appropriate. In these situations, nature works in our favor: The kinetics of particle formation and surface attachment of T-species conspire to give the optimal results.

Understanding of the intimate molecular level distribution of components within a multicomponent gel can be obtained using a number of approaches, among which solid-state NMR studies are widely applied. Dynamic techniques, usually involving measurements of Si—H cross-polarization and/or spin-lattice relaxation rates provide detailed information on the mobilities of silicon centers within materials. Large differences in site mobilities for Q- and T-type silicon centers can be interpreted in terms of phase separation to form separate domains of bulk-type silica and T-polymers, while similar dynamic behavior typically indicates a homogeneous distribution of precursors throughout the material structure. Examples of such applications of solid-state NMR techniques can be found in many places. The review by Lindner et al. (53) provides a very useful introduction to the techniques in context of a wide range of metal containing hybrid materials. Maciel and co-workers (52, 75, 196, 197) also contributed substantially to the application of NMR techniques to the study of silicon-based polymers. In studies of a number of admixtures of amine, thiol, and phosphine ligands with equimolar amounts of TEOS, these workers showed from NMR relaxation data that the mobilities of the components within the solid structures were essentially identical (52). Since the chemical environments of the various components are similar, the hybrid structures were inferred to be homogeneous copolymers.

2. *Postmodification Approaches*

Early studies involving postmodification of sol–gel materials [i.e., method (*c*)] demonstrated that many potential modification sites (e.g., immobilized ligands for metal binding) would be located in inaccessible regions of the material matrix. As a consequence, this approach, which was widely used in the modification of preformed silica, has been little used for sol–gel derived materials. The MTS materials, which are often prepared by sol–gel routes, have significantly more accessible pore surfaces and are therefore amenable to post-modification procedures. In fact, multistep construction sequences are emerging that demonstrate that assembly of TM binding ligands on the pore walls of MTS materials can be approached in essentially the same way as conventional ligand syntheses. Anchoring of the organosilyl tether onto the MTS surface can be accomplished by postmodification, treating the MCM as any preformed silica (37), but methods have also been developed in which appropriate alkylsilicates are included during the micelle-imprinting step to create surface-functionalized MTSs (198–200). Some examples of postmodification procedures that have been used to introduce potential metal-binding ligands into MTS materials include introduction of amino functional groups (201–203), thiols (199, 204), amphetamines (205), phosphines (203), and Schiff's bases (205). *In situ* imprinting of copper complexes has also been reported (40, 161). More extensive discussions of MTS modification procedures, and use of the resultant materials for catalytic purposes can be found elsewhere (205–207).

3. Issues Concerning Survival of Metal Centers

The conditions of the sol–gel reaction are quite innocuous relative to the extreme temperatures needed to make an ordinary glass, but may still present an environment that can interfere severely with coordination chemistry. To avoid disruption of coordination complexes, immobilization of the ligand can be performed prior to addition of a suitable metal complex, so that metal ion binding is accomplished as a postmodification step, as described in Section IV.C.1. This approach can result in inefficient metal binding, however, due to inaccessibility of many ligand sites, as discussed above (76, 178). Another significant potential drawback of this approach is that ligand immobilization in the absence of metal ions should afford a random distribution of the ligands throughout the gel. Such a random distribution is unlikely to result in a favorable spatial disposition for the majority of immobilized ligands when two or more are required for binding of a transition metal in a desired structural configuration (e.g., cis- vs *trans*-, or *mer*- vs *fac*-). Some reports suggest that specific structures can be obtained despite random immobilization of ligands, however, although this intriguing phenomenon has yet to be adequately explained (76, 169, 170). It is possible that in favorable cases, metal-binding thermodynamics may direct migration of the ligands through the silica matrix to afford a particularly stable metal-binding configuration: Such ligand migration across the silica surface has been implicated in deactivation of certain immobilized catalysts during turnover (82). Detailed studies that would explain the specific architectures obtained on metal treatment of randomly immobilized ligands do not appear to have been attempted, however. Despite the examples to the contrary, it is unlikely that self-assembly of prebound ligands into a desired metal-binding architecture is likely to occur in the majority of cases. Imprinting of the ligand architecture is therefore most commonly achieved by using the metal complex to direct the attachment of the ligands in the desired spatial arrangement. To obtain the desired imprinting effect, it is essential that the transition metal centers retain their structure during the immobilization step.

Survival of metal complex architectures during sol–gel immobilization requires either: (1) that there are no thermodynamically favorable interactions between the metal centers and any components of the sol–gel medium that will disrupt the desired coordination structure, or (2) that the ligand exchange kinetics at the metal centers are sufficiently slow that the structure will not be disrupted on the timescale of the attachment reaction. For our own studies, we have found that categorization of TM complexes according to rate behavior is useful in simplifying synthetic considerations.

Categorization of TM complexes as being "substitution inert" or "substitution labile" is an idea that originated in the 1950s and has been of inestimable value in the design of fundamentally important studies in inorganic mechanism (208). Application of the reactivity concept to studies of supported complexes requires only a small adaptation of these well-established ideas. Most textbooks will define a substitution-labile metal complex as one whose reactions are complete within the time of mixing, while inert metal complexes react on a longer timeframe. These definitions are appropriate for solution reactions, but it proves to be more convenient in the context of immobilized metal complexes if substitution rates are categorized based on immobilization rate as a benchmark.

The immobilization process can be considered in terms of two competing sets of pathways: Immobilization reactions, which stabilize the metal complex architecture, and ligand substitution reactions, which can disrupt the target structure.

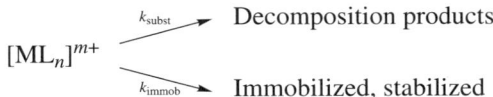

This model is in good agreement with data obtained during the immobilization of silylated $[Fe(phen)_3]^{2+}$ (ferroin) under decomposition conditions (209), and has afforded a conditional value for the immobilization rate ($\sim 10^{-4}$ s^{-1}). Immobilization rates are dependent on the rate of hydrolysis and condensation reactions of the silylated side chains, and hence on the reaction conditions, but to a first approximation are considered as being independent of the nature of the metal complex to which the tethers are attached.

Note that the kinetic model assumes that immobilization will stabilize the structures toward reactions with the sol–gel matrix, or with its components. Structural stabilization afforded by covalent attachment has not been studied extensively, but experiments with mononuclear Fe complexes (162, 172, 181), mononuclear copper complexes (40, 161), and mono- (162), di-, and trinuclear ruthenium complexes (186) suggest that the immobilization reactions do tend to have this stabilizing effect.

When the relative rate constants of the two competing paths are considered, immobilized metal complexes can be considered as belonging to one of three classes, according to Table II.

As far as immobilization behavior is concerned, synthetic considerations are very different for the three types of metal complex, and need to be addressed

TABLE II

Classification of Metal Complex Lability Relative to Immobilization Rate

Class	Designation	Kinetics
Class A	Substitution inert	$k_{sub} \ll k_{immob}$
Class B	Substitution labile	$k_{sub} \gg k_{immob}$
Class C	Intermediate	$k_{immob} \sim k_{sub}$

separately. For Class A complexes, the slow substitution rates will exert kinetic control over access to decomposition products, even if other coordination modes are thermodynamically favorable: Metal architectures are therefore likely to survive immobilization without disruption. Such behavior is typical of second- and third-row TM complexes—substitution rates are typically slow, and the softer nature of these ions also makes ICC interactions less favorable: Faithful imprinting with Class A complexes is quite well assured. Synthetic issues are more concerned with removal and replacement strategies than with the imprinting step (186).

For Class B (substitution labile) metal complexes, reequilibration to more thermodynamically favorable coordination modes will be very rapid relative to immobilization. Such behavior is typical of first-row TM complexes. In addition, these ions are usually very oxophilic, so the metal complexes are typically subject to ICC interactions with oxide materials. Since these metal ions are generally immobilized under conditions of thermodynamic control, all pertinent speciation equilibria, including ICC reactions (Section III.B), must be considered in order to understand or predict the outcome of immobilization reactions. It is essential to understand the relevant equilibria if direct imprinting of active site structures is to be successful. The studies of Klonkowski et al. (210–213), for example, underscore this point: Sol–gel immobilization of copper complexes bearing silylated amine and ethylenediamine ligands were shown by EPR to result in multiple copper environments, suggesting competition between immobilization and ICC reactions.

Class C (intermediate behavior: $k_{immob} \sim k_{subst}$) covers a range of possible behaviors, and specific considerations depend heavily on the nature of the metal ions under study. In most cases, conditions may be found that permit faithful imprinting, but structural integrity cannot generally be assumed. Because of their similar rate behavior, intermediate complexes can be used as convenient probes for the kinetics of immobilization processes. For example, ferrous tris-phenanthroline (ferroin) ($k_{subst} \sim 1 \times 10^{-4}\ s^{-1}$) has been used to obtain benchmark values for the immobilization rate ($\sim 10^{-6}$–$10^{-4}\ s^{-1}$) (Fig. 20) (209).

With labile (Class A) metal ions, any perturbations introduced by the sol–gel environment will result in rapid reequilibration of the metal complex to more

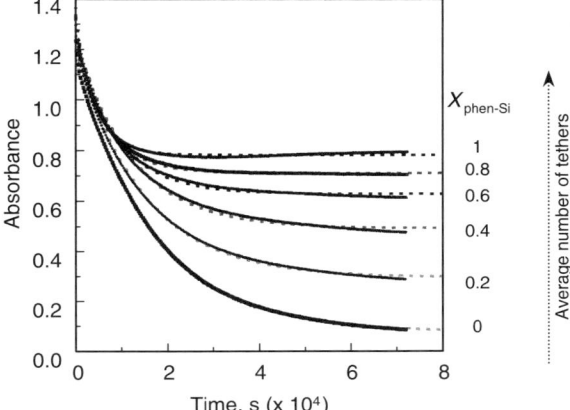

Figure 20. Ferroin immobilization rates versus metal substitution rates. Immobilization kinetics were measured by competition with ligand dissociation reactions, induced by the addition of excess Ni^{2+} (24 equiv) to sol–gel materials containing silylated ferroin complexes during the immobilization step. Treatment of the data according to a rate law based on competing parallel reactions afforded first-order immobilization rate constants in the range 3×10^{-6}–1×10^{-4} s^{-1}. The rate constants were measured in acidic (0.26 mM HCl) TMOS sols at 22.5°C. The parameter $X_{phen-Si}$ refers to the fraction of silylated phenanthroline ligand present, with the total phen/Fe ratio maintained at 3:1. [Adapted from (209).]

thermodynamically favorable forms. Many metal complexes will be prone to react either with the silica material (e.g., Sections III.B.2 and III.C), with the silicate intermediates that arise from hydrolysis of the orthosilicate precursor, or simply with the solvent under the conditions used for catalysis of the sol–gel reaction. A large number of coordination complexes, particularly those having complex polynuclear structures, are therefore unlikely to survive processing. Careful control of the conditions may make it possible to immobilize a particular cluster of labile ions, but this approach requires consideration of the interplay between solution speciation and surface-binding equilibria. Although some control over the structural features of the immobilized centers can be attained, creating silica or sol–gel based materials in which labile metal ions are bound in a single type of environment represents a considerable outstanding challenge (214).

Imprinting approaches using inert clusters to direct attachment of ligands at appropriate points on the silica matrix, followed by removal and replacement of the inert metal ions with more reactive metals present a possible alternative to direct immobilization of labile complexes. Such approaches are largely unexplored in sol–gel silica materials, generally because there are significant difficulties associated with synthesis of suitably modified (i.e., silylated) complexes of substitution-inert metals for purposes of imprinting. Inert metal complexes are much more likely to retain complex architectural features during immobilization (186), but difficulties with removing the inert template ions from the sol–gel hybrid materials must be overcome. The work of Hwang and Sasaki (162) on imprinting of oxidized silicon surfaces suggests a promising approach that may be applicable to sol–gel imprinted materials, although extension of that work to silica sol–gels has not yet been explored.

D. Characterization of Materials

To the classically trained coordination chemist entering the area of sol–gel immobilized complexes, by far the most exasperating aspects are the loss of crystallinity in the materials, which effectively precludes characterization of the metal complexes by diffraction methods, and the inability to obtain accurate elemental analysis data. Materials chemists instead have to rely on "sporting techniques"—that is, spectroscopic and physical methods, which typically lack the resolution and accuracy to which most chemists are accustomed. This loss of accuracy in determining structural details is an inevitable consequence of the transition into amorphous materials, and parallels to some extent the situation that existed in bioinorganic chemistry before the advent of readily accessible crystallographic methods for protein structure determination. A desirable property (e.g., catalytic activity) may be observed, but tracing this activity to a particular structural motif can be difficult. The literature contains much conjecture on the origin of unexpected behavior in sol–gel supported metal complexes, much of which passes into lore without confirmation through detailed physical study.

In the case of most sol–gel materials, there is (by definition) no hope of producing crystalline samples, and the presence of the silicate component will invariably interfere with efforts to obtain accurate analytical data. Despite these limitations, many techniques familiar to the coordination chemist have been successfully applied to the study of immobilized metal complexes, and new techniques are emerging that together provide—albeit at a lower resolution than is possible with X-ray crystallography—detailed information about the environment, homogeneity, and dynamics of TM complexes immobilized in silica materials. Many of the techniques used in the characterization of supported reagents of all types are discussed in detail in the book by Clark et al. (215). Techniques such as EXAFS, which are independent of the physical state of the sample, are widely applied and provide detailed structural information (57, 97, 216). The UV/vis and luminescence spectroscopies can often be used without any additional consideration, particularly when optically transparent gel samples are under study. Similarly, vibrational spectroscopies have been used extensively for the characterization of silica-supported metal complexes for many years, although these studies can be restricted by intense vibrational absorbances that arise from the silicate materials. Many of the considerations associated with application of these characterization techniques have been mentioned previously, with reference to the literature that provides access to more specific detail concerning the experimental nuances. Adaptation of many familiar techniques to the study of opaque solids is relatively straightforward. Cross-polarization–magic angle spinning (CP–MAS) NMR instrumentation, while less common than that used for liquids, is nevertheless quite widely available and many pulse sequences that provide detailed information on structure and dynamics within solids have been developed. Vibrational and electronic spectroscopies can often be performed using conventional instrumentation, by pressing the materials into self-supporting pellets, by use of semitransparent mulls, or by refractive index matching (217) to afford sufficiently transparent samples. Alternatively, diffuse reflectance methods can be used (218), which generally require the purchase of relatively inexpensive attachments that can be used with conventional instruments. Because of the inhomogeneity of light scattering from solid materials with randomly oriented particles, however, it can be considerably more difficult to draw good comparisons between data for different samples, and distortion of spectra can occur because of nonlinearity of scattering effects at different wavelengths. Despite some minor drawbacks, diffuse reflectance visible spectroscopy is widely applied and is a reliable means for characterizing supported compounds. Instead of the more familiar Beer–Lambert dependence of absorbance as a function of concentration used in transmittance work, diffuse reflectance spectra are linearly correlated with concentration of spectrally active species using the Kubelka–Munk function (Eq. 1) (Fig. 10).

E. Environment of Tethered TM Complexes

Although there is much indirect evidence that reactivity can be strongly influenced by entrapment and immobilization, surprisingly few studies have directly addressed the issue of environmental effects on the chemical reactivity of metal complexes. An indication of altered reactivity is seen in the photo-induced charge separation that was observed on reaction of an entrapped iridium complex with 1,4-dimethoxybenzene. Retardation of the back-electron-transfer reaction by matrix entrapment is clearly apparent and results in catalyzed generation of H_2 from H_2O (219). Meyer and co-workers (220) observed retardation of photochemically induced dissociation reactions by matrix entrapment in silica gel–polymer mixtures. Electrochemical studies of immobilized ferrocenes have also indicated that electron-transfer reactions are affected significantly relative to solution, although interpretations of electrochemical data are complicated by possible contributions from a number of different mechanisms that can affect electron transfer (159). Effects such as inhomogeneous surface distributions, and electrochemical inaccessibility of metal centers buried within the interior of the silicate structure were invoked for the ferrocene complexes.

Factors that can significantly affect the reactivity of tethered metal complexes include site isolation and effects arising from enforced proximity to the silica matrix—that is, medium effects. Influences of site–site interactions are often evident in catalytic applications, and are covered in more detail in Section V.C.4.

Quantifying medium effects would make a significant contribution to our understanding of the behavior of doped sol–gels used for many applications. To this end, we have examined the chemistry of ferroin, $[Fe(phen)_3]^{2+}$, as a probe of reactivity perturbation in silica sol–gels. Ferroin was primarily chosen for these studies because it is readily prepared and is very stable in solution but at the same time is subject to a number of inducible ligand exchange and redox reactions. The reactions of ferroin tend to occur on a convenient timeframe for kinetic measurements, with half-lives on the order of hours. The solution reactivity of ferroin has therefore been studied extensively, providing useful benchmark data for comparison with sol–gel results (221–224).

Initial experiments examined the influence of covalent attachment on the ligand dissociation reactivity in the presence of Ni^{2+} competitor ions (181). The effects of tethering were found to be quite dramatic. Immobilized ferroin complexes bearing various numbers of silylated phenanthroline ligands were challenged with Ni^{2+}, which competes with Fe^{2+} for the phenanthroline ligands. Loss of the first phen ligand from $[Fe(phen)_3]^{2+}$ results in a spin transition and bleaching of the intensely red-orange color of the complex. Under conditions where the color of ferroin is entirely bleached due to ligand exchange with Ni^{2+} in solution, covalent attachment to the silica matrix retards the ligand exchange reaction significantly, resulting in residual coloration after reaction had ceased. Although it was initially assumed that the inhibition of the reaction would show an incremental dependence on the number of covalent attachments, it was in fact found that the effect of one or two tethering ligands was indistinguishable. A third tethered ligand resulted in a considerable increase in stability of the complexes toward the competitor Ni^{2+} ions (Fig. 21). A similar dependence on the number of tethers was observed in the reaction with Ce^{4+} (172).

Interpretation of the results in terms of changes in molecular mechanism was ambiguous, since there were a large number of unknowns concerning the fundamental reactivity changes that might occur even in the absence of covalent attachments. Regardless of the origins of these reactivity changes, it was unexpected that the change from one to two tethers would have a negligible effect on stability of the ferroin complexes, since attachment of two separate ligands to the silica polymer would be expected to introduce a macrochelate effect. Only one ligand (i.e., the untethered ligand) needs to be lost to the Ni^{2+} competitor

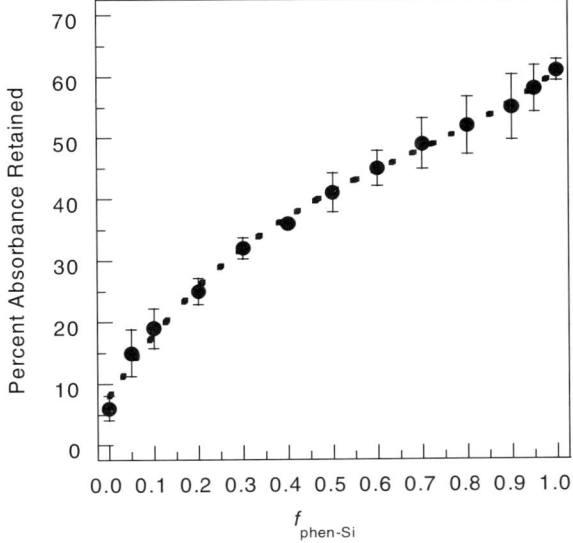

Figure 21. Plot of residual ferroin absorbance after reaction with excess (24 equiv) Ni^{2+} as a function of the mole fraction of silylated phenanthroline. [Adapted from (181).]

for the orange color of ferroin to be lost, however, so the apparent absence of this macrochelate effect can be justified. The strong stabilization of the ferroin complex, observed for the complexes having all three ligands attached most probably arises through macrochelate formation. Additional stabilization factors for any number of tethers may include steric inaccessibility of the complex to the competitor Ni^{2+} ions due to occlusion within constraining regions, or complete inhibition of the molecular motions that lead to ligand dissociation. The data that are currently available do not allow us to determine the relative importance of these potential contributing factors, but it is clear from these studies that immobilization of TM complexes can result in considerable perturbation of reactivity. Results from photophysical studies of structurally similar species (such as $[Ru(bpy)_3]^{2+}$) indicate that a fraction of the complexes spend a significant amount of time in an environment that alters their mobilities and photochemical behavior (14). Luminescence and Raman studies have also demonstrated that solvent ordering or constriction effects can limit ligand vibrational modes (119, 122). Although these latter effects were specifically documented for lanthanides with mixed-polypyridyl ligand systems such as $Eu(p\text{-}ABA)_3 \cdot phen$ and $Tb(p\text{-}ABA)_3 \cdot phen$ ($p\text{-}ABA = para$-aminobenzoate), it is plausible that such effects would also be significant for other polypyridyl TM complexes such as ferroin. Enforced proximity to the silica surface due to tethering would enhance such effects. Damping of vibrational modes, particularly those of the Fe—N bonds, would be expected to result in a substantial inhibition of ligand dissociation, accounting for the changes in the extent of the substitution reactions that were observed for the complexes with one or two tethering ligands. Untethered ferroin complexes undergo complete reaction with an excess of competitor Ni^{2+} ions, as might be expected for complexes that are not constrained to spend all of their time under the retarding influence of the silica surface. This result rules out the possibility that the Ni^{2+} competitor ions are in some way sequestered and prevented from reacting with the ferroin centers, through ICC or other effects. Although they do react to completion, sol–gel entrapped ferroins exhibit significantly diminished rate constants for the competition reaction relative to the analogous behavior in methanol–water mixtures having compositions similar to those expected for the sol–gel environment (~ 40–60% MeOH) (Fig. 22) (22). The observed reduction in rate constants must arise from a significant degree of interaction of the ferroin centers with the silica component of the sol–gel mixture. As will be described below, it is quite likely that ion pairing between the negatively charged silica surface and the ferroin molecules plays a considerable role in modulating the reactivity of ferroin toward ligand exchange.

Ion pairing effects are indicated by NMR studies of ferroin in methanol–water mixtures (Fig. 22). At low mole fractions of methanol, the chemical shift data are consistent with preferential solvation of the ferroin cation by methanol (222). Solvation of the cation by a distribution of methanol and water molecules representative of the bulk solvent composition would afford a linear dependence of chemical shift versus mole fraction of methanol. Deviation from linearity indicates that the cation environment is enriched in one component relative to the bulk composition (226). At higher methanol concentrations, however, the trend of the observed chemical shifts toward a maximum value, as would be expected for preferential solvation, is reversed before the 100% methanol composition is reached. Such behavior is inconsistent with a simple preferential solvation model, and suggests that another effect becomes important under these solvent compositions. The temperature dependence of the chemical shift values also undergoes a reversal that is inconsistent with the usual effects of temperature on solvent polarity. In the more polar solvent mixtures, an increase in chemical shift reflects decreasing solvent polarity as the temperature is raised (227). It is possible that ion pair formation between the ferroin cation and perchlorate counterions is responsible for the change in behavior at higher methanol concentrations. It is to be expected that close approach of the perchlorate counterions will result in an increase in the apparent polarity of the ferroin environment, consistent with the observed chemical shift trend. Further, ion pairing effects will be enhanced at higher temperatures (228),

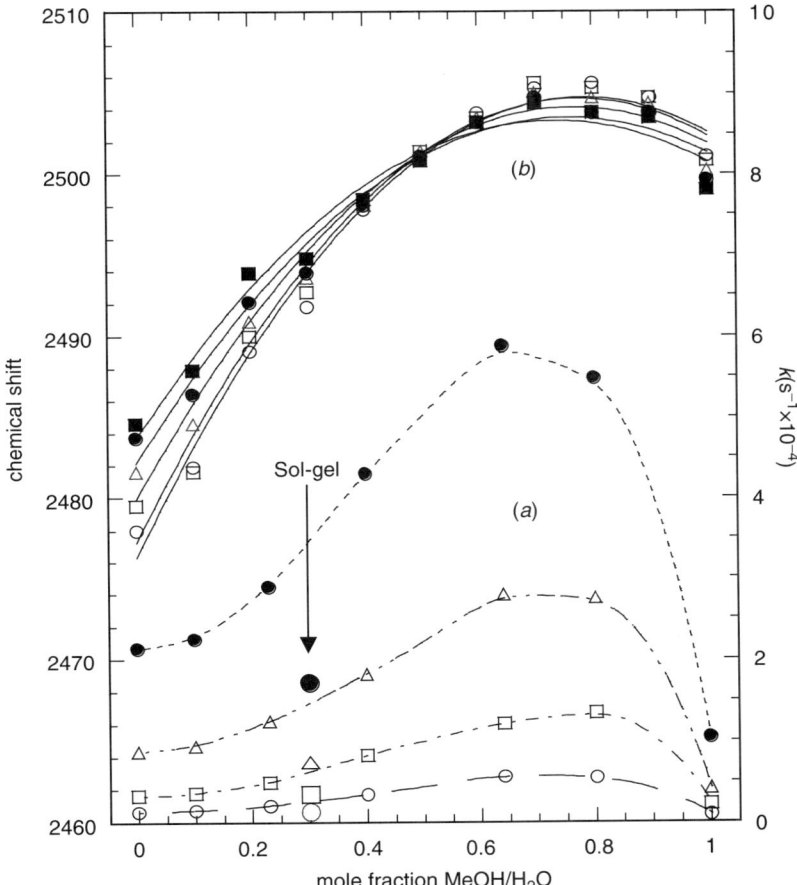

Figure 22. Kinetic data for the dissociation of ferroin in the presence of Ni^{2+} monitored by UV/vis (a) and chemical shift data for the 5,6-protons of phen ligands on ferroin (b). The dissociation of ferroin in sol–gel is indicated by the arrow at the experimental temperature described below. The placement of the sol–gel points for mole fraction methanol were based on entropic and ethalpic parameters and literature data. Each curve within regions (a) and (b) represent a different experiment temperature. Key: (a) open circle 15°C, open square 20°C, open triangle 25°C, closed circle 30°C (b) open circle 25°C, open square 30°C, open triangle 35°C, closed circle 40°C, closed triangle 25°C. [Adapted from (225).]

resulting in a decrease in chemical shift instead of the increase that would be expected due to reduced solvent polarity.

The NMR data show a clear parallel with rate data obtained for the ferroin–Ni^{2+} reaction in comparable solvent mixtures (Fig. 22), indicating that these ion pairing effects may contribute to the observed rate changes as a function of solvent composition. Ion pairing thus diminishes the rate of dissociation of the phenanthroline ligands from the Fe^{2+} ion, presumably because of reduced electrophilicity at the metal center. Assuming this model to be correct, the very low rates for the sol–gel reactions would indicate a high degree of ion pairing as might be expected in the presence of a large concentration of negatively charged silica particles and oligomers. The ion pairing model is consistent with all data obtained to date, although the chemical shift behavior of ferroin in the sol–gel medium has yet to be determined, due to a strong broadening of the proton resonances. Work is currently in progress to resolve this experimental difficulty, in order to gain improved insight into the environment of sol–gel entrapped ferroin molecules. Ion pairing effects also appear to be an important factor in determining the extent of changes in the luminescence behavior of europium complexes as a function of gel aging and drying (Section III.D.2.a) (130).

V. APPLICATIONS

A. Sensors

Sol–gels can be prepared in optically transparent form, and thus are suitable for the development of optical sensing devices for a range of species. However, the suitability of the sol–gel matrix for this application must be considered. Conditions for the synthesis of sol–gel derived materials can be adjusted to control the degree of porosity of the material (Section II.B), which can have important effects on the function of a sensing device. Problems can arise from a material that is not sufficiently porous to allow the diffusion of the analyte of interest to the indicating dopant or, alternatively, is too porous and does not retain the sensing molecule when exposed to solution media. The stability of the support material is also an important issue in the formation of an optical device. The matrix should be relatively unchanging, since the environment affects the response of a trapped reagent, which must be reproducible. In many cases, the device must be able to withstand environmental stresses such as high temperature and pressure to be used in certain settings. Furthermore, recyclability of the device is important for both cost effectiveness and environmental considerations. The incorporation of a sensor molecule that can be regenerated may offer an environmentally friendly alternative to solution phase detection but matrix effects on the incorporated molecule can change the solution-like behavior either for better or worse.

The majority of research involving optical sensing devices can be separated into four well-studied areas, involving proton, O_2, metal, and enzyme-based sensing. Many of these studies involve the use of organic molecules rather than metal complexes and thus will only be mentioned where historically significant. Readers are directed to more applications-oriented reviews for more information on organic molecules as sensors (22, 229).

Proton-sensitive organic dyes have been incorporated into silica materials for the purpose of creating optical pH sensors (230–234). As pointed out in the studies of Collinson et al. (146), with metal complexes in sol–gel films, leaching of small dopant molecules from sol–gel materials can occur if the device is used in an aqueous medium. Therefore, studies involving the use of pH probe molecules must address this issue. A pertinent discussion can be found in a paper by Butler et al. (233), where bromophenol blue was found to resist leaching to a certain extent by using acid catalysis and a water/alkoxide ratio (r-ratio, R_w) of at least 4. Rottman et al. (231), used two pH indicators (*para*-methyl red and methyl orange) and two solvatochromatic polarity indicators in a similar but more comprehensive study. In addition to determining the optimal r-ratio, they examined humidity effects on indicator performance. In a separate study, they showed that incorporation of a covalent attachment between the methyl red indicator and the sol–gel matrix or coentrapment of surfactant can improve the durability of the pH probe by reducing the leaching behavior (230). Subtle changes in the size of the indicator can also prevent leaching, as observed in comparisons of sensing behavior of entrapped thymolphthalein and phenolphthalein: the former molecule, although only slightly larger, did not leach from a sol–gel monolith while the latter did (232).

Some pH indicators have shown incredibly good response ranges and thermal, chemical, and mechanical stability. For example, methyl red on a sol–gel dipped coated sodalime glass slide showed a sensitivity range of 0.3–14 pH units by changes in λ_{max} with reversibility and no fatigue, thermal stability to $\sim$90°C, and chemical–mechanical resistance after treatment with a variety of harsh chemical solutions (234).

As discussed in Section III.D.2.b, O_2 sensing is an application of doped sol–gel materials. Much of the work associated with O_2 sensing has been performed using the quenching properties of fluorescing molecules such as pyrene and polypyridyl ruthenium complexes. Pyrene has been used extensively as an organic probe molecule because of its well-characterized excited states and well-resolved vibronic structure (14, 22, 235). The study of pyrene in sol–gel monoliths has made a significant contribution to the study of the sol–gel interior, however, its performance as a viable O_2 sensor is questionable. Dunbar et al. (235), for example, found the sensitivity of pyrene as an O_2 gas sensor deteriorates as sol–gel derived thin films were aged due to a decrease in the bimolecular quenching constant and fluorophore lifetime. Among its contributions as a probe molecule, pyrene has aided in detecting changes in solvent composition within sol–gel materials because the vibronic structure of its luminescence spectrum is solvent sensitive (22). Diffusion-controlled fluorescence quenching of excited state pyrene by Cu(II) ion during the sol-to-gel conversion has contributed to understanding of photoinduced electron-transfer mechanisms in sol–gel monoliths (236, 237). The study of the charge separation between pyrene and an electron acceptor molecule is more applicable to the discussion of photonics and will be discussed in Section V.B.

The use by Avnir and co-workers (21) of excited-state dynamics of pyrene-doped sol–gel materials to probe the interior of the sol–gel environment has helped pave the way for similar studies with luminescent probes such as $[Ru(bpy)_3]^{2+}$ (14). From an applications standpoint, the attributes of $[Ru(bpy)_3]^{2+}$ (Section III.D.2.b) make this metal complex a particularly good O_2 sensor. The matrix can have an enormous effect on O_2 diffusibility to the doped sensor depending on the conditions of synthesis. The sensing ability of $[Ru(bpy)_3]^{2+}$ is probably best suited for gas-phase O_2 as the molecule has been shown to leach from sol–gels under solution conditions (146). A typical optically based O_2 sensing device using such a system is shown in Fig. 23 (143).

Another promising example of a metal-based sensor can be seen in the work of Lee and Okura (239) who used entrapped platinum octaethylporphyrin complexes to form photostable devices. In their study, they found by adding the surfactant Triton X-100 to the sol–gel mixture, they could improve homogeneity and phosphorescence of the dye-containing glass. Encapsulated porphyrin molecules have also been used successfully in the sensing of nitrogen dioxide (240) and, as will be discussed below, can be used to sense metal ions.

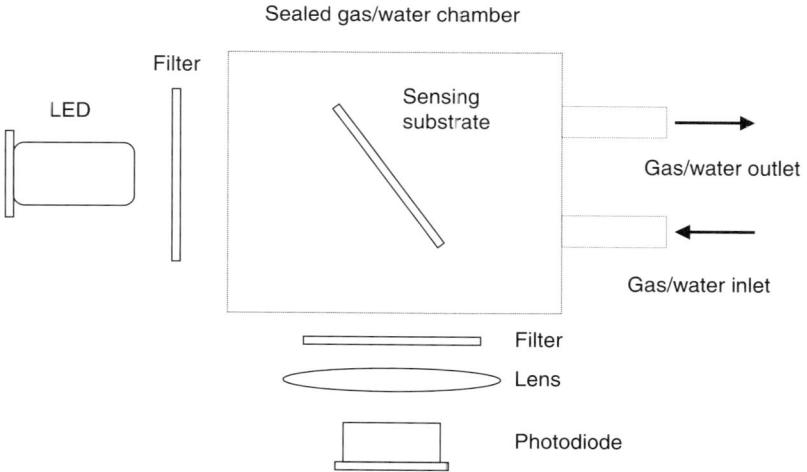

Figure 23. Schematic of all solid-state fluorescence intensity planar-based characterization system. [Adapted from (143).]

Many organic ligands have been incorporated for the purpose of sensing metal ions, although much of the earlier work tended to be more qualitative than quantitative. Zusman et al. (232) showed that many ligands are capable of sensing metal ions: These sensor ligands included phenanthroline for Fe(II), quinalizarin and pyridine for Al(II), α-nitroso-β-napthol for Co(II), dimethylglyoxime for Ni(II), rubeanic acid for Cu(II), and gallacianine for Pb(II). In the case of phenanthroline, the metal could also be incorporated as a dopant for sensing of the ligand, although we have observed that the diffusion of phenanthroline in Fe(II) doped sol–gels can be quite limited in aged monoliths (225). Zaitoun and Lin (241a), incorporated EDTA in silica sol–gels and showed the materials to have the ability to sense multiple metal ions. The researchers found the detecting response times to be better than those found in solution, and observed only small changes in the visible spectra of the complexes. Note, however, that the EDTA ligand appeared to lose its double zwitterion structure as a result of confinement in the silica matrix. In contrast to the EDTA study, Flamini et al. (241b), showed a decrease in response time compared to solution for the detection of Hg(II) by a polyvinylformyl-containing sol–gel material doped with 2-(5-amino-3,4-dicyano-2-H-pyrrol-2ylidene)-1,1,2-tricyanoethanide. Reisfeld and Shamrakov (242) described a cadmium sensor that used the effect of incorporating the 8-hydroxyquinoline-5-sulfonic acid into a matrix to advantage. The sulfonate group of the ligand interacts with the silica matrix and the cadmium ion binds weakly to the OH group of the ligand relative to solution. The cadmium could be displaced through exposure of the sensor–metal complex to water for quick regeneration of the sensors. What the weakly chelating ligand (relative to solution) gives up as far as detectability is compensated by quick regeneration of the sensing property that otherwise could not be obtained in solution.

An article by Plaschke et al. (243) using porphyrin-doped sol–gel films to detect Hg(II) by fluorimetric methods highlights the need to balance various sol–gel synthesis parameters to achieve an optimal and reproducible response. They found that acid-catalyzed sol–gel films retained the sensor molecule over long periods of time (days), while base-catalyzed films exhibited considerable leaching on the timeframe of hours. It was shown that, unfortunately, the same conditions that promoted retention of the porphyrin sensor components prevented the Hg(II) analyte from gaining access to them. The researchers solved this problem by covalently attaching the porphyrin sensor to aminodextran before incorporation into the sol–gel material. This procedure allowed them to use the base-catalyzed recipe to afford a more porous material while avoiding loss of the porphyrin sensor through leaching. The material also showed promise for simultaneous detection of other metal ions, using spectral subtraction methods. Delmarre et al. (244) demonstrated, however, that capping of pendant silanol groups on the silica surface may be necessary to prevent nonspecific metal ion adsorption on the silica. In their study they found roughly 90% of Hg(II), Pb(II), and Cd(II) salts trapped in the matrix rather than available for binding to the entrapped porphyrin molecules.

The use of enzymes in sol–gel materials as biosensors is a relatively recent advance that has rapidly developed into an expansive area that cannot be completely covered here. A review by Avnir et al. (26) provides a more detailed description of studies reported prior to 1994. The use of sol–gel encapsulated proteins to create biosensors was first established in the early 1990s (25, 27). The study of Braun et al. (25) used alkaline phosphatase as a dopant molecule and showed that, when encapsulating proteins, the sol–gel synthesis conditions must be considered carefully. They found only 30% retention of biological activity, and observed that the resultant material was an opaque powder containing both denatured and aggregated protein components. A major step forward was made by Ellerby et al. (27), who demonstrated that several proteins could be encapsulated within sol–gel materials without greatly perturbing their biological function. They used visible spectroscopy to monitor the demetallation of Cu/Zn superoxide dismutase, redox reactivity of ferricytochrome c, and reactivity of myoglobin with dioxygen or carbon monoxide, and showed that the metalloproteins exhibited solution-like reactivity in all cases. The key to maintaining biological activity and transparency of their materials was their sol–gel recipe in which buffered solutions of the enzymes were doped into sonicated sol with no added methanol for TMOS cosolvent.

Dave et al. (16) pointed out the advantages of sol–gel encapsulation, which include retention of biological activity in the aqueous environment within the solvent filled pores and prevention of aggregation by site isolation. For the sensor to be used to advantage, however, one must consider the effect of the matrix on diffusion of small molecules to the sensing protein. Shen and Kostic (245) give an excellent example of how this issue is relevant to the sensing of several molecules that quench cytochrome c, including $[Fe(CN)_6]^{3-}$, O_2, and p-benzoquinone. By considering UV/vis, circular dichroism, and resonance Raman spectra of a zinc-substituted protein, they found that the monophasic kinetics of ^{3}ZnCyt triplet-state quenching observed in solution became biphasic in the sol–gel medium. The change in kinetic behavior was attributed to limited diffusion, especially in the case of the charged quencher $[Fe(CN)_6]^{3-}$. This result is consistent with the findings of Collinson et al. (84), where negatively charged $[Fe(CN)_6]^{3-}$ resists diffusion through a negatively charged sol–gel surface interior (Section III.D.3). The uniform penetration of even small O_2 molecules into the matrix was diminished relative to solution.

Other studies with sol–gel materials containing encapsulated cytochrome c and glucose oxidase showed that despite diffusion limitations, the sensors could be very sensitive and their responses quite reproducible. For example, Blyth et al. (246), showed that a sensor fashioned from sol–gel and the enzyme cytochrome c could detect micromolar levels of NO: Comparable sensitivity could be obtained through at least seven cycles of regeneration of the NO free ferric state of the enzyme by purging with argon between NO sensing experiments. Similarly, Braun and co-worker (247) found that a glucose oxidase sensor could be used daily for six months. By using initial rates of reduction in the presence of various glucose concentrations they were able to achieve an analytical detection range of [Glu] between 1 and 100 mM.

Many groups have attempted to improve the sensing ability of glucose oxidase by modifying the synthesis or form of the sol–gel material (248–252). Kunzelmann and Bottcher (253) used this strategy and obtained good results for the electrochemical response of encapsulated glucose oxidase using an ORMOSIL material. Most enzyme containing sensors are made in monolithic, thin-film, or spin coated form. One of the more interesting modifications to the sol–gel form is the layering by spin coating or laying down successive thin films to make a so-called "sandwich configuration". Pandey et al. (248), used a layer of sol–gel followed by a layer of glucose oxidase and a final layer of sol–gel to fabricate an electrode. The electrode was then analyzed by cyclic voltammetry (CV) using ferrocene monocarboxylic acid in the presence and absence of glucose. Wolfbeis et al. (254), use a similar synthesis strategy with two thin-film layers doped with the luminescent oxygen probe, $[Ru(dpp)_3]^{2+}$, around a glucose oxidase layer to measure the consumption of oxygen.

Rather than the problem of leaching, encapsulated proteins are more frequently subject to macromolecular crowding. The rotation of a protein in a sol–gel material can be impeded and is highly dependent on sol–gel synthesis conditions and age of the material as was demonstrated in a study using a magnesium myoglobin doped sol–gel material (255). Time-resolved fluorescence anisotropy was used to show that the protein had a diminished decay rate, consistent with impeded rotational diffusion. Eggers and Valentine (28) used circular dichroism to examine the effects of confinement on the structure and stabilty of lysozyme, α-lactalbumin, and myoglobin in sol–gels. In general, they found that the thermal stability of the proteins was enhanced relative to solution and proteins, except for apomyoglobin, retained their native structures. After finding that apomyoglobin in particular becomes unfolded in sol–gel media, they expressed a need to understand the difference between possible solvent effects and steric constraints in the sol–gel pores.

Since enzymes are known for their selectivity and catalytic efficiency, they or their active site analogues make excellent candidates for heterogeneous catalysts when doped in sol–gel materials.

B. Photonics

Intense interest in photonic devices such as solar cells and photochromic materials has been spurred by their many potential applications. Such applications are reviewed in more detail elsewhere and will be discussed here only to provide a brief background (14, 256–259). The rationale behind using a solid support to create a solar device is to assist in the production of transient stored energy by physical separation of charged pairs. One component of the pair is converted to the excited state by vis or UV light. The second component accepts an electron from the excited-state chromophore. In solution, recombination of the charged pair to their ground-state counterparts is rapid but matrix isolation in a sol–gel environment has the potential to hinder this reaction. The chromophoric donors that have been studied most extensively are pyrene, and iridium and ruthenium polypyridyl complexes, all of which have attractive optical properties. In addition, lanthanide systems have been used to create photovoltaic cells.

Amorphous silicon, an inexpensive material for solar devices, can use visible light much more efficiently than UV light. Thus, from an applications perspective, the use of pyrene-doped materials for devices that utilize solar energy is diminished by the need to excite the molecule using UV light (14). The study of the quenching behavior of excited-state pyrene by methyl viologen (MV) has, however, made an invaluable contribution to understanding these processes in sol–gel materials (260, 261). The kinetics of quenching were quite complex and required at least a biexponential model to adequately account for the observed behavior. Modeling using two second-order processes showed the charge-separated pairs had an impressive increase in lifetime of five orders of magnitude relative to their behavior in methanol solution. The lifetimes could be further enhanced by the addition of a third component, N,N'-tetramethylene-$2,2'$-bipyridine, which acts as a shuttle between the two molecules. A four order of magnitude increase in fluorescence lifetime versus solution was achieved for ~25% of Ir(II) produced using the charge separated pair $[Ir(bpy)_2C^3,N'(bpy)]^{3+}$ and 1,4-dimethylbenzene (219).

Ruthenium polypyridyls incorporated into sol–gels have been studied widely (14, 138, 140, 217, 262, 263) and were recently reviewed in detail (14). As discussed in Section III.D.2.b, since these ruthenium complexes absorb visible light rather than UV, they offer an advantage over their pyrene counterpart for use in solar devices. Unfortunately, $[Ru(bpy)_3]^{2+}$ and MV pairs have not shown an increase in charge separation lifetime in sol–gels when compared to solution (260). This result is quite interesting as it was shown in other studies that ruthenium complexes tend to interact with sol–gel matrices to produce longer fluorescence lifetimes as the material becomes more rigid (140). Evidently, the interaction of the ruthenium complex with MV is not affected by this phenomenon to any appreciable extent.

In the lanthanide systems, the polypyridyl ligands absorb UV light that is transferred to the lanthanide ion, which then luminesces, emitting visible light. The function of the doped sol–gel component is to increase the efficiency of light harvesting via the wavelength-shifting properties of the inorganic dopant.

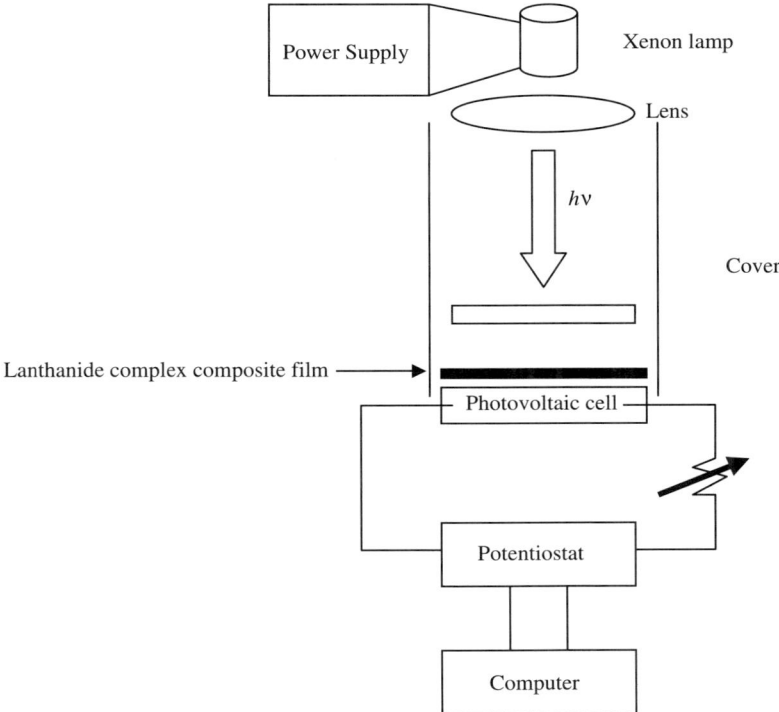

Figure 24. Typical apparatus for measurement of J–V characteristics using a photovoltaic cell. [Adapted from (123).]

Jin et al. (123, 264) recently reported on photovoltaic cells constructed using doped sol–gel coatings to utilize the fluorescence characteristics of [Eu(phen)$_2$]$^{3+}$ and [Tb(bpy)$_2$]$^{3+}$ (Fig. 24). The output of the photovoltaic devices fabricated with single crystal or amorphous silicon increases by 8–18%—a very significant improvement in efficiency.

Another area of interest is the application of metal-doped sol–gels in photochemical splitting of water into H_2 and O_2, which constitutes a valuable counterpart to photovoltaic power generation as a clean energy source. The catalysis of H_2 generation from H_2O was observed by Slama-Schwok et al. (219), as a result of charge separation of Ir(II) and DMB$^+$. Water splitting has the added advantage of producing H_2, a valuable fuel that can be transported readily. Sol–gels have promise for eliminating some of the problems encountered with water splitting by soluble complexes.

C. Catalysts

Supported catalysts have been used since the emergence of the chemical industry, with silica being one of the most widely used dispersants. Many excellent reviews (15, 19, 265–267) and books (34, 35, 46) are available that cover the topic extensively. Because of the vast extent of this literature topic, only some aspects that pertain to the behavior of supported catalytic metal complexes are covered herein.

1. Entrapment verus Immobilization

The principal advantage of using untethered complexes in sol–gel materials is simplicity of preparation; synthetic issues associated with preparation of tethered complexes (Section IV.B) are avoided completely. The approach has been applied to the catalysis of a wide range of reactions (19). Potential disadvantages that must be considered are loss of accessibility of the metal centers due to entrapment within constraining regions of the silicate matrix or conversely, mobility of the metal centers through the matrix, which may lead to inactivation by formation of dinuclear or polynuclear clusters, for example. These effects are not as well studied, and may be less significant contributors to catalyst deactivation than is leaching of the metal species into the reaction mixtures. Although difficult to quantify, migration effects are quite common, however, and have even been documented on occasion for complexes anchored to the silica materials (82).

2. Site Accessibility

Accessibility of active sites is of crucial importance to the operation of many hybrid materials—particularly those under study for sensing, separations and, in particular, catalysis. Site accessibility will naturally depend on material structure, which in turn is dependent on the composition of the material and the processing conditions. It is frequently found that the specific activities of metal complexes in catalytic reactions are considerably diminished when the complexes are supported on a silica gel material. Such reductions in activity may arise from changes in the coordination state of the complex or other environmental influences. In many cases, however, the reduced activity may simply result from a large fraction of supported metal complexes being rendered inaccessible to solvent or external reagents. Actual activities, expressed on a "per-accessible-site" basis may be considerably greater than are apparent based on the total concentration of active sites present. The observation of catalysis or any other activity from a supported metal complex implies at least some degree of site accessibility, but does not in itself provide quantitative information on the number of available sites. Determining the number of sites that are actually accessible to external solvent, or are capable of taking part in catalysis is a challenging problem that is rarely addressed directly.

Probing the accessibility of a site in the interior of a sol–gel is best accomplished by monitoring of a stoichiometric reaction that results in an observable spectroscopic change. Binding of metal complexes to sol–gel immobilized ligands illustrates this point. Parish et al. (75) showed that hydrosilylation of sol–gel immobilized vinyl groups resulted in only 10% inclusion of Ph_2PH functional groups. Similarly, reaction of chlorosilanes with aniline derivatives was 25% effective, whereas reaction with the larger Ph_2PH was only 10% successful. Hardee et al. (178) showed, using NMR studies, that only small proportion of sol–gel immobilized ligands ($\sim$5%) were accessible for binding to metal centers.

Reaction of small molecule probes such as H_2 or CO_2 can also be used to probe the accessibility of many organometallic complexes, using either vibrational or NMR spectroscopies to detect the binding event (53, 187, 268). Note, however, that accessibility of small molecule probes does not necessary guarantee access for larger substrate molecules.

3. Leaching

Leaching of the metal centers out of the material during catalysis is by far the most important problem associated with supported catalysts of all types, and is of major concern for untethered complexes in sol–gel media. Many studies of supported catalysts indicate a slow loss of catalytic activity due to loss of metal complexes into the reaction mixture. Leaching can give misleading results, because the observed catalytic activity may actually be due to the leached metal complexes in solution.

The recyclability of a catalyst is of major concern in the area of Green Chemistry (46, 269–271) and leaching is a major contributor to catalyst deactivation during repeated use. Contamination of the products with dissolved catalyst is also a significant concern since it may add costly purification steps. Because of these concerns, "leachproof" is currently one of the most common terms used to describe advantageous properties of a supported catalyst.

Although the approach must be pursued carefully, entrapment of metal complexes has afforded a range of active and leachproof catalysts (19, 272) that have been used in a variety of reactions. Engineered ion pairing to promote binding of charged metal complexes to silica surfaces is an interesting approach to leachproofing of catalysts that minimizes the synthetic concerns associated with the preparation of tethered complexes (19, 174, 273).

Evaluation of metal ion leaching from a supported catalyst is not a trivial undertaking. Most commonly, leaching is first detected as a loss of activity on recycling of a supported catalyst, although it can sometimes be apparent from the appearance of coloration in filtered reaction mixtures, particularly when transition metal-based catalysts are involved. Ion-selective methods such as atomic absorption or emission spectroscopies are frequently used to monitor metal complex concentrations in supernatant solutions after catalysis. Leaching can also be evaluated quantitatively by electrochemical techniques that, if generally adopted, may ultimately provide the most sensitive and reliable probes of leaching behavior in supported catalysts (146).

The effects of leaching on the behavior of silica-supported catalysts have recently been discussed in detail by Sheldon, whose discussion focuses on the widely applied MCM based supported titanium catalyst, titanium–silicalite (TS-1) (45). The review highlights the caution that must exercised when evaluating leaching behavior and determining its contribution to the apparent activities of supported catalysts, and provides useful insight into the design of appropriate control experiments. That review is an excellent source for anyone embarking on catalytic studies with supported complexes.

4. Site Isolation

One of the beneficial effects of immobilizing metal complexes on solid supports is that the metal centers are prevented from diffusing, which can eliminate formation of inactive polynuclear structures. Alternatively, enforced proximity of two or more metal centers may be essential to the successful operation of a catalytic cycle. The effects of site isolation (or absence thereof) on reactivity of metal catalysts supported on a variety of materials have been observed in a number of cases (217, 274, 275). For example, Collman et al. (217) demonstrated that the hydroformylation activity of silica-supported rhodium phosphine complexes increased as a function of rhodium loading, and were thus able to identify a dinuclear step in the catalytic mechanism. Jacobsen and co-worker (275) observed similar effects in the operation of polymer and dendrimer supported (276), Schiff's base complexes. Operation of many supported heme-type oxidation catalysts also relies critically on site isolation due to immobilization, since the high-valent heme-oxo intermediates implicated in the catalytic mechanisms are well known to undergo intermolecular decomposition reactions that inactivate the catalytic centers (277). Site isolation due to immobilization of the catalytic centers thus provides a convenient alternative to creation of sterically hindered porphyrins as an approach to stabilizing reactive intermediates (278). In enforcing either site isolation or site–site interactions, the support plays a similar role to the protein component of a metalloenzyme. Relative to the very large number of organometallic complexes that have been developed, very few supported catalysts have been prepared that mimic the behavior of metalloenzymes. With only a few exceptions (167, 169, 170, 214, 279–282), all of these complexes mimic hemoprotein active sites (283–291). In the absence of the structural definition afforded by the heme unit, immobilization of labile complexes of biologically relevant TM ions is a more challenging undertaking—complexes containing these reactive metal centers will be very prone to decomposition under sol–gel conditions (Section IV.C.3). Mimicry of metalloenzyme attributes, such as efficient and selective catalysis in neutral water under ambient conditions, in synthetic catalysts has long been a goal of bioinorganic chemistry. Coupled with the technical advantages of supported catalysts, these properties provide a strong driving force for further development of this class of hybrid materials—particularly as the need for environmentally benign chemical processes becomes more pressing. It seems likely that exploration at this frontier will rapidly gain momentum in the near future.

VI. SUMMARY TOPICS

A. Synthesis Issues

At this point, we would like to summarize some of the key issues discussed above in context of the choices confronting an inorganic chemist setting out for the first time to make a metal-doped silicate material. Some essential questions that must be addressed are

1. How does one design a synthesis for a particular doped silica gel so that the product material contains a uniform composition with regard to metal centers and at the same time has the necessary attributes to perform a desired function?
2. How can the synthetic considerations for the faithful immobilization of the metal complex be balanced with obtaining desirable material properties?
3. Importantly—How will you know when you have succeeded?

To approach at least some partial answer to these questions, we suggest the following:

First ask: What is the application and what material properties are particularly important? To answer this question requires a little knowledge of the types of material available, and which of their properties are important. Common materials properties that are desirable for specific functions have been covered in some detail, as well as some information on the relationship between synthetic conditions and material properties.

Next ask: Do I need to include covalent attachments? Entrapped complexes are much easier from a synthetic point of view, but are prone to leaching and may be subject to undesirable chemistry in the sol–gel interior. Immobilization may eliminate some of these problems but can involve a significant increase in synthetic effort, and may prove to be challenging. The approach, however, has some definite advantages, depending on the intended purpose for the material. Again, the basis for making this choice has much to do with the desired application. This question may not have a totally clear-cut answer, but the necessary considerations are easily understood, and inorganic chemists with different levels of interest in synthesis will be able to determine whether the additional effort is worthwhile.

Next ask: What conditions will I use for immobilization? Based on the literature presented here, the answer to this question basically boils down to whether high surface area or optical transparency is the most important property for the purpose. A reasonably clear-cut distinction can be made in terms of intended applications. It may be possible to find a compromise between the two properties by judicious juggling of conditions, but for most applications they are rarely optimized simultaneously.

Next ask: Will my complex survive the immobilization conditions? Consider the rates of typical reactions for the target complex, and the complexity of structure to be immobilized. For labile complexes, careful consideration of the metal chemistry under the immobilization conditions is essential, and may require significant study of solution chemistry to predict behavior in the sol–gel medium. If reactions (such as inner-sphere ICC) do occur, the particular form of the metal complex obtained may be adequate for the application, in which case this is a nonissue. Otherwise, if a particular structural motif is desired, finding the right balance between immobilization conditions and metal chemistry can be tricky. On the other hand, the emergence of ICC ideas points to another possibility: Can interactions with the material itself be used as an integral part of synthetic plan? Surface coordination reactions can in many cases introduce deleterious side reactions that will disrupt desired coordination structures. It has now been demonstrated conclusively, however, that some very specific surface coordination interactions can occur. If sufficiently well understood, these specific interactions may be put to use in creating a coordination environment for a supported metal complex. It cannot be overemphasized, however, that careful analysis of materials prepared by immobilizing labile metal complexes to assess structural identity and sample composition is essential. For inert complexes, where the immobilization reaction is rapid relative to any thermodynamically favorable side reactions, immobilization issues are of less importance, although careful characterization will still be required. It is assumed that immobilization "locks in" the structure, although few studies have unequivocally established that this is in fact the case, and the validity of the assumption is likely to vary depending on the specific complex under study.

Much as we like sol–gels, we note that an important question should be addressed while considering immobilization conditions and the fate of metal complexes: *Is a silica-based material the best or only option?* Silica and other oxide materials have many beneficial properties and have been used extensively to create metal-doped materials for a host of applications, so the answer will often be in the affirmative. It should be noted, however, that there exists an equally vast and diverse literature concerning polymer-supported complexes (292) with many areas of study paralleling the sol–gel literature discussed here. If organic polymers can be made that have properties appropriate to the application at hand, the conditions used for polymer synthesis may be more compatible with the particular metal chemistry being attempted. Some compromise between desirable properties may also be obtained by exploration of ORMOSILs, although the relationship between ultimate material properties and starting conditions is much more complicated, and significantly less well understood at present.

Finally ask: How will I know if I have made what I wanted to? It is essential for most applications to ensure that the constitution of the metal centers in the product material is uniform, and that the metal complexes have the intended structure. As discussed extensively above, sol–gel processing is a mild approach for preparation of silica glasses, but the sol–gel environment is far from chemically inert. It should not be assumed a priori that what you put in will be what you get out, although this assumption is frequently (and frustratingly) made without experimental corroboration. In the absence of crystallography and good elemental analysis data, establishing that a metal complex has survived immobilization intact is rarely a trivial matter. Many spectroscopic and physical methods are applicable for the characterization of supported metal complexes, but depending on the form of the material, some may require special equipment or techniques. There is often something of a learning curve associated with adapting to the specifics of studying materials instead of soluble or crystalline compounds, but in most cases it is relatively minor.

B. Conclusions and Future Prospects

A wide variety of metal complexes have been incorporated into sol–gel silica gels to create hybrid materials that have been applied in a number of technologically important contexts. A great deal of information about the behavior of entrapped and immobilized complexes is available from studies in several areas. In this chapter, we have categorized several subsets within this immense field with the inorganic chemist in mind. The most striking conclusion from our search though the literature is the difficulty in comparing various results between research groups using similar metal complexes but different sol–gel "recipes." There is, therefore, a need for consistency in preparative approaches that allow many researchers to combine thoughts, ideas and results. With the advent of new techniques such as micelle templating to afford quasi-regular material structures, whole new areas for exploration have been opened up, and further developments in these areas are expected to emerge rapidly. This approach promises to allow much more reliable comparisons between results from different research groups. However, even with its lack of consistency, the classic sol–gel technique still offers a range of possibilities for the coordination chemist. Hopefully, this chapter will provide some useful insights that will assist them in navigating the burgeoning literature.

ABBREVIATIONS

acac	Acetylacetonato (ligand)
BET	Brunauer–Emmett–Teller (adsorption isatherm)
Boc	*tert*-Butoxycarbonyl
bpy	2,2′-Bipyridyl
cod	1,4-Cyclooctadiene
CP–MAS	Cross-polarization–magic angle spinning
CV	Cyclic voltammetry
D centers	Dialkyl substituted silicate centers
DBM	Dibenzoylmethanide
DCCA	Drying chemical control agents
DMAP	N,N-Dimethylaminopyridine
dpp	4,7-Diphenyl-1,10-phenanthroline
DMF	N,N-Dimethylformamide
DMPE	1,2-Bis(dimethylphosphino)ethane
DMSO	Dimethyl sulfoxide
EDTA	Ethylenediaminetetraacetic acid
en	Ethylenediamine
EPR	Electron paramagnetic resonance
EXAFS	Extended X-ray absorption fine structure
Fc	Ferrocene
H_3Mptr	3-Methyl-5-(pyridin-2-yl)-1,2,4-triazole
HMPA	Hexamethylphosphoric triamide
ICC	Interfacial coordination chemistry
ITO	Indium tin oxide
IR	Infrared
LMT	Ligand-to-metal transfer
M centers	Trialkyl substituted silicate centers

MCM	Mobil composition of matter (see MTS)
MLCT	Metal-to-ligand charge transfer
MTS	Micelle templated silicas
MV	Methyl viologen
NaPSS	Sodium poly(styrenesulfonate) or Nafion
neo	neocuproine(2,9-Dimethyl-1,10-phenanthroline)
NLO	Nonlinear optics
NMR	Nuclear magnetic resonance
ORMOSIL	ORganically MOdified SILicates
p-ABA	$para$-aminobenzoate
PDMDAAC	Poly(dimethyldiallylammonium chloride)
phen	1,10-Phenanthroline
PTA	Phosphotungstic acid ($H_3PW_{12}O_{40}$)
PVA	Poly(vinyl alcohol)
PVSA	Poly(vinylsulfonic acid)
Qn	Q refers to a silicate center that can make up to "n" siloxane linkages
SEM	Scanning electron microscopy
STA	Silicotungstic acid ($H_3SiW_{12}O_{40}$)
T centers	Monoalkyl substituted silicate centers
TEOS	Tetraethylorthosilicates
TGA	Thermogravimetric analysis
THF	Tetrahydrofuran
TM	Transition metal
TMOS	Tetramethyl orthosilicate
TMS-Cl	Trimethylsilylchloride
TMS-OMe	Methoxytrimethylsilane
TS-1	Titanuim–silicalite
UV–vis	Ultraviolet–visible
XANES	X-Ray absorption near-edge spectroscopy
XPS	X-Ray photoelectron spectroscopy

ACKNOWLEDGMENTS

Acknowledgment is made to the Donors of The Petroleum Research Fund, administered by the American Chemical Society, for partial funding of our research. We are also grateful to Research Corporation and the Thomas F. and Kate Miller Jeffress Memorial Trust for support of our research program, as well as to the National Science Foundation for funding of REU students (Christina M. Moore, Dawn E. Zbell and Gerard Rowe) who have contributed significantly to our work. We are also grateful to Professors Laura E. Pence and Deborah C. Bebout for their helpful suggestions concerning the preparation of this manuscript, and for their ongoing scientific contributions. SPW thanks Professor Joan

S. Valentine, Professor Michael J. Maroney, Professor Daniel R. Gamelin, and Professor Vincent M. Rotello for many useful discussions. Dr. Stephen Paul Watton is a Cottrell Scholar of Research Corporation.

REFERENCES

1. A. Corma, *Chem. Rev.*, 97, 2373 (1997).
2. G. Cerveau and R. J. P. Corriu, *Coord. Chem. Rev.*, 180, 1051 (1998).
3. E. Lindner, M. Kemmler, T. Schneller, and H. A. Mayer, *Inorg. Chem.*, 34, 5489 (1995).
4. J. E. Mark, *Heterogeneous Chem. Rev.*, 3, 307 (1996).
5. A. B. Wojcik and L. C. Klein, *Appl. Organomet. Chem.*, 11, 129 (1997).
6. R. Gvishi, U. Narang, G. Ruland, D. N. Kumar, and P. N. Prasad, *Appl. Organomet. Chem.*, 11, 107 (1997).
7. T. Graham, *Ann. Chem.*, 121, 36 (1862).
8. M. Ebelman, *Ann. Chim. Phys.*, 16, 129 (1846).
9. C. J. Brinker and G. W. Scherer, *Sol–Gel Science: The Physics and Chemistry of Sol–Gel Processing*, Academic, New York, 1990.
10. R. K. Iler, *The Chemistry of Silica*, John Wiley & Sons, Inc., New York, 1979.
11. H. Brintzinger and W. Brintzinger, *Anorg. Allg. Chem.*, 181, 237 (1929).
12. H. Brintzinger and B. Troemer, *Anorg. Allg. Chem.*, 196, 44 (1931).
13. L. L. Hench and J. K. West, *Chem. Rev.*, 90, 33 (1990).
14. F. N. Castellano and G. J. Meyer, *Prog. Inorg. Chem.*, 44, 167 (1997).
15. U. Schubert, *New J. Chem.*, 18, 1049 (1994).
16. B. C. Dave, B. Dunn, J. S. Valentine, and J. I. Zink, *Anal. Chem.*, 66, 1120A (1994).
17. U. Schubert, N. Huesing, and A. Lorenz, *Chem. Mater.*, 7, 2010 (1995).
18. D. Avnir, S. Braun, O. Lev, and M. Ottolenghi, *Proc. SPIE-Int. Soc. Opt. Eng.*, 1758, 456 (1992).
19. J. Blum, D. Avnir, and H. Schumann, *Chemtech*, 2, 32 (1999).
20. D. Avnir, D. Levy, and R. Reisfeld, *J. Phys. Chem.*, 88, 5956 (1984).
21. D. Avnir, *Acc. Chem. Res.*, 28, 328 (1995).
22. B. Dunn and J. I. Zink, *Chem. Mater.*, 9, 2280 (1997).
23. D. Avnir, *Chem. Phys. Lett.*, 109, 593 (1984).
24. R. Reisfeld and C. K. Jorgensen, *Coordination Chemistry*, ACS Symposium Series, ACS, Washington, DC, 1994, p. 439.
25. S. Braun, S. Rappaport, R. Zusman, D. Avnir, and M. Ottolenghi, *Mater. Lett.*, 10, 1 (1990).
26. D. Avnir, S. Braun, O. Lev, and M. Ottolenghi, *Chem. Mater.*, 6, 1605 (1994).
27. L. M. Ellerby, C. R. Nishida, F. Nishida, S. A. Yamanaka, B. Dunn, J. S. Valentine, and J. I. Zink, *Science*, 255, 1113 (1992).
28. D. K. Eggers and J. S. Valentine, *Protein Sci.*, 10, 250 (2001).
29. C. Y. Shen and N. M. Kostic, *J. Electroanal. Chem.*, 438, 61 (1997).
30. A. Bronshtein, N. Aharonson, D. Avnir, A. Turniansky, and M. Altstein, *Chem. Mater.*, 9, 2632 (1997).

31. D. Shabat, F. Grynszpan, S. Saphier, A. Turniansky, D. Avnir, and E. Keinan, *Chem. Mater.*, 9, 2258 (1997).
32. E. J. A. Pope, *J. Sol–Gel Sci. Tech.*, 4, 225 (1995).
33. E. J. A. Pope, K. Braun, and C. M. Peterson, *J. Sol–Gel Sci. Tech.*, 8, 635 (1997).
34. F. R. Hartley, *Supported Metal Complexes*, Reidel, Dordrecht, The Netherlands, 1985.
35. Y. I. Yermakov, B. N. Kuznetsov, and V. A. Zacharov, *Catalysis by Supported Complexes*, Elsevier, Amsterdam, The Netherlands, 1981.
36. Z. C. Brzezinska and W. R. Cullen, *Can. J. Chem.*, 58, 744 (1980).
37. U. Deschler, P. Kleinschmit, and P. Panster, *Angew. Chem. Int. Ed. Engl.*, 25, 236 (1986).
38. S. Dai, Y. Shin, C. E. Barnes, and L. M. Toth, *Chem. Mater.*, 9, 2521 (1997).
39. S. Dai, Y. Shin, L. M. Toth, and C. E. Barnes, *J. Phys. Chem. B*, 101, 5521 (1997).
40. S. Dai, M. C. Burleigh, Y. Shin, C. E. Barnes, and Z. Xue, *Angew. Chem. Int. Ed. Engl.*, 38, 1235 (1999).
41. M. M. Collinson, *Crit. Rev. Anal. Chem.*, 29, 289 (1999).
42. S. Vidyasankar, M. Ru, and F. H. Arnold, *J. Chromatogr. A*, 775 (1997).
43. K. Kimura, T. Sunagawa, S. Yajima, S. Maiyake, and M. Yokoyama, *Anal. Chem.*, 70, 4309 (1998).
44. S. Bourg, J. C. Broudic, O. Conocar, J. J. E. Moreau, D. Meyer, and M. W. C. Man, *Chem. Mater.*, 13, 491 (2001).
45. R. A. Sheldon, M. Wallau, I. W. C. E. Arends, and Schuchardt, *Acc. Chem. Res.*, 31, 485 (1998).
46. J. H. Clark, *Catalysis of Organic Reactions by Supported Inorganic Reagents*, VCH, New York, 1994.
47. L. T. Zhurvalev, *Langmuir*, 3, 316 (1987).
48. J. Fricke, *Aerogels: Springer Proceedings in Physics*, Vol. 6, Springer-Verlag, Heidelberg, Germany, 1986.
49. J. Fricke and A. Emmerling, *Struct. Bonding (Berlin)*, 77, 37 (1992).
50. J. Fricke and A. Emmerling, *J. Sol–Gel Sci. Technol.*, 13, 299 (1998).
51. U. Schubert and N. Husing, *Synthesis of Inorganic Materials*, Wiley-VCH, Weinheim, 2000.
52. G. E. Maciel, *Solid-State NMR Spectroscopy of Inorganic Materials*, J. J. Fitzgerald, Ed., American Chemical Society Symposiun Series, Washington, DC, 1999, p. 326.
53. E. Lindner, T. Schneller, F. Auer, and H. A. Mayer, *Angew. Chem. IEE*, 38, 2155 (1999).
54. J. H. Clark and D. J. Macquarrie, *Chem. Commun.*, 853 (1998).
55. U. Schubert, C. Egger, K. Rose, and C. Alt, *J. Mol. Catal.*, 55, 330 (1989).
56. U. Schubert, K. Rose, and H. Schmidt, *J. Non-Cryst. Solids*, 105, 165 (1988).
57. C. M. Taylor and S. P. Watton, *Inorg. Chem.*, submitted for publication.
58. S. J. Gregg and K. S. W. Sing, *Adsorption, Surface Area and Porosity*, Academic, New York, 1982.
59. J. Rouquerol, D. Avnir, C. W. Fairbridge, D. H. Everett, J. H. Haynes, N. Pernicone, J. D. Ramsay, K. S. W. Sing, and K. K. Unger, *Pure Appl. Chem.*, 66, 1739 (1994).
60. E. P. Barrett, L. G. Joyner, and P. P. Halenda, *J. Am. Chem. Soc.*, 73, 373 (1951).
61. W. W. Lukens, P. Schmidt-Winkel, D. Zhao, J. Feng, and G. D. Stucky, *Langmuir*, 15, 5403 (1999).
62. J. P. Olivier, *J. Porous Mater.*, 2, 9 (1995).
63. D. W. Aksnes, K. Forland, and L. Kimtys, *Phys. Chem. Chem. Phys.*, 3, 3203 (2001).

64. B. D. Kay and R. A. Assink, *Mat. Res. Soc. Symp. Proc.*, Mat. Res. Soc., 157 (1986).
65. R. A. Assink and B. D. Kay, *J. Non-Cryst. Solids*, *107*, 35 (1988).
66. L. L. Hench and G. Orcel, *J. Non-Cryst. Solids*, *105*, 223 (1988).
67. S. H. Tolbert, A. Firouzi, G. D. Stucky, and B. F. Chmelka, *Science*, *278*, 264 (1997).
68. C. Rottman, G. Grader, Y. De Hazan, S. Melchior, and D. Avnir, *J. Am. Chem. Soc.*, *121*, 8533 (1999).
69. K. Matsui, T. Nakazawa, and H. Morisaki, *J. Phys. Chem.*, *95*, 976 (1988).
70. M. Schneider and A. Baiker, *Catal. Rev.-Sci. Eng.*, *37*, 515 (1995).
71. G. M. Pajonk, *Catal. Today*, *35*, 319 (1997).
72. D. R. Rolison and B. Dunn, *J. Mater. Chem.*, *11*, 963 (2001).
73. (a) E. Lindner, T. Schneller, H. A. Mayer, H. Bertagolli, T. S. Ertel, and W. Horner, *Chem. Mater.*, *9*, 1524 (1997); (b) R. J. P. Corriu and D. Leclercq, *Angew. Chem.-Inter. Ed. Engl.*, *35*, 1420 (1996).
74. (a) E. Lindner, T. Schneller, H. A. Mayer, H. Bertagnolli, T. S. Ertel, and W. Horner, *Chem. Mater.*, *9*, 1524 (1997); (b) E. Lindner, M. Kemmler, H. Mayer, and P. Wegner, *J. Am. Chem. Soc.*, *116*, 348 (1994).
75. I. M. El-Nahhal, J. J. Yang, I. S. Chuang, and G. E. Maciel, *J. Non-Cryst. Solids*, *208*, 105 (1996).
76. R. V. Parish, D. Habibi, and V. Mohammadi, *J. Organomet. Chem.*, *369*, 17 (1989).
77. H. Schumann, M. Hasan, F. Gelman, D. Avnir, and J. Blum, *Inorg. Chim. Acta*, *280*, 21 (1998).
78. G. M. Kloster, C. M. Taylor, and S. P. Watton, in preparation.
79. E. I. Mayo, D. D. Poore, and A. E. Stiegman, *Inorg. Chem.*, *39*, 899 (2000).
80. D. Avnir and V. R. Kaufman, *J. Non-Cryst. Solids*, *92*, 181 (1987).
81. R. J. P. Corriu, J. J. E. Moreau, P. Thepot, and M. Worg Chi Man, *J. Mater. Chem.*, *4*, 987 (1994).
82. R. S. Drago and D. C. Pribich, *Inorg. Chem.*, *24*, 1983 (1985).
83. J. D. Badjic and N. M. Kostic, *J. Mater. Chem.*, *11*, 408 (2001).
84. M. M. Collinson, P. J. Zambrano, H. M. Wang, and J. S. Taussig, *Langmuir*, *15*, 662 (1999).
85. C. Lepetit and M. Che, *J. Mol. Catal. A*, *100*, 147 (1995).
86. D. Avnir, S. Braun, and M. Ottolenghi, *ACS Symp. Ser.*, *499*, 384 (1992).
87. R. L. Burwell, R. G. Pearson, G. L. Haller, P. J. Tjok, and S. P. Chock, *Inorg. Chem.*, *4*, 1123 (1965).
88. L. D. Dugger, J. H. Stanton, B. N. Irby, B. L. McConnell, W. W. Cummings, and R. W. Maatman, *J. Phys. Chem.*, *68*, 757 (1964).
89. F. L. Tang, C. S. Zhu, and F. X. Gan, *J. Sol-Gel Sci. Tech.*, *9*, 279 (1997).
90. M. A. Garcia and A. Campero, *J. Sol-Gel Sci. Tecn.*, *13*, 651 (1998).
91. R. Litran, E. Blanco, M. RamirezdelSolar, A. Hierro, M. A. DiazGarcia, A. GarciaCabanes, and F. AgulloLopez, *Synt. Met.*, *83*, 273 (1996).
92. J.-F. Lambert, M. Hoogland, and M. Che, *J. Phys. Chem. B.*, *101*, 10347 (1997).
93. L. Bonneviot, D. Olivier, and M. Che, *J. Mol. Catal.*, *21*, 415 (1983).
94. A.-R. Badiei and L. Bonneviot, *Inorg. Chem.*, *37*, 4142 (1998).
95. M. Rogoviv and R. Neumann, *J. Molec. Cataly. A.* *138*, 315 (1999).
96. O. Clause, L. Bonnevoit, and M. Che, *J. Catal.*, *138*, 195 (1992).
97. L. Bonnevoit, O. Clause, M. Che, A. Manceau, and H. Dexpert, *Catal. Today*, *6*, 39 (1989).
98. P. Burrattin, M. Che, and C. Louis, *J. Phys. Chem. B*, *103*, 6171 (1999).

99. P. Burrattin, M. Che, and C. Louis, *J. Phys. Chem. B*, *101*, 7060 (1997).
100. P. Burrattin, M. Che, and C. Louis, *J. Phys. Chem. B*, *102*, 2722 (1998).
101. S. N. Towle, G. E. J. Brown, and G. A. Parks, *J. Coll. Interf. Sci.*, *217*, 299 (1999).
102. S. N. Towle, J. R. Bargar, G. E. J. Brown, and G. A. Parks, *J. Coll. Interf. Sci.*, *217*, 312 (1999).
103. N. Spanos and A. Lycourghiotis, *Langmuir*, *9*, 2250 (1993).
104. L. Vordonis, N. Spanos, P. G. Koutsoukos, and A. Lycourghiotis, *Langmuir*, *8*, 1736 (1992).
105. F. Beland, A. R. Badiei, M. Ronning, D. Nicholson, and L. Bonneviot, *PCCP Phys. Chem. Chem. Phys.*, *1*, 605 (1999).
106. Y. Iwasawa, *Catal. Today*, *6*, 27 (1989).
107. C. Louis and M. Che, *J. Phys. Chem.*, *91*, 2875 (1987).
108. C. Papadopoulou and A. Lycourghiotis, *Appl. Catal.*, *38*, 255 (1988).
109. L. Karakonstantis, C. Kordulis, and A. Lycourghiotis, *Langmuir*, *8*, 1318 (1992).
110. T. Lopez, J. Mendez-Vivar, and M. Asomoza, *Thermochem. Acta*, *216*, 279 (1993).
111. M. E. Raimondi, E. Gianotti, L. Marchese, G. Martra, T. Maschmeyer, J. M. Seddon, and S. Coluccia, *J. Phys. Chem. B.*, *104*, 7102 (2000).
112. N. Spanos, L. Vordonis, C. Kordullis, P. G. Koutsoukos, and A. Lycourghiotis, *J. Catal.*, *124*, 315 (1990).
113. N. Spanos, H. K. Matralis, C. Kordullis, and A. Lycourghiotis, *J. Catal.*, *138*, 432 (1992).
114. N. Spanos, L. Vordonis, C. Kordullis, P. G. Koutsoukos, and A. Lycourghiotis, *J. Catal.*, *124*, 301 (1990).
115. R. Reisfeld and C. K. Jorgensen, *Struct. Bonding (Berlin)*, *77*, 207 (1992).
116. A. A. G. Tomlinson, C. Bellitto, O. Piovesana, and C. Furlani, *J. Chem. Soc. Dalton Trans.*, 350 (1971).
117. S. P. Watton, M. I. Davis, L. E. Pence, J. Rebek, Jr., S. J. Lippard, *Inorg. Chim. Acta*, *1–2*, 195 (1995).
118. X. Fan, M. Wang, Z. Wang, and Z. Hong, *Mater. Res. Bull.*, *32*, 1119 (1997).
119. B. Yan, H. J. Zhang, and J. Z. Ni, *Mater. Sci. Eng. B-Solid State Mater. Adv. Technol.*, *52*, 123 (1998).
120. H. J. Zhang, L. S. Fu, S. B. Wang, Q. G. Meng, K. Y. Yang, and J. Z. Ni, *Mater. Lett.*, *38*, 260 (1999).
121. L. R. Matthews, X. J. Wang, and E. T. Knobbe, *J. Non-Cryst. Solids*, *178*, 44 (1994).
122. B. Yan, H. Zhang, S. Wang, and J. Ni, *Mater. Res. Bull.*, *33*, 1517 (1998).
123. T. Jin, S. Inoue, S. Tsutsumi, K. Machida, and G. Adachi, *Chem. Lett.*, 171 (1997).
124. L. S. Fu, H. J. Zhang, S. B. Wang, Q. G. Meng, K. Y. Yang, and J. Z. Ni, *J. Sol-Gel Sci. Tech.*, *15*, 49 (1999).
125. T. Jin, S. Tsutsumi, Y. Deguchi, K. Machida, and G. Adachi, *J. Electrochem. Soc.*, *143*, 3333 (1996).
126. T. Jin, S. Tsutsumi, Y. Deguchi, K. Machida, and G. Adachi, *J. Alloys Compounds*, *252*, 59 (1997).
127. B. Yan, H. J. Zhang, S. B. Wang, and J. Z. Ni, *J. Photochem. Photobiol. A-Chem.*, *112*, 231 (1998).
128. B. Yan, H. J. Zhang, S. B. Wang, and J. Z. Ni, *Mater. Chem. Phys.*, *51*, 92 (1997).
129. D. C. Lai, B. Dunn, and J. I. Zink, *Inorg. Chem.*, *35*, 2152 (1996).
130. M. J. Lochhead and K. L. Bray, *J. Non-Cryst. Solids*, *170*, 143 (1994).

131. W. V. Moreshead, J.-L. R. Nogues, and R. H. Krabill, *J. Non-Cryst. Solids*, *121*, 267 (1990).
132. R. Campostrini, G. Carturan, M. Ferrari, M. Montagna, and O. Pilla, *J. Mater. Res.*, *7*, 745 (1992).
133. L. S. Fu, H. J. Zhang, S. B. Wang, Q. G. Meng, H. Shao, and J. Z. Ni, *J. Mater. Sci. Technol.*, *15*, 187 (1999).
134. L. S. Fu, H. J. Zhang, S. B. Wang, Q. G. Meng, K. Y. Yang, and J. Z. Ni, *Chem. Res. Chinese Univ.*, *15*, 100 (1999).
135. H. D. Gafney, *Coord. Chem. Rev.*, *104*, 113 (1990).
136. J. McKiernan, J. C. Pouxviel, B. Dunne, and J. I. Zink, *J. Phys. Chem.*, *93*, 2129 (1989).
137. R. Reisfeld, *J. Non-Cryst. Solids*, *121*, 254 (1990).
138. K. F. Mongey, J. G. Vos, B. D. MacCraith, C. M. McDonagh, C. Coates, and J. J. McGarvey, *J. Mater. Chem.*, *7*, 1473 (1997).
139. K. F. Mongey, G. V. Johannes, B. D. MacCraith, and C. M. McDonagh, *Coord. Chem. Rev.*, *185–186*, 417 (1999).
140. P. Innocenzi, H. Kozuka, and T. Yoko, *J. Phys. Chem. B*, *101*, 2285 (1997).
141. K. Matsui, K. Sasaki, and N. Takahashi, *Langmuir*, *7*, 2866 (1991).
142. O. Katz, J. Samuel, D. Avnir, and M. Ottolenghi, *J. Phys. Chem.*, *99*, 14893 (1995).
143. C. McDonagh, B. D. MacCraith, and A. K. McEvoy, *Anal. Chem.*, *70*, 45 (1998).
144. U. Schubert, F. Schwertfeger, and C. Gorsmann, *Nanotechnology*, ACS Symposium Series, Washington, DC, 1996, p. 366.
145. O. Lev, Z. Wu, S. Bharathi, V. Glezer, A. Modestov, J. Gun, L. Rabinovich, and S. Sampath, *Chem. Mater.*, *9*, 2354 (1997).
146. M. M. Collinson, C. G. Rausch, and A. Voigt, *Langmuir*, *13*, 7245 (1997).
147. M. D. Petit-Dominguez, H. Shen, W. R. Heineman, and C. J. Seliskar, *Anal. Chem.*, *69*, 703 (1997).
148. Z. M. Hu, C. J. Seliskar, and W. R. Heineman, *Anal. Chem.*, *70*, 5230 (1998).
149. Z. M. Hu, A. F. Slaterbeck, C. J. Seliskar, T. H. Ridgway, and W. R. Heineman, *Langmuir*, *15*, 767 (1999).
150. G. Villemure and T. J. Pinnavaia, *Chem. Mat.*, *11*, 789 (1999).
151. O. Dvorak and M. K. Dearmond, *J. Phys. Chem.*, *97*, 2646 (1993).
152. S. D. Holmstrom, B. Karwowska, J. A. Cox, and P. J. Kulesza, *J. Electroanal. Chem.*, *456*, 239 (1998).
153. J. A. Cox, A. M. Wolkiewicz, and P. J. Kulesza, *J. Solid State Electrochem.*, *2*, 247 (1998).
154. G. Gun, M. Tsionsky, and O. Lev, *Anal. Chim. Acta*, *294*, 261 (1994).
155. J. Gun and O. Lev, *Anal. Chim. Acta*, *336*, 95 (1996).
156. J. Gun and O. Lev, *Anal. Lett.*, *29*, 1933 (1996).
157. K. Ogura, K. Nakaoka, M. Nakayama, M. Kobayashi, and A. Fujii, *Anal. Chim. Acta*, *384*, 219 (1999).
158. P. C. Pandey, S. Upadhyay, H. C. Pathak, and C. M. D. Pandey, *Electroanalysis*, *11*, 950 (1999).
159. P. Audebert, G. Cerveau, R. J. P. Corriu, and N. Costa, *J. Electroanal. Chem.*, *413*, 89 (1996).
160. J. X. Wang and M. M. Collinson, *J. Electroanal. Chem.*, *455*, 127 (1998).
161. M. C. Burleigh, S. Dai, E. W. Hagaman, and J. S. Lin, *Chem. Mater.*, *13*, 2537 (2001).
162. K. O. Hwang and T. Sasaki, *J. Mater. Chem.*, *8*, 2153 (1998).

163. M. A. Markowitz, P. R. Kust, G. Deng, P. E. Schoen, J. S. Dordick, D. S. Clark, and B. P. Gaber, *Langmuir*, *16*, 1759 (2000).
164. O. Ramström, L. I. Andersson, and Mosbach, K., *J. Org. Chem.*, *58*, 7562 (1993).
165. P. Tien and L. K. Chau, *Chem. Mater.*, *11*, 2141 (1999).
166. Y. Barness, O. Gershevitz, M. Sekar, and C. Sukenik, *Langmuir*, *16*, 247 (2000).
167. B. R. Bodsgard and J. N. Burstyn, *Chem. Commun.*, 647 (2001).
168. O. Krocher, R. A. Koppel, and A. Baiker, *J. Mol. Catal. A-Chem.*, *140*, 185 (1999).
169. M. Louloudi, Y. Deligiannakis, and N. Hadjiliadis, *Inorg. Chem.*, *37*, 6847 (1998).
170. M. Louloudi, Y. Deligiannakis, and N. Hadjiliadis, *J. Inorg. Biochem.*, *79*, 93 (2000).
171. H. H. Weetall, *Appl. Biochem. Biotechnol.*, *41*, 157 (1993).
172. G. M. Kloster and S. P. Watton, *Inorg. Chim. Acta*, *297*, 156 (2000).
173. A. Kraus, M. Schneider, A. Gugel, and K. Mullen, *J. Mater. Chem.*, *7*, 763 (1997).
174. J. Blum, A. Rosenfeld, N. Polak, O. Israelson, H. Schumann, and D. Avnir, *J. Mol. Catal. A-Chem.*, *107*, 217 (1996).
175. W. Urbaniak and U. Schubert, *Liebigs Ann. Chem.*, 1221 (1991).
176. D. J. Brondani, R. J. P. Corriu, S. E. Ayoubi, J. J. E. Moreau, and M. W. C. Man, *J. Organomet. Chem.*, *451*, C1 (1993).
177. C. Chuit, R. J. P. Corriu, G. Dubois, and C. Reye, *Chem. Comm.*, 723 (1999).
178. J. R. Hardee, S. E. Tunney, J. Frye, and J. K. Stille, *J. Poly. Sci.*, *28*, 3669 (1990).
179. C. Malins, S. Fanni, H. G. Glever, J. G. Vos, and B. D. MacCraith, *Anal. Commun.*, *36*, 3 (1999).
180. P. K. Ghosh and T. G. Spiro, *J. Am. Chem. Soc.*, *102*, 5543 (1980).
181. G. M. Kloster, C. M. Taylor, and S. P. Watton, *Inorg. Chem.*, *38*, 3954 (1999).
182. H. Irving and D. H. Mellor, *J. Chem. Soc.*, 5222 (1962).
183. E. Lindner, T. Schneller, F. Auer, P. Wegner, and H. A. Mayer, *Chem. Eur. J.*, *3*, 1833 (1997).
184. O. Krocher, R. A. Koppel, M. Froba, and A. Baiker, *J. Catal.*, *178*, 284 (1998).
185. A. Adima, J. J. E. Moreau, and M. Wong Chi Man, *J. Mater. Chem.*, *7*, 2331 (1997).
186. L. E. Pence, G. Rowe, and S. P. Watton, unpublished data.
187. U. Schubert and K. Rose, *Transition Met. Chem. (NY)*, *14*, 291 (1989).
188. M. Capka, U. Schubert, B. Heinrich, and J. Hjortkjaer, *Collect. Czech. Chem. Commun.*, *57*, 2615 (1992).
189. M. Capka, M. Czakoova, and U. Schubert, *Appl. Organomet. Chem.*, *7*, 369 (1993).
190. N. Husing, U. Schubert, K. Misof, and P. Fratzl, *Chem. Mater.*, *10*, 3024 (1998).
191. G. Cerveau, R. J. P. Corriu, and C. Lepeytre, *Chem. Mater.*, *9*, 2561 (1997).
192. B. K. Coltrain, W. T. Ferrar, C. J. T. Landry, T. R. Molaire, and N. Zumbulyadis, *Chem. Mater.*, *4*, 358 (1992).
193. C. A. Fyfe and P. P. Aroca, *J. Phys. Chem. B*, *101*, 9504 (1997).
194. E. Lindner, A. Bader, and H. A. Mayer, *Inorg. Chem.*, *30*, 3783 (1991).
195. N. Husing and U. Schubert, *Angew. Chem. Int. Ed. Eng.*, *37*, 23 (1998).
196. I.-S. Chuang and G. E. Maciel, *J. Phys. Chem. B.*, *101*, 3052 (1997).
197. J. J. Yang, I. M. El-Nahhal, I. S. Chuang, and G. E. Maciel, *J. Non-Cryst. Solids*, *212*, 281 (1997).
198. D. J. MacQuarrie, *Chem. Comm.*, 1961 (1996).
199. M. H. Lim, C. F. Blanford, and A. Stein, *Chem. Mater.*, *10*, 467 (1998).

200. S. L. Burkett, S. D. Sims, and S. Mann, *Chem. Comm.*, 1961 (1996).
201. A. Cauvel, G. Renard, and D. Brunel, *J. Org. Chem.*, 62, 749 (1997).
202. C. W. Lee, D. H. Ahn, B. Wang, J. S. Hwang, and S.-E. Park, *Micropor. Mesopor. Mater.*, 44–45, 587 (2001).
203. A. M. Liu, K. Hidajat, and S. Kawi, *J. Mol. Catal. A*, 168, 303 (2001).
204. X. Feng, G. E. Fryxell, L.-Q. Wang, A. Y. Kim, J. Liu, and K. M. Kemner, *Science*, 276, 923 (1997).
205. D. Brunel, *Micropor. Mesopor. Mater.*, 27, 329 (1999).
206. D. Brunel, N. Bellocq, P. Sutra, A. Cauvel, M. Laspéras, P. Moreau, F. Di Renzo, A. Galarneau, and F. Fajula, *Coord. Chem. Rev.*, 178–180, 1085 (1998).
207. A. Sayari and S. Hamoudi, *Chem. Mater.*, 13, 3151 (2001).
208. F. Basolo and R. G. Pearson, *The Mechanisms of Inorganic Reactions*, John Wiley & Sons, Inc., New York, NY, 1958.
209. C. M. Taylor, G. M. Kloster, and S. P. Watton, *Inorg. Chem.*, submitted for publication (2002).
210. A. M. Klonkowski and C. W. Schlaepfer, *J. Non-Cryst. Solids*, 129, 101 (1991).
211. A. M. Klonkowski, K. Koehler, T. Widernik, and B. Grobelna, *J. Mater. Chem.*, 6, 579 (1996).
212. A. M. Klonkowski, K. Koehler, and C. W. Schlaepfer, *J. Mater. Chem.*, 3, 105 (1993).
213. A. M. Klonkowski, B. Grobelna, T. Widernik, A. Jankowska-Frydel, and W. Mozgawa, *Langmuir*, 15, 5814 (1999).
214. S. C. Bowman and S. P. Watton, *J. Am. Chem. Soc.*, submitted for publication (2002).
215. J. H. Clark, A. P. Kybett, and D. J. Macquarrie, *Supported Reagents: Preparation, Analysis, and Applications*, VCH, New York, 1992.
216. E. Lindner, A. Jager, P. Wegner, H. A. Mayer, A. Benez, D. Adam, and E. Plies, *J. Non-Cryst. Solids*, 255, 208 (1999).
217. J. P. Collman, J. A. Belmont, and J. I. Brauman, *J. Am. Chem. Soc.*, 105, 7288 (1983).
218. M. M. Frodyma and V. T. Lieu, *Modern Aspects of Reflectance Spectroscopy*, W. W. Wendlandt, Ed., Plenum Press, New York, 1968, Chapter 6.
219. A. Slama-Schwok, D. Avnir, and M. Ottolenghi, *J. Phys. Chem.*, 93, 7544 (1989).
220. M. Adelt, M. Devenney, T. J. Meyer, D. W. Thompson, and J. A. Treadway, *Inorg. Chem.*, 37, 2616 (1998).
221. L. Seiden, F. Basolo, and H. M. Neumann, *J. Am. Chem. Soc.*, 81, 3809 (1959).
222. F. M. Van Meter, H. M. Neumann, M. Orban, K. Kurin-Csoergei, A. M. Zhabotinsky, and I. R. Epstein, *J. Am. Chem. Soc.*, 98, 1382 (1976).
223. F. M. Van Meter, H. M. Neumann, *J. Am. Chem. Soc.*, 98, 1388 (1976).
224. M. J. Blandamer, J. Burgess, and D. L. Roberts, *J. Chem. Soc., Dalton Trans.*, 1086 (1978).
225. C. M. Taylor and S. P. Watton, unpublished results.
226. C. Reichardt, *Solvents and Solvent Effects in Organic Chemistry*, VCH, Weinheim, 1988.
227. G. U. Bublitz and S. G. Boxer, *J. Am. Chem. Soc.*, 120, 3988 (1998).
228. N. Altounian, A. Glatfelter, S. Bai, and C. Dybowski, *J. Phys. Chem. B*, 104, 4723 (2000).
229. D. Levy, *New J. Chem.*, 18, 1073 (1994).
230. C. Rottman, A. Turniansky, and D. Avnir, *J. Sol-Gel Sci. Tech.*, 13, 17 (1998).
231. C. Rottman, M. Ottolenghi, R. Zusman, O. Lev, M. Smith, G. Gong, M. L. Kagan, and D. Avnir, *Mater. Lett.*, 13, 293 (1992).
232. R. Zusman, C. Rottman, M. Ottolenghi, and D. Avnir, *J. Non-Cryst. Solids*, 122, 107 (1990).

233. T. M. Butler, B. D. MacCraith, and C. McDonagh, *J. Non-Cryst. Solids*, *224*, 249 (1998).
234. M. A. Villegas and L. Pascual, *Thin Solid Films*, *351*, 103 (1999).
235. R. A. Dunbar, J. D. Jordan, and F. V. Bright, *Anal. Chem.*, *68*, 604 (1996).
236. A. M. Eremenko, V. P. Kondilenko, O. I. Lyuksutova, N. P. Smirnova, I. G. Tarasov, T. A. Kikteva, and V. M. Ogenko, *Proc. Indian Acad. Sci.-Chem. Sci.*, *107*, 779 (1995).
237. T. A. Kikteva, B. V. Zhmud, N. P. Smirnova, A. M. Eremenko, Y. Polevaya, and M. Ottolenghi, *J. Coll. Interf. Sci.*, *193*, 163 (1997).
238. C. M. McDonagh, A. M. Shields, A. K. McEvoy, B. D. Maccraith, and J. F. Gouin, *J. Sol-Gel Sci. Technol.*, *13*, 207 (1998).
239. S. K. Lee and I. Okura, *Analyst*, *122*, 81 (1997).
240. O. Worsfold, C. M. Dooling, T. H. Richardson, M. O. Vysotsky, R. Tregonning, C. A. Hunter, and C. Malins, *J. Mater. Chem.*, *11*, 399 (2001).
241. (a) M. A. Zaitoun and C. T. Lin, *J. Phys. Chem. B*, *101*, 1857 (1997); (b) A. Flamini and A. Panusa, *Sens. Actuators B-Chem.*, *42*, 39 (1997).
242. R. Reisfeld and D. Shamrakov, *Sens. Mater.*, *8*, 439 (1996).
243. M. Plaschke, R. Czolk, and H. J. Ache, *Anal. Chim. Acta*, *304*, 107 (1995).
244. D. Delmarre, R. Meallet1, C. Bied-Charreton, and R. B. Pansu, *J. Photochem. Photobiol. A*, *124*, 23 (1999).
245. C. Y. Shen and N. M. Kostic, *J. Am. Chem. Soc.*, *119*, 1304 (1997).
246. D. J. Blyth, J. W. Aylott, J. W. B. Moir, D. J. Richardson, and D. A. Russell, *Analyst*, *124*, 129 (1999).
247. S. Shtelzer and S. Braun, *Biotechnol. Appl. Biochem.*, *19*, 293 (1994).
248. P. C. Pandey, U. S., and H. C. Pathak, *Electroanalysis*, *11*, 59 (1999).
249. P. C. Pandey, S. Upadhyay, and H. C. Pathak, *Sensors Actuators B*, *60*, 83 (1999).
250. P. C. Pandey, S. Upadhyay, H. C. Pathak, I. Tiwari, and V. S. Tripathi, *Electroanalysis*, *11*, 1251 (1999).
251. Q. W. Li, G. A. Luo, Y. M. Wang, and X. R. Zhang, *Mater. Sci. Eng. C – Biomimetic Supramolecular Systems*, *11*, 67 (2000).
252. U. Georgi, H. Graebner, G. Roewer, and G. Wolf, *J. Sol-Gel Sci. Technol.*, *13*, 295 (1998).
253. U. Kunzelmann and H. Bottcher, *Sensors Actuators B-Chemical*, *39*, 222 (1997).
254. O. S. Wolfbeis, I. Oehme, N. Papkovskaya, and I. Klimant, *Biosens. Bioelectr.*, *15*, 69 (2000).
255. D. S. Gottfried, A. Kagan, B. M. Hoffman, and J. M. Friedman, *J. Phys. Chem. B*, *103*, 2803 (1999).
256. D. Levy and L. Esquivias, *Adv. Mater.*, *7*, 120 (1995).
257. D. Levy, *Chem. Mater.*, *9*, 2666 (1997).
258. D. Levy, *Mol. Cryst. Liq. Cryst. Sci. Technol. Sect. A-Mol. Cryst. Liq. Cryst.*, *297*, 31 (1997).
259. L. C. Klein, *Annu. Rev. Mater. Sci.*, *23*, 437 (1993).
260. A. Slama-Schwok, D. Avnir, and M. Ottolenghi, *J. Am. Chem. Soc.*, *113*, 3984 (1991).
261. A. Slama-Schwok, D. Avnir, and M. Ottolenghi, *Photochem. Photobiol.*, *54*, 52 (1991).
262. M. Sykora, K. A. Maxwell, and T. J. Meyer, *Inorg. Chem.*, *38*, 3596 (1999).
263. M. M. Collinson and S. A. Martin, *Chem. Commun.*, 899 (1999).
264. T. Jin, S. Inoue, K. Machida, and G. Adachi, *J. Electrochem. Soc.*, *144*, 4054 (1997).
265. I. F. J. Vankelecom and P. A. Jacobs, *Chiral Catalyst Immobilization and Recycling*, D. E. De Vos et al., Eds., Wiley-VCH, Weinheim, Germany, 2000, p. 19.

266. G. M. Pajonk, *Heterogeneous Chem. Rev.*, *2*, 129 (1995).
267. J. J. E. Moreau and M. W. C. Man, *Coord. Chem. Rev.*, *180*, 1073 (1998).
268. E. Lindner, M. Kemmler, T. Schneller, and H. A. Mayer, *Inorg. Chem.*, *34*, 5489 (1995).
269. J. H. Clark and D. J. Macquarrie, *Chem. Soc. Rev.*, *25*, 303 (1996).
270. P. T. Anastas and J. C. Warner, *Green Chemistry: Theory and Practice*, Oxford University Press, New York, 1998.
271. A. K. Williams and J. T. Hupp, *J. Am. Chem. Soc.*, *120*, 4366 (1998).
272. F. Gelman, D. Avnir, H. Schumann, and J. Blum, *J. Mol. Catal. A: Chem.*, *146*, 123 (1999).
273. A. Rosenfeld, J. Blum, and D. Avnir, *J. Catal.*, *164*, 363 (1996).
274. B. Pugin, *J. Mol. Catal. A: Chemical*, *107*, 273 (1996).
275. D. A. Annis and E. N. Jacobsen, *J. Am. Chem. Soc.*, *121*, 4147 (1999).
276. R. Breinbauer and E. N. Jacobsen, *Angew. Chem. IEE*, *39*, 3604 (2000).
277. G. B. Jameson and J. A. Ibers, *Bioinorganic Chemistry*, I. Bertini et al., Eds., University Science, Mill Valley, CA, 1994.
278. G. B. Jameson and J. P. Collmann, *J. Am. Chem. Soc.*, *102*, 3224 (1980).
279. J. V. Walker, M. Morey, H. Carlsson, A. Davidson, G. D. Stucky, and A. Butler, *J. Am. Chem. Soc.*, *119*, 6921 (1997).
280. J. Evans, A. B. Zaki, M. Y. El-Sheikh, and S. A. El-Safty, *J. Phys. Chem. B*, *104*, 10271 (2000).
281. K. Neimann, R. Neumann, A. Rabion, R. M. Buchanan, and R. H. Fish, *Inorg. Chem.*, *38*, 3575 (1999).
282. H. N. Choksi, A. Zippert, P. Berdahl, J. A. Bertrand, D. L. Perry, M. B. Mitchell, and M. G. White, *J. Molec. Catal. A*, *97*, 85 (1995).
283. K. J. Ciuffi, H. C. Sacco, J. B. Valim, C. M. C. P. Manso, O. A. Serra, O. R. Nascimento, E. A. Vidoto, and Y. Iamamoto, *J. Non-Cryst. Solids*, 146 (1999).
284. J. H. Wang, *J. Am. Chem. Soc.*, *80*, 3168 (1958).
285. K. Miki and Y. Sato, *Bull. Chem. Soc. Jpn.*, *66*, 2385 (1993).
286. M. das Dores Assis and J. R. Lindsay-Smith, *J. Chem. Soc. Perkin 2*, 2221 (1998).
287. Y. G. Akopyants, S. A. Borisenkova, O. L. Kaliya, V. M. Derkacheva, and E. A. Lukyanets, *J. Mol. Catal.*, *83*, 1 (1993).
288. M. A. Martinez-Lorente, P. Battioni, W. Kleemiss, J. F Bartoli, and D. Mansuy, *J. Mol. Catal. A.*, *113*, 343 (1996).
289. M. Sanchez, A. Hadasch, A. Rabion, and B. Meunier, *C. R. Acad. Sci. Paris, Ser. IIc*, 241 (1999).
290. O. Leal, D. L. Anderson, R. G. Bowman, F. Basolo, and J. R. L. Burwell, *J. Am. Chem. Soc.*, *97*, 5125 (1975).
291. Y. Iamamoto, Y. M. Idemori, and S. Nakagaki, *J. Mol. Catal. A: Chemical*, *99*, 197 (1995).
292. D. C. Sherrington, *J. Polym. Sci., Part A Polym. Chem.*, *39*, 2364 (2001).

Crystal Chemistry of Organically Templated Vanadium Phosphates and Organophosphonates

ROBERT C. FINN and JON ZUBIETA

Department of Chemistry
Syracuse University
Syracuse, NY

ROBERT C. HAUSHALTER

Parallel Synthesis Technologies
Menlo Park, CA

CONTENTS

I. INTRODUCTION

II. SYNTHESIS

III. INORGANIC VANADIUM PHOSPHATE PHASES

    A. General Structural Characteristics
    B. Vanadium Phosphates with Charge Neutral V—P—O Covalent Linkages
    C. Quaternary M′/V/P/O Phases with Negatively Charged V—P—O Covalent Linkages and Charge Compensating M′ Cations
        1. Infinite Chains of Vanadium Polyhedra
        2. Isolated Vanadium Polyhedra
        3. Binuclear Vanadium Clusters
        4. Binuclear Vanadium Clusters and Vanadium Chains
        5. Tetranuclear Vanadium Clusters
        6. Pentanuclear Vanadium Clusters
        7. Nonanuclear Vanadium Clusters
    D. Inorganic Vanadium Phosphates with d-Block Elements as Additional Framework Constituents
    E. Vanadium Silicophosphates, Aluminophosphates, Gallophosphates, Borophosphates, and Phosphate Fluorides

---

Progress in Inorganic Chemistry, Vol. 51, Edited by Kenneth D. Karlin.
ISBN 0-471-26534-9 © 2003 John Wiley & Sons, Inc.

1. General Comments
2. Vanadium Silicophosphates
3. Vanadium Aluminophosphates and Gallophosphates
4. Vanadium Borophosphates
5. Vanadium Phosphate Fluorides

IV. VANADIUM PHOSPHATE PHASES WITH CHARGE-COMPENSATING ORGANIC CATIONS

    A. One-Dimensional Structures
    B. Two-Dimensional Structures
    C. Three-Dimensional Structures

V. VANADIUM ORGANOPHOSPHONATE PHASES

    A. General Characteristics
    B. Oxovanadium Monophosphonate Solids
    C. Oxovanadium Diphosphonate Solids

VI. OXOVANADIUM PHOSPHATES AND ORGANOPHOSPHONATES WITH SECONDARY META–LIGAND SUBUNITS INTEGRATED INTO COVALENTLY LINKED V/O/M′/P SCAFFOLDINGS

    A. General Considerations
    B. Influence of M^{II}–Organodiimine Subunits on Oxovanadium Phosphate Structures
    C. Influence of M^{II}–Organodiimine Subunits on Oxovanadium Organophosphonate Structures

VII. CONCLUSIONS

ABBREVIATIONS

REFERENCES

I. INTRODUCTION

Because solid-state oxides exist for a majority of the elements, these materials are ubiquitous in our world. The huge range of solid-state properties is a result of the diversity of chemical composition and structure types, and inorganic oxides occur across the geosphere, biosphere, and noosphere (1–3). In addition to the hydrogen oxide that forms the basis of life, the land on which we live is largely made up of silicon and aluminum oxides. While many naturally occurring oxides and minerals possess complex crystal structures, the majority are of simple composition and have highly symmetrical structures with rather small unit cells. Most silicates, important ores, gems, many rocks and soils are examples of these materials. Although these simple oxides can have unique and specific properties, such as piezoelectricity, ferromagnetism, or catalytic activity, as a general rule there is a correlation between the complexity of the structure of a material and its functionality (4). For example, the $CaCO_3$ based shell of a mollusc or a calcium phosphate tooth or bone from an animal, have much more functionality than crystals of $CaCO_3$ or $Ca_3(PO_4)_2$ due to their incorporation into a complex structure. In these cases, the inorganic oxide contributes to the increased functionality via assimilation as one component in a hierarchical structure where there is a synergistic interaction between an organic and inorganic component (5, 6). This interaction within these hybrid organic–inorganic materials must ultimately derive from the nature of the interface between the organic material and the inorganic oxide. Therefore, synthetic studies of materials possessing such an interface, coupled with the acquisition of the appropriate structural information, should contribute to the development of an increased understanding on how to control the structure–property relationships within these hybrid materials.

Prior to the publication of the first microporous octahedral–tetrahedral framework transition metal phosphate (TMPO) material with entrained organic cations (7), there were three primary classes of materials in which organic materials played an important structural role. These materials were zeolites (8), mesoporous oxides of which MCM-41 was the first example (9), and biomineralized materials. The important field of biomineralization, which includes studies of the nature of the structure directing role of the organic component, is receiving increasing attention (5, 6) and will not be discussed here. However, one very important and relevant fact is that the biologically based organic material appears to interact with inorganic oxide primarily via $O-H \cdots O$ and $N-H \cdots O$ hydrogen bonds.

The first synthetic materials in which extensive organic molecules were used to influence the growth of an inorganic oxide were the synthetic zeolites made during the pioneering work of Barrer (10). The incorporation of various organic molecules, especially amines and ammonium salts, into aluminosilicate frameworks has allowed this technologically extremely important class of materials to be readily prepared. In this case, not only does the organic molecule often provide charge compensation for the Al^{3+} substituting for Si^{4+} within the zeolite framework, but by virtue of its size, charge, polarity, and shape, directs the formation and crystallization of the silicate precursors into the zeolite lattice. Since zeolites are prepared under highly basic conditions to facilitate the dissolution of the SiO_2 starting material, only quaternary ammonium salts remain cationic under these conditions. Therefore, the nature of the template-framework interaction depends primarily on van der Waals type interactions. This type of interaction of the relatively low-charged zeolite framework with hydrophobic organic molecules is confirmed by the exothermic sorption of hydrocarbons into many zeolite frameworks.

The third major class of inorganic oxides in which organic materials play an important structure directing role is the rapidly expanding field of the mesoporous oxides, which emanate from the pioneering work on the MCM-41 type of materials derived by Kresge and co-workers (9, 11) at Mobil. These oxides are prepared by the polymerization of soluble silica precursors around various organic assemblies derived from the associative behavior of amphiphilic organic molecules in aqueous solution. By altering the size, shape, and charge of organic surfactants in water, by changing concentration, pH, temperature, and shear forces, the resulting oxide structures derived from ensuing silicate polymerization take on the form of the organic aggregate, which persist as entrained assemblies within the silicate. Although many morphologies have now been realized, the most important contain periodic arrays of voids on the 30–100-Å size scale (or larger) within the amorphous silica matrix after removal of the organics. Based on recent advances in extending this basic methodology to the preparation of oxide frameworks other than Si based, as well as studies of the grafting or ion exchange of additional organic or inorganic species into the internal pore volume of the host framework, this area is expected to receive considerable future attention.

In 1989, we reported the first example of the synthesis of non-silicate, open framework inorganic oxide with entrained organic cations in the molybdenum phosphate (MoPO) system. This compound, $(Me_4N)(H_3O)[Mo_4O_4(PO_4)] \cdot xH_2O$ (12, 13) contained an octahedral–tetrahedral framework built up from Mo_4-oxoclusters connected by PO_4 tetrahedra into a three-dimensional (3D) covalent framework. Significantly, it was possible to remove the organic counterions from the material and render it microporous as demonstrated by sorption measurements. Subsequently, other open framework MoPO materials were prepared and characterized and they contributed the only crystalline organic–inorganic materials of this type, in addition to the zeolites, at that time.

One of the important questions formulated during the initial investigations of the TMPO materials was that of the possibility of observing any sort of shape selective reactivity between the framework and small molecules in a fashion analogous to zeolites. It might seem that the incorporation of stoichiometric amounts of transition elements into a crystalline microporous framework, assuming the transition element is in direct contact with the voids that constitute the micropores, might produce materials differing greatly in selectivity and reactivity from the aluminosilicates. Reactions involving oxidation–reduction, atom abstraction, and photochemistry normally associated with the d-block elements could thus be combined with the shape selective absorptivity inherent in the open framework. Based on assessments of the known reactivities and potential extensions of the system to hydrothermal treatment, organically templated vanadium phosphates and phosphonates were chosen as the initial study area. These materials constitute the main subject matter of this chapter.

The commonalities linking the structurally diverse biomineralized structures, zeolites, mesoporous oxides, and TMPO materials are two of the phenomena that critically influence the organization of biological structures, namely, hydrogen bonding and hydrophobic–hydrophilic interactions. These two factors are the main structure directing influences in the formation of the organically templated vanadium phosphates (VOPO) and vanadium organophosphonates (VOPRO). In the case of the vanadium phosphates, no materials incorporating the relatively nonpolar tetraalkylammonium cations have been observed thus far, but only polar organic ammonium cations containing N—H groups, inorganic cations (primarily alkali metal cations), or their mixtures have been found to serve as charge compensation for the anionic VOPO framework. Thus, both the relatively higher charge per unit volume of lower alkylated ammonium cations as compared to tetraalkyammonium cations, and the prevalence of higher hydrates indicate that the interior surface of the organic vanadium phosphates are much more polar, and thus complement the correspondingly less polar micropores in the more highly siliceous zeolites. The important structure directing role of the entrained polar ammonium cation manifests itself in the strong, geometrically very specific hydrogen bonding between the N—H groups of the cation and the O or O—H groups of the framework. As an example, Figure 1 shows the variety of ways that a single organic ammonium cation, in this case the $H_3NCH_2CH_2NH_3^{2+}$ ethylenediammonium cation, interacts through hydrogen bonds to an oxide framework. As discussed below, it seems likely that the large number of hydrigen bonds found in the organic VOPO, which constitute the strongest interactive interactions after the covalent bonds of the oxide framework, may render the specific material isolated the kinetically least soluble species present in the hydrothermal reaction medium. It is obvious that one could speculate concerning the utilization of preformed organic cations, which could present a geometrically well-defined array of hydrogen-bond donors or acceptors, to influence the crystallization of the oxide framework in a geometrically complicated manner. The organic cation serves in this role to imprint structural information onto the inorganic framework. Such organization of inorganic components by organic molecules via numerous noncovalent interactions that determine the kinetic pathway of the crystallization process has been extensively exploited in zeolite synthesis (14).

The influences of hydrophobic–hydrophilic interactions are important in the structures of the VOPRO, some of which are shown in Figure 2. Two chemical features appear to strongly influence the structure, namely, the association of the organic groups from the neighboring phosphonate groups and the incorporation of more hydrophobic ammonium cations of lower charge–volume than those found in the organic VOPO materials. Thus, the energetically favorable situation of reducing the interaction of the less polar organic moiety with the polar oxide framework often results in the formation of a 2D oxide layer with each surface covered by organic groups that interdigitate with the adjacent layer. While much less common one-dimensional (1D) and 3D organophosphonate materials are known, most of the materials form 2D sheets or molecular clusters whose surfaces are terminated by organic groups.

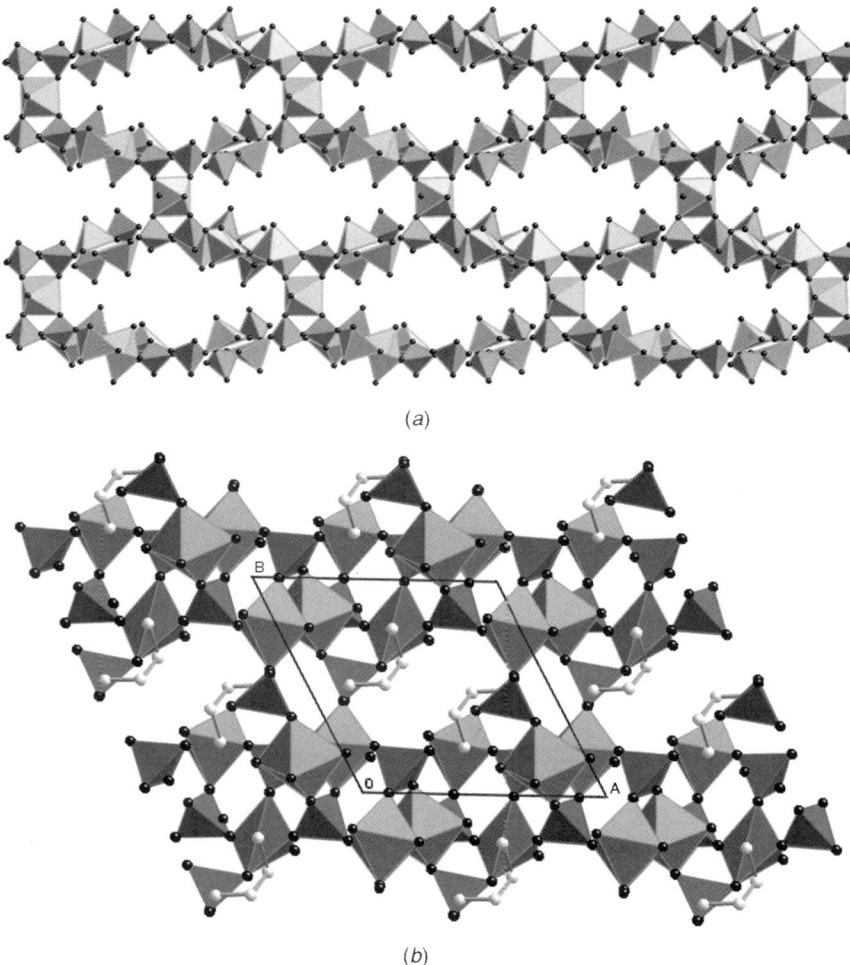

Figure 1. Views of two V/P/O phases containing the ethylenediammonium cations. Polyhedral representations of the structures of $[H_3NCH_2CH_2NH_3]_2[H_3NCH_2CH_2NH_2][V(H_2O)_2(VO)_8(OH)_4(HPO_4)_4(PO_4)_4] \cdot 2H_2O$, viewed parallel to the crystallographic c axis (a) and of $[H_3NCH_2CH_2NH_3](VO)_3(H_2O)_2(PO_4)_2(HPO_4)]$, viewed parallel to the b axis (b).

CRYSTAL CHEMISTRY OF ORGANICALLY TEMPLATED VANADIUM PHOSPHATES 427

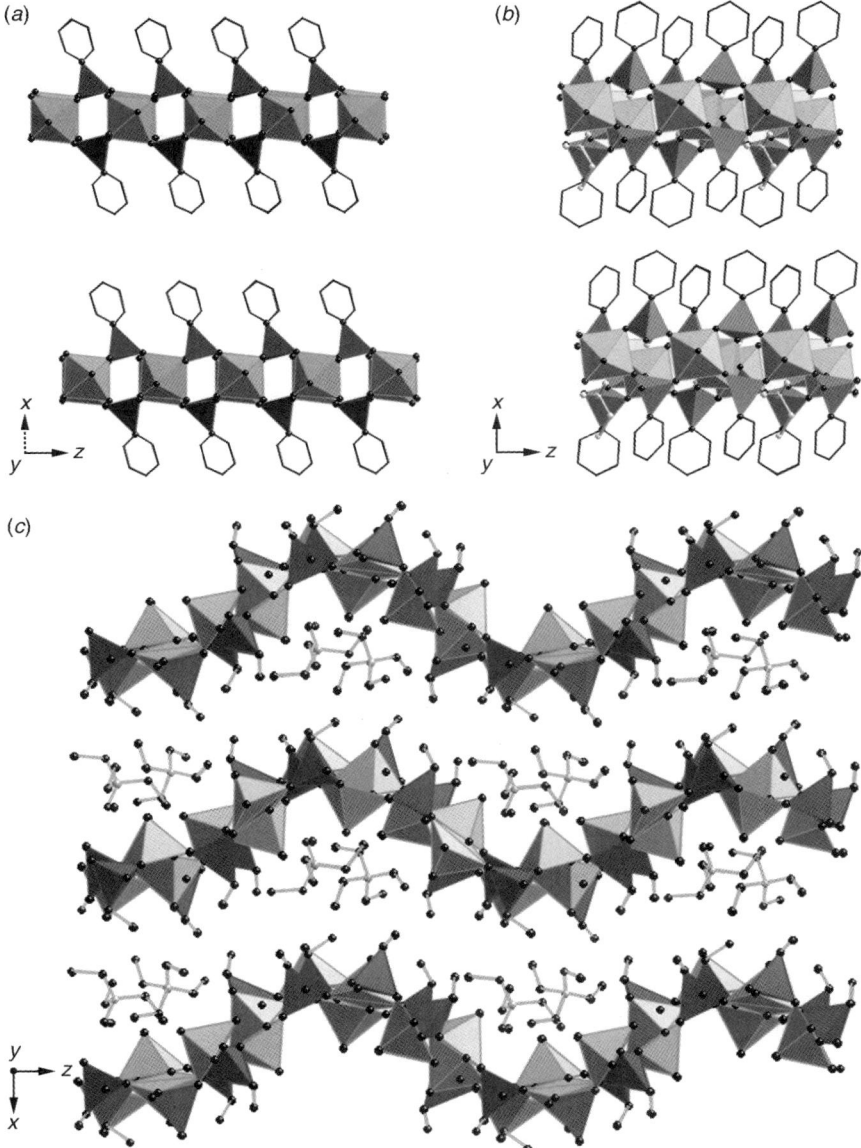

Figure 2. Polyhedral views of representative two-dimensional (2D) oxovanadium organophosphonate structures: [(VO)(O$_3$PPh)(H$_2$O)] (a); (H$_3$NEt)[(VO)$_3$(H$_2$O)(O$_3$PPh)$_4$] (b); (Et$_4$N)[(VO)$_3$(OH)(H$_2$O)(O$_3$PEt)$_3$ • H$_2$O] (c).

II. SYNTHESIS

The synthesis of these organically templated oxides is not complicated to execute, but the large number of interacting variables renders the observations of many trends difficult. In the case of many oxide materials, the synthesis must be performed at high temperatures (800–1200°C) because the starting materials are solid-state compounds and the high temperatures are mandatory in order to achieve diffusion rates adequate for product formation within a reasonable time period. Obviously, the ideas discussed above, which could utilize the rational synthetic procedures of organic chemistry to synthesize a template that could imprint or transfer structural information to the inorganic oxide framework, would not be possible at temperature much above ∼300°C because of the lack of thermal stability of the organic component. However, based on a large number of mineralogical examples, and more importantly the successful preparation of numerous aluminosilicates, it appeared that the hydrothermal technique would be the synthetic method of choice (15). This fact has been verified by the subsequent experimentation that has lead to the materials that are the subject of this chapter.

The hydrothermal experiments to prepare the VOPO and VOPRO materials are quite simple. An aqueous solution of the organic template, the vanadium source and the phosphate or phosphonate are treated at 150–200°C for ∼3 days. It appears that in most cases, the crystals form at the reaction temperature and not upon cooling the reaction mixture and, in fact, with reaction temperatures much <150°C, that the formation of well-crystallized products occurs slowly. These well-crystallized samples are imperative to any study of these systems as single crystal X-ray diffraction is the primary characterization tool for these materials.

While it is very clear that small changes in one or few of the reactions variables of time, temperature, concentrations, V/P ratio, the amount of template, the chemical structure of the template, the nature of the starting materials, and pH can have very dramatic effects on the outcome of the reaction, the number and degree of coupling between these numerous variables renders quantitative predictions of reaction products extremely difficult. The potential qualitative trends discussed below may represent real constraints on this system or may simply be a reflection of the fact that there have been relatively few materials characterized compared to the number of variables and their possible permutations.

The V/P ratios for the vanadium phosphates with inorganic and organic cations and for the vanadium organophosphonates are plotted against frequency in Figure 3. Not unexpectedly, the V/P ratio of 1:1 predominates, reflecting charge balance considerations for cations and negatively charge oxide frameworks. The most common constituents of such phases are vanadyl VO^{2+} and

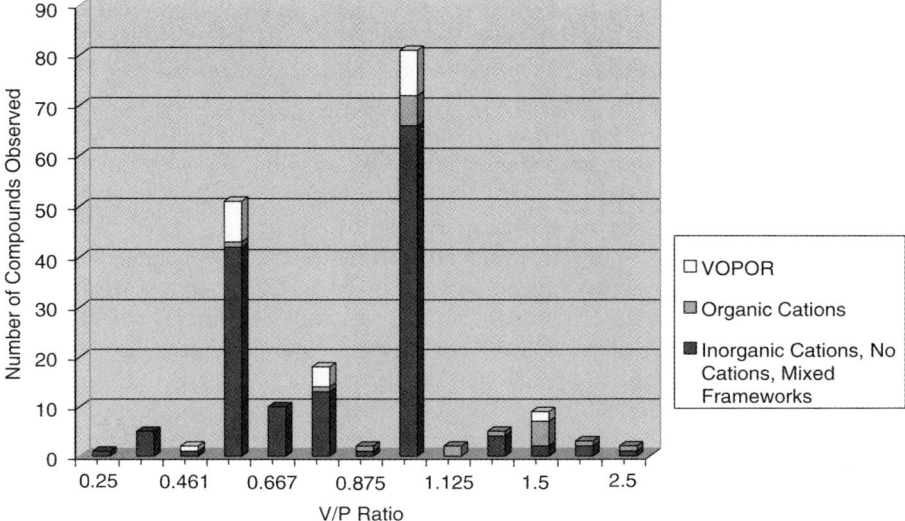

Figure 3. Frequencies of V/P ratios for the types of V/P/O phases.

PO_4^{3-}, which at a 1:1 ratio generate the negatively charged oxide framework and require a charge-compensating inorganic or organic cation. Vanadium/ phosphorus ratios in the range $0.5 \leq V/P < 1.0$ are also common, since at the acidic pH of the majority of the syntheses, protonation of the phosphate groups to give $H_nPO_4^{3-n}(n = 1,2)$ is not unusual. Likewise, divalent and trivalent cations as charge compensating groups can accommodate somewhat smaller V/P ratios. Consequently, for the ~150 V/P/O structures with no cation, inorganic cations, and mixed frameworks, 94% exhibit V/P ratios of 1:1 or less. A similar trend is observed for the oxovanadium organophosphonate phases that exhibit V/P ratios of 1:1 or less for ~92% of reported structures. Curiously, the introduction of organic cations has one unexpected consequence: V/P ratios >1:1 are quite common, representing 50% of the reported structures. While conclusions based on a relatively meager data set (19 structures) may be premature, it is tempting to speculate that this observation is related to charge balance considerations. Replacement of large charge/volume ratio inorganic cations by organic cations where the charge is considerably more diffuse may necessitate a compensatory incorporation of additional vanadium sites into the V/P/O scaffolding to balance charge. Furthermore, the increased hydrophobicity of the organic cations may encourage anionic scaffoldings with augmented vanadyl sites, which are considerably less hydrophilic than phosphate oxygen atoms.

However, V/P ratio <0.5 are extremely unusual, an observation that reflects the inaccessibility of P—O—P bond formation in aqueous media. Similarly, V/P ratios >1.0 are observed at low frequency. The requirement that the V/P/O inorganic framework be anionic or neutral would tend to rule out structures based on V/P ratios >1.5 and containing isolated VO^{2+} units. To preserve the negatively charged inorganic framework, large vanadium concentrations can only be achieved if these sites are introduced as neutral VO_2 units or as clusters $\{V_xO_y\}^{n-}$, that is, by incorporating additional oxo groups.

As mentioned above, the reactions proceed in the 150–200°C range and one would expect that many organic materials could begin to show thermal instability above ~250°C. Furthermore, the reactions are carried out in polytetrafluoroethylene containers that begins to soften considerably above these temperatures. Most of these reactions were carried out below a pH value of ~8, above which the probability of V—O—P bond hydrolysis would occur, but V—O—V bonds can be formed at higher pH [cf. $VO_2 \cdot \frac{1}{2}H_2O$ (16)]. Also, if the presence of $\{N-H\}^+$ groups is desirable as discussed above, then the pK_a of the $\{N-H\}^+$ moieties must be taken into account. The duration of the reaction is usually ~2 or 3 days. Although we have observed cases where the product appears within the first few hours and then transforms to another product, the product and crystal size formed after ~3 days change much more slowly with longer reaction times. Since the products are nearly always quite insoluble in water at 20°C, the absolute amount of water present does not seem to be a critical variable; although the *relative* concentrations of all of the reactants is a very strong influence on the outcome of the reaction. The chemical nature of the starting materials (e.g., for vanadium: V_2O_5, $NaVO_3$, VCl_4, etc.) can have a profound influence as the solubility (amount of V in solution at any given time), the presence or absence of different counterions and the rate of reaction all depend on the chemical properties of the V or P sources.

III. INORGANIC VANADIUM PHOSPHATE PHASES

A. General Structural Characteristics

The solid-phase inorganic phosphates constitute a vast family of materials of considerable technological importance. Orthophosphate salts of most elements of the periodic table are known and may include minerals, synthetic products, acid salts, hydrates, and polymorphic varieties (17). The continuing scientific interest in phosphate materials reflects their practical applications as fertilizers, catalysts, phosphors, ion conductors, piezoelectric materials, biotechnological materials, and sorbents (18).

The oxovanadium phosphate system represents an important subclass of the family of inorganic phosphates. The dramatic expansion of the chemistry of the V/P/O system derives its impetus from the observation that vanadyl pyrophosphate $(VO)_2P_2O_7$ (19, 20) and vanadyl phosphite (21, 22) are effective in the selective air oxidation of butane to maleic anhydride, a remarkable transformation involving 14 electrons and 7 oxygen atoms. In addition to studies directed at the elucidation of the catalytic reaction mechanism, the synthesis and characterization of related materials has provided an expansive chemical literature. The rich structural chemistry associated with the V/P/O phases derives from several noteworthy characteristics of the vanadium oxide (23) and phosphate chemistries:

1. The accessibility of the V(III), V(IV), and V(V) oxidation states under the synthetic conditions employed to prepare these phases.
2. The occurrence of different vanadium coordination polyhedra, including tetrahedral [V(V)], square pyramidal [V(IV) and V(V)], trigonal bipyramidal [V(IV)], distorted octahedral [V(IV) and V(V)], and regular octahedral [V(III)] geometries.
3. The formation of larger polyhedral aggregates $\{V_xO_y\}$ or even 1D chains through condensation of vanadium polyhedra through shared oxo groups.
4. Variable connectivities between vanadium polyhedra of the $\{V_xO_y\}$ aggregates, which may share common corners, edges, or faces and combinations of these modes.
5. Water molecule and hydroxyl group participation in the vanadium ligation.
6. Variability in the tetrahedral phosphate building block, which may be present as PO_4^{3-}, HPO_4^{2-}, $H_2PO_4^-$, or $P_2O_7^{4-}$.
7. Variations in the number of vertices of the phosphate tetrahedral unit linked to neighboring metal polyhedra, allowing η^2, η^3, or η^4- coordination modes and combinations.
8. The ability of phosphate oxygen donor to bond to more than one vanadium center, providing the possibility of μ or μ^2 coordination modes at a given tetrahedral vertex.
9. The conformational flexibility of the common bridging unit, the eight-membered ring motif $\{(O=V)(OPO)_2(V=O)\}$ which may adopt various orientations of the vanadyl oxygens relative to each other and several different ring conformations.
10. The introduction of additional metal cations, which may function as charge-compensating units located in interlamellar regions, tunnels or cages provided by the V—P—O covalent backbone or as an integral part of the covalent structure.

The polyhedral geometries associated with the oxidation states observed for vanadium in V/P/O phases exhibit several noteworthy characteristics. As noted above, three oxidation states are encountered in the synthetic phases, as well as in minerals (23) : +3, +4, and +5. Trivalent vanadium, with the electronic configuration [Ar] $3d^2$, exhibits octahedral geometry with V—O bond distances generally falling in the 1.98–2.04-Å range (23a). On the other hand, tetravalent vanadium ([Ar]$3d^1$) may occur in five- or six-coordinate geometries. However, in contrast to the small range of V—O bonds encountered for V(III), V(IV) geometries are characterized by the presence of a short vanadyl bond in the range of ~1.57–1.68 Å. Consequently, the five-coordinate geometry may be described as 1 + 4 distorted square pyramidal, with one short apical vanadyl bond and four equatorial bonds in the 1.80–2.15 Å range. The distorted six-coordinate geometry adopts 1 + 4 + 1 bond distributions with a short axial vanadyl bond, four longer equatorial bonds, and long axial bond trans to the vanadyl group, generally in the 2.20–2.30-Å range.

Pentavalent vanadium may adopt four-, five-, or six-coordinate geometries. Four-coordinate V(V) species exhibit distorted tetrahedral geometries. However, five-coordinate V(V) in V/P/O phases may be found in 1 + 4 square pyramidal geometry or, less frequently, in 2 + 3 trigonal bipyramidal geometry. The 1 + 4 geometry is characterized by the short vanadyl bond in the apical position and four longer equatorial bonds. The trigonal bipyramidal geometry in general exhibits two short equatorial vanadyl bonds (1.55–1.70 Å), a longer V—O equatorial bond distance (1.90–2.00 Å), and two long axial bonds.

Six-coordinate V(V) species feature 1 + 4 + 1 and 2 + 2 + 2 bond distributions. The 1 + 4 + 1 geometry is similar to that adopted by V(IV): one short axial vanadyl bond (1.55–1.70 Å), one long axial bond (trans to the vanadyl, 2.00–2.60 Å), and four intermediate equatorial bonds (1.80–2.25 Å). The 2 + 2 + 2 geometry is similar to that observed for Mo(VI) species with {MoO_6} polyhedra: two short vanadyl groups in a cis orientation (1.55–1.70 Å), two long V—O bonds trans to the vanadyl groups (2.10–2.50 Å), and two intermediate bonds (~1.90–2.05 Å). Polyhedra containing two vanadyl groups invariably exhibit a cis arrangement of these multiple bonded oxo groups, which is the preferred geometry since the strong π-bonding oxo groups interact more effectively with the metal t_{2g} orbitals than in the trans orientation.

Since many of the structural principles encountered for the inorganic V/P/O phases extend to the organic–inorganic composite materials, a brief excursion into this structural chemistry provides a basis for the development of the structural systematics of these families of materials. As a gross structural classification, the inorganic V/P/O phases have been grouped into three types: (1) phosphates with neutral V/P/O backbones, that is, materials containing only vanadium, phosphorus, oxygen, and hydrogen; (2) phosphates with negatively charged V/P/O motifs and charge-compensating inorganic cations or materials of the ionic salt category; (3) phosphate phases with other framework constituents in addition to vanadium, phosphorus, and oxygen.

(Text continued on page – 459)

TABLE I

Selected Structural Information for Vanadium Phosphates with Neutral V/P/O Backbones

Compound	V/P/O Framework Structure	Vanadium Coordination and Linkages	Cell Parameters[a]	Space Group	References
$VO(H_2PO_4)_2 \cdot H_2O$	Chain	Isolated V(IV) octahedra	12.046(3) 8.147(1) 7.548(2) $\beta = 121.83(2)$	Cc	24
$\alpha\text{-}VO(PO_4)$	Layered	Isolated V(V) square pyramids with long (2.5–2.6 Å) interaction to the site *trans* to the (V=O) unit linking layers.	6.20(1) 4.11(1) 6.014(7) 4.434(2)	$P4/n$	25–27
$VO(PO_4) \cdot 2H_2O$	Layered	Isolated V(V) octahedra	6.202(1) 7.410(1)	$P4/mmm$ or $P4/n$	28 29
$VO(HPO_4) \cdot 0.5H_2O$	Layered	Dimers of face-sharing octahedra	7.416(1) 9.592(2) 5.689(1)	$Pmmn$	30
$\alpha\text{-}VO(HPO_4) \cdot 2H_2O$	Layered	Chains of corner-sharing octahedra	7.613(5) 7.431(5) 9.482(7) $\beta = 95.44(8)$	$P2_1/c$	31
$\beta\text{-}VO(HPO_4) \cdot 2H_2O$	Layered	Isolated V(IV) octahedra	5.659(2) 7.578(4) 12.623(5) $\alpha = 89.66(2)$ $\beta = 102.14(2)$ $\gamma = 92.23(2)$	$P\bar{1}$	32

(*continues*)

TABLE I (Continued)

Compound	V/P/O Framework Structure	Vanadium Coordination and Linkages	Cell Parameters[a]	Space Group	References
$VO(HPO_4) \cdot 4H_2O$	Layered	Isolated V(IV) octahedra	6.379(2) 8.921(2) 13.462(3) $\alpha = 79.95(2)$ $\beta = 76.33(3)$ $\gamma = 71.03(3)$	$P\bar{1}$	33
$\beta\text{-}VO(PO_4)$	3D	Chains of corner sharing V(IV)octahedra	7.770(3) 6.143(3) 6.965(3)	$Pnma$	34
$VO(H_2PO_4)_2$	3D	Chains of corner-sharing V(IV) octahedra	8.953(2) 8.953(2) 7.965(2)	$P4/ncc$	35
$(VO)_2(P_2O_7)$	3D	Double chains of corner sharing V(IV) octahedra, linked into edge-sharing pairs	7.73808(7) 9.58698(8) 16.5895(1)	$Pca2_1$ (orthorhombic form)	36
			7.7276(3) 16.5885(4) 9.5796(5) $\beta = 89.975(3)$	$P2_1$ (monoclinic form)	37a
			7.5801(2) 9.5458(3) 8.3629(3)	$Pnab$ (high pressure form)	37b
$V_4O(OH)_2(PO_4)_3$	3D	Strings of face-sharing V(III)/V(IV) octahedra, linked by oxo-bridges into a double chain of such dimers	19.528(5) 7.404(2) 7.383(2) $\beta = 101.94(2)$	$C2/c$	38

Compound	Dim.	Structural features	Unit cell parameters (Å, °)	Space group	Ref.
$V_{5.12}(PO_4)_4(OH)_{3.36}(H_2O)_{0.64} \cdot 0.84 H_2O$	3D	Chains of face-sharing V(III) octahedra	5.1646(1); 12.9217(4)	$I4_1/amd$	39
$V_{4.92}(PO_4)_4(OH)_{2.76}(H_2O)_{1.24} \cdot 1.32 H_2O$	3D	Chains of face-sharing V(III) octahedra	5.1811(2); 12.9329(4)	$I4_1/amd$	40
$V(PO_3)_3$	3D	Isolated V(III) octahedra; infinite polyphosphate chains	13.223(2); 6.380(3); 10.643(3); $\beta = 134.95(3)$	Cc	41
$V(PO_4) \cdot 2H_2O$	3D	Isolated V(III) octahedra	5.3112(2); 9.8447(2); 8.6536(2); $\beta = 91.23(1)$	$P2_1/n$	39
$V_2(VO)(P_2O_7)_2$	3D	Linear edge-sharing trinuclear units of $2 \times V(III)$ octahedra and $1 \times V(IV)$ square pyramid	17.459(3); 12.185(2); 5.24311(7)	$Pnma$	42
$V_4(P_2O_7)_3$	3D	Binuclear units of face-sharing V(III) octahedra	7.443(1); 9.560(2); 21.347(4)	$Pmcn$	43

[a] The estimated standard deviation values for unit cell parameters are given in parentheses.

TABLE II

Selected Structural Information for Vanadium Phosphates with Negatively Charged V/P/O Motifs and Charge-Compensating Inorganic Cations

Compound	V/P/O Framework Structure	Vanadium Coordination and Linkages	Cell Parameters	Space Group	References
$(NH_4)(VO)(PO_4) \cdot H_2O$	Chain	Isolated V(IV) square pyramids with H_2O in one basal position	6.7756(5) 4.9147(7) 8.4539(8) $\beta = 91.092(7)$	$P2_1$	44
$(NH_4)_2(VO)(HPO_4)_2 \cdot H_2O$	Chain	Chains of corner-sharing V(IV) octahedra	7.2499(8) 8.1244(9) 8.6529(9) $\alpha = 73.472(2)$ $\beta = 80.085(2)$ $\gamma = 85.077(2)$	$P1$	45
$Sr_2(VO)(PO_4)_2$	Chain	Chains of trans corner-sharing V(IV) octahedra	6.744(4) 15.866(4) 7.032(2) $\beta = 115.41(2)$	$I2/a$	46
$Ba_2(VO)(PO_4)_2$	Chain	Isolated V(IV) square pyramids	9.471(2) 5.443(1) 16.972(4) $\beta = 101.65(2)$	$I2$	47
$Ba_2(VO)(PO_4)_2 \cdot H_2O$	Chain	Isolated V(IV) octahedra	12.420(3) 5.219(2) 6.941(2) $\beta = 104.52(2)$	$C2/m$	48

Compound	Structure	Lattice parameters	Space group	Ref.
$Ba_8(VO)_6(PO_4)_2(HPO_4)_{11} \cdot 3H_2O$	Chain	Isolated V(IV) octahedra	31.685(11) 5.208(2) 7.784(3) $\beta = 90.59(3)$	$C2/m$	49
$\alpha\text{-}(NH_4)(VO_2)(HPO_4)$	Chain	Chains of corner-sharing V(V) square pyramids	6.830(1) 9.233(2) 8.817(2)	$Pb2_1a$	50
$A(VO_2)(HPO_4)$ ($A = K^+, Rb^+, NH_4^+$ – β phase)	Chain	Isolated V(V) square pyramids	6.7550(3) 9.1026(4) 17.0808(6) (for K^+)	$Pbca$	51
$Ba_2(VO_2)(PO_4)(HPO_4) \cdot H_2O$	Chain	Isolated V(V) trigonal bipyramids	5.0772(5) 8.724(2) 10.806(1) $\beta = 90.795(8)$	$P2_1$	52
$Tl(VO_2)(HPO_4)$	Chain	Chains of corner-sharing V(IV) square pyramids	9.257(2) 17.518(4) 6.810(2)	$Pbca$	53
$Ba_3(V_2O_3)(PO_4)_3$	Ribbons	Binuclear units of corner-sharing V(IV) square pyramids and V(V) octahedra	5.137(1) 12.418(1) 20.724(3)	$P2_12_12_1$	54
$Ca_2V(PO_4)(HPO_4)_2 \cdot H_2O$	Chain	Chains of corner-sharing V(III) octahedra	7.531(2) 15.522(4) 9.149(2) $\beta = 113.52(2)$	$C2/c$	55
$(NH_4)(VO)(PO_4) \cdot 1.5H_2O$	Layered	Isolated V(IV) square pyramids	6.3160(5) 13.540(2)	$I4/mmm$	56

(*continues*)

TABLE II (Continued)

Compound	V/P/O Framework Structure	Vanadium Coordination and Linkages	Cell Parameters	Space Group	References
$(NH_4)(VO)_2(PO_4)_2 \cdot 3H_2O$	Layered	Isolated V(IV)/V(V) octahedra	6.9669(6) 17.663(2) 8.9304(8) $\beta = 105.347(1)$	$P2_1/m$	56
$Li(VO)(PO_4) \cdot 0.5H_2O$	Layered	Binuclear units of face-sharing V(IV) octahedra	7.4651(6) 9.4167(8) 6.0762(6)	$P2_12_12$	57
$Na_{0.5}(VO)(PO_4) \cdot 2H_2O$	Layered	Isolated V(IV) and V(V) octahedra	6.2851(7) 6.284(1) 13.262(2) $\alpha = 80.30(1)$ $\beta = 87.434(9)$ $\gamma = 89.94(1)$	$P\bar{1}$	58
$Na_{0.5}(VO)(PO_4) \cdot 2H_2O$	Layered	Isolated V(IV) and V(V) octahedra	6.293(3) 6.294(3) 6.844(5) $\alpha = 107.04(5)$ $\beta = 92.34(5)$ $\gamma = 90.13(5)$	$P1$	59
$K_{0.5}(VO)(PO_4) \cdot 1.5H_2O$	Layered	Isolated V(IV) and V(V) octahedra	6.282(2) 6.285(1) 6.679(2) $\alpha = 89.11$ $\beta = 72.84$ $\gamma = 89.98$	$P\bar{1}$	58

Compound	Structure	Description	Parameters	Space group	Ref
$Rb_{0.5}(VO)(PO_4) \cdot 1.5H_2O$	Layered	Isolated V(IV) and V(V) octahedra	6.285(1) 6.2908(6) 6.849(2) $\alpha = 89.73$ $\beta = 107.79(2)$ $\gamma = 90.147(13)$	$P\bar{1}$	60
$M_x(VO)(PO_4) \cdot yH_2O$ ($M = Ag_{0.43}$, $Cu_{0.16}$, $Zn_{0.11}$) ($y = 2$ for Ag, 2.5 for Cu, Zn)	Layered	Isolated V(V)/V(IV) octahedra	6.287(1) 6.283(1) 13.240(1) $\alpha = 80.98(1)$ $\beta = 86.58(1)$ $\gamma = 90.00(1)$ (M = Ag)	$P\bar{1}$	61
$Ba(VO)(PO_4)(H_2PO_4) \cdot H_2O$	Layered	Isolated V(IV) octahedra	10.848(3) 6.408(3) 11.376(2) $\beta = 102.32(2)$	$P2_1/n$	62
$Li_2(VO_2)(PO_4)$	Layered	Isolated V(V) octahedra	10.3219(3) 4.6355(1) 8.5620(4)	$Pna2_1$	63
$K_2(VO)_2(HPO_4)_3 \cdot 1.125H_2O$	Layered	Isolated V(IV) octahedra	6.404(2) 8.9512(9) 12.070(1) $\alpha = 106.234(9)$ $\beta = 98.61(2)$ $\gamma = 103.33(2)$	$P\bar{1}$	64

(*continues*)

TABLE II (Continued)

Compound	V/P/O Framework Structure	Vanadium Coordination and Linkages	Cell Parameters	Space Group	References
$K_2(VO_2)V(PO_4)_2(HPO_4)(H_2PO_4)(H_2O_2)$	Layered stair-step ribbons, seven polyhedra wide	Isolated V(IV) square pyramids and isolated V(III) octahedra	10.571(2) 12.815(3) 6.407(2) $\alpha = 102.73(2)$ $\beta = 97.43(2)$ $\gamma = 105.26(2)$	$P\bar{1}$	65
$Ca(VO)_2(PO_4)_2 \cdot 4H_2O$	Layered	Isolated V(IV) octahedra	6.3484(8) 6.350(1) 6.597(1) $\alpha = 106.81(2)$ $\beta = 94.09(1)$ $\gamma = 90.02(1)$	$P1$	66
$Co(VO)_2(PO_4)_2 \cdot 4H_2O$	Layered	Isolated V(IV) square pyramids	6.264(1) 13.428(4)	$I4/mmm$	66
$Ag_2(VO_2)(PO_4)$	Layered	Binuclear units of edge-sharing V(V) octahedra	12.431(3) 6.298(1) 6.300(2) $\beta = 90.38(2)$	$C2/m$	67
$A(VO_2)(PO_4)$ $A = Ba, Sr$	Layered	Binuclear units of edge-sharing V(V) octahedra	5.616(2) 10.062(1) 8.727(1) $\beta = 90.90(2)$	$P2_1/c$ (Ba^{2+})	68
$\alpha\text{-}A(VO_2)_2(PO_4)_2 \cdot 4H_2O$ $A = Sr, Pb$	Layered	Isolated V(IV) octahedra	9.026(2) 9.010(3) 12.841(3)	Cc	66

Compound	Type	Structure detail	Cell parameters	Space group	Ref.
Ba(VO)$_2$(PO$_4$)$_2$·4H$_2$O	Layered	Isolated V(IV) octahedra	6.3860(3) 12.7796(9) 6.3870(5) $\beta = 90.172(6)$ $\beta = 100.19(2)$ (A = Sr)	Pn	69
β-Pb(VO)$_2$(PO$_4$)$_2$·4H$_2$O	Layered	Isolated V(IV) square pyramids	6.377(9) 6.384(1) 25.357(11)	$P2_12_12_1$	70
(NH$_4$)$_2$(VO)(A$_2$O$_7$) A = V, P	Layered	Isolated V(IV) square pyramids; isolated V(V) tetrahedra	8.6233(6) 5.6384(7)	$P4bm$	71
			8.629(2) 5.648(4)	$P4$	72
M$_2$(VO)(P$_2$O$_7$) (M=Na, Ag)	Layered	Isolated V(IV) octahedra	7.7178(6) 13.3233(8) 6.2871(3) $\beta = 99.49(1)$ (for Na)	$P2_1/c$	73, 74
K$_2$(VO)(P$_2$O$_7$)	Layered	Isolated V(IV) square pyramids	8.277(3) 5.420(2)	$P4bm$	75
K$_2$(VO)(V$_2$O$_7$)	Layered	Isolated V(IV) square pyramids and {V$_2$O$_7$} dimers	8.870 5.215	$P4bm$	76
β-K$_2$(VO)$_3$(P$_2$O$_7$)$_2$	Layered	Isolated V(IV) square pyramids	9.298(3) 4.879(2) 17.998(9) $\beta = 114.98(3)$	$P2_1/c$	77

(*continues*)

TABLE II (Continued)

Compound	V/P/O Framework Structure	Vanadium Coordination and Linkages	Cell Parameters	Space Group	References
$M_2(VO)(P_2O_7)$ M = Rb	Layered	Isolated V(IV) square pyramids	7.101(1) 9.172(2) 12.801(4)	$P2_12_12_1$	78
M = Cs	Layered	Isolated V(IV) square pyramids	13.280(3) 7.247(2) 9.518(2)	$Pnma$	78
$Cd_2(VPO_7)$	$[Cd_2O_6]_\infty$ layers	Isolated V(V) tetrahedra in $\{O_3VOPO_3\}$ units	4.712(1) 10.791(1) 5.620(1) $\beta = 97.34(1)$	$P2_1/c$	79
$(NH_4)(VO)(PO_4)$	3D	Chains of corner-sharing V(IV) octahedra	10.5119(2) 12.917(3) 6.462(1)	$Pnn2$	80, 81
$(NH_4)_2[(VO)_5(PO_4)_4(H_2O)_7]\cdot 3H_2O$	3D	Chains of corner-sharing V(IV) octahedra; isolated V(IV) octahedra	10.2523(9) 12.263(1) 12.362(1) 69.041(2) 65.653(2) 87.789(2)	$P\bar{1}$	56
$\alpha\text{-Li}(VO)(PO_4)$	3D	Chains of corner-sharing V(IV) octahedra	6.748(1) 7.206(1) 7.922(1) $\alpha = 89.84(1)$ $\beta = 91.32(1)$ $\gamma = 116.99(1)$	$P1$	82

Compound	Dim.	Structural description	Cell parameters	Space group	Ref.
β-Li(VO)(PO$_4$)	3D	Chains of corner-sharing V(IV) octahedra	7.444(2) 6.300(1) 7.174(2)	$Pnma$	83
Na(VO)(PO$_4$)	3D	Chains of corner-sharing V(IV) octahedra	6.524(1) 8.455(4) 7.119(1) $\beta = 115.29(1)$	$P2_1/c$	84a 84b
K(VO)(PO$_4$)	3D 6.3648(4)	Chains of corner-sharing V(IV) octahedra	12.7640(8) 10.5153(9)	$Pn2_1a$	85, 86
K$_2$(VO)$_2$(P$_4$O$_{13}$)	3D	Corner-sharing V(IV) octahedra and {P$_4$O$_{13}$}$^{6-}$ clusters	22.181(2) 11.564(1) 9.948(1)	$Pbca$	87
M(VO)(PO$_4$) (M = Rb, Cs)	3D	Isolated V(IV) square pyramids	7.8669(9) 7.5848(8) 8.3771(10) (Rb$^+$) 7.367(2) 7.615(2) 9.804(2) (Cs$^+$)	$P2_12_12_1$	88
M(VO)$_2$(PO$_4$)$_2$ (M = Ca, Sr)	3D	Chains of trans corner-sharing V(IV) octahedra	11.795(4) 15.784(6) 7.190(4) (for Ca)	$Fdd2$	89, 90
Ag(VO)$_2$(PO$_4$)$_2$	3D	Tetranuclear units of edge- and corner-sharing V(IV)/V(V) octahedra	5.256(1) 8.117(1) 16.966(1) $\beta = 91.46(1)$	$P2_1/c$	91

(continues)

TABLE II (Continued)

Compound	V/P/O Framework Structure	Vanadium Coordination and Linkages	Cell Parameters	Space Group	References
α-Cd(VO)$_2$(PO$_4$)$_2$	3D	Chains of edge- and corner-sharing V(IV) octahedra	5.187(1) 7.959(2) 17.187(2) β = 92.74(1)	$P2_1/c$	92
β-Cd(VO)$_2$(PO$_4$)$_2$	3D	Chains of trans corner-sharing V(IV) octahedra	11.571(1) 15.880(2) 7.138(1)	$Fdd2$	93
Ba(VO)$_2$(PO$_4$)$_2$	3D	Binuclear units of corner-sharing V(IV) square pyramid and octahedra	5.2204(3) 9.1701(7) 16.3247(9) β = 92.757(5)	$P2_1/c$	93
M(VOPO$_4$)$_2$ · 4H$_2$O (M = Ca, Ba, Cd)	3D	Isolated V(IV) octahedra	6.376(1) 12.754(1) 9.006(1) β = 135.03(2)	Pc	95
α-Pb(VO)$_2$(PO$_4$)$_2$	3D	Binuclear units of corner-sharing V(IV) square pyramids and octahedra	5.2306(4) 8.5805(9) 16.790(1) β = 91.01(1)	$P2_1/c$	96
Pb(VO)$_2$(PO$_4$)$_2$ · 3H$_2$O	3D	Isolated V(IV) octahedra	10.343(4) 7.580(2) 12.596(6) β = 90.22(3)	$P2_1/c$	97

Compound	Structure	Lattice parameters	Space group	Ref.
$Na_3(VO)(PO_4)(HPO_4)$	3D, consisting of {$VO(PO_4)$} layers linked through HPO_4^{2-} groups	9.6402(7) 6.4075(5) 6.2637(6) $\beta = 104.861(7)$	$P2_1/m$	98
	Isolated V(IV) octahedra			
$K(VO)V(HPO_4)_3(H_2O)_2$	3D	6.399(3) 10.275(2) 17.810(2) $\beta = 96.12(2)$	$P2_1/n$	99
	Isolated V(IV) and V(III) octahedra			
$K_2(VO)_3(HPO_4)_4$	3D	10.756(1) 9.147(3) 8.662(1) $\beta = 112.842(9)$	$P2/c$	100
	Chains of corner-sharing V(IV) octahedra			
$K_3(VO)(V_2O_3)(PO_4)_2(HPO_4)$	3D	7.023(4) 13.309(7) 14.294(7)	$Pnma$	101
	Chains of corner-sharing octahedra; binuclear units of corner-sharing square pyramids; mixed valence V(V)/V(IV)			
$K_3(VO)(HV_2O_3)(PO_4)_2(HPO_4)$	3D	6.975(3) 13.559(7) 14.130(7)	$Pnma$	101
	Chains of corner-sharing V(IV) octahedra; binuclear units of corner-sharing V(IV) square pyramids			
$M_3(VO)(V_2O_3)(PO_4)_2(HPO_4)$ (M = Tl, NH_4)	3D	7.160(2) 13.378(2) 14.422(1) (for Tl)	$Pnma$	53, 81
	Chains of corner-sharing V(IV) octahedra; binuclear units of corner-sharing V(IV)/V(V) square pyramids			
$Rb_6(V_6O_7)(PO_4)_6$	3D	7.0656(4) 13.4988(8) 14.4198(9)	$Pnma$	102
	Chains of corner-sharing V(IV) octahedra; binuclear units of corner-sharing V(V) square pyramids			

(*continues*)

TABLE II (Continued)

Compound	V/P/O Framework Structure	Vanadium Coordination and Linkages	Cell Parameters	Space Group	References
$Cd_5(VO)V_2(PO_4)_6$	3D	Isolated V(IV) square pyramids, isolated V(III) octahedra	15.861(1) 4.710(1) 24.160(2) $\beta = 103.07$	$C2/c$	103
$Pb_2V_2(VO)(PO_4)_4$	3D	V(III) octahedra and V(IV) square pyramids. Corner sharing trinuclear units and isolated V(III) octahedra	17.747(2) 18.051(2) 9.344(1) $\beta = 117.03(1)$	$C2/c$	104
$Na_{2.44}(V_4O)(OH)(PO_4)_4$	3D	Isolated octahedra, chains of edge-sharing octahedra. Mixed-valent V(III)/V(IV)	13.723(5) 6.314(2) 16.139(4)	$Pnma$	105
$M(V_2O)(PO_4)_2$ (M = Ca, Cd, Sr)	3D	Chains of edge-sharing V(III) octahedra with corner-sharing V(III) octahedra as "outriders"	14.308(2) 6.318(1) 7.248(1) (M = Cd)	$Pnma$	106–108
$Ba_2(V_5\text{-}O_8)(PO_4)_4$	3D	Tetranuclear clusters of edge-sharing V(IV)/V(V) octahedra and V(V) square pyramids; isolated V(V) tetrahedra	7.6405(1) 22.777(1) 11.6157(7) $\beta = 103.38(1)$	$I1m1$	109
$(Ba_{0.38}Ca_{0.20}K_{0.06}Na_{0.02})$ $[(V_{3.44}Al_{0.46})(OH)_{5.66}O_{2.34}$ $(PO_4)_2] \cdot 12H_2O$	3D	Tetranuclear clusters of edge-sharing V(IV) octahedra	15.470(4)	$I\bar{4}3m$	110
$HK_4(VO)_{10}(OH)_4(PO_4)_7$ $(H_2O)_2 \cdot 9H_2O$	3D	Pentanuclear unit of corner-sharing octahedra and tetrahedra	12.136(1) 30.581(2)	$I4_1/a$	111

Compound	Dim.	Structure	Cell parameters	Space group	Ref.
$Cs_3(V_5O_9)(PO_4)_2 \cdot xH_2O$	3D	Cross-shaped pentanuclear units of edge-sharing V(IV)/V(V) square pyramids	32.306(4)	$Fd\bar{3}m$	112
$K_{11}(V_{15}O)(PO_4)_{18}$	3D	Nonanuclear unit of corner, edge, and face-sharing V(III) octahedra	9.896(1)	$P2_13$	113
$\alpha\text{-}(NH_4)V(HPO_4)_2$	3D	Isolated V(III) octahedra	7.173(2) 8.841(2) 9.458(2) $\alpha = 65.08(2)$ $\beta = 70.68(2)$ $\gamma = 69.59(2)$	$P\bar{1}$	114
$\beta\text{-}NH_4V(HPO_4)_2$	3D	Isolated V(III) octahedra	5.201(2) 8.738(2) 14.398(3) $\beta = 94.83(1)$	$P2_1/c$	115
$NaV(HPO_4)_2$	3D	Isolated V(III) octahedra	8.430(2) 9.712(2) 22.133(4) $\beta = 90.07(3)$	Cc	84
$Na_3V_2(PO_4)_3$	3D	Isolated V(III) octahedra	8.67(2) 21.71(3)	$R\bar{3}c$	116,117
$NaV_3(PO_4)_3$	3D	Isolated octahedra; binuclear units of edge-sharing octahedra. Mixed-valent V(II)/V(III)	10.488(2) 13.213(3) 6.455(1)	$Imma$	118

(*continues*)

TABLE II (Continued)

Compound	V/P/O Framework Structure	Vanadium Coordination and Linkages	Cell Parameters	Space Group	References
$K_6V_2(PO_4)_4$	3D	Isolated V(III) octahedra	9.578(1) 11.097(1) 18.127(2) $\beta = 121.67(1)$	$P2_1/c$	119
α-$RbV(HPO_4)_2$	3D	Isolated V(III) octahedra	8.831(1) 9.450(2) 7.188(2) $\alpha = 109.55(2)$ $\beta = 110.26(1)$ $\gamma = 65.34(1)$	$P\bar{1}$	115
β-$RbV(HPO_4)_2$	3D	Isolated V(III) octahedra	5.211(1) 8.789(8) 14.330(5) $\beta = 94.39(4)$	$P2_1/c$	115
$Sr_2V(PO_4)_2(H_2PO_4)$	3D	Isolated V(III) octahedra	11.078(3) 11.742(3) 8.951(3) $\beta = 125.35(2)$	$C2/c$	120
$Cd_3V_4(PO_4)_6$	3D	Binuclear units of edge-sharing V(III) octahedra	12.446(1) 12.547(1) 6.487(1) $\beta = 115.66(1)$	$C2/c$	121
$BaV_2(HPO_4)_4 \cdot H_2O$	3D	Isolated V(III) octahedra	9.441(2) 7.913(2) 9.521(2) $\beta = 117.91(2)$	$P2_1$	122

Compound	Dim.	Structural features	Cell parameters (Å, °)	Space group	Ref.
$CsV_2(PO_4)(HPO_4)_2$ $(H_2O)_2$	3D	Isolated V(III) octahedra	10.004(1) 17.812(3) 6.370(2) $\beta = 97.02(2)$	$P2_1/a$	99
$KV_4(PO_4)(P_2O_7)(P_4O_{13})$	3D	Binuclear units of edge-sharing V(III) octahedra	10.0846(7) 10.2309(6) 10.8283(9) $\alpha = 112.757(5)$ $\beta = 109.226(7)$ $\gamma = 104.675(5)$	$P\bar{1}$	123
$Ba_2V_3H(PO_4) \cdot {}_2(P_2O_7)_2$	3D	Isolated V(III) octahedra	4.858(2) 7.908(2) 11.300(3) $\alpha = 89.10(2)$ $\beta = 89.28(2)$ $\gamma = 87.25(2)$	$P\bar{1}$	124
$Ca_2V(PO_4)(P_2O_7)$	3D	Isolated V(III) octahedra	6.391(2) 6.6362(9) 19.071(2) $\beta = 99.26(2)$	$P2_1/c$	125
$(NH_4)_2(VO)(P_2O_7)$	3D	Isolated V(IV) square pyramids	8.3039(2) 5.7658(2)	$P4bm$	126
$\alpha\text{-}(NH_4)_2(VO)_3(P_2O_7)_2$	3D	Chains of edge-sharing V(IV) octahedra; binuclear units of edge-sharing V(IV) octahedra and square pyramids	17.4973(4) 11.3655(3) 7.2769(2)	$Pnam$	126

(*continues*)

TABLE II (Continued)

Compound	V/P/O Framework Structure	Vanadium Coordination and Linkages	Cell Parameters	Space Group	References
α-K$_2$(VO)$_3$(P$_2$O$_7$)$_2$	3D	Chains of corner-sharing V(IV) octahedra; binuclear units of corner sharing V(IV) square pyramids and octahedra	17.407(1) 11.3438(7) 7.2964(15)	$Pna2_1$	127
β-K(VO)$_3$(P$_2$O$_7$)$_2$	3D	Chains of corner-sharing octahedra: $2 \times$ V(IV); $1 \times$ V(V)	7.8654(5) 10.0104(7) 16.2715(8)	$P2_12_12_1$	128
Rb(V$_3$O$_{3+x}$)(P$_2$O$_7$)$_2$ ($x = 0.14$)	3D	Chains of corner-sharing V(V) and V(IV) octahedra	13.651(2) 7.289(2)	$P4_2/mmm$	129
Rb$_2$(VO)$_3$(P$_2$O$_7$)$_2$	3D	As for Cs$_2$(VO)$_3$(P$_2$O$_7$)$_2$	17.502(7) 7.292(2) 11.399(6)	$Pnma$	130
M(VO)$_3$(P$_2$O$_7$)$_2$ M = Sr, Pb, Ba	3D	Isolated V(III) octahedra	7.594(1) 10.075(1) 16.291(1) (for Sr^{2+}) 7.6871(8) 10.0898(7) 16.746(3) $\gamma = 90.036(7)$ (for Ba^{2+})	$P2_12_12_1$ $P11b$ (intergrowth)	131

Compound	Dim.	Structural features	Cell parameters	Space group	Ref.
$Cs_2(VO)_3(P_2O_7)_2$	3D	Infinite chains of corner-sharing V(IV) octahedra; binuclear units of a V(IV) square pyramid sharing a corner with a V(IV) octahedron	17.613(5) 7.328(2) 11.600(4)	$Pnma$	132
$BiBa_2V_2PO_{11}$ [based on $MBa_2(XO_4)(X_2O_7)$ type]	3D	Isolated V(V) tetrahedra	12.266(2) 7.615(2) 11.312(2) $\beta = 103.32(2)$	$P2_1/c$	133
$(H_3O)V(P_2O_7)$	3D	Isolated V(III) octahedra	7.519(2) 10.054(3) 8.244(3) $\beta = 105.86(3)$	$P2_1/c$	134
$(NH_4)V(P_2O_7)$	3D	Isolated V(III) octahedra	7.5149(2) 10.0384(3) 8.2422(2) $\beta = 105.998(3)$	$P2_1/c$	126
$LiV(P_2O_7)$	3D	Isolated V(III) octahedra	4.8048(6) 8.113(1) 6.9393(9) $\beta = 109.01(1)$	$P2_1$	135
$NaV(P_2O_7)$	3D	Isolated V(III) octahedra	7.324(5) 7.930(4) 9.586(6) $\beta = 111.96(4)$	$P2_1/c$	136

(continues)

TABLE II (Continued)

Compound	V/P/O Framework Structure	Vanadium Coordination and Linkages	Cell Parameters	Space Group	References
KV(P$_2$O$_7$)	3D	Isolated V(III) octahedra	7.3686(2) 10.0527(5) 8.1874(4) $\beta = 106.580(3)$	$P2_1/c$	137
RbV(P$_2$O$_7$)	3D	Isolated V(III) octahedra	7.511(4) 10.035(2) 8.254(2) $\beta = 105.74(2)$	$P2_1/c$	138
CsV(P$_2$O$_7$)	3D	Isolated V(III) octahedra	7.701(3) 9.997(2) 8.341(4) $\beta = 104.82(4)$	$P2_1/c$	139
α-BaV$_2$(P$_2$O$_7$)$_2$	3D	Isolated V(III) octahedra	10.6213(8) 10.4685(7) 9.7063(13) $\beta = 103.074(9)$	$C2/c$	140
β-BaV$_2$(P$_2$O$_7$)$_2$	3D	Isolated V(III) octahedra	6.269(1) 7.864(3) 6.1592(9) $\alpha = 101.34(2)$ $\beta = 105.84(1)$ $\gamma = 96.51(2)$	$P\bar{1}$	141

TABLE III

Selected Structural Information for Vanadium Phosphates with Additional Framework Constituents

Compound	V/P/O Framework	Vanadium Coordination and Linkages	Cell Parameters	Space Group	References
A. d-Block Elements as Framework Constituents					
[Mn(H$_2$O)]$_x$(VO)$_{1-x}$(PO$_4$) · 2H$_2$O	2D, derived from VO(PO$_4$) · 2H$_2$O	Isolated V(V) octahedra	6.203(1) 13.814(1)	$P4/nmm$ or $P4/n$	142
[Cu$_{0.5}$(OH)$_{0.5}$[(VO)(PO$_4$) · 2H$_2$O	2D, consisting of {VO(PO$_4$)(H$_2$O)} layers, decorated with {Cu(H$_2$O)$_3$}$^{2+}$ units	Isolated {VO$_6$} octahedra bridging through an oxo group to the "4 + 2" {CuO$_6$} site	6.614(2) 8.930(2) 9.071(2) β = 103.79(2)	$P2_1/m$	143
Zn$_3$(XO$_4$)$_2$ X = 0.25 V, 0.75 P	Layered	Isolated V(V) tetrahedra	15.941(6) 5.314(2) 8.265(2) β = 106.96(3)	$C2/c$	144
Na$_3$TiV(PO$_4$)$_3$	3D NASICON	Isolated V(III) octahedra	8.759(3) 21.699(4)	$R\bar{3}c$	145
K$_2$TiV(PO$_4$)$_3$	3D Langbeinite	Isolated V(III) octahedra	9.855(3)	$P2_13$	145
Mn$_2$(VO)(PO$_4$)$_2$ · H$_2$O	3D	2 × V(IV) octahedra and 2 × Mn(II) polyhedra in tetranuclear units	8.957(2) 8.806(2) 9.329(2) β = 100.95(1)	$P2_1/c$	146

(*continues*)

TABLE III (Continued)

Compound	V/P/O Framework	Vanadium Coordination and Linkages	Cell Parameters	Space Group	References
$[Co(H_2O)_4][(VO)(PO_4)]_2$	3D consisting of 2D $\{VO(PO_4)\}$ layers linked by $[Co(H_2O)_4]^{2+}$ units	Isolated V(IV) square pyramids	6.264(1) 13.428(4)	$I4/mmm$	147
$K[M_{0.5}(V_{0.5}O)(PO_4)]$ $M = Nb, Ta$	3D	Chains of corner-sharing V(III) and Ti(V) octahedra	12.949(6) 6.431(8) 10.686(4)	$Pna2_1$	148
$Ni_{0.5}(VO)(PO_4) \cdot 1.5H_2O$	3D	Trinuclear unit of 2 V(IV) octahedra and one Ni octahedron sharing a common corner	10.344(1) 9.350(3) 9.731(1)	$Pnma$	149
$[Ni(H_2O)_4][(VO)(PO_4)]_2$	3D consisting of 2-D $\{VO(PO_4)\}$ layers linked by $[Ni(H_2O)_4]^{2+}$ units	Isolated V(IV) square pyramids	6.251(1) 13.338(3)	$I4/mmm$	150, 151
$[M(H_2O)_4][(VO)(PO_4)]_2$ $M = Mg, Zn$	3D	Isolated V(IV) square pyramids	6.251(3) 6.258(3) 8.065(4) $\alpha = 112.85(1)$ $\beta = 112.86(1)$ $\gamma = 89.93(1)$	$P\bar{1}$	152
$[Cu_{0.5}(OH)_{0.5}][(VO)(PO_4)] \cdot 2H_2O$	3D, consisting of $\{VO(PO_4)\}$ layers linked through $\{Cu_2(OH)_2H_2O_4\}^{4+}$ binuclear units	$\{VO_5\}$ square pyramids	7.0124(8) 12.6474(9) 6.3022(6) $\alpha = 90.079(8)$ $\beta = 96.076(9)$ $\gamma = 74.002(7)$	$P\bar{1}$	143

$Zn_2(VO)(PO_4)_2$	3D	Chains of corner-sharing V(IV) octahedra; binuclear units of edge-sharing Zn(II) square pyramids	8.9227(13) 9.039(3)	$I4cm$	153
$BaZn(VO)(PO_4)_2$	3D, consisting of {$VO(PO_4)$} layers linked by {$Zn(PO_4)$} layers	Isolated V(IV) square pyramids; Zn(II) tetrahedra	8.814(1) 9.039(1) 18.538(1)	$Pbca$	154
$Zn_3V_4(PO_4)_6$	3D	Binuclear units of edge-sharing V(III) octahedra, linked into infinite chains by Zn(II) trigonal bipyramids and cross-linked by Zn(II) octahedra	6.349(2) 7.869(1) 9.324(2) $\alpha = 105.32(1)$ $\gamma = 108.66(2)$ $\gamma = 101.23(2)$	$P\bar{1}$	155
$(NH_4)_3Zn_2V(PO_4)_2(HPO_4)_2$	3D	Isolated V(III) octahedra	20.977(2) 5.274(2) 15.406(2) $\beta = 121.422(7)$	$C2/c$	156

B. *Phosphosilicates*

$(VO)Si(PO_4)_2$	3D	Chains of corner-sharing V(IV) octahedra	8.747(2) 8.167(2)	$P4/ncc$	157
$V_3SiP_5O_{19}$	3D	Binuclear units of edge-sharing V(III) octahedra	14.4671(11) 7.4605(2)	$P6_3$	158

(*continues*)

TABLE III (Continued)

Compound	V/P/O Framework	Vanadium Coordination and Linkages	Cell Parameters	Space Group	References
\multicolumn{6}{c}{C. Vanadium Aluminophosphates and Gallophosphates}					
Cs(VO)Al(PO$_4$)$_2$ · H$_2$O	3D	Isolated {V(IV)O$_6$} octahedra	8.014(2) 8.088(2) 14.141(2) $\beta = 105.98(1)$	$P2_1/n$	159
Rb(VO)Al(PO$_4$)$_2$ · H$_2$O	3D	Isolated {V(IV)O$_6$} octahedra	7.880(2) 8.063(2) 14.062(2) $\beta = 105.83(1)$	$P2_1/n$	159
[H$_3$NCH$_2$CH$_2$NH$_2$] [(VO)Al(PO$_4$)$_2$]	3D	Isolated {VO$_5$N} octahedra	8.172(2) 14.232(2) 8.873(1) $\beta = 109.88(1)$	$P2_1/n$	159
(MeNH$_3$)[(VO)Al(PO$_4$)$_2$]	3D	Isolated {VO$_5$} square pyramid	8.151(1) 8.784(2) 7.842(2) $\alpha = 116.22(2)$ $\beta = 95.88(2)$ $\gamma = 69.16(1)$	$P\bar{1}$	159
Rb(VO)(H$_2$O)Ga(PO$_4$)$_2$	3D	Isolated {V(IV)O$_6$} octahedra	7.928(2) 8.049(2) 13.983(3) $\beta = 104.274(5)$	$P2_1/c$	160

Formula	Dimensionality	Description	Cell parameters	Space group	Ref.
$Cs(VO)(H_2O)Ga(PO_4)_2$	3D	Isolated $\{V(IV)O_6\}$ octahedra	8.0423(5) 8.0661(5) 14.1284(5) $\beta = 105.094(1)$	$P2_1/c$	160
$[M(H_2O)_1]_x H(VO)_{1-x}(PO_4)] \cdot 2H_2O$ M = Al, Ga and also Cr, Mn, Fe	Layered	Isolated V(V) square pyramids	17.498 11.365 7.277	$Pnam$	161
$[H_3N(CH_2)_4NH_3]_2[Ga_{4-x}V_x(HPO_4)_2(PO_4)_3(OH)_3] \cdot yH_2O$ ($x = 0.4, y = 0.6$)	3D	Isolated V(III) octahedra	15.2344(4) 28.908(9)	$I4_1/a$	162
$[NH_3(CH_2)_2NH_3]_4[Ga_{4-x}V_x(HPO_4)_5(PO_4)_3H(OH)_2]$ ($x = 1.65$)	2D	Ga/V(III) tetramers of edge and corner sharing MO_6 octahedra	9.991(1) 12.367(1) 15.082(1) 90.751(6) 91.720(7) 91.449(8)	$P\bar{1}$	163
D. Vanadium Borophosphates					
$[H_3NCH_2CH_2NH_3]_2[(VO)_5(H_2O){O_3POB(O)_2OPO_3}_2] \cdot 1.5H_2O$	3D	Binuclear units of edge-sharing V(IV) square pyramids; isolated V(IV) octahedra	14.206(3) 33.308(4) 11.587(5)	$Fdd2$	164
E. Vanadium Phosphate Fluorides					
$[(H_3NCH_2CH_2NH_2)VF(PO_4)]$	Layered	Corner-sharing chains of $\{VF_2O_3N\}$ octahedra	9.2272(3) 7.3532(2) 9.8496(2) $\beta = 101.315(1)$	$P2_1/c$	165

(continues)

TABLE III (Continued)

Compound	V/P/O Framework	Vanadium Coordination and Linkages	Cell Parameters	Space Group	References
[H$_3$NCH$_2$CH$_2$NH$_3$][(VO$_2$)$_2$F(PO$_4$)]	3D	Zigzag chain of corner-sharing {VO$_5$F} octahedra and {VO$_4$F} square pyramids	8.2939(6) 9.9260(8) 12.5021(8)	$P2_12_12_1$	166
[H$_2$N(C$_2$H$_2$)$_2$NH$_2$]$_{0.5}$[(VO)$_4$V(HPO$_4$)$_2$(PO$_4$)$_2$F$_2$(H$_2$O)$_4$]·2H$_2$O	3D	Binuclear units of face-sharing {V(IV)O$_5$F} octahedra; isolated V(III) octahedra	18.425(4) 7.417(1) 8.954(2) $\beta = 93.69(2)$	$C2/m$	167
K$_2$[(VO)$_3$(PO$_4$)$_2$F$_2$(H$_2$O)]·H$_2$O	3D	Binuclear units of face-sharing {V(IV)O$_5$F} octahedra; binuclear units of edge-sharing {V(IV)O$_4$F$_2$} octahedra	7.298(1) 8.929(2) 10.090(2) $\alpha = 104.50(3)$ $\beta = 100.39(3)$ $\gamma = 92.13(3)$	$P\bar{1}$	167

Tables I–III present a structural summary of the three categories of V/P/O solids. In addition to the unit cell and space group information, the members of each category are subdivided into the type of extended motif adopted: 1D, 2D, or 3D. A brief description of the vanadium coordination polyhedra and linkages is also provided. Several representative members of each category and subtype are discussed and illustrated in the following sections.

B. Vanadium Phosphates with Charge Neutral V–P–O Covalent Linkages

While compositionally the simplest of the inorganic phases, the vanadium phosphates with neutral V–P–O covalent structures display many of the structural motifs common to the larger family. Thus, 1D, 2D, and 3D structures with a variety of polyhedral connectivities have been described. The layered neutral V/P/O phases have received considerable attention as precursors to the catalytically active $(VO)_2(P_2O_7)$ phase (36).

Vanadyl phosphate $VO(PO_4)$ has five polymorphic forms: α_I and α_{II}, which are layered, β, γ and δ (26, 27, 34, 168, 169). The structure of the α_{II} form, shown in Figure 4, is characteristic of the $MO(PO_4)$ family of phases (170) and consists of corner-sharing $\{VO_6\}$ octahedra and $\{PO_4\}$ tetrahedra. The four equatorial oxygens of each V(V) octahedron are shared with four different $\{PO_4\}$ tetrahedra in such a way as to form a tetragonal layer. These tetragonal layers are stacked so as to allow corner sharing of the axial trans vertices of the vanadium octahedra. In the characteristic fashion of metal oxides (171, 172), the vanadium site is displaced from the centroid of the octahedron along the $\{V=O\}$ axis, to produce an alternation of long and short V–O bond distances of 2.85 and 1.58 Å, respectively. Alternatively, the structure may be described as $\{VO_5\}$ square pyramids and $\{PO_4\}$ tetrahedra in a layered structure, with weak V–O interactions between layers. The weak interlayer connectivity allows α-$VO(PO_4)$ to act as a host for intercalation of neutral guest molecules (173) and in the redox intercalation of alkali metal and alkaline earth cations (174).

Intercalation of water into the $VO(PO_4)$ host yields $VO(PO_4) \cdot 2H_2O$, in which one aquo group coordinates to the vanadium atom of the layer while the second occupies the interlayer space. Reductive intercalation of $VO(PO_4) \cdot 2H_2O$ leads not to simple intercalation but to the formation of a new layer structure V(IV)/hydrogen phosphate $VO(HPO_4) \cdot 0.5H_2O$ (30), a member of the oxovanadium (IV) hydrogen phosphate phases, which constitute a series of thermochemically related materials (175), which serve as precursors for the synthesis of the catalytically active $(VO)_2P_2O_7$ (37, 176).

While structural analyses are available for $VO(HPO_4) \cdot xH_2O$ with $x = 0.5$, 1, 2 (α and β), 3 and 4 (30–33, 172, 175, 177), the structures of VO $(HPO_4) \cdot 0.5H_2O$,

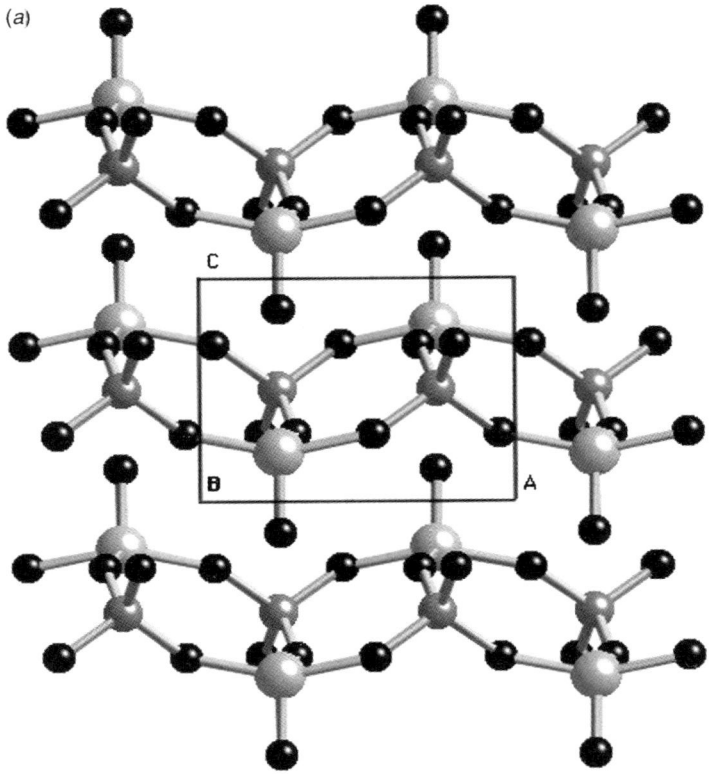

Figure 4. Ball-and-stick-representations of the structure of α_{II}-VO(PO$_4$) parallel to the V–P–O plane.

α-VO(HPO$_4$) • 2H$_2$O and β-VO(HPO$_4$) • 2H$_2$O serve to illustrate the relevant structural characteristics of these phases. Although the three phases share layered structural motifs and the {VO$_6$} octahedron as the fundamental building block, the polyhedral connectivities within the layers are dramatically different.

As shown in Figure 5, the structure of VO(HPO$_4$) • 0.5H$_2$O consists of layers constructed from face-sharing {VO$_6$} dimers and η^3-{HPO$_4$} tetrahedra. The pendant {P–OH} units of the hydrophosphate groups project from either surface of the layer and provide hydrogen bonding to the adjacent layers. The aqua ligand is trans to the multiply bonded vanadyl oxygen {V=O} and bridges the vanadium sites of the binuclear unit. The equatorial oxygen atoms of the vanadium sites are provided by the hydrogen phosphates and each binuclear unit links to six hydrogen phosphate tetrahedra. Consequently, each {HPO$_4$} tetrahedron adopts the η^3, μ^4 modality.

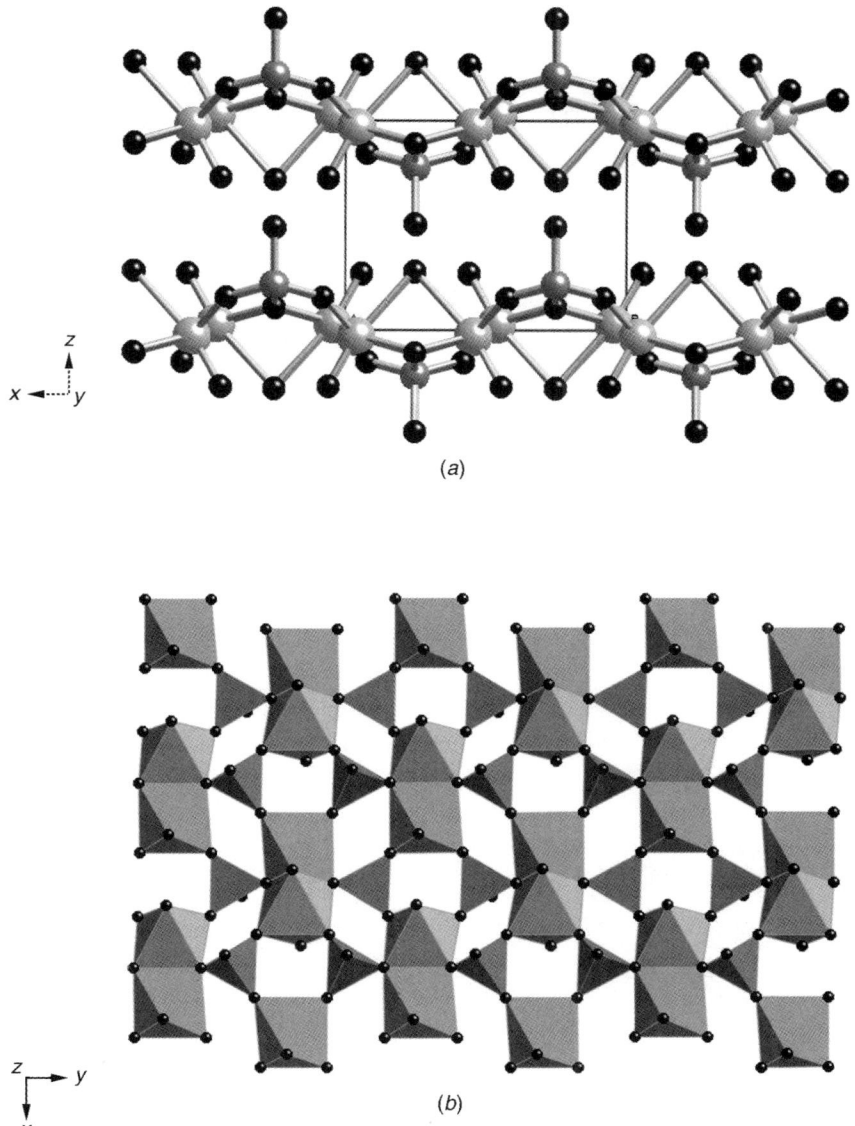

Figure 5. (a) Ball-and-stick representation of the structure of VO(HPO$_4$) · 0.5H$_2$O viewed parallel to the layer. (b) Polyhedral view of the layer connectivity.

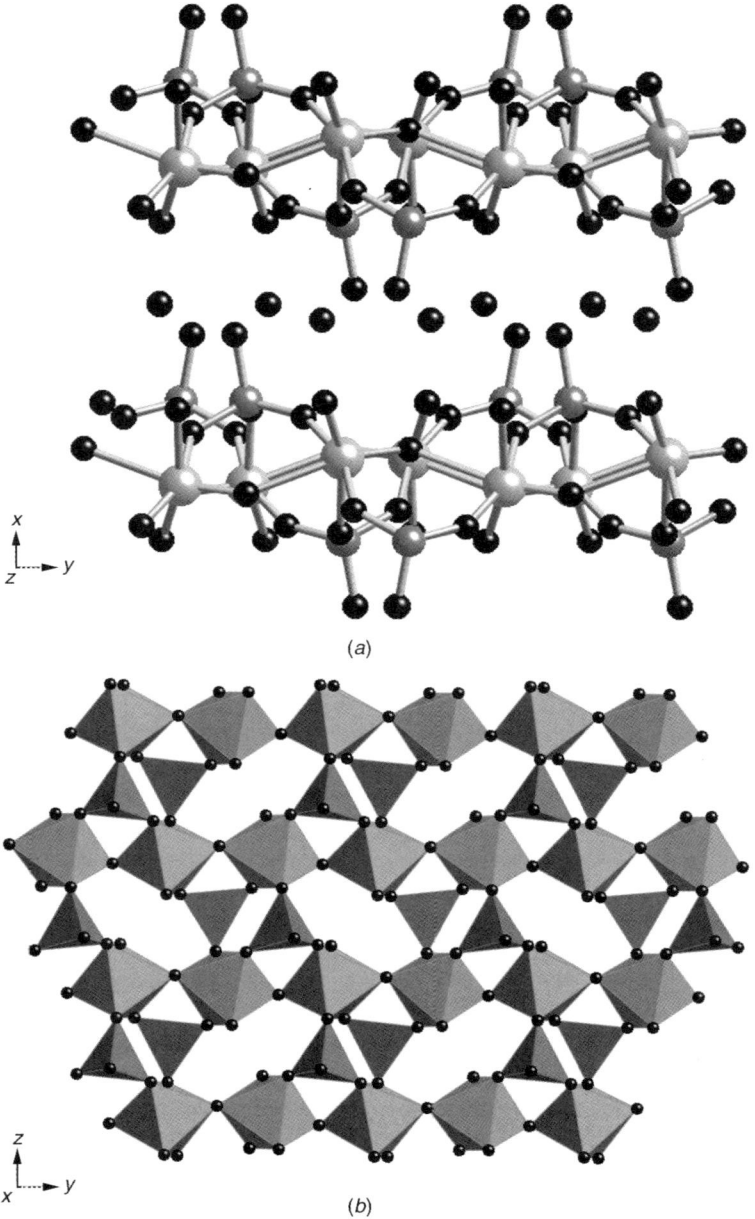

Figure 6. A view of the structure of α-VO(HPO$_4$) · 2H$_2$O parallel to the crystallographic c axis (a). Polyhedral representation of the structure, viewed normal to the layer plane (b).

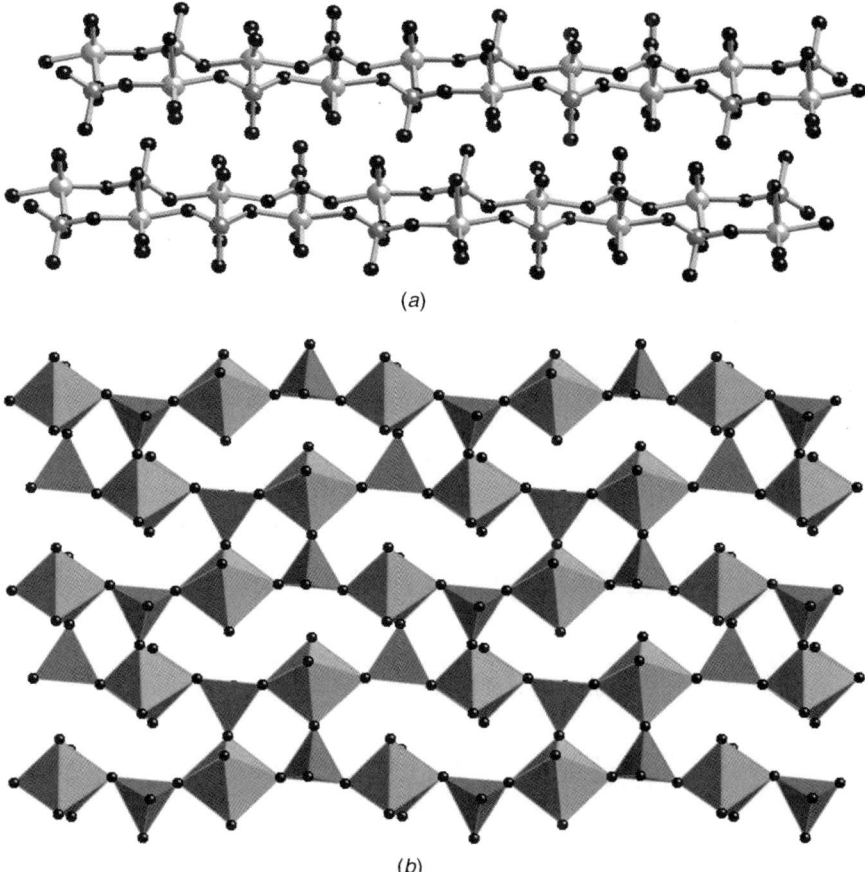

Figure 7. (a) A view of the structure of β-VO(HPO$_4$) • 2H$_2$O parallel to the b axis. (b) Polyhedral view of the layer parallel to the a axis.

The layer structure of α-VO(HPO$_4$) • 2H$_2$O, shown in Figure 6, is quite distinct from that of the hemihydrate. The vanadium polyhedral motif consists of chains of trans- corner-sharing {VO$_6$} octahedra. The chains exhibit the common alternation of long and short V—O distances and are linked by {HPO$_4$} tetrahedra adopting the η^3, μ^3 coordination mode. Each vanadium site is coordinated to two oxo groups, three phosphate oxygens, and an aqua ligand. The second water molecule occupies the interlamellar region.

In contrast to the structure of α-VO(HPO$_4$) • 2H$_2$O, that of β-VO(HPO$_4$) • 2H$_2$O, shown in Figure 7, exhibits no interlamellar water but rather two aqua ligands on each vanadium site, one trans to the terminal oxo group. The

{VO_6} octahedra are "isolated" in the sense that no common vertices are shared. Each {HPO_4} unit connects three {VO_6} polyhedra and projects a {P—OH} group into the interlayer space to provide hydrogen bonding between layers.

Although it is 3D, the structure of $(VO)_2P_2O_7$ is closely related to that of the hemihydrate, $VO(HPO_4) \cdot 0.5H_2O$ (30). As shown in Figure 8, the structure consists of double chains of corner-sharing vanadium octahedral, linked into edge-sharing pairs. The chains exhibit the common alternation of long and short trans vanadium–oxo interactions. Oxygen atoms from two pyrophosphate groups provide the vertices for the shared edges of the binuclear pairs. When viewed down a, the structure of $(VO)_2(P_2O_7)$ reveals a layer motif similar to that of $VO(HPO_4) \cdot 0.5H_2O$. However, the layers of $(VO)_2(P_2O_7)$ are linked through the {V—O=V} bonds of the edge-sharing octahedral double chains and the P—O—P bonds of the pyrophosphate. In converting the hemihydrate to the pyrophosphate, the vanadium octahedra change from face- to edge-shared binuclear units with a concomitant expansion of one axis from 7.43–8.288 Å (a and $b/2$, respectively). The transformation is topotactic and occurs by condensation of the {P—OH} groups between adjacent layers and elimination of the coordinated water molecules. This topotaxy has been established by detailed electron microscopy and X-ray diffraction studies and supported by MNDO self-consistent field–molecular orbital (SCF–MO) calculations (37b).

Examination of the structures of these vanadium phosphates with neutral V/P/O covalent networks reveals three general structural types based on the nuclearity of the vanadium sites: phases with mononuclear vanadium polyhedra, those with binuclear vanadium sites, and species exhibiting 1D chains or double chains. There can be considerable structural variability within these broad categories. Thus, the binuclear core may exhibit corner-, edge-, or face-sharing of vanadium octahedra. Likewise, the chain structures may be further distinguished by rotations or inclinations of the relative positions of the basal planes.

Additional structural characteristics of some significance are afforded by bridging of the fundamental vanadium polyhedral types by phosphate tetrahedra. Two common structural motifs linking vanadium centers of mononuclear, binuclear or chain sites are the $\mu(O, O')$ phosphate and the di-$\mu(O, O')$ phosphate bridges illustrated in Figure 9. In fact, these structural building blocks exhibit considerable geometric flexibility, reflecting the relative orientations available to the terminal oxo groups of the vanadium sites (syn, anti) and the boat, chair or skew conformations that may be adopted by the eight-membered {$V_2P_2O_4$} ring of the di-$\mu(O, O')$ phosphate bridged unit. This latter unit is nearly ubiquitous in the structural chemistry of metal phosphates and metal organophosphonate phases in general.

A notable feature of the {$V_2P_2O_4$} ring motif is the correlation with the magnetic properties exhibited by the V(IV) phases (178, 179). A parallel

Figure 8. A view of the structure of $(VO)_2(P_2O_7)$ parallel to the a axis (a). A polyhedral representation parallel to the c axis (b).

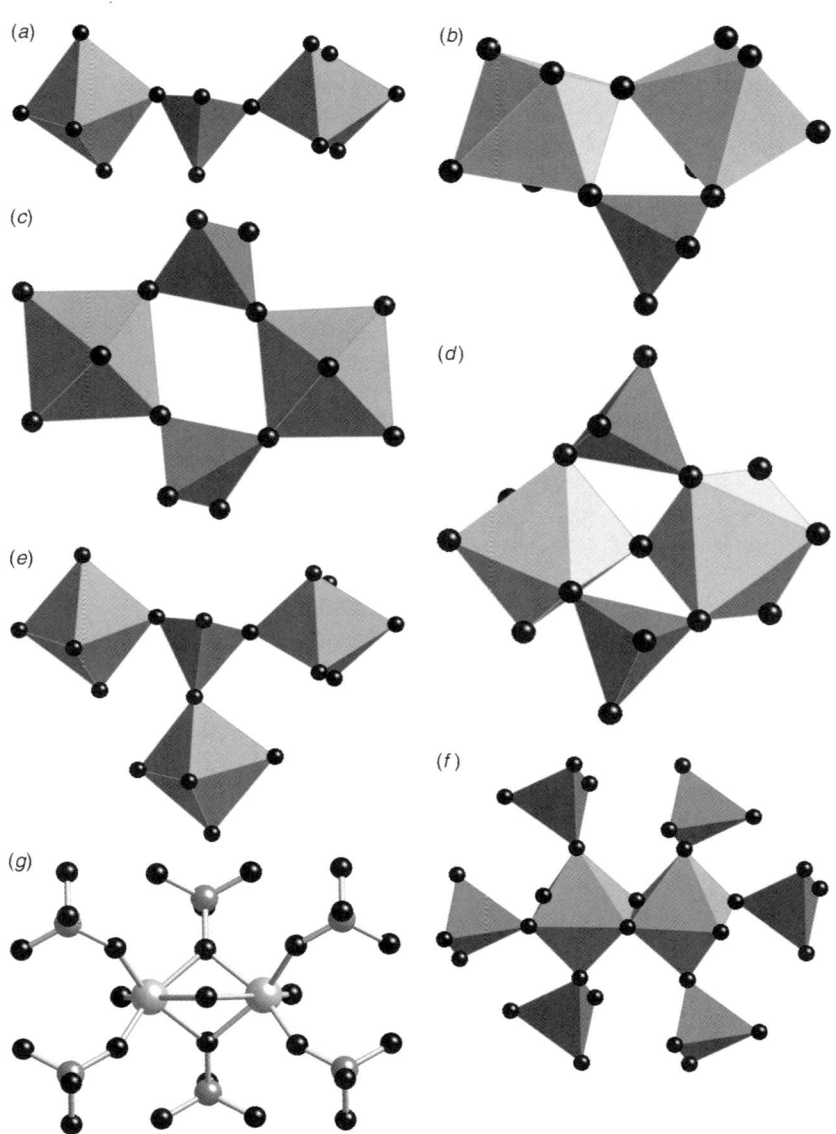

Figure 9. Ball-and-stick and polyhedral representations of common structural building blocks for V/P/O phases. From top to bottom: (a) the μ(O,O') phosphate group linking isolated vanadium sites; (b) the μ(O,O') phosphate unit linking corner-sharing vanadium sites, (c) di-μ(O,O') phosphate groups linking isolated vanadium centers; (d) di-μ(O,O') groups linking corner-sharing vanadium sites; (e) the μ(O,O'O'') phosphate group linking three isolated vanadium centers; (f) edge-sharing binuclear vanadium sites; (g) face-sharing binuclear vanadium sites.

arrangement of the basal planes of the $\{(VO)O_4\}$ square pyramids is optimal for antiferromagnetic coupling. This preference indicates that through-space interactions are not dominant, since coupling would then be expected between the d_{xy} magnetic orbitals of the V(IV) sites. Superexchange through the phosphate bridges must carry the magnetic coupling. The same arguments can be applied to ferromagnetic coupling in the mixed-valence materials $M_xVO(PO_4) \cdot nH_2O$ (176).

As illustrated by the structure of the mixed-valence V(III)/V(IV) phase $V_2(VO)(P_2O_7)_2$ (42), shown in Figure 10, structure-types based on the incorporation into the V—P—O covalent network of vanadium subunits of nuclearity other than one, two, or infinite chain are not unusual. The linear trinuclear cluster consists of a central square pyramidal V(IV) site sharing opposite basal edges with V(III) octahedra. The basal plane of the V(IV) square pyramid is defined by the oxygens from chelating pyrophosphate groups. The structure is constructed from the trinuclear building blocks linked into a 3D framework through P—O—V bonds. Each pyrophosphate group adopts the η^6, μ^8 coordination mode in connecting five neighboring trinuclear units.

The structure of $V_2(VO)(P_2O_7)$ reveals several structural features common to the V/P/O class of materials in general: the incorporation of V(IV) in square pyramidal geometry and the distinctive geometry of the V(III) site in comparison to V(V) and V(IV) sites. Square pyramidal V(IV) geometry is a recurrent structural theme of the V/P/O phases. The square pyramidal structures conform to the general classifications developed for the octahedrally based geometries, with the exception that the weakly coordinated site trans to the terminal oxo group is absent.

A noteworthy feature of V(III) geometry is the absolute absence of strongly multiply bonded oxo groups. Thus, in contrast to the highly distorted octahedral or square pyramidal geometries adopted by V(IV) and V(V) sites, which exhibit the vanadium center displaced from the centroid of the polyhedron in the direction of the short $\{V=O\}$ bond, V(III) sites exhibit more regular octahedral geometries with nearly identical bond lengths. This characteristic also serves to enhance the connectivity of the V(III) polyhedron to adjacent vanadium and phosphorus polyhedra, a feature consistent with the greater tendency of V(III) containing phases to adopt 3D structures, as noted in Table I.

Both charge neutral and cation incorporating V/P/O phases exhibit as a recurring theme the isolation of high-temperature, dense-phase materials based on vanadium polyhedra and pyrophosphate $\{P_2O_7\}^{4-}$ subunits. Such materials have attracted considerable attention as components or models for industrial catalysts. While considerable controversy surrounds the nature of the active component of such catalysts (180–186), in the case of vanadyl pyrophosphate the active phase involved in the redox cycle for the organic oxidation appears to be structurally related to δ-VO(PO)$_4$ (187).

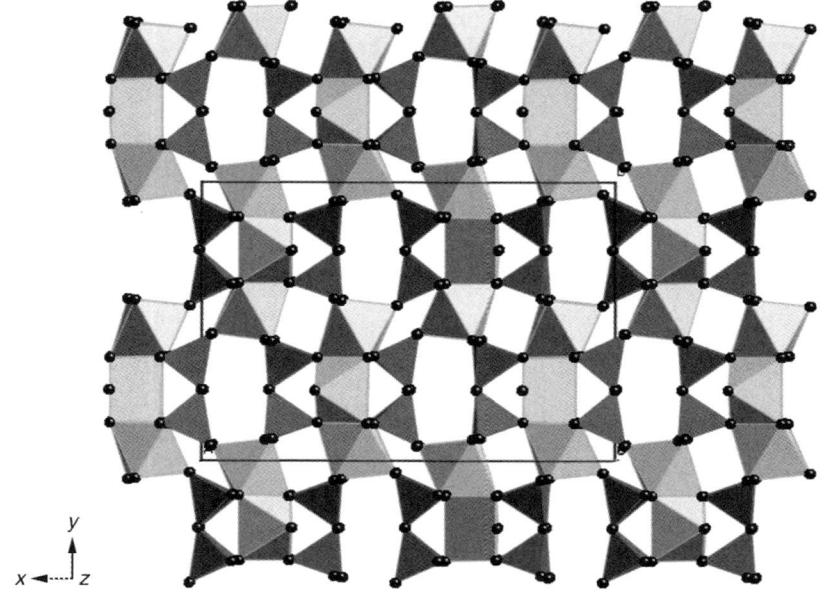

Figure 10. Polyhedral representation of the V/P/O network of $V_2(VO)(P_2O_7)_2$, viewed parallel to the c axis.

C. Quaternary M′/V/P/O Phases with Negatively Charged V–P–O Covalent Linkages and Charge Compensating M′ Cations

A variety of metal cations may be accommodated by negatively charged V/P/O structures as charge compensating and space-filling units. Such quaternary materials, of general type M′/V/P/O, may adopt chain or layer structures with the M'^{n+} cations serving to separate the chains or residing in the interlamellar region. As illustrated by Table II, 3D frameworks are most commonly encountered with the M'^{n+} cation located in tunnels or cages provided by the V/P/O framework.

In contrast to the charge neutral phases discussed previously, which may intercalate neutral guest molecules, the M'^{n+} cations of the quaternary phases are intimate parts of the overall structure of what is essentially an ionic salt. Thus, while many of these quaternary phases exhibit open framework V/P/O structures, the cavities are occupied by the M'^{n+} cations which cannot be removed without causing framework collapse.

Many of the structural elements identified for the charge neutral materials extend to the quaternary phases. However, additional structural elements are introduced through the incorporation of structural motifs based on vanadium oligomers of nuclearity greater than two and less than infinite chain or from the presence of more than one vanadium motif in the V—P—O covalent network.

The quaternary phases with negatively charged V/P/O structures may be classified according to their dimensionality and further subcategorized in terms of the nuclearity of the vanadium motifs incorporated into the anionic chain, layer, or framework. The following sections discuss a representative sampling of the general class of quaternary phases $M'/V/P/O$ in which the M'^{n+} cations serve as charge-compensating units. The structures of the remaining members of this class exhibit the common structural motifs identified previously for the charge neutral V/P/O materials or possess features similar to those of these representative quaternary phases.

1. Infinite Chains of Vanadium Polyhedra

The structures of α-$(NH_4)(VO_2)(HPO_4)$ (56), $(NH_4)(VO)(PO_4)$ (80), and $LiVO(PO_4) \cdot 0.5H_2O$ (57) serve as prototypes for the numerous quaternary phases constructed from this motif. The structure of α-$(NH_4)(VO_2)(HPO_4)$, shown in Figure 11, consists of 1D chains of trans-corner sharing $\{VO_5\}$ square pyramids, bridged by $\mu(O, O')$ hydrogen phosphate groups, with NH_4^+ cations occupying positions between the strands and strongly hydrogen bonded to three neighboring chains. The vanadium polyhedra are defined by two trans bridging oxo groups and two oxygen donors from two hydrogen phosphate groups in the basal plane and an apical $\{V=O\}$ group. The $\{V=O-V\}$ chain exhibits the common motif of alternating short and long V—O bonds. The square pyramid is consequently severely distorted by the presence of a cis-dioxo $\{VO_2\}$ unit. The $\{HPO_4\}$ unit adopts η^2, μ^2 coordination and consequently projects pendant $\{P=O\}$ and $\{P-OH\}$ units into the interchain region.

In contrast to the 1D structure based on trans oxo-linked vanadium square pyramids, the structure of $(NH_4)(VO)(PO_4)$ consists of a 3D framework constructed from $\{VO_6\}$ octahedra fused through alternating cis- and trans-O—V—O linkages. As shown in Figure 12, the vanadium octahedra are defined by two bridging oxo groups and four oxygen donors from four *neighboring* $\{PO_4\}$ tetrahedra. Neighboring vanadium sites exhibit alternating cis and trans-bridging oxo groups producing an undulating $\{V=O-V\}$ chain. This $\{V=O-V\}$ backbone exhibits the common alternation of short and long $\{V-O\}$ distances. These chains are linked through η^4, μ^4 phosphate groups, each of which bridges two vanadium sites in each of two adjacent strands. The polyhedral connectivity generates six polyhedral connects forming 12-membered $\{V_4P_2O_6\}$ rings that form the walls of the cavities encapsulating the NH_4^+ cations. While $KVOPO_4$ (169) exhibits a similar 3D structure based on linkage of cis–trans $\{VO_6\}$ octahedra, the location of the cations within the tunnels is quite distinct from that observed for $(NH_4)(VO)(PO_4)$.

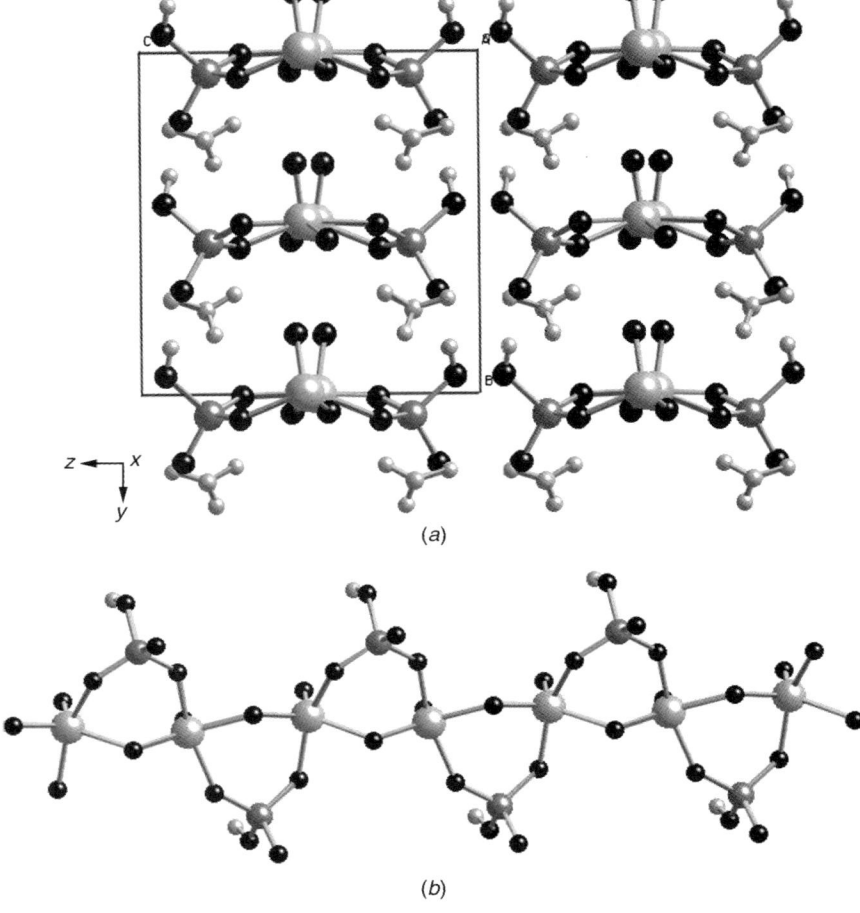

Figure 11. The structure of α-(NH$_4$)(VO$_2$)(HPO$_4$) viewed parallel to the *a* axis. A view of the structure showing the infinite [O–V···O–V···] chain.

2. Isolated Vanadium Polyhedra

The structure of Li$_2$(VO$_2$)(PO$_4$) (63), shown in Figure 13, consists of layers of {VO$_6$} octahedra and phosphate tetrahedra with Li$^+$ cations occupying the interlamellar region. The octahedral geometry about the vanadium sites is defined by two cis terminal oxo groups and four oxygen donors from three neighboring phosphate tetrahedra. The phosphates adopt the η^4, μ^3 coordination mode, such that edge sharing between the phosphate tetrahedra and vanadium

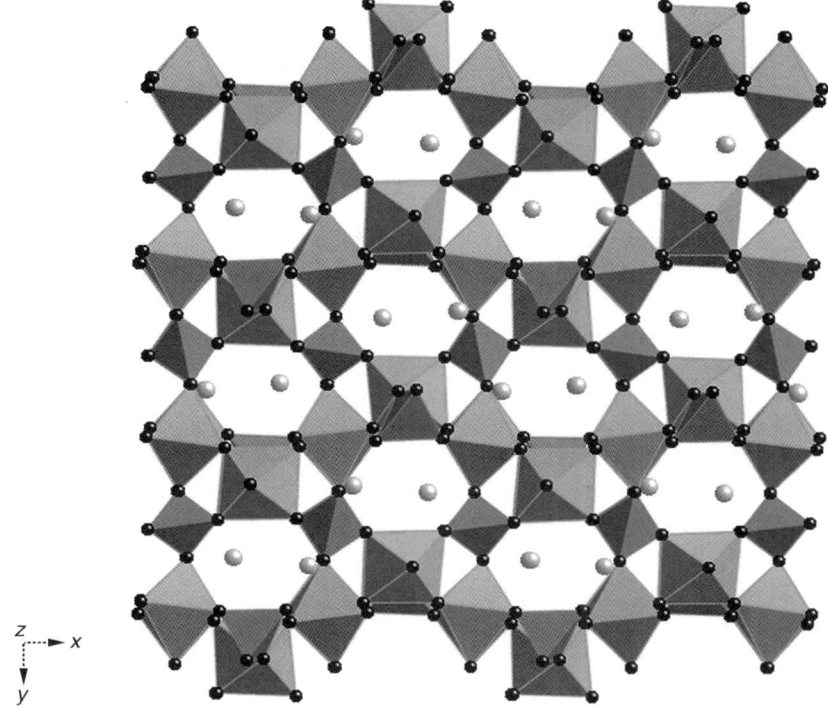

Figure 12. Polyhedral representation of the structure of $(NH_4)(VO)(PO_4)$ viewed parallel to the c axis and showing the location of NH_4^+ cation in the cavities formed by the 3D V/P/O connect.

octahedra leads to a strained four-membered ring. The presence of the edge-shared phosphate and the cis-dioxo unit precludes the formation of the common di-µ (O, O′) phosphate linkage and the eight-membered ring geometry $\{V_2P_2O_4\}$ that is common to the majority of V/P/O phases.

The structure of the mixed-valence V(IV)/V(V) species $K_{0.5}(VO)(PO_4) \cdot 1.5H_2O$ (58), shown in Figure 14, is closely related to that of $VO(PO_4) \cdot 2H_2O$ and is constructed from layers of corner-sharing $\{VO_6\}$ octahedra and phosphate tetrahedra with interlamellar K^+ cations and water molecules. Each vanadium site shares its four equatorial oxygens with four neighboring phosphorus centers. The axial sites are defined by a terminal oxo group and an aqua ligand with a long V—O bond length as a consequence of the trans influence of the multiply bonded oxo group. The vanadyl oxygen atoms are directed alternately up and down with respect to the V/P/O layer. The layer stacking, in contrast to that in $VOPO_4 \cdot 2H_2O$, is displaced to provide large cavities occupied by the K^+ cation and the interlamellar H_2O. The K^+ site

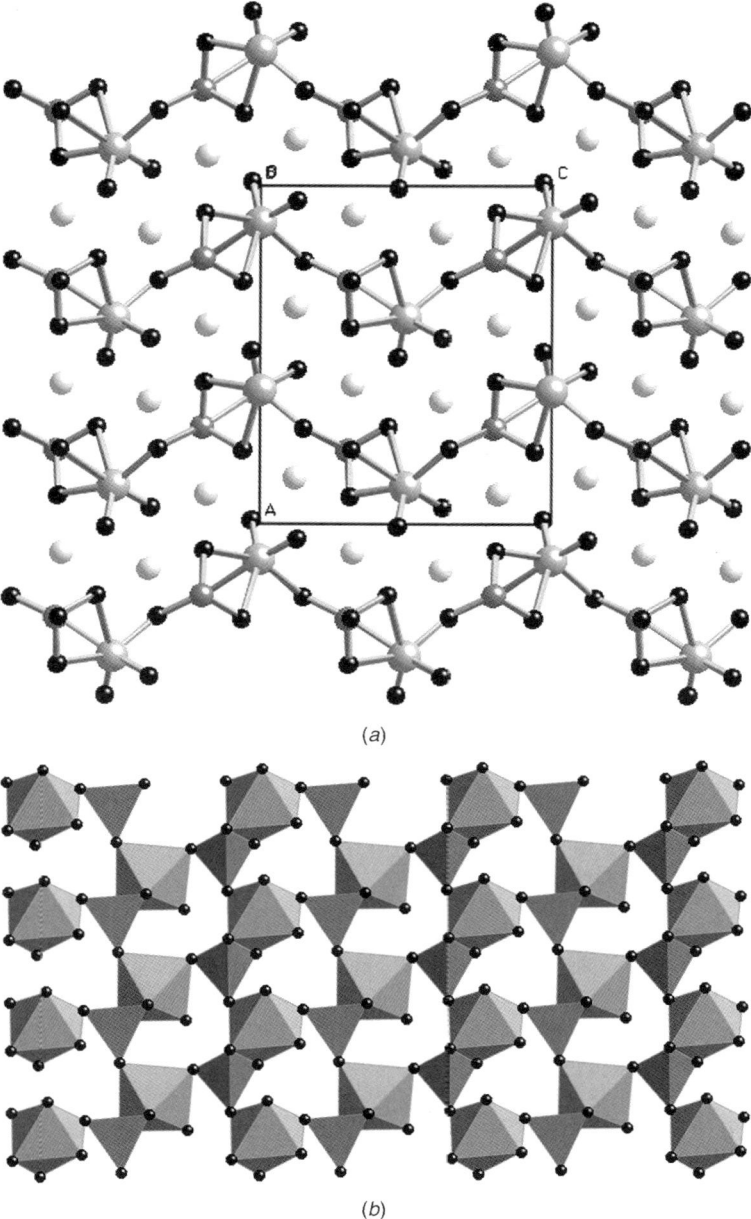

Figure 13. (a) A ball-and-stick representation of the stacking of layers in $Li_2(VO_2)(PO_4)$, viewed parallel to the crystallographic b axis. (b) A polyhedral view of the V/P/O network.

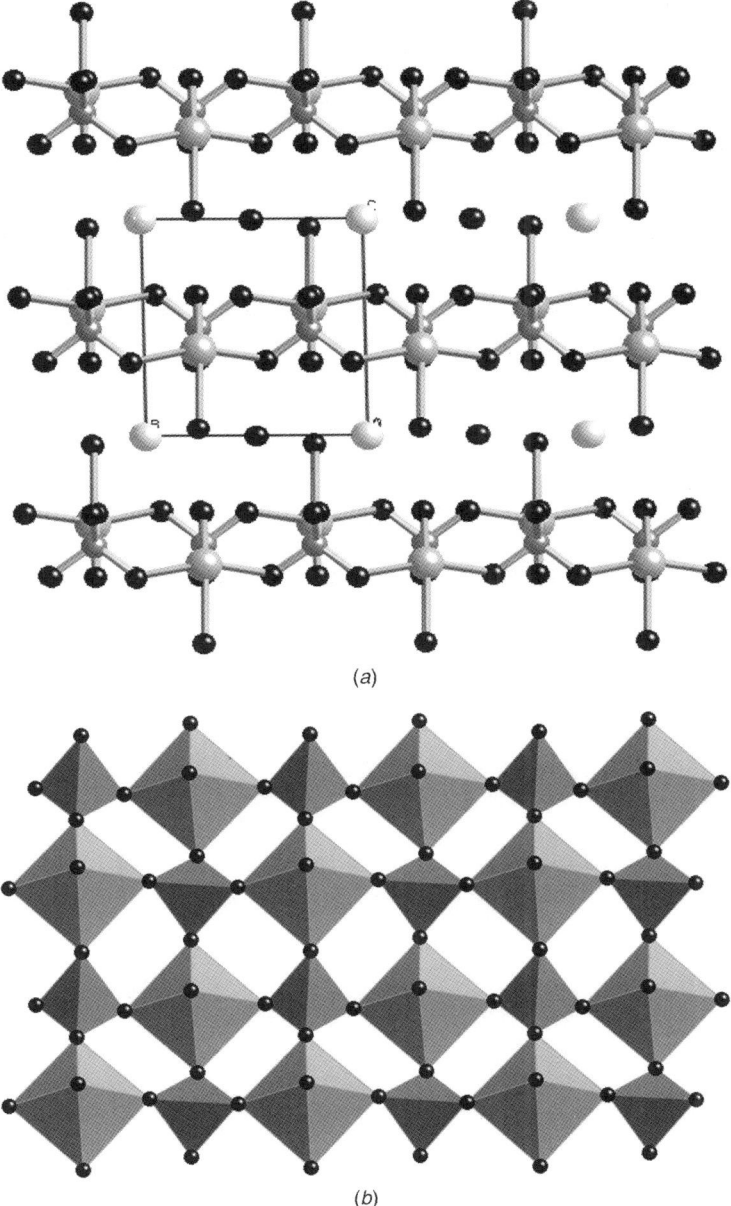

Figure 14. The structure of $K_{0.5}(VO)(PO)_4 \cdot 1.5H_2O$: parallel to the a axis and showing the layer stacking (a) and showing the V/P/O network (b).

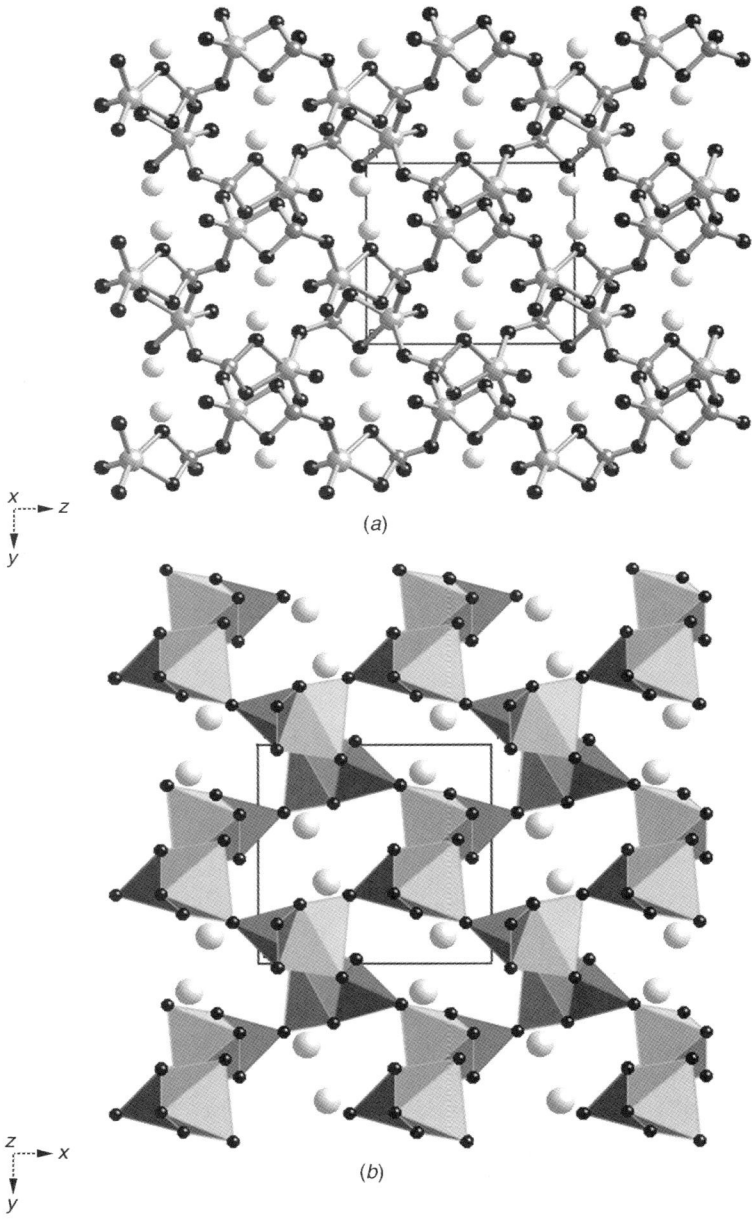

Figure 15. Ball-and-stick (*a*) and polyhedral (*b*) views of the structure of Rb(VO)(PO$_4$), showing the cavities occupied by the Rb$^+$ cations.

CRYSTAL CHEMISTRY OF ORGANICALLY TEMPLATED VANADIUM PHOSPHATES   475

coordinates to twelve oxygen atoms including both coordinated and interlamellar water and oxo groups and phosphate oxygens from adjacent layers.

The structure of the V(IV) species $Rb(VO)(PO_4)$ (88) is quite distinct from that of the mixed valence materials $M_{0.5}VO(PO_4) \cdot nH_2O$ and of the parent compound $VO(PO_4) \cdot 2H_2O$. As shown in Figure 15, the structure consists of $\{VO_5\}$ square pyramids edge- and corner-sharing with phosphate tetrahedra to produce a 3D framework. The four basal oxygen atoms of the vanadium square pyramid are shared with four phosphate tetrahedra. The phosphate groups adopt the η^4, μ^3 coordination mode. The polyhedral connectivity produces rings formed by eight polyhedra $\{V_4P_4O_8\}$ that provide channels for the Rb^+ cations. Curiously, the $Cs(VO)(PO_4)$ derivative is not isomorphous with $Rb(VO)(PO_4)$ but differs in the orientation of the vanadyl oxygens.

3. Binuclear Vanadium Clusters

Binuclear units with a variety of connectivities are common structural features of the V/P/O phases. A member of the charge-neutral class of materials, $VO(HPO_4) \cdot 0.5H_2O$, exhibited binuclear units constructed from face-sharing $\{VO_6\}$ octahedra. In contrast, the structure of $Ag_2(VO_2)(PO_4)$ (67) is based on dimers of edge-sharing vanadium octahedra. As shown in Figure 16, the layers are constructed from binuclear units each linked through six $\{PO_4\}$ tetrahedra in the η^3, μ^3 coordination mode to six adjacent binuclear units. The octahedral geometry about the vanadium sites is defined by two cis bridging oxo groups, a terminal oxo group and three oxygen donors from three adjacent $\{PO_4\}$ tetrahedra. The oxo groups are in the meridional orientation. The pendant $\{P=O\}$ groups of the phosphate units project into the interlamellar region and alternate up and down with respect to the layer. These units interact strongly with the interlamellar Ag^+ cations, which also bond to the oxo groups and other phosphate oxygens of adjacent layers.

As illustrated in Figure 17, the structure of $Li(VO)(PO_4) \cdot 0.5H_2O$ (57) also consists of V/P/O layers constructed from binuclear vanadium octahedra with interlamellar metal cations. However, the binuclear units exhibit the face-sharing motif previously described for $VO(HPO_4) \cdot 0.5H_2O$. The overall structure of $Li(VO)(PO_4) \cdot 0.5H_2O$ is similar to that of $VO(HPO_4) \cdot 0.5H_2O$, with the notable differences of Li^+ cations in the interlamellar region and deprotonation of the pendant {P–OH} groups of latter. The bond distance from the Li^+ cation to the oxygen of the pendant $\{P=O\}$ group is relatively short, as is the Li^+–OH_2 interaction.

4. Binuclear Vanadium Clusters and Vanadium Chains

The 3D structures of the quaternary phases may exhibit quite complex polyhedral connectivities, at times combining several vanadium building blocks

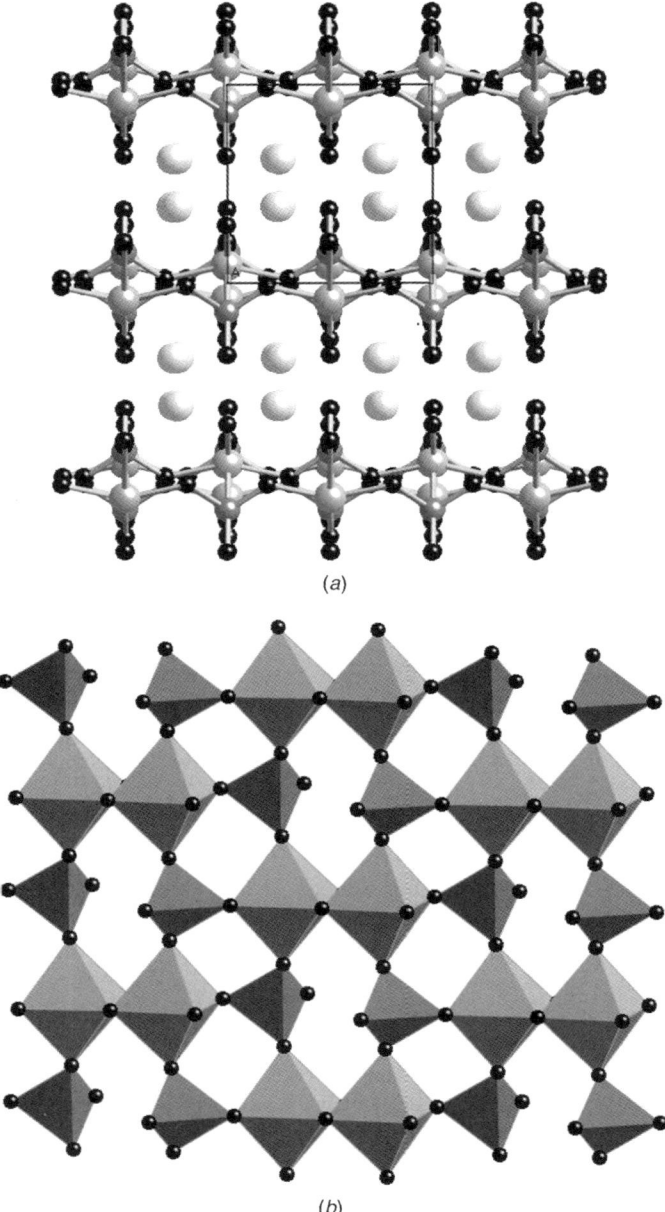

Figure 16. The structure of $Ag_2(VO_2)(PO_4)$. (a) View parallel to the a axis showing the stacking of layers. (b) Polyhedral representation of the V/P/O network.

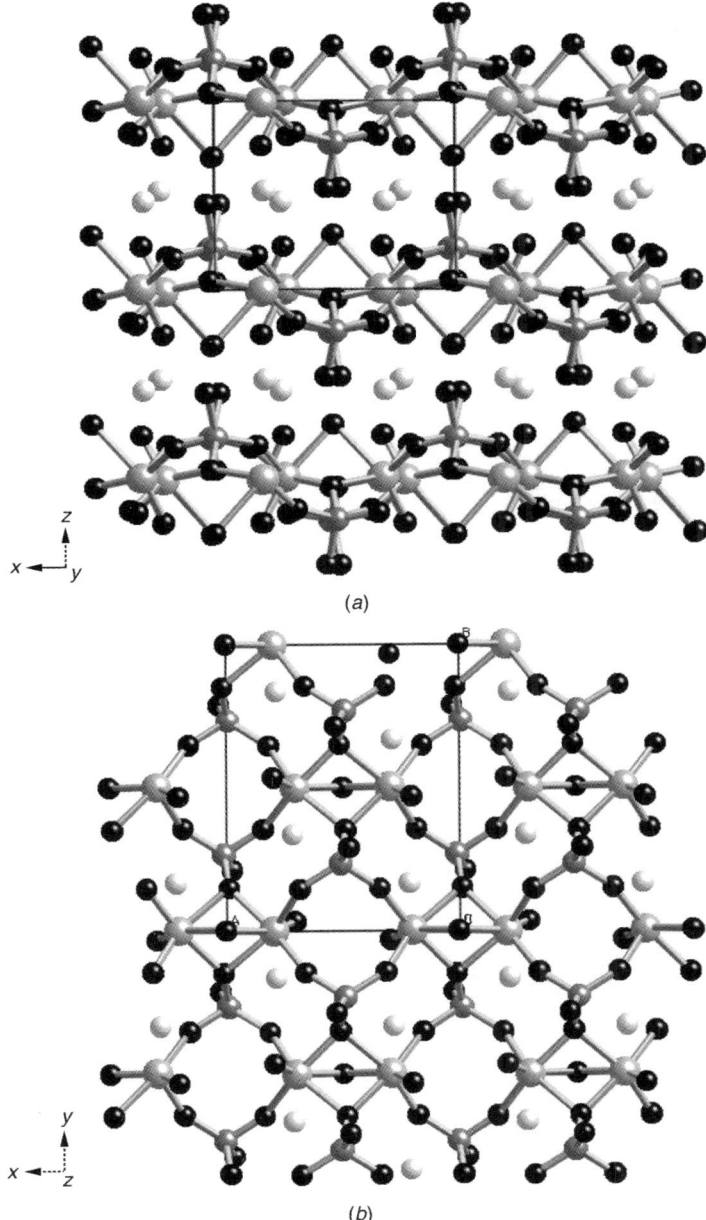

Figure 17. Views of the structure of $Li_2VOPO_4 \cdot 0.5H_2O$: showing the stacking of layers and the locations of the Li^+ cations (a) and looking normal to the V/P/O plane (b).

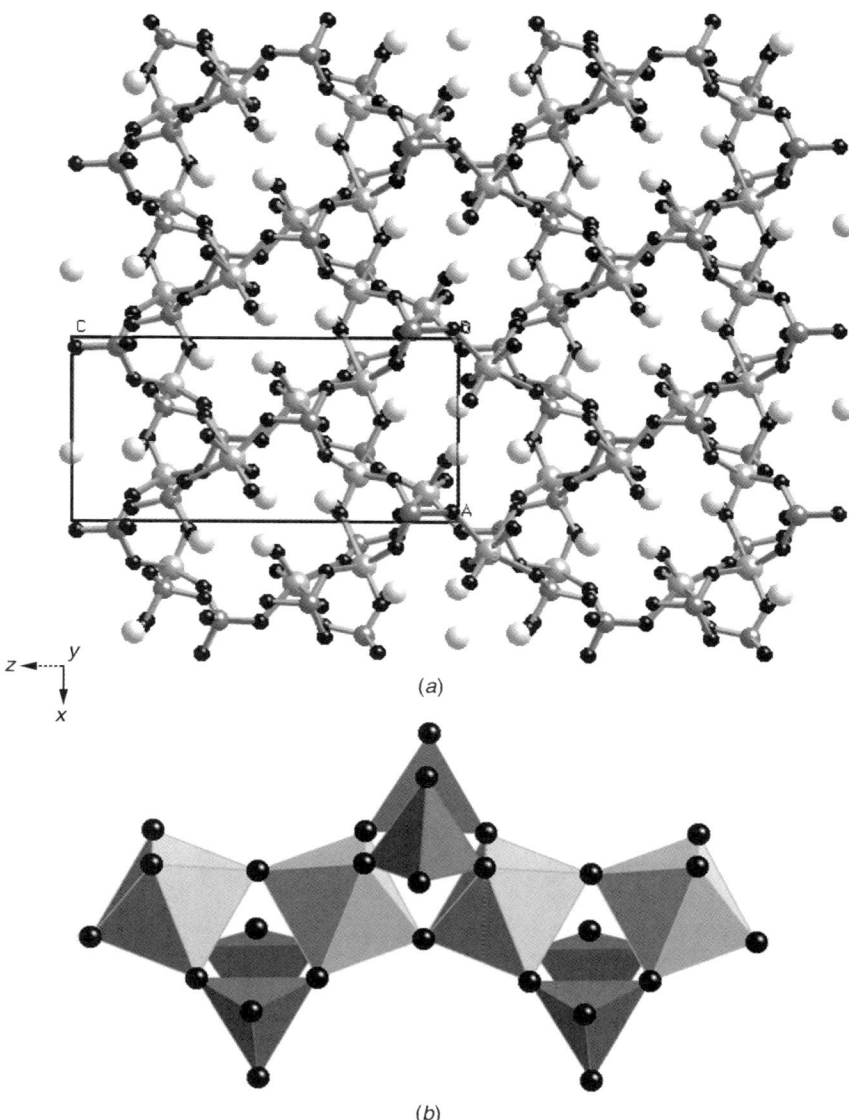

Figure 18. (a) a view of the structure of $K_3(VO)(HV_2O_3)(PO_4)_2(HPO_4)$ viewed parallel to the b axis. (b) polyhedral representation of the chain of corner-sharing vanadium octahedra.

in the framework construction. The phases $K_3(VO)(HV_2O_3)(PO_4)_2(HPO_4)$ (101) and $\alpha\text{-}K_2(VO)_3(P_2O_7)_2$ (127) are representative of structures containing both vanadium binuclear units and 1D chains as submotifs of the 3D framework.

The structure of $K_3(VO)(HV_2O_3)(PO_4)_2(HPO_4)$, shown in Figure 18, possesses chains of corner-sharing $\{VO_6\}$ octahedra with the common pattern of alternating short-long $\{V=O \cdots V\}$ interactions along the vanadium–oxide chain. Each vanadium of the chain bonds to four oxygens of four phosphate groups in equatorial dispositions and to two trans oxo groups. These chains are linked in turn to a binuclear unit of corner-sharing $\{VO_5\}$ square pyramids. The vanadium geometry within the binuclear units is defined by a terminal oxo group in the apical position and a basal plane defined by the bridging hydroxy group and three oxygens from three phosphate groups. The phosphate group adopts the η^4, μ^4 coordination mode linking two binuclear units to two adjacent vanadium sites of the neighboring chain. The $\{HPO_4\}^{2-}$ group exhibits η^2, μ^2 coordination, serving to bridge the vanadium sites of a binuclear unit and projecting pendant $\{P=O\}$ and $\{P-OH\}$ groups into the channel occupied by the K^+ cation. The deprotonated, V(V)/V(IV) mixed-valence analogue $K_3(VO)(V_2O_3)(PO_4)_2(HPO_4)$ is isomorphous.

While $\alpha\text{-}K_2(VO)_3(P_2O_7)_2$ also exhibits 1D chains and binuclear vanadium motifs, the 3D structure, shown in Figure 19, is unrelated to that of $K_3(VO)(HV_2O_3)(PO_4)(HPO_4)$. Once again, there are 1D chains of trans corner-sharing $\{VO_6\}$ octahedra, with the common alternating short-long $\{V=O \cdots V\}$ interactions. These chains are linked through pyrophosphate groups to binuclear units from corner sharing of a vanadium square pyramid and a vanadium octahedron. The square pyramidal geometry is defined by an apical bridging oxo group and four basal oxygens from the pyrophosphate groups, while the octahedral site consists of the bridging oxo group, a trans terminal oxo group, and four equatorial pyrophosphate oxygen donors. Each pyrophosphate bonds through two oxygen donors, one from each $\{PO_3\}$ group, to adjacent vanadium sites of the chain, and employs the remaining four oxygen donors to connect to two adjacent binuclear units. The polyhedral connectivity is such that intersecting tunnels occupied by K^+ cations are formed. The structure of $Cs_2(VO)_3(P_2O_7)$ (132) is analogous to that of $K_2(VO)_3(P_2O_7)$ with Cs^+ cations residing in the tunnels, as shown in Figure 20.

5. Tetranuclear Vanadium Clusters

The structure of $Ba_2(V_5O_8)(PO_4)_4$ (109) exhibits octahedral, square pyramidal, and tetrahedral vanadium sites. As shown in Figure 21, one structural motif consists of a tetranuclear cluster of V(IV)/V(V) octahedra and V(IV) square pyramids in a corner-sharing $\{V_4O_4\}$ ring. These clusters are linked through phosphate tetrahedra into layers two polyhedra in thickness. The layers are interconnected through bridging vanadium tetrahedra into a 3D framework. The

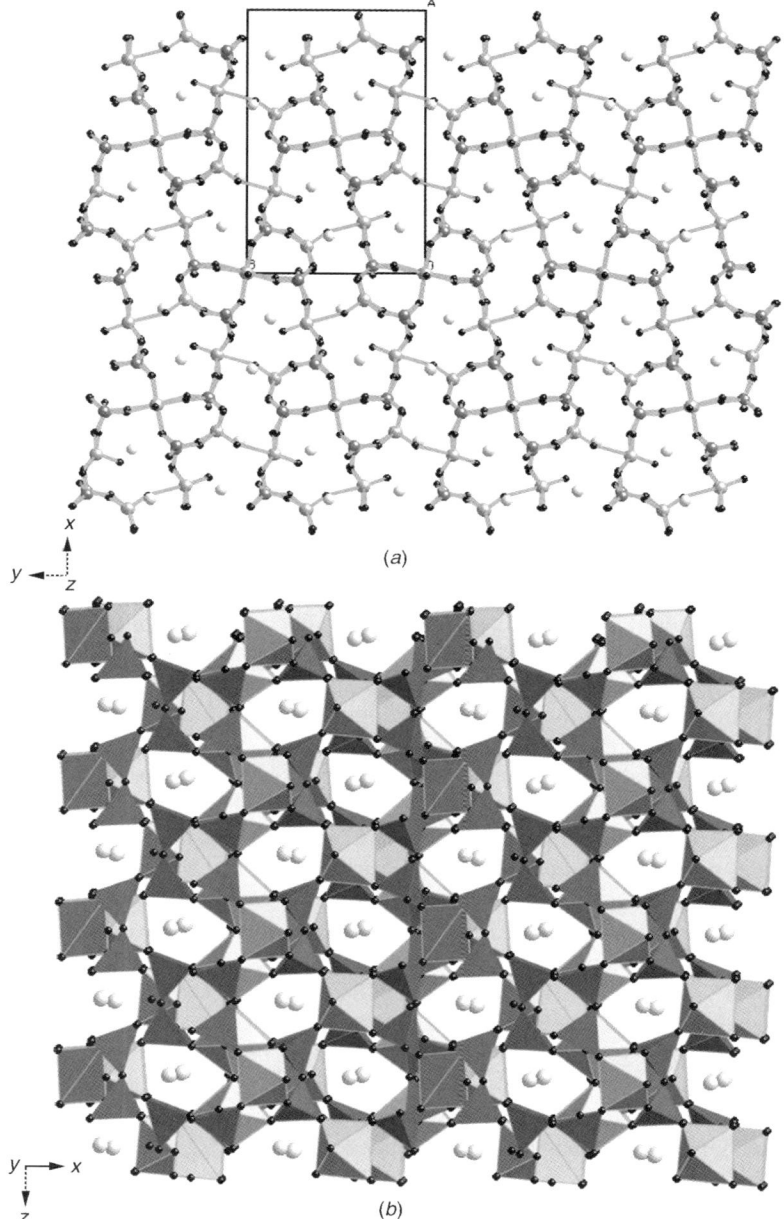

Figure 19. (a) Ball-and-stick representation of the structure of α-$K_2(VO)_3(P_2O_7)_2$, viewed parallel to the crystallographic c axis. (b) A polyhedral representation viewed parallel to the b axis showing the cavities occupied by the K^+ cations.

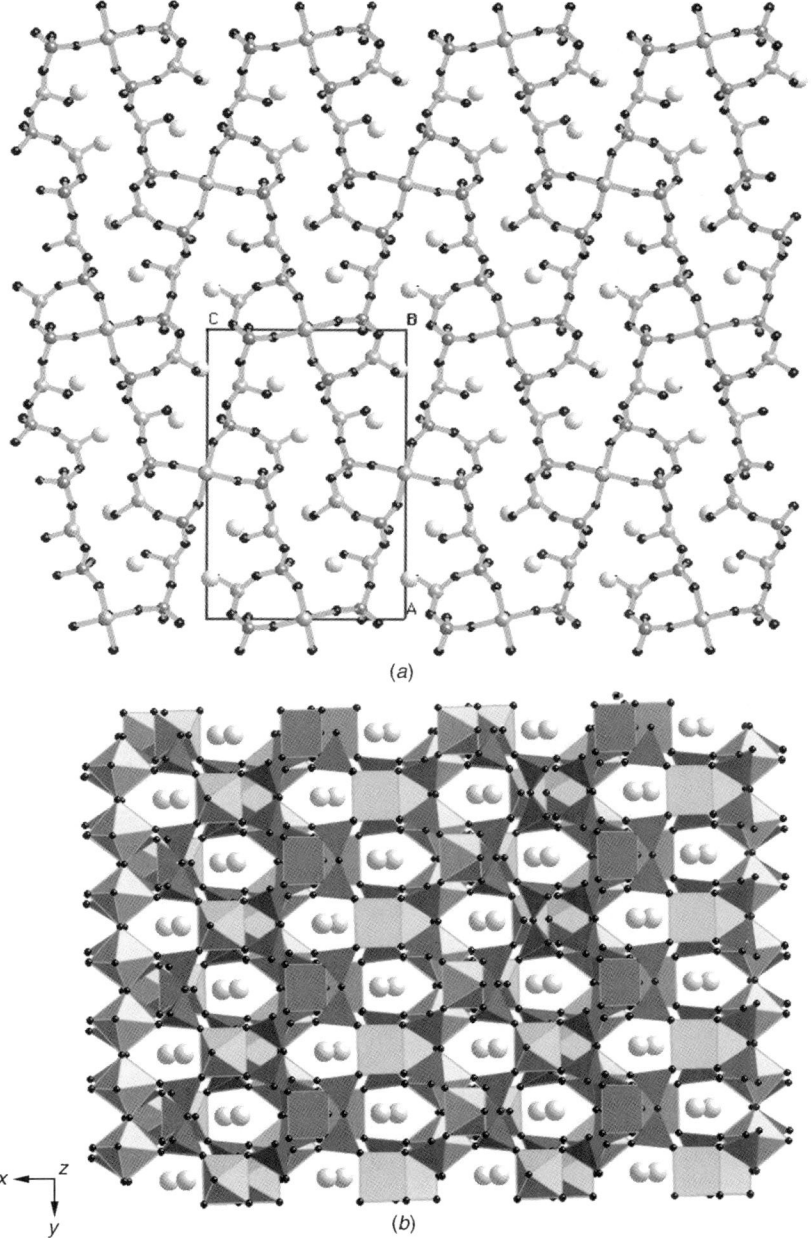

Figure 20. The structure of $Cs_2(VO)_3(P_2O_7)$ viewed parallel to the crystallographic b axis (a), and c axis (b) to illustrate the cavities occupied by the Cs^+ cations.

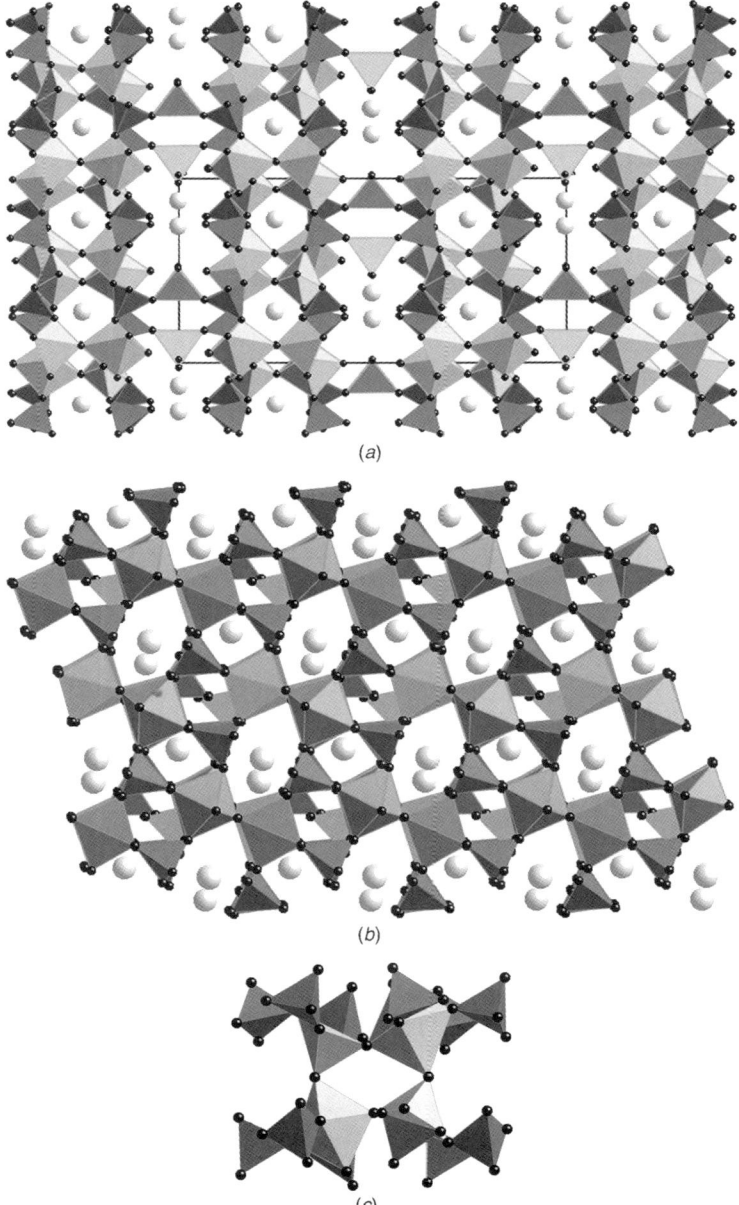

Figure 21. Polyhedral representations of the structure of $Ba_2[V_5O_8(PO_4)_4]$, viewed parallel to the a and b axis (a and b, respectively). (c) The tetranuclear core of the structure, with the attendant phosphate groups.

Ba^{2+} cations occupy the considerable void space produced by the V/O/P framework.

An unusual example of a tetranuclear cluster constructed from edge sharing octahedra is provided by the mineral phosphovanadylite $(Ba_{0.38}Ca_{0.20}K_{0.06}Na_{0.02})$ $[(V_{3.44}Al_{0.46})(OH)_{5.66}O_{2.34}(PO_4)_2] \cdot 12H_2O$ (110). The core of the cluster consists of a $\{V_4O_4\}$ cube; consequently each $\{VO_6\}$ octahedron shares an edge with the three adjacent octahedra (Fig. 22). Phosphate tetrahedra link these tetranuclear clusters into an open-framework zeolitic structure. The resultant 7.0- and 5.5-Å diameter cavities are occupied by cations and water of crystallization.

6. Pentanuclear Vanadium Clusters

The structure of $Pb_2V_2(VO)(PO_4)_4$ (104) is uniquely constructed from mixed valence V(III)/V(IV) pentanuclear units and isolated V(III) octahedra providing the metal oxide motifs. The pentanuclear unit consists of two V(IV) square pyramids sharing cis basal vertices with V(III) octahedra, as shown in Figure 23. The apical position of the V(IV) sites is defined by a terminal oxo group while two positions on the equatorial plane involve corner sharing with $\{PO_4\}$ tetrahedra; the remaining two positions arise from edge sharing with a $\{PO_4\}$ tetrahedron, which also serves to link two vertex sharing V(III) octahedra to the V(IV) center and to bridge to a V(III) site of a neighboring pentanuclear unit. The terminal V(III) sites of the clusters are each linked to six phosphate tetrahedra, one of which provides the common vertex with the V(IV) site. The central V(III) site is also linked to six phosphate tetrahedra, two of which each provide a shared vertex with a V(IV) square pyramid. Adjacent pentanuclear units are linked through phosphate tetrahedra, which also provide connectivity to isolated V(III) octahedra. The complex polyhedral arrangement generates tunnels occupied by the Pb^{2+} cations, which exhibit distorted geometries reflecting the stereoactivity of the lone pair. While $(Me_2NH_2)K_4[V_{10}O_{10}(H_2O)_2(OH)_4(PO_4)_7] \cdot 4H_2O$ (7) and $[HN(CH_2CH_2)_3NH]K_{1.35}[V_5O_9(PO_4)_2] \cdot xH_2O$ (112) are also constructed from pentanuclear vanadium building blocks, the structures of the clusters and consequently of the 3D frameworks are quite distinct, as discussed in Section IV.C.

7. Nonanuclear Vanadium Clusters

The structure of $K_{11}(V_{15}O)(PO_4)_{18}$ (113) (Fig. 24) contains a most unusual nonanuclear vanadium (III) cluster as building block. The central core of the cluster consists of three $\{VO_6\}$ octahedra sharing a common oxo group. This group of vanadium octahedra is in turn sandwiched between two rings of six polyhedra, each composed of an alternating arrangement of corner-sharing $\{VO_6\}$ octahedra and $\{PO_4\}$ tetrahedra. Each vanadium site of the trinuclear

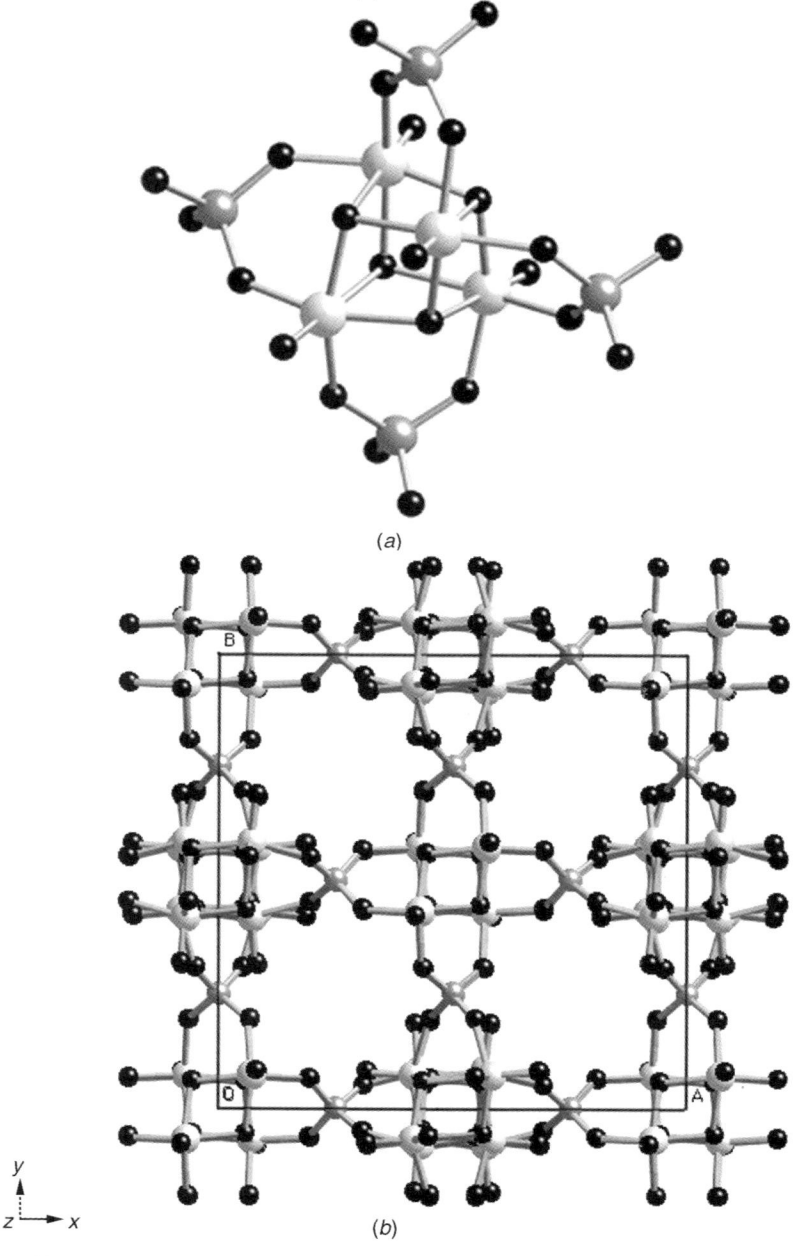

Figure 22. (a) The tetranuclear cluster embedded in the framework structure of $(Ba_{0.38}Ca_{0.20}K_{0.06}Na_{0.02})[(V_{3.44}Al_{0.46})(OH)_{5.66}O_{2.34}(PO_4)_2] \cdot 12H_2O$. (b) A view of the V/P/O framework.

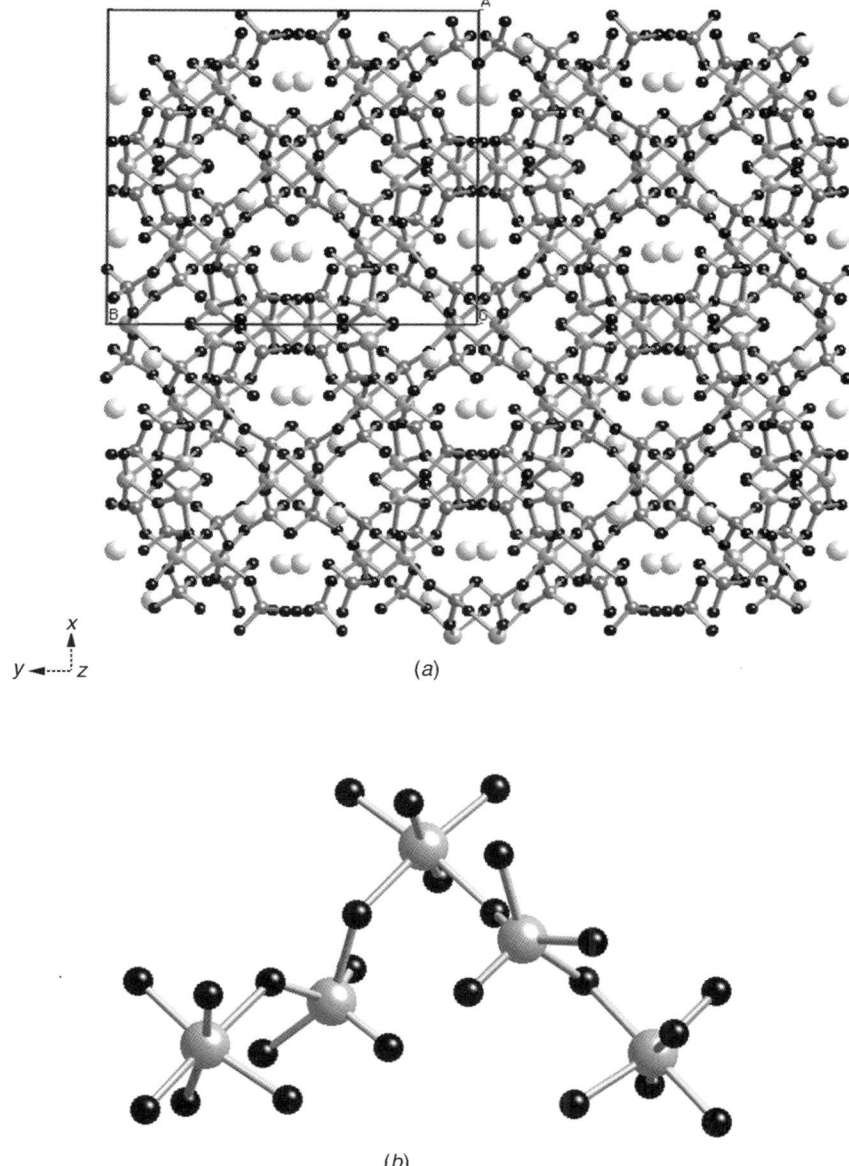

Figure 23. (a) A view of the structure of $Pb_2V_2(VO)(PO_4)_4$, viewed parallel to the c axis. (b) The pentanuclear cluster embedded in the 3D framework.

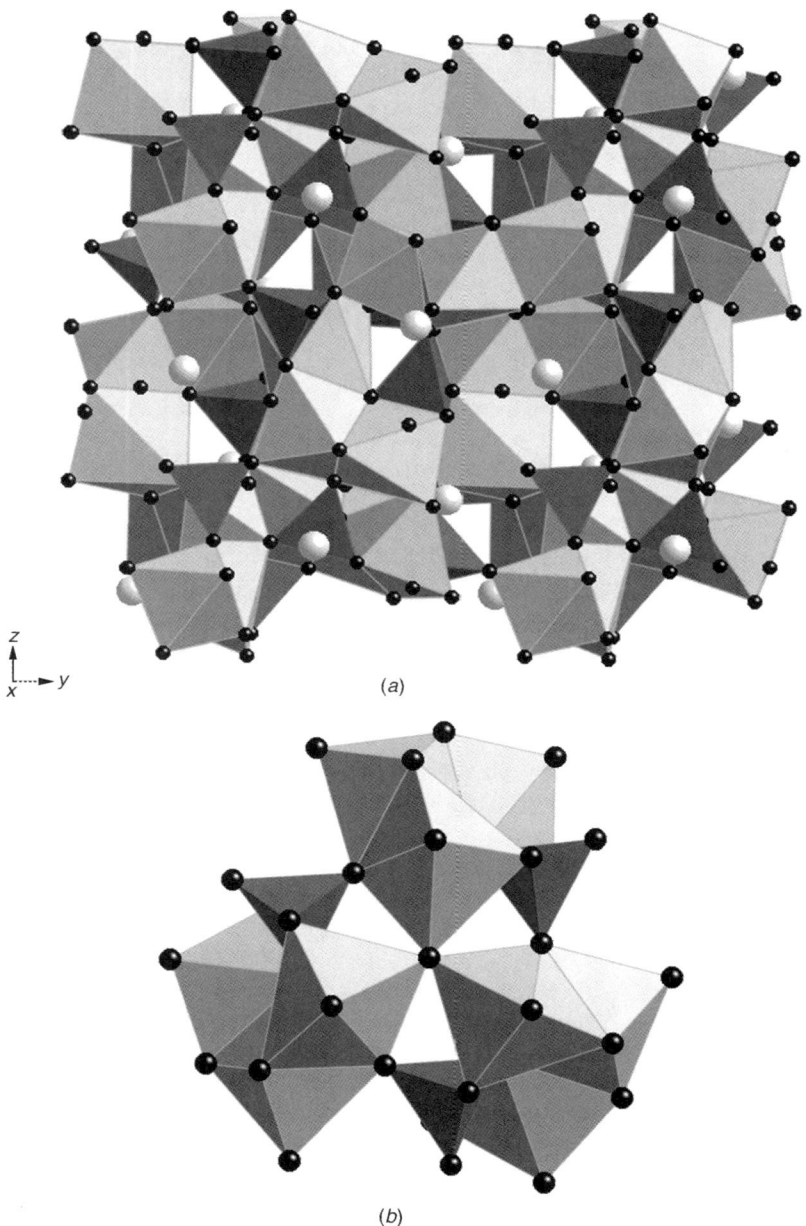

Figure 24. A view of the structure of $K_{11}V_{15}O(PO_4)_{18}$ parallel to the crystallographic a axis (a). The nonanuclear vanadium cluster embedded in the structure (b).

core shares a face with one vanadium of a six polyhedral connect ring and an edge with a vanadium of the other six polyhedral connect ring. These unusual clusters are linked through phosphate tetrahedra into a complex 3D framework that delimits cages that encapsulate the K^+ cations.

D. Inorganic Vanadium Phosphates with d-Block Elements as Additional Framework Constituents

The introduction of additional d-block elements as integral, covalently attached framework constitutes results in a considerable expansion of the structural diversity of the quaternary phases $M'/V/P/O$. While such materials remain relatively unexplored, the few examples described to date suggest an expansive chemistry.

The compound $[Ni(H_2O)_4][(VO)(PO_4)]_2$ (150) represents a direct structural analogue of a parent V/P/O phase. The structure consists of $(VO)(PO_4)$ layers constructed from isolated $\{VO_5\}$ square pyramids and $\{PO_4\}$ tetrahedra, linked by $\{NiO_6\}$ octahedra. The layer structure is essentially that of $VO(PO_4)$, absent the long range long V $\cdots$ O interactions between neighboring layers, and with parallel shearing to alter the registry of layers so as to point the $\{V=O\}$ vertices of adjacent layers at one another. These $\{V=O\}$ units link to the interlamellar Ni^+ site, which consequently coordinates to two trans oxo groups and four equatorial aqua ligands. The spacing between $\{VO(PO_4)\}$ planes has expanded in $Ni(H_2O)_4[VO(PO_4)]_2$ to 4.84 Å, compared to 3.91 Å in $VO(PO_4)$.

Curiously, the structure of $[Ni(H_2O)_4][(VO)(PO_4)]_2$ is quite distinct from that of $Ni_{0.5}(VO)(PO_4) \cdot 1.5\ H_2O$ (149), shown in Figure 25. In contrast to the structure of $[Ni(H_2O)_4[(VO)(PO_4)]_2$, $Ni_{0.5}(VO)(PO_4) \cdot 1.5H_2O$ exhibits 3D architecture. The structural motif consists of a trinuclear unit of one Ni(II) and two V(IV) octahedra sharing a common vertex. One vanadium octahedron is defined by the oxo group common to the trinuclear unit, a trans aqua ligand and an equatorial plane constructed from oxygen atoms from four phosphate tetrahedra. The second vanadium site exhibits a trans dioxo motif, defined by the common vertex, and an oxo group bridging to the Ni(II) site of a neighboring trinuclear unit; the basal plane results from corner sharing with four phosphate tetrahedra. The Ni(II) site is defined by a trans dioxo unit, again involving the common vertex and an oxo group bridging to a vanadium site of a second adjacent trinuclear unit. The equatorial plane consists of oxygen atoms from two phosphate groups and two aqua ligands. The complex polyhedral connectivity provides channels into which the aqua groups project. The phosphate adopts the η^4, μ^4 mode, bridging two vanadium sites of one trinuclear unit and a vanadium site and nickel center of an adjacent cluster.

The Zn(II) containing structures $Zn_3V_4(PO_4)_6$ (155) and $Zn_2(VO)(PO_4)_2$ (153) demonstrate the structural variability which may be achieved by

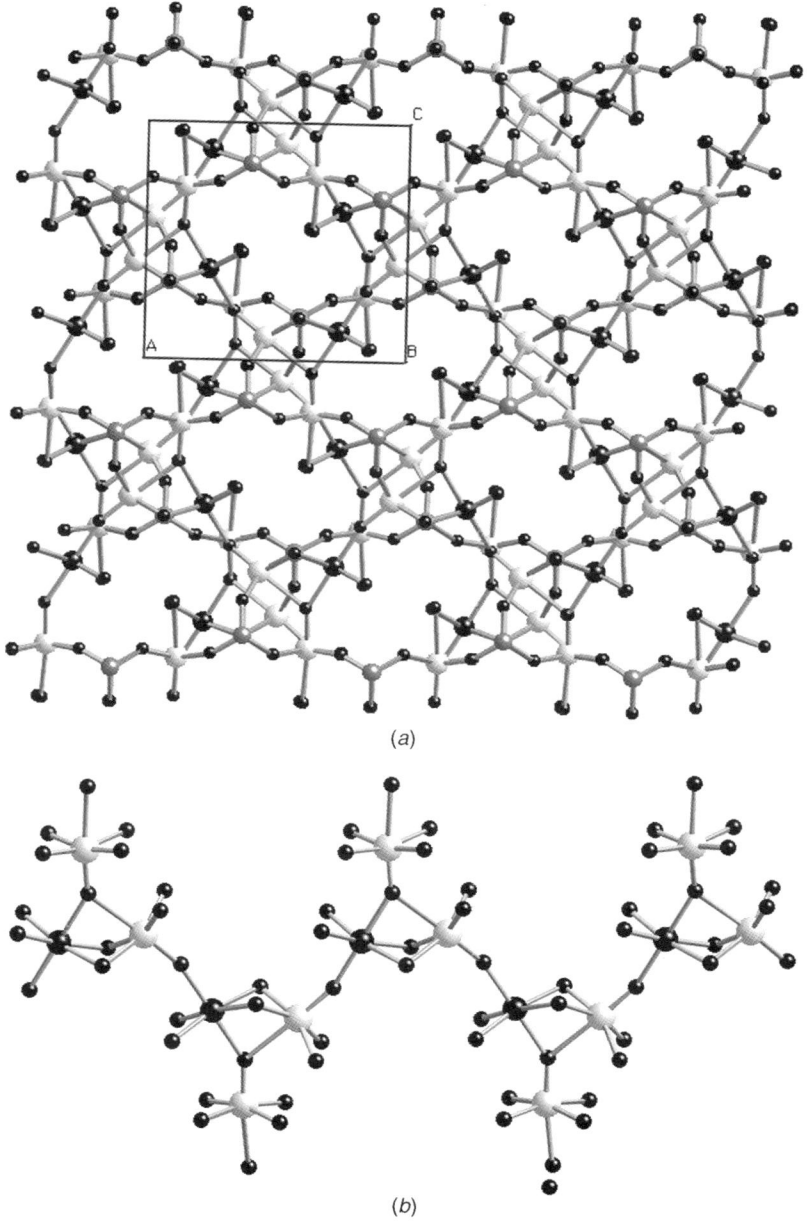

Figure 25. (a) A polyhedral representation of the structure of $Ni_{0.5}VO(PO_4) \cdot 1.5H_2O$, viewed parallel to the b axis. (b) A view of the chain of trinuclear {NiV_2} clusters.

introducing additional d-block elements with variable coordination requirements. The structure of $Zn_3V_4(PO_4)_6$, shown in Figure 26, is constructed from V(III) octahedra, Zn(II) octahedra, and Zn(II) trigonal bipyramids. Each V(III) octahedron engages in corner sharing with six phosphate tetrahedra and shares an edge with an adjacent V(III) site to form a binuclear unit. The {ZnO_5} trigonal bipyramids share two edges with two neighboring V(III) binuclear units to form an infinite corrugated ribbon. The octahedral Zn(II) site shares two trans vertices with trigonal bipyramidal Zn sites from adjacent ribbons and with phosphate tetrahedra; two vertices with V(III) sites of adjacent ribbons and with phosphate tetrahedra; and two vertices with phosphate tetrahedra only. A

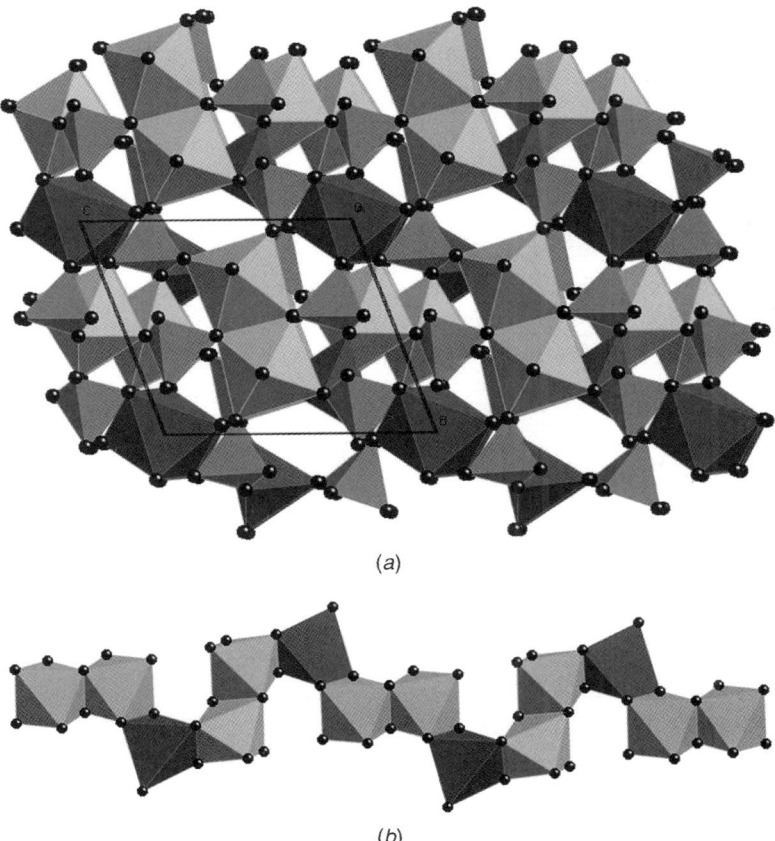

Figure 26. The structure of $Zn_3V_4(PO_4)_6$. (a) Viewed parallel to the crystallographic a axis. (b) The 1D substructure of edge-sharing nickel and vanadium polyhedra.

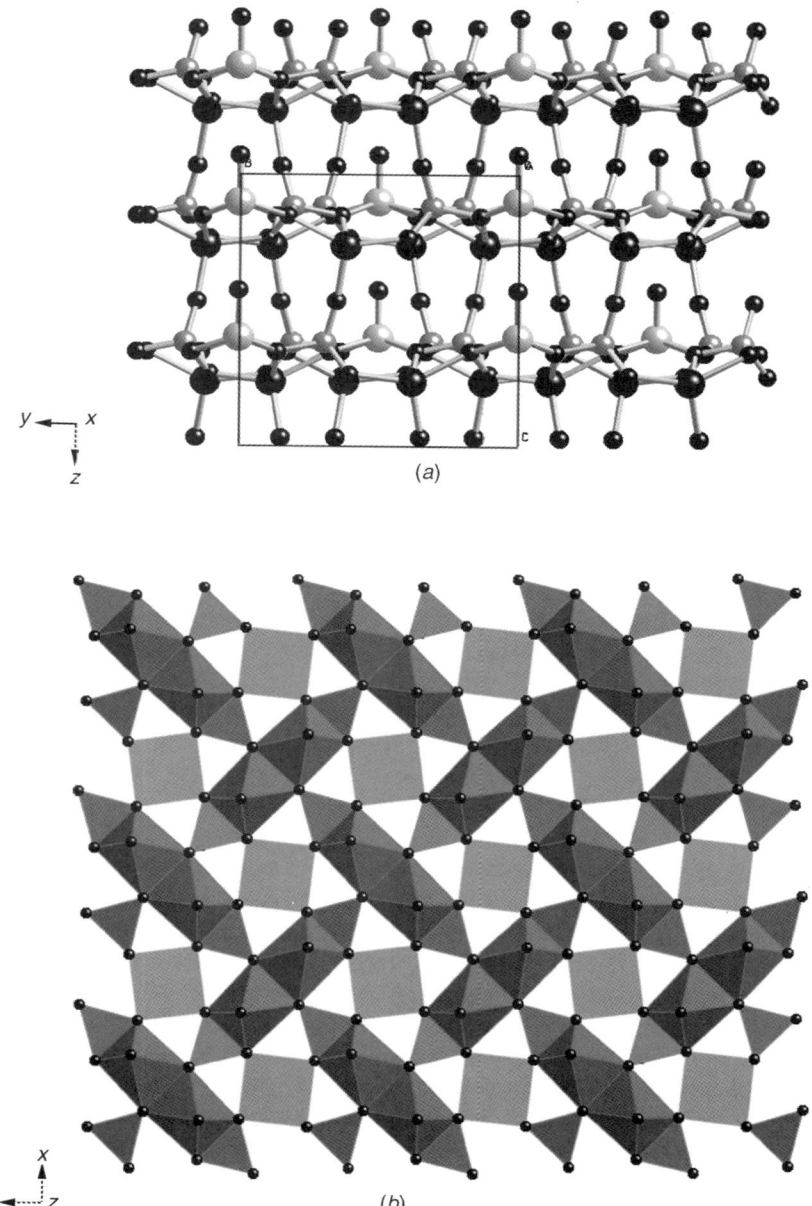

Figure 27. Two views of the structure of $Zn_2VO(PO_4)$. Down [100] (a) and down [001] (b).

curious feature of the structure is the large percentage of μ^2-bridging phosphate oxygen atoms, resulting in η^4, μ^6 and η^4, μ^7 coordination modes for the phosphate tetrahedra.

The structure of $Zn_2(VO)(PO_4)_2$, shown in Figure 27, is constructed from 1D chains of trans corner sharing $\{VO_6\}$ octahedra and edge-sharing $\{ZnO_5\}$ square pyramids. The V(IV) octahedra are defined by an apical oxo group and a trans oxo group from a neighboring V(IV) site at 2.95 Å, and by four equatorial vertices each shared with a Zn(II) square pyramid and a phosphate tetrahedron. Alternatively, the structure may be described as isolated $\{VO_5\}$ square pyramids corner sharing with four adjacent Zn(II) binuclear units. Each Zn(II) site bonds to five oxygens from five phosphate groups; two vertices are also shared with the adjacent zinc site of the binuclear unit and two with two adjacent V(IV) square pyramids; the apical vertex is shared only with a phosphorus tetrahedron. The phosphate tetrahedra are in the η^4, μ^7 coordination mode, suggesting that multiple bridging may be a common feature of such quaternary covalent structures.

Although some structural correspondence might be anticipated between $Zn_2(VO)(PO_4)_2$ and $Mn_2(VO)(PO_4)_2 \cdot H_2O$ (146), the structures are unrelated. The metal oxide core of the latter consists of a unique tetranuclear cluster, constructed from corner-sharing $\{VO_6\}$ octahedra and $\{MnO_5\}$ distorted polyhedra. The vanadium octahedra are defined by a terminal oxo group and five phosphate oxygens. As shown in Figure 28, the vanadium shares the oxo group with a neighboring $\{MnO_5\}$ polyhedron. Two such binuclear units are linked through common phosphate oxygens to generate a cyclic tetranuclear cluster. A second manganese site with distorted octahedral geometry enjoys a face-sharing interaction with the V(IV) center to generate an unusual hexanuclear core. Once again, the phosphate tetrahedra exhibit unusually high consideration number, adopting the η^4, μ^7 coordination mode.

E. Vanadium Silicophosphates, Aluminophosphates, Gallophosphates, Borophosphates, and Phosphate Fluorides

1. General Comments

Additional elements or functional groups may be incorporated into the anionic framework of the VOPO class in order to increase framework stability or to provide additional structural flexibility. The presence of aluminum in the framework, for example, results in enhanced thermal stability of the phases. Borate serves not only as a framework constituent but as an effective solubilizer of the metal oxide components. Addition of fluoride to silicates and aluminophosphates and gallophosphates (188–191) results in enhanced mineralization and induces crystallization in neutral and acidic pH. While VOPO phases incorporating such additional constituents are relatively unexplored, the few

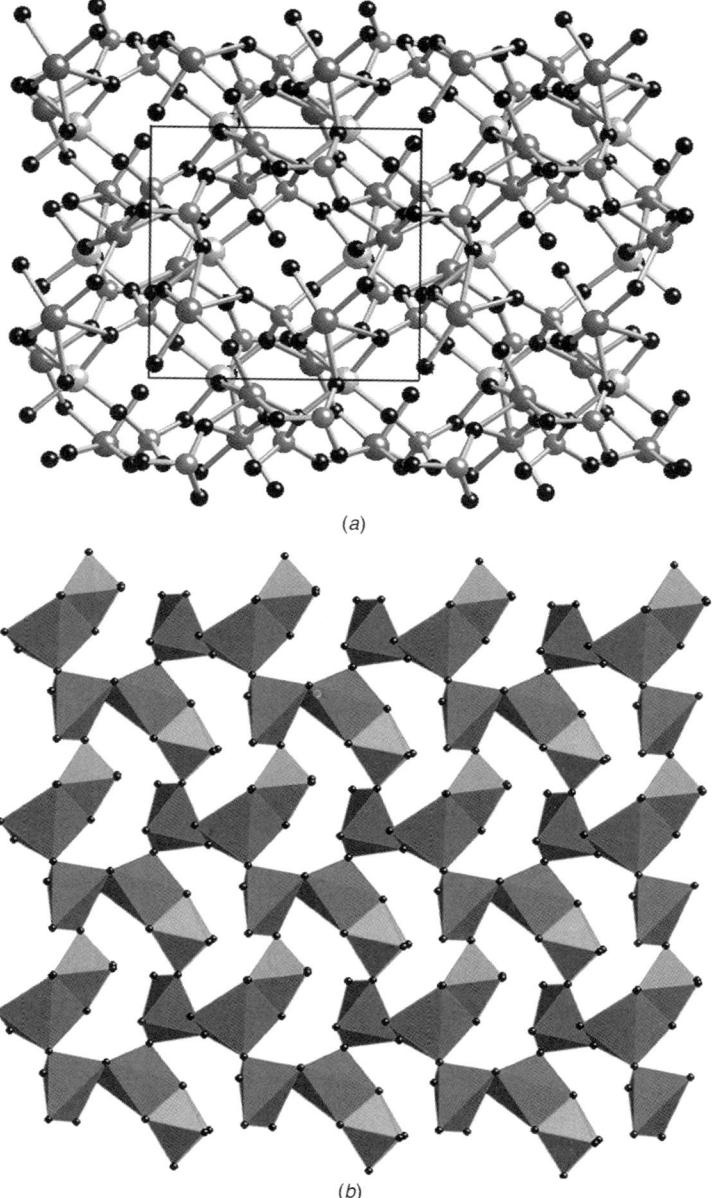

Figure 28. Views of the structure of $Mn_2VO(PO_4)_2 \cdot H_2O$. (a) Ball-and stick-representation, parallel to the a axis. (b) Polyhedral representation of the Mn–V–O framework (phosphorus tetrahedra omitted), parallel to the c axis.

examples identified to date suggest a diverse structural chemistry of materials with modified physical properties.

2. Vanadium Silicophosphates

Since the substitution of P^{5+} for Si^{4+} in silicates and of Si^{4+} for P^{5+} in phosphate phases is not common in minerals (192), it is not surprising that vanadium silicophosphate materials are relatively unexplored. The two prominent examples are dense phase materials with distinctly different structures.

The oxovanadium(IV) diphosphatomonosilicate $(VO)Si(PO_4)_2$ (157) exhibits a 3D structure constructed from chains of corner-sharing $\{VO_6\}$ octahedra linked through $-Si(PO_4)_2Si(PO_4)_2Si-$ chains. Both types of chains are connected to four chains of the other type by corner sharing of phosphate tetrahedra, as shown in Figure 29. The phosphosilicate chain consists of four polyhedral rings $\{P_2Si_2O_{12}\}$ linked through apical Si atoms. Thus, each Si center bridges four phosphorus sites from two adjacent rings.

In contrast, the 3D structure of $V_3P_5SiO_{19}$ (158) is constructed from $\{PO_4\}$ tetrahedra, $\{Si_2O_7\}$ disilicate units and binuclear units of face-sharing $\{VO_6\}$ octahedra, with vanadium in the +3 oxidation state. Each disilicate unit employs three oxygen atoms at each Si center not involved in the Si—O—Si linkage to bridge to a phosphorus site, to give a $\{Si_2P_6O_{25}\}$ moiety. The six $\{PO_4\}$ tetrahedra of this unit share three corners with three different binuclear vanadium sites. A second substructural unit is formed by a diphosphate group similarly linked to six phosphate tetrahedra to produce a $\{P_8O_{25}\}$ unit that again bridges six different vanadium binuclear sites. The $\{SiP_6O_{25}\}$ structural motif is similar to that observed for materials of the type $MMo_3P_6Si_2O_{25}$ (193).

3. Vanadium Aluminophosphates and Gallophosphates

The first attempts to introduce $\{AlO_4\}$ tetrahedra into vanadium phosphate structures were predicated on the assumption that such incorporation would increase the thermal stability of the V/P/O phases. Furthermore, the diverse coordination environments and structural connectivities of both aluminophosphates and vanadium phosphate systems suggest that introduction of aluminum structural units into the V/P/O phase would lead to unusual materials.

The structures of $Cs(VO)(H_2O)Al(PO_4)_2$, $Rb(VO)(H_2O)Al(PO_4)_2$, and $[H_3NCH_2CH_2NH_2][(VO)Al(PO_4)_2]$ (159) are comparable. As shown in Figure 30, all are constructed from vanadium octahedra, phosphorus tetrahedra, and aluminum tetrahedra. In the Cs and Rb phases, the vanadium octahedron is defined by four oxygen donors from four phosphate groups, a terminal oxo group and an aqua ligand. Each Al site shares vertices with four $\{PO_4\}$ tetrahedra, while each P site shares vertices with two vanadium octahedra and two aluminum tetrahedra. In the case of $[H_3NCH_2CH_2NH_2][(VO)Al(PO_4)_2]$, the aqua ligand is displaced by the 1-aminoethane-2-ammonium ligand that

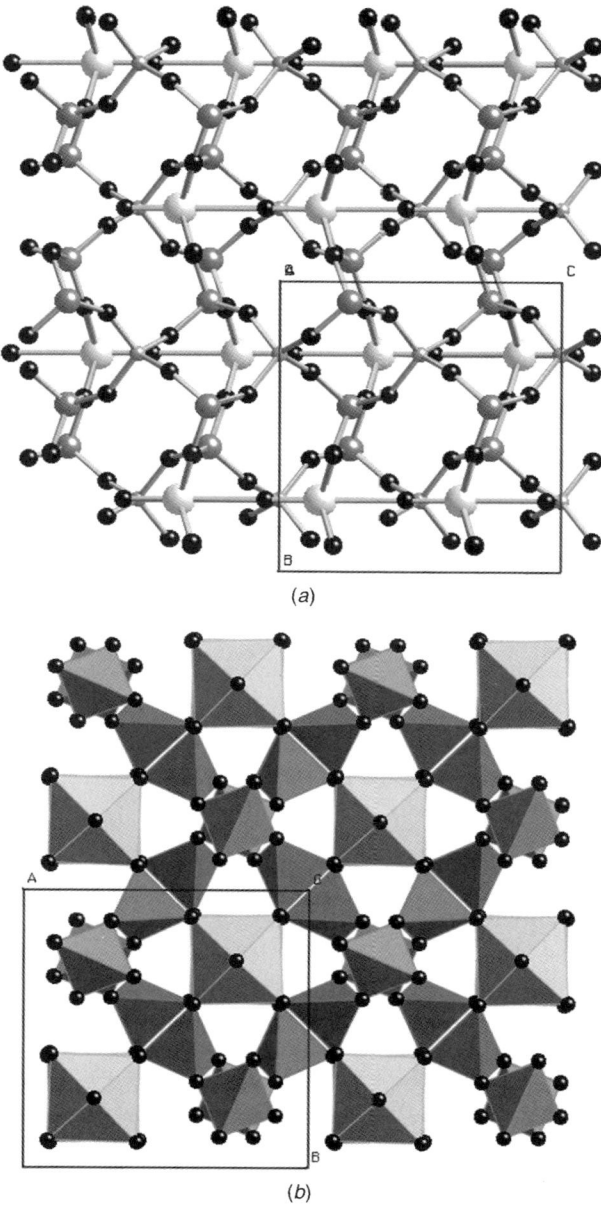

Figure 29. The structure of [(VO)Si(PO$_4$)$_2$]. (a) Ball-and-stick representation, viewed parallel to the a axis. (b) Polyhedral view parallel to the c axis.

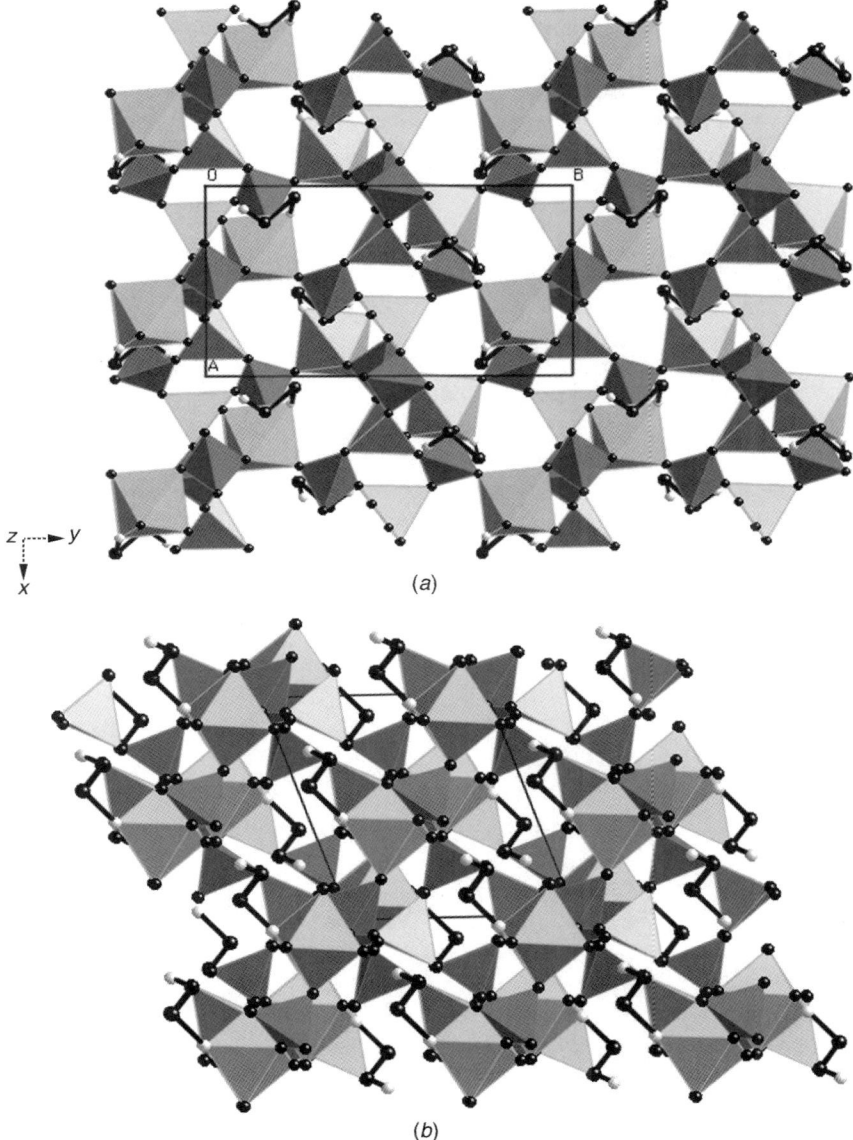

Figure 30. Two views of the structure of [H₃NCH₂CH₂NH₂][(VO)Al(PO₄)]. (*a*) Parallel to the *c* axis. (*b*) parallel to the *b* axis.

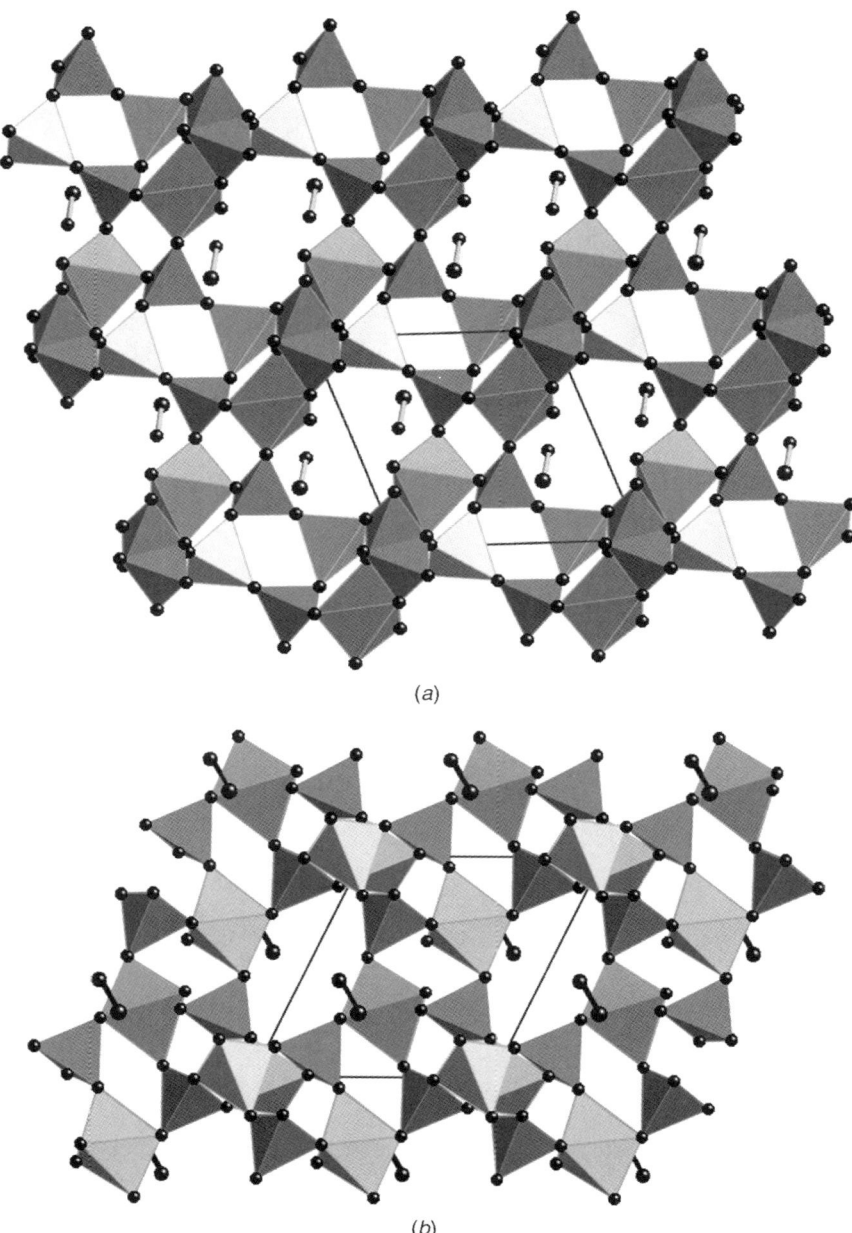

Figure 31. The structure of (MeNH$_3$)[(VO)Al(PO$_4$)$_2$] projected down the a (a) and c axes (b).

provides an integral connection to the framework as well as acting as a cavity occupant, allowing for the commonly diprotonated cation to be singly protonated and able to fit into the same sized cavity that the smaller cations occupy in the analogous structures.

The structure of $(MeNH_3)[(VO)Al(PO_4)_2]$ may be differentiated from the other members of this family by the absence of a six donor group to the V(IV) site. Consequently, the vanadium centers in this case exhibit square pyramidal geometry, rather than octahedral. However, the similar topology of the Al/V/P/O phases is evident in Figure 31, which is a view projected down [100]. The structures can be envisioned as having conceptual "layers" that lie in planes parallel to (100). Views perpendicular to these layers show the corner-sharing connection between the vanadium octahedra, aluminum tetrahedra, and phosphorus tetrahedra in Cs^+, Rb^+, and $H_2NCH_2CH_2NH_3^+$ species and the corner-sharing connection between the square pyramidal vanadium atoms, tetrahedral aluminum, and phosphorus atoms in the $MeNH_3^+$ phase.

The vanadium gallophosphates are isostructural with the vanadium aluminophosphates, $Rb(VO)(H_2O)Al(PO_4)_2$ and $Cs(VO)(H_2O)Al(PO_4)_2$. The replacement of the Al^{3+} ion with the larger Ga^{3+} ion does not perturb the structure, and results in a minimal increase in volume per oxygen atom. Vanadium gallophosphate phases have been shown to possess heterogeneous catalytic activity with respect to ammoxidation of toluene and benzonitrile (161).

4. Vanadium Borophosphates

While a large number of borate mineral structures are known (194), only two mineral borophosphates have been characterized, $Mg_3B_2P_2O_8(OH)_6 \cdot 6H_2O$ (195), and $Mn_3BPO_4(OH)_6$ (196). Synthetically prepared examples of metal borophosphates (197–207) are rare. One example of an open framework metal borophosphate $[H_3NCH_2CH_2NH_3][Co(OH)B_2P_3O_{12}]$ has been reported recently (208).

Although the cluster chemistry of oxovanadium borophosphates has witnessed recent expansion (209–211), the only example of an extended vanadium borophosphate structure is $[H_3NCH_2CH_2NH_3]_2[(VO)_5(H_2O)\{O_3POB(O)_2OPO_3\}_2] \cdot 1.5H_2O$ (164). The structure of this material consists of an open 3D framework constructed from vanadium square pyramids and octahedra, in combination with phosphorus and boron tetrahedra, providing channels occupied by the diammonium cations, $(H_3NCH_2CH_2NH_3)^{2+}$, as shown in Figure 32. The framework structure is constructed from three simple building blocks: binuclear units of edge-sharing vanadium square pyramids, isolated vanadium octahedra, and $\{O_3POB(O)_2OPO_3\}^{7-}$ borophosphate units. The binuclear units adopt an anti configuration of the vanadyl groups with respect to the $\{V_2O_2\}$ bridging group, with both vanadium centers in the V(IV) oxidation state. As illustrated in Figure 33, adjacent binuclear units are linked through

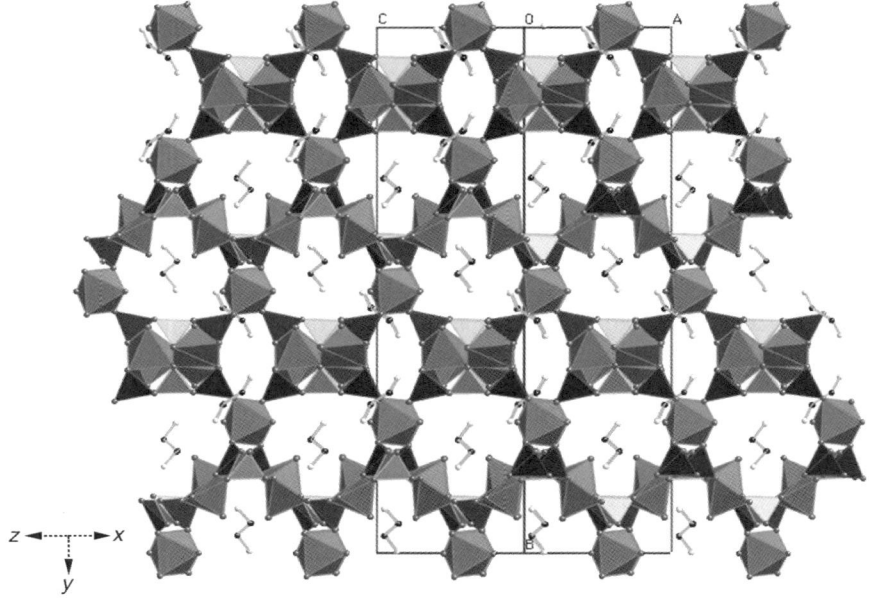

Figure 32. A polyhedral view of the structure of [H$_3$NCH$_2$CH$_2$NH$_3$]$_2$[(VO)$_5$(H$_2$O){O$_3$POB(O)$_2$OPO$_3$}$_2$] • 1.5H$_2$O, showing the cavities occupied by the H$_3$NCH$_2$CH$_2$NH$_3^{2+}$ cations.

{O$_3$POB(O)$_2$OPO$_3$}$^{7-}$ units, which consist of a central {BO$_4$} tetrahedron corner sharing with two {PO$_4$} tetrahedra. The central {BO$_4$} group contributes a bridging oxygen to each of two neighboring binuclear {V$_2$O$_{10}$} moieties, while each phosphate bonds to a vanadium site on each of two neighboring binuclear units. In this fashion, each {O$_3$POB(O)$_2$OPO$_3$}$^{7-}$ group exhibits three point attachments to each of two binuclear {V$_2$O$_{10}$} sites. This core motif

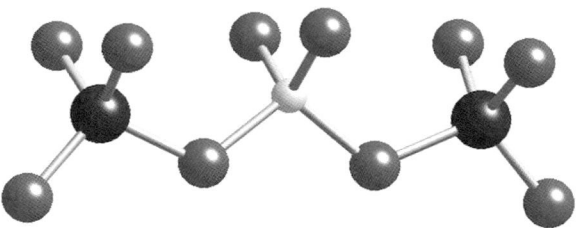

Figure 33. The {O$_3$POB(O)$_2$OPO$_3$}$^{7-}$ unit of [H$_3$NCH$_2$CH$_2$NH$_3$]$_2$[(VO)$_5$(H$_2$O){O$_3$POB(O)$_2$OPO$_3$}$_2$] • 1.5H$_2$O.

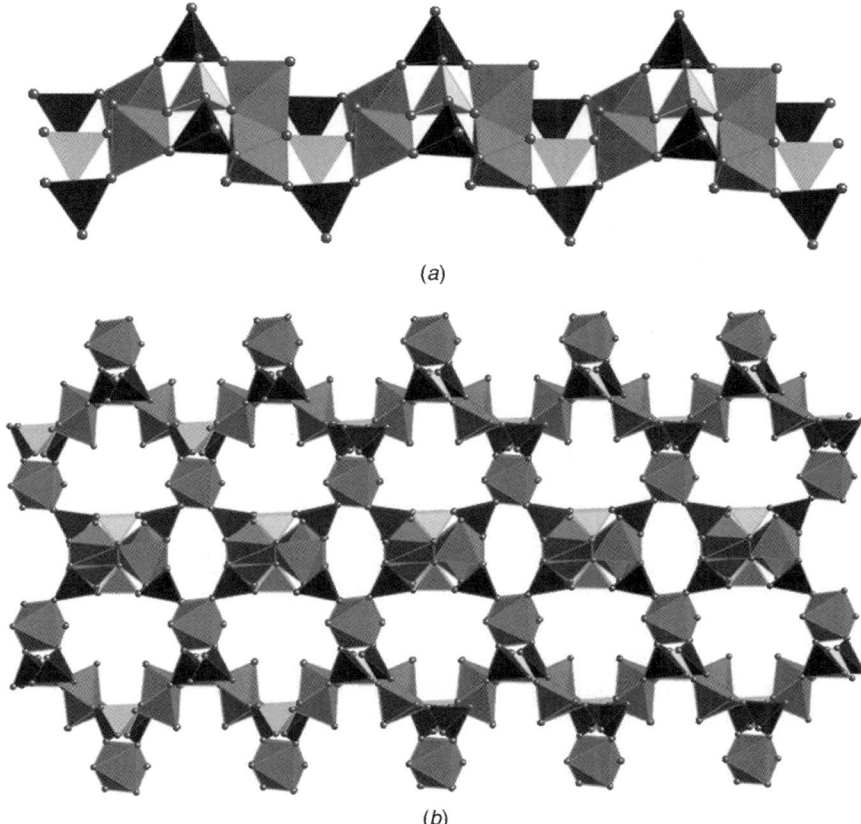

Figure 34. (a) The $\{(V_2O_{10})_2[O_3POB(O)_2OPO_3]\}_m^{n-}$ ribbons of $[H_3NCH_2CH_2NH_3]_2[(VO)_5(H_2O)\{O_3POB(O)_2OPO_3\}_2]\cdot 1.5H_2O$. (b) The linking of $\{(VO)_2[O_3POB(O)_2OPO_3]\}$ ribbons by $\{VO_6\}$ sites.

propagates as ribbons parallel to the *ab* plane, as shown in Figure 34. It is noteworthy that the structural unit of Figure 34 is identical to that observed for the *molecular anionic cluster* of $[H_3NCH_2CH_2NH_3]_2[Na(VO)_{10}\{HO_3POB(O)_2OPO_3H\}_5]\cdot 22.5H_2O$ (209), suggesting that the borophosphate cluster $\{H_nO_3POB(O)_2OPO_3H_n\}$ may represent a common structural motif in the chemistry of metal borophosphates and manifesting a topological relationship between the molecular and solid-state chemistries of these species, reminiscent of that of polyoxoanions and metal oxides (212).

The remaining oxygen donor on each phosphate of the $\{O_3POB(O)_2OPO_3\}^{7-}$ groupings bonds to the mononuclear octahedral vanadium centers. As shown in Figure 34, these {VO$_6$} sites serve to link adjacent {(VO)$_2$[O$_3$POB(O)$_2$OPO$_3$]} ribbons into layers running parallel to the *ac* plane. Furthermore, these octahedral centers link adjacent layers such that the ribbons of neighboring layers propagate at right angles to each other. In this fashion, each octahedral site provides connectivity through four phosphate oxygens to four ribbons, two in each of two adjacent layers. The remaining two coordination sites are occupied by an aqua ligand and a trans oxo group.

This complex polyhedral connectivity generates channels parallel to the *ac* plane occupied by the diammonium cations and water molecules of crystallization. The aqua ligands of one vanadium site also project into these channels. The cations are locked into position by strong multipoint hydrogen bonding to oxygen atoms of the framework, as indicated by N $\cdots$ O distances in the 2.75–2.95-Å range.

5. Vanadium Phosphate Fluorides

Fluoride may adopt a number of roles in the hydrothermal synthesis of phosphate-based materials. In addition to its mineralizing effect, fluoride may have a catalytic role, as manifested in the synthesis of AlPO$_4$-14A, which requires fluoride, although it is not incorporated into the framework (213). However, the most extensive use of fluoride has been in the synthesis of new aluminophosphate and gallophosphate architectures that directly incorporate the fluoride into the framework (214, 215). While fluoride incorporation into vanadium phosphate structures remains relatively unexplored, the phases studied to date reveal profound structural influences concomitant to incorporation of fluoride into the anionic scaffolding.

The prototypical structure of this family is provided by (H$_2$en)[(VO$_2$)$_2$(PO$_4$)F] (en = ethylenediamine)(166). The structure is constructed from zigzag chains of {VO$_5$F} octahedra and {VO$_4$F} square pyramids linked through {PO$_4$} tetrahedra into an open-framework 3D structure. The fluoride serves to bridge pairs of vanadium sites in the chains. The connectivity pattern produces 10-polyhedral connect rings {V$_6$P$_4$O$_{10}$} and smaller 8-polyhedral connect tunnels {V$_4$P$_4$O$_8$} to produce the 3D pore structure of the material.

The material [H$_2$N(CH$_2$CH$_2$)$_2$NH$_2$]$_{0.5}$[(VO)$_4$V(HPO$_4$)$_2$(PO$_4$)$_2$F$_2$(H$_2$O)$_4$] · 2H$_2$O (167) has a complex 3D structure with cavities occupied by the organic cations and water molecules of crystallization. The overall structure of the anionic framework may be described as V/P/F/O layers, containing octahedral V(IV) binuclear units as the fundamental building blocks, linked through V(III) octahedra, as shown in Figure 35. The V(IV) binuclear motifs are present as face-sharing octahedra shown in Figure 36, rather than the more common edge- or corner-sharing units. The coordination geometry about each V(IV) center is defined by a terminal oxo group, a bridging fluoride ligand, two phosphate oxygen donors in a bridging ligation mode, and two terminal phosphate oxygen

CRYSTAL CHEMISTRY OF ORGANICALLY TEMPLATED VANADIUM PHOSPHATES 501

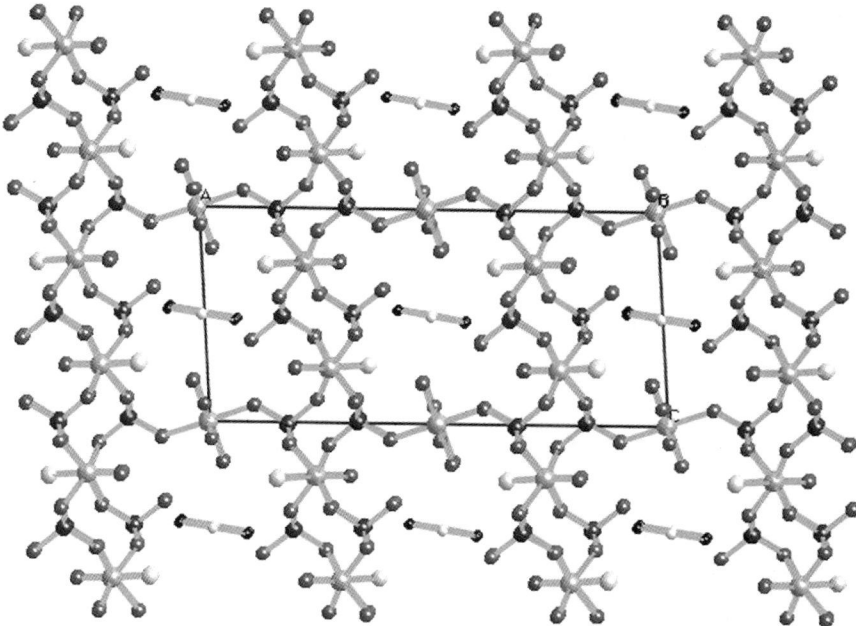

Figure 35. A view of the structure of $[H_2N(CH_2CH_2)_2NH_2]_{0.5}[(VO)_4V(HPO_4)_2(PO_4)_2F_2(H_2O)_4] \cdot 2H_2O$.

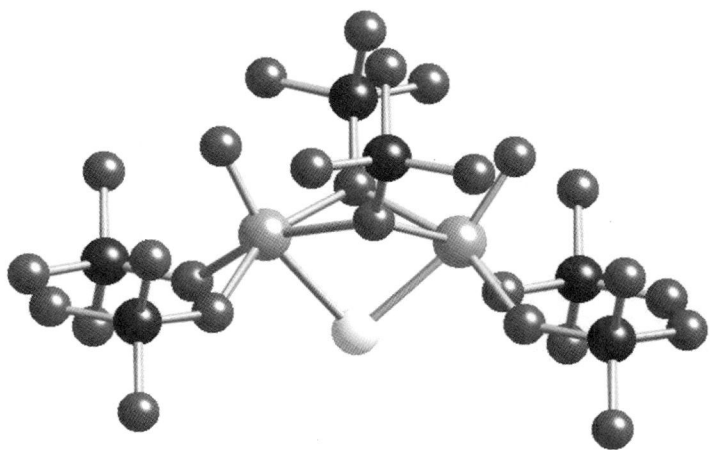

Figure 36. The confacial bioctahedral unit of $[H_2N(CH_2CH_2)_2NH_2]_{0.5}[(VO)_4V(HPO_4)_2(PO_4)_2F_2(H_2O)_4] \cdot 2H_2O$ and its environment.

donors. The bridging fluoride adopts as orientation trans to the terminal oxogroup. Each $\{V_2F_2O_8\}$ confacial bioctahedral unit is linked to four adjacent binuclear units in the layer through bridging phosphate groups.

There are two distinct phosphate units. One site contributes one oxygen donor to the bridge between two V(IV) sites of a binuclear unit. Two oxygen donors serve to bridge to each of two adjacent binuclear units in a monodentate fashion, while the fourth oxygen is protonated and pendant. The second site is present on a $(PO_4)^{3-}$ unit, which contributes one oxygen donor to the bridge between V(IV) centers of a binuclear unit and utilizes two oxygen donors to bridge to each of two adjacent vanadium motifs in the layer. However, the fourth oxygen is used to bond to the V(III) center that serves to bridge the V/P/F/O layers.

The V(III) sites adopt a tetragonally distorted, axially elongated octahedral geometry. Each V(III) site bonds to two phosphate oxygens, one from each of two adjacent layers, in a trans orientation. The equatorial plane is defined by four aqua ligands. The V(III) oxidation state assignment is consistent with valence sum calculations and with the charge requirements of the material. Thus, the gross structure is reminiscent of that of $(H_3NCH_2CH_2NH_3)_{2.5}$ $[V(VO)_8(OH)_4(HPO_4)_4(PO_4)_4(H_2O)_4] \cdot 2H_2O$, which exhibits a structure constructed from V/P/O layers linked by V(III) octahedra (see below).

The piperazinium cations and water molecules of crystallization occupy the void spaces between the layers and the V(III) buttresses. There is considerable hydrogen bonding interaction between the cations, waters of crystallization, V(III) bound aqua ligands, and the pendant {P–OH} groups of the layers. The presence of potentially labile aqua ligands suggests incipient vacant coordination sites within a micropore, whose properties appear to be dictated by a balance of hydrophilic–hydrophobic interactions that determine the manner in which organic–inorganic solids can crystallize.

As shown in Figure 37, the phosphate $K_2[(VO)_3(PO_4)_2F_2(H_2O)] \cdot H_2O$ adopts a similar 3D framework constructed from a layer motif identical to that of $[H_2N(C_2H_2)_2NH_2]_{0.5}[(VO)_4V(HPO_4)_2(PO_4)_2F_2(H_2O)_4] \cdot 2H_2O$ but with binuclear $\{(V^{IV}O)_2F_2O_6\}$ bridges replacing the $\{V^{III}O_6\}$ groups of VOPOF-1. The overall structure may be described in terms of edge- and face-sharing V(IV) binuclear units linked through phosphate tetrahedra into a 3D anionic framework, providing interconnecting tunnels occupied by K^+ cations and water molecules of crystallization.

An unusual feature of the structure is the presence of two distinct V(IV) binuclear units and of three different V(IV) environments. One binuclear motif, Type 2a, is similar to that observed for $[H_2N(C_2H_2)_2NH_2]_{0.5}[(VO)_4V(HPO_4)_2(PO_4)_2F_2(H_2O)_4] \cdot 2H_2O$, consisting of a confacial bioctahedral unit. Each V(IV) site of this unit coordinates to a terminal oxo group, a bridging fluoride ligand, two bridging phosphate oxygen atoms from each of two phosphate

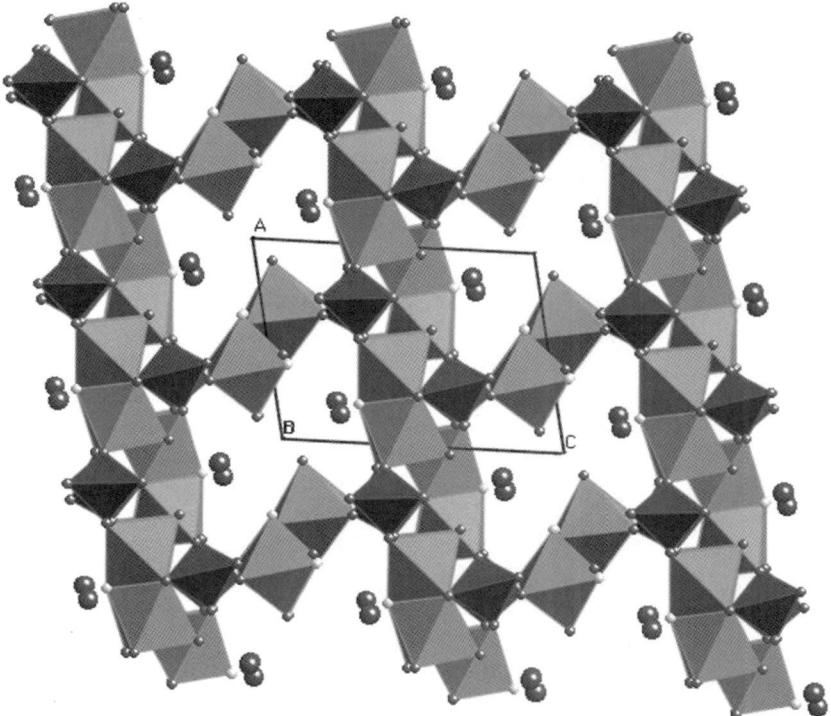

Figure 37. A polyhedral representation of the structure of $K_2[(VO)_3(PO_4)_2F_2(H_2O)] \cdot H_2O$, viewed parallel to the crystallographic b axis. The spheres represent the K^+ locations.

groups, and two oxygen donors in a terminally bonding mode from two additional phosphate groups, as illustrated in Figure 38. The second binuclear motif, Type 2b, consists of an edge-sharing unit, with each V(IV) geometry defined by two bridging fluorides, a terminal oxo group, two terminal phosphate oxygen donors, and an aqua ligand.

Note that the fluoride of the confacial bioctahedral unit is trans to both terminal oxo-groups, resulting in a symmetrical bridging interaction. In contrast, the fluoride bridges of the second binuclear site are trans to a single oxo group, producing an unsymmetrical bridge with V—F distances of 1.983(3) and 2.180(3) Å. The structural versatility exhibited by this limited set of V/P/F/O materials is quite remarkable.

Each phosphate group provides an oxygen atom to bridge the vanadium sites of the confacial bioctahedral unit. Of the remaining three oxygen atoms, one is deployed as a donor to a neighboring corner-sharing binuclear unit and two link

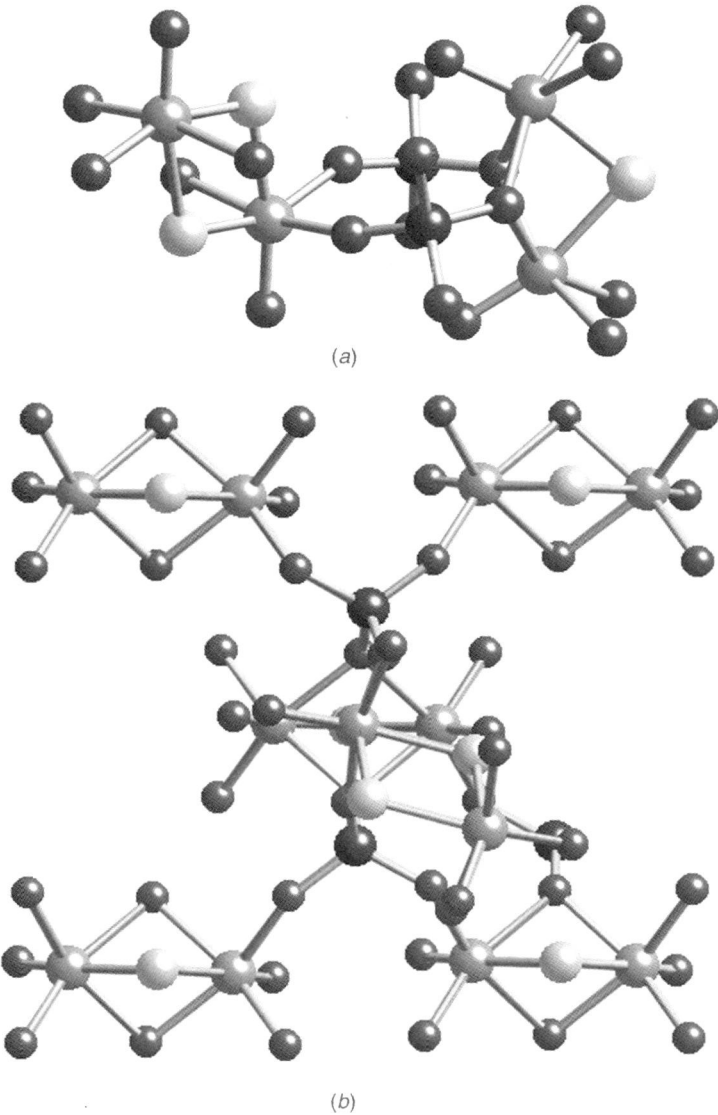

Figure 38. (a) A ball-and-stick representation of the confacial bioctahedral Type 2a subunit linked through bridging phosphate groups to the edge-sharing Type 2b subunit of $K_2[(VO)_3(PO_4)_2 F_2(H_2O)] \cdot H_2O$. (b) The linking of a Type 2a subunit to an adjacent Type 2b subunit and four Type 2a building blocks.

to each of two adjacent Type 2a units. In this fashion, as illustrated in Figure 38, the phosphorus tetrahedra serve to bridge each Type 2a binuclear unit to one Type 2b and four Type 2a neighbors.

When viewed along the crystallographic b axis as in Figure 37, this complex connectivity pattern produces channels defined by a 10 polyhedron ring. The water molecules of crystallization occupy these channels, and the aqua ligands of the Type 2b sites project into the cavity, resulting in significant hydrogen bonding between the various water molecules. When the structure is projected down the b axis, a second set of channels is revealed. These intersect the first set and the location of K^+ cations within these tunnels is clearly observed. A 12 polyhedral ring defines the perimeter of this latter set of tunnels. There is significant hydrogen bonding between the phosphate oxygens, aqua ligands and water molecules of crystallization. It is also noteworthy that the interiors of the channels in $K_2(VO)_3(PO_4)_2F_2(H_2O)] \bullet H_2O$ contain chains of fluoride bridged K^+ cations with K—F distances of 2.78 Å; this distance may be compared to that of 2.67 Å in KF.

IV. VANADIUM PHOSPHATE PHASES WITH CHARGE-COMPENSATING ORGANIC CATIONS

It has been amply demonstrated in the chemistry of zeolites that novel structures may be derived from the use of organocations as guests in the crystallizing product. The organic component effectively stabilizes the "void" regions within the inorganic structure by adopting charge compensating and space-filling roles (216). Such guest species are often referred to as templates or as structure-directing agents (SDA) (217). The successful exploitation of organic cations as structure-directing components of zeolitic structures suggested that similar roles may be adopted in the design of open-framework materials based on transition metal–oxide polyhedra and appropriate negatively charged subunits, in place of aluminum and silicon tetrahedra. Such materials would then conflate the size and shape selectivity of zeolites with the reactivity, electronic and magnetic properties of transition elements (218).

A general observation is relevant in the specific instance of the vanadium–phosphate–organic cation family of materials. In contrast to zeolitic systems, there are no examples to date of incorporation of a quaternary organoammonium cation into a V/P/O scaffolding. This observation is consistent with the importance of multipoint hydrogen bonding in directing the structures of these organic–inorganic composite materials and suggests that strongly hydrophobic components will not be readily entrained in this family of materials.

It is also noteworthy that the same organocation guest may be capable of crystallizing more than one V/P/O phase, an observation common to zeolitic

TABLE IV

Selected Structural Information for Vanadium Phosphates with Negatively Charged V/P/O Motifs and Change-Compensating Organic Cations

Compound	V/P/O Framework Structure	Vanadium Coordination and Linkages	Cell Parameters	Space Group	References
[(VO)(PO$_4$)(H$_2$NCH$_2$CH$_2$NH$_3$)]	Chain	Zigzag chain of corner-sharing {VO$_5$N} octahedra	17.593(4) 4.798(1) 9.037(2) $\beta = 114.21(3)$	$C2$	220
(CN$_3$H$_6$)[(VO)(HPO$_4$)(H$_2$PO$_4$)(H$_2$O)]·2H$_2$O	Chain	Isolated V(IV) octahedra	13.956(2) 11.717(2) 13.961(2) $\beta = 94.97(1)$	$C2/c$	221
[H$_3$NCH$_2$CH$_2$NH$_3$][V(H$_2$PO$_4$)(P$_2$O$_7$)]	Chain	Isolated V(III) octahedra	8.6139(4) 14.929(1) 9.3109(4) $\beta = 106.541(4)$	Cc	222
[H$_2$N(C$_2$H$_4$)$_2$NH$_2$][(VO)$_2$(PO$_4$)$_2$]	Layered	Isolated V(IV) trigonal bipyramids	8.786(2) 8.257(2) 8.566(2) $\beta = 111.07(3)$	$P2_1/c$	223
[H$_2$N(C$_2$H$_4$)$_2$NH$_2$]$_2$[(VO)$_3$(HPO$_4$)$_2$(PO$_4$)$_2$]·H$_2$O	Layered	Isolated V(IV) octahedra and square pyramids	14.631(3) 8.706(2) 17.635(4)	$Pna2_1$	223
[HNEt$_3$NH][(VO)$_3$(OH)$_2$(PO$_4$)$_2$]	Layered	Ribbons constructed from binuclear units of face sharing {VO$_6$} octahedra linked by corner sharing into infinite chains. Each binuclear unit corner-shares to a {VO$_5$} square pyramid	12.048(2) 6.3470(10) 20.249(4) $\beta = 105.30(3)$	$P2_1/n$	224

Compound	Structure	Description	Lattice parameters	Space group	Ref.
$[H_2NEt_2NH_2][(VO)_4(OH)_4(PO_4)_2]$	Layered	Binuclear units of edge-sharing $\{VO_5\}$ square pyramids	10.682(2) 8.991(2) 8.951(2) $\beta = 110.41(3)$	$P2_1/c$	224
$[HNEt_3NH]_2[(VO)_8(HPO_4)_3(PO_4)_4(OH)_2] \cdot 2H_2O$	Layered	One tetranuclear unit constructed of two edge-sharing V(IV) octahedra, each corner sharing to a square pyramid. Another tetranuclear unit of two binuclear corner sharing square pyramids linked by bridging phosphates	9.559(2) 8.840(2) 24.309(5) $\beta = 100.07(2)$	$P2/n$	224
$[(H_3NPrNHEt_2NH(PrNH_3)]$ $[(VO)_5(OH)_2(PO_4)_4] \cdot 2H_2O$	Layered	Binuclear units of edge-sharing square pyramids; isolated square pyramids	9.433(3) 17.799(3) 9.356(1) $\alpha = 103.83(1)$ $\beta = 91.80(2)$ $\gamma = 95.90(2)$	$P\bar{1}$	224
$(CN_3H_6)_2[(VO_2)_3(PO_4)(HPO_4)]$	Layered	Corner-sharing V(V) octahedra, giving hexagonal tungsten oxide-like layers	12.446(3) 7.287(2) 17.819(5) $\beta = 97.23(3)$	$C2/c$	225
$[(VO_2)(terpy)][(VO_2)_2(PO_4)]$	Layered	Chains of corner-sharing V(V) square pyramids; isolated $\{VO_2(terpy)\}^{1+}$ square pyramids as cations	12.315(1) 10.835(1) 29.181(2) $\beta = 101.62(2)$	$C2/c$	226
$[H_3N(CH_2)_3NH_3][(VO)_3(OH)_2(H_2O)_2(PO_4)_2]$	3D	Trinuclear unit of central $\{VO_5\}$ square pyramid corner-sharing with two $\{VO_6\}$ octahedra	10.507(2) 17.136(3) 8.451(2)	$Pnma$	227

(continues)

TABLE IV (Continued)

Compound	V/P/O Framework Structure	Vanadium Coordination and Linkages	Cell Parameters	Space Group	References
[H$_3$NCH$_2$CH$_2$NH$_3$][(VO)$_3$(H$_2$O)$_2$ (PO$_4$)$_2$(HPO$_4$)]	3D	Binuclear units of corner-sharing V(IV) octahedra; isolated V(IV) square pyramids	10.187(1) 10.241(1) 8.2137(7) $\alpha = 90.398(8)$ $\beta = 95.930(9)$ $\gamma = 117.325(7)$	$P\bar{1}$	228
[H$_3$N(CH$_2$)$_3$NH$_3$][(V$_3$O$_5$)(PO$_4$)$_2$(H$_2$O)$_2$]	3D	Trinuclear unit of central V(IV) square pyramid, corner sharing to two V(V) octahedra	10.567(1) 16.970(2) 8.413(1)	$Pnma$	229
[H$_3$N(CH$_2$)$_3$NH$_3$]K[(VO)$_3$(PO$_4$)$_3$]	3D	Isolated square pyramids	9.047(2) 9.747(2) 10.288(2) $\alpha = 109.68(3)$ $\beta = 101.78(3)$ $\gamma = 98.11(3)$	$P\bar{1}$	230
[H$_2$NEt$_2$NH$_2$][(VO)$_4$(H$_2$O)$_4$ (HPO$_4$)$_2$(PO$_4$)$_2$]	3D	One-dimensional chains of trans corner-sharing octahedra; isolated {VO$_6$} octahedra	7.025(1) 9.470(2) 16.570(3) $\beta = 96.03(3)$	Im	231, 232
[HNEt$_3$NH]K$_{1.35}$[(V$_5$O$_9$) (PO$_4$)$_2$] · xH$_2$O	3D	Cross-shaped pentanuclear units of edge-sharing V(IV)/V(V) square pyramids	26.247(3)	$I\bar{4}3m$	112
[H$_3$NCH$_2$CH$_2$NH$_3$]$_2$[H$_3$NCH$_2$CH$_2$NH$_2$] [V(H$_2$O)$_2$(VO)$_8$(OH)$_4$(HPO$_4$)$_4$ (PO$_4$)$_4$] · 2H$_2$O	3D	Binuclear units of corner-sharing V(IV) square pyramids and isolated V(III) octahedra	14.313(3) 10.151(2) 18.374(4) $\beta = 90.39(2)$	$P2_1/n$	233

[H$_3$NCH$_2$CH$_2$NH$_3$]$_4$[V(H$_2$O)$_2$(VO)$_6$ (OH)$_2$ (HPO$_4$)$_3$(PO$_4$)$_5$] · 3H$_2$O	3D	Binuclear units of corner-sharing V(IV) square pyramids and isolated V(IV) square pyramids and V(III) octahedra	20.674(2) 9.956(2) 23.694(1) $\beta = 101.154$	$C2/c$	234
[Me$_2$NH$_2$]K$_4$[(VO)$_{10}$(H$_2$O)$_2$(OH)$_4$ (PO$_4$)$_7$] · 4H$_2$O	3D	Pentanuclear unit of corner-sharing octahedra and tetrahedra	12.130 30.555	$P4_3$	7
[H$_3$NCH$_2$CH$_2$NH$_3$][(VO)$_2$(PO$_4$)$_2$ (H$_2$PO$_4$)]	3D	Isolated V(IV) and V(V) octahedra	8.891(3) 15.971(4) 18.037(5)	$Pc2_1n$	235

systems also (219). Consequently, a systematic evaluation of favorable guest–host interactions must be undertaken in order to elaborate a more rational approach to the design of these materials.

In discussing the structural chemistry of the V/P/O organocation family of materials, we will proceed in order of increasing dimensionality of the V/P/O substructure. Within a subfamily, the materials will be further classified according to the nuclearity of the vanadium sites or clusters embedded within the extended structure (Table IV).

A. One-Dimensional Structures

The paucity of 1D structures of the V/P/O family is not surprising in view of the tendency of vanadium sites to bridge via V—O—V bonds and of phosphate polyhedra to adopt multiple bridging roles. These factors encourage spatial extension of the structure in 2D or 3D, rather than low dimensionality chain or ribbon structures.

An exception to this common trend is the 1D phase [(VO)(PO$_4$)(H$_2$NCH$_2$CH$_2$NH$_3$)] (168), shown in Figure 39. The structure consists of 1D ribbons constructed from V(IV) octahedra and PO$_4^{3-}$ tetrahedra. The coordination about the vanadium centers is defined by a terminal oxo group, four phosphate oxygen donors from three PO$_4^{3-}$ groups, and a monodentate

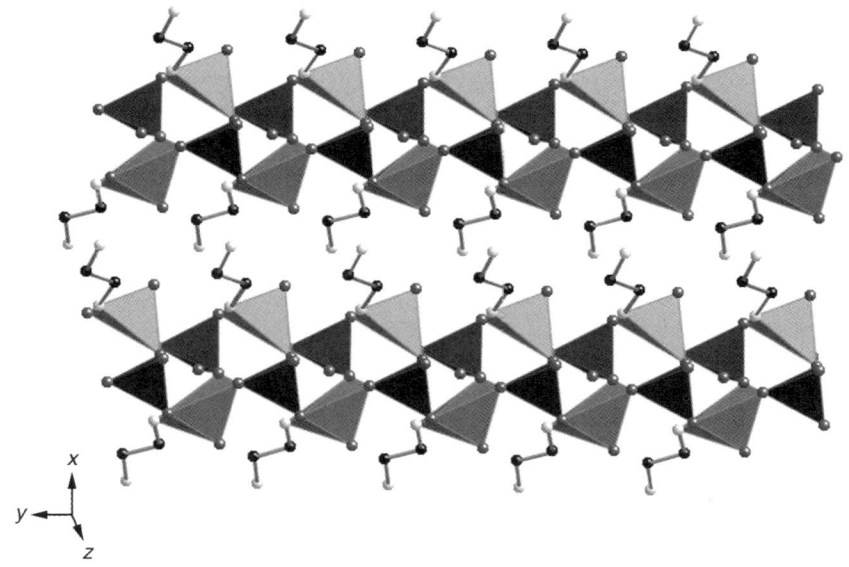

Figure 39. The structure of [(VO)(PO$_4$)(H$_2$NCH$_2$CH$_2$NH$_3$)].

$(H_2NCH_2CH_2NH_3)^+$ ligand. Each vanadium octahedron shares corners in a cis disposition with two adjacent vanadium octahedra to produce a zigzag polyhedral chain, which when viewed along the direction of propagation reveals a ribbon two octahedra in width. Each phosphate bridges three vanadium centers through three of four oxygen atoms. While two of the coordinated phosphate oxygen atoms chelate to a single vanadium center, the third bridges two vanadium sites. Curiously, the pendant site is unprotonated, {P=O}. The coordinated $(H_2NCH_2CH_2NH_3)^{1+}$ ligands project outward from the V/P/O ribbon, with the $-NH_3^+$ groups participating in strong hydrogen bonding to adjacent chains. The noteworthy feature of the structure is the presence of the organic component as both ligand and charge-compensating cation. It would appear that direct incorporation of the organic component into the covalent scaffolding of the solid can passivate the additional metal site with respect to forming additional V—O—V or V—O—P linkages and consequently higher dimensionality materials. Such passivation may reflect the blocking of coordination sites through coordinate covalent bonding or through the steric requirements of the ligated component. This observation suggests that the organic subunit may be introduced as a ligand, rather than as a charge-compensating group, and that the overall structure will reflect the geometric constraints of the ligand. Consequently, the ligand may be tailored to effect the design of the solid-phase material.

A second example of a 1D structure is provided by $(CN_3H_6)[(VO)(HPO_4)(H_2PO_4)(H_2O)] \cdot 2H_2O$ (221), whose structure is shown in Figure 40. In contrast to the structure of $[(VO)(PO_4)(H_2NCH_2CH_2NH_3)]$, which exhibits a V—O—V linked chain, the vanadium octahedra of $(MeH_6)[(VO)(HPO_4)(H_2PO_4)(H_2O)] \cdot H_2O$ are isolated. Each vanadium site is coordinated to a terminal oxo group, an aqua ligand trans to the oxo group, and four oxygen donors from four η^2, μ^2-bridging $(HPO_4)^{2-}$ or $(H_2PO_4)^-$ groups. The role of strong multipoint

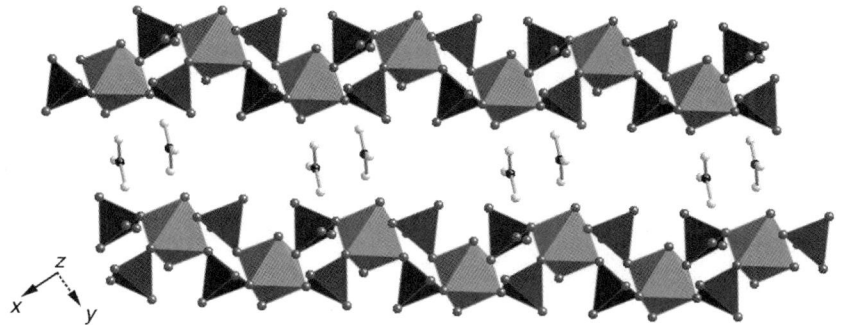

Figure 40. A view of the structure of $(CN_3H_6)[(VO)(HPO_4)(H_2PO_4)(H_2O)] \cdot H_2O$.

hydrogen bonding is evident in the networks of contacts between guanidium protons, coordinated water, phosphate, and water of crystallization.

B. Two-Dimensional Structures

The prototypical 2D network structure for the organically templated vanadium phosphates is provided by $[H_2N(CH_2CH_2)_2NH_2][(VO)_2(PO_4)_2]$ (223). The structure contains a $\{VOPO_4\}$ framework of corner-sharing $\{VO_5\}$ polyhedra and $\{PO_4\}$ tetrahedra. As shown in Figure 41, the vanadium polyhedra adopt a distorted triangular bipyramidal geometry in which each vanadium center is connected to four phosphate tetrahedra by corner sharing, while the fifth points toward the interlamellar region. Four out of the five V—O distances in the $\{VO_5\}$ polyhedra are similar (1.92–2.03 Å) and typical of V—O bond lengths, whereas this fifth one is much shorter (1.60 Å), consistent with a V=O bond. The presence of these triangular bipyramidal centers, as opposed to octahedral centers, causes distortions or puckerings of the $VOPO_4$ layer, resulting in the unusual layered structure of $[H_2N(CH_2CH_2)_2NH_2][(VO)_2(PO_4)_2]$. Figure 41 also shows the interlayer region, which exhibits an atom to atom (O $\cdots$ O) distance of 7.8 Å and is occupied by piperazinium dications.

Prepared under conditions analogous to those used to isolate $[H_2N(CH_2CH_2)_2NH_2][(VO)_2(PO_4)_2]$, the phosphate $[H_2N(CH_2CH_2)_2NH_2][(VO)_3(HPO_4)_2(PO_4)_2] \cdot H_2O$ (223) is also a layered compound containing alternating organic–inorganic regions. In contrast, however, to the $\{VO_5\}$ and $\{PO_4\}$ building blocks of $[H_2N(CH_2CH_2)_2NH_2][(VO)_2(PO_4)_2]$, $[H_2N(CH_2CH_2)_2NH_2][(VO)_3(HPO_4)_2(PO_4)_2] \cdot H_2O$ has V/O/P layers constructed from $\{VO_6\}$ octahedra, $\{VO_5\}$ square pyramids, and $\{PO_4\}$, and $\{HPO_4\}$ tetrahedra. The five-coordinate vanadium site exhibits common square pyramidal geometry with its basal positions defined by oxygen donors from each of four adjacent phosphate groups and its apical position by a terminal oxo group, while the six-coordinate vanadium sites associated with it are extremely distorted. This observation is a consequence of the polyhedral environment of the $\{VO_6\}$ sites that are defined by three oxygen donors from each of three adjacent phosphates (the terminal oxide group and two oxygen donors from a bidentate phosphate unit) giving rise to an edge-sharing interaction between the $\{VO_6\}$ and $\{PO_4\}$ polyhedra that is unusual in the chemistry of solid V/O/P phases. The layered structure of phosphate $[H_2N(CH_2CH_2)_2NH_2]_2[(VO)_3(HPO_4)_2(PO_4)_2] \cdot H_2O$ is depicted in Figure 42.

The V/P/O layers of $[H_2N(CH_2CH_2)_2NH_2][(VO)_2(PO_4)_2]$ and $[H_2N(CH_2CH_2)_2NH_2][(VO)_3(HPO_4)_2(PO_4)_2]$ are structurally related to the layer section of $MOXO_4$ (M = V, Nb, Ta, Mo; X = P, As, S, Mo) type compounds. The isotypic $MoOPO_4$ (236) and α_{II}-$VOPO_4$ provide the structural prototypes. However, $[H_2N(CH_2CH_2)_2NH_2][(VO)_2(PO_4)_2]$ exhibits a distorted network compared to

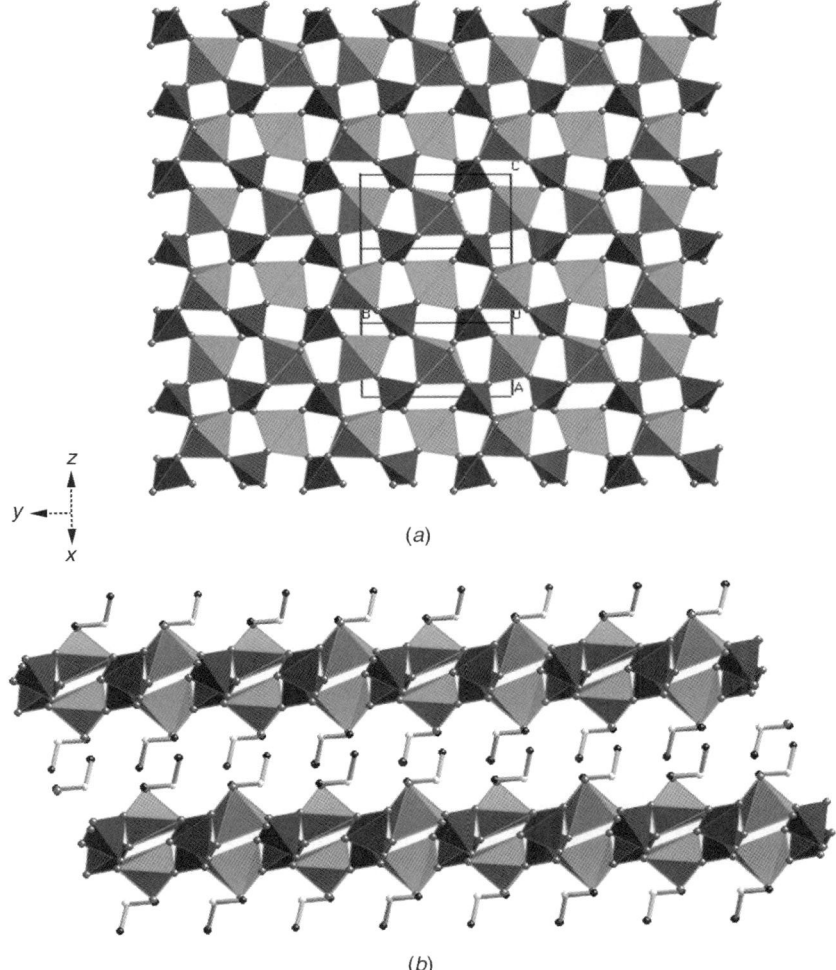

Figure 41. (a) A polyhedral representation of the V–P–O network of [H$_2$N(CH$_2$CH$_2$)$_2$NH$_2$][(VO)$_2$(PO$_4$)$_2$]. (b) The stacking of layers in [H$_2$N(CH$_2$CH$_2$)$_2$NH$_2$][(VO)$_2$(PO$_4$)$_2$], showing the interlamellar location of the cations.

[VO(PO$_4$)], as a consequence of the trigonal bipyramidal geometry at the vanadium centers, in contrast to the more common square pyramidal structure observed in [VO(PO$_4$)]. In the case of the {(VO)$_3$(HPO$_4$)$_2$(PO$_4$)$_2$} layer of the second phase, the V/P ratio of 3:4 for this composition, rather than the V/P

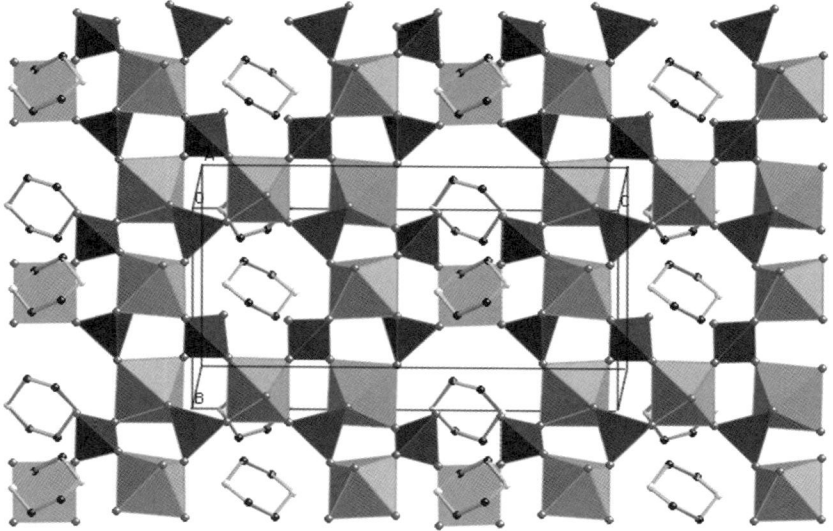

Figure 42. A polyhedral representation of the V/P/O network of $[H_2N(CH_2CH_2)_2NH_2]_2$ $[(VO)_3(HPO_4)_2(PO_4)_2] \cdot H_2O$.

ratio of 1:1 in the prototypical $[VOPO_4]$ results in a missing octahedral site in the layer giving rise to a "defect" structure.

It is apparent that minor changes in reaction conditions for the synthesis of $[H_2N(CH_2CH_2)_2NH_2][(VO)_2(PO_4)_2]$ and $[H_2N(CH_2CH_2)_2NH_2][(VO)_3(HPO_4)_2(PO_4)_2]$ result in distinct structures incorporating the same organic structure-directing components. This structural complexity reflects the variability of coordination numbers and polyhedral type associated with the vanadium center, as well as the flexibility of connectivity patterns to various phosphate polyhedra in different protonation states, $\{H_nPO_4\}^{(3-n)-}$.

This structural versatility is clearly manifested in phases exhibiting vanadium oxide clusters of varying nuclearities imbedded within the layer structures of the phosphate phases. An example of such structural submotifs is provided by $[H_2N(CH_2CH_2)_2NH_2][(VO)_4(OH)_4(PO_4)_2]$ (172). The fundamental structural motif associated with the V/P/O planes in this layered vanadium phosphate is a binuclear unit of edge-sharing $\{VO_5\}$ square pyramids (shown in Figure 43). The coordination at each vanadium center in this unit is defined by a terminal oxo group, two oxygen donors from each of two μ_3-phosphate groups, and two bridging hydroxyl groups. This binuclear unit is propagated in the layer construction by linkage through a μ_4-PO group, which serves to connect four adjacent $\{V_2O_{10}\}$ moieties. The structure of $[H_2N(CH_2CH_2)_2NH_2][(VO)_4(OH)_4(PO_4)_2]$

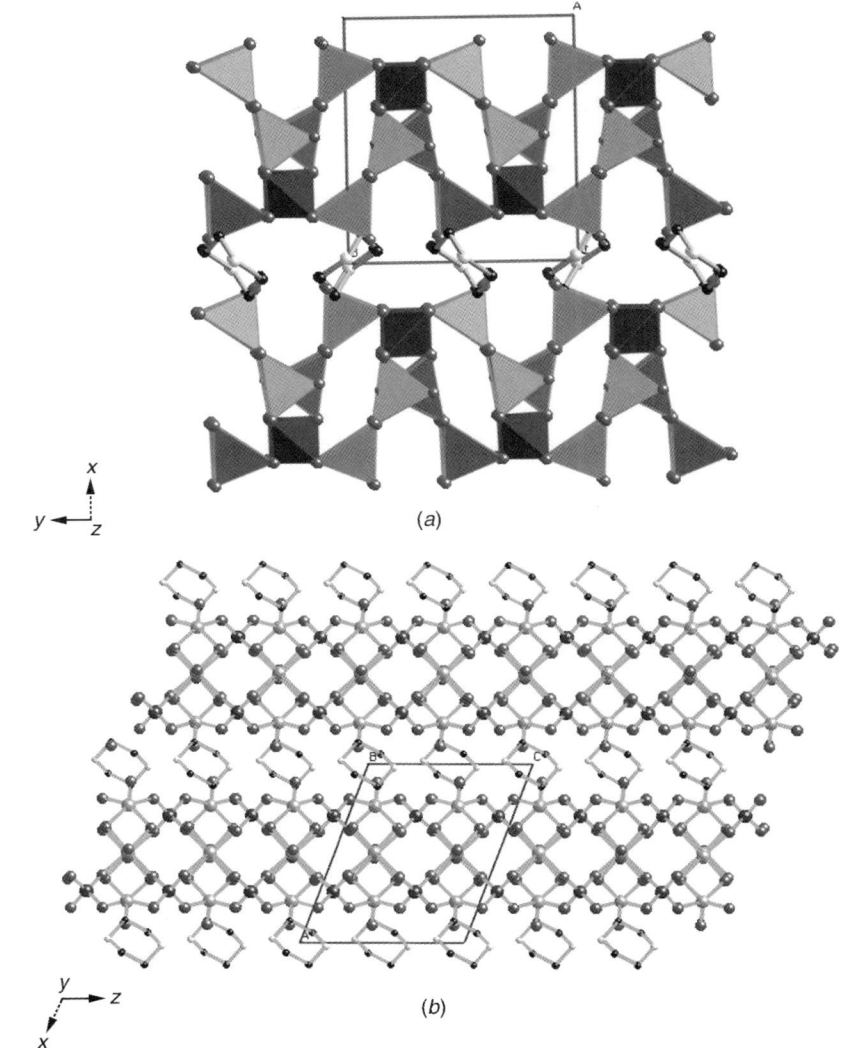

Figure 43. Two views of the structure of $[H_2N(CH_2CH_2)_2NH_2][(VO)_4(OH)_4(PO_4)_2]$. (a) The V–P–O layer. (b) the stacking of layers showing the interlayer occupancy by the organic cations.

is reminiscent of the layered $VOPO_4$ prototype and may be constructed from that of $VO(PO)_4$ by expansion of the plane through insertion of an $\{VO(OH)_2\}$ unit into a cis-$\{V(OPO_3)_2\}$ motif at each vanadium site in the layer. In addition to this expanded layer motif, $[H_2N(CH_2CH_2)_2NH_2][(VO)_4(OH)_4(PO_4)_2]$

contrasts with VOPO$_4$ through the introduction of organic cations that occupy the interlamellar regions as shown in Figure 43.

Unexpected structural complexity is exhibited by [(NH$_3$C$_3$H$_6$)NH(C$_2$H$_4$)$_2$NH (C$_3$H$_6$NH$_3$)][(VO)$_5$(OH)$_2$(PO$_4$)$_4$] • 2H$_2$O (224), which exhibits V/P/O layers defined by structural motifs based both on two distinct binuclear vanadium sites and a unique mononuclear center that serves as the hinge for the connectivity of polyhedra within the plane Figure 44. The first of the binuclear vanadium sites comprises a {V$_2$O$_{10}$} unit of edge-sharing square pyramids. Each vanadium atom in this unit is coordinated to a terminal oxo group, two oxygen donors form each of two phosphate groups, and two bridging hydroxyl groups. The second binuclear site consists of two {VO$_5$} square pyramids symmetrically

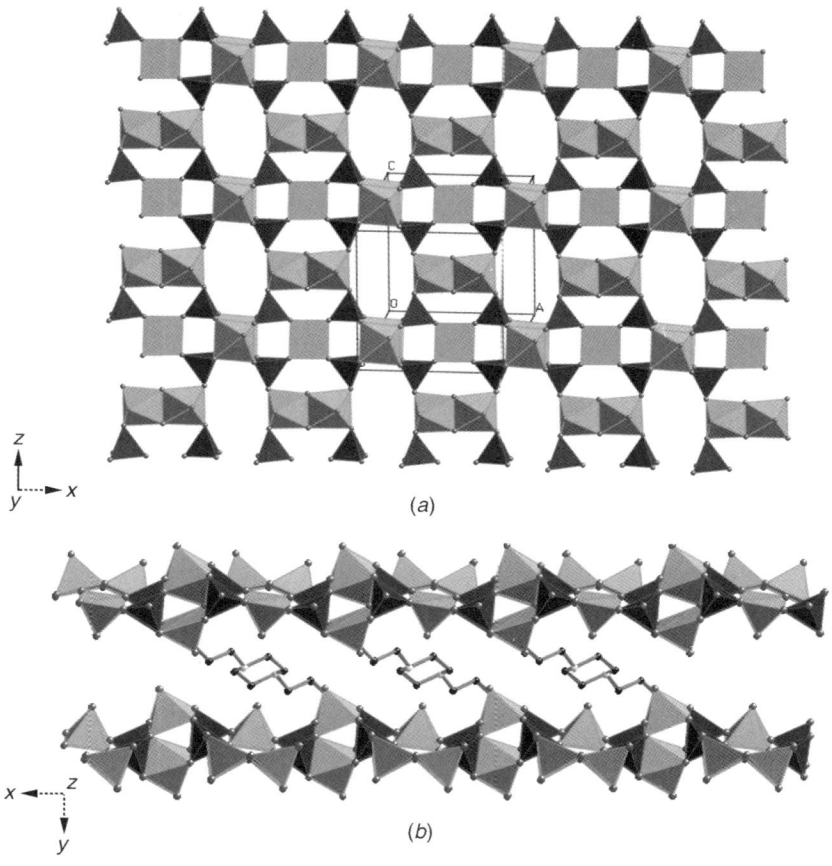

Figure 44. (a) A polyhedral representation of the V/P/O network of [(H$_3$NC$_3$H$_6$)NH(C$_2$H$_4$)$_2$NH(C$_3$H$_6$NH$_3$)] [(VO)$_5$(OH)$_2$(PO$_4$)$_4$] • 2H$_2$O. (b) A view of the layer stacking.

bridged in the bidentate mode by four {PO$_4$} groups to produce a copper acetate-type core {V$_2$(μ_2-PO$_4$)$_4$}. The final vanadium site consists of an isolated square pyramid, defined by a terminal oxo group and four oxygen donors from each of the four phosphate groups. The polyhedral view of Figure 44 demonstrates the complexity of this layered structure, with the organic cations and water molecules (not shown) occupying the interlamellar regions.

The phase [HN(CH$_2$CH$_2$)$_3$NH]$_2$[(VO)$_8$(HPO$_4$)$_3$(PO$_4$)$_4$(OH)$_2$] • 2H$_2$O (224) exhibits tetranuclear vanadium oxide clusters as structural motifs within the V/P/O network. These are best described as dimers of binuclear units. The first binuclear unit Figure 45 consists of two face-sharing {VO$_6$} octahedra, each in

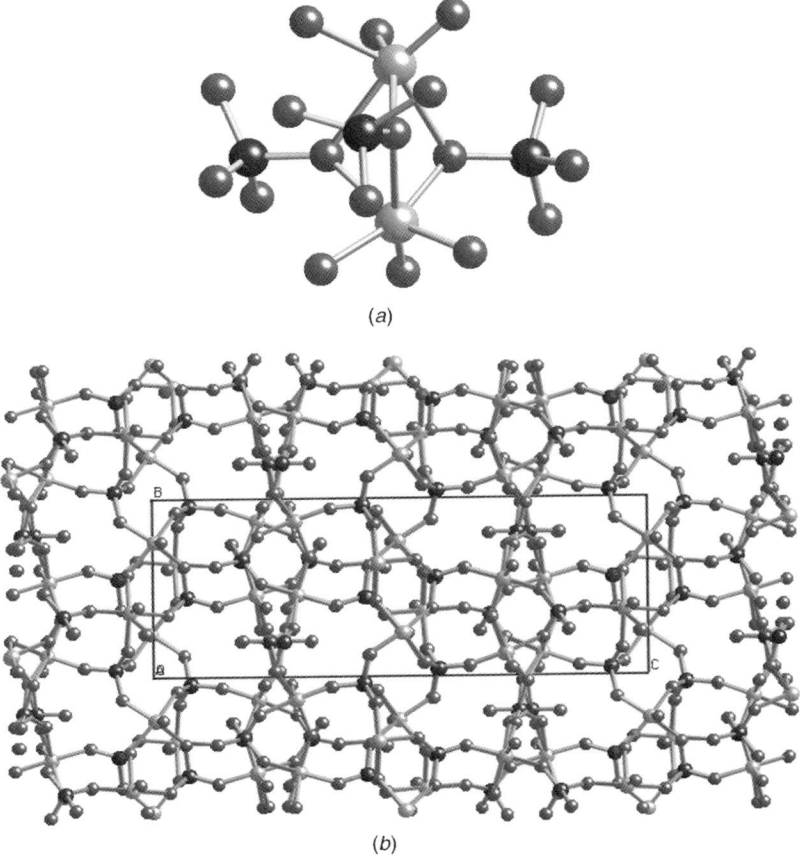

Figure 45. (a) The confacial bioctahedral subunit of [HN(CH$_2$CH$_2$)$_3$NH]$_2$[(VO)$_8$(HPO$_4$)$_3$(PO$_4$)$_4$(OH)$_2$] • 2H$_2$O. (b) A ball-and-stick representation of the layer.

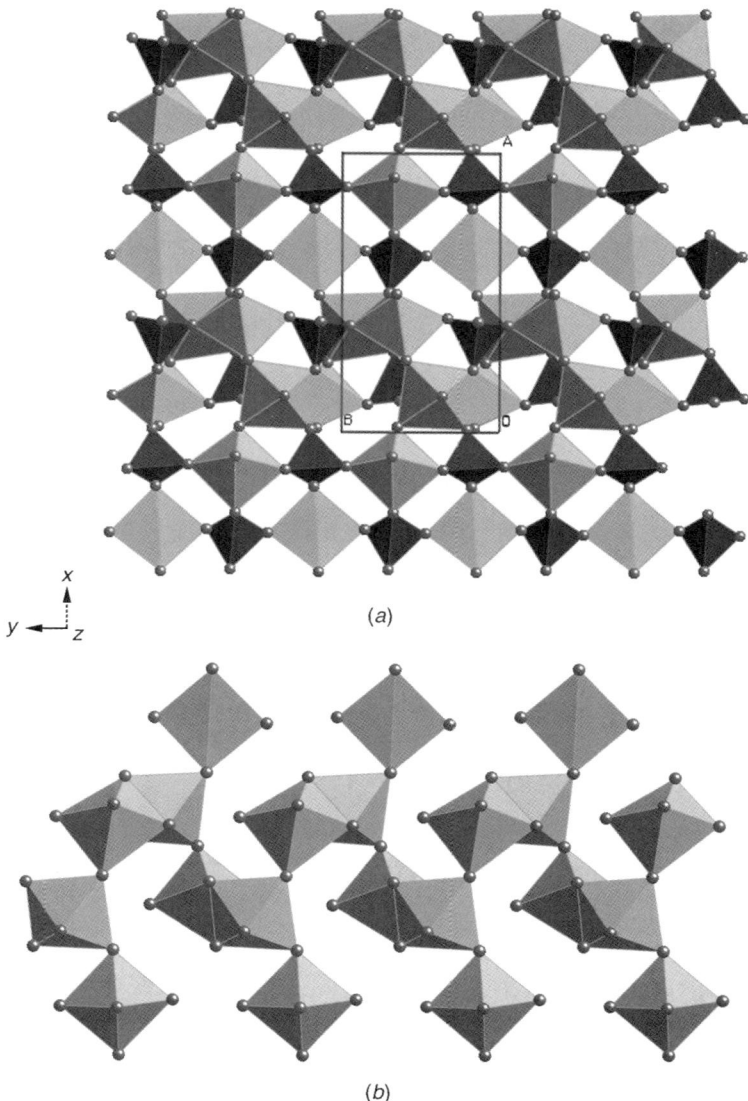

Figure 46. (a) A polyhedral representation of the V/P/O network of [HN(CH$_2$CH$_2$)$_3$NH][(VO)$_3$(OH)$_2$(PO$_4$)$_2$]. (b) The chain constructed of trinuclear units of face- and corner-sharing vanadium polyhedra.

turn linked via corner sharing through a bridging —OH group to a square pyramidal {VO_5} site. The common face of the vanadium octahedra in this unit consists of three oxygen donors from each of three neighboring phosphate groups, while the remaining octahedral sites are defined by a terminal oxo group, the bridging hydroxyl group, and the oxygen donor of a fourth phosphate group. The square-pyramidal center is coordinated by a terminal oxo group, a bridging hydroxyl group, and oxygen donors from each of three phosphate ligands. The second binuclear unit, consists of corner-sharing {VO_5} square pyramids linked by two symmetrically bridging phosphate groups. The unit comprises a {$V_4P_2O_6$} ring in which the vanadium sites exhibit square pyramidal geometry exclusively, with each vanadium atom bonded to a terminal oxo group, the bridging oxygen donor of an associated phosphate group, and three oxygen donors from the other neighboring phosphate groups. Infinite alternating ribbons of these binuclear units are linked together to form the layered structure shown in Figure 45, with the organic cations (not shown) occupying the interlamellar regions.

The structure of [$HN(CH_2CH_2)_3NH$][$(VO)_3(OH)_2(PO_4)$] is unique in exhibiting infinite chains of {V—O(H)—V} linked vanadium octahedra and {VO_5} square pyramids that are connected through phosphate tetrahedra into sheets, shown in Figure 46, that are separated into layers by organic cations (not shown). The fundamental structural motif in this phosphate consists of a trinuclear unit of face-sharing {VO_6} octahedra linked by corner sharing to a {VO_5} square pyramid (Fig. 46). The two six-coordinated vanadium sites exhibit distorted octahedral geometries, with three oxygen donors from each of three neighboring phosphate groups providing the common face. These binuclear units are connected through a hydroxyl linkage via corner sharing to produce an infinite chain of {VO_6} octahedra, with alternating corner-sharing and face-sharing connectivities, and fused at each of the vanadium sites to a pendant {VO_5} square pyramid.

An unusual example of a V/P/O network constructed from a vanadium oxide network substructure is provided by $(CN_3H_6)_2[(VO_2)_3(PO_4)(HPO_4)]$ (225). As shown in Figure 47, the vanadium oxide substructure consists of vertex-sharing {VO_6} octahedra linked into triangles and six polyhedra ring windows. The V–O–V bridges all exhibit a short V=O and a long V—O bond. Three adjacent V—O(P) bonds in each three ring are capped by a P—O(H) unit. Thus, the triply bridging {PO_4} and {HPO_4} groups project one vertex into the interlamellar region to form strong multipoint hydrogen bonds to the guanidinium cations. The octahedral linkage pattern of the vanadium oxide motif is identical to that observed for the hexagonal tungsten oxide type $M(VO_2)_3(SiO_3)_2$ (237) and to $M[(VO_2)_3(MePO_3)_2]$, discussed in Section V.B.

A novel application of the organic component is provided by the structure of [$(VO)_2$(terpy)][$(VO_2)_2(PO_4)$] (226). Whereas in all other cases discussed in this

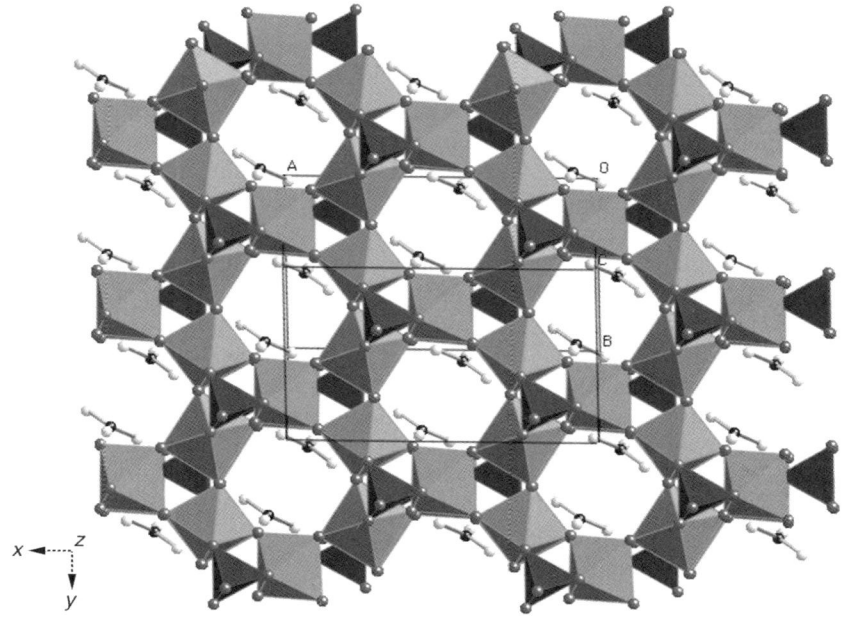

Figure 47. A polyhedral representation of the V/P/O network of $(CN_3H_6)_2[(VO_2)_3(PO_4)(HPO_4)]$, viewed parallel to the a axis. The organic cations are shown as ball-and-stick models occupying interlamellar positions above and below the six polyhedral rings of the layer.

section the organic moiety is present in its protonated form as the charge compensating cation, terpyridine functions as a ligand to the $\{V(V)O_2\}^{1+}$ cations that occupy the interlamellar regions between the $[(VO_2)_2(PO_4)]_n^{n-}$ layers of the phase. As shown in Figure 48, the 2D structure of the $[(VO_2)_2(PO_4)]_n^{n-}$ subunit consists of undulating 1D chains of corner-sharing $\{VO_5\}$ square pyramids linked by η^4, μ^4-$\{PO_4\}^{3-}$ tetrahedra into a covalently linked network. This connectivity generates eight polyhedra rings $\{V_6P_2O_8\}$, and the $\{VO_2(terpy)\}^{1+}$ cations are aligned with these cavities.

The structure of $[(VO_2)(terpy)][(VO_2)_2(PO_4)]$ suggests that the role of the organic component in solid-state vanadium phosphates may be expanded from that of charge compensating and space filling to ligation, either directly to the V/P/O skeleton or to a secondary transition metal center. This theme will be expanded upon in Section VI.

C. Three-Dimensional Structures

While layered V/P/O phases are common and often exhibit structural similarities to the network motif of $MOPO_4$ prototypes, V/P/O skeletal frameworks

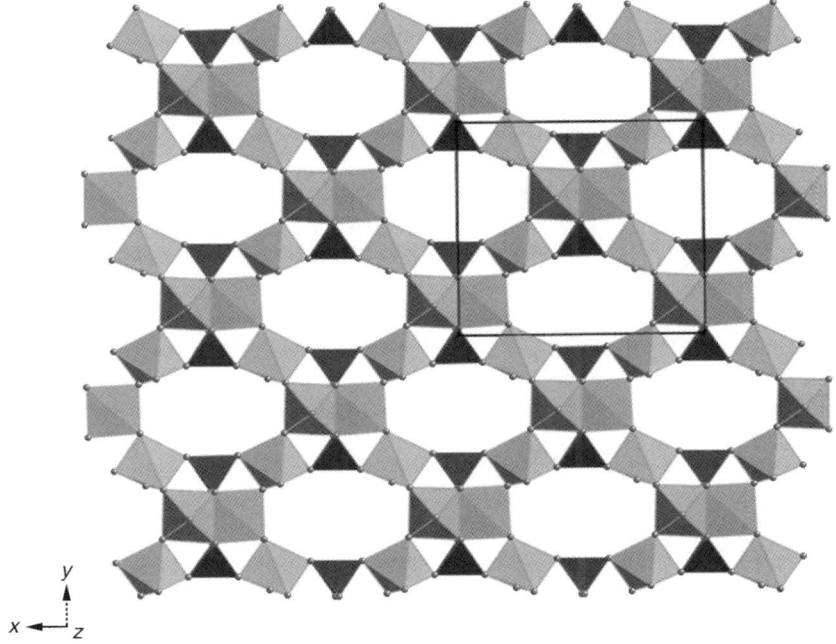

Figure 48. A polyhedral representation of the $[(VO_2)_2(PO_4)]_n^{n-}$ network of [(VO$_2$)(terpy)] [(VO$_2$)$_2$(PO$_4$)] viewed parallel to the crystallographic c axis. (terpy = terpyridine)

exhibit the structural flexibility to achieve 3D connectivities and thereby to entrain organic structure-directing cations within an anionic inorganic scaffolding.

Once again, a variety of vanadium oxide subunit nuclearities are encountered as building blocks in these framework structures. The phosphate [H$_3$NCH$_2$CH$_2$NH$_3$]K[(VO)$_3$(PO$_4$)$_3$] (178), is constructed from isolated corner-sharing {VO$_5$} square pyramids and phosphate tetrahedra, and employs {(VO)$_2$(μ^2-PO$_4$)$_2$} building blocks that aggregate in such a fashion as to produce rings containing 12 polyhedra: 6 vanadium square pyramids and 6 phosphorus tetrahedra in an alternating alignment. The structure of this phosphate viewed down the [111] cell diagonal is shown in Figure 49. In this figure, the vanadyl oxo groups of a given ring can be seen oriented in an alternating endocyclic–exocyclic pattern that is reversed in the next ring of the stack, resulting in the illusion of six vanadyl oxo groups directed toward the interior of the cavity. Adjacent rings are fused through bridging phosphate groups to produce a network of parallel tunnels in which a given ring shares structural elements with six adjacent rings.

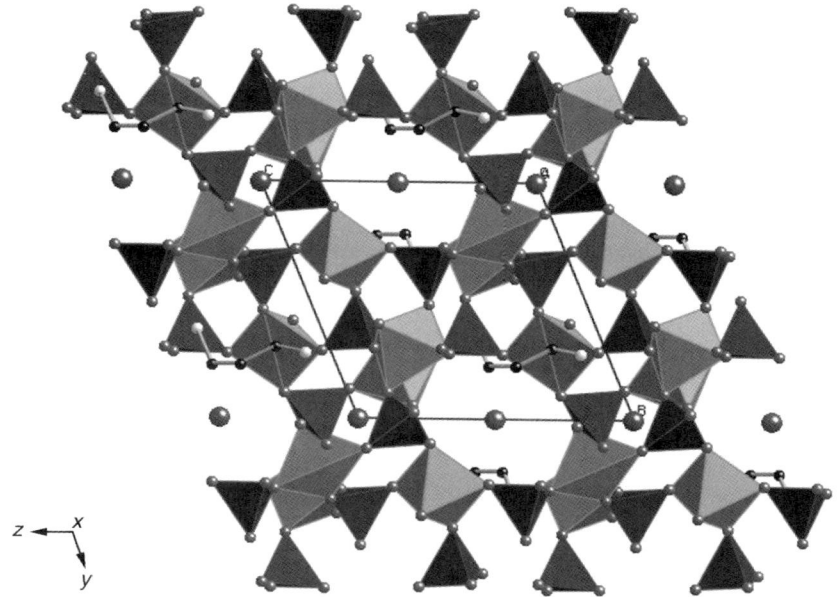

Figure 49. A view of the structure of [H$_3$NCH$_2$CH$_2$NH$_3$]K[(VO)$_3$(PO$_4$)$_3$].

This connectivity generates distinct tunnels which contain the (H$_3$NCH$_2$CH$_2$CH$_2$NH$_3$)$^{2+}$ and K$^+$ cations.

The mixed-valence phosphate phase [(H$_2$en)$_2$(Hen)][V(H$_2$O)$_2$(VO)$_8$(OH)$_4$(HPO$_4$)$_4$(PO$_4$)$_4$] • 2H$_2$O (233) exhibits both mononuclear and binuclear vanadium sites. The 18.4 × 5.6-Å elliptical channels are formed from undulating layers of corner sharing (through µ^2-OH groups) {VO$_5$} square pyramids and {HPO$_4$} and {PO$_4$} tetrahedra connected by {VO$_6$} octahedra, as seen in Figure 50. The fundamental building blocks of these layers are domed rings made up of eight polyhedra: four {VO$_5$} square pyramids, and two {HPO$_4$} and two {PO$_4$} tetrahedra. Each ring exhibits V=O and P—OH groups in an anti orientation and is surrounded by six other rings. In two of the neighboring rings, the V=O groups are oriented up and the P—OH groups down along the b axis (V=O up/P—OH down) with the other four rings having the arrangement V=O down/P—OH up. Within each layer, strips formed from rings with V=O up/P—OH down alternate with parallel strips with V=O down/P—OH up. This up–down connectivity of the domed rings gives rise to the undulations and large holes (6.2–7.2 Å) in the layers. The 3D structure of this phosphate with its unusual elliptical channels is formed by the connection of these layers through {V^{III}(H$_2$O)$_2$} centers. Other M(III) cations of the appropriate radius may be

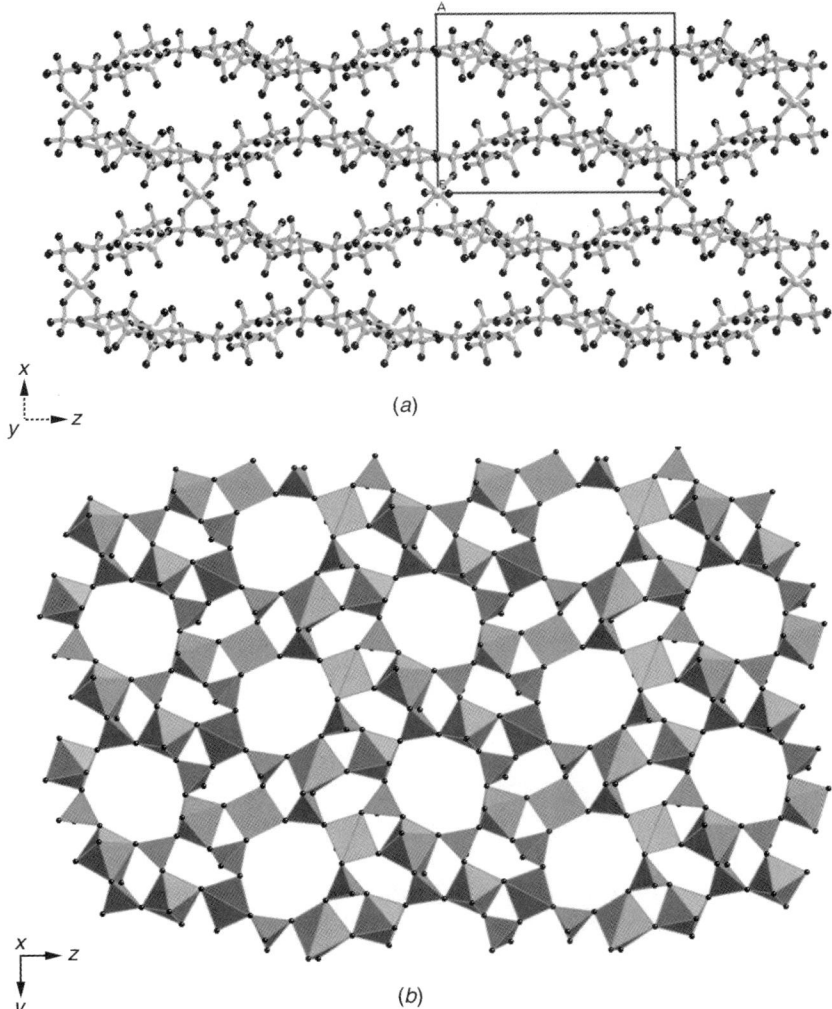

Figure 50. (a) A view of the structure of [(H$_2$en)$_2$(Hen)][V(H$_2$O)$_2$(VO)$_8$(OH)$_4$(HPO$_4$)$_4$(PO$_4$)$_4$] • 2H$_2$O parallel to the crystallographic b axis. The organic cations are not shown. (b) Polyhedral representation of the {(VO)$_8$(OH)$_4$(HPO$_4$)$_4$(PO$_4$)$_4$(H$_2$O)$_2$}$^{8-}$ network, viewed parallel to the a axis.

substituted for the layer connecting V(III) as demonstrated in [(H$_2$en)$_2$(Hen)][Fe(H$_2$O)$_2$(VO)$_8$(OH)$_4$ (HPO$_4$)$_4$(PO$_4$)$_4$] • 4H$_2$O (238).

The material [H$_3$NCH$_2$CH$_2$NH$_3$][(VO)$_3$(H$_2$O)$_2$(PO$_4$)$_2$(HPO$_4$)] (228) also exhibits both mononuclear and binuclear vanadium substructures. The structure

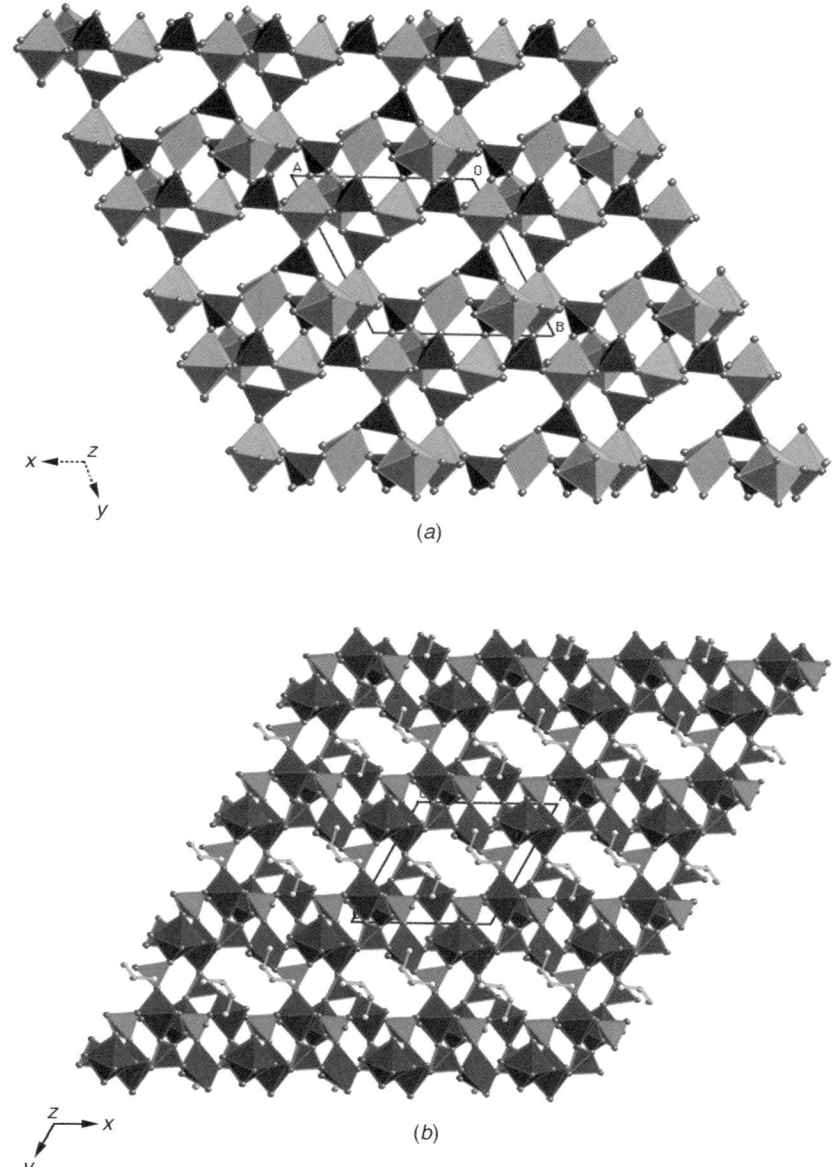

Figure 51. (a) Polyhedral representation of the structure of [H$_3$NCH$_2$CH$_2$NH$_3$] [(VO)$_3$(H$_2$O)$_2$ (PO$_4$)$_2$(HPO$_4$)], viewed along the c axis. (b) A view of the structure showing the locations of the organic cations.

consists of binuclear units of corner-sharing {VO$_6$} octahedra and square pyramidal {VO$_5$} units linked by phosphate tetrahedra, forming double extended polyhedral slabs in the (010) plane with these slabs connected by corner-sharing {HPO$_4$} tetrahedra to form a corner-sharing 3D cage structure about the diammonium cations. The structure, depicted in Figure 51, contains three crystallographically independent vanadium atoms, all in the +4 oxidation state. The coordination environment about the octahedral vanadium centers comprises either four phosphate oxygens, a vanadyl oxygen and a water molecule, or three phosphate oxygens, a water molecule, a remote bridging oxo group and a terminal oxygen. The coordination environment around the square pyramidal {VO$_5$} units consists of an apical vanadyl oxygen and four basal phosphate oxygens. These structural motifs combine to form an octahedral corner-sharing {VO$_5$—O—VO$_5$} dimer and an isolated square pyramidal {VO$_5$} unit connected by two tetrahedral {PO$_4$} groups and one tetrahedral {HPO$_4$} group through corner sharing. The connectivity of the {VO$_5$—O—VO$_5$} dimer and the isolated {VO$_5$} square pyramid forms double extended polyhedral layers in the (101) plane. The {HPO$_4$} tetrahedral units are then used to link the complex double extended polyhedra layers into a 3D structure through corner sharing with the {VO$_5$—O—VO$_5$} dimer and square pyramidal {VO$_5$} units along the b axis. As seen in Figure 51, the ethylenediammonium cations are inserted between the {HPO$_4$} tetrahedra in the [100] direction and between the extended double polyhedra layers, in a cage defined by a ring of eight tetrahedra, four octahedra, and four square pyramids all connected through corner sharing of the polyhedra.

Higher nuclearity {V$_x$O$_y$} subunits are quite common in the V/P/O framework structures. For example, the basic structural motif found in [H$_3$N(CH$_2$)$_3$NH$_3$][(VO)$_3$(OH)$_2$(H$_2$O)$_2$(PO$_4$)$_2$] (227) is built up from vanadium trimers and a single crystallographically unique phosphate group that create an anionic framework encapsulating the propanediammonium (PDA) cations. This motif is depicted in Figure 52 which shows the unit cell contents projected down [001]. The vanadium trimers, which consist of two distorted {VO$_6$} octahedra on the ends of the trimer and a square pyramidal {VO$_5$} located in the center, lie on the crystallographic mirror planes that are perpendicular to the b axis at $b = \frac{1}{4}$ and $\frac{3}{4}$. The central {VO$_5$} square is bridged to the {VO$_6$} octahedra via μ^2-OH groups, and the vanadium trimers are connected by phosphate groups into stacks that run parallel to [100].

The phase [Me$_2$NH$_2$]K$_4$[(VO)$_{10}$(H$_2$O)$_2$(OH)$_4$(PO$_4$)$_7$] • 4H$_2$O (7) consists of a 3D covalently bonded framework built up from {VO$_6$} octahedra, {VO$_5$} square pyramids, and {PO$_4$} tetrahedra. It crystallizes in the space group $P4_3$ (or its enantiomorph $P4_1$), and therefore the crystals are enantiomorphic and the unit cell contents are chiral. The fundamental building blocks of this phosphate are two structurally similar, crystallographically independent vanadium oxide

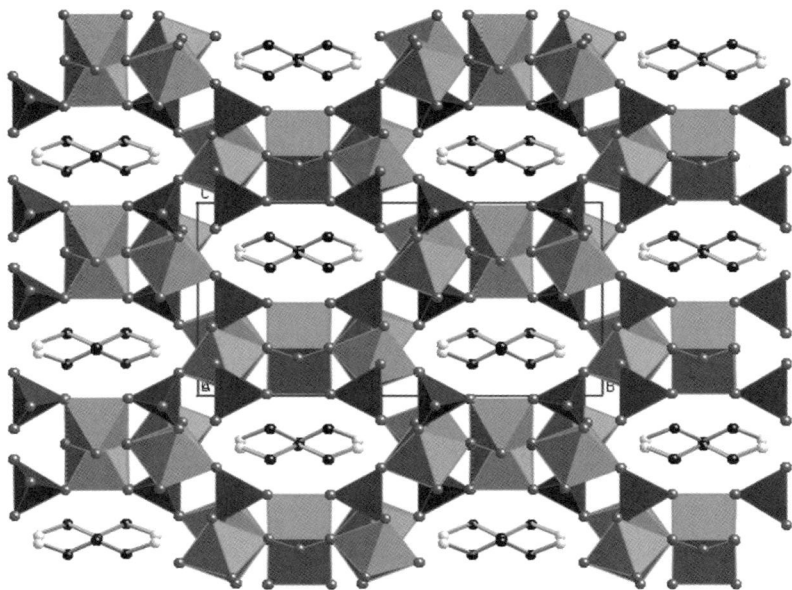

Figure 52. A view of the structure of [H$_3$N(CH$_2$)$_3$NH$_3$][(VO)$_3$(OH)$_2$(H$_2$O)$_2$(PO$_4$)$_2$], parallel to the crystallographic a axis.

pentanuclear clusters, one of which is shown in Figure 53. These clusters have a V—O—V backbone containing four V—O—V and two V—OH—V bonds. The connectivity is such that there is a central trimer of three {VO$_6$} octahedra, with the central octahedron sharing trans corners with the two outer octahedra. Each of the two outer octahedra of the trimer then share an edge with two {VO$_5$} square pyramids forming the clusters that are arranged so as to form spirals, with four clusters per spiral of unit cell length along [001]. The spirals, in turn, are intertwined to give the two strands of a double helix. Seven types of P^{5+} cations are also present in the unit cell of this phase, some serving to join the pentamers, some to connect the strands to one another to form the helix, and some to bond one double helix to another. The resulting connectivity generates cavities and a topologically unusual array of tunnels that contain the K$^+$ and Me$_2$NH$_2^+$ cations (Fig. 53).

A pentanuclear building block is also a prominent feature of the structure of [HNEt$_3$NH]K$_{1.35}$[(V$_5$O$_9$)(PO$_4$)$_2$] • xH$_2$O (112) (Fig. 54). However, the structure of the {V$_5$O$_9$(PO$_4$)$_{4/2}$} unit of this material is quite distinct from the pentanuclear unit of [Me$_2$NH$_2$]K$_4$ [(VO)$_{10}$(H$_2$O)$_2$(OH)$_4$(PO$_4$)$_7$] • 4H$_2$O. As shown in Figure 55, the vanadium square pyramids form an unusual cross-shaped V$_5$ pentanuclear unit, consisting of a central {VO$_5$} square pyramid sharing each of

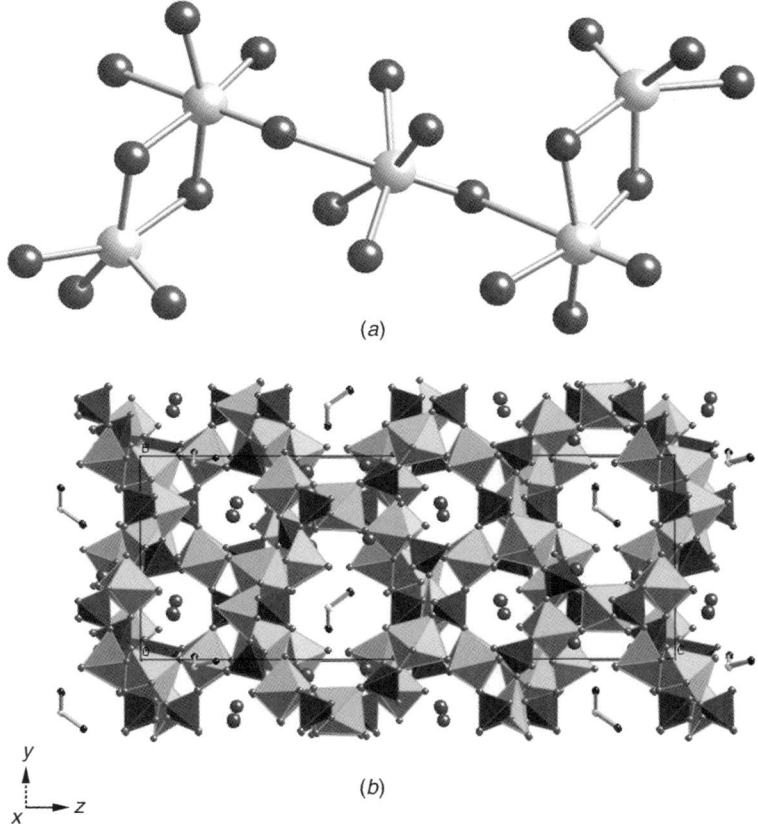

Figure 53. (a) The pentanuclear vanadium cluster embedded in the structure of $[Me_2NH_2]K_4$ $[(VO)_{10}(H_2O)_2(OH)_4(PO_4)_7] \cdot 4H_2O$. (b) A view of the 3D structure of $[Me_2NH_2]K_4[(VO)_{10}$ $(H_2O)_2(OH)_4(PO_4)_7] \cdot 4H_2O$, showing the cavities occupied by K^+ and $[Me_2NH_2]^+$ cations.

its four edges with edges from four neighboring square pyramidal vanadium centers. Each pentamer is connected to four other V_5 units by $\frac{4}{2}$ phosphate tetrahedra. The most prominent feature of the structure is the presence of extremely large cavities in the V/P/O framework, which are filled in a remarkably complicated, but highly symmetric, fashion with a mixture of organic and inorganic cations. The center of the cavity contains an aggregate of 12 $[HN Et_3NH]^{2+}$ (diprotonated 1,4-diazabicyclooctane) cations, surrounded by water and 32 K^+ cations. The organic and K^+ cations, and the water molecules of crystallization, are in turn surrounded by a cavity formed from 12

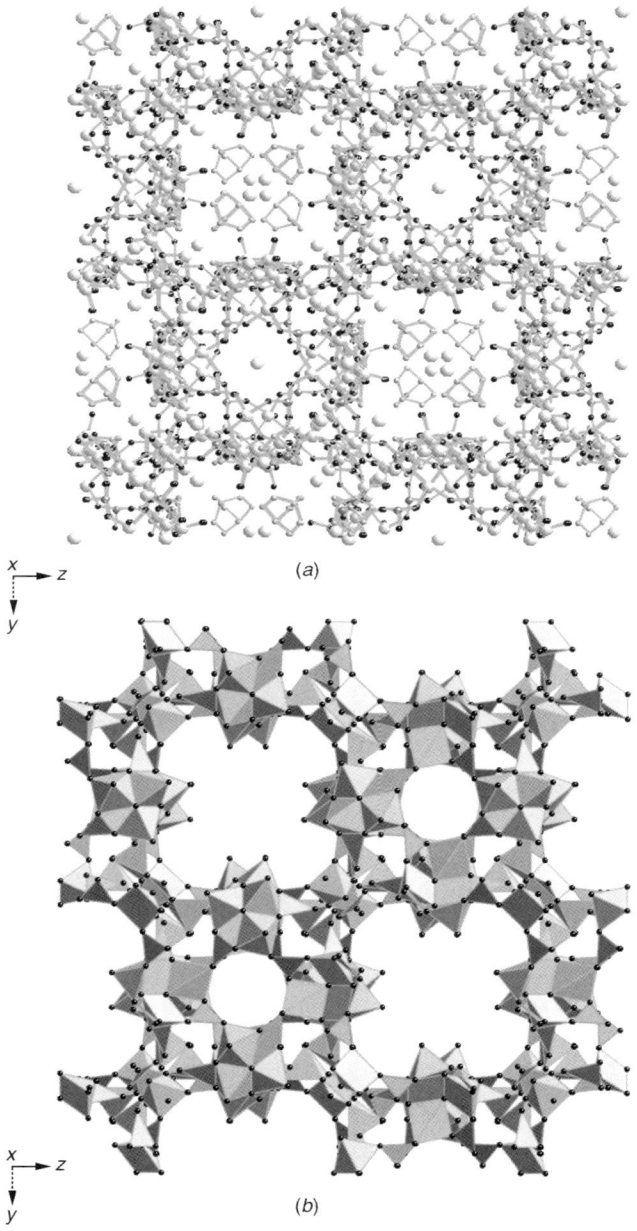

Figure 54. (a) A ball-and-stick representation of the structure of [HNEt$_3$NH]K$_{1.35}$[(V$_5$O$_9$)(PO$_4$)$_2$] · xH$_2$O, showing the cavities occupied by the HNEt$_3$NH^{2+} and K$^+$ cations. (b) A polyhedral view of the structure with the cations removed.

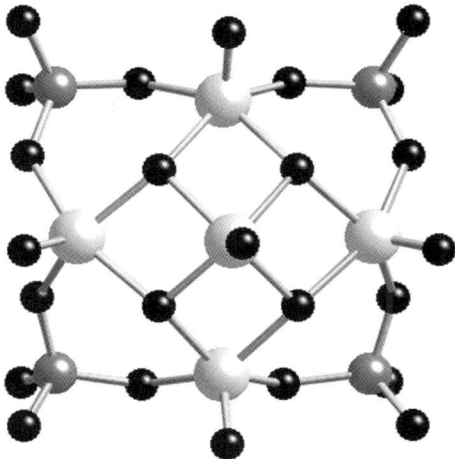

Figure 55. The pentanuclear subunit of $[HN(CH_2CH_2)_3NH]K_{1.35}[(V_5O_9)(PO_4)_2] \cdot xH_2O$.

vanadium pentanuclear units, connected by phosphate groups and corners of other adjacent pentamers.

The pronounced curvature of the $\{V_5O_9(PO_4)_{4/2}\}$ units, together with the tetrahedral geometric requirements of the phosphorus sites, the distances between $\frac{4}{2}PO_4$ groups and the relatively low charge per volume in the pentanuclear unit, all encourage metrically large structures which would have difficulty filling space in a characteristically dense fashion. The surprisingly large volume of the cavities that result is reflected in a very low framework metal atom density of 9.3 M atoms (M = V, P)/1000 Å^3 compared to 12.7 atoms/1000 Å^3 faujasite (M = Si). The cross-section of the cavity reveals a 32 polyhedral connect ring encircling the maximum diameter of the cavity. Each cavity communicates with six symmetry equivalent cavities through six 16-ring windows.

The structure of $[HN(CH_2CH_2)_3NH]K_{1.35}[(V_5O_9)(PO_4)_2]$ exhibits the same fundamental building blocks as those of the previously described $Cs_3[(V_5O_9)(PO_4)_2] \cdot xH_2O$. However, in this later instance, the polyhedral connectivities link each super cage to four, rather than six, adjacent cages. It has been noted (239), that the structure of $[HN(CH_2CH_2)_3NH]K_{1.35}[(V_5O_9)(PO_4)_2]$ is related to the sodalite net (240), while those of $Cs_5[(V_5O_9)(PO_4)_2] \cdot xH_2O$ (112) and its disordered analogue $Na_x[(V_5O_9)(PO_4)_2](PO_4)_y(OH)_z$ (241) exhibit the zeolite rho (242) structure.

Fusing of the vanadium oxide subunits into infinite $\{V_xO_y\}_n$ chains is also a feature of the V/P/O framework structures. The 3D structure of

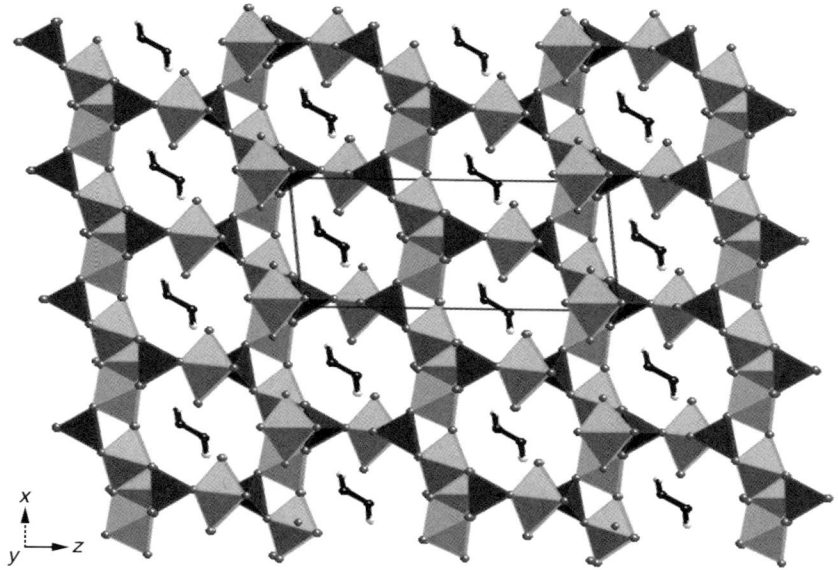

Figure 56. A view of the structure of [H$_2$N(CH$_2$CH$_2$)$_2$NH$_2$][(VO)$_4$(H$_2$O)$_4$(HPO$_4$)$_2$(PO$_4$)$_2$], parallel to the crystallographic b axis.

[H$_2$N(CH$_2$CH$_2$)$_2$ NH$_2$][(VO)$_4$(H$_2$O)$_4$(HPO$_4$)$_2$(PO$_4$)$_2$] (231, 232) can best be understood and qualitatively described as a severely defected {VOPO$_4$} structure type, containing several different {VO$_6$} environments that are related to the infinite {O=V ··· O}$_n$ strings present in VOPO$_4$. As seen in Figure 56, the framework is built up from distorted V(IV) octahedra combined with {PO$_4$} or {HPO$_4$} tetrahedra, and has a V/P ratio of 1. The polar nature of this phosphate is also demonstrated in this figure (which is viewed down [010]), with the 1D ··· O–V–O–V ··· chains running parallel to the crystallographic a direction. Similar chains exist in the structure of VOPO$_4$, but in VOPO$_4$ these 1D chains are bonded to adjacent parallel chains via {PO$_{4/2}$} tetrahedra, while in [H$_2$N(CH$_2$CH$_2$)$_2$NH$_2$][(VO)$_4$(H$_2$O)$_4$(HPO$_4$)$_2$(PO$_4$)$_2$] they are bonded to two {PO$_4$} and two {HPO$_4$} tetrahedra. Furthermore, two vertices of the {PO$_4$} groups are bound to adjacent octahedra in the 1D chain, and one vertex is coordinated to the vanadium that bears an oxo group trans to an aqua ligand while the remaining vertex is bonded to the vanadium with an oxo group and three water ligands. Two vertices of the {HPO$_4$} tetrahedra bridge adjacent octahedra in the 1D chain, in a similar fashion to the {PO$_4$} tetrahedra, and a third vertex forms an equatorial contact to the monoprotonated vanadium while the remaining oxygen is present as a P–OH group. The most obvious difference between the structures of this phosphate and $VOPO_4$, however, is that an octahedral coordination site in $VOPO_4$ is replaced by a piperazinium cation, which gives rise to the "severely defected" description of $[H_2N(CH_2CH_2)_2NH_2]$ $[(VO)_4(H_2O)_4(HPO_4)_2(PO_4)_2]$.

V. VANADIUM ORGANOPHOSPHONATE PHASES

A. General Characteristics

The contemporary interest in metal organophosphonate coordination chemistry has received considerable impetus from the applications of such materials as sorbents, catalysts and catalyst supports [243, 244]. The structural chemistry of the metal organophosphonate system is extremely rich [245, 246] and is represented by mononuclear coordination complexes [247], molecular clusters [248–252], 1D materials [253], and layered phases [254]. Extensive investigations have been reported on layered phosphonates of divalent [255–258], trivalent [259, 260], tetravalent [261–265], and, recently, hexavalent elements [266]. Several examples of 3D metal organophosphonate frameworks have also been described since the first report by Bujoli [267–270]. A comprehensive review of metal organophosphonate chemistry by Clearfield is also available [271].

Oxovanadium organophosphonate chemistry [272] includes solid materials with 1D, 2D, and 3D structures, incorporating either single valence V(IV) or V(V) sites or mixed-valence V(IV)/V(V) sites [273–278], as well as molecular species exhibiting a remarkable structural diversity [248–257, 279–286]. A noteworthy common feature of these clusters is the organization of the V/P/O framework about a template, generally present as a guest or captive species. The range of templates is quite diverse, including neutral molecules, anions and even cations. Template effects are, of course, also a persistent theme of the solid-state chemistry of the $V-O-RPO_3^{2-}$ system.

For the purposes of the structural discussion, two subclasses of the oxovanadium organophosphonate phases are identified: the oxovanadium monophosphonate group, $V/O/RPO_3^{2-}$ and the oxovanadium diphosphonate solids, $V/O/\{O_3P/R'/PO_3\}^{4-}$ Table V. The first type is characterized by layer structures with the organic substituents projecting from the two surfaces of the layer to give the common alternating pattern of inorganic layers separated by organic bilayers. The latter group exhibits a diverse structural chemistry including chain, layer, and framework (3D) structures.

B. Oxovanadium Monophosphonate Solids

The structural prototype for the solid phases of the $V/O/RPO_3^{2-}$ system is the V(IV) species $[(VO)(PhPO_3)(H_2O)]$ (262), whose structure is shown in

TABLE V

Comparison of Structural Features of Vanadium Organophosphate and Vanadium Organodiphosphonate Phases

Compound	V/P/O Framework Structure	Vanadium Coordination and Linkages	Cell Parameters	Space Group	$V=O_t^{a-c}$	$V-O_{phosphonate}^{d}$	$V-O_{other}$ (type)	References
$(H_2pip)[(VO)(O_3PCH_2PO_3)]$	Chain	Isolated V(IV) Square pyramids	8.012(2) 18.191(4) 7.744(2)	$Pnma$	1.595(3)	1.980(2) (x4)		276
$Cs[(VO)(HO_3PCH_2PO_3)]$	Chain	Chains of corner-sharing V(IV) octahedra	10.212(2) 10.556(2) 14.699(3) $\beta = 94.57(2)$	$C2$	1.60(1)	2.00(1) (x4)	2.40(1) (bridging oxo)	287
$(NH_4)_2(VO)(O_3PCH_2PO_3)$	Chain	Chains of corner-sharing V(IV) octahedra	7.3182(1) 16.5633(1) 7.5225(2)	$Pnma$	1.623(1)	1.984(1) (x4)	2.371(2) (bridging oxo)	288
$[(VO)(O_3PPh)(H_2O)]$	Layer	Chains of corner-sharing V(IV) octahedra	28.50(3) 7.18(2) 9.42(2) $\beta = 97.1(2)$	$C2/c$	1.606(3)	1.96(1) (x3)	2.11(1) (aqu) 2.14(1) $(V \cdots O=V)$	262
$[(VO)_2(O_3PCH_2PO_3)_2(H_2O)_4]$	Layer	Isolated V(IV) octahedra	12.805(4) 10.592(3) 15.037(5)	$Pbca$	1.606(3) (V1) 1.619(2) (V2)	1.993(3) (x2) 2.021(3) (x4)	2.103(4) (aqua (x3) 2.215(2) (aqua)	263
$M[(VO_2)_3(O_3PMe)_2]$ (M = K$^+$, NH$_4$, Rb, Tl)	Layer	3 ring/6 ring units of corner-sharing V(V) octahedra	7.139(3) 19.109(5) (K$^+$)	$R32$	1.658(4) (x2)	1.961(2) (x2)	2.174(4) (x2, $V \cdots O=V$)	270
$(H_2en)[(VO)(O_3PCH_2CH_2PO_3)]$	Layer	Isolated V(IV) square pyramids	5.023(1) 16.322(3) 13.314(3) $\beta = 92.54(3)$	$P2_1/n$	1.586(6)	1.977(7) (x4)		276

Compound	Structure	Unit	Cell parameters	Space group	V=O	V-O	Other	Ref
(H₃NEt)[(VO)₃(H₂O)(O₃PPh)₄]	Layer	Trinuclear units of corner-sharing V(IV) octahedra	14.708(3), 12.406(2), 10.275(2), β = 93.66(1)	$P2_1/c$	1.62(1)(V1) 1.638(9)(V2)	1.981(8) (x4) 2.006(8) (x4)	2.276(7) (V⋯O=V) 2.162(9) (V⋯O=V)	273
(H₂NEt₂)(H₂NMe₂)[(VO)₄(OH)₂(O₃PPh)₄]	Layer	Binuclear units of corner-sharing V(IV) square pyramids	21.022(4), 19.034(3), 10.771(2), β = 99.63(2)	$P2_1/c$	1.53(1)	1.958(9) (x3)	1.96(1) (hydroxy bridge)	274
(Et₄N)[(VO)₃(OH)(H₂O)(O₃PEt₃)]·H₂O	Layer	Binuclear units of corner-sharing V(IV) square pyramids; isolated V(IV) square pyramids	21.461(4), 13.912(3), 19.247(4)	$Pca2_1$	1.54(1) (VI) 1.53(1) (V3)	1.98(1) (x3) 1.97(1) (x3)	1.96(1) (hydroxy bridge) 2.01(1) (aqua)	275
Cs[(VO)₂V(O₃PCH₂PO₃)₂(H₂O)₂]	Layer	Trinuclear units of corner-sharing V(IV) and V(III) octahedra	9.724(2), 8.136(2), 10.268(2), β = 103.75(3)	$C2/m$		1.97(1) (x4)ᵉ 1.99(1) (x6)ᶠ	1.783(9) (bridging oxo) 2.14(1)(aqua)	287
Cs[(VO)₂(OH)(O₃PCH₂CH₂PO₃)]	Layer	Binuclear units of corner-sharing V(IV) square pyramids	11.798(2), 11.056(2), 8.682(2), β = 104.92(3)	$P2_1/c$	1.586(9)	1.971(11) (x4)	2.017(10) (bridging OH)	289
(H₂pip)[(VO)₂(O₃PCH₂CH₂PO₃H)₂]	"Stair-stepped" layer	Isolated V(IV) octahedra	6.244(1), 8.687(2), 11.156(2), α = 88.92(3), β = 74.63(3), γ = 74.32(3)	$P\bar{1}$	1.585(6)	1.964(7) (x4)		277

(continues)

TABLE V (Continued)

Compound	V/P/O Framework Structure	Vanadium Coordination and Linkages	Cell Parameters	Space Group	$V=O_t^{a-c}$	$V-O_{phosphonate}{}^d$	$V-O_{other}$ (type)	References
[(VO){O$_3$PCH$_2$NH(C$_2$H$_4$)$_2$NHCH$_2$PO$_3$}(H$_2$O)]	"Pillared" layer	Isolated V(IV) octahedra	7.999(2) 8.009(2) 9.201(2) $\beta = 100.24(3)$	$P2_1/c$	1.578(4)	2.008(3) (x4)		277
(H$_2$en)[(VO)$_4$(OH)$_2$(H$_2$O)$_2$(O$_3$PCH$_2$CH$_2$CH$_2$PO$_3$)$_2$]·4H$_2$O	"Pillared" layer	Binuclear units of corner-sharing V(IV) square pyramids and octahedra	14.870(3) 10.245(2) 18.868(4) $\beta = 99.50(3)$	$C2/c$	1.583(7)^g 1.615(8)^h	1.971(8) (x3)^g 1.993(8) (x3)^h	1.930(7)^g (bridging hydroxy) 1.9995(7) (bridging hydroxy) 2.388(9)(aqua)	276
(H$_3$O)$_2$[(VO)$_2$(OH)$_2$(O$_3$PCH$_2$CH$_2$PO$_3$)$_2$]·H$_2$O	"Pillared" layer	Chains of corner-sharing V(III) octahedra and isolated V(IV) square pyramids	7.150(1) 7.809(2) 9.996(2) 76.55(2) 70.17(2) 88.91(2)	$P\bar{1}$	1.53(2)^j	1.981(7) (x4)^i 1.95(1) (x4)^j	1.924(9)^i (hydroxy bridge x2)	278
[(VO)$_2$(OH)(O$_3$PCH$_2$CH$_2$PO$_3$)]·H$_2$O	"Pillared" layer	Binuclear units of face-sharing V(IV)/V(V) {VO$_5$(OH)} octahedra	7.903(1) 9.355(1) 7.485(1) $\beta = 118.25(1)$	$P2_1/m$	1.646(2) 1.55(3)	2.03(3)	2.33(2) (bridging hydroxy)	290
(H$_3$O)[V$_3$(O$_3$PCH$_2$CH$_2$PO$_3$)(HO$_3$PCH$_2$CH$_2$PO$_3$H)$_3$]	3D	Isolated V(III) octahedra	9.863(1) 46.403(9)	$R\bar{3}c$		2.00(1) (x6)		278
(NH$_4$)[(V$_2$O$_3$)(O$_3$PCH$_2$CH$_2$PO$_3$)(H$_2$O)]	"Pillared" layer	V(IV) octahedra and V(V) square pyramids	7.4548(3) 8.0825(3) 10.1660(4) $\alpha = 75.244(1)$ $\beta = 68.883(1)$	$P\bar{1}$	1.612(2)	1.910(2) (x2)	1.644(2) (bridging oxo) 2.118(2) (aqua)	291

534

Compound		Structure	Space group	Lattice parameters			Ref.
						(bridging oxo)	
$(NH_4)[(VO)_2(OH)(O_3PCH_2CH_2CH_2PO_3)(H_2O)] \cdot 2H_2O$	"Pillared" layer	Binuclear units of corner-sharing $V(IV)$ octahedra and square pyramids	$C2/c$	14.2928(2) 10.2440(2) 18.9901(1) $\beta = 96.658(1)$	1.599(2)	1.960–2.011(2) (x3)	1.977(3) (bridging hydroxy) 2.480(1) (aqua)	292
$(NH_4)[(V_2O_3)(O_3PCH_2CH_2PO_3)(H_2O)] \cdot 2(H_2O)$	"Pillared" layer	Binuclear units of $V(IV)/V(V)$ octahedra and square pyramids	$C2/c$	14.8998(8) 10.2903(6) 18.515(1) $\beta = 101.079(1)$	1.587(4)	1.941–1.990(3) (x3)	1.753(3), 1.947(3) (bridging oxo); 2.388(3) (aqua)	292
$[V(HO_3PCH_2PO_3)(H_2O)]$	3D	Isolated $V(III)$ octahedra	$P2_1/n$	5.341(1) 11.516(2) 10.558(2) $\beta = 99.89(1)$		1.982(6) (x4)(V1) 2.015(6) (x6)(V2)	2.056(6) (aqua x2)	287
$(NH_4)_4[(VO)_6(O_3PCH_2PO_3)_4](H_2O)_4] \cdot 4H_2O$	3D	$V(IV)$ octahedra and $V(IV)$ square pyramids	$Pmmn$	13.512(7) 16.194(8) 4.677(2)	1.591(7) 1.649(3)	1.969(3)x4 2.013x4	2.251(3) (H_2O)	291
$Na[(VO)(O_3PCH_2CH_2CO_2)] \cdot 2H_2O$	3D	Isolated $V(IV)$ octahedra	$P2_1$	6.4041(3) 17.6252(8) 7.7271(3) $\beta = 95.586(2)$	1.59(1)	2.00(1) (x3)	2.22(1)(x2) (carboxylate)	293

[a] The symbol O_t refers to the multiply bonded terminal oxo groups.
[b] where there is more than one crystallographically unique vanadium site, this is noted in parentheses.
[c] where there is more than one vanadium–organophosphate bond the number of bonds is given in parentheses (xn) and the average vanadium–oxygen distance is reported.
[d] Organophosphate refers to the oxygen donors of the organophosphate ligand.
[e] This V–O average distance is associated with the $V(IV)$ site.
[f] This V–O average distance is associated with the $V(VIII)$ site.
[g] Values for the square pyramidal site.
[h] Values for the octahedral site.
[i] Values for the $V(VIII)$ site.
[j] Values for the $V(IV)$ site.

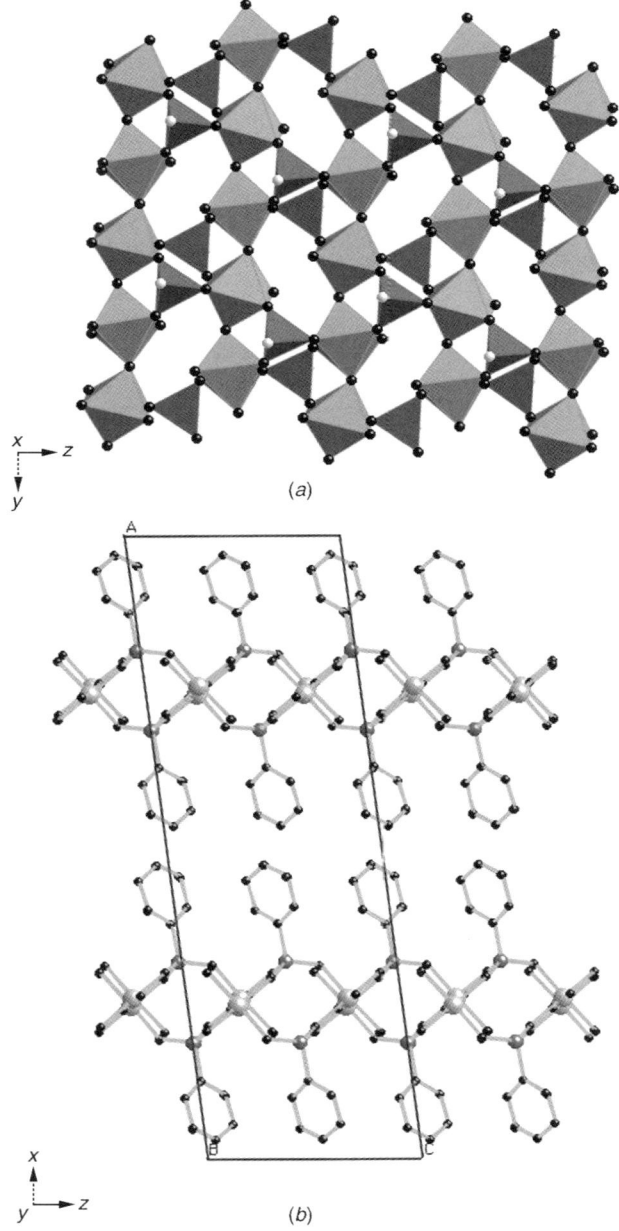

Figure 57. (a) A polyhedral representation of the V/P/O network of [(VO)(PhPO$_3$)(H$_2$O)] viewed parallel to the a axis. (b) The stacking of layers in [(VO)(PhPO$_3$)(H$_2$O)].

Figure 57. The structure consists of layers of corner-sharing {VO$_6$} octahedra linked through {CPO$_3$} tetrahedra. Phenyl groups extend from both faces of the inorganic V/O/P layer, to produce a repeating motif of alternating inorganic layers and organic bilayers with a layer repeat distance of 14.14 Å between V/O/P layers. Within the inorganic layer, the vanadium octahedra share an axial oxygen to form infinite [—V=O ⋯ V—] chains with alternating short–long V—O distances. The chains are separated by phosphate tetrahedra linking two adjacent vanadium sites in one chain and one vanadium center in the neighboring chain. The interchain interactions produce the common {(VO)$_2$(μ$_2$-O$_3$PC)$_2$} structural motif. The network of vanadium octahedra and phosphorus tetrahedra also results in a 12-membered interchain heterocycle [P—O—V—O—P—O]$_2$ and an intrastrand motif [V$_2$O$_3$(μ$_2$-O$_3$PC)] characterized by the presence of the {V$_2$O$_3$}$^{2+}$ unit and a symmetrically bridging {μ$_2$-O$_3$PC} group.

The chemistry of the V/O/RPO$_3^{2-}$ phases may be expanded dramatically by introducing organic cations, which may not only occupy interlamellar voids, thus increasing the separation between planes, but also influence the connectivities between vanadium polyhedra and organophosphonate tetrahedra. Hydrothermal synthesis provides an expedient method for introducing a variety of potential templates to direct the organization of new phases and for adequate crystal growth. The optimal synthetic conditions reflect the vastness of the hydrothermal parameter space and certainly provide no sense of rational design of solid phases by well-defined reaction routes. On the other hand, by exploiting hydrothermal techniques and judicious choice of organic cations, it is evident that major structural modifications may be accomplished and that a significant structural chemistry of these phases may be developed. Furthermore, since the functional group of the organophosphonate is also subject to variation, presynthesis and/or postsynthesis modification may afford routes to more rational synthetic design. This feature of the chemistry will be revisited in the discussion of the diphosphonate, V/O/R(PO$_3$)$_2^{4-}$ phases in Section II.B.

The phase most closely related to the prototype structure [VO(PhPO$_3$)(H$_2$O)] is the organically modified (EtNH$_3$)[(VO)$_3$(H$_2$O)(PhPO$_3$)$_4$] (273). As shown in Figure 58, the structure of (EtNH$_3$)[(VO)$_3$(H$_2$O)(PhPO$_3$)$_4$] may be described as layers of corner-sharing {VO$_6$} octahedra and {PO$_3$C} tetrahedra, with the phenyl groups extending from both sides of the V/O/P layer. The pattern of alternating organic and inorganic layers is reminiscent of the structure of [(VO)(PhPO$_3$)(H$_2$O)]. However, the structural similarities do not extend to the detailed structure of the oxide layers, which distort in (EtNH$_3$)[(VO)$_3$(H$_2$O)(PhPO$_3$)$_4$] to accommodate the presence of the organic cationic templates. Whereas the structure of [(VO)(PhPO$_3$)(H$_2$O)] exhibits an oxide layer with a vanadium-oxo/phenylphosphonate composition of 1:1 and exhibiting infinite

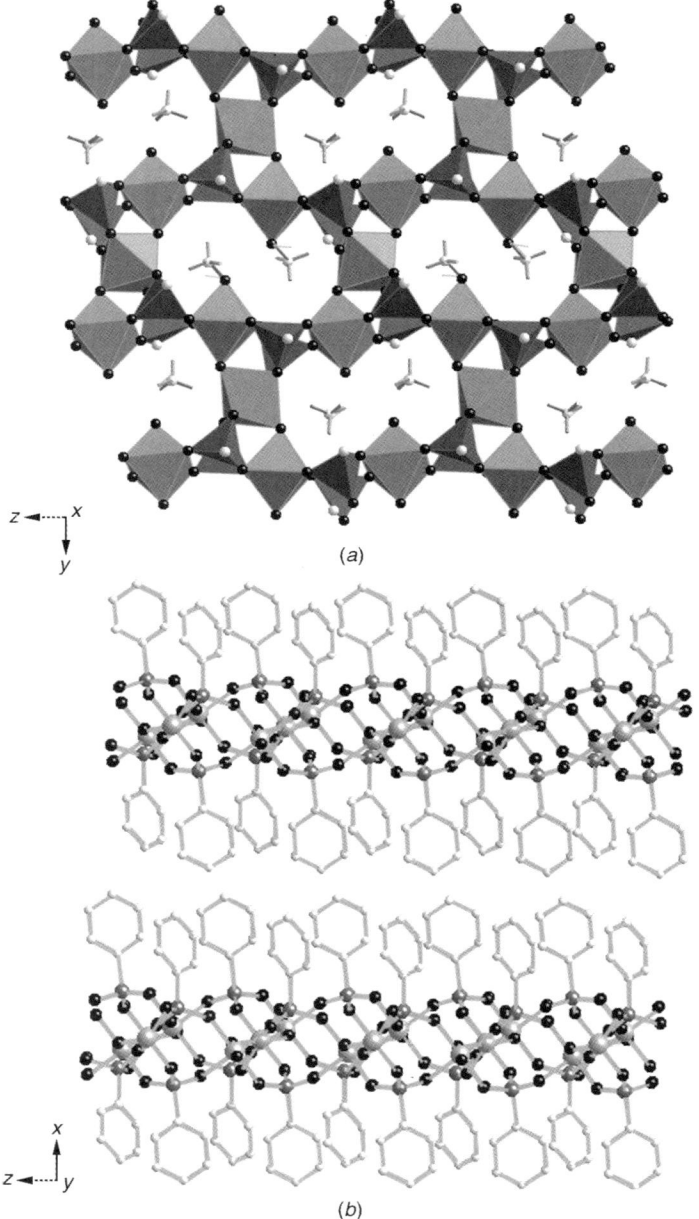

Figure 58. Views of the layer structure of $(EtNH_3)[(VO)_3(H_2O)(PhPO_3)_4]$. (a) Normal to the plane of the V/P/O network; (b) Showing the layer stacking.

{−V=O−V−} chains with alternating short and long V−O bonds, [(VO)$_3$(H$_2$O)(PhPO$_3$)$_4$] features a {VO}/(RPO$_3$)$^{2-}$ composition of 3:4 with discrete trinuclear {V$_3$O$_3$(H$_2$O)} units bridged through {PhPO$_3$}$^{2-}$ groups. The {V$_3$O$_3$(H$_2$O)} trinuclear units bridge through {PhPO$_3$}$^{2-}$ groups. In addition, the trinuclear units exhibit zigzag {−V=O−V−} chains with short–long alternation of V−O bonds, such that each {(VO)$_3$(H$_2$O)} chain terminates in an oxo-group atom at one end and an aqua ligand at the other. The kinks in the {−V=O−V−} chains are a consequence of edge sharing between the three vanadium octahedra of the trinuclear unit and four bridging (RPO$_3$)$^{2-}$ groups. Thus, the central vanadium octahedron of the unit coordinates to four bridging organophosphonates, while each terminal vanadium octahedron shares two of these bridging (RPO)$_3^{2-}$ tetrahedra with the central vanadium and forms two additional edge-sharing interactions to phosphate tetrahedra that bridge to an adjacent trinuclear unit.

The trinuclear structural motifs fuse in such a fashion as to generate cavities in the V/P/O layer, defined by rings constructed from the edge sharing of six vanadium octahedra and four (RPO$_3$)$^{2-}$ tetrahedra. These cavities are polar in character with aqua groups and vanadyl oxygen atoms projecting into the void space. The V/P/O layers are stacked to produce tunnels parallel to the cell a axis, which are occupied by the (EtNH$_3$)$^+$ cations. The walls of these tunnels are defined by the organophosphonate phenyl groups that project above and below the V/P/O planes. The organic cations are oriented in the cavities with the ammonium group, −NH$_3^+$, directed toward the hydrophilic holes in the V/P/O layer, while the −Et group trails into the hydrophobic region generated by the phenyl substituents. The incorporation of the organic template into the V/O/RPO$_3^{2-}$ system results in dramatic structural rearrangement within the V/P/O layer of (Et$_3$N)[(VO)$_3$(H$_2$O)(PhPO$_3$)$_4$] as compared to [(VO)(PhPO$_3$)(H$_2$O)], in contrast to a relatively minor distortion of the gross alternating pattern of inorganic and organic layers. The structure may be described as amphiphilic and rationalized on the basis of partitioning into hydrophilic and hydrophobic domains. While the templating mechanism remains enigmatic, it is clear that charge-compensating and space-filling effects contribute to the process and that different template molecules will provide different distortions of the V/P/O layer, as in [Et$_2$NH$_2$][Me$_2$NH$_2$][(VO)$_4$(OH)$_2$(PhPO$_3$)$_4$] (274) and [Et$_4$N][(VO)$_3$(OH)(H$_2$O)(EtPO$_3$)$_3$] • H$_2$O (275).

Additional template modification forces more severe structural modification, as observed for [Et$_2$NH$_2$][Me$_2$NH$_2$][(VO)$_4$(OH)$_2$(PhPO$_3$)$_4$] (274) shown in Figure 59. The introduction of the intercalating organic cations has disrupted the registry of layers such that the structure of [Et$_2$NH$_2$][Me$_2$NH$_2$][(VO)$_4$(OH)$_2$(PhPO$_3$)$_4$] exhibits a trilayer repeat represented by a layer of Et$_2$NH$_2^+$ cations sandwiched between inorganic V/P/O layers that are in turn bounded by organic bilayers consisting of phenyl group from adjacent V/O/PhPO$_3$ slabs.

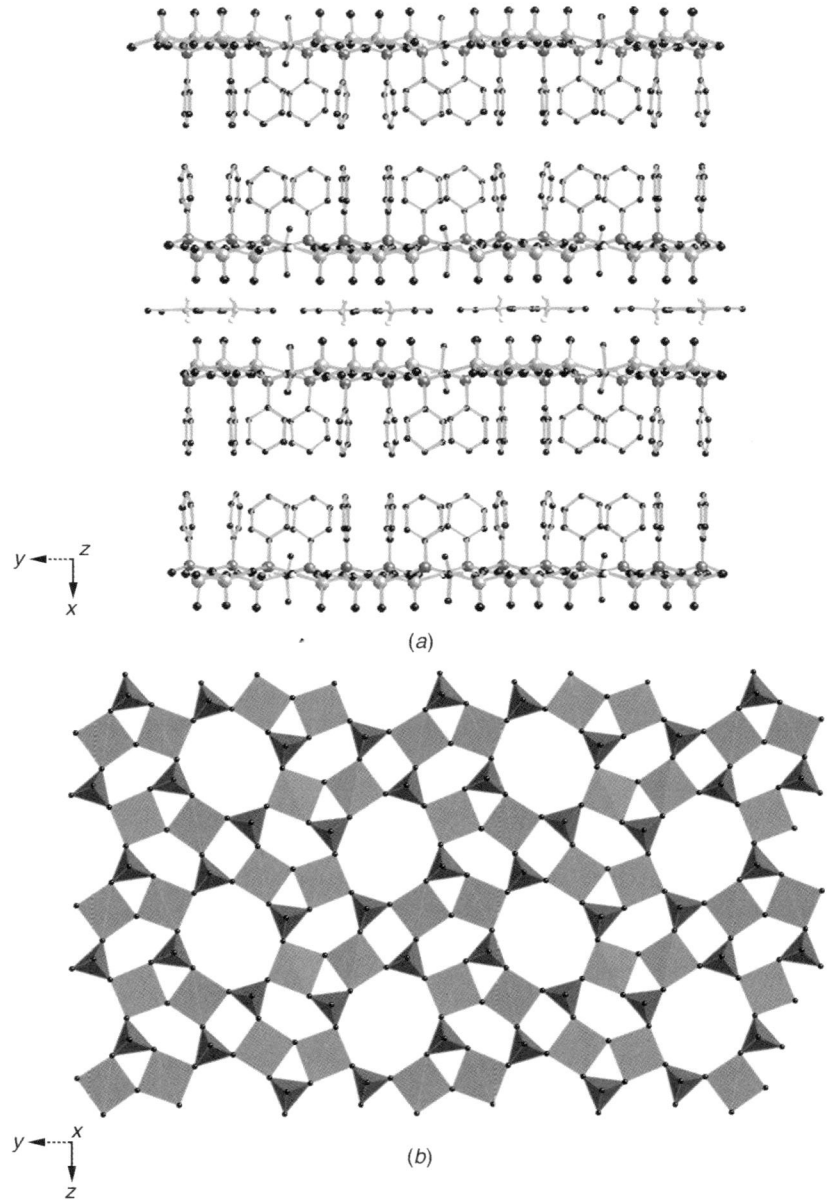

Figure 59. (a) A view of the structure of [Et$_2$NH$_2$][Me$_2$NH$_2$][(VO)$_4$(OH)$_2$(PhPO$_3$)$_4$], parallel to the crystallographic c axis. (b) A polyhedral representation of the V–P–O network.

The structure of [(VO)(PhPO$_3$)(H$_2$O)] is constructed from layers of corner-sharing {VO$_6$} octahedra and {CPO$_3$} tetrahedra, with *phenyl groups extending from both sides of the metal oxide layers*. In contrast, the structure of [Et$_2$NH$_2$][Me$_2$NH$_2$][(VO)$_4$(OH)$_2$(PhPO$_3$)$_4$] consists of corner-sharing {VO$_5$} square pyramids and {CPO$_3$} tetrahedra, with *phenyl groups directed exclusively to one face of the layer* while the {V=O} groups are directed to the other. Since the {V=O} groups of the adjacent layer are directed toward the vanadyl face of the neighboring layer, a highly hydrophilic region is produced that accommodates the layer of Et$_2$NH$_2^+$ cationic templates. Whereas the structure of [VO(PhPO$_3$)(H$_2$O)] is defined by an interlayer repeat of 14.14 Å, the registry of layers in [Et$_2$NH$_2$][Me$_2$NH$_2$] [(VO)$_4$(OH)$_2$(PhPO$_3$)$_4$] is such as to produce two repeat distances: 8.89 Å between V/P/O layers sandwiching the Et$_2$NH$_2^+$ *intercalators* and 12.29 Å between V/P/O layers sandwiching phenyl group bilayers. The partitioning of the structure into polar and nonpolar domains is evident, and [Et$_2$NH$_2$][Me$_2$NH$_2$][(VO)$_4$(OH)$_2$(PhPO$_3$)$_4$] appears to be amphiphilic with hydrophobic interactions determining the layer packing.

The structural differences between [(VO)(PhPO$_3$)(H$_2$O)] and [Et$_2$NH$_2$][Me$_2$NH$_2$][(VO)$_4$(OH)$_2$(PhPO$_3$)$_4$] are further accentuated in the oxide layer, shown in Figure 59. The vanadium octahedra of [VO(PhPO$_3$)(H$_2$O)] share axial oxygen atoms, forming infinite {—V=O—V=O—} chains with alternating short–long interactions. In [Et$_2$NH$_2$][Me$_2$NH$_2$][(VO)$_4$(OH)$_2$(PhPO$_3$)$_4$], however, discrete {V$_2$O$_2$(OH)}$^{3+}$ binuclear units are bridged through phosphonate tetrahedra to form the layer motif. The vanadium centers of each binuclear unit are bridged by a phosphonate group that directs the third oxygen donor to an adjacent binuclear unit. The coordination at each vanadium site of the binuclear unit is completed by bonding to two oxygen donors from two phosphonate groups each of which in turn bridges to an adjacent binuclear unit. Thus, while the largest cavity in the V/P/O layer of [(VO)(PhPO$_3$)(H$_2$O)] is constructed from the corner sharing of four vanadium octahedra and two phosphonate tetrahedra, producing a 12-membered ring, the cavity size in [Et$_2$NH$_2$][Me$_2$NH$_2$] [(VO)$_4$(OH)$_2$(PhPO$_3$)$_4$] is expanded by corner sharing of four vanadium square pyramids and four phosphonate tetrahedra, resulting in a 16-membered ring with a V—V diagonal distance of ∼10.5 Å. The distortion of the V/P/O layer in [Et$_2$NH$_2$][Me$_2$NH$_2$][(VO)$_4$(OH)$_2$(PhPO$_3$)$_4$] from that in [(VO)(PhPO$_3$)(H$_2$O)] reflects the necessity of accommodating the Me$_2$NH$_2^+$ cation, which projects through the V/P/O inorganic layer. The structure of [Et$_2$NH$_2$][Me$_2$NH$_2$][(VO)$_4$(OH)$_2$ (PhPO$_3$)$_4$] demonstrates the dramatic topological flexibility of vanadium polyhedra and phosphonate tetrahedra in incorporating a variety of substrates. The structure of [Et$_2$NH$_2$][Me$_2$NH$_2$][(VO)$_4$(OH)$_2$ (PhPO$_3$)$_4$] is thus able to accommodate intercalation of organic cations both between layers and within layers, by suitable modification of the [(VO)(PhPO$_3$)(H$_2$O)] parent structure.

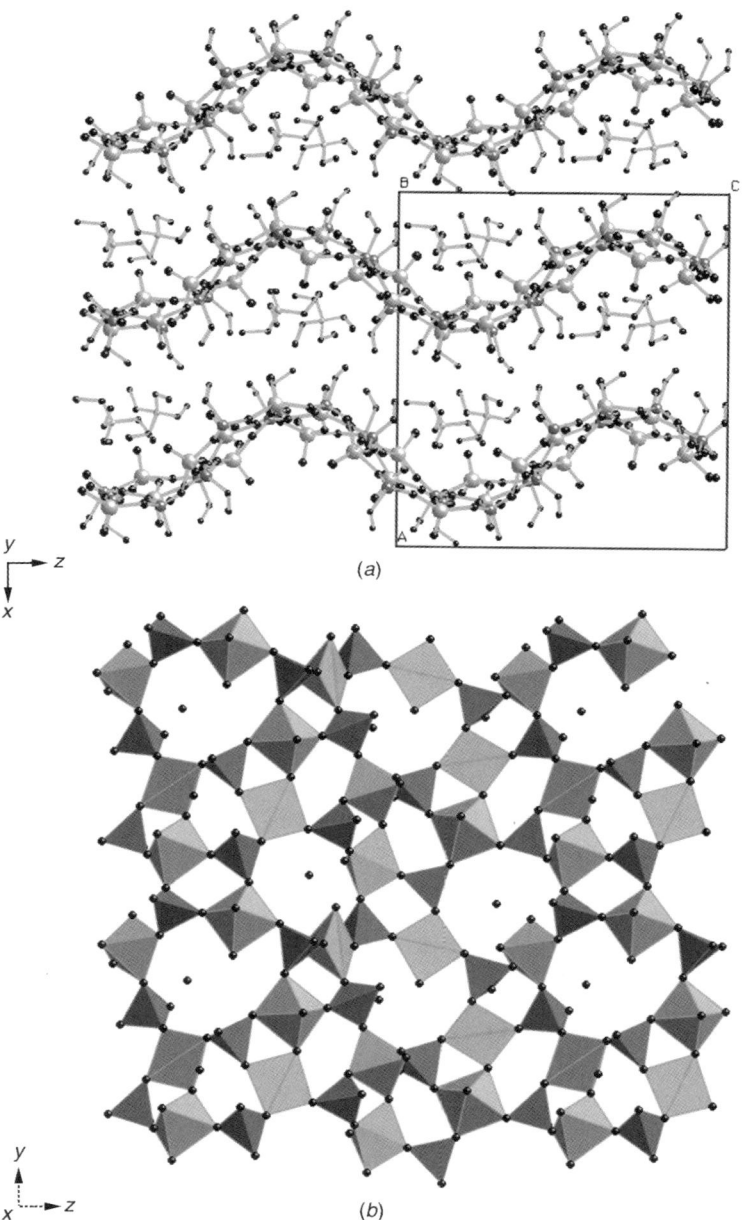

Figure 60. (a) A view of the layer stacking for $(Et_4N)[(VO)_3(OH)(H_2O)(EtPO_3)_3]$. (b) A polyhedral representation of the V/P/O layer.

The steric constraints of bulky organic cations will also cause structural reorganization relative to the prototype structure. The structure of (Et_4N) $[(VO)_3(OH)(H_2O)(EtPO_3)_3] \cdot H_2O$ (275) exhibits layers of corner-sharing V(IV) pyramids and $\{PO_3C\}$ tetrahedra, with the phosphate ethyl groups and the vanadyl $\{V=O\}$ moieties projecting from both surfaces of the V/P/O layer, as shown in Figure 60. In contrast to the prototypical structure of $[(VO)(PhPO_3)(H_2O)]$, which exhibits planar V/P/O layers and alternating inorganic layers and phenyl bilayers, the V/P/O layers of $(Et_4N)[(VO)_3(OH)(H_2O)(EtPO_3)_3]$ undulate in such a fashion as to produce interlamellar cavities occupied by the organic cationic templates. The organic cations produce channels defined by the crests and troughs of adjacent V/P/O layers, which exhibit an interlayer repeat distance of 10.73 Å. These channels are bounded on two flanks by the ethyl substituents of the organophosphonate groups, on a third side by the concave surface of the V–P–O layer, which projects both ethyl groups and $\{V-O\}$ units into the interlamellar region, and on the remaining side by the convex surface of the adjacent layer, which projects vanadyl groups exclusively. Thus, through the interplay of hydrophobic–hydrophilic interactions, nonpolar interlamellar regions bounded by organic substituents and relatively nonpolar vanadyl groups have been created.

As shown in Figure 60, there are two distinct vanadium environments associated with the V/P/O planes of $(Et_4N)[(VO)_3(OH)(H_2O)(EtPO_3)_3]$. Two vanadium centers form a hydroxy-bridged binuclear unit with the terminal oxo groups adopting the anticonfiguration relative to the $\{V_2O_7\}$ plane. On the other hand, the third site consists of isolated $\{VO_5\}$ square pyramids with the coordination geometry at the vanadium defined by a terminal oxo group, three oxygen donors from each of three phosphonate units, and an aqua liquid. The geometries adopted by the vanadium centers of $[Et_2NH_2][Me_2NH_2]$ $[(VO)_4(OH)_2(PhPO_3)_4]$ or $[Et_4N][(VO)_3(OH)(H_2O)(EtPO_3)_3]$ may be contrasted to those observed for $[VO(PhPO_3)(H_2O)]$, which contains infinite 1D $\{-V=O-V=O-\}$ chains and for $(Et_3NH)[(VO)_3(H_2O)(PhPO_3)_4]$, which is characterized by isolated $\{VO_6\}$ octahedra and $\{RPO_3\}$ tetrahedra in a corner-sharing array.

Ethylphosphonate groups in three distinct environments serve to bridge the vanadium centers and to link the structure into its unique 2D array. Each organophosphonate group associated with two of the phosphonate sites adopts the symmetrical bridge mode between V atoms of the $\{V_2(\mu^2\text{-}OH)(\mu^2\text{-}O_2P(Et)O)\}$ binuclear units, as well as serving as a monodentate ligand to an adjacent binuclear unit. The second type of organophosphonate group bridges vanadium centers from each of two adjacent binuclear units to a mononuclear vanadium site, while the third type links a vanadium site of a binuclear unit to two adjacent mononuclear vanadium centers. The complexity of the connectivity pattern is reflected in the presence of 6-membered $\{V_2PO_3\}$ rings,

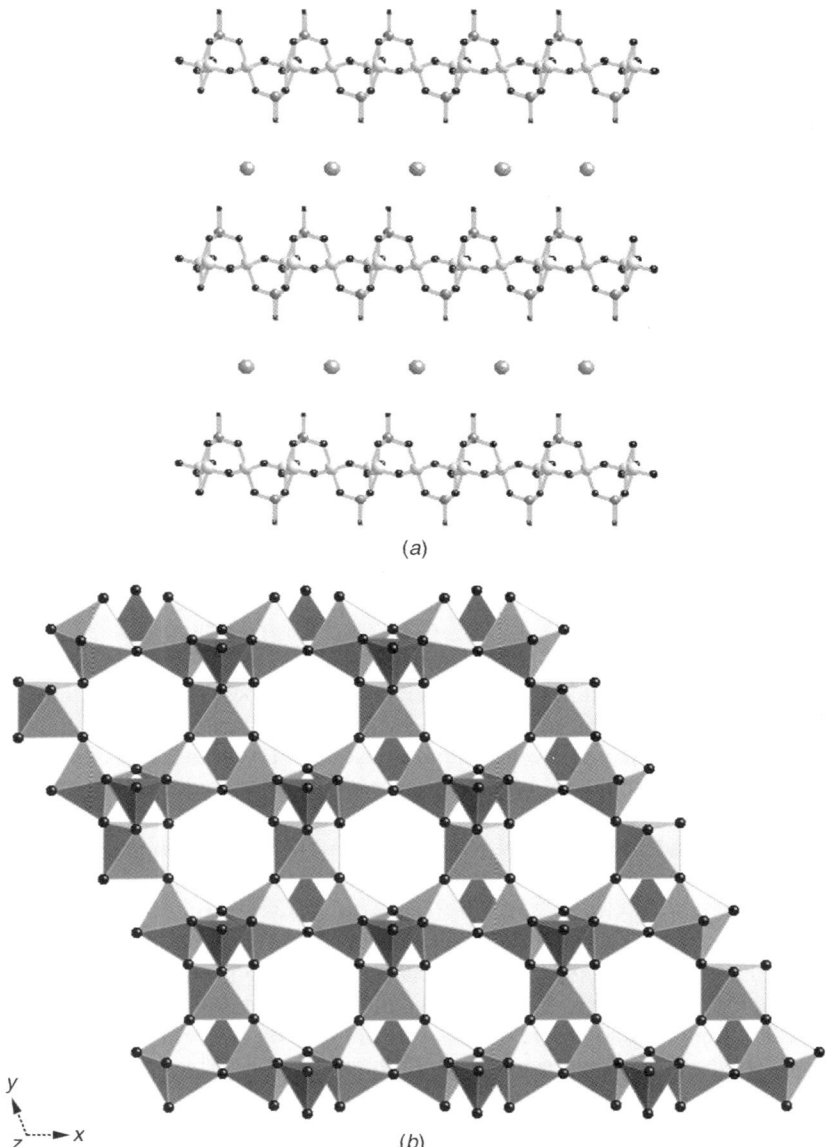

Figure 61. The structure of K[(VO$_2$)$_3$(MePO$_3$)$_2$] showing the layer stacking (a) and the V–P–O network (b).

8-membered $\{V_2P_2O_4\}$ rings, 10-membered $\{V_3P_2O_5\}$ rings, and 16-membered $\{V_4P_4O_8\}$ rings formed by the corner sharing of the $\{VO_5\}$ square pyramids and the $\{O_3PC\}$ tetrahedra.

The family of phosphates $M[(VO_2)_3(MePO_3)_2]$ (M = K, NH$_4$, Rb, Tl) (270) represents the unique example of a V–O–RPO$_3^{2-}$ phase containing exclusively V(V) sites. As illustrated in Figure 61 for $(NH_4)[(VO_2)_3(MePO_3)_2]$, the structure consists of NH$_4^+$ cations and layers constructed from corner-sharing $\{VO_6\}$ octahedra and $(RPO_3)^{2-}$ tetrahedra. The V(V) octahedra each share four $\{V–O\}$ vertices with four adjacent $\{VO_6\}$ units to produce a hexagonal oxide type layer with a 3 ring/6 ring motif (294–296). The $(MePO_3)^{2-}$ groups cap both faces of the $\{VO_2\}_n^{n+}$ layer by ligation to the three vanadium sites of a 3 ring. The methyl groups project into the interlamellar region, so as to produce the characteristic alternation of inorganic V/P/O layers and organic bilayers.

C. Oxovanadium Diphosphonate Solids

In contrast to the oxovanadium monophosphonate solids that possess layer structures exclusively, the solids of the vanadium diphosphonate subclass exhibit 1D, 2D, and 3D dimensional covalent connectivities. Not unexpectedly, the most common structural type of the V/O/R′(PO$_3$)$_2^{4-}$ subclass is a variant of the alternating inorganic layer–organic bilayer motif adopted by the oxovanadium monophosphonate solids, with structures constructed from inorganic V/P/O layers pillared by the covalently attached organic backbone of the diphosphonate groups. However, the length and steric demands of the tether are significant determinants of structure, allowing the isolation of chain structures, of 2D materials with the organic backbone forming part of the 2D network rather than pillaring adjacent layers, and complex 3D frameworks with no discernible layer substructure.

In contrast to the pronounced tendency of most organodiphosphonate ligands to bridge metal sites in the construction of complex oligomeric or solid species, the methylene–diphosphonate group will adopt a simple bidentate coordination mode, as shown in Figure 62 for Cs$_2$ $[(VO)(HO_3PCH_2PO_3H)_2(H_2O)]$ (287). The V(IV) site of this species exhibits distorted octahedral coordination, defined by a terminal oxo group, an aqua ligand and four oxygen donors from two methylenediphosphonate ligands, each in the bidentate ligation mode. One pendant oxygen of each phosphorus site is protonated. The pendant $\{P–OH\}$ and $\{P=O\}$ groups are appropriately oriented to adopt a bisbidentate bridging geometry to adjacent metal centers in the formation of larger oligomers or solids. Thus, the complex $[(VO)(HO_3PCH_2PO_3H)_2(H_2O)]^{2-}$ serves as a potential synthetic precursor and structural prototype for vanadium methylenediphosphonate solids.

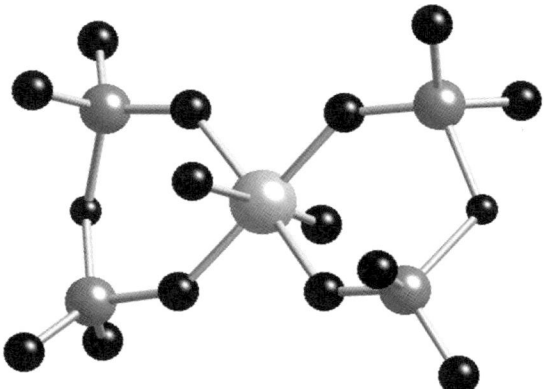

Figure 62. A view of the structure of the molecular anion of $Cs_2[(VO)(HO_3PCH_2PO_3H)_2(H_2O)]$.

The conservation of this structural core is illustrated in the structure of $Cs[(VO)(HO_3PCH_2PO_3)]$ (287), which consists of Cs^+ cations occupying the interstrand regions between undulating negatively charged $[(VO)(HO_3PCH_2PO_3)]^-$ chains. As shown in Figure 63, the vanadium sites adopt distorted octahedral geometry, defined by two trans oxo groups and four oxygen donors from two bidentate methylene diphosphonate groups. The chain is constructed from corner-sharing $\{VO_6\}$ octahedra through $\{V=O-V\}$ linkages with alternating short–long $\{V-O\}$ distances. The bisbidentate diphosphonate ligands bridge adjacent vanadium octahedra.

The influence of reaction conditions and organic cation templating effects are manifested in the structure of $[H_2N(CH_2)_4NH_2][(VO)(O_3PCH_2PO_3)]$ (276), shown in Figure 64, which consists of infinite puckered chains of $[(VO)(O_3PCH_2PO_3)]^{2-}$ units with piperazinium cations occupying the interstrand regions. The V(IV) sites exhibit square pyramidal coordination geometry with the apical position occupied by a terminal oxo group and the four equatorial positions defined by phosphonate oxygen atoms, two from each of

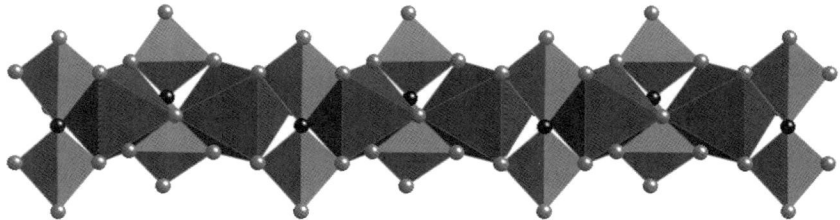

Figure 63. The 1D V/P/O chain of $Cs[(VO)(O_3PCH_2PO_3H)]$.

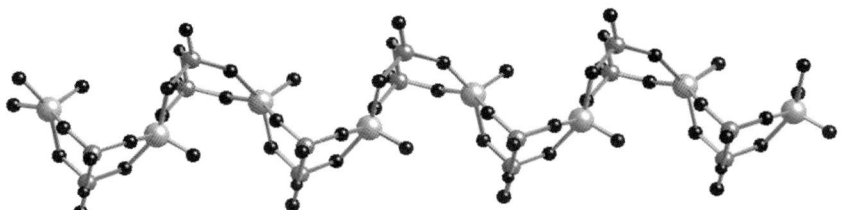

Figure 64. The 1D chain of [H$_2$N(CH$_2$)$_4$NH$_2$][(VO){CH$_2$(PO$_3$)$_2$}].

two diphosphonate ligands. Each diphosphonate bridges two vanadium centers, acting as a symmetrically bridging bidentate ligand to each metal site. Each phosphonate unit of the ligand retains a pendant {P=O} group, confirmed by the short P—O distance of 1.518(6) Å.

The structure relationship between Cs[(VO)(HO$_3$PCH$_2$PO$_3$)] (287) and (H$_2$pip)[(VO)(O$_3$PCH$_2$PO$_3$)] (276) is apparent in that the anionic chain [VO(HO$_3$PCH$_2$PO$_3$)]$^-$ may be derived from the [(VO)(O$_3$PCH$_2$PO$_3$)]$^{2-}$ chain by folding along the O···O vectors of the oxygen donors of the methylenediphosphonate ligand, so as to contract the V—O—P angle and concomitantly the angles between O$_4$(phosphonate) planes of adjacent vanadium polyhedra. It is likewise evident that the pendant {P=O} or {P—OH} groups of the 1D structures may serve as sites for further condensation in the construction of higher dimensionality solids.

The ability of the methylenediphosphonate group to function as an η^6-ligand is realized in the structures of Cs[(VO)$_2$V(O$_3$PCH$_2$PO$_3$)$_2$(H$_2$O)$_2$] (287) and [(VO)$_2$(O$_3$PCH$_2$PO$_3$)$_2$(H$_2$O)$_4$] (263). As shown in Figure 65, the structure of Cs[(VO)$_2$V(O$_3$PCH$_2$PO$_3$)$_2$(H$_2$O)$_2$] consists of Cs$^+$ cation encapsulated within channels of a 2D vanadium–oxygen–methylenediphosphonate network. This 2D network is constructed from {(VO)$_2$V(O$_3$PCH$_2$PO$_3$)$_2$(H$_2$O)$_2$}$^-$ building blocks, containing an unusual mixed-valence V(IV)/V(III) trinuclear unit of corner-sharing {VO$_6$} octahedra. The two V(IV) sites each direct an oxo group into the interior of the layer to bond to the central V(III) site and project an aqua ligand from the top and bottom surfaces of the layer. Within the building block, the methylenediphosphonate group adopts the common bisbidentate bridging mode. The remaining oxygen donors serve to bridge the trinuclear building block to two adjacent trinuclear units in the construction of the 2D network.

While the structure of [(VO)$_2$(O$_3$PCH$_2$PO$_3$)(H$_2$O)$_4$], shown in Figure 66, is quite different from that of Cs [(VO)$_2$V(O$_3$PCH$_2$PO$_3$)$_2$(H$_2$O)$_2$], it shares the common structural motif of a bisbidentate methylenediphosphonate group bridging two vanadium sites and employing the remaining oxygen donors to bridge this unit to adjacent vanadium centers. The structure possesses undulating layers of corner-sharing {VO$_6$} octahedra and diphosphonate polyhedra

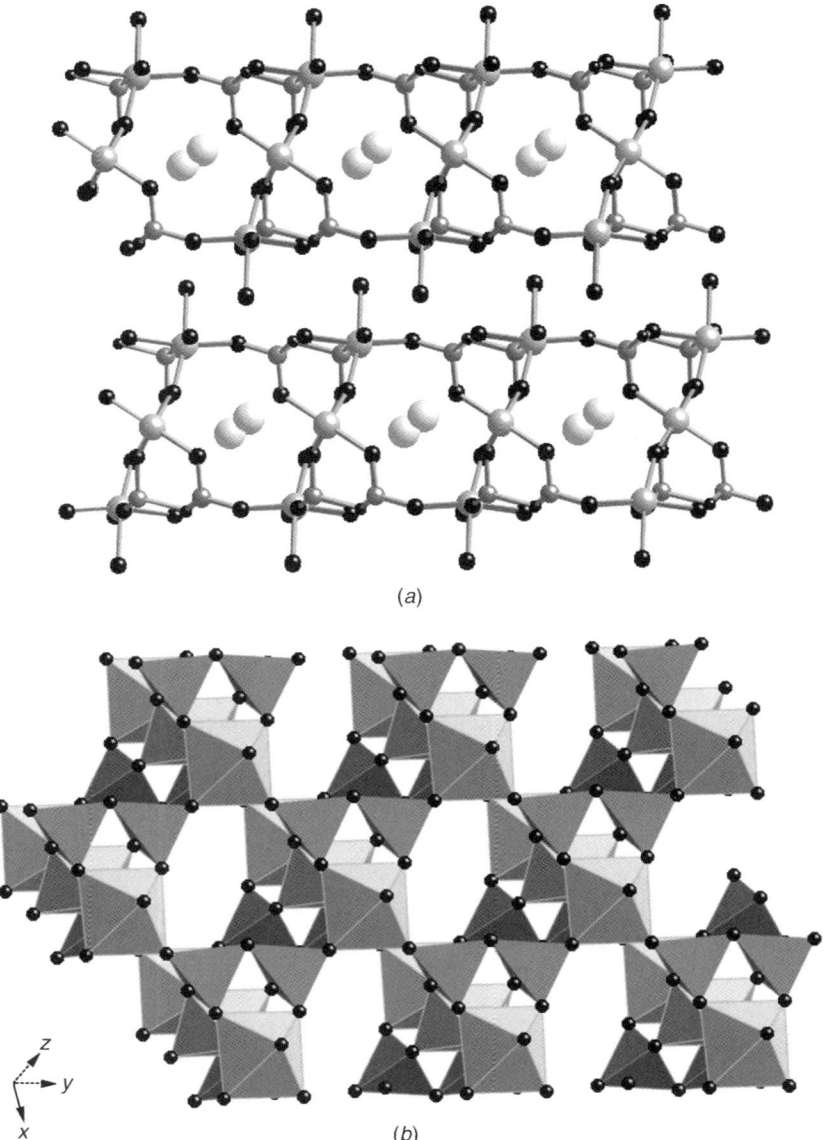

Figure 65. (a) A view of the layer stacking in $Cs[(VO)_2V(O_3PCH_2PO_3)_2(H_2O)_2]$. (b) A polyhedral representation of the trinuclear units of corner-sharing vanadium polyhedra from which the structure is constructed.

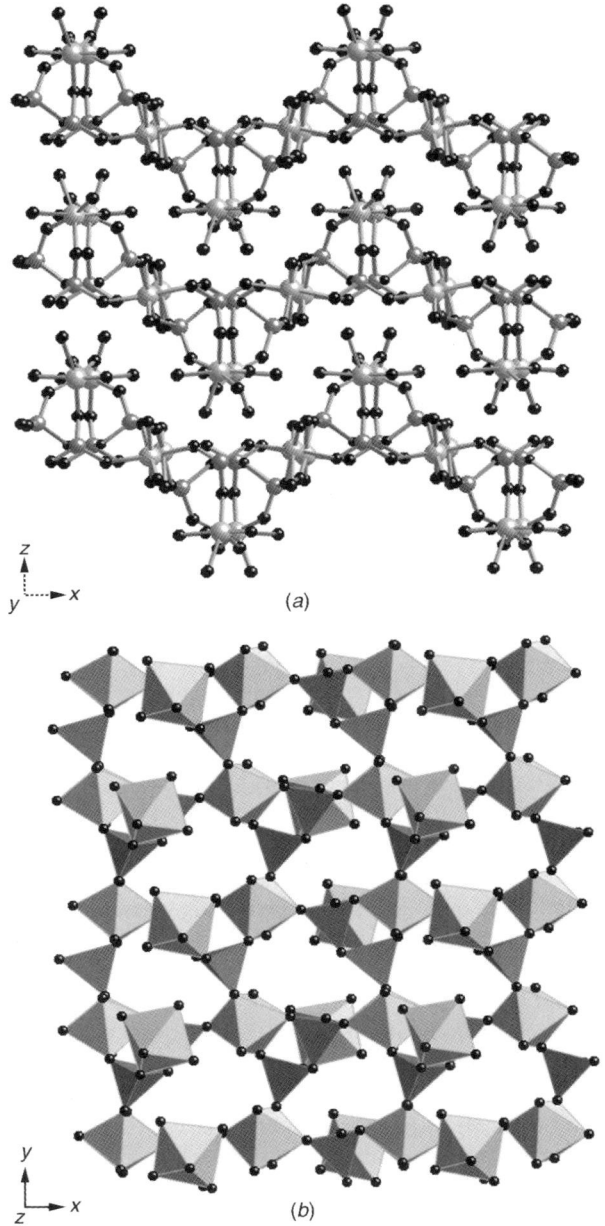

Figure 66. (a) The stacking of layers in [(VO)$_2$(O$_3$PCH$_2$PO$_3$)(H$_2$O)$_4$], viewed parallel to the b axis. (b) A polyhedral representation of the V/P/O network.

stacked with a repeat distance of 15.08 Å. There are four coordinated water molecules, three associated with one vanadium center and one with the second unique vanadium site; these all project into the interlamellar void. The diphosphonate ligand employs all six oxygen donors in coordination to vanadium centers, acting as a bidentate ligand toward two vanadium sites and a monodentate ligand for two additional vanadium centers. One consequence of this coordination mode and the presence of extensive aqua ligation is the absence of direct interactions between $\{VO_6\}$ octahedra through $\{V-O-V\}$ linkages.

Since metal diphosphonate phases in general tend to adopt the common structure type based on M—O—P sheets pillared by the organic backbone, the chain structures adopted by $Cs[(VO)(HO_3PCH_2PO_3)]$ and (H_2pip) $[(VO)(O_3PCH_2PO_3)]$ and the isolated layer structure, incorporating the —CH_2— bridge into the covalent network of the V/P/O/C composite layer, presented by $[(VO)_2(O_3PCH_2PO_3)_2(H_2O)_4]$ and $Cs[(VO)_2V(O_3PCH_2PO_3)_2 H_2O)_2]$ are atypical. This observation suggests that the methylene bridge backs the spatial extension to provide an effective pillar between V/P/O layers and displays rather a marked propensity to form bisbidentate bridge units that provide building blocks most suitable for the construction of chain or layer structures, with the methylene groups locked into the covalent network of the layer.

Lengthening of the organic tether between the $\{O_3P-\}$ units of the diphosphonate by one methylene in ethylenediphosphonate provides a tether of intermediate extension that adopts the role either of pillar or of layer network constituent, depending on the subtleties of reaction conditions and template involvement. The structures of $(H_2en)[(VO)(O_3PCH_2CH_2PO_3)]$ (276), $Cs[(VO)_2 (OH)(O_3PCH_2CH_2PO_3)]$ (288), and $(H_3O)_2[(VO)V_2(OH)_2(O_3PCH_2CH_2PO_3)_2]$ • H_2O (278) illustrate these observations.

As shown in Figure 67, the structure of $[H_2en][(VO)(O_3PCH_2CH_2PO_3)]$ may be described as layers of V(IV) square pyramids corner sharing with diphosphonate tetrahedra to produce $[(VO)(O_3OCH_2CH_2PO_3)]_n^{n-}$ layers with ethylenediammonium cations occupying the interlamellar regions. In addition to the apical oxo group, each V(IV) site is bonded in the equatorial positions to four diphosphonate oxygen donors, one from each of four diphosphonate groups. Hence, each diphosphonate ligand bridges four vanadium centers in a monodentate fashion through two oxygen donors of each $\{PO_3\}$ unit. In a fashion akin to $(H_2pip)[(VO)(O_3PCH_2PO_3)]$, each phosphate group exhibits a pendant $\{P=O\}$ group, which projects into the interlamellar space, as do the vanadyl oxo groups. In comparison to the prototypical layered structure $[(VO)(O_3PPh) (H_2O)]$ the V/O/RPO_3 layer of $[H_2en][(VO)(O_3PCH_2CH_2PO_3)]$ exhibits a more open net as a consequence of the presence of 14-membered $\{V-O-P-C-C-P-O\}_2$ rings. The layer also exhibits the 8-membered

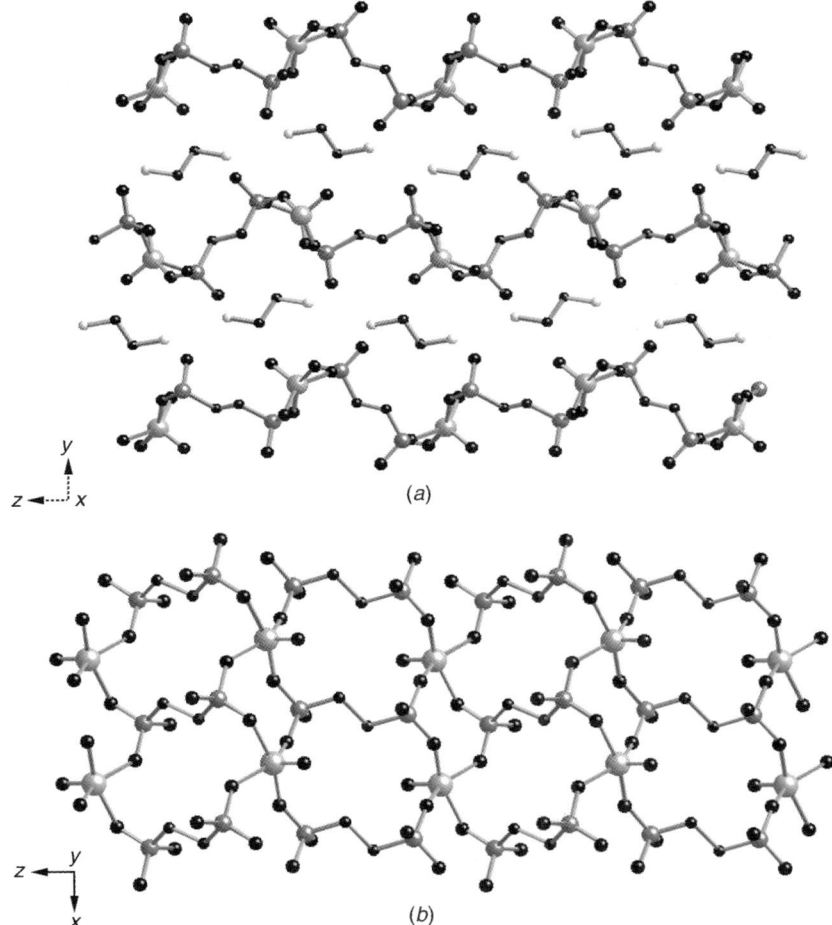

Figure 67. (a) A view parallel to the crystallographic a axis of the stacking of layers in [H$_2$en][(VO)(O$_3$PCH$_2$CH$_2$PO$_3$)]. (b) The V/P/O network.

{V—O—P—O—V—O—P—O—} ring, characteristic of both molecular and solid-phase structures of the V—O—RPO$_3^{2-}$ system.

The structure of Cs[(VO)$_2$(OH)(O$_3$PCH$_2$CH$_2$PO$_3$)], shown in Figure 68, manifests the ethylene diphosphonate ligand in a pillaring role, albeit an atypical type. The structure consists of a 2D anionic framework constructed from complementary V/P/O layers linked by the organic backbone of the diphosphonate ligand. The overall structure may be described as bilayers constructed of two inorganic V/P/O surfaces connected by the organic backbone with an

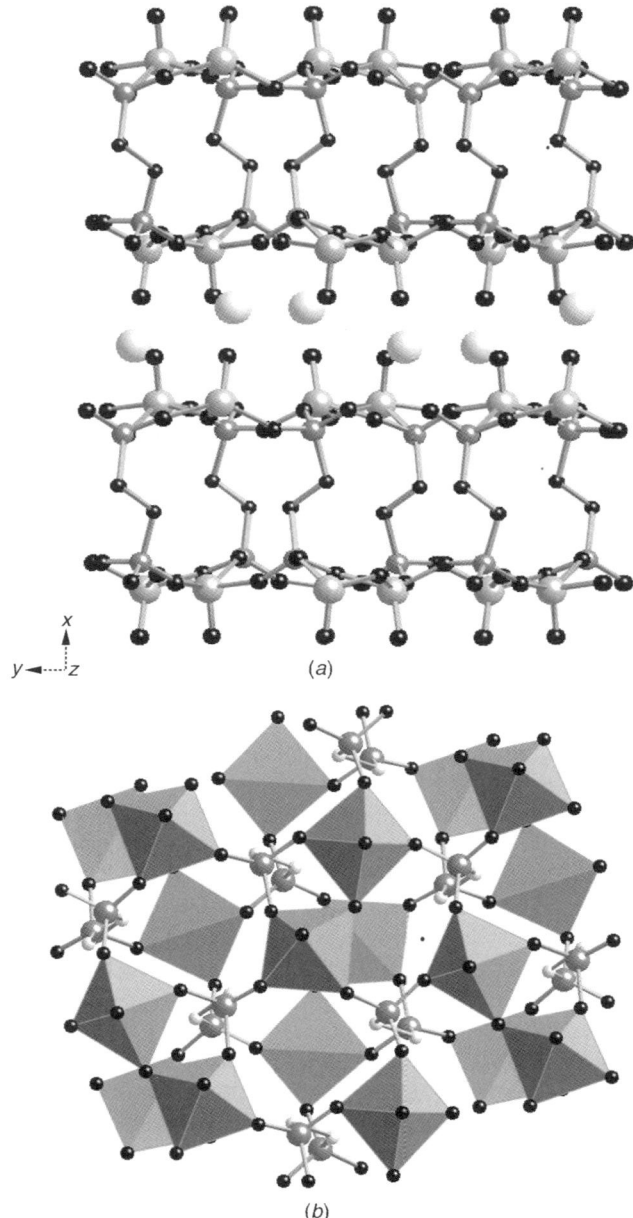

Figure 68. Views of the structure of $Cs[(VO)_2(OH)(O_3PCH_2CH_2PO_3)]$. (a) Parallel to the layers. (b) Normal to the layers.

interlamellar region populated by Cs^+ cations. As noted, the organic tethers of the diphosphonate groups generally project outward from both faces of the inorganic V/P/O sheet so as to produce a 3D covalent framework. The isolation of the organic component in the less polar interior of the bilayer may reflect the templating influence of the Cs^+ cations, which interact with the vanadyl oxygens of the surfaces of the bilayer. Within the inorganic sheets, there are two distinct $\{VO_5\}$ square pyramids, which are linked through a hydroxyl bridge to form $\{(VO)_2(OH)\}$ binuclear units. Each diphosphonate ligand coordinates to six vanadium sites, three in each of the sheets forming the surfaces of the bilayer.

The prototypical pillaring motif is apparent in the structure of $(H_3O)_2[(VO)V_2(OH)_2(O_3PCH_2CH_2PO_3)_2]\cdot H_2O$, whose structure is shown in Figure 69. The overall structure may be characterized as inorganic V/P/O layers buttressed by the ethylene groups of the diphosphonate at a distance of 7.20 Å. Since the organic pillars project from both faces of the V/P/O layers, the effect is to produce 3D V–P–O–C connectivity. It is noteworthy that the polyhedral connectivity within the V/P/O layers is unique. There are both V(III) and V(IV) sites contributing to the unusual layer structure. The V(III) sites form 1D chains of hydroxyl bridged $\{VO_6\}$ octahedra, which are linked through phosphonate tetrahedra to the isolated square planar V(IV) sites. The layer consists of parallel chains of corner sharing V(III) octahedra linked through $\{(VO)(O_3PR)_4\}$ units. This unusual polyhedral connectivity serves to generate cavities within the planes, defined by corner sharing of four V(III) octahedra, two V(IV) square pyramids, and eight phosphonate tetrahedra. The hydronium cations occupy positions above and below these cavities and exhibit significant hydrogen-bonding interactions to the bridging —OH groups and to the vanadyl oxygens.

The structure of $(H_3O)[V_3(O_3PCH_2CH_2PO_3)(HO_3PCH_2CH_2PO_3H)_3]$ (278) confirms the observation that the common pillaring motif can accommodate a variety of polyhedral connectivities within the V–P–O layer and oxidation states V(III), V(IV), and/or V(V). The structure of the anionic network of $(H_3O)[V_3(O_3PCH_2CH_2PO_3)(HO_3PCH_2CH_2PO_3H)_3]$ consists of V/O/P layers of corner-sharing $\{VO_6\}$ octahedra and phosphorus tetrahedra pillared by the organic backbone to produce a 3D V/P/O/C framework (Fig. 70). The distance between V/P/O layers is 7.75 Å. There are two distinct diphosphonate environments: one adopts η^6 modality bonding to three V(III) sites of each of two adjacent layers while the second exhibits a protonated and pendant {P—OH} unit on each phosphorus, resulting in η^4 connectivity. The pattern of polyhedral connectivity within the V/P/O planes results in the formation of cavities circumscribed by six phosphate tetrahedra and six vanadium octahedra. The six pendant —OH groups of the phosphate tetrahedra provide the boundary of a hydrophilic cavity occupied by the hydronium cation.

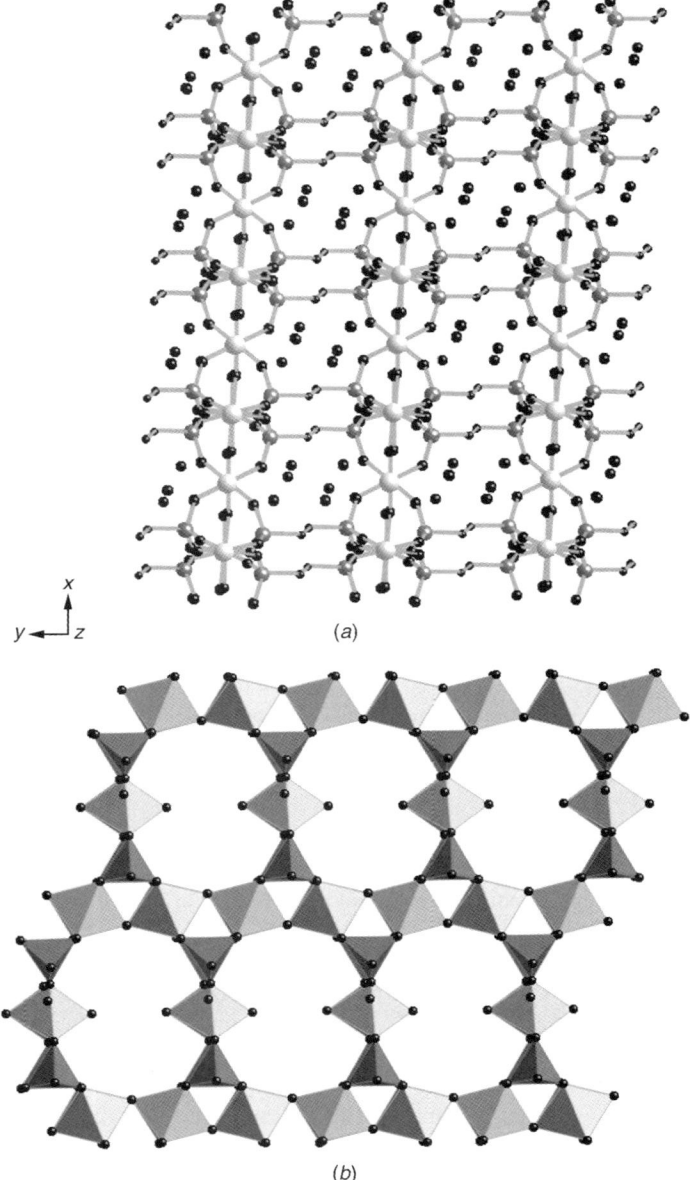

Figure 69. (a) A view of the structure of $(H_3O)_2[(VO)V_2(OH)_2(O_3PCH_2CH_2PO_3)_2] \cdot H_2O$ parallel to the crystallographic c axis. (b) A polyhedral representation of the structure showing the substructure of corner-sharing vanadium octahedra linked through $\{(VO)(RPO_3)_4\}$ subunits.

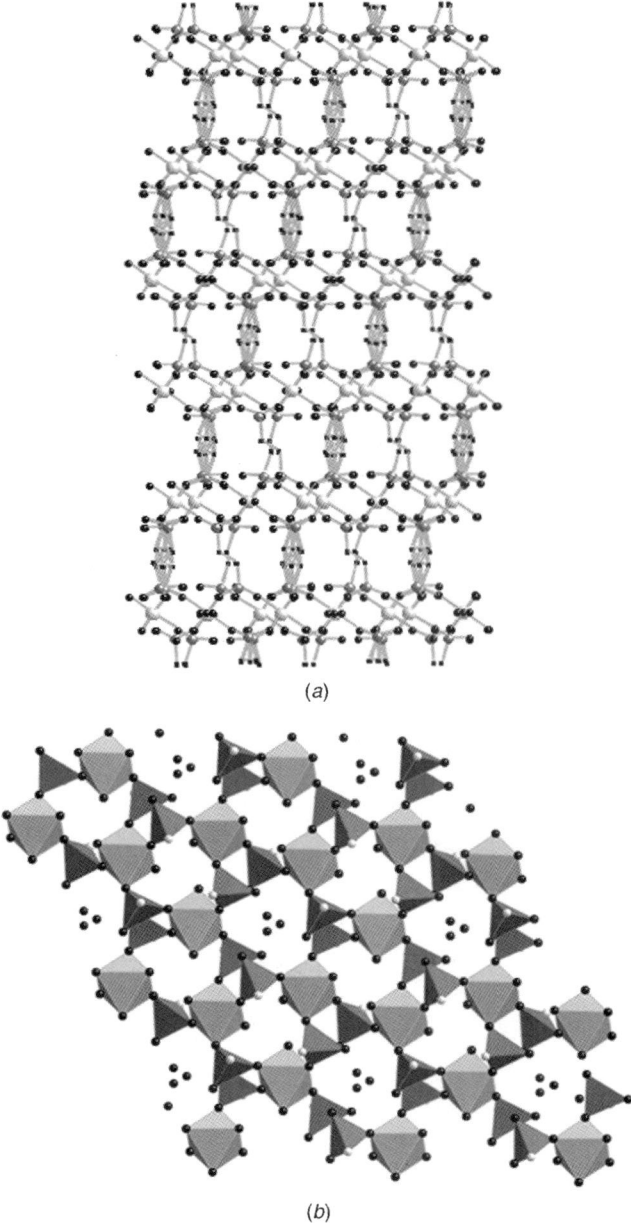

Figure 70. (a) A view of the framework structure of $(H_3O)[V_3(O_3PCH_2CH_2PO_3)(HO_3PCH_2CH_2PO_3H)_3]$, parallel to the b axis. (b) A polyhedral representation of the structure, parallel to the c axis, showing the locations of the H_3O^+ cations and H_2O molecules of crystallization.

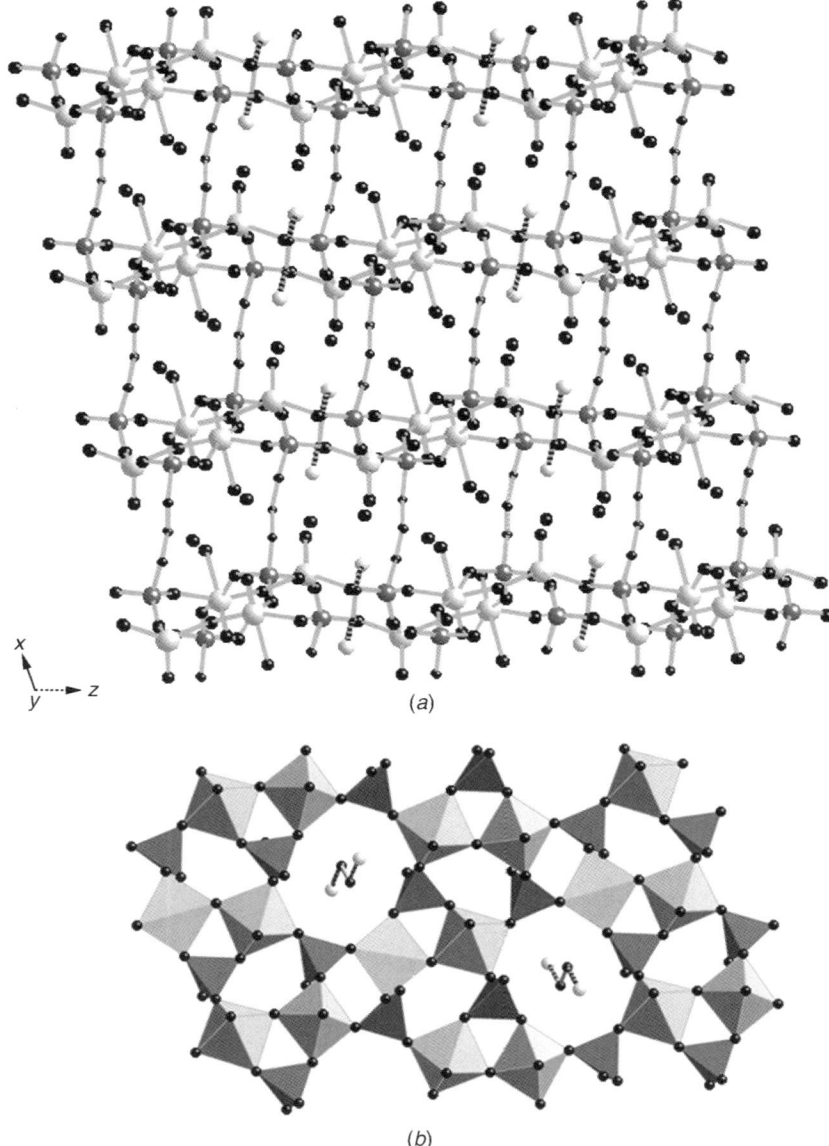

Figure 71. The structure of $(H_2en)[(VO)_4(OH)_2(H_2O)_2(O_3PCH_2CH_2CH_2PO_3)_2] \cdot 4H_2O$. (a) Parallel to the b axis. (b) A polyhedral representative showing the location of the $H_3NCH_2CH_2NH_3^{2+}$ cations.

The expectation that expansion of the tether length of the diphosphonate group would result in the common pillared motif was realized in the structure of $(H_2en)[(VO)_4(OH)_2(H_2O)_2(O_3PCH_2CH_2CH_2PO_3)_2] \cdot 4H_2O$ (276). As shown in Figure 71, the structure may be described in terms of inorganic V/P/O layers joined covalently by the trimethylene bridges of the diphosphonate groups. Since the covalently linked framework includes the ligand carbon framework, the solid may alternatively be described as a three dimensional V/P/O/C framework with ethylenediammonium cations occupying the channels parallel to the crystallographic a axis. There are two distinct V(IV) environments: the first is square pyramidal with bonds to an apical oxo group, to three phosphonate oxygen donors, and to bridging hydroxy group, while the second site is distorted octahedral with bonding to an apical oxo group, three phosphonate oxygen atoms, the bridging hydroxy group, and a terminal aqua ligand. The binuclear $\{(VO)_2(OH)\}$ unit is similar to that observed for $Cs[(VO)_2(OH)(O_3PCH_2CH_2PO_3)]$, illustrating that common structural motifs may be conserved in materials with different cations, different tether lengths and even with dissimilar overall structures: the pillared layer motif for $(H_2en)[(VO)_4(OH)_2(H_2O)_2(O_3PCH_2CH_2CH_2PO_3)_2]$ and the pillared bilayer for $Cs[(VO)_2(OH)(O_3PCH_2CH_2PO_3)_2]$.

The role of the organic cation as structural determinant is demonstrated by the structure of the "stepped" layer phase $[H_2pip][(VO)_2(O_3PCH_2CH_2CH_2PO_3H)_2]$ (277). The structure, as seen in Figure 72, consists of stepped layers constructed of V(IV) square pyramids and diphosphonate polyhedra. Adjacent vanadium sites are bridged by phosphonate oxygen atoms from two different diphosphonate ligands to give the recurrent motif $\{(VO)_2(\mu_2-O_3PC-)_2\}$. The $\{V_2P_2O_4\}$ eight-membered rings formed by this connectivity pattern fuse along common edges to produce ribbons of linked vanadium square pyramids and phosphonate tetrahedra. Each propylenediphosphonate ligand bridges adjacent strips. However, while one $\{-PO_3\}$ group employs all three oxygen donors in bonding to the vanadium centers of one strip, the other $\{-PO_3\}$ terminus of the diphosphonate group bonds through a single oxygen donor, leaving two pendant $\{P=O\}$ units on the same phosphorus. This coordination mode has the effect of terminating the connectivity and producing ribbons two vanadium and four phosphate polyhedra in width. The propylene bridges are disposed in pairs above and below the plane of each strip, so as to connect neighboring ribbons in a "stepped" orientation. The vanadyl oxygen atoms and the terminal $\{P=O\}$ groups project into the interlamellar region such that adjacent "stepped" layers generate well-defined tunnels occupied by the organic cations.

The structure of $[H_2pip][(VO)_2(O_3PCH_2CH_2CH_2PO_3H)_2]$ is analogous to that of $Cs[(VO)_2(OH)(O_3PCH_2CH_2PO_3)]$ in the sense that the organic backbones of the diphosphonate groups serves to buttress the two V/P/O sheets that sandwich the alkyl bridge. However, despite these common characteristics, the

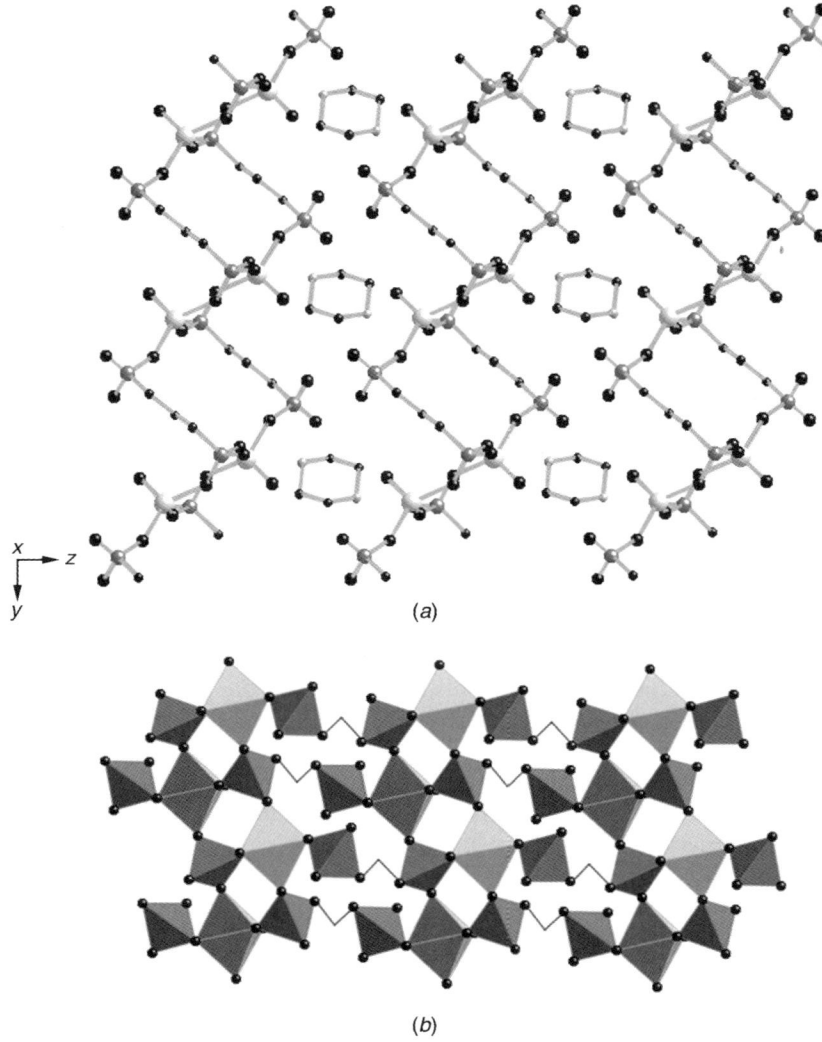

Figure 72. (a) A view of the "stepped" layer structure of [H$_2$pip][(VO)$_2$(O$_3$PCH$_2$CH$_2$CH$_2$PO$_3$H)$_2$], viewed parallel to the crystallographic a axis. (b) A polyhedral representation of the layer connectivity.

structures of the V/P/O sheets are distinct, binuclear units of corner-sharing V(IV) square pyramids for Cs[(VO)$_2$(OH)(O$_3$PCH$_2$CH$_2$PO$_3$)] and isolated V(IV) square pyramids for (H$_2$pip)[(VO)$_2$(O$_3$PCH$_2$CH$_2$CH$_2$PO$_3$H)$_2$], as well as the extended sheet structures that are planar for the former and distinctly stepped for the latter.

The organic tether of the diphosphonate unit is amendable to considerable variation. The organic backbone of N,N'-piperazinebis(methylenephosphonic acid) possesses the unusual characteristic of acting both as linker between V/P/O layers and protonation site at the nitrogen atoms, so as to serve as the counterion. As illustrated in Figure 73 for [(VO){$O_3PCH_2NH (C_2H_4)_2 NHCH_2PO_3$} ($H_2O$)] (277), the structure consists of corner-sharing vanadium octahedra and phosphorus tetrahedra fused to generate V/P/O layers covalently joined by the {$-CH_2NH(C_2H_4)_2NHCH_2-$}$^{2+}$ backbones of the diphosphonate ligands. The resultant structure adopts the common "pillared" layer motif. The vanadium site exhibits octahedral geometry defined by four oxygen donors from each of four phosphonate groups in an equatorial arrangement, with a terminal oxo ligand and the oxygen donor of an aqua group occupying the axial positions. While (H_2en)[(VO)($O_3PCH_2CH_2PO_3$)] and (H_2pip)[(VO)$_2$($O_3PCH_2CH_2CH_2PO_3H$)] also exhibit layer motifs consisting of a V(IV) site with four equatorial phosphonate oxygen donors and an apical oxo group, the details of the layer structures are quite distinct. Thus, in [(VO){$O_3PCH_2NH(C_2H_4)_2NHCH_2PO_3$}($H_2O$)], the octahedral V(IV) sites and tetrahedral phosphorus sites form 16-membered {$V_4P_4O_8$} rings as the consequence of one {P=O} group at each phosphorus adopting a pendant modality. In contrast, (H_2pip) [(VO)$_2$($O_3PCH_2CH_2CH_2PO_3H$)$_2$] exhibits V(IV) square pyramids and phosphorus tetrahedra in 8-membered {$V_2P_2O_4$} rings, somewhat reminiscent of such rings in VO(PO_4), while the intralayer location of the ethylene bridges in (H_2en)[VO($O_3PCH_2CH_2PO_3$)] results in 14-membered {$V_2P_4O_4C_4$} and 8-membered {$V_2P_2O_4$} rings.

In contrast to other "pillared" oxovanadium diphosphonates, (H_2pip) [(VO)$_2$($O_3PCH_2CH_2CH_2PO_3$)$_2$] and (H_2en)[(VO)$_4$(OH)$_2$(H_2O)$_2$($O_3PCH_2CH_2CH_2PO_3$)$_2$] which exhibit alkyl tethers at right angles to the inorganic layers, [(VO){$O_3PCH_2NH(C_2H_4)_2NHCH_2PO_3$}($H_2O$)] exhibits an organic tether at a 60° angle with respect to the V/P/O layers. Consequently, the interlayer separation for the latter is 8.0 Å, while the former two examples exhibit layer separations of 11.2 and 14.9 Å, respectively.

The V(III) phase [V(HO$_3PCH_2PO_3$)(H_2O)] is unique among the family of V–O–organophosphonates in exhibiting a structure with no V/P/O layer submotif. The structure exhibits two distinct octahedral V(III) sites: one defined by a pair of trans aqua ligands and four equatorial oxygen donors from each of four diphosphonate groups, while the record sites exhibits coordination to two methylenediphosphonate groups in the common bidentate coordination mode and to two oxygens to each of two diphosphonate ligands. Each methylenediphosphonate group bonds to four V(III) sites in η^5 modality, leaving one pendant {P—OH} site. The connectivity pattern results in a complex 3D structure, illustrated in Figure 74.

The structures of the members of the subgroup of 2D oxovanadium organophosphonates may be described in terms of heteronuclear ring motifs

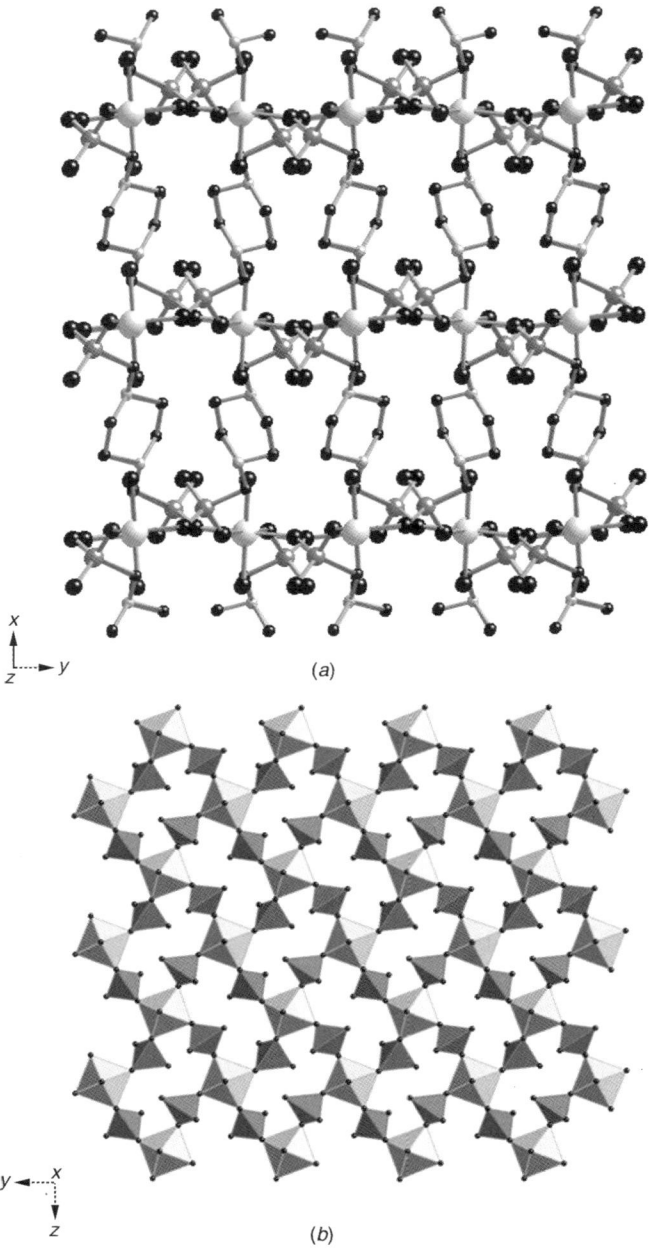

Figure 73. (a) A view of the "pillared" layer adopted by [(VO){O$_3$PCH$_2$NH(C$_2$H$_4$)$_2$NHCH$_2$PO$_3$}(H$_2$O)]. (b) A polyhedral representation of the V/P/O network.

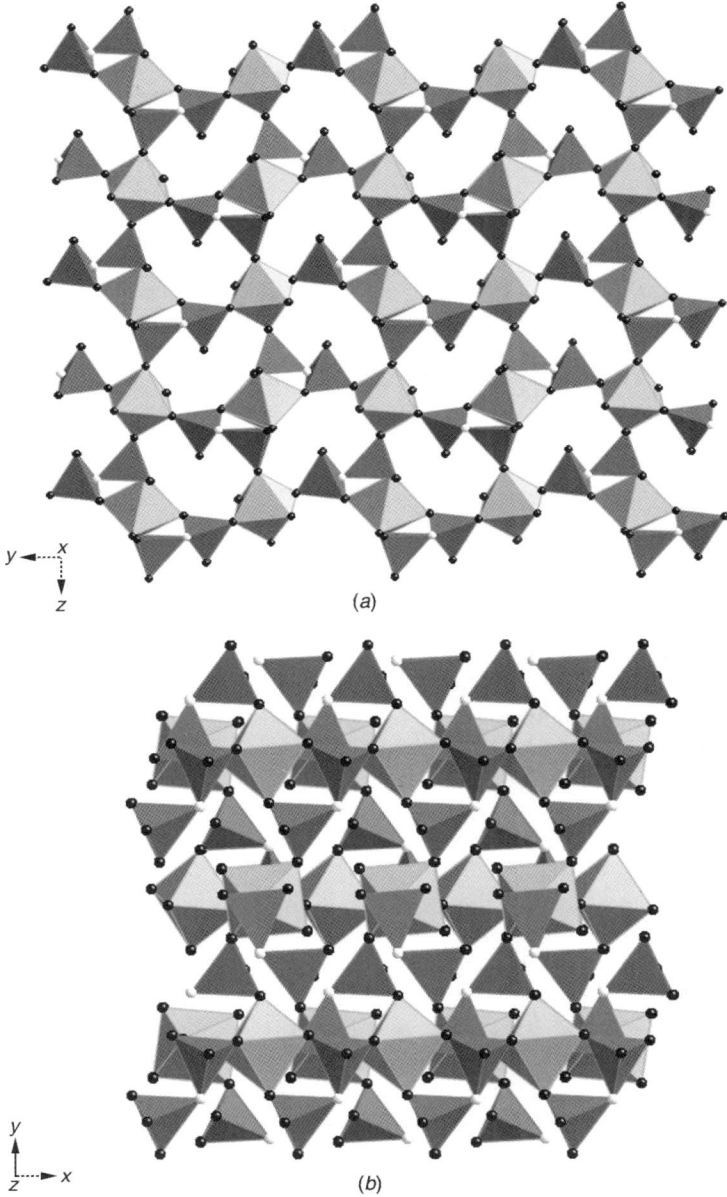

Figure 74. Polyhedral representations of the 3D structure of [V(HO$_3$PCH$_2$PO$_3$)(H$_2$O)]. (*a*) Parallel to the *a* axis. (*b*) Parallel to the *c* axis.

TABLE VI

Fundamental Building Blocks for Two-Dimensional Oxovanadium Organophosphonate and Oxovanadium Organodiphosphonate Structures

Ring Connectivity	Ring Size	Designator	Frequency
I. Class A			
{V–O–V–O–P–O–}	6	Type 1	13
{V–O–P–O–}$_2$	8	Type 2	16
{V–O–V–O–P–O–V–O–P–O–}	10	Type 3	6
{V–O–V–O–P–O–}$_2$	12	Type 4	1
{V–O–P–O–}$_3$	12	Type 4a	2
{V–O–P–O–}$_4$	16	Type 5	8
II. Class B			
{V–O–V–O–V–O–}	6	Type 1a	3
{V–O–V–O–}$_3$	12	Type 4b	1
{V–O–V–O–P–O–V–O–P–O–}$_2$	20	Type 6	1
III. Class C			
{V–O–P–C–P–O–}	6	Type 1b	4
{V–O–P–C–P–O–V–O–}	8	Type 2a	1
{V–O–P–C–P–O–}$_2$	12	Type 4c	1
{V–O–P–C–C–P–O–}$_2$	14	Type 7	5
{V–O–P–O–V–O–P–O–V–O–P–C–P–O–}	14	Type 7a	2
{V–O–P–C–C–C–P–O–}$_2$	16	Type 5a	5
{V–O–V–O–P–C–P–O–}	16	Type 5b	1
{V–O–V–O–P–C–C–C–P–O–V–O–P–C–C–C–P–O–}	18	Type 8	3
{V–O–V–O–P–C–C–C–P–O–}$_2$	20	Type 9	4
{V–O–P–C–N–C–C–N–P–O–}$_2$	22	Type 10	1
IV. Class D			
{M–O–P–C–P–O–}	6	Type 1c	5
{M–O–V–O–P–O–}	6	Type 1d	4
{M–O–P–O–}$_2$	8	Type 2b	1
{M–O–V–O–V–O–P–O–}	8	Type 2c	1
{M–O–V–O–P–O–V–O–P–O–}	10	Type 3a	2
{M–O–P–O–H–O–P–O–V–O–P–O–V–O–P–O–}	16	Type 5b	1
{M–O–V–O–P–C–P–O–V–O–M–O–P–O–V–O–V–O–}	18	Type 8b	1
{M–O–P–C–C–C–P–O–V–O–V–O–}$_2$	24	Type 11	1

that constitute the fundamental building blocks of the layers. Inspection of Figures 56–74 reveals a number of ring types common to both oxovanadium organophosphonate and oxovanadium organodiphosphonate layered compounds. These include 6-, 8-, 10-, 12-, and 16-membered rings as fundamental building blocks (FBBs) of Class A, which are identified in Table VI. Several ring types (Class B) are unique to the oxovanadium-organophosphonate (V—O—RPO$_3^{2-}$) subclass, while Class C types occur only as a consequence of the tethering of {PO$_3$} groups through the organic frameworks of the organodiphosphonate family of ligands. The fundamental building blocks of Class C may form part of the V/P/O layer, so as to introduce the organic tether into the inorganic–organic composite network, or may serve as buttresses between adjacent inorganic layers.

As summarized in Table VII, the structures of the 2D oxovanadium organophosphonates and oxovanadium organodiphosphonates are constructed from several recurring fundamental building blocks. However, we note that the unique example of a V(V) species for this family, (NH$_4$)[(VO$_2$)$_3$(O$_3$PMe)$_2$], exhibits unusual ring motifs. Similarly, the unusual 2D structures of [(VO)$_2$(O$_3$PCH$_2$PO$_3$)], (H$_3$NCH$_2$CH$_2$NH$_3$)[(VO)$_4$(OH)$_2$(H$_2$O)$_2$(O$_3$PCH$_2$CH$_2$ CH$_2$PO$_3$)$_2$], and [(VO)(H$_2$O)(O$_3$PCH$_2$NH(C$_2$H$_4$)$_2$NHCH$_2$PO$_3$)(H$_2$O)] are also associated with unique ring geometries.

Not unexpectedly, three- and four polyhedral connect rings, Types 1 and 2, are the most commonly encountered building blocks for the layer structures, occurring in 9 of 13 structures listed in Table VI. Such motifs are also common features of oxovanadium organophosphonate polyanion clusters and reflect the coordination requirements of the metal polyhedra and the ligand geometries. However, it is noteworthy that the three polyhedral connect ring Type 1 requires corner sharing between vanadium polyhedra and consequently cannot constitute a building block of those structures constructed exclusively from isolated vanadium polyhedra, namely, [(VO)$_2$(O$_3$PCH$_2$PO$_3$)$_2$(H$_2$O)$_4$], (H$_2$en) [(VO)(O$_3$PCH$_2$CH$_2$PO$_3$)], (H$_2$pip)[(VO)$_2$(O$_3$PCH$_2$CH$_2$CH$_2$PO$_3$H)$_2$], and [(VO) {O$_3$PCH$_2$NH(C$_2$H$_4$)$_2$NHCH$_2$PO$_3$}(H$_2$O)]. The Type 1 FBB is ubiquitous in those layer structures exhibiting {V—O—V} linkages.

While polyhedral connect rings >4 are less common, the Type 3 and Type 5 FBBs occur with frequencies of four and six, respectively. Since the five polyhedral connect ring also requires {V—O—V} bonding, the frequency is limited to such structure types. The 16-membered FBB Type 5, which is constructed of four vanadium polyhedra alternating with four phosphorus tetrahedra, represents an unique example of an FBB of ring dimensions > 10 that occur with frequency >2. There is no counterpart to this FBB in the polyanion structural chemistry, suggesting that extension of {V—O—RPO$_3$} structures into 2D requires some ring expansion to alleviate steric interactions between the ligand substituents. Consequently, all structures of Table VII exhibit

TABLE VII

Occurrence of Fundamental Building Blocks in the Structures of Two-Dimensional Oxovanadium Organophosphonates and Oxovanadium Organodiphosphonates

Compound	Fundamental Building Blocks
I. Oxovanadium Organophosphonates (V—O—RPO$_3$)	
[(VO)(H$_2$O)(O$_3$PPh)]	Type 1
	Type 2
	Type 4
(EtNH$_3$)$_2$[(VO)$_3$(H$_2$O)(O$_3$PPh)$_4$]	Type 1
	Type 2
	Type 6
(Et$_2$NH$_2$)(Me$_2$NH$_2$)[(VO)$_4$(OH)$_2$(O$_3$PPh)$_4$]	Type 1
	Type 2
	Type 3
	Type 5
(Et$_4$N)[(VO)$_3$(OH)(H$_2$O)(O$_3$PEt)$_3$]	Type 1
	Type 2
	Type 3
	Type 5
(NH$_4$)[(VO$_2$)$_3$(O$_3$PMe)$_2$]	Type 1
	Type 1a
	Type 4b
II. Oxovanadium Organodiphosphonates (V—O—O$_3$PRPO$_3$)	
[(VO)$_2$(O$_3$PCH$_2$PO$_3$)]	Type 1b
	Type 2a
	Type 7a
Cs[(VO)$_2$V(H$_2$O)(O$_3$PCH$_2$PO$_3$)]	Type 1
	Type 2
	Type 1b
	Type 5
(H$_2$en)[(VO)(O$_3$PCH$_2$CH$_2$PO$_3$)]	Type 2
	Type 7
[(VO)$_2$(OH)(O$_3$PCH$_2$CH$_2$PO$_3$)] • H$_2$O	Type 2
	Type 7
(H$_3$O)$_2$[(VO)V$_2$(OH)$_2$(O$_3$PCH$_2$CH$_2$CH$_2$PO$_3$)$_2$]	Type 1
	Type 2
	Type 5
	Type 7
Cs[(VO)$_2$(OH)(O$_3$PCH$_2$CH$_2$PO$_3$)]	Type 1
	Type 3
	Type 7
	Type 4a

TABLE VII (Continued)

Compound	Fundamental Building Blocks
$(H_2en)[(VO)_4(OH)_2(H_2O)_2(O_3PCH_2CH_2CH_2PO_3)_2]$	Type 1
	Type 2
	Type 3
	Type 5
	Type 5a
	Type 8
	Type 9
$(NH_4)[(VO)_2(OH)(H_2O)\{O_3P(CH_2)_3PO_3\}] \cdot 2H_2O$	Type 1
	Type 2
	Type 3
	Type 5
	Type 5a
	Type 8
	Type 9
$(NH_4)[V_2O_3(H_2O)\{O_3P(CH_2)_3PO_3\}] \cdot H_2O$	Type 1
	Type 2
	Type 3
	Type 5
	Type 5a
	Type 8
	Type 9
$(H_2pip)[(VO)_2(O_3PCH_2CH_2CH_2PO_3)_2]$	Type 2
	Type 5a
$[(VO)(H_2O)(O_3PCH_2NHEt_2NHCH_2PO_3)]$	Type 5
	Type 10
III. Oxovanadium Organodiphosphonates with M^{II}–Organoimine Subunits	
$[Cu\text{-}bpy)(VO)(O_3PCH_2PO_3)]$	Type 1b
	Type 1c
	Type 7a
$[Cu(bpy)(VO)(O_3PCH_2CH_2PO_3)]$	Type 1c
	Type 2
	Type 7
$[Cu(bpy)(VO)(O_3PCH_2CH_2CH_2PO_3)]$	Type 1d
	Type 2
	Type 3a
	Type 5a
$[Cu(phen)(VO)(H_2O)(O_3PCH_2PO_3)]$	Type 1b
	Type 1c
	Type 5b
$[\{Cu(phen)\}_2(V_2O_5)(O_3PCH_2CH_2PO_3)]$	Type 1d
	Type 2b
	Type 11
	(*continues*)

TABLE VII (Continued)

Compound	Fundamental Building Blocks
[Cu(phen)(V$_2$O$_3$)(O$_3$PCH$_2$CH$_2$CH$_2$PO$_3$)]	Type 1
	Type 1d
	Type 2
	Type 2c
	Type 5a
	Type 9
[{Cu(bpydicarbH$_2$)}$_2$(V$_3$O$_5$)(O$_3$PCH$_2$PO$_3$)$_2$ (H$_2$O)] • 2H$_2$O	Type 1
	Type 1c
	Type 2
	Type 4c
	Type 5b
[Cu(phen)(V$_2$O$_3$)(O$_3$PCH$_2$PO$_3$)]	Type 1b
	Type 1c
	Type 3a
	Type 8b

FBBs of Type 4 (12-membered rings) or larger; in fact, all but two exhibit FBBs of Type 5 (16-membered rings) or larger.

While FBBs larger than Type 5 are common for the oxovanadium organodiphosphonate subclass of solids, such rings span V/P/O layers rather than provide part of the covalent connectivity within the layer. The only example of an FBB larger than Type 5 that does not bridge adjacent layers occurs in (NH$_4$)[(VO$_2$)$_3$(O$_3$PMe)$_2$], which exhibits a 20-membered ring structure, Type 6. However, this species is best described as a vanadium oxide layer with {O$_3$PMe} ligands attached above and below the plane, rather than the more common {V–P–O} layer with the organic substituents occupying interlamellar positions. This V(V) species exhibits an unusually high density of vanadium sites within the layer, in contrast to the more open structures adopted by the V(IV) species. Whether such high density is a characteristic of V(V) organophosphate materials is an issue requiring further development of the V(V) chemistry of the V/O/RPO$_3$ system.

VI. Oxovanadium Phosphates and Organophosphonates with Secondary Metal–Ligand Subunits Integrated into Covalently Linked V/O/M′/P Scaffoldings

A. General Considerations

While organoamine constituents have been conventionally introduced as charge-balancing counterions in zeolite synthesis and in the preparations of

V/P/O phases discussed above, in this application the organic component serves as a ligand to a secondary metal site, a first-row transition or posttransition metal cation (Tables VIII and IX). Consequently, a coordination complex cation is assembled that serves to provide charge-compensation, space-filling, and structure-directing roles.

The structure of the organoamine–secondary metal complex cation is derived, of course, from the geometrical requirements of the ligand as well as the coordination preferences of the metal. The ligand set may include chelating agents that coordinate to a single metal center or bridging ligands of various extensions that may provide a polymeric cationic scaffolding for the entraining of the V/P/O substructure. One strategy adopts appropriate stoichiometric control to form mononuclear metal–ligand chelate complex cations $\{M(N\text{--}N)_2\}^{n+}$ that are coordinatively unsaturated, and hence capable of bonding to the oxo groups of the oxide substructure, so as to provide linkage between the oxide subunits.

The properties of this cationic component may be tuned by exploiting the preferred coordination modes of various transition and posttransition metal cations. For example, while a Cu^{II}–organoamine fragment will likely exhibit $4+2$ or $4+1$ coordination geometries, the Cu(I) counterpart will result in low coordination numbers with tetrahedral or trigonal-planar geometries predominating. Similarly, a Ni(II) based cation will adopt more regular octahedral coordination, while Zn(II) species may adopt various coordination modes.

B. Influence of M^{II}–Organodiimine Subunits on Oxovanadium Phosphate Structures

The prototypical 1D structure is provided by [Cu(bpy)(VO$_2$)(PO$_4$)], shown in Figure 75. The structure consists of two parallel chains of alternating and corner-sharing V(V) tetrahedra and phosphate tetrahedra. The chains are linked by "$4+1$" square pyramidal {Cu(bpy)O$_3$} units into a ribbon two polyhedra in thickness. Each copper bonds to the imine nitrogen donors of the bpy ligand, a vanadium oxo group and a phosphate oxygen from one chain and a phosphate oxygen from the second chain.

The ligand influences can be quite unanticipated, as illustrated by the structure of α-[Cu(phen)(VO$_2$)(PO$_4$)] (297), which exhibits the identical Cu–organoimine–VO$_x$–PO$_4$ stoichiometry as [Cu(bpy)(VO$_2$)(PO$_4$)]. However, in this instance the V/P/O substructure is not based on oxovanadium–phosphate chains, but rather on {V$_2$P$_2$O$_4$} rings constructed from four corner-sharing tetrahedra. These rings are in turn linked through pairs of {Cu(phen)O$_3$} square pyramids into a 1D chain (Fig. 76). Both [Cu(bpy)(VO$_2$)(PO$_4$)] and [Cu(phen)(VO$_2$)(PO$_4$)] share the {Cu$_2$P$_2$O$_4$} ring motif as a common building block. Curiously, the {Zn(bpy)}$^{2+}$ analogue, [Zn(bpy)(VO$_2$)(PO$_4$)] exhibits a

TABLE VIII

Selected Structural Features of Vanadium Phosphate Phases with Secondary Metal–Ligand Subunits

Compound	V–P–O Framework Structure	Vanadium Coordination and Linkage Parameters	Cell	Space Group	$V=O_t^{a,b}$	$V-O_{phosphate}^{c}$	V–O (other)
[Cu(phen)(VO$_2$)(PO$_4$)]	Chain	Isolated V(V) tetrahedra	20.213(1) 11.600(1) 14.526(1) $\beta = 126.92(1)$	$C2/c$	1.612(2)	1.852(2) × 2	1.635(2) (V–O–Cu)
[Zn(bpy)(VO$_2$)(PO$_4$)]	Chain	Isolated V(V) tetrahedra	7.786(1) 17.290(1) 9.790(2) $\beta = 104.74(1)$	$P2_1/c$	1.608(2)	1.850(2) × 2	1.646(2) (V–O–Zn)
[Zn(terpy)(VO$_2$)(PO$_4$)]^d	Chain	Isolated V(V) tetrahedra	10.454(1) 11.887(1) 13.998(1) $\alpha = 93.034(1)$ $\beta = 93.776(1)$ $\gamma = 105.468(1)$	$P\bar{1}$	V1, 1.626(3) × 2 V2, 1.620(2) × 2	V1, 1.875(3) × 2 V2, 1.863(3) × 2	
[Cu(bpy)(VO$_2$)(PO$_4$)]	Chain	Isolated V(V) tetrahedra	5.553(1) 10.226(1) 11.736(1) $\alpha = 73.606(2)$ $\beta = 79.982(2)$ $\gamma = 86.438(2)$	$P\bar{1}$	1.622(2)	1.851(2) × 2	1.635(3) (V–O–Cu)

Compound					
[{Cu(terpy)}$_2$(VO$_2$)$_3$(PO$_4$)]	Chain	Isolated V(V) tetrahedra, two corner-sharing V(V) square pyramids	8.991(1) 11.301(1) 19.451(2) $\alpha = 106.762(1)$ $\beta = 91.208(1)$ $\gamma = 103.984$	$P\bar{1}$	V1, 1.623(4) × 2 V2, 1.594(4) V3, 1.620(3)	V1, 1.850(3) × 2 V2, 1.956(3) × 3 V3, 1.985(4) × 2	1.702(3) (V–O–V) 1.647(3) (V–O–Cu) 1.987(2) (V–O–V)
[{Cu(bpy)}$_2$(VO)$_3$(PO$_4$)$_2$](HPO$_4$)$_2$]·2H$_2$O	Layer	Isolated V(IV) square pyramids, isolated V(V) octahedra	8.150(1) 9.721(1) 11.947(1) $\alpha = 106.811(1)$ $\beta = 98.943$ $\gamma = 110.429$	$P\bar{1}$	1.607(2) 1.586(4)	1.963(2) × 4 2.017(2) × 4	2.105(2) (V–O–Cu)
β-[Cu(phen)(VO$_2$)(PO$_4$)]	Layer	Isolated V(IV) square pyramids, V(V) tetrahedra	8.055(1) 9.186(1) 10.238(1) $\alpha = 97.375(2)$ $\beta = 100.673(2)$ $\gamma = 104.269(2)$	$P\bar{1}$	1.614(3)	1.852(4) × 2 1.866(4) × 2	1.618(4) (V–O–Cu)

[a] O$_t$ is the terminal, multiply bonded oxo group.
[b] Where there are more than one oxo group associated with a metal site (× n), the average value of the bond length is given.
[c] Where there is more than one phosphate oxygen bound to a vanadium site (× n) the average value is given.
[d] 2,2′:6′,2″-Terpyridine = terpy.

TABLE IX

Selected Structural Features of Vanadium Organophosphonate Phases with Secondary Metal–Ligand Subunits

Compound	V–P–O Framework Structure	Vanadium Coordination and Linkage Parameters	Cell	Space Group	V=O$_t^{a,b}$	V–O$_{phosphonate}$^{c}	V–O (other)
[{Cu(bpy)H$_2$O)}(VO(PO$_3$CH$_2$PO$_3$)]	Layer	Isolated V(IV) square pyramids	11.616(1) 10.755(2) 12.300(1)	$Pca2_1$	1.593(1)	1.987(1) × 4	
[Cu(bpy)(VO(PO$_3$CH$_2$CH$_2$PO$_3$)]	Layer	Isolated V(IV) square pyramids	8.200(1) 9.405(2) 19.948(1) β = 96.603(1)	$P2_1/n$	1.611(2)	1.962(2) × 4	
[Cu(bpy)(VO(O$_3$PCH$_2$CH$_2$PO$_3$)]·H$_2$O	Layer	Isolated V(IV) square pyramids	9.914(1) 17.678(2) 20.876(2) β = 103.167(1)	$P2_1/c$	1.610(2)	1.967(3) × 4	
[{Cu(bpy{COOH}$_2$)}$_2$ (V$_3$O$_5$)$_2$(PO$_3$CH$_2$CH$_2$PO$_3$)]·H$_2$O	Layer	Corner sharing of one V(IV) and two V(V) square pyramids	7.899(1) 11.042(2) 11.858(1) α = 100.845(1) β = 98.400(2) γ = 105.290(1)	$P\bar{1}$	V1, 1.659(4) V2, 1.619(4) × 2	V1, 1.973(2) × 2 V2, 1.962(4) × 3	1.926(1) (V–O–V) 1.619(2) (V–O–Cu)
[{Cu(phen)}$_2$(V$_3$O$_5$) PO$_3$CH$_2$CH$_2$CH$_2$PO$_3$)]	Layer	Distorted corner-sharing V(IV) square pyramids	9.436(1) 10.656(1) 11.035(1) 118.187(1) 91.416(2) 107.821(2)	$P\bar{1}$	V1, 1.608(7) V2, 1.597(2)	V1, 1.943(2) × 3 V2, 1.949(3) × 3	1.607(1) (V–O–Cu) 1.831(2) (V–O–V)

Compound	Structure	Cell parameters	Space group	V=O_t	V–O–P / V–O–V	Other
[Cu(phen)(VO)(H₂O)(PO₃CH₂CH₂PO₃)]	Isolated V(IV) square pyramids	9.066(1) 8.658(2) 20.935(1) β = 97.306	$P2_1/c$	1.604(3)	3 × 1.954(2)	2.033(2) (water)
[{Cu(phen)}₂VO₂VO₃(PO₃CH₂CH₂PO₃)]	Two corner-sharing V(V) tetrahedra	10.601(2) 11.695(1) 13.180(1) α = 71.369(1) β = 70.790(2) γ = 80.738(1)	$P\bar{1}$	1.618(2)	1.871(2) 1.875(2)	1.644(2) (V–O–Cu)
[{Cu(phen)}₂(V₃O₅)(O₃PCH₂PO₃)₂(H₂O)]	3D	Isolated V(IV) tetrahedra and square pyramids	8.395(1) 16.840(2) 11.914(1) β = 93.903	$P2_1/n$	V1, 1.636(2) V2, 1.611(2)	V1, 2.004(4) × 4 V2, 1.863(4) × 2	V2, 1.635(2) (V–O–Cu) V1, 2.314(1) (water)

[a] O_t is the terminal, multiply bonded oxo group.
[b] Where there are more than one oxo group associated with a metal site (×n) the average value of the bond length is given.
[c] Where more than one organophosphate oxygen is bound to the vanadium site (×n) the average value is given.

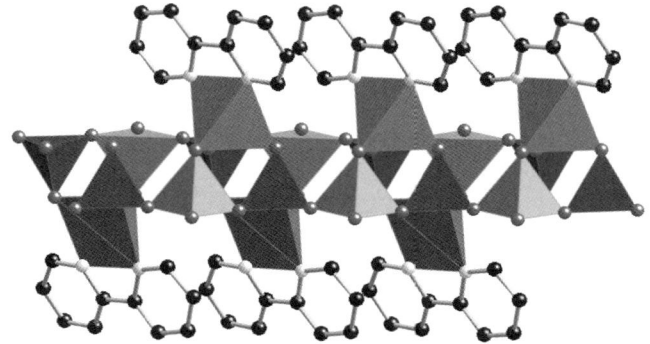

Figure 75. A view of the 1D structure of [Cu(bpy)(VO$_2$)(PO$_4$)].

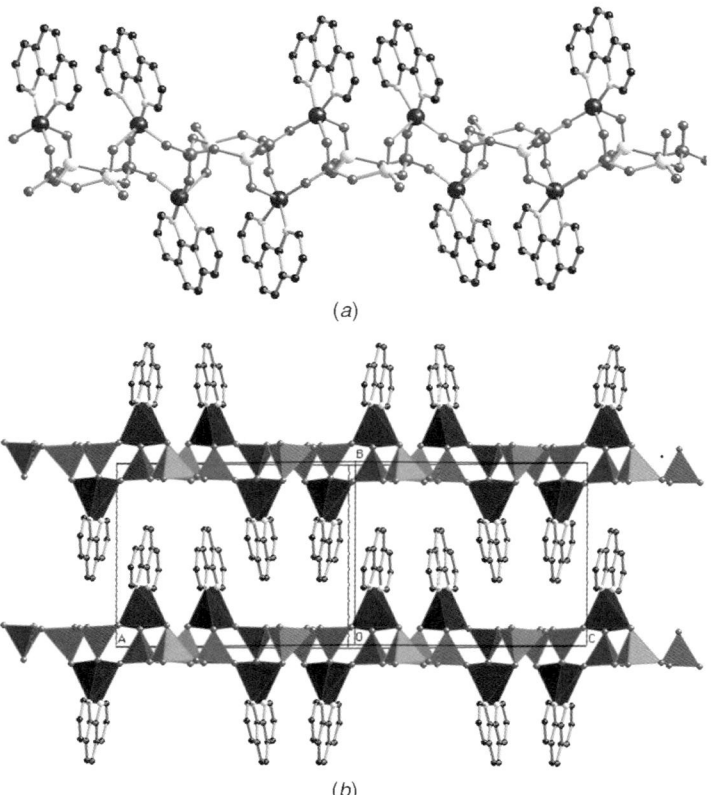

Figure 76. (a) The 1D structure of α-[Cu(phen)(VO$_2$)(PO$_4$)]. (b) A polyhedral representation of the cell contents.

chain structure identical to the copper–phenanthroline species [Cu(phen)(VO$_2$)(PO$_4$)] rather than to the copper–bipyridine [Cu(bpy)(VO$_2$)(PO$_4$)].

Substitution of the more sterically demanding and tridentate ligand terpyridine also results in a 1D structure, [Zn(terpy)(VO$_2$)(PO$_4$)], shown in Figure 77. Once again the {V$_2$P$_2$O$_4$} ring motif of corner-sharing tetrahedra is observed. The rings are linked into a chain through pairs of {Zn(terpy)O$_3$} square pyramids. In contrast to the structures of [Cu(phen)(VO$_2$)(PO$_4$)] and [Zn(bpy)(VO$_2$)(PO$_4$)], where the M(II) site bonds to vanadium oxo group and a phosphate oxygen from the ring and a phosphate oxygen from a neighboring ring, in [Zn(terpy)(VO)$_2$(PO$_4$)] the Zn(II) sites bridge through two phosphate

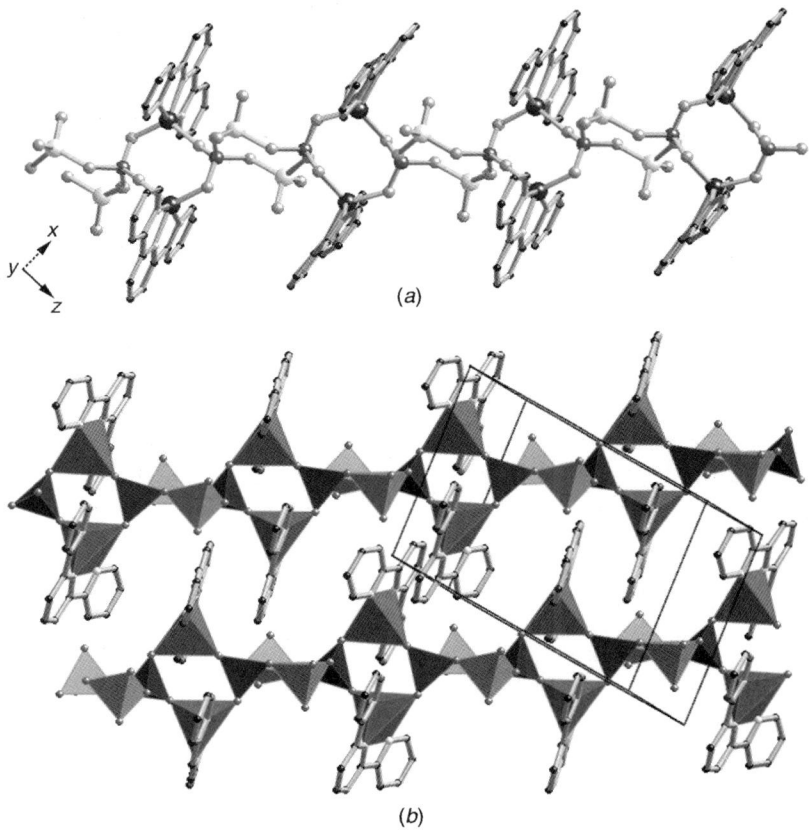

Figure 77. (a) An isolated chain of [Zn(terpy)(VO$_2$)(PO$_4$)]. (b) A polyhedral view of adjacent chains.

oxygen atoms only. Consequently, there are no V—O—Zn bonds, and each vanadium site exhibits two terminal oxo groups.

The structural versatility of this family is further demonstrated by [{Cu(terpy)}$_2$(VO$_2$)$_3$(PO$_4$)(HPO$_4$)$_2$], shown in Figure 78. The chain consists of exclusively corner-sharing V(V) tetrahedra, square pyramids and trigonal bipyramids, copper square pyramids, and {PO$_4$}$^{3-}$ and {HPO$_4$}$^{2-}$ tetrahedra. There are three distinct vanadium centers: an "isolated" {VO$_4$} tetrahedra site (i.e., no V—O—V linkages) and a binuclear site consisting of an oxo-bridged {VO$_5$} trigonal bipyramid and a {VO$_5$} square pyramid. The {PO$_4$}$^{3-}$ group participates in corner sharing to bridge the vanadium sites of the binuclear unit, the tetrahedral vanadium site, and the copper. The remaining phosphate sites each bridge three metal centers and exhibit a pendant and protonated oxygen site. The terpy ligand occupies three coordination sites on the copper and introduces significant steric constraints, leaving two coordination sites available for linking to the oxovanadium phosphate chain. One copper site coordinates to two oxygen donors from two phosphorus tetrahedra, while the second bonds to a phosphorus oxygen and a bridging oxo group of the trigonal bipyramidal vanadium site. The presence of three distinct vanadium polyhedra is a structurally unique feature for the oxovanadium phosphate family of materials.

This family of 1D structures exhibit V/P/O substructures decorated with M^{II}–diimine subunits, suggesting that the ligands may serve to passivate the surface of the growing oxide so as to limit the dimensionality of the phase. While this speculation is reasonable, expansion into 2D structures is not precluded, as demonstrated by the structure of β-[Cu(phen)(VO$_2$)(PO$_4$)], shown

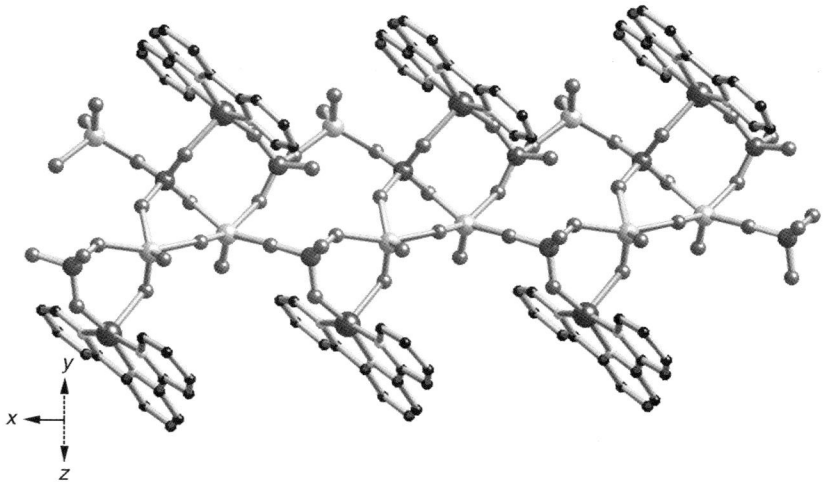

Figure 78. The structure of [{Cu(terpy)}$_2$(VO$_2$)$_3$(PO$_4$)(HPO$_4$)$_2$].

in Figure 79. The overall structure consists of V—P—O chains, constructed from {$V_2P_2O_4$} rings of corner-sharing tetrahedra linked through {VO_5} square pyramids; these chains are in turn linked through binuclear units of edge-sharing Cu(II) square pyramids into a 2D network. An unusual substructure is provided by the chains of binuclear copper clusters linked through corner-sharing vanadium square pyramids. It is also noteworthy that the structure

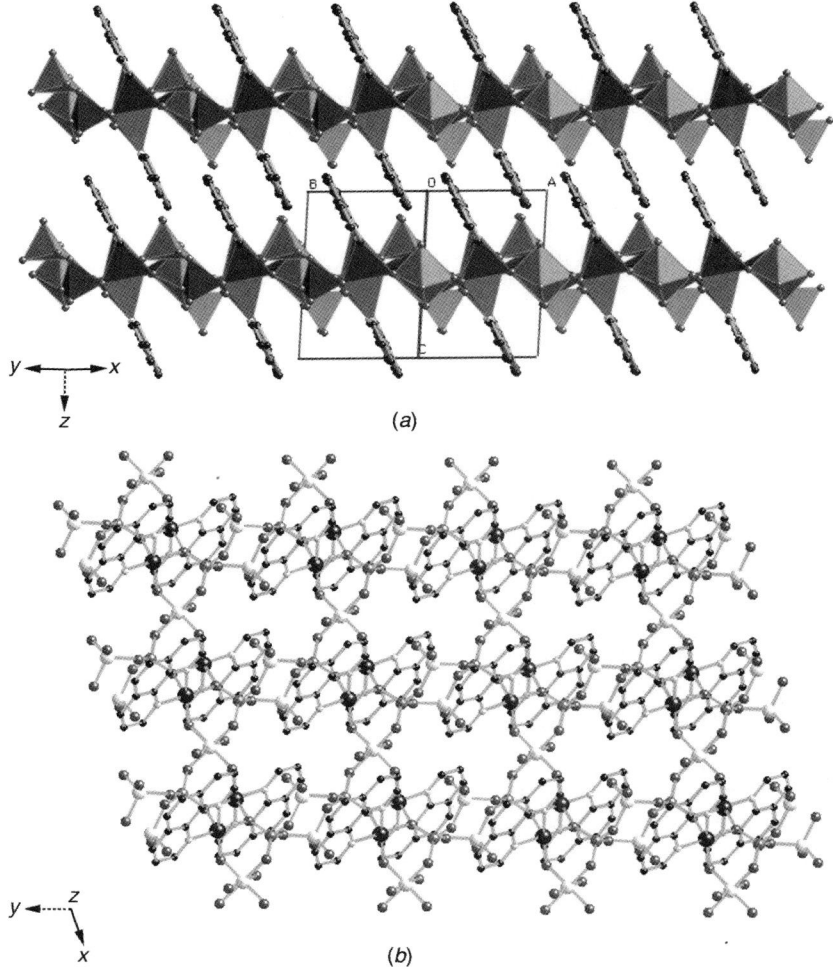

Figure 79. A polyhedral representation of the structure of β-[Cu(phen)(VO$_2$)(PO$_4$)] (a); ball-and-stick representation of the layer structure (b).

possesses cavities defined by 10 polyhedral connects, that is 20-membered $\{Cu_2V_4P_4O_{10}\}$ rings.

The structure of $[\{Cu(bpy)\}_2(VO_2)_3(PO_4)(HPO_4)_2]$, shown in Figure 80, may be best described as $\{Cu_2(bpy)_2(VO)_2(PO_4)_2(HPO_4)_2\}$ chains linked through square pyramidal $\{VO_5\}$ units into a 2D network. Embedded within the chain are $\{Cu_2(bpy)_2(VO)_2(PO_4)_2\}$ clusters, constructed of a central ring of edge- and corner-sharing $\{VO_6\}$ octahedra and $\{PO_4\}$ tetrahedra capped by two

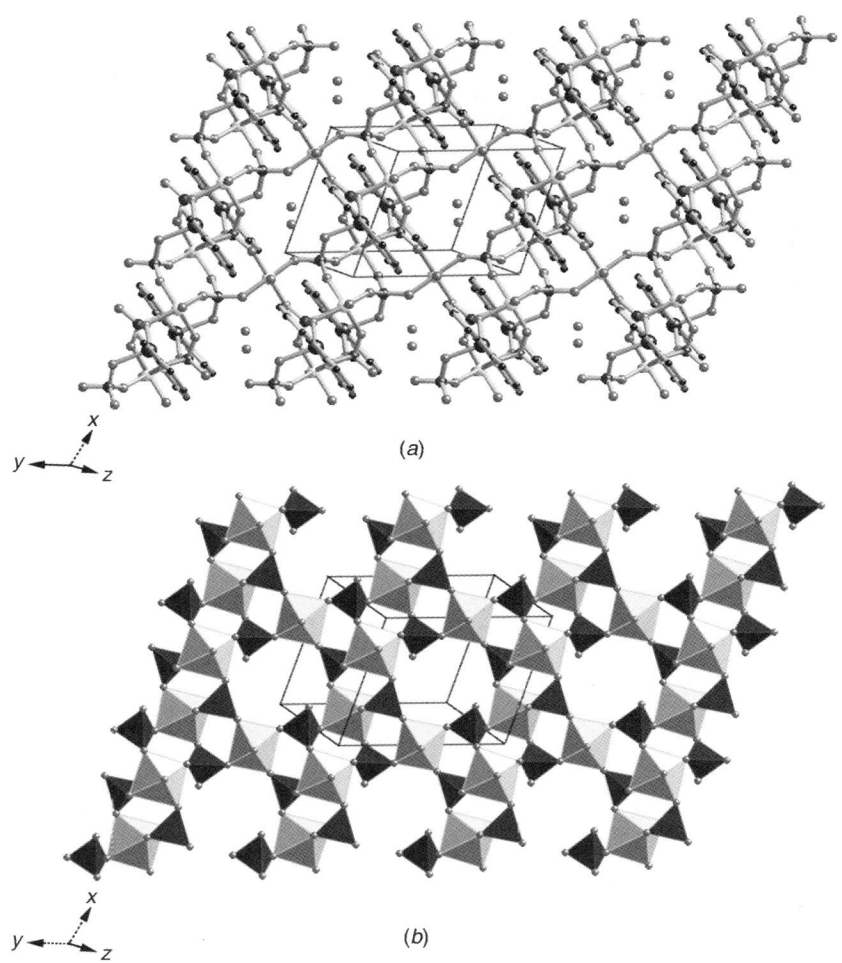

Figure 80. (a) A ball-and-stick representation of $[\{Cu(bpy)\}_2(VO)_3(PO_4)_2(HPO_4)_2]\cdot H_2O$, showing the location of the water molecules of crystallization. (b) A polyhedral view of the V/P/O network.

{CuO$_3$N$_2$} square pyramids. The phosphate groups of this cluster motif each link two copper and two vanadium sites, participating in edge-sharing with a vanadium polyhedron and contributing two triply bridging oxygen donors. The fourth vertex of each of these phosphate groups serves to bridge to the square pyramidal vanadium site. The individual clusters are linked into the chain motif through four corner-sharing {PO$_4$} tetrahedra. The fourth vertex of each of these phosphate sites again bridges to the square pyramidal vanadium center. One consequence of this unprecedented polyhedral connectivity is to generate eight polyhedral connect rings {V$_4$P$_4$O$_8$}, which create network cavities. The water molecules of crystallization are situated within these cavities and strongly hydrogen bonded to the phosphate oxygens and the protonated phosphate oxygen. Since these intralamellar cavities are contained within the hydrophobic regions defined by the bpy groups, there is no facile pathway for removal of H$_2$O from the crystal.

C. Influence of M^{II}-Organodiimine Subunits on Oxovanadium Organophosphonate Structures

As noted above, oxometal organophosphonate phases are a well-documented family of materials for which structural modification through incorporation of an organic subunit is effected through variations in the steric demands of the organic, the tether length, and/or the presence of additional functional groups. The structural demands of the organophosphonate may be combined with the influences of secondary metal–organoimine components to exploit both the spatial transmission of structural information by the diphosphonate ligand and the surface modification of the oxide material by a multidentate organonitrogen ligand.

Since structural modifications are consequences of both the tether length of the diphosphonate ligand {O$_3$P(CH$_2$)$_x$PO$_3$}$^{4-}$ and of the nature of the M^{II}–diimine subunit, one approach to the development of the chemistry is to vary the spatial extension of the organodiphosphonate for a variety of M^{II}–organodiimine subunits. Two series have been studied to date, Cu^{II}(LL)–O$_3$P(CH$_2$)$_x$PO$_3$ with LL = bpy, x = 1, 2, and 3 (298) and with LL = phen, x = 1, 2, and 3.

The structure of [{Cu(bpy)(H$_2$O)}(VO)(O$_3$PCH$_2$PO$_3$)] (Fig. 81) consists of 2D layers constructed from corner-sharing Cu(II) square pyramids, V(V) square pyramids, and phosphorus tetrahedra. The coordination sphere of the Cu(II) site is defined by the nitrogen donors of a bpy ligand and two oxygen donors of a chelating methylenediphosphonate ligand in the basal plane, and an aqua ligand in the apical position. The basal plane of the vanadium center consists of two oxygen donors from a chelating methylenediphosphonate ligand and two oxygen donors from each of two monodentate methylenediphosphonate groups; the

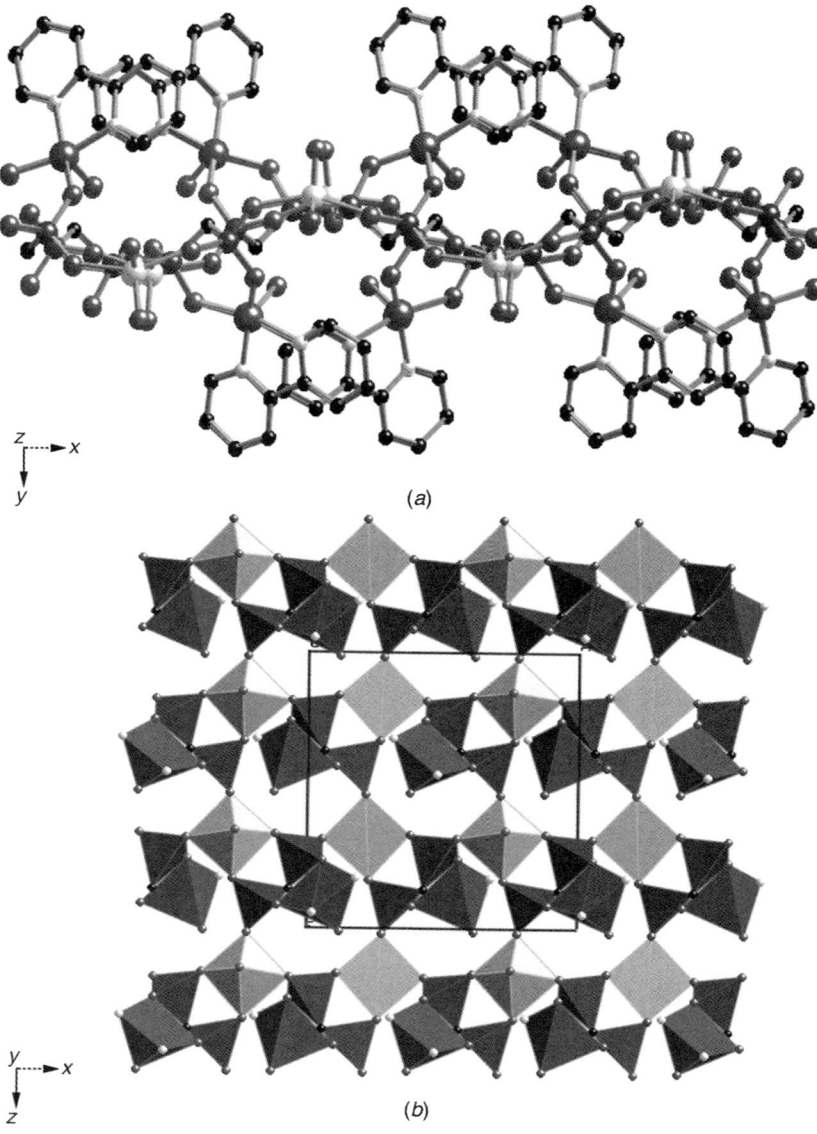

Figure 81. (a) A view of the structure of [{Cu(bpy)(H$_2$O)}(VO)(O$_3$PCH$_2$PO$_3$)] parallel to the crystallographic c axis. (b) A polyhedral representation of the V/P/O network.

apical position is occupied by the terminal oxo group. Each methylenediphosphonate ligand links one copper and three vanadium sites.

The layer structure of [{Cu(bpy)(H$_2$O)}(VO)(O$_3$PCH$_2$PO$_3$)] may be described as undulating {(VO)(O$_3$PCH$_2$PO$_3$)}$^{2-}$ networks decorated with {Cu(bpy)H$_2$O)}$^{2+}$ groups that project into the interlamellar regions. The {(VO)(O$_3$PCH$_2$PO$_3$)}$^{2-}$ network, shown in Figure 81, is constructed from seven polyhedral connect rings {V$_3$P$_4$O$_6$C} linked through corner-sharing interactions into a 2D covalently linked layer. The {Cu(bpy)(H$_2$O)}$^{2+}$ groups are sited above and below the cyclic cavities.

As shown in Figure 82, the structure of [Cu(bpy)(VO)(O$_3$PCH$_2$CH$_2$PO$_3$)] consists of a 2D network again constructed from corner-sharing vanadium {VO$_5$} square pyramids, phosphate tetrahedra, and copper {CuO$_3$N$_2$} square pyramids. The V(IV) site is defined by four oxygen donors from each of four phosphonate groups in the basal plane and an apical oxo group that bridges to the Cu(II) center. The copper geometry consists of the nitrogen donors of the bpy ligand and two oxygen donors from two phosphonate ligands in the basal

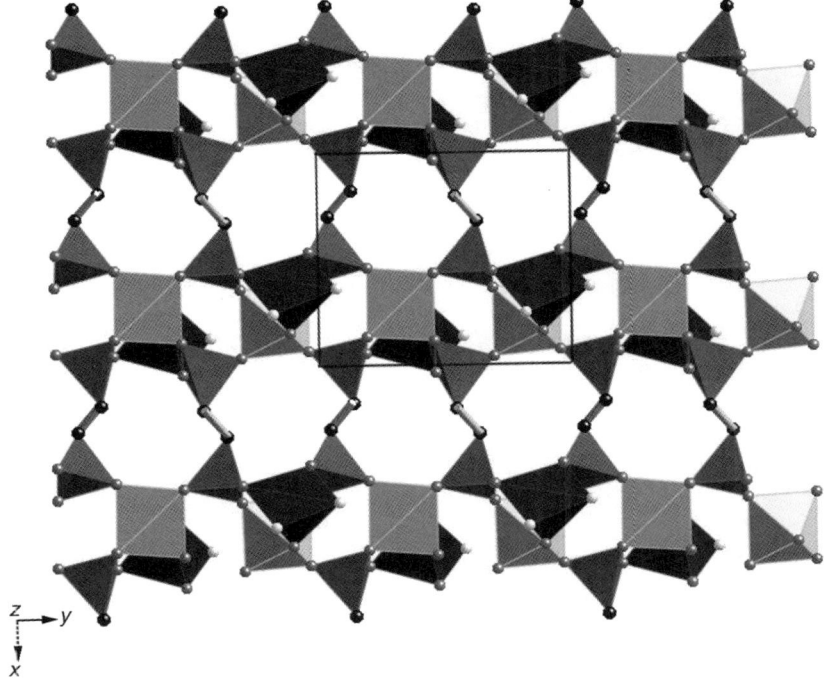

Figure 82. A polyhedral representation of the structure of [Cu(bpy)(VO)(O$_3$PCH$_2$CH$_2$PO$_3$)], showing the network connectivity. The bpy rings have been omitted for clarity.

plane and the bridging oxo group in the apical position. The oxo bridge is asymmetric with a V—O bond distance of 1.611(2) Å and a Cu—O distance of 2.394(2) Å. Each {PO_3} terminus of the diphosphonate ligand bridges one copper and two vanadium sites. Consequently, the structure may be described as {Cu(bpy)(VO)($O_3PCH_2PO_3$)} chains, tethered through the ethylene bridges of the diphosphonate ligands into a 2D covalently linked network. Within the 1D oxide substructure, the common cyclic motif {$V_2P_2O_4$} of corner-sharing {VO_5} square pyramids and {O_3PR} tetrahedra is observed. These rings are linked through the vanadium sites into a chain, which is decorated by the {Cu(bpy)}$^{2+}$ moieties, linked to the vanadium oxo group and the oxygen donors of the diphosphonate ligands not involved in the {$V_2P_2O_4$} ring formation. This results in the additional ring motifs {$VCuP_2O_4$} and {$VCuPO_3$}. The connectivity gives rise to a distinctive partitioning of the layer structure into inorganic chain substructures and tethering organic domains. The {Cu(bpy)}$^{2+}$ subunits project into the interlamellar regions above and below the plane.

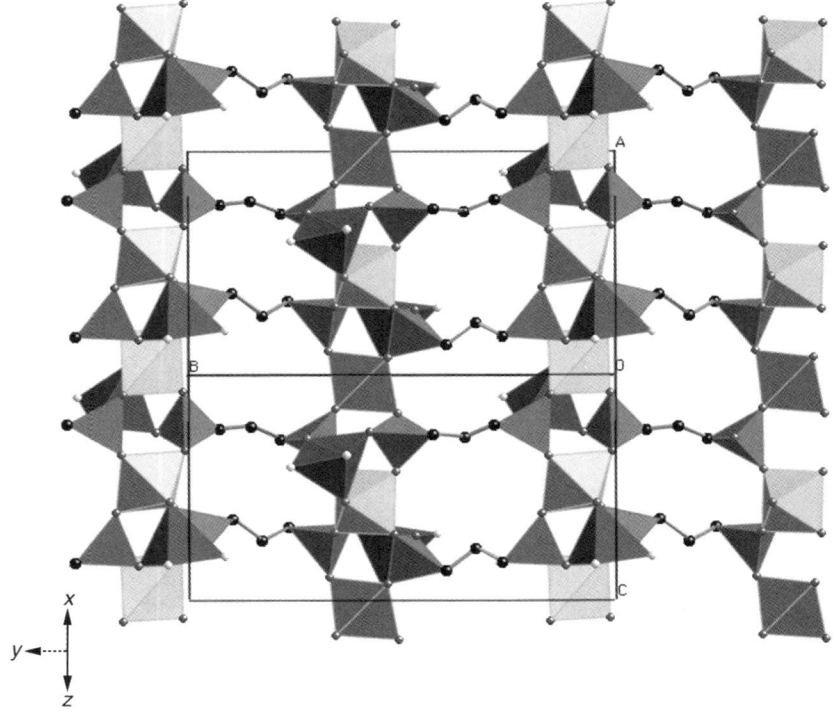

Figure 83. A polyhedral representation of the structure of [Cu(bpy)(VO)($O_3PCH_2CH_2CH_2PO_3$)] • H_2O. The bpy rings have been omitted.

The structure of the propylenediphosphonate derivative [Cu(bpy)(VO) ($O_3PCH_2CH_2CH_2PO_3$)] • H_2O is analogous to that of ethylene-bridged analogue, as shown in Figure 83. The structures share the common structural motifs of {Cu(bpy)(VO)($O_3PCH_2PO_3$)} chains and cyclic {$V_2P_2O_4$} rings within the chains. The propylene groups bridge chains into a 2D network. However, the consequences of tether extension are apparent in the relative orientations of the {Cu(bpy)}$^{2+}$ moieties. In contrast to [Cu(bpy)(VO)($O_3PCH_2CH_2PO_3$)] in which the {Cu(bpy)}$^{2+}$ units link to phosphonate oxygen donors on the same edge of the oxovanadium phosphonate chain and consequently orient the bpy plane approximately parallel to the V—V vectors of the chain, the {Cu(bpy)}$^{2+}$ groups of [Cu(bpy)(VO)($O_3PCH_2CH_2CH_2PO_3$)] coordinate oxygen donors from phosphonate groups on opposite edges of the chain, with a concomitant orientation of the bpy plane perpendicular to the V—V vectors of the chain. It would appear that chain lengthening in the propylene analogue relieves the steric strain that would result from the perpendicular orientation of bpy units in a shorter tether system.

The structural consequences of tether lengthening in [Cu(bpy)(VO) ($O_3PCH_2CH_2PO_3$)] are significant. While structure [Cu(bpy)(VO)($O_3PCH_2PO_3$)] contains vanadium centers with terminal oxo groups, the oxo groups in the ethylene analogue bridge to the copper sites. Consequently, although the copper geometries in both the methylene and the ethylene analogues are square pyramidal {CuN_2O_3}, one oxygen donor in the methylene analogue is provided by an aqua ligand, in contrast to the bridging oxo group in the ethylene case. The [(VO){$O_3P(CH_2)_nPO_3$}]$^{2-}$ networks, shown in Figures 81 and 82, are quite distinct, with the former displaying 14-membered [$V_3P_4O_6C$] rings and the latter 14-membered [$V_2P_4O_4C_4$] rings.

While the structure of the {(VO)($O_3PCH_2PO_3$)}$^{2-}$ network of [Cu(bpy)(VO) ($O_3PCH_2PO_3$)] is quite distinct from other examples of oxovanadium methylenediphosphonate layered materials, the {(VO)($O_3PCH_2CH_2PO_3$)}$^{2-}$ network of the ethylene species is reminiscent of the oxovanadium phosphonate layer of [$H_3NCH_2CH_2NH_3$][(VO)($O_3PCH_2CH_2PO_3$)]. The covalent connectivities of the oxovanadium phosphonate network structures of [Cu(bpy)(VO)($O_3PCH_2PO_3$)] and this latter material are identical, and the terminal oxo group and pendant P—O groups observed in [$H_3NCH_2CH_2NH_3$][(VO)($O_3PCH_2CH_2PO_3$)] serve to anchor the {Cu(bpy)}$^{2+}$ moiety to the layer structure in [Cu(bpy)(VO) ($O_3PCH_2PO_3$)].

Note that substitution of phen for bpy results in a structurally unique series of materials. For example, the network structure of [Cu(phen)(VO)(H_2O) ($O_3PCH_2PO_3$)], illustrated in Figure 84, contrasts with that of [{Cu(bpy) (H_2O)}(VO)($O_3PCH_2PO_3$)] in several important features. The aqua ligand in the latter material resides on the Cu(II) site. Consequently, the copper links to the V/P/O network only through two phosphonate oxygen donors. The

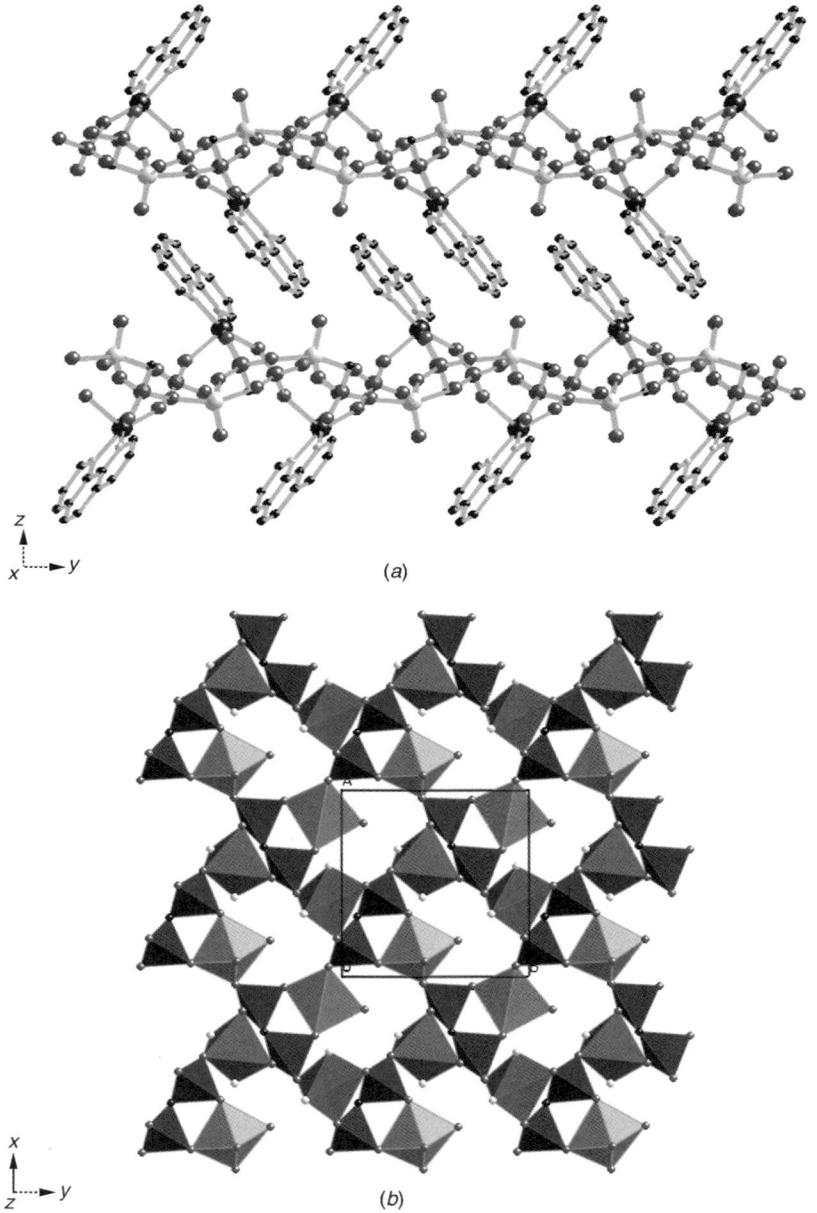

Figure 84. (a) The stacking of layers in [Cu(phen)(VO)(H$_2$O)(O$_3$PCH$_2$PO$_3$)]. (b) A polyhedral view of the Cu/V/P/O network connectivity.

vanadium center coordinates to four phosphonate oxygen donors. However, the aqua ligand of [Cu(phen)(VO)(H$_2$O)(O$_3$PCH$_2$PO$_3$)] is situated on the V(IV) center that consequently links to only three phosphate oxygen donors. Whereas tracing the covalent connectivity of the oxovanadium diphosphonate substructure of [{Cu(bpy)(H$_2$O)}(VO)(O$_3$PCH$_2$PO$_3$)] reveals a network substructure, the V–O–diphosphonate substructure of [Cu(phen)(VO)(H$_2$O)(O$_3$PCH$_2$PO$_3$)] consists of 1D chains. These chains are connected through the {Cu(phen)O$_3$} square pyramids into the 2D structure. It is evident that while the {Cu(bpy)(H$_2$O)O$_2$} subunit of [{Cu(bpy)(H$_2$O)}(VO)(O$_3$PCH$_2$PO$_3$)] serves to decorate the surface of the oxovanadium diphosphonate network, the {Cu(phen)O$_3$} subunit of [Cu(phen)(VO)(H$_2$O)(O$_3$PCH$_2$PO$_3$)] is an integral component of the network.

The structure of the ethylene bridged analogue [{Cu(phen)}$_2$(V$_2$O$_5$)(O$_3$PCH$_2$CH$_2$PO$_3$)] is also quite distinct from that of [Cu(bpy) (VO)(O$_3$PCH$_2$CH$_2$PO$_3$)]. As shown in Figure 85, the structure of the former is constructed from chains of corner-sharing V(V) tetrahedra and diphosphonate groups, linked through copper square pyramids. Several features of the structure are note

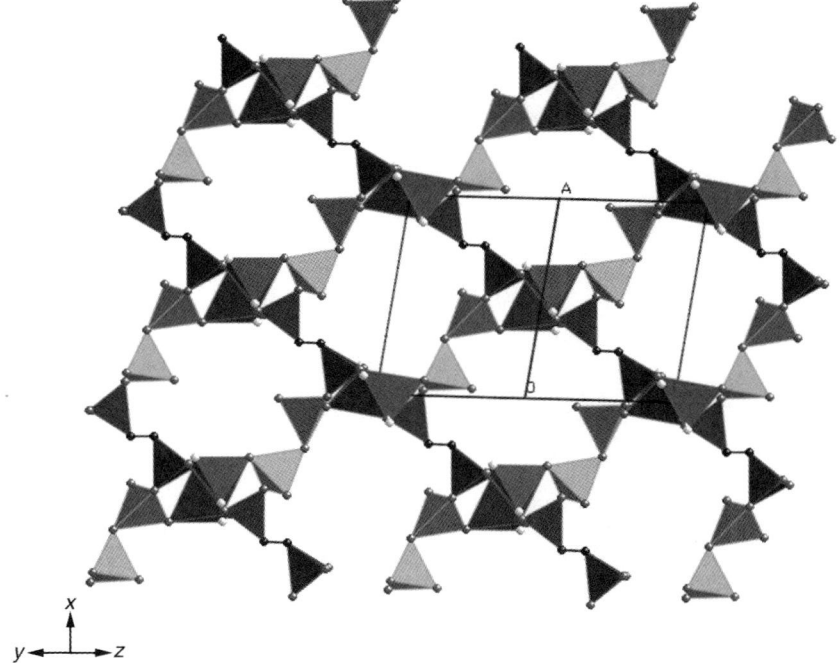

Figure 85. A polyhedral representation of the network connectivity in [{Cu(phen)}$_2$(V$_2$O$_5$)(O$_3$PCH$_2$CH$_2$PO$_3$)]. The phen rings have been omitted for clarity.

worthy. The presence of binuclear units of corner-sharing V(V) tetrahedra is unique, and the +5 oxidation state is unusual for oxovanadium organophosphonate structures in general.

The structural complexity of this family of materials is also manifested in the structure of [Cu(phen)(V$_2$O$_3$)(O$_3$PCH$_2$CH$_2$CH$_2$PO$_3$)]. As shown in Figure 86, the network structure of this propylene diphosphonate analogue, like that of the

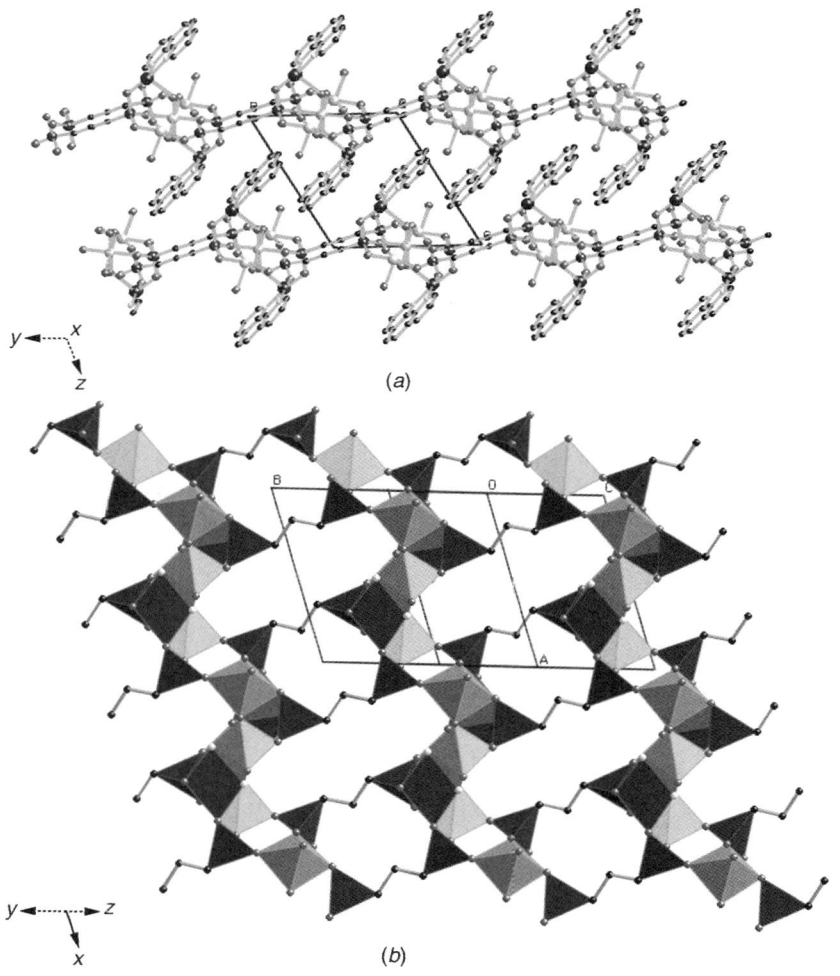

Figure 86. (a) The layer stacking in [{Cu(phen)}$_2$(V$_3$O$_5$)(O$_3$PCH$_2$CH$_2$CH$_2$PO$_3$)$_2$(H$_2$O)]. (b) A polyhedral representation of the network connectivity.

previously described [Cu(bpy)(VO)(O$_3$PCH$_2$CH$_2$CH$_2$PO$_3$)], is constructed from V—P—O—Cu chains linked through the propylene bridges of the diphosphonate into a 2D network. However, the detailed connectivity within the inorganic chains of [Cu(phen)(V$_2$O$_3$)(O$_3$PCH$_2$CH$_2$CH$_2$PO$_3$)] is quite distinct from that of [Cu(bpy)(VO)(O$_3$PCH$_2$CH$_2$CH$_2$PO$_3$)]. The oxide chain of the *o*-phen derivative contains a trinuclear submotif of two V(IV) square pyramids corner sharing with a central V(IV) octahedron. The trinuclear units are bridged by {O$_3$PC} tetrahedra to form the puckered 1D chain. In addition, the central vanadium site of the trinuclear unit engages in a corner-sharing interaction through a bridging oxo group with the {Cu(phen)}$^{2+}$ subunit, which serves to decorate the exterior of the chain.

In comparing the three bpy phases discussed above to the three phen containing materials, the striking feature is the contrast in the nuclearities of the vanadium oxide substructures of these materials. While the bpy family exhibits exclusively mononuclear V(IV) sites as submotifs, that is, no V—O—V bridging, the instances of the *o*-phen family contain higher oligomers, binuclear V(V) sites in [{Cu(phen)}$_2$(V$_2$O$_5$)(O$_3$PCH$_2$CH$_2$PO$_3$)] and trinuclear V(IV) sites in [Cu(phen)(V$_2$O$_3$)(O$_3$PCH$_2$CH$_2$CH$_2$PO$_3$)]. Similarly, the presence of V(V) sites in the ethylenediphosphonate derivative is unusual as all other examples of this class of compounds contain V(IV) sites, exclusively, with the exception of the mixed-valence [{Cu(bpydicarbH$_2$)}$_2$(V$_3$O$_5$) (O$_3$PCH$_2$PO$_3$)$_2$(H$_2$O)] • 2H$_2$O, described below where bpydicarb = 2,2'-bipridyl-4,4'-dicarboxylic acid. We can only conclude that under reactions conditions favoring the isolation of metastable species, the complications inherent in the variable oxidation states, coordination polyhedra and degree of aggregation of vanadium oxides are fully realized.

Another curious feature of the Cu—phenvanadate series is the isolation of a 3D phase with [O$_3$PCH$_2$PO$_3$]$^{4-}$, in addition to the 2D [Cu(phen)(VO) (H$_2$O)(O$_3$PCH$_2$PO$_3$)], discussed above. As shown in Figure 87, the structure of [Cu(phen)(V$_2$O$_3$)(O$_3$PCH$_2$PO$_3$)] consists of {Cu(phen)(VO)$_2$(O$_3$PCH$_2$PO$_3$)} layers linked through vanadium square pyramids into a covalently linked framework. The layers are constructed from chains of corner-sharing V(V) and (O$_3$PC} tetrahedra, linked through Cu(II) square pyramids. Each diphosphonate ligand projects two oxygen vertices into the interlamellar region, which bond to the square pyramidal V(IV) sites which serve to link the V/P/O networks.

Attempts to modify the structures by introducing steric bulk and/or additional functionality on the organoimine ligand are represented by [{Cu(bpydicarbH$_2$)}$_2$(V$_3$O$_5$)(O$_3$PCH$_2$PO$_3$)$_2$(H$_2$O)] • 2H$_2$O. As shown in Figure 88, the V/P/O network is constructed from corner-sharing V(IV) and V(V) square pyramids and {O$_3$PC} tetrahedra. The square pyramidal {Cu(bpydicarbH$_2$)O$_3$} subunits decorate the surfaces of the layer. The {V$_3$O$_5$) motif consists of a

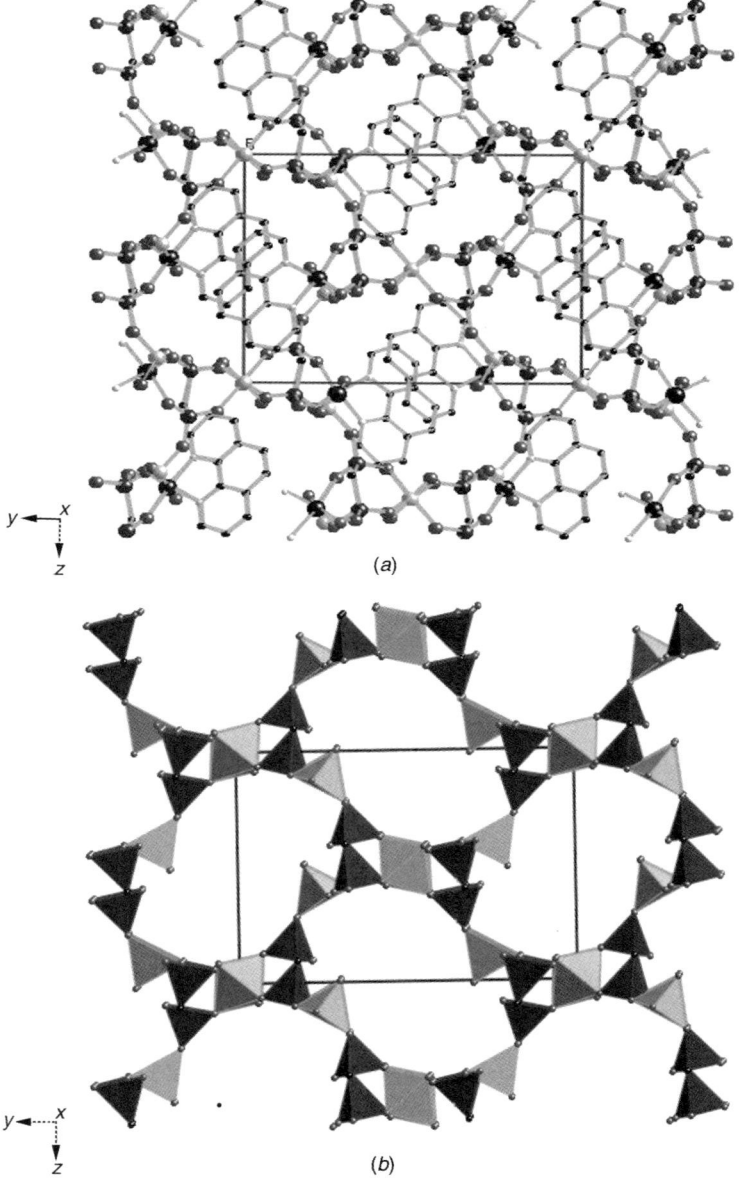

Figure 87. (a) A view of the 3D framework structure of [Cu(phen)[{Cu(phen)}$_2$(V$_3$O$_5$)(O$_3$PCH$_2$PO$_3$)$_2$(H$_2$O)], parallel to the crystallographic a axis. (b) A polyhedral representation parallel to the a axis showing the unusual ring motifs.

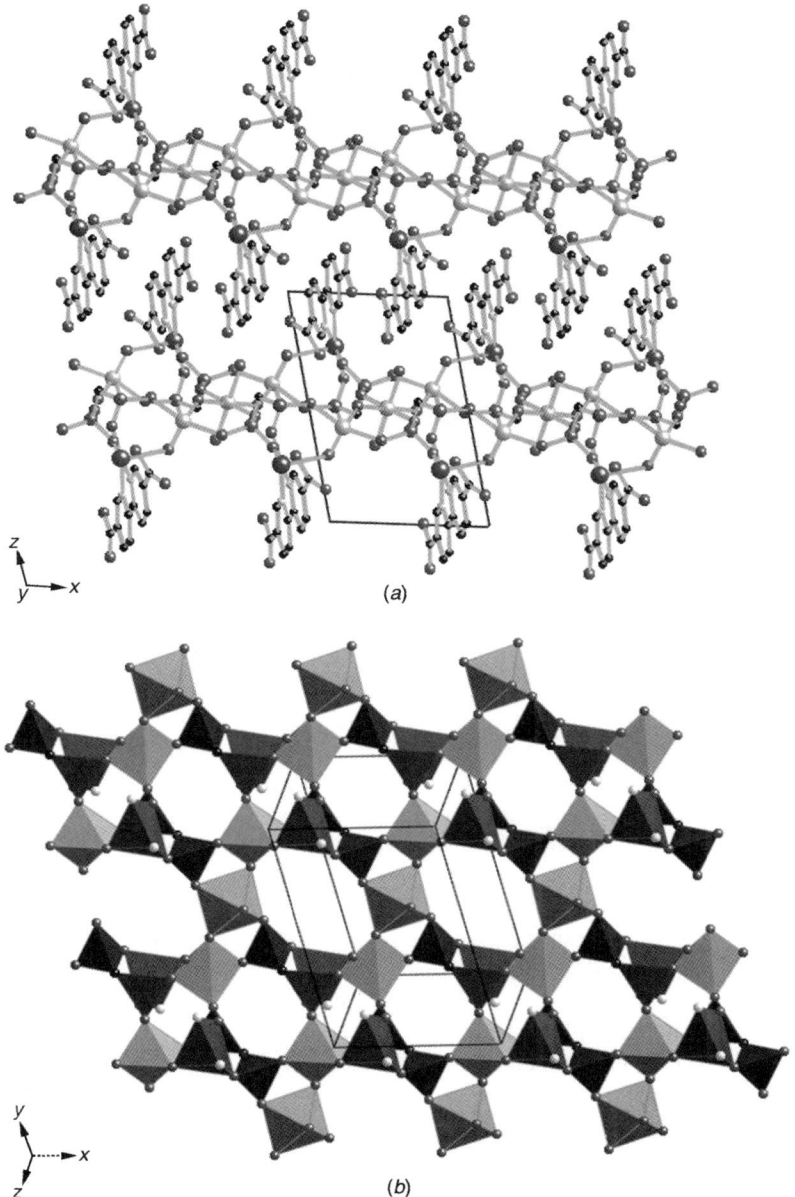

Figure 88. (a) A view of the stacking of layers in [{Cu(bpydicarbH$_2$)}$_2$(V$_3$O$_5$)(O$_3$PCH$_2$PO$_3$)$_2$(H$_2$O)]·2H$_2$O. (b) A view of the polyhedral connectivity in the layer.

trinuclear unit of corner-sharing polyhedra. The central vanadium site exhibits a terminal oxo group and bonds to the two peripheral vanadium centers through oxo groups and to two phosphonate oxygen donors. Each peripheral vanadium site is additionally coordinated to three phosphonate oxygen atoms and to a copper site through a V—O—Cu bridging oxo group. Each methylenediphosphonate ligand acts as a bidentate chelator of a Cu(II) site and bridges four vanadium centers from three trinuclear units. The unusual connectivity pattern generates 16-membered rings $\{V_4P_4O_8\}$, resulting in considerable void volume, which is occupied by water molecules of crystallization.

VII. CONCLUSIONS

The evolution of the chemistry of TMPOs and TMPROs parallels the contemporary interest in properties of inorganic solid state materials. These useful physicochemical properties, derived from molecular recognition, endow such materials with applications to sorption, catalysis, molecular electronics, and optical systems (299–303). Molecular recognition, in this sense of appropriate juxtapositions of functional groups for reactivity and cooperative electronic effects, is often related to the periodicity of the crystalline state. Such considerations render the goals of "crystal engineering" quite attractive (301).

However, it would appear premature to classify the synthetic routes to TMPOs and TMPROs in terms of "designed" syntheses or of "engineering" of crystalline architecture. Organic–inorganic materials of the type discussed here are structurally complex and represent metastable phases that are kinetically, not thermodynamically, stable. The vast reaction parameter space, involving conditions such as pH, stoichiometry, temperature, pressure, fill volume, together with the variable speciation of the transition metal sites reflected in coordination polyhedra, coordination numbers, oxidation states, and nuclearity and with the flexible coordination modalities of the phosphate or organophosphonate groups, endow the crystal chemistry of these hybrid materials with an unusual diversity. Concomitant with the complexity of such materials is a natural limit on the degree of predictability. While it has been argued that such complexity renders these materials undesignable, this is true only if design requires total predictability (304). By exploiting the reciprocity of structure–function relationships of a class of materials, the expectation of control of desired properties should be achieved as the chemistry of the products of empirical development are unraveled. For new classes of materials, such as TMPOs and TMPROs, whose crystal chemistry is as yet represented by relatively few examples, such utilitarian goals must await further evolution on the phenomenological level.

Despite this practical caveat, it is abundantly clear that organic molecules can alter inorganic microstructures, offering a powerful tool for the synthesis at low temperature of new organic–inorganic hybrid materials. The preparation of solid materials capable of molecular recognition was pioneered by Pauling nearly one-half of a century ago in studies on the imprinting of structural information from an organic molecule on silica gels (305). In the past 20 years, templating via noncovalent interactions has witnessed significant development. The structure directing role of organic cations on anionic metal oxide frameworks has been extensively documented and discussed in terms of organic–inorganic interactions whose geometric correspondence gives rise to the architecture of the framework. In addition to hydrophobic–hydrophilic interactions, the organoammonium family of cations engages in strong multipoint hydrogen bonding in its structure-directing role as the oxide framework. Moreover, in addition to this noncovalent role, the organic component may participate in the structure of the hybrid as tethers for functional groups anchoring metal oxide sites, as ligands covalently bound to the oxide backbone, and as donors that passivate the metal coordination sphere so as to favor low-dimensional materials.

The materials that constitute the topic of this chapter manifest the structural diversity associated with the organic component. The largest subclass of the VOPO family of materials incorporates the organic component as an organoammonium cation adopting a noncovalent structure-directing role with respect to the inorganic backbone, providing a general representation $(cat)_n[V_xO_yH_m(PO_4)_z]$ (cat = cation). For a small subclass of the VOPO family, the organoamine is covalently attached to the metal oxide backbone through a nitrogen-donor group. A second family of composites is represented by the VOPROs that incorporate organophosphonate groups. The organic components of this latter family may adopt the roles of substituents on the phosphorus oxide moiety that serve to eliminate potential metal ligation at one vertex of the phosphorus tetrahedron and of tethers linking the phosphorus oxide constituents of the V/P/O inorganic networks, thereby promoting 2D structures. In all cases, the introduction of organic components serves to partition the complex structures of these materials into organic and inorganic domains through hydrophobic–hydrophilic interactions. The structural diversity achieved is represented both by low-dimensional materials with variable interstrand or interlamellar distances and shapes and by 3D frameworks displaying mixed-valence and open pore structures.

Despite the absence of detailed structural predictability, the combination of hydrothermal syntheses and the incorporation of organic components into oxide phases have provided a facile, rapid and inexpensive method for the isolation of crystalline phases whose dimensionalities, oxidation states, and void volumes can be to some extent manipulated. The technique has provided a variety of lamellar and zeolitic materials incorporating nontetrahedral sites. In fact, the supercage material $(H_2DABCO)K_{1.35}[V_5O_9(PO_4)_2]$ (DABCO = diaminobicyclooctane) represents the only example of demonstrated microporosity in the TMPO family of materials and confirms that zeolitic materials with transition metals as major constituents of the polyhedral framework are not only readily accessible but may also exhibit potentially useful properties.

It is noteworthy that the introduction of d-block elements into zeolitic materials renders the frameworks capable of absorbing visible light and magnetically active. The variable oxidation states associated with d-block elements allows control of the charge/volume ratio of the translational repeats of the framework; variations in the oxidation state of the framework may also influence the contents of the cavities.

Prefabricated organic components can also be introduced with substituents whose geometries serve to mold interlamellar regions or framework cavities or that provide redox or optically active sites. The inorganic backbone can also serve to immobilize organic functional groups or to provide appropriate geometrical orientations for reactivity or cooperative electronic effects.

While low-temperature routes for metastable organic–inorganic hybrid materials are as yet relatively unexplored, the potential for developing a breadth of novel crystal chemistry is exceptional. As structure–function relationships for these materials evolve, elements of designed synthesis based on reciprocity may emerge and allow a certain predictability in modification of crystalline architectures.

ABBREVIATIONS

bpy	2, 2′-Bipyridine
bpydicarb	2, 2′-Bipyridyl-4, 4′-dicarboxylic acid
cat	Cation
DABCO	Diaminobicyclooctane
en	Ethylenediamine
Et	Ethyl or ethylene
FBB	Fundamental building block
LL	Bidentate ligand
Me	Methyl
MNDO	Modified neglect of diatomic overlap
MoPO	Molybdenum phosphate
PDA	Propanediammonium
Ph	Phenyl
phen	1, 10-Phenanthroline
pip	Piperazine
Pr	Propylene

SCF   Self-consistent field
SDA   Structure-directing agents
terpy   2, 2': 6', 2''-Terpyridine
TMPO   Transition metal phosphate
VOPO   Oxovanadium phosphate
VOPRO   Oxovanadium organophosphonate

REFERENCES

1. A. F. Wells, *Structural Inorganic Chemistry*, 4th ed., Oxford University Press, Oxford, 1975.
2. N. N. Greenwood and A. Earnshaw, *Chemistry of the Elements*, Pergamon Press, New York, 1984.
3. T. de Chardin, *The Phenomenon of Man*, Harper Collins, New York, 1980.
4. S. I. Stupp and P. V. Braun, *Science*, 277, 1242 (1997).
5. (a) S. Mann, *Nature (London)*, 365, 499 (1993). (b) S. Mann, *J. Chem. Soc. Dalton Trans.*, 21, 3953 (1997). (c) S. Mann, S. L. Burkett, S. A. Davis, C. E. Fowler, N. H. Mendelson, S. D. Sims, D. Walsh, and N. T. Whilton, *Chem. Mater.*, 9, 2300 (1997). (d) M. Li, K. W. Wong, and S. Mann, *Chem. Mater.*, 11, 23 (1999).
6. L. L. Hench, *Inorganic Biomaterials*, in *Materials Chemistry, An Emerging Discipline*, L. V. Interannte, L. A. Casper, and A. B. Ellis, Eds., ACS Series 245, Washington, DC, 1995, Chapter 21, pp. 523–547.
7. V. Soghomonian, Q. Chen, R. C. Haushalter, J. Zubieta, and C. J. O'Connor, *Science*, 259, 1596 (1993).
8. J. V. Smith, *Chem. Rev.* 88, 149 (1988); M. L. Occelli and H. C. Robson, *Zeolite Synthesis*, American Chemical Society, Washington, DC, 1989.
9. C. T. Kresge, M. E. Leonowicz, W. J. Roth, J. C. Vartuli, and J. S. Beck, *Nature (London)*, 359, 710 (1992).
10. R. M. Barrer, *Hydrothermal Chemistry of Zeolites*, Academic, New York, 1982.
11. J. S. Beck, J. C. Vartuli, W. J. Roth, M. E. Leonowicz, C. T. Kresge, K. D. Schmitt, C. T.-W. Chu, D. H. Olson, E. W. Sheppard, S. B. McCullen, J. B. Higgins, and J. L. Schlenker, *J. Am. Chem. Soc.* 114, 10834 (1992).
12. R. C. Haushalter, K. G. Strohmaier, and F. W. Lai, *Science*, 246, 1289 (1989).
13. R. C. Haushalter and L. A. Mundi, *Chem. Mater.*, 4, 31 (1992).
14. M. E. Davis, A. Katz, and W. R. Ahmad, *Chem. Mater.*, 8, 1820 (1996) and references cited therein.
15. A. Rabenau, *Angew. Chem., Int. Ed. Engl.*, 97, 1017 (1985).
16. D. Hagrman, J. Zubieta, C. J. Warren, L. M. Myer, M. M. J. Treacy, and R. C. Haushalter, *J. Sol. State Chem.*, 138, 178 (1998).
17. D. E. C. Corbridge, *Phosphorus*, Elsevier, Amsterdam, The Netherlands, 1978.
18. T. Kanazawa, *Inorganic Phosphate Materials*, Materials Science Monographs, 52, Elsevier, Amsterdam, The Netherlands, 1989.
19. G. Centi, F. Trifiro, J. R. Ebner, and V. M. Franchetti, *Chem. Rev.*, 88, 55 (1988).
20. G. Centi, Ed., *Catalysis Today*, Elsevier, Amsterdam, The Netherlands, Vol. 16, 1993.

21. M. T. Sananes, G. J. Hutchings, and J.-C. Volta, *J. Chem. Soc. Chem. Comm.*, 243 (1995).
22. M. T. Sananes, G. J. Hutchings, and J.-C. Volta, *J. Catal.*, 154, 253 (1995).
23. For general discussions of the structural chemistry of compounds containing $\{V_xO_y\}$ units see: (a) M. Schindler, F. C. Hawthorne, and W. H. Baur, *Chem. Mater.*, 12, 1248 (2000). (b) P. Y. Zavalig. and M. S. Whittingham, *Acta Crystallogr.*, B55, 627 (1999). (c) H. T. Evans, Jr., and J. S. White, Jr., *Mineral Rec.*, 18, 333 (1987).
24. A. LeBail, M. D. Marcos, and P. Amoros, *Inorg. Chem.*, 33, 2607 (1994).
25. M. Tachez, F. Theobald, and E. Bordes, *J. Solid State Chem.*, 40, 280 (1981).
26. E. Bordes, P. Courtine, and G. Pannetier, *Ann. Chim.*, 8, 105 (1973).
27. B. D. Jordan and C. Calvo, *Can. J. Chem.*, 51, 2621 (1973).
28. H. R. Tietze, *Aust. J. Chem.*, 34, 2035 (1981).
29. M. Tachez, F. Theobald, J. Bernard, A. W. Hewat, *Rev. Chim. Miner.*, 19, 291 (1982).
30. C. C. Torardi and J. C. Calabrese, IC 23, 1308 (1984).
31. A. LeBail, G. Ferey, P. Amoros, and D. Beltran-Porter, *Eur. J. Solid State Inorg. Chem.*, 26, 419 (1989).
32. A. LeBail, G. Ferey, P. Amoros, D. Beltran-Porter, and G. Villeneuve, *J. Solid State Chem.*, 79, 169 (1989).
33. M. E. Leonowicz, J. W. Johnson, J. F. Brody, H. F., Shannon, Jr., and J. M. Newsam, *J. Solid State Chem.*, 56, 370 (1985).
34. R. Gopal and C. Calvo, *J. Solid State Chem.*, 5, 432 (1972).
35. S. A. Linde, Y. E. Gorbunova, A. V. Lavrov, and Y. G. Kustnetsov, *Dokl. Akad. Nauk SSSR* 244, 1411 (1979).
36. (a) N. E. Middlemiss, Ph. D. Thesis, "Crystallographic Studies in the V/P/As/O Systems", Department of Chemistry, McMaster University, Hamilton, ON, Canada, 1978. (b) Yu. E. Gorbunova and S. A. Linde, *Sov. Phys.-Dokl. (Engl. Transl.)* 1979, 24, 138. (c) J. R. Ebner and M. R. Thompson, *Studies in Surface Sciences and Catalysis*, R. K. Graselli, A. W. Sleight, Eds., Elsevier, France, 1992. (d) C. C. Torardi, Z. G. Li, H. S. Horowitz, W. Liang, and M.-H. Whangbo, *J. Solid State Chem.*, 119, 349 (1995). (e) Z. Hiroi, M. Azuma, Y. Fujishiro, T. Saito, M. Takano, F. Izumi, T. Kamiyama, and T. Ikeda, *J. Solid State Chem.*, 146, 369 (1999).
37. (a) P. T. Nguyen, R. D. Hoffman, and A. W. Sleight, *Mat. Res. Bull.*, 30, 1058 (1995). (b) M. Azuma, T. Saito, Y. Fujishiro, Z. Hiroi, M. Takano, F. Izumi, T. Ikida, Y. Narumi, and K. Kindo, *Phys. Rev. B* 60, 10145 (1999).
38. C. C. Torardi, W. M. Reiff, and L. Takacs, *J. Solid State Chem.*, 82, 203 (1989).
39. M. Schindler, W. Joswig, and W. H. Baur, *Eur. J. Solid State Inorg. Chem.*, 32, 109 (1995).
40. J. T. Vaughey, W. T. A. Harrison, and A. J. Jacobson, *Inorg. Chem.*, 33, 2481 (1994).
41. S. A. Linde, Yu. E. Gorbunova, and A. V. Lavrov, *Russian J. Inorg. Chem.*, 28, 16 (1983).
42. J. W. Johnson, D. C. Johnston, H. E., King, Jr., T. R. Halbert, J. F. Brody, and D. P. Goshorn, *Inorg. Chem.*, 27, 1646 (1988).
43. K. K. Palkina, S. I. Maksimova, N. T. Chibiskova, K. Schlesinger, and G. Ladwig, *Z. Anog. Allg. Chem.*, 529, 89 (1985).
44. Z. Bircsak and W. T. A. Harrison, *Inorg. Chem.* 37, 5387 (1998).
45. L. Liu, X. Wang, R. Bontchev, K. Ross, and A. J. Jacobson, *J. Mater. Chem.*, 9, 1585 (1999).
46. C. Wadewitz and Hk. Müller-Buschbaum, *Z. Naturforsch.*, 51b, 929 (1996).
47. C. Wadewitz and Hk. Müller-Buschbaum, *Z. Naturforsch.*, 51b, 1290 (1996).

48. (a) W. T. A. Harrison, S. C. Lim, J. T. Vaughey, A. J. Jacobson, D. P. Goshorn, and J. W. Johnson, *J. Solid State Chem.*, *113*, 444 (1994). (b) G. Bagnasco, L. Benes, P. Galli, M. A. Massucci, P. Patrono, M. Turco, and V. Zima, *J. Therm. Anal. Calorim.*, *52*, 615 (1998).
49. W. T. A. Harrison, J. T. Vaughey, A. J. Jacobson, D. J. Goshorn, and J. W. Johnson, *J. Solid State Chem.*, *116*, 77 (1995).
50. P. Amoros and A. LeBail, *J. Solid State Chem.*, *97*, 283 (1992).
51. P. Amoros, D. Beltran-Porter, A. LeBail, G. Ferey, and G. Villeneuve, *Eur. J. Solid State Chem.*, *25*, 599 (1989).
52. Z. Bircsak and W. T. A. Harrison, *J. Sol. State Chem.*, *140*, 272 (1998).
53. G. Huan, J. W. Johnson, A. J. Jacobson, E. W. Corcoran, and D. P. Goshorn, *J. Solid State Chem.*, *93*, 514 (1991).
54. M. M. Borel, A. Leclaire, J. Chardon, C. Michel, J. Provost, and B. Raveau, *J. Sol. State Chem.*, *135*, 302 (1998).
55. K. H. Lii, N. S. Wen, C. C. Su, and B. R. Chueh, *Inorg. Chem.*, *31*, 439 (1992).
56. J. Do, R. P. Bontchev, and A. J. Jacobson, *Inorg. Chem.*, *39*, 3230 (2000).
57. D. Lozano-Calero, S. Bruque, M. A. G. Aranda, M. Martinez-Lara, and L. Moreno, *J. Solid State Chem.*, *103*, 481 (1993).
58. S. L. Wang, H. Y. Kang, C. Y. Cheng, and K. H. Lii, *Inorg. Chem.*, *30*, 3496 (1991).
59. T. Yamase and H. Makino, *J. Chem. Soc., Dalton Trans.*, 1143 (2000).
60. D. Papoutsakis, J. E. Jackson, and D. G. Nocera, *Inorg. Chem.*, *35*, 800 (1996).
61. P. Ayyappan, A. Ramanan, and C. C. Torardi, *Inorg. Chem.*, *37*, 3628 (1998).
62. W. T. A. Harrison, S. C. Lim, L. L. Dussack, A. J. Jacobson, D. P. Goshorn, and J. W. Johnson, *J. Solid State Chem.*, *118*, 241 (1995).
63. V. C. Korthuis, R.-D. Hoffmann, J. Huang, and A. W. Sleight, *J. Solid State Chem.*, *105*, 294 (1993).
64. K. H. Lii and H. J. Tsai, *Inorg. Chem.*, *30*, 446 (1991).
65. R. C. Haushalter, Z. Wang, M. E. Thompson, J. Zubieta, and C. O'Connor, *J. Inorg. Chem.*, *32*, 3966 (1993).
66. H. Y. Kang, W. C. Lee, S. L. Wang, and K. H. Lii, *Inorg. Chem.*, *31*, 4743 (1992).
67. H.-Y. Kang, S.-L. Wang, P.-P. Tsai, and K. H. Lii, *J. Chem. Soc. Dalton Trans.*, 1525 (1993).
68. H. Y. Kang, S. L. Wang, and K. H. Lii, *Acta Crystallogr.*, *C48*, 975 (1992).
69. M. Roca, M. D. Marcos, P. Amoros, J. Alamo, A. Beltran-Porter, and D. Beltran-Porter, *Inorg. Chem.*, *36*, 3414 (1997).
70. E. LeFur and J. Y. Pivan, *J. Mater. Chem.*, *9*, 2589 (1999).
71. S. Boudin, A. Grandin, P. Labbé, and B. Raveau, *Acta Crystallogr.*, *C52*, 2668 (1996).
72. C. Ninclaus, R. Retoux, D. Riou, and G. Ferey, *J. Solid State Chem.*, *122*, 139 (1996).
73. L. Benhamada, A. Grandin, M. M. Borel, A. Leclaire, and B. Raveau, *J. Solid State Chem.*, *101*, 154 (1992).
74. A. Daidouh, M. L. Veiga, and C. Pico, *J. Solid State Chem.*, *130*, 28 (1997).
75. Yu. E. Gobunova, S. A. Linde, A. V. Lavrov, and I. V. Tananaev, *Dokl. Akad. Nauk SSSR 250*, 350 (1980); *Sov. Phys. Dokl.*, *25*, 9 (1980).
76. J. Galy and A. Carpy, *Acta Crystallogr.*, *B31*, 1794 (1975).
77. K. H. Lii, H. J. Tsai, and S. L. Wang, *J. Solid State Chem.*, *87*, 396 (1990).
78. K. H. Lii and S. L. Wang, *J. Solid State Chem.*, *82*, 239 (1989).

79. S. Boudin, A. Grandin, A. Leclaire, M. M. Borel, and B. Raveau, *J. Solid State Chem.*, *111*, 365 (1994).
80. R. C. Haushalter, Q. Chen, V. Soghomonian, J. Zubieta, and C. J. O'Connor, *J. Solid State Chem.*, *108*, 128 (1994).
81. M. Schindler, W. Joswig, and W. M. Baur, *J. Solid State Chem.*, *134*, 286 (1997).
82. A. V. Lavrov, V. P. Nikolaev, G. G. Sadikov, and M. A. Porai-Koshits, *Sov. Phys. Dokl. Engl. Trans.*, *27*, 680 (1982).
83. K. H. Lii, C. H. Li, C. Y. Cheng, and S. L. Wang, *J. Solid State Chem.*, *95*, 352 (1991).
84. (a) K. H. Lii, C. H. Li, T. M. Chen, and S. L. Wang, *Z. Kristallogr.*, *197*, 67 (1999). (b) L. Benhamada, A. Grandin, M. M. Borel, A. Leclaire, and B. Raveau, *C. R. Acad. Sci. Paris, Ser. II*, *314*, 585 (1992).
85. M. L. F. Phillips, W. T. A. Harrison, E. G. Thurman, G. Stucky, G. V. Kulkarni, J. K. Burdett, *Inorg. Chem.*, *29*, 2158 (1990).
86. L. Benhamada, A. Grandin, M. M. Borel, A. Leclaire, and B. Raveau, *Acta Crystallogr.*, *C47*, 1138 (1991).
87. M. M. Borel, A. Leclaire, J. Chardon, J. Provost, H. Rebbah, and B. Raveau, *J. Solid State Chem.*, *132*, 41 (1997).
88. K.-H. Lii and W.-C. Liu, *J. Solid State Chem.*, *103*, 38 (1993).
89. K. H. Lii, B. R. Chueh, H. Y. Kang, and S. L. Wang, *J. Solid State Chem.*, *99*, 72 (1992).
90. F. Berrah, A. Leclaire, M.-M. Borel, A. Guesdon, and B. Raveau, *Acta Crystallogr.*, *C55*, 288 (1999).
91. A. Grandin, J. Chardon, M. M. Borel, A. Leclaire, and B. Raveau, *J. Solid State Chem.*, *104*, 226 (1993).
92. A. Leclaire, Q. Grandin, J. Chardon, M. M. Borel, and B. Raveau, *Eur. J. Solid State Inorg. Chem.*, *30*, 393 (1993).
93. A. Leclaire, J. Chardon, A. Grandin, M. M. Borel, and B. Raveau, *Eur. J. Solid State Inorg. Chem.*, *30*, 461 (1993).
94. A. Grandin, J. Chardon, M. M. Borel, A. Leclaire, and B. Raveau, *J. Solid State Chem.*, *99*, 297 (1992).
95. E. Le Fur, O. Peña, and J. Y. Pivan, *J. Alloys Compd.*, *285*, 89 (1999).
96. A. Grandin, J. Chardon, M. M. Borel, A. LeClaire, and B. Raveau, *Acta Crystallogr.*, *C48*, 1913 (1992).
97. K. H. Lii, *J. Chin. Chem. Soc.*, *39*, 569 (1992).
98. M. Schindler, W. Josvig, and W. H. Baur, *J. Solid State Chem.*, *145*, 15 (1999).
99. R. C. Haushalter, Z. Wang, M. E. Thompson, and J. Zubieta, *Inorg. Chem.*, *32*, 3700 (1993).
100. K. H. Lii and H. J. Tsai, *J. Solid State Chem.*, *91*, 331 (1991).
101. J. T. Vaughey, W. T. A. Harrison, and A. J. Jacobson, *J. Solid State Chem.*, *110*, 305 (1994).
102. L. Benhamada, A. Grandin, M. M. Borel, A. Leclaire, and B. Raveau, *J. Solid State Chem.*, *94*, 274 (1991).
103. P. Crespoa, A. Grandin, M. M. Borel, A. Leclaire, and B. Raveau, *J. Solid State Chem.*, *105*, 307 (1993).
104. A. Leclaire, J. Chardon, A. Grandin, M. M. Borel, and B. Raveau, *J. Solid State Chem.*, *108*, 291 (1994).
105. A. LeBail, M. LeBlanc, and P. Amoros, *J. Solid State Chem.*, *87*, 178 (1990).

106. S. Boudin, A. Grandin, M. M. Borel, A. Leclaire, and B. Raveau, *J. Solid State Chem.*, *111*, 380 (1994).
107. S. Boudin, A. Grandin, A. Leclaire, and M. M. Borel, B. Raveau, *Acta Crystallogr. Sect. C.*, *51*, 796 (1995).
108. S. Boudin, A. Grandin, Ph. Labbé, J. Provost, and B. Raveau, *J. Solid State Chem.*, *127*, 325 (1996).
109. M. M. Borel, A. Leclaire, J. Chardon, and B. Raveau, *J. Mater. Chem.*, *8*, 693 (1998).
110. M. D. Medrano, H. T., Evans, Jr., H.-R. Wenk, and D. Z. Piper, *Am. Mineral.*, *83*, 889 (1998).
111. F. Berrah, A. Guesdon, A. Leclaire, M. M. Borel, J. Provost, and B. Raveau, *J. Solid State Chem.*, *148*, 189 (1999).
112. M. I. Khan, L. M. Meyer, R. C. Haushalter, A. L. Schweitzer, and J. Zubieta, J. L. Dye, *Chem. Mater.*, *8*, 43 (1996).
113. A. Benmoussa, M. M. Borel, A. Grandin, E. A. Leclaire, and B. Raveau, *J. Solid State Chem.*, *97*, 314 (1992).
114. Z. Bircsak and W. T. A. Harrison, *Acta Crystallogr.*, *C54*, 1195 (1998).
115. R. C. Haushalter, Z. Wang, M. E. Thompson, and J. Zubieta, *Inorg. Chim. Acta*, *232*, 83 (1995).
116. C. Delmas, R. Olazcuaga, F. Cherkaoui, R. Brochu, and G. C. LeFlem, *C. R. Acad. Sci., Ser. C* *287*, 169 (1978).
117. L. O. Hagman, and P. Kierkegaard, *Acta Chim. Scand.*, *22*, 1822 (1968).
118. N. Kinomura, N. Matsui, N. Kumada, and F. Muto, *J. Solid State Chem.*, *79*, 232 (1989).
119. L. Benhamada, A. Grandin, M. M. Borel, A. Leclaire, and B. Raveau, *J. Solid State Chem.*, *91*, 264 (1991).
120. K. H. Lii, T. C. Lee, S. N. Liu, and S. L. Wang, *J. Chem. Soc. Dalton Trans* 1051 (1993).
121. S. Boudin, A. Grandin, M. M. Borel, A. Leclaire, and B. Raveau, *J. Solid State Chem.*, *110*, 43 (1994).
122. Z. Wang, R. C. Haushalter, M. E. Thompson, and J. Zubieta, *Materials Chem. Phys.*, *35*, 205 (1993).
123. L. Benhamada, A. Grandin, M. M. Borel, A. Leclaire, and B. Raveau, *J. Solid State Chem.*, *104*, 193 (1993).
124. E. Dvoncova, K.-H. Lii, C.-H. Li, and T.-M. Chen, *J. Solid State Chem.*, *106*, 485 (1993).
125. K. H. Lii, N. S. Wen, C. C. Su, and B. R. Chueh, *Inorg. Chem.*, *31*, 439 (1992).
126. J. Trommer, H. Worzala, S. Rabe, and M. Schneider, *J. Solid State Chem.*, *136*, 181 (1998).
127. A. Leclaire, H. Chahboun, D. Groult, and B Raveau, *J. Solid State Chem.*, *77*, 170 (1988).
128. L. Benhamada, A. Grandin, M. M. Borel, A. Leclaire, and B. Raveau, *J. Solid State Chem.*, *97*, 131 (1992).
129. K. H. Lii, and C. S. Lee, *Inorg. Chem.*, *29*, 3298 (1990).
130. K. H. Lii, Y. P. Wang, C. Y. Cheng, S. L. Wang, and H. C. Ku, *J. Chin. Chem. Soc.*, *37*, 141 (1990).
131. M. M. Borel, M. Hervieu, A. Leclaire, C. Michel, J. Chardon, J. Provost, and B. Raveau, *Chem. Mater.*, *11*, 3655 (1999).
132. K. H. Lii, Y. P. Wang, and S. L. Wang, *J. Solid State Chem.*, *80*, 127 (1989).
133. J. Huang, Q. Gu, and A. W. Sleight, *J. Solid State Chem.*, *110*, 226 (1994).
134. A. El Badraoui, J. Y. Pivan, M. Maunaye, M. Louer, and D. Louer, *J. Alloys Compounds 245*, 47 (1996).

135. K. H. Lii, Y. P. Wang, Y. B. Chen, and S. L. Wang, *J. Solid State Chem.*, 86, 143 (1990).
136. Y. P. Wang and K. H. Lii, *Acta Crystallogr.*, C45, 1417 (1989).
137. L. Benhamada, A. Grandin, M. M. Borel, A. Leclaire, and B. Raveau, *Acta Crystallogr.*, C47, 424 (1991).
138. U. Flörke, *Z. Kristallogr.*, 191, 137 (1990).
139. Y. P. Wang and K. H. Lii, *Acta Crystallogr.*, C45, 1210 (1989).
140. L. Benhamada, A. Grandin, M. M. Borel, A. Leclaire, and B. Raveau, *Acta Crystallogr.*, C47, 2437 (1991).
141. S. J. Hwu, R. I. Carroll, and D. L. Serra, *J. Solid State Chem.*, 110, 290 (1994).
142. K. Richtrová, J. Votinsky, J. Kalousová, L. Beneš, and V. Zima, *J. Solid State Chem.*, 116, 400 (1995).
143. Y. Zhang, A. Clearfield, and R. C. Haushalter, *J. Sol. State Chem.*, 117, 157 (1995).
144. K. L. Idler and C. Calvo, *Can. J. Chem.*, 53, 3665 (1975).
145. (a) K. K. Rangan and J. Gopalakirshnan, *J. Solid State Chem.*, 109, 116 (1994). (b) Langbeinite structure: A. Zeemann and J. Zeemann, *Acta Crystallogr.*, 10, 409 (1957).
146. A. Mar, F. Leroux, D. Cuyomard, A. Verbaere, and Y. Piffard, *J. Solid State Chem.*, 115, 76 (1995).
147. H. Y. Kang, W. C. Lee, S. L. Wang, and K. H. Lii, *Inorg. Chem.*, 31, 4743 (1992).
148. J. Gopalakrishnan, K. K. Rangan, B. R. Prasad, and C. K. Subramanian, *J. Solid State Chem.*, 111, 41 (1994).
149. K. H. Lii and L. F. Mao, *J. Solid State Chem.*, 96, 436 (1992).
150. R. C. Haushalter, V. Soghomonian, Q. Chen, and J. Zubieta, *J. Solid State Chem.*, 105, 512 (1993).
151. K.-H. Lii, L.-S. Wu, and H.-M. Gau, *Inorg. Chem.*, 32, 4153 (1993).
152. E. Le Fur and J. Y. Pivan, *Mater. Res. Bull.*, 34, 1117 (1999).
153. K. H. Lii and H. J. Tsai, *J. Solid State Chem.*, 90, 291 (1991).
154. S. Meyer and Hk. Müller-Buschbaum, *Z. Naturforsch.*, 52b, 367 (1997).
155. S. Boudin, A. Grandin, A. LeClaire, M. M. Borel, and B. Raveau, *J. Solid State Chem.*, 115, 140 (1995).
156. Z. Bircsak and W. T. A. Harrison, *Acta Crystallogr.*, C54, 1197 (1998).
157. C. E. Rice, W. R. Robinson, and B. C. Tofield, *Inorg. Chem.*, 15, 345 (1976).
158. A. Leclaire, H. Chahboun, D. Groult, and B. Raveau, *J. Solid State Chem.*, 65, 168 (1986).
159. L. M. Meyer, R. C. Haushalter, and J. Zubieta, *J. Solid State Chem.*, 125, 200 (1996).
160. R. P. Hammond and J. Zubieta, *J. Solid State Chem.*, in press.
161. A. Martin, U. Steinike, K. Melanova, and V. Zima, *J. Mater. Chem.*, 9, 2523 (1999).
162. A. M. Chippindale, K. J. Peacock, and A. R. Cowley, *J. Solid State Chem.*, 145, 379 (1999).
163. A. M. Chippindale and A. R. Cowley, *J. Solid State Chem.*, 159, 59 (2001).
164. C. J. Warren, R. C. Haushalter, D. J. Rose, and J. Zubieta, *Chem. Mater.*, 9, 2694 (1997).
165. M. Riou-Cavellec, C. Serre, and G. Ferey, *C. R. Acad. Sci. Paris, Ser. IIc* 2, 147 (1999).
166. D. Riou and G. Férey, *J. Solid State Chem.*, 111, 422 (1994).
167. G. Bonavia, R. C. Haushalter, and J. Zubieta, *J. Solid State Chem.*, 126, 292 (1996).
168. E. Bordes and P. Courtine, *J. Chem. Soc. Chem. Commun.*, 294 (1985).

169. J. W. Johnson, D. C. Johnston, and A. J. Jacobson, *Preparation of Catalysts, IV*, 181 (1987).
170. See, for example: (a) J. M. Longo and P. Kierkegaard, *Acta Chem. Scand.*, 20, 72 (1966). (b) J. M. Longo, J. W. Pierce, and J. A. Kafalas, *Mat. Res. Bull.*, 6, 1157 (1971). (c) H. A. Eick and L. Kihlborg, *Acta Chem. Scand.*, 1, 394 (1970). (d) P. Kierkegaard and M. Westerlund, *Acta Chem. Scand.*, 18, 2217 (1964).
171. A. Kilhborg, *Ark. Kemi*, 21, 357 (1963).
172. P. Day, in *High Temperature Superconducting Materials*, W. E. Hatfield and J. H. Miller, Eds; Dekhu, New York; 1988.
173. See, for example, J. W. Johnson, A. J. Jacobson, J. F. Brody, and S. M. Rich, *Inorg. Chem.*, 21, 3820 (1982).
174. A. J. Jacobson, J. W. Johnson, J. F. Brody, J. C. Scanlon, and J. T. Lewandowski, *Inorg. Chem.*, 24, 1782 (1985).
175. D. Beltran-Porter, A. Beltran-Porter, P. Amoros, R. Ibañez, E. Martinez, A. LeBail, G. Ferey, and G. Villeneuve, *Eur. J. Solid State Inorg. Chem.*, 28, 131 (1991).
176. J. W. Johnson, D. C. Johnston, A. J. Jacobson, and J. F. Brody, *J. Am. Chem. Soc.*, 106, 8123 (1984).
177. R. Amoros, R. Ibañez, E. Martinez-Tamayo, A. Beltran-Porter, D. Beltran-Porter, and G. Villeneuve, *Mat. Res. Bull.*, 24, 1347 (1989).
178. P. Amoros, A. Beltran, and D. Beltran, *J. Alloys Compounds*, 188, 123 (1992).
179. D. Beltran, P. Amoros, R. Ibañez, E. Martinez, A. Beltran, A. LeBail, G. Ferey, and G. Villeneuve, *Solid State Ionics 32*, 57 (1989).
180. T. Shimoda, T. Okuhara, and M. Misono, *Bull. Chem. Soc. Jpn.*, 58, 2163 (1985).
181. F. B. Abdelouahab, R. Olier, N. Guilhaume, F. Lefebvre, and J. C. Volta, *J. Catal.*, 134, 151 (1992).
182. T. P. Moser and G. L. Schrader, *J. Catal.*, 92, 216 (1985).
183. G. Busca, F. Cavani, G. Centi, and F. Trifiro, *J. Catal.*, 99, 400 (1986).
184. E. Bordes, *Catal. Today 1*, 499 (1987).
185. H. Morishige, J. Tamaki, N. Miura, and N. Yamazoe, *Chem. Lett.*, 1513 (1990).
186. S. Albonetti, F. Cavani, F. Trifiro, P. Venturoli, G. Calestani, M. L. Granados, and J. L. G. Fierro, *J. Catal.*, 160, 52 (1996).
187. G. Koyano, T. Okuhara, and M. Misono, *J. Am. Chem. Soc.*, 120, 767 (1998).
188. J. C. Guth, H. Kessler, J. M. Higel, J. M. Lamblin, J. Patarin, A. Seive, and J. M. Chezeau, and R. Wey, *ACS Symp. Ser.*, 398, 176 (1989).
189. H. Kessler, *MRS Symp. Proc.*, 233, 47 (1991).
190. P. Caullet, J. L. Guth, J. Hazm, J. M. Lamblin, and H. Gies, *Eur. J. Solid State Inorg. Chem.*, 28, 345 (1991).
191. M. Estermann, C. B. McCusker, C. Baerlocher, G. Merroude, and H. Kessler, *Nature (London)* 352, 320 (1995).
192. B. C. Tofield, G. R. Crane, P. M. Bridenbaugh, and R. C. Sherwood, *Nature (London)* 253, 722 (1975).
193. A. Leclaire, M. M. Borel, A. Grandin, and B. Raveau, *Mater. Chem. Phys.*, 12, 537 (1985).
194. P. C. Burns, J. D. Grice, and F. C. Hawthorne, *Can. Mineral 33*, 1131 (1995) and references cited therein.
195. P. K. S. Gupta, G. H. Swihart, R. Dimitrijevich, and M. B. Hossain, *Am. Mineral*, 76, 1400 (1991).

196. P. B. Moore and S. Ghose, *Am. Mineral*, *56*, 1527 (1971).
197. R. Kniep, G. Gözel, B. Eisenmann, C. Rohr, M. Asbrand, and M. Kizilyalli, *Angew. Chem. Int. Ed. Engl.*, *33*, 749 (1994).
198. C. H. Park and K. Z. Bluhm, *Naturforsch.*, *50b*, 1617 (1995).
199. C. Hauf, T. Friedrich, and R. Z. Kniep, *Kristallografiya*, *210*, 446 (1995).
200. R. Kniep, H. G. Will, I. Boy, and C. Rohr, *Angew. Chem. Int. Ed. Engl.*, *36*, 1013 (1997) and references cited therein.
201. K. Bluhm and C. H Park, *Z. Naturforsch.*, *52b*, 102 (1997).
202. C. Hauf and R. Kniep, *Z. Naturforsch 54b*, 102 (1999).
203. R. Kniep and H. Engelhardt, *Z. Anorg. Allg. Chem.*, *624*, 1291 (1998).
204. I. Boy and R. Kniep, *Z. Naturforsch.*, *54b*, 895 (1999).
205. I. Boy, G. Cordier, B. Eisenmann, and R. Kniep, *Z. Naturforsch.*, *53b*, 165 (1998).
206. Y. Shi, J. Liang, H. Zhong, J. Yang, W. Zhuang, and G. Rao, *J. Solid State Chem.*, *129*, 45 (1997).
207. R. P. Bontchev and S. C. Sevov, *Inorg. Chem.*, *35*, 6910 (1996).
208. S. C. Sevov, *Angew. Chem. Int. Ed. Engl.*, *35*, 2630 (1996).
209. C. J. Warren, R. C. Haushalter, D. J. Rose, and J. Zubieta, *Inorg. Chem. Commun.*, *1*, 4 (1998).
210. R. P. Bontchev, J. Do, and A. J. Jacobson, *Inorg. Chem.*, *38*, 2231 (1999).
211. R. P. Bontchev, J. Do, and A. J. Jacobson, *Angew. Chem. Int. Ed. Engl.*, 38, 1937 (1999).
212. W. G. Klemperer, T. A. Marquart, and O. M. Yaghi, *Angew. Chem. Int. Ed. Engl.*, *31*, 49 (1992).
213. M. Goepper and J. L. Guth, *Zeolites*, *11*, 477 (1991).
214. H. Kessler, *Mater. Res. Soc. Symp. Proc.*, *233*, 47 (1991).
215. M. Estermann, L. B. McCusker, C. Baerlocher, A. Merrouche, and A. Kessler, *Nature (London)* *352*, 320 (1991).
216. P. Wagner, Y. Nakagawa, G. S. Lee, M. E. Davis, S. Elomari, R. C. Medrud, and S. I. Zones, *J. Am. Chem. Soc.*, *122*, 263 (2000).
217. R. F. Lolio, S. I. Zones and M. G. Davis, in *Inclusion Chemistry with Zeolites, Nanoscale Materials by Design*; N. Herron and D. Corlim, Eds.: Kluwer Academic Press Dordrecht, The Netherlands, 1995, Vol. 21, p. 47.
218. A. K. Cheetham, G. Férey, and T. Loiseau, *Angew. Chem. Int. Ed. Engl.*, *38*, 3268 (1999).
219. P. Wagner, M. Yoshikawa, M. Lovallo, K. Tsuji, M. Tsapatsis, and M. E. Davis, *Chem. Commun.*, 2179 (1997).
220. M. I. Khan, R. C. Haushalter, C. J. O'Connor, C. Tao, and J. Zubieta, *Chem. Mater.*, *7*, 593 (1995).
221. Z. Bircsak, A. K. Hall, and W. T. A. Harrison, *J. Solid State Chem.*, *142*, 168 (1999).
222. A. M. Chippindale, *Chem. Mater.*, *12*, 818 (2000).
223. (a) V. Soghomonian, R. C. Haushalter, Q. Chen, and J. Zubieta, *Inorg. Chem.*, *33*, 1700 (1994).
     (b) D. Riou and G. Ferey, *Eur. J. Solid State Inorg. Chem.*, *31*, 25 (1994).
224. V. Soghomonian, Q. Chen, Y. Zhang, R. C. Haushalter, C. J. O'Connor, C. Tao, and J. Zubieta, *Inorg. Chem.*, *34*, 3509 (1995).
225. Z. Bircsak and W. T. A. Harrison, *Inorg. Chem.*, *37*, 3204 (1998).
226. R. C. Finn and J. Zubieta, unpublished results.
227. V. Soghomonian, Q. Chen, R. C. Haushalter, J. Zubieta, C. J. O'Connor, and Y.-S. Lee, *Chem. Mater.*, *5*, 1690 (1993).
228. Y. Lu, R. C. Haushalter, and J. Zubieta, *Inorg. Chim. Acta 257*, 268 (1997).

229. T. Loiseau and G. Férey, *J. Solid State Chem.*, *111*, 416 (1994).
230. V. Soghomonian, Q. Chen, R. C. Haushalter, and J. Zubieta, *Chem. Mater.*, *5*, 1595 (1993).
231. V. Soghomonian, R. C. Haushalter, J. Zubieta, and C. J. O'Connor, *Inorg. Chem.*, *35*, 2826 (1996).
232. X. Bu, P. Feng, and G. D. Stucky, *Chem. Commun.*, 1337 (1995).
233. V. Soghomonian, Q. Chen, R. C. Haushalter, and J. Zubieta, *Angew. Chem., Int. Ed. Engl.*, *32*, 610 (1993).
234. Y. Zhang, A. Clearfield, and R. C. Haushalter, *Chem. Mater.*, *7*, 1221 (1995).
235. W. T. A. Harrison, K. Hsu, and A. J. Jacobson, *Chem. Mater.*, *7*, 2004 (1995).
236. P. Kierkegaard and M. Westerlund, *Acta Chem. Scand.*, *18*, 2217 (1964).
237. J. T. Vaughey, W. T. A. Harrison, L. L. Dussack, and A. J. Jacobson, *Inorg. Chem.*, *33*, 4370 (1994).
238. M. Roca, M. D. Marcos, P. Amoros, A. Beltran-Porter, A. J. Edwards, and D. Beltran-Porter, *Inorg. Chem.*, *35*, 5613 (1996).
239. M. Schindler and W. H. Baur, *Angew. Chem., Int. Ed. Engl.*, *36*, 91 (1997).
240. W. H. Meier, D. H. Olsen, and C. Baerlocher, *Atlas of Zeolite Structure Types*, 4th ed., Elsevier, London, 1996.
241. M. Schindler, W. Joswig, and W. H. Baur, *Z. Anorg. Allg. Chem.*, *623*, 45 (1997).
242. W. H. Baur, A. Bieniok, R. D. Shannon, and E. Prince, *Z. Kristallogr.*, *187*, 253 (1989).
243. D. A. Burwell, K. G. Valentine, J. H. Timmermans, and M. E. Thompson, *J. Am. Chem. Soc.*, *114*, 4144 (1992).
244. A. Clearfield, *Chem. Rev.*, *88*, 125 (1988).
245. Y. Zhang and A. Clearfield, *Inorg. Chem.*, *31*, 2821 (1992) and references cited therein.
246. D. Rong, H.-G. Hong, Y. I. Kim, J. S. Kreuger, J. E. Mayer, and T. E. Mallouk, *Coord. Chem. Rev.*, *97*, 237 (1990).
247. E. T. Clark, P. R. Rudolf, A. E. Martell, and A. Clearfield, *Inorg. Chim. Acta 164*, 59 (1989).
248. G. Huan, V. W. Day, A. J. Jacobson, and D. P. Goshorn, *J. Am. Chem. Soc.*, *113*, 3188 (1991).
249. G. Huan, A. J. Jacobson, and V. W. Day, *Angew. Chem. Int. Ed. Engl.*, *30*, 422 (1991).
250. A. Müller, K. Hovemeier, and R. Rohlfing, *Angew. Chem. Int. Ed. Engl.*, *31*, 1192 (1992).
251. A. Müller, K. Hovemeier, E. Krickemeyer, and H. Boegge, *Angew. Chem. Int. Ed. Engl.*, *34*, 779 (1995).
252. (a) Y.-D. Chang and J. Zubieta, *Inorg. Chim. Acta*, *245*, 177 (1996); (b) M. I. Khan, Q. Chen, J. Zubieta, *Inorg. Chim. Acta*, *235*, 135 (1995).
253. B. Bujoli, P. Palvadeau, and J. Rouxel, *Chem. Mater.*, *2*, 582 (1990).
254. G. Cao, H. Lee, V. M. Lynch, and T. E. Mallouk., *Inorg. Chem.*, *27*, 2781 (1988).
255. K. Martin, P. J. Squattrito, and A. Clearfield, *Inorg. Chim. Acta*, *155*, 7 (1989).
256. G. Cao and T. E. Mallouk, *Inorg. Chem.*, 30, 1434 (1991).
257. J. LeBideau, B. Bujoli, A. Jouanneaux, C. Payen, P. Palvadeau, and J. Rouxel, *Inorg. Chem.*, *32*, 4017 (1993).
258. C Bhardwaj, H. Hu, and A. Clearfield, *Inorg. Chem.*, *32*, 4294 (1993).
259. D. M. Poojary and A. Clearfield, *J. Am. Chem. Soc.*, *117*, 11278 (1995) and references cited therein.
260. G. Cao, V. M. Lynch, J. S. Swinnea, and T. E. Mallouk, *Inorg. Chem.*, *29*, 2112 (1990).

261. G. Alberti, M. Casciola, and R. K. Biswas, *Inorg. Chim. Acta 201*, 207 (1992) and references cited therein.
262. G. Huan, A. J. Jacobson, J. W. Johnson, E. W. Corcoran, Jr., *Chem. Mater.*, 2, 91 (1990).
263. G. Huan, J. W. Johnson, A. J. Jacobson, and J. S. Merola, *J. Solid State Chem.*, 89, 220 (1990).
264. J. L. Snover and M. E. Thompson, *J. Am. Chem. Soc.*, 116, 765 (1994).
265. C. Y. Ortiz-Avila, C. Bhardwaj, and A. Clearfield, *Inorg. Chem.*, 33, 2499 (1994).
266. W. T. A. Harrison, L. L. Dussack, and A. J. Jacobson, *Inorg. Chem.*, 34, 4774 (1995).
267. J. LeBideau, C. Payen, P. Polvadeau, and B. Bujoli, *Inorg. Chem.*, 33, 4885 (1994); S. Drumel, P. Janvier, D. Deniaud, and B. Bujoli, *J. Chem. Soc. Chem. Commun.*, 1051 (1995); S. Drumel, P. Janvier, P. Barboux, M. Bujoli-Doeuff, and B. Bujoli, *Inorg. Chem.*, 34, 148 (1995).
268. M. Poojary, D. Grohol, and A. Clearfield, *Angew. Chem. Int. Ed. Engl.*, 34, 1508 (1995); M. Poojary, A. Cabeza, M. A. G. Aranda, S. Bruque, and A. Clearfield, *Inorg. Chem.*, 35, 1468 (1996).
269. K. Maeda, J. Akimoto, Y. Kiyozumi, and F. Mizukami, *Angew. Chem. Int. Ed. Engl.*, 34, 1199 (1995); K. Maeda, J. Akimoto, Y. Kiyozumi, and F. Mizukami, *J. Chem. Soc. Chem. Commun.*, 1033 (1995).
270. W. T. A. Harrison, L. L. Dussack, and A. J. Jacobson, *Inorg. Chem.*, 35, 1461 (1996).
271. A. Clearfield, *Prog. Inorg. Chem.*, 47, 371 (1998).
272. M. I. Khan and J. Zubieta, *Prog. Inorg. Chem.*, 43, 1 (1995).
273. M. I. Khan, Y.-S. Lee, C. J. O'Connor, R. S. Haushalter, and J. Zubieta, *Inorg. Chem.*, 33, 3855 (1994).
274. M. I. Khan, Y.-S. Lee, C. J. O'Connor, R. S. Haushalter, and J. Zubieta, *J. Am. Chem. Soc.*, 116, 4525 (1994).
275. M. I. Khan, Y.-S. Lee, C. J. O'Connor, R. S. Haushalter, and J. Zubieta, *Chem. Mater.*, 6, 721 (1994).
276. V. Soghomonian, Q. Chen, R. C. Haushalter, and J. Zubieta, *Angew. Chem. Int. Ed. Engl.*, 34, 223 (1995).
277. V. Soghomonian, R. Diaz, R. C. Haushalter, C. J. O'Connor, and J. Zubieta, *Inorg. Chem.*, 34, 4460 (1995).
278. V. Soghomonian, R. C. Haushalter, and J. Zubieta, *Chem. Mater.*, 7, 1648 (1995).
279. Q. Chen, J. Salta, and J. Zubieta, *Inorg. Chem.*, 32, 4485 (1993).
280. Q. Chen and J. Zubieta, *J. Chem. Soc. Chem. Commun.*, 2663 (1994).
281. Q. Chen and J. Zubieta, *Angew. Chem. Int. Ed. Engl.*, 32, 610 (1993).
282. J. Salta, Q. Chen, Y.-D. Chang, and J. Zubieta, *Angew. Chem. Int. Ed. Engl.*, 33, 757 (1994).
283. Y.-D. Chang, J. Salta, and J. Zubieta, *Angew. Chem. Int. Ed. Engl.*, 33, 325 (1994).
284. Q. Chen and J. Zubieta, *J. Chem. Soc. Chem. Commun.*, 1635 (1994).
285. M. I. Khan and J. Zubieta, *Angew. Chem. Int. Ed. Engl.*, 33, 760 (1994).
286. J. Salta and J. Zubieta, *J. Cluster Sci.*, 7, 531 (1996).
287. G. Bonavia, R. C. Haushalter, C. J. O'Connor, and J. Zubieta, *Inorg. Chem.*, 35, 5603 (1996).
288. C. Ninclaus, C. Serre, D. Riou, and G. Ferey, *C. R. Acad. Sci. Paris Ser IIc 1*, 551 (1998).
289. G. Bonavia, R. C. Haushalter, S. Lu, R. C. O'Connor, and J. Zubieta, *J. Solid State Chem.*, 132, 144 (1997).
290. D. Riou, C. Serre, and G. Férey, *J. Solid State Chem.*, 141, 89 (1998).
291. D. Riou, O. Roubeau, and G. Férey, *Microporous and Mesoporous Materials*, 23, 23 (1998).

292. D. Riou and G. Ferey, *J. Mater. Chem.*, *8*, 2733 (1998).
293. M. Riou-Cavellec, M. Sanselme, and G. Ferey, *J. Mater. Chem.*, *10*, 745 (2000).
294. M. Tournoux, M. Ganne, and Y. Piffard, *J. Solid State. Chem.*, *96*, 141 (1992).
295. F. Jona, G. Shirane, and R. Pepinsky, *Phys. Rev.*, *98*, 903 (1955).
296. B. Gérand, G. Nowogrocki, and M. Figlarz, *J. Solid State Chem.*, *38*, 312 (1981).
297. R. C. Finn and J. Zubieta, *Chem. Commun.*, 2000, in press.
298. R. C. Finn and J. Zubieta, *J. Chem. Soc., Dalton Trans.*, 1821 (2000).
299. A. K. Cheetham, *Science*, *264*, 794 (1994) and references cited therein.
300. *Inorganic Materials*, D. W. Bruce and D. O'Hare, Eds., Wiley, Chichester, UK, 1992.
301. M. Fujita, Y. J. Kwon, O. Sasaki, K. Yamaguchi, and K. Ogura, *J. Am. Chem. Soc.*, *117*, 7287 (1995) and references cited therein.
302. *Introduction to Molecular Electronics*, M. C. Petty, M. R. Bryce, and D. Bloor, Eds., Edward Arnold, London, 1995.
303. *Magnetic Molecular Materials*, D. Gatteschi, O. Kahn, J. S. Miller, and F. Palacio, Eds., NATO ASI Ser. Ser. E, 1991.
304. G. B. Olson, *Science*, *277*, 1237 (1997).
305. *Chem. Eng. News*, *27* (13), 913 (1949).

Subject Index

Absorption mechanisms, in vivo chromium(III), 206–207
Absorption spectroscopy, lead(II) compounds, 16–24
  electronic transitions, 16
  optical electronegativities, 16–19
  solid-state spectra, 19–24
Acid catalysis, sol-gel processing, initial stages, 346–347
Aging process, sol-gel processing, 347–349
Air, lead concentrations in, 77–78
Air-water interface, cryptand-based amphiphiles, 311–319
  Langmuir-Blodgett film, 311–317
  molecular recognition, 317–319
Alkoxysilanes:
  sol-gel covalently attached complexes, silylated precursor preparation, 378–739
  sol-gel processing, initial stages, 345–347
Alkyl silicon compounds, sol-gel surface modification, 352
Aluminum ions:
  entrapped sol-gel complexes, 355
  vanadium aluminophosphates, 493, 495–497
  vanadium phosphates (VOPO), research overview, 491, 493
Amino acids, chromium-chromodulin interaction, 211–214
δ-Aminolevulinic acid dehydratase (ALAD): lead(II) compounds:
  EXAFS studies, 51–52
  heavy atom derivatives, 50
  protein-lead interactions, 65
  lead poisoning, heme biosynthesis, 117–118
  zinc-lead interactions, 104–110
δ-Aminolevulinic acid synthase (ALAS), lead molecular targets, 111
Amphiphiles, laterally asymmetric aza-cryptands, 304–319
  air-water interface, 311–319
    Langmuir-Blodgett film, 311–317
    molecular recognition, 317–319
  vesicular aggregation, 305–311

Anhydrous alcohols, sol-gel covalently attached complexes, silylated precursor preparation, 378–379
Anthracene fluorescent sensor:
  cryptand-based fluorescent sensors, 289–295
  lead biodistribution, 96–99
Antioxidants, chromium toxicity, oxidative stress and, 178–180
Apoptosis, chromium toxicity, promotion of, 181–182
Aquatic systems, lead mineral distribution in, 84–87
  environmental contamination measurements, 87–89
Aqueous solution, protonation-lead(II) binding constants, 67–69
Associative mechanism:
  lead-chelate dissociation/substitution reactions, 59–60
  lead-ligand interactions, 53–54
Atomic absorption spectroscopy (AAS):
  chromium uptake, distribution, and metabolism, 163–167
  lead poisoning:
    blood lead levels (BLLs), 115
    urine, teeth, hair, and nail levels, 119

Backbone structures:
  vanadium organophosphates (VOPRO), oxovanadium diphosphonate solids, 550–566
  vanadium phosphates, 431–459
    charge-compensating cations, three-dimensional structures, 525–531
Ball-and-stick structure:
  vanadium phosphate fluorides, 503–505
  vanadium phosphates:
    binuclear clusters and chains, 479–480
    charge-compensating cations, three-dimensional structures, 527–531
    charge neutral V-P-O covalent linkages, 460–468
    isolated vanadium polyhedra, 471–475

Ball-and-stick structure: (*Continued*)
  vanadium phosphates/organophosphates (VOPO/VOPRO), M^{II}-organodiimine subunits, V/O/M'/P scaffoldings, 576–577
  vanadium silicophosphates, 493–494
Barium compounds, vanadium phosphates, tetranuclear cluster structures, 479, 482–483
Base catalysis, sol-gel processing, initial stages, 346–347
Beer-Lambert relationship, ICC to ISSC conversion, 358
BI crown fluorescent sensor, lead biodistribution, 96–99
Binding constant, thermodynamics, lead-ligand interactions, 65–66
Binuclear vanadium sites:
  vanadium phosphate fluorides, 502–505
  vanadium phosphates/organophosphates (VOPO/VOPRO), M^{II}-organodiimine subunits, V/O/M'/P scaffoldings, 575–577
  vanadium phosphates (VOPO), 464–468
    charge-compensating cations:
      three-dimensional structures, 523–531
      two-dimensional structures, 516–520
    cluster and chain structures, 475, 478–481
    quaternary M'/V/P/O phases, 475–477
Biodistribution mechanics, lead compounds:
  body burden and uptake, 92–93
  body distribution, 93
  fluorescent sensors, 93–99
  toxicokinetics, 99–101
Biological chemistry:
  chromium(III), 203–225
    chromodulin, 209–214
    food supplement efficacy and safety, 214–220
    glucose tolerance factor, 207–208
    low molecular weight chromium-binding substance, 208–209
    molecular mechanisms, 220–225
  lead compounds:
    biodistribution, 92–101
    biomarkers, lead poisoning, 111–119
      blood lead levels (BLLs), 113–115
      bone lead levels, 118–119
      heme biosynthesis alterations, 115–118
    body burden and uptake, 92–93
    chelation therapy, 120–123
    distribution in body, 93
    fluorescent sensors, 93–99
    lead poisoning, 90–92
    molecular targets, 101–111
      calcium protein interactions, 102–103
      RNA/DNA interactions, 110–111
      zinc protein interactions, 103–110
    toxicokinetics, 99–101
Biomarkers:
  chromium toxicity, DNA damage, chromium(VI) exposure, 175
  lead poisoning, 111–119
    blood lead levels (BLLs), 113–115
    bone lead levels, 118–119
    heme biosynthesis alterations, 115–118
Biomolecules:
  chromium, 147–159
    electronic absorption spectroscopy, 158–159
    electrospray mass spectrometry, 158
    EPR spectroscopy, 148–156
      chromium(III) and chromium(IV), 153–154
      chromium(V), 148–153
      organic radicals, 154
      oxygen radicals and singlet oxygen, 154–156
    in vivo studies, 156
    XRD/XAS structural characterization, 156–158
    XRAS spectroscopy, 157–158
    x-ray structures, 156–157
  lead(II) compounds, x-ray crystal structures, 47–50
    heavy atom derivatives, proteins and molecules, 49–50
2,2'-Bipyridine:
  cryptand-based fluorescent sensors, 283–295
  sol-gel noncoordinated interactions, 359–360
  vanadium phosphates/organophosphates (VOPO/VOPRO), M^{II}-organodiimine subunits, V/O/M'/P scaffoldings, 581–588
Bis(acetylacetonate) complexes, sol-gel catalysis, 351
3,3'-Bis(isoquinoline) (biqn), cryptand-based fluorescent sensors, 283–295
Blood-brain barrier, lead toxicokinetics, 101

SUBJECT INDEX

Blood lead levels (BLLs), lead poisoning, 90–92
   biomarkers, 113–115
Blue shift, sol-gel luminescent probes, ruthenium polypyridyl complexes, 362–366
Bone lead levels, lead poisoning biomarkers, 118–119
Borate:
   vanadium borophosphates, 497–500
   vanadium phosphates (VOPO), research overview, 491, 493
Bottom-up technology, laterally asymmetric aza-cryptands, chemical logic, fluorescent signaling, 277–278
Bovine serum albumin (BSA), chromium(VI)+reductant systems, 191–193
British anti-Lewisite (BAL):
   lead(II) compound small molecules, 43–46
   lead poisoning, chelation therapy with, 120–122
terti-Butoxycarbonyl (Boc) group, cryptand-based amphiphiles, vesicular aggregation, 309–311

Cadmium(II), cryptand metal complexation, 266–274
Calcium proteins:
   lead toxicokinetics, 100–101
   molecular targets, lead interactions, 102–103
Cancer-related genes, chromium toxicity and, 177–178
Capping, sol-gel surface modification, 352
Carbonyl groups, cryptand-based fluorescent sensors, 288–295
Carcinogenicity studies:
   chromium(III) toxicity, 161–163
   chromium toxicity, direct/indirect pathways, 182–183
   chromium(VI) toxicity, 160–163
Catalysts:
   sol-gel materials, 404–407
      entrapment *vs.* immobilization, 404
      leaching, 405–406
      site accessibility, 404–405
      site isolation, 406–407
   sol-gel processing, 350
Catalytic DNA fluorescent sensor, lead biodistribution, 99

Catecholamines, chromium(VI)+reductant systems, 189–193
Cation input. *See also* Charge-compensating cations
   cryptand-based fluorescent sensors, 283, 286–295
Cell membrane-wall interactions, chromium toxicity, 171–173
Cell signaling mechanisms, chromium toxicity, 180–182
Cellular toxicity mechanisms, chromium toxicology, 163–183
   carcinogenicity, direct and indirect pathways, 182–183
   DNA and transcription factor damage, 174–178
   genotoxicity assays, 167–171
   membrane and wall interactions, 171–173
   oxidative stress and antioxidants, 178–180
   signaling and aopotosis, 180–182
   uptake, distribution, and metabolism, 163–167
Cerium(IV), sol-gel catalysis, 350–351
Ceruse (white lead), structure and composition, 79, 81
Cerussite ($PbCO_3$), structure and composition, 79, 81
Cesium compounds:
   vanadium aluminophosphates, 493, 495–497
   vanadium organophosphates (VOPRO), oxovanadium diphosphonate solids, 547–566
Chalcogenides, lead(II) compounds, photoelectron spectroscopy, 27–28
Charge compensating cations:
   vanadium organophosphates (VOPRO), oxovanadium diphosphonate solids, 557–566
   vanadium phosphates (VOPO):
      organic cations, 505–531
      one-dimensional structures, 510–512
      three-dimensional structures, 520–531
      two-dimensional structures, 512–520
      quaternary $M'/V/P/O$ phases, covalent linkages and charge compensating M' cations, 468–487
Charge neutral V-P-O covalent linkages, vanadium phosphates, 459–468

Charge-transfer reactions:
  cryptands, nonlinear optical effects, 296–297
  lead(II) compounds, 16
    optical electronegativities, 16–19
    solid-state absorption spectra, 19–24
Chelation therapy agents:
  lead-chelate dissociation/substitution reactions, 59–63
  lead(II) compounds:
    ring size, 73
    structural studies, 41–46
  lead poisoning, 120–123
  protonation-lead(II) binding constants, 67, 70, 72
Chemical logic, laterally asymmetric aza-cryptands, fluorescent signaling, 277–279
Chemical shifts, lead-207 NMR spectroscopy, 30–33
Chloropropyltrialkoxysilane derivatives, sol-gel covalently attached complexes, 375–376
Chromium:
  biologically active complexes, 147–159
    electronic absorption spectroscopy, 158–159
    electrospray mass spectrometry, 158
    EPR spectroscopy, 148–156
      chromium(III) and chromium(IV), 153–154
      chromium(V), 148–153
      organic radicals, 154
      oxygen radicals and singlet oxygen, 154–156
      in vivo studies, 156
    molecular mechanisms, 220–225
    XRD/XAS structural characterization, 156–158
    XRAS spectroscopy, 157–158
    x-ray structures, 156–157
  nutrient properties, 203–225
    chromium(III) biological activity, 203–206
    chromodulin, 209–214
    food supplement efficacy and safety, 214–220
    glucose tolerance factor, 207–208
    low molecular weight chromium-binding substance, 208–209
    molecular mechanisms, 220–225
    chromium(III) in vivo absorption, metabolism, and speciation, 206–207
  structure and chemistry, 146–147
  toxicology, 159–203
    cellular/subcellular toxicity mechanisms, 163–183
      carcinogenicity, direct and indirect pathways, 182–183
      DNA and transcription factor damage, 174–178
      genotoxicity assays, 167–171
      membrane and wall interactions, 171–173
      oxidative stress and antioxidants, 178–180
      signaling and aopotosis, 180–182
      uptake, distribution, and metabolism, 163–167
    epidemiological/experimental evidence, 159–163
    molecular toxicity mechanisms, 183–203
      chromium(III)-DNA interactions, 195–196
      chromium(III)+oxidant systems, 193–195
      chromium(V) DNA damage mechanisms, 196–203
      chromium(VI)+reductant systems, 184–193
Chromium(III):
  biological activity, molecular mechanisms, 220–225
  cellular genotoxicity assays, 169–171
  cellular uptake, distribution, and metabolism, 163–167
  entrapped sol-gel complexes, 355
  EPR spectroscopic chemical studies, 153–154
  as nutrient:
    biological activity, 203–206
    chromodulin, 209–214
    food supplement efficacy and safety, 214–220
    glucose tolerance factor, 207–208
    low molecular weight chromium-binding substance, 208–209
    molecular mechanisms, 220–225
    in vivo absorption, metabolism, and speciation, 206–207
  toxicity, 161
    DNA adducts, 174–175
    DNA damage mechanisms, 196–203
    oxidant systems, 193–195

shuttle vector mutation, 177
x-ray structures, 156–157
Chromium(III) tris-picolinate, efficacy and safety, 215–220
Chromium(IV):
  EPR spectroscopic chemical studies, 153–154
  toxicity:
    chromium(III)-oxidant systems, 194–195
    DNA damage mechanisms, 196–203
    x-ray structures, 157
Chromium(V):
  cellular genotoxicity assays, 169–171
  cellular uptake, distribution, and metabolism, 164–167
  EPR spectroscopic chemical studies, 148–153
  toxicity:
    cell membrane/wall interactions, 172–173
    chromium(III)-oxidant systems, 194–195
    chromium(VI)+reductant systems, 185–193
    DNA damage mechanisms, 196–203
    x-ray structures, 157
Chromium(VI):
  biological activity, molecular mechanisms, 221–225
  cellular genotoxicity assays, 167–171
  cellular uptake, distribution, and metabolism, 163–167
  chromodulin interaction, 211–214
  organic radicals, 154
  toxicity:
    cancer-related genes, 177–178
    cell membrane/wall interactions, 171–173
    cell signaling and apoptosis mechanisms, 180–182
    DNA biomarkers, 175
    epidemiological/experimental evidence, 159–163
    oxidative DNA damage, 174
    oxidative stress and antioxidants, 178–180
    reductant systems, 184–193
    shuttle vector mutation, 177
    transcription factor and gene expression changes, 176
  x-ray structures, 157
Chromodulin, chromium(III) biological activity, 209–214
Chromophore formation, cryptand derivatives, D-π-A nonlinear systems, 197–203

Cobalt(III) ions, cryptands:
  amphiphiles, 304–319
  metal complexation, 260–265
  tripodal capping, 257–260
Cobalt(II) ions:
  cryptands:
    cryptand-based fluorescent sensors, 281–295
    tripodal capping, 257–260
  sol-gel processing:
    entrapped sol-gel complexes, 356
    ICC to ISSC conversion, 357–358
Collision-induced energy conversion, laterally asymmetric aza-cryptands, fluorescent signaling, 279–280
Concentration effects, sol-gel processing, luminescent probes, lanthanide complexes, 361–362
Condensation reactions, sol-gel processing, initial stages, 345–347
Coordination numbers, lead(II) structures, 36–40
  thermodynamic preferences, 69–71, 72–73
Copper complexes, vanadium phosphates/organophosphates (VOPO/VOPRO), M^{II}-organodiimine subunits, V/O/M′/P scaffoldings, 575–588
Copper(II):
  cryptand-based amphiphiles, 305–311
    air-water interface, Langmuir-Blodgett film, 311–317
  cryptand-based fluorescent sensors, 281–295
  cryptand metal complexation, 265–266
Coupling constants, lead-207 NMR spectroscopy, 30–33
Covalent bonds:
  entrapped sol-gel complexes, 355
  vanadium phosphates:
    charge neutral V-P-O linkages, 459–468
    quaternary M′/V/P/O phases, covalent linkages and charge compensating M′ cations, 468–487
V/O/M′/P scaffoldings, vanadium phosphates/organophosphates (VOPO/VOPRO):
  M^{II}-organodiimine subunits:
    oxovanadium organophosphate structures, 577–588
    oxovanadium phosphate structures, 567–577
    research background, 566–567

Covalently attached sol-gel complexes, 372–396
 future research issues, 408–409
 material characterization, 390–391
 material preparation, 380–390
  immobilization techniques, 380–385
  metal center survival, 386–390
  postmodification techniques, 385–386
 precursor preparation, 372–380
  purification, 379–380
  silylated ligands, 372–376
  silylated metal complexes, 376–379
 research background, 372
 tethered TM complex environment, 391–396
C2 protein domain, lead interactions, 103
Cross-polarization-magic angle spinning (CP-MAS) NMR, sol-gel materials characterization, 391
Cryptands. See also Laterally asymmetric aza-cryptands
 amphiphiles, 304–319
  air-water interface, 311–319
   Langmuir-Blodgett film, 311–317
   molecular recognition, 317–319
  vesicular aggregation, 305–311
 fluorescent senors, 280–295
 lateral asymmetry:
  research background, 252–253
  synthesis, 253–260
 nonlinear optical materials, 295–304
  D-π-A nonlinear systems, 297–304
  nonlinear optical effect, 295–297
 structure and properties, 252–253
 synthesis, 253–260
  methodology, 253–255
  tripodal capping, 257–260
  tripodal coupling, 255–257
Cryptate effect, cryptand-based fluorescent sensors, 283–295
Crystallinity, sol-gel materials, 390–391
Cystein-rich sites, zinc-lead interactions, 107–110
Cytochrome c, sol-gel sensors, 400–401

d-block elements, vanadium phosphates, 487–491
Density, silica materials, 342–343
Diazatrithia-15-crown-5 fluorescent sensor, lead biodistribution, 96–98
Diet, lead concentrations in, 77–78

Differential pulse polarography (DPP), lead poisoning, blood lead levels (BLLs), 115
"Diffusion-limited" reaction, sol-gel processing, luminescent probes, ruthenium polypyridyl complexes, 366
$meso$-2,3-Dimercaptosuccinic acid (DMSA):
 lead(II) compounds:
  lead-protein thermodynamic stability, 76
  small molecules, 43–46
 lead poisoning, chelation therapy with, 120–122
N,N-dimethylaminopyridine (DMAP), sol-gel processing, 350
5-5-Dimethylpyrroline-N-oxide (DMPO):
 chromium oxygen radicals and singlet oxygen, 154–156
 chromium(VI) organic radicals, 154
Dimethyl sulfoxide (DMSO), chromium toxicity, DNA damage mechanisms, 200–203
Dirac equation, lead(II) spectroscopic analysis, 14–15
Dissociation constant, thermodynamics:
 lead-ligand interactions, 65–66
 lead-protein interactions, 74–76
Dissociative mechanism, lead-ligand interactions, 53–54
Distribution mechanisms, chromium toxicity, 163–167
DNA:
 chromium toxicity:
  cancer-related genes, 177–178
  carcinogenicity pathways, 183
  chromium(III)-DNA adducts, 174–175
  chromium(III) interactions, 195–196
  chromium(III)-oxidant systems, 193–195
  chromium(V) complexes, 196–203
  chromium(VI) exposure biomarkers, 175
  chromium(VI)+reductant systems, 184–193
  enzyme activity changes, 175–176
  oxidative damage, 174
  shuttle vector mutations, 177
 lead interactions, 110–111
Donor groups:
 cryptand metal complexation:
  N and O donors, 265–274
  N and S donors, 260–265
 lead(II) structures, 36
  thermodynamic preferences, 68–69, 72

Double chain structures, vanadium phosphates, 464–468
D-π-A nonlinear systems, cryptand derivatives, 297–304
Drying chemical control agents (DCCAs), sol-gel processing, 349
Drying process, sol-gel processing, 347–349
Dynamic light scattering (DLS), cryptand-based amphiphiles, vesicular aggregation, 306–311

EDTA:
 chromium-chromodulin interaction, 213–214
 chromium toxicity, chromium(III)-DNA adducts, 174–175
 lead(II) compounds:
  chelation therapy agents, 43–46
  lead-protein thermodynamic stability, 76
  small molecule complexes, 39–40
  vibrational spectroscopy, 28–29
 lead minerals, soil distribution, 82–84
 lead poisoning, chelation therapy with, 120–122
 sol-gel sensors, 399–401
Electric field-induced second harmonic generation (EFISH), 297–304
Electrochemical studies, chromium complexes, 159
Electrode structure, sol-gel entrapped complexes, redox active probes, 366
Electronegativity:
 lead compounds, 11
 lead(II) compounds, optical electronegativities, 16–19
Electronic absorption spectroscopy, chromium complexes, 158–159
Electron nuclear double resonance (ENDOR), chromium(V) complexes, 152–153
Electron pairs, lead(II) compounds, small molecule structural effects, 40–41
Electron paramagnetic resonance (EPR) spectroscopy:
 chromium biomolecules, 148–156
  chromium(III) and chromium(IV), 153–154
  chromium(V), 148–153
  organic radicals, 154
  oxygen radicals and singlet oxygen, 154–156
  uptake, distribution, and metabolism, 165–167
  in vivo studies, 156
 chromium toxicity, chromium(VI)+reductant systems, 185–193
 cryptand metal complexation, 269–274
Electrospray mass spectrometry, chromium complexes, 158
Entrapped sol-gel complexes, 353–371
 catalysts, 404
 inner-sphere sol-gel chemistry (ICC to ISSC), 357–358
 noncoordinated interactions, 358–371
  luminescent probes, 360–366
   lanthanide complexes, 360–362
   ruthenium polypyridyl complexes, 362–366
  redox active probes, 366–371
   monolithic studies, 369–370
   thin-film studies, 366–369
   research background, 358–360
 research background, 353–354
 silica chemistry, 354–356
Environmental distribution, lead compounds:
 aquatic systems, 84–87
 contamination history and geochemistry, 87–88
 mineral distribution, 78–81
 research background, 76–78
 soil distribution, 81–84
Enzyme activity:
 chromium toxicity, DNA-related changes, 175–176
 sol-gel noncoordinated interactions, 371
 sol-gel sensors, 400–401
EP/ZPP levels, lead poisoning, heme biosynthesis, 116–118
Europium(III):
 cryptand-based fluorescent sensors, 283–295
 sol-gel processing, luminescent probes, 360–362
EXAFS studies:
 lead(II) compounds, 51–52
 sol-gel materials characterization, 390–391
Excimeric emissions, cryptand-based fluorescent sensors, 292–295
Exciplex emissions, cryptand-based fluorescent sensors, 292–295

Ferrocenes:
 sol-gel covalently attached complexes, immobilization techniques, 383–385
 sol-gel noncoordinated interactions, 371

Ferroin derivatives, sol-gel covalently attached complexes:
  silylated precursor preparation, 377–379
  tethered TM complexes, 392–396
Flexibility, silica materials, 342–343
Fluorescent sensors:
  laterally asymmetric aza-cryptands, 275–295
    chemical logics, 277–278
    cryptand-based sensors, 280–295
    design principles, 275–276
    metal ion sensing, 277
    transition metal ions, 279–280
  lead compounds, biodistribution, 93–99
  sol-gel processing, luminescent probes, lanthanide complexes, 360–362
Fluorides:
  sol-gel catalysts, 350
  vanadium phosphate fluorides, 500–505
  vanadium phosphates (VOPO), research overview, 491, 493
"Fluorophore-spacer-receptor" model:
  cryptand-based fluorescent sensors, 280–295
  laterally asymmetric aza-cryptands, fluorescent signaling, 277
Food supplements, chromium(III) in, 214–220
Fundamental building blocks (FBBs), vanadium organophosphates (VOPRO), oxovanadium diphosphonate solids, 562–566

Galena (PbS), structure and composition, 79
Gas-liquid chromatography (GLC), chromium uptake, distribution, and metabolism, 163–167
Gelation process, sol-gel processing, 347–349
Gene expression, chromium toxicity, transcription factors, 176
Genotoxicity assays, chromium toxicity, 167–171
  food supplement efficacy and safety, 218–220
Geochemistry, environmental lead contamination, 87–89
Geometries:
  lead(II) small molecule complexes, 36–40
  vanadium phosphate structures, 431–459
Glucose metabolism, chromium(III) biological activity, 224–225
  chromodulin, 210–214

Glucose oxidase, sol-gel sensors, 401
Glucose tolerance factor (GTF), chromium(III), 207–208
Glutathione, chromium-chromodulin interaction, 211–214
Glycoproteins, chromium toxicity:
  cell membrane/wall interactions, 173
  chromium(VI)+reductant systems, 191–193
Graphite furnace AAS, chromium uptake, distribution, and metabolism, 164–167
Grignard reagents, sol-gel covalently attached complexes, silylated ligand preparation, 375–376
Group 14 (VIA) analogues, lead(II) compounds, 46–47

Hair analysis, lead poisoning biomarkers, 119
Halides, lead(II) compounds, photoelectron spectroscopy, 27
Heat shock proteins, molecular targets, 111
Heavy atom derivatives:
  laterally asymmetric aza-cryptands, fluorescent signaling, 279–280
  lead(II) biomolecules, 49–50
Heme biosynthesis, lead poisoning, alterations in, 115–118
Heme proteins, lead interactions, 111
Hemihydrate structures, vanadium phosphates, charge neutral V-P-O covalent linkages, 460–468
HEPES (4-(2-hydroxyethyl)-1-piperazeneethanesulfuric acid), chromium toxicity, DNA damage mechanisms, 200–203
Hexamethylphosphoric triamide (HMPA), sol-gel processing, 350
High dilution synthesis, cryptands, 253–255
Highest occupied molecular orbital (HOMO), cryptands:
  fluorescent sensors, 281–295
  nonlinear optical effects, 296–297
Human tissue, lead levels, 93–94
Hybrid-organic inorganic materials, sol-gel materials, 335
Hydrogen bonding, cryptand-based amphiphiles, air-water interface, 317–319
Hydrolysis, sol-gel processing, 336
  catalyst contamination, 350
  initial stages, 345–347
Hydrophilicity, silica materials, 342–343

Hydrophobic-hydrophilic interactions, vanadium organophosphates (VOPRO), 425–427
Hydrosilylation, sol-gel covalently attached complexes, silylated ligand preparation, 373–376
Hyperfine coupling, chromium(V), EPR spectroscopy, 150–153
Hyperpolarizability, cryptands, nonlinear optical effect (NLO), 296–297
Hyper-Rayleigh scattering (HRS) technique, cryptand derivatives, D-π-A nonlinear systems, 297–304

ICC interactions:
  conversion to ISSC (inner-sphere sol-gel chemistry), 357–358
  entrapped sol-gel complexes, 356
  future research issues, 408–409
  transition-metal complexes, metal center survival, 388–390
Immobilization techniques:
  lead minerals, soil distribution, 83–84
  sol-gel catalysts, 404
    site isolation, 406–407
  sol-gel covalently attached complexes, 380–385
    future research issues, 408–409
    metal center survival, 386–390
Indium-tin oxide (ITO):
  cryptand-based amphiphiles, air-water interface, Langmuir-Blodgett film, 315–317
  sol-gel entrapped complexes, redox active probes, 366–369
Indo-1 fluorescent sensor, lead biodistribution, 95–99
Inductively coupled plasma-mass spectrometry (ICP-MS):
  lead chemical analysis, 10
  lead poisoning:
    blood lead levels (BLLs), 115
    urine, teeth, hair, and nail levels, 119
"Inert" ($6s^2$) electron pairs, photoelectron spectroscopy, lead(II) spectroscopic analysis, 25, 123
Infinite chain structures, vanadium phosphates (VOPO):
  charge-compensating cations, two-dimensional structures, 519–520
  polyhedral structures, 469–470
Infrared (IR) spectroscopy, lead(II) compounds, 28–29
Inner-sphere sol-gel chemistry (ISSC), ICC conversion to, 357–358
Insulin receptors, chromium(III) biological activity, chromodulin, 209–214
Intramolecular reactions, chromium(III) biological activity, 223–225
Intrinsic active sites, cryptand metal complexation, 273–274
In vivo studies:
  chromium EPR spectroscopic analysis, 156
  chromium(III) absorption, metabolism, and speciation, 206–207
Ionophores, lead toxicokinetics, 101–102
Ion-pairing effects, sol-gel processing, tethered TM complexes, 394–396
Iron(III):
  cryptand-based fluorescent sensors, 281–295
  entrapped sol-gel complexes, 355
Iron uptake, lead toxicokinetics, 101
Isocyanates, sol-gel covalently attached complexes, 373–376
Isotope dilution mass spectrometry (IDMS), lead isotopes, 10
Isotopes, lead compounds, 9–10

Jahn-Teller effect, solid-state absorption spectra, lead-thiolate complexes, 22–24

Kinetics:
  lead-ligand interactions, 55–65
    lead-chelate dissociation and substitution, 59–63
    lead(II)-protein interactions, 64–65
    macrocyclic ligands, 64
    simple ligand-exchange constants, 55–58
  sol-gel processing:
    immobilization techniques, 384–385
    initial stages, 346–347
Kubelka-Munk function, sol-gel materials, ICC to ISSC conversion, 357–358, 360

Langmuir-Blodgett film, cryptand-based amphiphiles, air-water interface, 311–317
Lanthanide ions:
  cryptand-based fluorescent sensors, 283–295
  sol-gel processing:

Lanthanide ions: (*Continued*)
  catalysis, 350–351
  luminescent probes, 360–362
  photonics, 402–403
  tethered TM complexes, 393–396
Laterally asymmetric aza-cryptands:
  amphiphiles, 304–319
    air-water interface, 311–319
      Langmuir-Blodgett film, 311–317
      molecular recognition, 317–319
    vesicular aggregation, 305–311
  fluorescent signaling, 275–295
    chemical logics, 277–278
    cryptand-based sensors, 280–295
    design principles, 275–276
    metal ion sensing, 277
    transition metal ions, 279–280
  metal complexation, 260–274
    N and O donors, 265–274
    N and S donors, 260–265
  nonlinear optical materials, 295–304
    D-π-A nonlinear systems, 297–304
    nonlinear optical effect, 295–297
  research background, 252–253
  synthesis, 253–260
    methodology, 253–255
    tripodal capping, 257–260
    tripodal coupling, 255–257
Layered structures:
  vanadium organophosphates (VOPRO):
    oxovanadium diphosphonate solids, 547–566
    oxovanadium monophosphonate solids, 537–545
  vanadium phosphates/organophosphates (VOPO/VOPRO), M^{II}-organodiimine subunits, V/O/M'/P scaffoldings, 581–588
Leaching techniques, sol-gel catalysts, 405–406
Lead:
  biological chemistry:
    biodistribution, 92–101
    biomarkers, lead poisoning, 111–119
      blood lead levels (BLLs), 113–115
      bone lead levels, 118–119
      heme biosynthesis alterations, 115–118
    body burden and uptake, 92–93
    chelation therapy, 120–123
    distribution in body, 93
    fluorescent sensors, 93–99

lead poisoning, 90–92
  molecular targets, 101–111
    calcium protein interactions, 102–103
    RNA/DNA interactions, 110–111
    zinc protein interactions, 103–110
  toxicokinetics, 99–101
electronic properties, 11
environmental distribution:
  aquatic systems, 84–87
  contamination history and geochemistry, 87–88
  mineral distribution, 78–81
  research background, 76–78
  soil distribution, 81–84
history of emissions, 4–6
isotopes, 9–10
ligand interactions, 52–76
  kinetics, 55–65
    lead-chelate dissociation and substitution, 59–63
    lead(II)-protein interactions, 64–65
    macrocyclic ligands, 64
    simple ligand-exchange constants, 55–58
  research background, 52–54
  thermodynamics, 65–76
    lead(II)-ligand binding constants, 66–67
    lead-protein interactions, 73–76
    lead-small molecule interactions, 67–73
mineral composition, 79–81
  cerussite ($PbCO_3$), 79, 81
  galena (PbS), 79
  litharge and massicot (PbO), 81
  minium (Pb_3O_4), 81
  schulterite and other phosphates, 81
mining production, 78–79
oxidation states, 10
research background, 3–9
structure and properties, 9–12
Lead-binding proteins, molecular targets, 111
Lead(II):
  cryptand-based fluorescent sensors, 283–295
  cryptand metal complexation, 266–274
  electronic properties, 9–10
  isotopes, 9–10
  oxidation states, 10
  solubility and water-exchange, 11–12
  spectroscopic studies, 13–35
    absorption spectroscopy, 16–24
    electronic transitions, 16
    optical electronegativities, 16–19

SUBJECT INDEX

solid-state spectra, 19–24
nuclear magnetic resonance (NMR), 29–35
   lead-207 parameters, 30–33
   proteins, lead-207 studies, 33–34
photoelectron spectroscopy, 24–28
   chalcogenides, 27–28
   halides, 27
   oxides, 26
   relativistic effects, 25
relativistic effects, 14–15
vibrational spectroscopy (IR and Raman), 28–29
structural studies:
   EXAFS spectroscopy, 51–52
   research background, 35–36
   X-ray crystals, biomolecules, 47–50
      heavy atom derivatives, proteins and molecules, 49–50
   X-ray crystals, small molecule complex:
      chelation therapy agents, 41–46
      electron pairs ($6s^2$), 40–41
      new compounds, 46–47
   X-ray crystals, small molecule complexes, 36–47
      coordination numbers and geometries, 36–40
      donor groups, 36
Lead-207 NMR spectroscopy, parameters, 30–33
   protein studies, 33–34
Lead phosphates, structure and composition, 81
Lead poisoning:
   biomarkers, 111–119
      blood lead levels (BLLs), 113–115
      bone lead levels, 118–119
      heme biosynthesis alterations, 115–118
   history of, 4–7
   symptoms, 90–92
Lead-thiolate complexes, solid-state absorption spectra, 19–24
Ligand-exchange constants, lead-ligand kinetics, 55–58
Ligand interactions:
   lead compounds, 52–76
      kinetics, 55–65
         lead-chelate dissociation and substitution, 59–63
         lead(II)-protein interactions, 64–65
         macrocyclic ligands, 64
         simple ligand-exchange constants, 55–58
      research background, 52–54
      thermodynamics, 65–76
         lead(II)-ligand binding constants, 66–67
         lead-protein interactions, 73–76
         lead-small molecule interactions, 67–73
lead poisoning, chelation therapy with, 120, 123
sol-gel covalently attached complexes, silylated ligands, 372–376
sol-gel sensors, 399–401
Ligand-to-metal charge-transfer (LMCT):
   lead(II) compounds, 16
   optical electronegativities, 16–19
   sol-gel processing, luminescent probes, lanthanide complexes, 361–362
   solid-state absorption spectra, lead-thiolate complexes, 23
Lime treatment, lead compounds, aquatic distribution, 86–87
Linear combination of atomic orbitals (LCAO), solid-state absorption spectra, lead-thiolate complexes, 21–24
Litharge (PbO), structure and composition, 81
Logarithmis relationships, entrapped sol-gel complexes, silica compounds, 355
Lowest unoccupied molecular orbital (LUMO), cryptands, nonlinear optical effects, 296–297
Low molecular weight substances, chromium(III), 208–209
Low-temperature synthesis, cryptands, 254–255
Luminescent probes, sol-gel noncoordinated interactions, 360–366
   lanthanide complexes, 360–362
   ruthenium polypyridyl complexes, 362–366

Macrocyclic ligands:
   kinetic studies, 64
   lead(II) compounds, non-macrocyclic ligands vs., 73
   protonation-lead(II) binding constants, 67, 71–72
Macromolecules:
   chromium carcinogenicity pathways, 183
   crowding, sol-gel sensors, 401
Magnetic circular dichroism (MCD), solid-state absorption spectra, lead-thiolate complexes, 22
Manganese compounds, vanadium phosphates, d-block elements, 491–492

Manganese(II):
  chromium(VI)+reductant systems, 188–193
  cryptand-based fluorescent sensors, 281–295
Massicot (β-PbO), structure and composition, 81
MCM-41 materials, research background, 423–427
Mercury(II) compounds, sol-gel sensors, 399–401
Metabolism reactions:
  chromium toxicity, 163–167
  in vivo chromium(III), 206–207
Metal center survival:
  sol-gel catalysts, leaching, 405–406
  sol-gel processes, 386–390
Metal complexation, laterally asymmetric aza-cryptands, 260–274
  N and O donors, 265–274
  N and S donors, 260–265
Metal-ion binding, laterally asymmetric aza-cryptands, fluorescent signaling, 275–276
Metal ion-fluorophore (M-F) communication:
  cryptand-based fluorescent sensors, 280–295
  laterally asymmetric aza-cryptands, fluorescent signaling, 279–280
Metal ion-receptor (M-R) interaction:
  cryptand-based fluorescent sensors, 280–295
  laterally asymmetric aza-cryptands, fluorescent signaling, 279–280
Metal ion sensing, laterally asymmetric aza-cryptands, fluorescent signaling, 277
Metal-ligand subunits, V/O/M'/P scaffoldings, vanadium phosphates/organophosphates (VOPO/VOPRO), M^{II}-organodiimine subunits:
  oxovanadium organophosphate structures, 577–588
  oxovanadium phosphate structures, 567–577
Metal substitution reactions, multidentate lead-ligand interactions, 61–63
Metal-to-ligand charge-transfer (MLCT):
  lead(II) compounds, 16
    optical electronegativities, 16–19
  sol-gel processing, luminescent probes, ruthenium polypyridyl complexes, 362–366
N-Methylimidazole, sol-gel processing, 350
Methyl viologen (MV), sol-gel photonics, 402–403

Micelle templated silicas (MTMs), sol-gel materials, 338
M^{II}-organodiimine subunits, V/O/M'/P scaffoldings, vanadium phosphates/organophosphates (VOPO/VOPRO):
  oxovanadium organophosphate structures, 577–588
  oxovanadium phosphate structures, 567–577
Minerals, lead concentrations in, 77–78
Minimum (Pb_3O_4), structure and composition, 81
Mitogen-activated protein (MAP) kinases, chromium toxicity, 182
Mixed-valence phase:
  vanadium phosphates, isolated vanadium polyhedra, 471–475
  vanadium phosphate structures, 467–468
    charge-compensating cations, three-dimensional structures, 522–531
M'^{n+} cations, vanadium phosphates, quaternary $M'/V/P/O$ phases, covalent linkages and charge compensating M' cations, 468–487
Mobile composition of matter (MCM):
  silica gel structures, 339–340
  sol-gel materials, 335, 338
    ICC to ISSC conversion, 357–358
    postmodification techniques, 385–386
Molecular mechanisms, chromium(III) biological activity, 220–225
Molecular orbital (MO) models, solid-state absorption spectra, lead-thiolate complexes, 21–24
Molecular recognition, cryptand-based amphiphiles, air-water interface, 317–319
Molecular targets, lead biological chemistry, 101–111
  calcium protein interactions, 102–103
  RNA/DNA interactions, 110–111
  zinc protein interactions, 103–110
Molybdenum phosphate (MoPO) system, research background, 424–427
Molybdenum(VI) complexes, entrapped sol-gel complexes, 356
Monolayer lipid membranes (MLMs), cryptand-based amphiphiles, vesicular aggregation, 309–311
Monolithic studies, sol-gel processing, redox active probes, 369–370

SUBJECT INDEX

Monomeric emissions, cryptand-based
 fluorescent sensors, 292–295
Mononuclear polyhedra:
 vanadium phosphates, 464–468
 vanadium phosphate structures, charge-
  compensating cations, three-dimensional
  structures, 523–531
MTS materials, sol-gel processing,
 postmodification techniques, 385–386
Multidentate ligands, lead(II) substitution
 reactions, 60–63

Nail analysis, lead poisoning biomarkers, 119
Neurotransmission, calcium protein-lead
 interactions, 103
Neutron activation analysis, chromium uptake,
 distribution, and metabolism, 166–167
Nickel compounds, vanadium phosphates,
 d-block elements, 487–491
Nickel(II):
 chromium carcinogenicity pathways, 183
 cryptand-based fluorescent sensors,
  281–295
 cryptand metal complexation, 266–274
 entrapped sol-gel complexes, 356
 sol-gel covalently attached complexes,
  tethered TM complexes, 392–396
Nicotinate, efficacy and safety, 216–220
Nonanuclear vanadium clusters, structural
 properties, 483, 486–487
Nonbonding distances, cryptand metal
 complexation, 272–274
Noncoordinated interactions, sol-gel entrapped
 complexes, 358–371
 luminescent probes, 360–366
  lanthanide complexes, 360–362
  ruthenium polypyridyl complexes, 362–366
 redox active probes, 366–371
  monolithic studies, 369–370
  thin-film studies, 366–369
 research background, 358–360
Nonlinear optical effect (NLO), cryptands,
 295–297
Nonlinear optical materials:
 laterally asymmetric aza-cryptands,
  295–304
 D-π-A nonlinear systems, 297–304
 nonlinear optical effect, 295–297
 sol-gel processing, 348–349
Nuclear magnetic resonance (NMR):

chromium complexes, 159
lead(II) compounds, 29–35
 lead-207 parameters, 30–33
 proteins, lead-207 studies, 33–34
lead(II)-ligand binding constants, 67
lead(II) oxides, 26
silica structural analysis, 340–341
sol-gel processing:
 catalyst contamination, 350
 immobilization techniques, 385
Nucleic acids, lead(II) biomolecules, heavy atom
 derivatives, 49–50
Nucleophilic substitution reactions, sol-gel
 covalently attached complexes, silylated
 ligand preparation, 373–376
Nutrients, chromium, 203–225
 chromium(III) biological activity, 203–206
 chromodulin, 209–214
 food supplement efficacy and safety,
  214–220
 glucose tolerance factor, 207–208
 low molecular weight chromium-binding
  substance, 208–209
 molecular mechanisms, 220–225
 chromium(III) in vivo absorption,
  metabolism, and speciation, 206–207

One-dimensional chain structures:
 vanadium organophosphates (VOPRO):
  oxovanadium diphosphonate solids,
   546–566
  oxovanadium monophosphonate solids,
   543–545
 vanadium phosphates, 464–468
  charge-compensating cations, 510–512
  charge-compensating cations, three-
   dimensional structures, 529–531
  infinite chain polyhedra, 469–470
 vanadium phosphates/organophosphates
  (VOPO/VOPRO), M^{II}-organodiimine
  subunits, V/O/M'/P scaffoldings, 567,
  572–577
Optical electronegativities, lead(II) compounds,
 16–19
Optical pH sensors, sol-gel preparations,
 397–401
Optical transparency, silica materials,
 342–343
Organic radicals, chromium (VI) chemical
 studies, 154

ORMOSILs (ORganically MOdified SILicates):
  sol-gel materials:
    classification, 335
    sensors, 401
  sol-gel processing, luminescent probes, lanthanide complexes, 361–362
  structural chemistry, 341
Oxidant systems, chromium(III) toxicity, 193–195
Oxidation states, lead compounds, 10
Oxidative damage, chromium toxicity:
  antioxidants, 178–180
  chromium(VI)+reductant systems, 185–193
  DNA damage, 174
Oxides:
  entrapped sol-gel complexes, 355–356
  inorganic compounds, research background, 422–427
  lead(II), photoelectron spectroscopy, 26
  sol-gel sensors, 397–401
Oxo-bridged clusters:
  sol-gel covalently attached complexes, silylated precursor preparation, 378–739
  vanadium phosphate structures, 467–468
Oxovanadium phosphate systems:
  diphosphonate solids, 545–566
  M^{II}-organodiimine subunits, V/O/M'/P scaffoldings, vanadium phosphates/organophosphates (VOPO/VOPRO):
    oxovanadium organophosphate structures, 577–588
    oxovanadium phosphate structures, 567–577
  monophosphonate solids, 531, 536–545
  structural characteristics, 430–459
Oxygen radicals:
  chromium EPR spectroscopic analysis, 154–156
  chromium toxicity, chromium(VI)+reductant systems, 189–193

Paramagnetic metal ions, laterally asymmetric aza-cryptands, fluorescent signaling, 279–280
Penicillamine:
  lead(II) compound small molecules, 43–46
  lead poisoning, chelation therapy with, 120–122
Pentanuclear vanadium clusters:

charge-compensating cations, three-dimensional structures, 526–531
  structural properties, 483, 485
Peptide-based fluorescent sensor, lead biodistribution, 99
Peptide systems, zinc-lead interactions, 107–110
Permissible exposure limit (PEL), chromium(VI) toxicity, 160–163
Phagocytosis, chromium uptake, distribution, and metabolism, 167
1,10-Phenanthroline (phen):
  cryptand-based fluorescent sensors, 283–295
  sol-gel noncoordinated interactions, 359–360
pH levels:
  lead compounds, soil distribution, 85–87
  sol-gel processing, 336
    entrapped sol-gel complexes, 355
    luminescent probes, ruthenium polypyridyl complexes, 364–366
Phosphorylation, chromium(III) molecular mechanisms, 223–225
Phosphovanadylite, tetranuclear cluster structure, 483–484
Photoelectron spectroscopy, lead(II) compounds, 24–28
  chalcogenides, 27–28
  halides, 27
  oxides, 26
  relativistic effects, 25
Photoinduced electron transfer (PET):
  cryptand-based fluorescent sensors, 280–295
  laterally asymmetric aza-cryptands, fluorescent signaling, 275–276
  transition metal ions, 279–280
Photonics, sol-gel materials, 401–403
Pillared structures, vanadium organophosphates (VOPRO), oxovanadium diphosphonate solids, 553–566
Platinum complexes, sol-gel sensors, 398–401
Polarizability, cryptands, nonlinear optical effect (NLO), 296–297
Polarography, lead(II)-ligand binding constants, 67
Polyhedral structures:
  vanadium borophosphates, 497–500
  vanadium organophosphates (VOPRO):
    oxovanadium diphosphonate solids, 550–566
    oxovanadium monophosphonate solids, 531, 536–545

vanadium phosphate fluorides, 500–505
vanadium phosphates/organophosphates (VOPO/VOPRO), M^{II}-organodiimine subunits, V/O/M'/P scaffoldings, 579–588
vanadium phosphates (VOPO):
   charge-compensating cations:
     three-dimensional structures, 521–531
     two-dimensional structures, 512–520
   infinite chain polyhedra, 469–470
   isolated vanadium polyhedra, 470–475
Polymer classification, sol-gel materials, 335
Porosity, silica materials, 341–342
Porphyrin complexes, sol-gel sensors, 398–401
Postmodification techniques, sol-gel processing, 385–386
Postprocessing, sol-gel materials, 352
Potassium ions, vanadium phosphates:
   binuclear vanadium clusters and chains, 475, 478–481
   isolated vanadium polyhedra, 471–475
   nonanuclear vanadium clusters, 483, 486–487
Potentiometry, lead(II)-ligand binding constants, 66–67
Protein kinase C, lead interactions, 103
Proteins:
   lead(II) biomolecules, heavy atom derivatives, 49–50
   lead(II)-ligand interactions, kinetic studies, 64–65
   lead-207 NMR spectroscopy, 33–34
   lead-protein interactions:
     calcium proteins, 102–103
     thermodynamic stability, 73–76
Proton-induced X-ray emission (PIXE), chromium uptake, distribution, and metabolism, 163–167
Proton sensors, sol-gel preparations, 396–401
Pseudo-threefold networks, cryptands, D-π-A nonlinear systems, 301–304
p53 tumor suppressor gene, chromium toxicity, 177–178
Purification methods, sol-gel covalently attached complexes, 379–380
α-(4-Pyrdil-1-oxide)-N-$tert$-butylnitrone (POBN), chromium(VI) organic radicals, 154
Pyrene, sol-gel sensors, 397–401

Q-T systems:

ORMOSIL structure, 341
sol-gel processing, immobilization techniques, 383–385
Quaternary M'/V/P/O phases, covalent linkages and charge compensating M' cations, vanadium phosphates (VOPO), 468–487
   binuclear clusters and chains, 475–479
   binuclear vanadium clusters, 475
   infinite chains, vanadium polyhedra, 469–470
   isolated vanadium polyhedra, 470–475
   nonanuclear clusters, 483–487
   pentanuclear clusters, 483
   tetranuclear vanadium clusters, 479–483

Racah B parameter, cryptands, metal complexation, 260–265
Raman spectroscopy, lead(II) compounds, 28–29
Rare earth salts, sol-gel processing, luminescent probes, lanthanide complexes, 361–362
Reactive oxygen species (ROS), chromium toxicity:
   carcinogenicity pathways, 182–183
   chromium(VI) organic radicals, 154
   food supplement efficacy and safety, 219–220
Reactivity concept, sol-gel covalently attached complexes:
   metal center survival, 387–390
   tethered TM complexes, 391–396
Red lead. See Minimum (Pb_3O_4)
Redox active probes, sol-gel entrapped complexes, 366–371
   monolithic studies, 369–370
   thin-film studies, 366–369
Redox cycle:
   chromium(VI)+reductant systems, 189–193
   laterally asymmetric aza-cryptands, fluorescent signaling, 279–280
Reductant systems, chromium(VI) toxicity, 184–193
Relativistic effects, lead(II) spectroscopic analysis, 14–15
   photoelectron spectroscopy, 25
Rhodium complex catalysts, sol-gel processing, 350
Rigid-group principle, cryptand synthesis, 254–255
"Rigidochromism," sol-gel luminescent probes, 362–366

Ring structures:
  vanadium organophosphates (VOPRO), oxovanadium diphosphonate solids, 562–566
  vanadium phosphates, 464–468
RNA, lead interactions, 110–111
Roasting, of lead, 4
Rubidium complexes:
  vanadium aluminophosphates, 493, 495–497
  vanadium phosphates, isolated vanadium polyhedra, 475
Ruthenium(II), cryptand metal complexation, 265
Ruthenium polypyridyl complexes, sol-gel processing:
  luminescent probes, 362–366
  photonics, 402–403
  sensor applications, 397–401

S. typhimurium, chromium genotoxicity assays, 170–171
Scandium, environmental lead contamination measurements, 87–89
Scanning tunneling microscopy (STM), lead(II) chalcogenides, 27–28
Schiff base condensation, tripodal coupling, cryptand synthesis, 255–257
Schrödinger equation, lead(II) spectroscopic analysis, 14–15
Schulterite, structure and composition, 81
Second harmonic generation (SHG), cryptands:
  D-π-A nonlinear systems, 297–304
  nonlinear optical effect (NLO), 296–297
Sediment analysis, environmental lead contamination measurements, 87–89
Seitz substitutional model, solid-state absorption spectra, lead-thiolate complexes, 20–24
Sensors:
  fluorescent sensors:
    laterally asymmetric aza-cryptands, 275–295
      chemical logics, 277–278
      cryptand-based sensors, 280–295
      design principles, 275–276
      metal ion sensing, 277
      transition metal ions, 279–280
    lead compounds, biodistribution, 93–99
    sol-gel processing, luminescent probes, lanthanide complexes, 360–362
  sol-gel materials, 396–401

Shuttle vectors, chromium-induced mutation, 177
Signal transduction, laterally asymmetric aza-cryptands, fluorescent signaling, 275–276
Silica compounds. *See also* Sol-gel materials
  entrapped sol-gel complexes, 354–356
  material properties, 341–344
  structural chemistry, 339–341
  vanadium phosphates (VOPO), vanadium silicophosphates, 493–494
Silver, lead codeposits with, 4–5
Silver(II), cryptand metal complexation, 269–274
Silylated ligands, sol-gel covalently attached complexes, 372–376
  precursor preparation, 372–380
  purification, 379–380
  silylated ligands, 372–376
  silylated metal complexes, 376–379
Silylated metal complexes, sol-gel covalently attached complexes, precursor preparation, 376–379
Silyl tethered amines, sol-gel covalently attached complexes, 373–376
Single oxygen, chromium EPR spectroscopic analysis, 154–156
Sintering, sol-gel materials, 352
Site accessibility, sol-gel catalysts, 404–405
Site isolation, sol-gel catalysts, 406–407
Small molecule complexes:
  biomolecules, lead(II) compounds, x-ray crystal structures, 47–50
  lead biodistribution, fluorescent sensors, 96–99
  lead(II) compounds, 36–47
    chelation therapy agents, 41–46
    coordination numbers and geometries, 36–40
    donor groups, 36
    electron pairs ($6s^2$), 40–41
    new compounds, 46–47
  lead-ligand thermodynamics, 67–73
  sol-gel catalysts, 405
Smelting, of lead, 4
$SnBU_2(OAc)_2$, sol-gel processing, 350
Soil distribution, lead compounds, 81–84
  environmental contamination measurements, 87–89
Sol-gel materials:

applications:
  catalysts, 404–407
    entrapment vs. immobilization, 404
    leaching, 405–406
    site accessibility, 404–405
    site isolation, 406–407
  photonics, 401–403
  sensors, 396–401
covalently attached complexes, 372–396
  material characterization, 390–391
  material preparation, 380–390
    immobilization techniques, 380–385
    metal center survival, 386–390
    postmodification techniques, 385–386
  precursor preparation, 372–380
    purification, 379–380
    silylated ligands, 372–376
    silylated metal complexes, 376–379
  research background, 372
  tethered TM complex environment, 391–396
dopant environment, 352–353
entrapped complexes, 353–371
  inner-sphere sol-gel chemistry (ICC to ISSC), 357–358
  noncoordinated interactions, 358–371
    luminescent probes, 360–366
      lanthanide complexes, 360–362
      ruthenium polypyridyl complexes, 362–366
    redox active probes, 366–371
      monolithic studies, 369–370
      thin-film studies, 366–369
    research background, 358–360
  research background, 353–354
  silica chemistry, 354–356
future research issues, 409–410
historical development, 337–338
postprocessing procedures, 352
processing, 336
  catalysts, 350
  gelation, aging and drying, 347–349
  initial stages, 345–347
  material formation, 344–345
  solvents, 351
  transition metal ion catalysis, 350–351
silica gels:
  material properties, 341–343
  ORMOSILs, 341
  structural chemistry, 339–341

structure and properties, 334–336
synthesis issues, 407–409
Solid-state absorption spectra:
  lead(II) compounds, 19–24
  research background, 422–427
Solubility, lead compounds, 11–12
Solution structures, chromium(V) complexes, 151–153
Solvents, sol-gel process, 351
Sonification, sol-gel processing, solvents, 351
Speciation studies:
  lead compounds:
    aquatic distribution, 84–87
    soil distribution, 82–84
  in vivo chromium(III), 206–207
Spectrophotometry, lead(II)-ligand binding constants, 67
Spectroscopic studies, lead(II), 13–35, 123
  absorption spectroscopy, 16–24
    electronic transitions, 16
    optical electronegativities, 16–19
    solid-state spectra, 19–24
  nuclear magnetic resonance (NMR), 29–35
    lead-207 parameters, 30–33
    proteins, lead-207 studies, 33–34
  photoelectron spectroscopy, 24–28
    chalcogenides, 27–28
    halides, 27
    oxides, 26
    relativistic effects, 25
  relativistic effects, 14–15
  vibrational spectroscopy (IR and Raman), 28–29
Spin-orbit interaction, lead(II) spectroscopic analysis, 14–15
Spin trapping, chromium(VI) organic radicals, 154
"Sporting techniques," sol-gel materials characterization, 390–391
Stability constant, thermodynamics, lead-ligand interactions, 65–66
Stearic acid (SA), cryptand-based amphiphiles, air-water interface, Langmuir-Blodgett film, 313–317
"Stepped" layer structure, vanadium organophosphates (VOPRO), oxovanadium diphosphonate solids, 557–566
Stepwise equilibria, thermodynamics, lead-ligand interactions, 65–66

Stepwise strategy, cryptand synthesis, 255
Stereochemistry, lead(II) compound
    thermodynamics, 73
Steric effects:
    chromium(V), EPR spectroscopy, 150–153
    macrocyclic ligands, 64
    multidentate lead-ligand interactions,
        62–63
Structural studies, lead(II) compounds:
    EXAFS spectroscopy, 51–52
    research background, 35–36
    X-ray crystals, biomolecules, 47–50
        heavy atom derivatives, proteins and
            molecules, 49–50
    X-ray crystals, small molecule complexes,
        36–47
        chelation therapy agents, 41–46
        coordination numbers and geometries,
            36–40
        donor groups, 36
        electron pairs ($6s^2$), 40–41
        new compounds, 46–47
Structure-directing agents (SDA), charge-
    compensating cations, vanadium
    phosphate phases, 505–510
Subcellular toxicity mechanisms, chromium
    toxicology, 163–183
    carcinogenicity, direct and indirect pathways,
        182–183
    DNA and transcription factor damage,
        174–178
    genotoxicity assays, 167–171
    membrane and wall interactions, 171–173
    oxidative stress and antioxidants, 178–180
    signaling and aopotosis, 180–182
    uptake, distribution, and metabolism,
        163–167
Substitution reactions:
    multidentate lead-ligand reactions, 60–63
    transition-metal complexes, "substitution
        inert"/"substitution labile" complexes,
        387–390
Supercritical drying, sol-gel processing, 349
Superhyperfine coupling, chromium(V)
    complexes, 152–153
Surface anchoring, sol-gel covalently attached
    complexes, immobilization techniques,
    380–385
Surface area, silica materials, 342–343
Surface modification, sol-gel materials, 352

Synchrotron radiation-induced X-ray emission
    (SRIXE):
    chromium complexes, 157–158
    chromium uptake, distribution, and
        metabolism, 166–167
Synthesis research:
    sol-gel materials, 407–409
    vanadium organophosphates (VOPRO)/
        vanadium phosphates (VOPO),
        428–430

Tb(III):
    cryptand-based fluorescent sensors,
        283–295
    sol-gel processing, luminescent probes,
        360–362
Teeth, lead poisoning biomarkers, 119
Template effect:
    charge-compensating cations, vanadium
        phosphate phases, 505–510
    cryptand synthesis, 254–255
    vanadium organophosphates (VOPRO):
        oxovanadium diphosphonate solids,
            546–566
        oxovanadium monophosphonate solids,
            539–545
Tethered TM complexes, sol-gel processing,
    391–396
Tetraethylorthosilicates (TEOS):
    ORMOSIL structure, 341
    sol-gel formation, 344–345
    sol-gel processing:
        immobilization techniques, 383–385
        luminescent probes:
            lanthanide complexes, 361–362
            ruthenium polypyridyl complexes,
                363–366
Tetrahydrofuran (THF):
    cryptand-based fluorescent sensors,
        281–295
    cryptand synthesis, tripodal coupling, 257
    sol-gel solvent, 351
Tetramethylorthosilicate (TMOS):
    ORMOSIL structure, 341
    sol-gel formation, 344–345
    sol-gel processing:
        immobilization techniques, 383–385
        luminescent probes, lanthanide complexes,
            361–362
        redox active probes, 367–369

SUBJECT INDEX

Tetranuclear vanadium clusters:
  charge-compensating cations, two-dimensional structures, 516–520
  structure and properties, 479, 482–483
Thermodynamics:
  entrapped sol-gel complexes, silica compounds, 354–356
  lead-ligand interactions, 65–76
    lead(II)-ligand binding constants, 66–67
    lead-protein interactions, 73–76
    lead-small molecule interactions, 67–73
Thin-film studies, sol-gel entrapped complexes, redox active probes, 366–369
Thiohydroxamic acids, lead(II) compound small molecules, 45–46
Three-dimensional structures:
  vanadium phosphates/organophosphates (VOPO/VOPRO), M^{II}-organodiimine subunits, V/O/M'/P scaffoldings, 585–588
  vanadium phosphates (VOPO), 520–531
Top-down technology, laterally asymmetric aza-cryptands, chemical logic, fluorescent signaling, 277–278
Toxicology, chromium, 159–203
  cellular/subcellular toxicity mechanisms, 163–183
    carcinogenicity, direct and indirect pathways, 182–183
    DNA and transcription factor damage, 174–178
    genotoxicity assays, 167–171
    membrane and wall interactions, 171–173
    oxidative stress and antioxidants, 178–180
    signaling and aopotosis, 180–182
    uptake, distribution, and metabolism, 163–167
  epidemiological/experimental evidence, 159–163
  molecular toxicity mechanisms, 183–203
    chromium(III)-DNA interactions, 195–196
    chromium(III)+oxidant systems, 193–195
    chromium(V) DNA damage mechanisms, 196–203
    chromium(VI)+reductant systems, 184–193
Toxicology studies, lead toxicokinetics, 99–101
Transcription factors:
  chromium toxicity:
    cell signaling and apoptosis, 180–181
    inducible gene expression, 176
  zinc-lead interactions, 110
Transfer RNA (tRNA), lead(II) biomolecules, heavy atom derivatives, 50
Transition metal ions. *See also* specific metals, e.g. Cobalt
  cryptand-based fluorescent sensors, 281–295
  laterally asymmetric aza-cryptands, fluorescent signaling, 279–280
  sol-gel processing, 337–338
    catalysis, 350–351
    metal center survival, 386–390
    noncoordinated interactions, 359–360
    tethered TM complexes, 391–396
Transition metal phosphate (TMPO):
  research background, 423–427
  vanadium structures, 588–591
Trialkoxysilyl functional groups, sol-gel covalently attached complexes, silylated ligand purification, 379–380
Trinuclear propionate, efficacy and safety, 217–220
Trinuclear structures, vanadium organophosphates (VOPRO), oxovanadium monophosphonate solids, 539–545
Tripodal capping, cryptand synthesis, 257–260
Tripodal coupling, cryptand synthesis, 255–257
Tris(3-aminopropyl)amine (trpn), cryptand metal complexation, 271–274
Tungsten complexes, sol-gel processing, redox active probes, 370
Tunneling electron microscopy (TEM), cryptand-based amphiphiles, vesicular aggregation, 306–311
Two-catalyst procedures, sol-gel processing, 350
Two-dimensional structures:
  vanadium organophosphates (VOPRO):
    oxovanadium diphosphonate solids, 550–566
    fundamental building blocks, 562–566
    oxovanadium monophosphonate solids, 543–545
  vanadium phosphates, charge-compensating cations, 512–520
  vanadium phosphates/organophosphates (VOPO/VOPRO), M^{II}-organodiimine subunits, V/O/M'/P scaffoldings, 579–588

Ultraviolet-visible (UV-vis) absorption:
  entrapped sol-gel complexes, 356
  sol-gel materials characterization, 390–391
  photonics, 402–403
Uptake mechanisms, chromium toxicity, 163–167
Urine analysis, lead poisoning biomarkers, 119
Urokinase-type plasminogen activator (uPA), chromium toxicity, cell membrane/wall interactions, 172–173

Valence shell electron-pair repulsion (VSEPR), lead(II) compounds, small molecule structural effects, 40–41
Vanadium borophosphates, structural properties, 497–500
Vanadium gallophosphates, structure and function, 497
Vanadium organophosphates (VOPRO):
  general characteristics, 531
  monophosphonate/diphosphonate comparisons, 531–535
  oxovanadium diphosphonate solids, 545–566
  oxovanadium monophosphonate solids, 531, 536–545
  research background, 425–427
  secondary metal-ligand subunits, V/O/M′/P scaffoldings:
    M^{II}-organodiimine subunits:
      oxovanadium organophosphate structures, 577–588
      oxovanadium phosphate structures, 567–577
    research background, 566–567
  synthesis, 428–430
Vanadium phosphate fluorides, structural properties, 500–505
Vanadium phosphates (VOPO):
  additional elements, research overview, 491, 493
  charge-compensating organic cations, 505–531
    one-dimensional structures, 510–512
    three-dimensional structures, 520–531
    two-dimensional structures, 512–520
  charge neutral V-P-O covalent linkages, 459–468
  d-block elements, 487–492
  inorganic phases, structural characteristics, 430–459

quaternary M′/V/P/O phases, covalent linkages and charge compensating M′ cations, 468–487
  binuclear clusters and chains, 475–479
  binuclear vanadium clusters, 475
  infinite chains, vanadium polyhedra, 469–470
  isolated vanadium polyhedra, 470–475
  nonanuclear clusters, 483–487
  pentanuclear clusters, 483
  tetranuclear vanadium clusters, 479–483
research background, 425–427
secondary metal-ligand subunits, V/O/M′/P scaffoldings:
  M^{II}-organodiimine subunits:
    oxovanadium organophosphate structures, 577–588
    oxovanadium phosphate structures, 567–577
  research background, 566–567
synthesis, 428–430
Vanadium silicophosphates, structure and properties, 493–494
Vanadyl phosphate, charge neutral V-P-O covalent linkages, 459–468
Vesicular aggregation, cryptand-based amphiphiles, 305–311
Vibrational spectroscopy (IR and Raman), lead(II) compounds, 28–29
Vibronic coupling, sol-gel processing, luminescent probes, lanthanide complexes, 361–362
V/O/M′/P scaffoldings, vanadium phosphates/organophosphates (VOPO/VOPRO):
  M^{II}-organodiimine subunits:
    oxovanadium organophosphate structures, 577–588
    oxovanadium phosphate structures, 567–577
  research background, 566–567
V/P/O phases:
  vanadium organophosphates (VOPRO):
    oxovanadium diphosphonate solids, 545–566
    oxovanadium monophosphonate solids, 531, 536–545
  vanadium organophosphates (VOPRO)/ vanadium phosphates (VOPO), 428–430
  vanadium phosphates (VOPO):
    charge-compensating cations, 505–531

one-dimensional structures, 510–512
three-dimensional structures, 520–531
two-dimensional structures, 512–520
charge neutral covalent linkages, 459–468
d-block elements, 487–492
quaternary M′/V/P/O phases, covalent linkages and charge compensating M′ cations, 468–487
structural characteristics, 430–459
V/P ratios, vanadium organophosphates (VOPRO)/vanadium phosphates (VOPO), 428–430

Water:
lead concentrations in, 77–78
sol-gel photochemical splitting, 403
Water-exchange rate:
lead(II) compounds, 11–12
lead-ligand kinetics, 55–58
White lead. See Ceruse (white lead)
Wine, environmental lead contamination measurements, 89

X-ray absorption fine structure (XAFS), chromium complexes, 157–158
X-ray absorption near edge spectroscopy (XANES):
chromium complexes, 157–158
lead(II) oxides, 26
X-ray absorption spectroscopy (XAR), chromium complexes, 157–158

X-ray crystals:
lead(II) compounds:
biomolecules, 47–50
heavy atom derivatives, proteins and molecules, 49–50
small molecule complexes, 36–47
chelation therapy agents, 41–46
coordination numbers and geometries, 36–40
donor groups, 36
electron pairs ($6s^2$), 40–41
new compounds, 46–47
sol-gel materials characterization, 390–391
zinc-lead interactions, 105–110
X-ray diffraction (XRD):
chromium complexes, 156–157
chromium(V) chemical studies, 149–153
X-ray fluorescence spectroscopy (XRF):
bone lead levels, 118–119
lead body burden and uptake, 92–93

Zeolites, research background, 423–427
Zinc-binding domains:
lead(II)-ligand interactions, 64–65
lead-protein interactions, thermodynamic stability, 74–76
Zinc compounds, vanadium phosphates, d-block elements, 487–491
Zinc(II):
cryptand-based fluorescent sensors, 281–295
cryptand metal complexation, 266–274
Zinc proteins, lead interactions, 103–110

Cumulative Index, Volumes 1–51

	VOL.	PAGE
Abel, Edward W., Orrell, Keith G., and Bhargava, Suresh K., *The Stereodynamics of Metal Complexes of Sulfur-, Selenium and Tellurium-Containing Ligands*	32	1
Adams, Richard D, and Horváth, Istváns, T., *Novel Reactions of Metal Carbonyl Cluster Compounds*	33	127
Adamson, A. W., *see* Fleischauer, P. D.		
Addison, C. C. and Sutton, D., *Complexes Containing the Nitrate Ion*	8	195
Albin, Michael, *see* Horrocks, William DeW., Jr.		
Allen, G. C. and Hush, N. S., *Intervalence-Transfer Absorption, Part I Qualitative Evidence for Intervalence Transfer Absorption in Inorganic Systems in Solution and in the Solid State*	8	357
Allison, John, *The Gas-Phase Chemistry of Transition-Metal Ions with Organic Molecules*	34	627
Ardizzoia, G. Attilio, *see* La Monica, Girolamo		
Arnold, John, *The Chemistry of Metal Complexes with Selenolate and Tellurolate Ligands*	43	353
Asprey, L. B. and Cunningham, B. B., *Unusual Oxidation States of Some Actinide and Lanthanide Elements*	2	267
Baird, Michael C., *Metal–Metal Bonds in Transition Metal Compounds*	9	1
Bakac, Andreja, *Mechanistic and Kinetic Aspects of Transition Metal Oxygen Chemistry*	43	267
Balch, Alan L., *Construction of Small Polynuclear Complexes with Trifunctional Phosphin-Based Ligands as Backbones*	41	239
Balhausen, C. J., *Intensities of Spectral Bands in Transition Metal Complexes*	2	251
Balkus, Kenneth J., Jr., *Synthesis of Large Pore Zeolites and Molecular Sieves*	50	217
Barton, Jacqueline K., *see* Pyle, Anna Marie		
Barwinski, Almut, *see* Pecoraro, Vincent L.		
Barrett, Anthony G. M., *see* Michel, Sarah L. J.		
Basolo, Fred and Pearson, Ralph G., *The Trans Effect in Metal Complexes*	4	381
Bastos, Cecilia M., *see* Mayr, Andreas		
Baum, Sven M., *see* Michel, Sarah L. J.		
Beattie, I. R., *Dinitrogen Trioxide*	5	1
Beattie, J. K. and Haight, G. P., Jr., *Chromium (IV) Oxidation of Inorganic Substrates*	17	93
Becke-Goehring, *Von Margot, Uber Schwefel Stickstoff Verbindungen*	1	207
Becker, K. A., Plieth, K., and Stranski, I. N., *The Polymorphic Modifications of Arsenic Trioxide*	4	1
Beer, Paul D. and Smith, David K., *Anion Binding and Recognition by Inorganic Based Receptors*	46	1
Bennett, L. F., *Metalloprotein Redox Reactions*	18	1
Beno, Mark A., *see* Williams, Jack M.		

	VOL.	PAGE
Berg, Jeremy M., *Metal-Binding Domains in Nucleic Acid-Binding and Gene-Regulatory Proteins*	37	143
Bertrand, J. A. and Eller, P. G., *Polynuclear Complexes with Aminoalcohols and Iminoalcohols as Ligands: Oxygen-Bridged and Hydrogen-Bonded Species*	21	29
Bharadwaj, Parimal K., *Laterally Nonsymmetric Aza-Cryptands*	51	251
Bhargava, Suresh K., *see* Abel, Edward W.		
Bickley, D. G., *see* Serpone, N.		
Bignozzi, C. A., Schoonover, J. R., and Scandola, F., *A Supramolecular Approach to Light Harvesting and Sensitization of Wide-Bandgap Semiconductors: Antenna Effects and Charge Separation*	44	1
Bodwin, Jeffery J., *see* Pecoraro, Vincent L.		
Bowler, Bruce E., Raphael, Adrienne L., and Gray, Harry B., *Long-Range Electron Transfer in Donor (Spacer) Acceptor Molecules and Proteins*	38	259
Bowman, Stephanie, *see* Watton, Stephen P.		
Bradley, D. C., *Metal Alkoxides*	2	303
Bridgeman, Adam J. and Gerloch, Malcolm. *The Interpretation of Ligand Field Parameters*	45	179
Brookhart, Maurice, Green, Malcom L. H., and Wong, Luet-Lok, *Carbon-Hydrogen-Transition Metal Bonds*	36	1
Brothers, Penelope, J., *Heterolytic Activation of Hydrogen by Transition Metal Complexes*	28	1
Brown, Dennis G., *The Chemistry of Vitamin B12 and Related Inorganic Model Systems*	18	177
Brown, Frederick J., *Stoichiometric Reactions of Transition Metal Carbene Complexes*	27	1
Brown, S. B., Jones, Peter, and Suggett, A., *Recent Developments in the Redox Chemistry of Peroxides*	13	159
Brudvig, Gary W. and Crabtree, Robert H., *Bioinorganic Chemistry of Manganese Related to Photosynthesis Oxygen Evolution*	37	99
Bruhn, Suzanne L., Toney, Jeffrey H., and Lippard, Stephen J., *Biological Processing of DNA Modified by Platinum Compounds*	38	477
Brusten, Bruce E. and Green, Michael, R., *Ligand Additivity in the Vibrational Spectroscopy, Electrochemistry, and Photoelectron Spectroscopy of Metal Carbonyl Derivatives*	36	393
Busch, Daryle H., *see* Meade, Thomas J.		
Canary, James W. and Gibb, Bruce C., *Selective Recognition of Organic Molecules by Metallohosts*	45	1
Caneschi, A., Gatteschi, D., and Rey, P., *The Chemistry and Magnetic Properties of Metal Nitronyl Nitroxide Complexes*	39	331
Cannon, Roderick D., White, Ross P., *Chemical and Physical Properties of Triangular Bridged Metal Complexes*	36	195
Carlson, K. Douglas, *see* Williams, Jack M.		
Carty, A., *see* Tuck, D. G.		
Carty, Arthur J., *see* Sappa, Enrico		
Castellano, Felix N. and Meyer, Gerald J., *Light-Induced Processes in Molecular Gel Materials*	44	167
Catlow, C. R. A., *see* Thomas, J. M.		

	VOL.	PAGE
Cattalini, L., *The Intimate Mechanism of Replacement in d^5 Square-Planar Complexes*	13	263
Chaffee, Eleanor and Edwards, John O., *Replacement as a Prerequisite to Redox Processes*	13	205
Chakravorty, A., *see* Holm, R. H.		
Chang, Hsuan-Chen, *see* Lagow, Richard J.		
Chapelle, Stella, *see* Verchère, Jean-François		
Chaudhuri, Phalguni and Wieghardt, Karl, *The Chemistry of 1,4,7-Triazacyclononane and Related Tridentate Macrocyclic Compounds*	35	329
Chaudhuri, Phalguni, and Wieghardt, Karl, *Phenoxyl Radical Complexes*	50	151
Chisholm, M. H. and Godleski, S., *Applications of Carbon-13 NMR in Inorganic Chemistry*	20	299
Chisholm, Malcolm H. and Rothwell, Ian P., *Chemical Reactions of Metal–Metal Bonded Compounds of Transition Elements*	29	1
Chock, P. B. and Titus, E. O., *Alkali Metal Ions Transport and Biochemical Activity*	18	287
Chow, S. T. and McAuliffe, C. A., *Transition Metal Complexes Containing Tridentate Amino Acids*	19	51
Churchill, Melvyn R., *Transition Metal Complexes of Azulene and Related Ligands*	11	53
Ciurli, A., *see* Holm, Richard M.		
Claudio, Elizabeth S., Godwin, Hilary Arnold, and Magyar, John S., *Fundamental Coordination Chemistry, Environmental Chemistry and Biochemistry of Lead (II)*	51	1
Clearfield, Abraham, *Metal-Phosphonate Chemistry*	47	371
Codd, Rachel, *see* Levina, Aviva		
Constable, Edwin C., *Higher Oligopyridines as a Structural Motif in Metal-Iosupramolecular Chemistry*	42	67
Corbett, John D., *Homopolyatomic Ions of the Post-Transition Elements-Synthesis, Structure, and Bonding*	21	129
Cotton, F. A., *Metal Carbonyls: Some New Observations in an Old Field*	21	1
Cotton, F. A., *see* Wilkinson, G.		
Cotton F. A. and Hong, Bo, *Polydentate Phosphines: Their Syntheses, Structural Aspects, and Selected Applicators*	40	179
Cotton, F. A. and Lukehart, C. M., *Transition Metall Complexes Containing Carbonoid Ligands*	16	487
Coucouvanis, Dimitri, *see* Malinak, Steven M.		
Coucouvanis, Dimitri, *The Chemistry of the Dithioacid and 1,1-Dithiolate Complexes*	11	233
Coucouvanis, Dimitri, *The Chemistry of the Dithioacid and 1,1-Dithiolate Complexes, 1968–1977*	26	301
Cowley, Alan H., *UV Photoelectron Spectroscopy in Transition Metal Chemistry*	26	45
Cowley, Alan H. and Norman, Nicholas C., *The Synthesis, Properties, and Reactivities of Stable Compounds Featuring Double Bonding Between Heavier Group 14 and 15 Elements*	34	1
Crabtree, Robert H., *see* Brudvig, Gary W.		
Cramer, Stephen P. and Hodgson, Keith O., *X-Ray Absorption Spectroscopy: A New Structural Method and Its Applications to Bioinorganic Chemistry*	25	1
Crans, Debbie C., *see* Verchère, Jean-François		

	VOL.	PAGE
Creutz, Carol, *Mixed Valence Complexes of d^5–d^6 Metal Centers*	30	1
Cummins, Christopher C., *Three-Coordinate Complexes of "Hard" Ligands: Advances in Synthesis, Structure and Reactivity*	47	685
Cunningham, B. B., *see* Asprey, L. B.		
Dance, Ian and Fisher, Keith, *Metal Chalcogenide Cluster Chemistry*	41	637
Darensbourg, Marcetta York, *Ion Pairing Effects on Metal Carbonyl Anions*	33	221
Daub, G. William, *Oxidatively Induced Cleavage of Transition Metal-Carbon Bonds*	22	375
Dean, P. A. W., *The Coordination Chemistry of the Mercuric Halides*	24	109
DeArmond, M. Keith and Fried, Glenn, *Langmuir-Blodgett Films of Transition Metal Complexes*	44	97
Dechter, James J., *NMR of Metal Nuclides, Part I: The Main Group Metals*	29	285
Dechter, James J., *NMR of Metal Nuclides, Part II: The Transition Metals*	33	393
De Los Rios, Issac, *see* Peruzzini, Maurizio		
Deutsch, Edward, Libson, Karen, Jurisson, Silvia, and Lindoy, Leonard F., *Technetium Chemistry and Technetium Radiopharmaceuticals*	30	75
Diamond, R. M. and Tuck, D. G., *Extraction of Inorganic Compounds into Organic Solvents*	2	109
DiBenedetto, John, *see* Ford, Peter C.		
Dillon, Carolyn T., *see* Levina, Aviva		
Doedens, Robert J., *Structure and Metal-Metal Interactions in Copper (II) Carboxylate Complexes*	21	209
Donaldson, J. D., *The Chemistry of Bivalent Tin*	8	287
Donini, J. C., Hollebone, B. R., and Lever, A. B. P., *The Derivation and Application of Normalized Spherical Harmonic Hamiltonians*	22	225
Dori, Zvi, *The Coordination Chemistry of Tungsten*	28	239
Doyle, Michael P. and Ren, Tong, *The Influence of Ligands on Dirhodium (II) on Reactivity and Selectivity in Metal Carbene Reactions*	49	113
Drago, R. S. and Purcell, D. F., *The Coordination Model for Non-Aqueous Solvent Behavior*	6	271
Drew, Michael G. B., *Seven-Coordination Chemistry*	23	67
Dunbar, Kim R. and Heintz, Robert A., *Chemistry of Transition Metal Cyanide Compounds: Modern Perspectives*	45	283
Dutta, Prabir K. and Ledney, Michael, *Charge-Transfer Processes in Zeolites: Toward Better Artificial Photosynthetic Models*	44	209
Dye, James L., *Electrides, Negatively Charged Metal Ions, and Related Phenomena*	32	327
Earley, Joseph E., *Nonbridging Ligands in Electron-Transfer Reactions*	13	243
Edwards, John O. and Plumb, Robert C., *The Chemistry of Peroxonitrites*	41	599
Edwards, John O., *see* Chaffee, Eleanor		
Eichorn, Bryan W., *Ternary Transition Metal Sulfides*	42	139
Eisenberg, Richard, *Structural Systematics of 1,1- and 1,2-Dithiolate Chelates*	12	295
Eller, P. G., *see* Bertand, J. A.		
Emge, Thomas J., *see* Williams, Jack M.		
Endicott, John F., Kumar, Krishan, Ramasami, T., and Rotzinger, François P., *Structural and Photochemical Probes of Electron Transfer Reactivity*	30	141
Epstein, Arthur J., *see* Miller, Joel S.		
Espenson, James H., *Homolytic and Free Radical Pathways in the Reactions of Organochromium Complexes*	30	189

	VOL.	PAGE

Evans, David A., *see* Rovis, Tomislav
Everett, G. W., *see* Holm. R. H.

	VOL.	PAGE
Fackler, John P., Jr., *Metal B-Ketoenolate Complexes*	7	361
Fackler, John P., Jr., *Multinuclear d^5–d^{10} Metal Ion Complexes with Sulfur-Containing Ligands*	21	55
Favas, M. C. and Kepert, D. L., *Aspects of the Stereochemistry of Four-Coordination and Five-Coordination*	27	325
Favas, M. C. and Kepert, D. L., *Aspects of the Stereochemistry of Nine-Coordination, Ten-Coordination, and Twelve-Coordination*	28	309
Feldman, Jerald and Schrock, Richard R., *Recent Advances in the Chemistry of "d^0" Alkylidene and Metallacyclobutane Complexes*	39	1
Felthouse, Timothy R., *The Chemistry, Structure, and Metal-Metal Bonding in Compounds of Rhodium (II)*	29	73
Fenske, Richard F., *Molecular Orbital Theory, Chemical Bonding, and Photoelectron Spectroscopy for Transition Metal Complexes*	21	179
Ferguson, J., *Spectroscopy of 3d Complexes*	12	159

Ferguson, James, *see* Krausz, Elmars

	VOL.	PAGE
Figgis, B. N. and Lewis, J., *The Magnetic Properties of Transition Metal Complexes*	6	37
Finn, Robert C., Haushalter, Robert C., and Zubieta, Jon, *Crystal Chemistry of Organically Templated Vanadium Phosphates and Organophosphonates*	51	421

Fisher, Keith, *see* Dance, Ian

	VOL.	PAGE
Fisher, Keith J., *Gas-Phase Coordination Chemistry of Transition Metal Ions*	50	343
Fleischauer, P. D., Adamson, A. W., and Sartori G., *Excited States of Metal Complexes and Their Reactions*	17	1

Floriani, Carlo, *see* Piarulli, Umberto

	VOL.	PAGE
Ford, Peter C., Wink, David, and DiBenedetto, John. *Mechanistic Aspects of the Photosubstitution and Photoisomerization Reactions of d^6 Metal Complexes*	30	213
Fowles, G. W. A., *Reaction by Metal Hallides with Ammonia and Aliphatic Amines*	6	1
Fratiello, A., *Nuclear Magnetic Resonance Cation Solvation Studies*	17	57

Frenking, Gernot, *see* Lupinetti, Anthony J.
Fried, Glenn, *see* DeArmond, M. Keith
Friedman, H. L., *see* Hunt, J. P.
Fu, Lei, *see* Mody, Tarak D.

Gatteschi, D., *see* Caneschi, A.

	VOL.	PAGE
Geiger, William E., *Structural Changes Accompanying Metal Complex Electrode Reactions*	33	275

Geiser, Urs, *see* Williams, Jack M.

	VOL.	PAGE
Geoffroy, George, L., *Photochemistry of Transition Metal Hydride Complexes*	27	123
George, J. W., *Halides and Oxyhalides of the Elements of Groups Vb and VIb*	2	33
George, Philip and McClure, Donald S., *The Effect of Inner Orbital Splitting on the Thermodynamic Properties of Transition Metal Compounds, and Coordination Complexes*	1	381
Gerfin, T., Grätzel, M., and Walder, L., *Molecular and Supramolecular Surface Modification of Nanocrystalline TiO_2 Films: Charge-Separating and Charge-Injecting Devices*	44	345
Gerloch, M., *A Local View in Magnetochemistry*	26	1
Gerloch, M. and Miller, J. R., *Covalence and the Orbital Reduction*	10	1

	VOL.	PAGE
Gerloch, Malcolm, *see* Bridgeman, Adam J.		
Gerloch, Malcolm and Woolley, R. Guy, *The Functional Group in Ligand Field Studies: The Empirical and Theoretical Status of the Angular Overlap Model*	31	371
Gibb, Bruce C., *see* Canary, James W.		
Gibb, Thomas, R. P., Jr., *Primary Solid Hydrides*	3	315
Gilbertson, Scott R., *Combinatorial-Parallel Approaches to Catalyst Discovery and Development*	50	433
Gibney, Brian, R., *see* Pecoraro, Vincent L.		
Gillard, R. C., *The Cotton Effect in Coordination Compounds*	7	215
Gillespie, Ronald J., *see* Sawyer, Jeffery F.		
Glasel, Jay A., *Lanthanide Ions as Nuclear Magnetic Resonance Chemical Shift Probes in Biological Systems*	18	383
Glick, Milton. D. and Lintvedt, Richard L., *Structural and Magnetic Studies of Polynuclear Transition Metal β-Polyketonates*	21	233
Godleski, S., *see* Chisholm, M. H.		
Godwin, Hilary Arnold, *see* Claudio, Elizabeth S.		
Gordon, Gilbert, *The Chemistry of Chlorine Dioxide*	15	201
Gratzel, M., *see* Gerfin, T.		
Gray, Harry B., *see* Bowler, Bruce E.		
Green, Malcom L. H., *see* Brookhart, Maurice		
Green, Michael R., *see* Burstein, Bruce E.		
Grove, David M., *see* Janssen, Maurits D.		
Grubbs, Robert H., *The Olefin Metathesis Reaction*	24	1
Gruen, D. M., *Electronic Spectroscopy of High Temperature Open-Shell Polyatomic Molecules*	14	119
Gultneh, Yilma, *see* Karlin, Kenneth D.		
Hahn, James E., *Transition Metal Complexes Containing Bridging Alkylidene Ligands*	31	205
Haight, G. P., Jr., *see* Beattie, J. K.		
Haim, Albert. *Mechanisms of Electron Transfer Reactions: The Bridged Activated Complex*	30	273
Hall, Kevin P. and Mingos, D. Michael P., *Homo- and Heteronuclear Cluster Compounds of Gold*	32	237
Hall, Tracy H., *High Pressure Inorganic Chemistry*		
Hancock, Robert D., *Molecular Mechanics Calculations as a Tool in Coordination Chemistry*	37	187
Haushalter, Robert C., *see* Finn, Robert C.		
Hayaishi, Osamu, Takikawa, Osamu, and Yoshida, Ryotaro, *Indoleamine 2,3-Dioxygenase, Properties and Functions of a Superoxide Utilizing Enzyme*	38	75
Heintz, Robert A., *see* Dunbar, Kim R.		
Hendry, Philip, and Sargeson, Alan M., *Metal Ion Promoted Reactions of Phosphate Derivatives*	38	201
Hennig, Gerhart R., *Interstitial Compounds of Graphite*	1	125
Henrick, Kim, Tasker, Peter A., and Lindoy, Leonard F., *The Specification of Bonding Cavities in Macrocyclic Ligands*	33	1
Herbert, Rolle H., *Chemical Applications of Mössbauer Spectroscopy*	8	1
Heumann, Andreas, Jens, Klaus-Joachim, and Réglier, Marius, *Palladium Complex Catalyzed Oxidation Reactions*	42	483

	VOL.	PAGE
Hobbs, R. J. M., see Hush, N. S.		
Hodgson, D. J., *The Structural and Magnetic Properties of First-Row Transition Metal Dimers Containing Hydroxo, Substituted Hydroxo, and Halogen Bridges*	19	173
Hodgson, Derek J., *The Stereochemistry of Metal Complexes of Nucleic Acid Constituents*	23	211
Hodgson, Keith O., see Cramer, Stephen P.		
Hoff, Carl, D., *Thermodynamics of Ligand Binding and Exchange in Organometallic Reactions*	40	503
Hoffman, Brian E., see Michel, Sarah L. J.		
Hollebone, B. R., see Domini, J. C.		
Holloway, John H., *Reactions of the Noble Gases*	6	241
Holm, R. H., Everett, G. W., and Chakravorty, A., *Metal Complexes of Schiff Bases and B-Ketoamines*	7	83
Holm, R. H. and O'Connor, M. J., *The Stereochemistry of Bis-Chelate Metal (II) Complexes*	14	241
Holm, Richard M., Ciurli, Stefano, and Weigel, John A., *Subsite-Specific Structures and Reactions in Native and Synthetic (4Fe-4-S) Cubane-Type Clusters*	38	1
Holmes, Robert R., *Five-Coordinated Structures*	32	119
Hong, Bo, see Cotton, F. A.		
Hope, Hakon, *X-Ray Crystallography: A Fast, First-Resort Analytical Tool*	41	1
Horrocks, William DeW., Jr. and Albin, Michael, *Lanthanide Ion Luminescence in Coordination Chemistry and Biochemistry*	31	1
Horváth, István T., see Adams, Richard D.		
Humphries, A. P. and Kaesz, H. D., *The Hydrido-Transition Metal Cluster Complexes*	25	145
Hunt, J. P. and Friedman, H. L., *Aquo Complexes of Metal Ions*	30	359
Hush, N. S., *Intervalence Transfer Absorption Part 2. Theoretical Considerations and Spectroscopic Data*	8	391
Hush, N. S., see Allen, G. C.		
Hush, N. S. and Hobbs, R. J. M., *Absorption-Spectra of Crystals Containing Transition Metal Ions*	10	259
Isied, Stephan S., *Long-Range Electron Transfer in Peptides and Proteins*	32	443
Isied, Stephan S., see Kuehn, Christa		
Jagirdar, Balaji R., *Organometallic Fluorides of the Main Group Metals Containing the C-M-F Fragment*	48	351
James, B. D. and Wallbridge, M. G. H., *Metal Tetrahydroborates*	11	99
James, David W., *Spectroscopic Studies of Ion-Ion Solvent Interaction in Solutions Containing Oxyanions*	33	353
James, David W. and Nolan, M. J., *Vibrational Spectra of Transition Metal Complexes and the Nature of the Metal-Ligand Bond*	9	195
Janssen, Maurits D., Grove, David M., and Koten, Gerard van, *Copper(I) Lithium and Magnesium Thiolate Complexes: An Overview with Due Mention of Selenolate and Tellurolate Analogues and Related Silver(I) and Gold(I) Species*	46	97
Jardine, F. H., *The Chemical and Catalytic Reactions of Dichlorotris(triphenylphosphine(II) and Its Major Derivatives*	31	265

	VOL.	PAGE
Jardine, F. H., *Chlrotris(triphenylphosphine)rhodium(I): Its Chemical and Catalytic Reactions*	28	63
Jeffrey, G. A. and McMullan, R. K., *The Clathrate Hydrates*	8	43
Jens, Klaus-Joachim, *see* Heumann, Andreas		
Johnson, B. F. G. and McCleverty, J. A., *Nitric Oxide Compounds of Transition Metals*	7	277
Jolly, William L., *Metal-Ammonia Solution*	1	235
Jones, Peter, *see* Brown, S. B.		
Jorgensen, Chr., Klixbull, *Electron Transfer Spectra*	12	101
Jorgensen, Chr., Klixbull, *The Nephelauxetic Series*	4	73
Jurisson, Silvia, *see* Deutsch, Edward		
Kadish, Karl M., *The Electrochemistry of Metalloporphyrins in Nonaqueous Media*	34	435
Kaesz, H. D., *see* Humphries, A. P.		
Kahn, M. Ishaque and Zubieta, Jon, *Oxovanadium and Oxomolybdenum Clusters and Solids Incorporating Oxygen-Donor Ligands*	43	1
Kamat, Prashant V., *Native and Surface Modified Semiconductor Nanoclusters*	44	273
Kampf, Jeff W., *see* Pecoraro, Vincent L.		
Kanatzidis, Mercouri G. and Sutorik, Anthony C., *The Application of Polychalcogenide Salts to the Exploratory Synthesis of Solid-State Multinary Chalogenides at Intermediate Temperatures*	43	151
Karlin, Kenneth D. and Gultneh, Yilma, *Binding and Activation of Molecular Oxygen by Copper Complexes*	35	219
Kennedy, John D., *The Polyhedral Metallaboranes, Part I: Metallaborane Clusters with Seven Vertices and Fewer*	32	519
Kennedy, John D., *The Polyhedral Metallaboranes, Part II: Metallaborane Clusters with Eight Vertices and More*	34	211
Kepert, D. L., *Aspects of the Stereochemistry of Eight-Coordination*	24	179
Kepert, D. L., *Aspects of the Stereochemistry of Seven-Coordination*	25	41
Kepert, D. L., *Aspects of the Stereochemistry of Six-Coordination*	23	1
Kepert, D. L., *Isopolytungstates*	4	199
Kepert, D. L., *see* Favas, M. C.		
Kesselman, Janet M., *see* Tan, Ming X.		
Kice, J. L., *Nucleophilic Substitution at Different Oxidation-States of Sulfur*	17	147
Kimura, Eiichi, *Macrocylic Polyamine Zinc(II) Complexes as Advanced Models for Zinc(II) Enzymes*	41	443
King, R. B., *Transition Metal Cluster Compounds*	15	287
Kingsborough, Richard P., *Transition Metals in Polymeric π-Conjugated Organic Frameworks*	48	123
Kitagawa, Teizo and Ogura, Takashi, *Oxygen Activation Mechanism at the Binuclear Site of Heme-Copper Oxidase Superfamily as Revealed by Time-Resolved Resonance Raman Spectroscopy*	45	431
Klingler, R. J. and Rathke, J. W., *Homogeneous Catalytic Hydrogenation of Carbon Monoxide*	39	113
Kloster, Grant M., *see* Watton, Stephen P.		
Kolodziej, Andrew F., *The Chemistry of Nickel-Containing Enzymes*	41	493

	VOL.	PAGE
Konig, Edgar, *Structural Changes Accompanying Continuous and Discontinuous Spin-State Transitions*	35	527
Koten, Gerard van, *see* Janssen, Maurits D.		
Kramarz, K. W. and Norton, J. R., *Slow Proton-Transfer Reactions in Organometallic and Bioinorganic Chemistry*	42	1
Krausz, Elmars and Ferguson, James, *The Spectroscopy of the $[Ru(bpy)_3]^{2+}$ System*	37	293
Kubas, Gregory J., *see* Vergamini, Philip J.		
Kuehn, Christa and Isied, Stephan S., *Some Aspects of the Reactivity of Metal Ion-Sulfur Bonds*	27	153
Kumar, Krishan, *see* Endicott, John F.		
Kustin, Kenneth and Swinehart, James, *Fast Metal Complex Reactions*	13	107
Laane, Jaan and Ohlsen, James R., *Characterization of Nitrogen Oxides by Vibrational Spectroscopy*	27	465
Lagow, Richard J. and Margrave, John L., *Direct Fluorination: A "New" Approach to Fluorine Chemistry*	26	161
Lagow, Richard J., and Chang, Hsuan-Chen, *High-Performance Pure Calcium Phosphate Bioceramics: The First Weight Bearing Completely Resorbable Synthetic Bone Replacement Materials*	50	317
Laibinis, Paul E., *see* Tan, Ming, X.		
La Monica, Girolamo, *The Role of the Pyrazolate Ligand in Building Polynuclear Transition Metal Systems*	46	151
Lange, Christopher W., *see* Pierpont, Cortlandt G.		
Laudise, R. A., *Hydrothermal Synthesis of Single Crystals*	3	1
Laure, B. L. and Schmulbach, C. D., *Inorganic Electrosynthesis in Nonaqueous Solvents*	14	65
Lay, Peter A., *see* Levina, Aviva		
Ledney, Michael, *see* Dutta, Prabir K.		
Le Floch, Pascal, *see* Mezaillies, Nicolas		
Lentz, Dieter, *see* Seppelt, Konrad		
Leung, Peter C. W., *see* Williams, Jack M.		
Lever, A. B. P., *see* Donini, J. C.		
Levina, Aviva, Codd, Rachel, Dillon, Carolyn T., and Lay, Peter A., *Chromium in Biology: Toxicology and Nutritional Aspects*	51	145
Lewis, J., *see* Figgis, B. N.		
Lewis, Nathan S., *see* Tan, Ming, X.		
Libson, Karen, *see* Deutsch, Edward		
Lieber, Charles M., *see* Wu, Xian Liang		
Liehr, Andrew D., *The Coupling of Vibrational and Electronic Motions in Degenerate Electronic States of Inorganic Complexes. Part I. States of Double Degeneracy*	3	281
Liehr, Andrew D., *The Coupling of Vibrational and Electronic Motions in Degenerate Electronic States of Inorganic Complexes. Part II. States of Triple Degeneracy and Systems of Lower Symmetry*	4	455
Liehr, Andrew D., *The Coupling of Vibrational and Electronic Motions in Degenerate and Nondegenerate Electronic States of Inorganic and Organic Molecules. Part III. Nondegenerate Electronic States*	5	385
Lindoy, Leonard F., *see* Deutsch, Edward		

	VOL.	PAGE
Lindoy, Leonard F., *see* Henrick, Kim		
Lintvedt, Richard L., *see* Glick, Milton D.		
Lippard, Stephen J., *see* Bruhn, Suzanne L.		
Lippard, Stephen J., *Eight-Coordination Chemistry*	8	109
Lippard, Stephen J., *Seven and Eight Coordinate Molybdenum Complexes and Related Molybdenum (IV) Oxo Complexes, with Cyanide and Isocyanide Ligands*	21	91
Lippen, Bernhard, *Platinum Nucleobase Chemistry*	37	1
Lobana, Tarlok, S., *Structure and Bonding of Metal Complexes of Tertiaryphosphine-Arsine Chalcogenides Including Analytical, Catalytic, and Other Applications of the Complexes*	37	495
Lockyer, Trevor N. and Manin, Raymond L., *Dithiolium Salts and Dithio-β-diketone Complexes of the Transition Metals*	27	223
Long, K. H., *Recent Studies of Diborane*	15	1
Lorand, J. P., *The Cage Effect*	17	207
Lukehart, C. M., *see* Cotton, F. A.		
Lupinetti, Anthony J., Strauss, Steven H., and Frenking, Gernot, *Nonclassical Metal Carbonyl*	49	1
McAuliffe, C. A., *see* Chow, S. T.		
McCleverty, J. A., *Metal 1,2-Dithiolene and Related Complexes*	10	49
McCleverty, J. A., *see* Johnson, B. F. G.		
McClure, Donald S., *see* George, Philip		
MacDonnell, Frederick M., *see* Wright, Jeffrey G.		
McMullan, R. K., *see* Jeffrey, G. A.		
Magyar, John S., *see* Claudia, Elizabeth S.		
Maier, L., *Preparation and Properties of Primary, Secondary and Tertiary Phosphines*	5	27
Malatesta, Lamberto, *Isocyanide Complexes of Metals*	1	283
Malinak, Steven M. and Coucouvanis, Dimitri, *The Chemistry of Synthetic Fe-Mo-S Clusters and Their Relevance to the Structure and Function of the Fe-Mo-S Center Nitrogenase*	49	599
Manoharan, P. T., *see* Venkatesh, B.		
Margrave, John L., *see* Lagow, Richard J.		
Marks, Tobin J., *Chemistry and Spectroscopy of f-Element Organometallics Part I: The Lanthanides*	24	51
Marks, Tobin J., *Chemistry and Spectroscopy of f-Element Organometallics Part II: The Actinides*	25	223
Martin, Raymond L., *see* Lockyer, Trevor N.		
Marzilli, Lulgi G., *Metal-ion Interactions with Nucleic Acids and Nucleic Acid Derivatives*	23	225
Marzilli, Luigi G., *see* Toscano, Paul J.		
Mathey, Francois, *see* Mezaillies, Nicolas		
Mayr, Andreas and Bastos, Cecilia M., *Coupling Reactions of Terminal Two-Faced π Ligands and Related Cleavage Reaction*	40	1
McKee, Vickie, *see* Nelson, Jane		
Meade, Thomas J. and Busch, Daryle H., *Inclusion Complexes of Molecular Transition Metal Hosts*	33	59

	VOL.	PAGE
Mehrotra, Ram C. and Singh, Anirudh, *Recent Trends in Metal Alkoxide Chemistry*	46	239
Meyer, Gerald J., *see* Castellano, Felix N.		
Meyer, Thomas J., *Excited-State Electron Transfer*	30	389
Meyer, T. J., *Oxidation-Reduction and Related Reactions of Metal-Metal Bonds*	19	1
Mézaillies, Nicolas, Mathey, Francois, and Le Floch, Pascal, *The Coordination Chemistry of Phosphinines: Their Polydentate and Macrocyclic Derivatives*	49	455
Michel, Sarah L. J., Hoffman, Brian M., Baum, Sven M., and Barrett, Anthony G. M., *Peripherally Functionalized Porphyrazines: Novel Metallomacrocycles with Broad Untapped Potential*	50	473
Miller, J. R., *see* Gerloch, M.		
Miller, Joel S. and Epstein, Anhur, J., *One-Dimensional Inorganic Complexes*	20	1
Mingos, D. Michael P., *see* Hall, Kevin P.		
Mirkin, Chad A., *see* Slone, Caroline S.		
Mitra, S., *Chemical Applications of Magnetic Anisotropy Studies on Transition Metal Complexes*	22	309
Mitzi, David B., *Synthesis, Structure and Properties of Organic-Inorganic Perovskites and Related Materials*	48	1
Mody, Tarak D., Fu, Lei, and Sessler, Jonathan L., *Texaphyrins: Synthesis and Development of a Novel Class of Therapeutic Agents*	49	551
Morgan, Grace, *see* Nelson, Jane		
Muetterties, E. L., *see* Tachikawa, Mamoru		
Murphy, Eamonn F., *see* Jugirdar, Balayi R.		
Natan, Michael J., *see* Wright, Jeffrey G.		
Natan, Michael J. and Wrighton, Mark S., *Chemically Modified Microelectrode Arrays*	37	391
Nelson, Jane, McKee, V. and Morgan, G. *Coordination Chemistry of Azacryptands*	47	167
Neumann, Ronny, *Polyoxometallate Complexes in Organic Oxidation Chemistry*	47	317
Nguyen, Sonbinh T., *see* Tan, Ming X.		
Nolan, M. J., *see* James, David W.		
Norman, Nicholas, C., *see* Cowley, Alan H.		
Norton, J. R., *see* Kramarz, K. W.		
Oakley, Richard T., *Cyclic and Heterocyclic Thiazines*	36	299
O'Connor, Charles J., *Magnetochemistry—Advances in Theory and Experimentation*	29	203
O'Connor, M. J., *see* Holm, R. H.		
Ogura, Takashi, *see* Kitagawa, Teizo		
O'Halloran, Thomas V., *see* Wright, Jeffrey G.		
Ohlsen, James R., *see* Laane, Jaan		
Oldham, C., *Complexes of Simple Carboxylic Acids*	10	223
Orrell, Keith, G., *see* Abel, Edward W.		
Ozin, G. A., *Single Crystal and Cas Phase Raman Spectroscopy in Inorganic Chemistry*	14	173
Ozin, G. A. and Vandèr Voet, A., *Cryogenic Inorganic Chemistry*	19	105

	VOL.	PAGE
Pandey, Krishna K., *Coordination Chemistry of Thionitrosyl (NS), Thiazate (NSO⁻), Disulfidothionitrate (S₃N⁻), Sulfur Monoxide (SO), and Disulfur Monoxide (S₂O) Ligands*	40	445
Parish, R. V., *The Interpretation of 119 Sn-Mössbauer Spectra*	15	101
Parkin, General, *Terminal Chalcogenido Complexes of the Transition Metals*	47	1
Paul, Purtha P., *Coordination Complex Impregnated Molecular Sieves-Synthesis, Characterization, Reactivity and Catalysis*	48	457
Peacock, R. D., *Some Fluorine Compounds of the Transition Metals*	2	193
Pearson, Ralph G., *see* Basolo, Fred		
Pecoraro, Vincent L., Stemmler, Ann J., Gibney, Brian R., Bodwin, Jeffrey J., Wang, Hsin, Kampf, Jeff W., and Barwinski, Almut, *Metallacrowns: A New Class of Molecular Recognition Agents*	45	83
Perlmutter-Hayman, Berta. *The Temperature-Dependence of the Apparent Energy of Activation*	20	229
Peruzzini, Maurizio, De Los Rios, Issac, and Romerosa, Antonio, *Coordination Chemistry of Transition Metals and Hydrogen Chalogenide and Hydrochalcogenido Ligands*	49	169
Pethybridge, A. D. and Prue, J. E., *Kinetic Salt Effects and the Specific Influence of Ions on Rate Constants*	17	327
Piarulli, Umberto and Floriani, Carlo, *Assembling Sugars and Metals: Novel Architectures and Reactivities in Transition Metal Chemistry*	45	393
Pierpont, Conlandt G. and Lange, Christopher W., *The Chemistry of Transition Metal Complexes Containing Catechol and Semiquinone Ligands*	41	331
Plieth, K., *see* Becker, K. A.		
Plumb, Robert C., *see* Edwards, John O.		
Pope, Michael T., *Molybdenum Oxygen Chemistry: Oxides, Oxo Complexes, and Polyoxoanions*	39	181
Power, Philip P., *The Structures of Organocuprates and Heteroorganocuprates and Related Species in Solution in the Solid State*	39	75
Prue, J. E., *see* Pethybridge, A. D.		
Purcell, D. F., *see* Drago, R. S.		
Pyle, Anna Marie and Banon, Jacqueline K. Banon, *Probing Nuclei Acids with Transition Metal Complexes*	38	413
Que, Lawrence, Jr., and True, Anne E., *Dinuclear Iron- and Manganese-Oxo Sites in Biology*	38	97
Ralston, Diana M., *see* Wright, Jeffrey G.		
Ramasami, T., *see* Endicott, John F.		
Raphael, Adrienne L., *see* Bowler, Bruce E.		
Rathke, J. W., *see* Klingler, R. J.		
Rauchfuss, Thomas B., *The Coordination Chemistry of Thiophenes*	39	259
Réglier, Marius, *see* Heumann, Andreas		
Ren, Tong, *see* Doyle, Michael P.		
Rey, P. *see* Caneschi, A.		
Reynolds, Warren L., *Dimethyl Sulfoxide in Inorganic Chemistry*	12	1
Rifkind, J. M., *see* Venkatesh, B.		
Roesky, Herbert W., *see* Jagirdar, Balaji R.		
Roesky, Herbert W., *see* Witt, Michael		

	VOL.	PAGE
Romerosa, Antonio, *see* Peruzzini, Maurizio		
Rothwell, Ian P. *see* Chisholm, Malcolm H.		
Rotzinger, Francois P., *see* Endicott, John F.		
Roundhill, D. Max. *Metal Complexes of Calixarenes*	43	533
Rovis, Tomislav, and Evans, David A., *Structural and Mechanistic Investigations in Asymmetric Copper(I) and Copper(II) Catalyzed Reactions*	50	1
Sappa, Enrico, Tiripicchio, Antonio, Carty, Anhur J., and Toogood, Gerald E., *Butterfly Cluster Complexes of the Group VIII Transition Metals*	35	437
Sargeson, Alan M., *see* Hendry, Philip		
Sanon, G., *see* Fleischauer, P. D.		
Sawyer, Donald T., *see* Sobkowiak, Andrzej		
Sawyer, Jeffery F., and Gillespie, Ronald J., *The Stereochemistry of SB(III) Halides and Some Related Compounds*	34	65
Scandola, F., *see* Bignozzi, C. A.		
Schatz, P. N., *see* Wong, K. Y.		
Schmulbach, C. D., *Phosphonitrile Polymers*	4	275
Schmulbach, C. D., *see* Laure, B. L.		
Schoonover, J. R., *see* Bignozzi, C. A.		
Schrock, Richard R., *see* Feldman, Jerald		
Schultz, Arthur J., *see* Williams, Jack M.		
Searcy, Alan W., *High-Temperature Inorganic Chemistry*	3	49
Seppelt, Konrad and Lentz, Dieter, *Novel Developments in Noble Gas Chemistry*	29	167
Serpone, N. and Bickley, D. G., *Kinetics and Mechanisms of Isomerization and Racemization Processes of Six-Coordinate Chelate Complexes*	17	391
Sessler, Jonhathan L., *see* Mody, Tarak D.		
Seyferth, Dietmar, *Vinyl Compounds of Metals*	3	129
Singh, Anirudh, *see* Mehrotra, Ram C.		
Slone, Caroline S., *The Transition Metal Coordination Chemistry of Hemilabile Ligands*	48	233
Smith, David K., *see* Beer, Paul D.		
Smith III, Milton R., *Advances in Metal Boryl and Metal-Mediated B-X Activation Chemistry*	48	505
Sobkowiak, Andrzej, Tung, Hui-Chan, and Sawyer, Donald T., *Iron- and Cobalt-Induced Activation of Hydrogen Peroxide and Dioxygen for the Selective Oxidation-Dehydrogenation and Oxygenation of Organic Molecules*	40	291
Spencer, James, T., *Chemical Vapor Deposition of Metal-Containing Thin-Film Materials from Organometallic Compounds*	41	145
Spiro, Thomas G., *Vibrational Spectra and Metal-Metal Bonds*	11	1
Stanbury, David M., *Oxidation of Hydrazine in Aqueous Solution*	47	511
Stanton, Colby E., *see* Tan, Ming X.		
Stemmler, Ann J., *see* Pecoraro, Vincent L.		
Stiefel, Edward I., *The Coordination and Bioinorganic Chemistry of Molybdenum*	22	1
Stranski, I. N., *see* Becker, K. A.		
Strauss, Steven H., *see* Lupinetti, Anthony J.		
Strouse, Charles E., *Structural Studies Related to Photosynthesis: A Model for Chlorophyll Aggregates in Photosynthetic Organisms*	21	159
Stucky, Galen D., *The Interface of Nanoscale Inclusion Chemistry*	40	99

	VOL.	PAGE
Suggett, A., see Brown, S. B.		
Sutin, Norman, *Theory of Electron Transfer Reactions: Insights and Hindsights*	30	441
Sutorik, Anthony C., see Kanatzidis, Mercouri G.		
Sutton, D., see Addison, C. C.		
Swager, Timothy M., see Kingsborough, Richard P.		
Swinehart, James, see Kustin, Kenneth		
Sykes, A. G. and Weil, J. A., *The Formation, Structure, and Reactions of Binuclear Complexes of Cobalt*	13	1
Tachikawa, Mamoru and Muetterties, E. L., *Metal Carbide Clusters*	28	203
Takikawa, Osamu, see Hayaishi, Osamu		
Tan, Ming X., Laibinis, Paul E., Nguyen, Sonbinh T., Kesselman, Janet M., Stanton, Colby E., and Lewis, Nathan S., *Principles and Applications of Semiconductor Photochemistry*	41	21
Tasker, Peter A., see Henrick, Kim		
Taube, Henry, *Interaction of Dioxygen Species and Metal Ions—Equilibrium Aspects*	34	607
Taylor, Colleen M., see Watton, Stephen P.		
Templeton, Joseph L., *Metal-Metal Bonds of Order Four*	26	211
Tenne, R., *Inorganic Nanoclusters with Fullerene-Like Structure and Nanotubes*	50	269
Thomas, J. M. and Callow, C. R. A., *New Light on the Structures of Aluminosilicate Catalysts*	35	1
Thorn, Robert J., see Williams, Jack M.		
Tiripicchio, Antonio, see Sappa, Enrico		
Titus, E. O., see Chock, P. B.		
Tofield, B. C., *The Study of Electron Distributions in Inorganic Solids: A Survey of Techniques and Results*	20	153
Tolman, William B., see Kitajima, Nobumasa		
Toney, Jeffrey. H., see Bruhn, Suzanne L.		
Toogood, Gerald E., see Sappa, Enrico		
Toscano, Paul J. and Marzilli, Luigi G., *B_{12} and Related Organocobalt Chemistry: Formation and Cleavage of Cobalt Carbon Bonds*	31	105
Trofimenko, S., *The Coordination Chemistry of Pyrazole-Derived Ligands*	34	115
True, Anne E., see Que, Lawrence Jr.		
Tuck, D. G., *Structures and Properties of Hx_2 and HXY Anions*	9	161
Tuck, D. G., see Diamond, R. M.		
Tuck, D. G. and Carty, A., *Coordination Chemistry of Indium*	19	243
Tung, Hui-Chan, see Sobkowiak, Andrzej		
Tyler, David R., *Mechanic Aspects of Organometallic Radical Reactions*	36	125
Vander Voet, A., see Ozin, G. A.		
van Koten, see Janssen, Maurits D.		
van Leeuwen, P. W. N. M., see Vrieze, K.		
Vannerberg, Nils-Gosta, *Peroxides, Superoxides, and Ozonides of the Metals of Groups Ia, IIa, and IIb*	4	125
Venkatesh, B., Rifkind, J. M., and Manoharan, P. T. *Metal Iron Reconstituted Hybrid Hemoglobins*	47	563
Verchère, Jean-Francois, Chapelle, S., Xin, F., and Crans, D. C., *Metal-Carboxyhydrate Complexes in Solution*	47	837

	VOL.	PAGE
Vergamini, Phillip J. and Kubas, Gregory J., *Synthesis, Structure, and Properties of Some Organometallic Sulfur Cluster Compounds*	21	261
Vermeulen, Lori A., *Layered Metal Phosphonates as Potential Materials for the Design and Construction of Molecular Photosynthesis Systems*	44	143
Vlek, Antonin A., *Polarographic Behavior of Coordination Compounds*	5	211
Vrieze, K. and van Leeuwen, P. W. N. M., *Studies of Dynamic Organometallic Compounds of the Transition Metals by Means of Nuclear Magnetic Resonance*	14	1

Walder, L., *see* Gerfin, T.
Wallbridge, M. G. H., *see* James, B. D.

Walton, R., *Halides and Oxyhalides of the Early Transition Series and Their Stability and Reactivity in Nonaqueous Media*	16	1
Walton, R. A., *Ligand-Induced Redox Reactions of Low Oxidation State Rhenium Halides and Related Systems in Nonaqueous Solvents*	21	105

Wang, Hsin, *see* Pecoraro, Vincent L.
Wang, Hua H., *see* Williams, Jack M.

Ward, Roland, *The Structure and Properties of Mixed Metal Oxides*	1	465
Watton, Stephen P., Taylor, Colleen M., Kloster, Grant M., and Bowman, Stephanie C., *Coordination Complexes in Sol–Gel Silica Materials*	51	333

Weigel, A., *see* Holm, Richard M.
Weil, J. A., *see* Sykes, A. G.
Weinberger, Dana A., *see* Slone, Caroline S.
Whangbo, Myung-Hwan, *see* Williams, Jack M.
White, Ross R. *see* Cannon, Roderick D.
Wieghardt, Karl, *see* Chaudhuri, Phalguni
Wieghardt, Karl, *see* Chaudhuri, Phalguni

Wigley, David E., *Organoimido Complexes of the Transition Metals*	42	239
Wilkinson, G. and Cotton, F. A., *Cyclopentadienyl and Arene Metal Compounds*	1	1
Williams, Jack M., *Organic Superconductors*	33	183
Williams, Jack M., Wang, Hau H., Emge, Thomas J., Geiser, Urs, Beno, Mark A., Leung, Peter C. W., Carlson, K. Douglas, Thorn, Robert J., Schultz, Arthur J., and Whangbo, Myung-Hwan, *Rational Design of Synthetic Metal Superconductors*	35	51
Williamson, Stanley M., *Recent Progress in Sulfur-Fluorine Chemistry*	7	39
Winchester, John W., *Radioactivation Analysis in Inorganic Geochemistry*	2	1

Wink, David, *see* Ford, Peter C.

Witt, Michael and Roseky, Herbert W., *Sterically Demanding Fluorinated Substituents and Metal Fluorides with Bulky Ligands*	40	353

Wong, Luet-Lok, *see* Brookhart, Maurice

Wong, K. Y. and Schatz, P. N., *A Dynamic Model for Mixed-Valence Compounds*	28	369
Wood, John S., *Stereochemical Electronic Structural Aspects of Five-Coordination*	16	227

Woolley, R. Guy, *see* Gerloch, Malcolm

Wright, Jeffrey G., Natan, Michael J., MacDonnell, Frederick M., Ralston, Diana, M., and O'Halloran, Thomas V. *Mercury(II)-Thiolate Chemistry and the Mechanism of the Heavy Metal Biosensor MerR*	38	323

Wrighton, Mark S., *see* Natan, Michael J.

	VOL.	PAGE
Wu, Xian Liang and Lieber, Charles M., *Applications of Scanning Tunneling Microscopy to Inorganic Chemistry*	39	431

Xin, Feibo, *see* Verchère, Jean-Francois

Yoshida, Ryotaro, *see* Hayaishi, Osamu

| Zubieta, J. A. and Zuckerman, J. J., *Structural Tin Chemistry t-Coordination* | 24 | 251 |

Zubieta, Jon, *see* Kahn, M. Ishaque
Zubieta, Jon, *see* Finn, Robert C.
Zuckerman, J. J., *see* Zubieta, J. A.